AF360788

Motion Control and Path Planning of Marine Vehicles

Motion Control and Path Planning of Marine Vehicles

Motion Control and Path Planning of Marine Vehicles

Editors

Bowen Xing
Bing Li

Basel • Beijing • Wuhan • Barcelona • Belgrade • Novi Sad • Cluj • Manchester

Editors
Bowen Xing
Shanghai Ocean University
Shanghai, China

Bing Li
Harbin Engineering University
Harbin, China

Editorial Office
MDPI
St. Alban-Anlage 66
4052 Basel, Switzerland

This is a reprint of articles from the Special Issue published online in the open access journal *Journal of Marine Science and Engineering* (ISSN 2077-1312) (available at: https://www.mdpi.com/journal/jmse/special_issues/2W81IG43WS).

For citation purposes, cite each article independently as indicated on the article page online and as indicated below:

Lastname, A.A.; Lastname, B.B. Article Title. *Journal Name* **Year**, *Volume Number*, Page Range.

ISBN 978-3-0365-9780-5 (Hbk)
ISBN 978-3-0365-9781-2 (PDF)
doi.org/10.3390/books978-3-0365-9781-2

Contents

Journal of
Marine Science and Engineering

Article

Smooth Sliding Mode Control for Path Following of Underactuated Surface Vehicles Based on LOS Guidance

Yuchao Wang [1], Yinsong Qu [1], Shiquan Zhao [1,*], Ricardo Cajo [2] and Huixuan Fu [1]

[1] College of Intelligent Systems Science and Engineering, Harbin Engineering University, Harbin 150001, China; wangyuchao@hrbeu.edu.cn (Y.W.); quyinsong@hrbeu.edu.cn (Y.Q.); fuhuixuan@hrbeu.edu.cn (H.F.)
[2] Facultad de Ingeniería en Electricidad y Computación, Escuela Superior Politécnica del Litoral, ESPOL, Campus Gustavo Galindo Km 30.5 Vía Perimetral, P.O. Box 09-01-5863, Guayaquil 090150, Ecuador; rcajo@espol.edu.ec
* Correspondence: zhaoshiquan@hrbeu.edu.cn

Abstract: In this paper, a solution to the problem of following a curved path for underactuated unmanned surface vehicles (USVs) with unknown sideslip angle and model uncertainties is studied. A novel smooth sliding mode control (SSMC) based on a finite-time extended state observer (FTESO) for heading control is proposed. Firstly, the model of a USV with rudderless double thrusters is established. Secondly, the path-following error dynamics of a USV is established in a path-tangential reference frame. Thirdly, a finite-time observer is introduced to estimate the unidentified sideslip angle, and the line-of-sight (LOS) guidance law is applied to produce the desired heading angle. Finally, an SSMC controller is proposed to force USV tracking at the desired heading angle and surge speed, in which FTESO is used to estimate and compensate the unknown disturbance in sliding mode dynamics. The theoretical analysis for FTESO-SSMC verifies that the controller can provide finite-time convergence to and stability on the sliding surface. Simulation studies and contrast test are conducted to demonstrate the robustness and rapidity of the proposed FTESO-SSMC controller.

Keywords: path following; LOS guidance; smooth sliding mode control; finite-time extended state observer

Citation: Wang, Y.; Qu, Y.; Zhao, S.; Cajo, R.; Fu, H. Smooth Sliding Mode Control for Path Following of Underactuated Surface Vehicles Based on LOS Guidance. *J. Mar. Sci. Eng.* **2022**, *11*, 2214. https://doi.org/10.3390/jmse11122214

Academic Editors: Mohamed Benbouzid and Yassine Amirat

Received: 21 September 2023
Revised: 18 November 2023
Accepted: 18 November 2023
Published: 22 November 2023

1. Introduction

Unmanned surface vehicles (USVs) provide unique capabilities for military and security applications, including harbor patrol, maritime interdiction, and riverine operations [1]. Therefore, motion control of such autonomous vehicles is considered an important area within the marine control research community. The problems related to motion control of USVs can be classified into three basic groups, such as objective tracking, path following, and trajectory tracking. In the literature, among most of the scenarios including stabilization, trajectory tracking, and path following, it is of practical importance to follow a predefined path [2]. For the path-following problem, the vehicle needs to converge to and follow a predefined path without any temporal constraints. Compared with the trajectory tracking and target tracking, the path following controller can provide a smoother trajectory for USVs to move to and follow the desired path so that it is generally not possible to saturate the actuators of USVs [3]. Therefore, it is extremely important for a ship to perform tasks at sea, such as maritime search, resource exploration, and nautical charting.

Conventional USVs are commonly equipped with one central propeller for surge speed control and one rudder for heading control [4]. However, the mechanical structure of a combination of thrusters and rudders is sophisticated, and the rudders can be damaged easily due to frequent steering processes in curved path-following tasks [5]. To overcome this problem, the structure of double propellers without a rudder is used. In this paper, the curved path-following problem for an underactuated USV with rudderless double thrusters is investigated. The model uncertainties and unknown exterior disturbance are also considered in this paper. The demand for underactuated USVs to move on a desired path under severe unknown disturbances and model uncertainties creates heavy challenges for robust controller design. For the path following, there are two main research directions: advanced guidance strategies and advanced control methods. They are introduced as follows.

For the guidance system, a widely used method for path following is the line-of-sight (LOS) approach. The main advantages of a LOS guidance law are simplicity and a small computational footprint [4]. However, the conventional LOS guidance has limitations when the vehicle is under the unknown drift forces that are generated by ocean waves, ocean wind, ocean currents, or other exterior disturbances. In [6], the sideslip angle was measured by accelerometers. However, the sensor is expensive and the measured data may be noisy and distorted. Therefore, plenty of research work has been carried out to improve LOS. The methods to improve LOS guidance can be divided into two types. The first type is based on adaptive law to compensate unknown sideslip angle [7–12]. The second type is based on an estimator to estimate and compensate the sideslip angle. A multifarious sideslip observer was proposed to strengthen the robustness of the path-following controller in [13–19]. In [13], a predictor-based LOS (PLOS) was developed for the estimation of vehicle sideslip under the assumption of small sideslip angle. In [14], the proposed observer could estimate large sideslip angle, and its estimation error was asymptotically convergent. However, it will be singular when the difference between heading angle and path tangential angle is $\pi/2$. In [16], a finite-time predictor-based LOS (FPLOS) guidance law was presented for the problem of path following. Compared to PLOS, the FPLOS can make sideslip estimation error convergent to zeros in a finite time and speed up the convergence process. In [19], a fixed-time predictor-based LOS (FTPLOS) was designed to ensure effective convergence of tracking sideslip angle. In this paper, the FPLOS is introduced to estimate the unknown sideslip angle and produce desired heading angle for the control system.

For the control system, lots of methods, such as PID with extended Kalman filtering techniques [20,21], adaptive disturbance rejection control (ADRC) [22], model predictive control (MPC) [23,24], deep reinforcement learning control [25,26], and sliding mode control (SMC) [27–29], are used to force USVs to follow the desired heading angle produced by the LOS guidance law. Compared to other control methods, SMC has attracted a significant interest due to its simplicity, high robustness to external disturbances, and low sensitivity to the system parameter variations [30]. In [26,27], a policy based on adaptive sliding-mode control was proposed for the path-following control of USVs. In [28], sliding-mode dynamic surface and adaptive techniques were employed to compensate for the uncertainty from parameters varying and constant bias caused by the exterior disturbances. In [29], a disturbance-observer-based sliding mode control was designed to reach good tracking performance, where the observer was designed to estimate and compensate for the modeling uncertainties and external disturbance. In the above SMC techniques, a high frequent switch function is often used to obtain a high robustness. In [31,32], a finite-time extended state observer (FTESO) is proposed to estimate system state and disturbances, and FTESO is verified to be effective in disturbances estimation. Different from the above methods, in this paper, the FTESO and SMC are combined to the design controller to realize the heading and surge control, which is called the smooth sliding mode controller (SSMC). The unknown term in the sliding manifold differential equation is taken as total disturbance, and extended state observer (ESO) is applied to estimate and compensate for the disturbance, then, a second-order control law is designed to satisfy the existence condition of the sliding mode.

The main contributions of this paper are given as follows.

(1) In the kinematic task, FPLOS is introduced to calculate the desired heading angle with sideslip angle compensation. Compared to the traditional LOS guidance law, the FPLOS can improve the performance of path-following control of USVs under the unknown ocean currents. The finite-time sideslip angle observer can make es-

timation error of sideslip angle convergent to zeros in finite time and speed up the convergence process.

(2)   In the kinetic task, a novel smooth second-order control law is proposed to satisfy the existence condition of the sliding mode. For the disturbance in sliding mode, FTESO is applied to estimate and compensate for it instead of using a strong discontinuous control signal, which will cause a strong chattering phenomenon.

The rest of this paper is organized as follows. Section 2 formulates preliminaries and the path-following problem of an USV under unknown sideslip angle. Section 3 derives the LOS guidance law based on sideslip angle observer. Then, the SSMC algorithm is designed for surge and heading control in kinetics in Section 4. The theoretical analysis verifies that the controller is semi-globally asymptotically stable in Section 5. Simulation studies are conducted in Section 6. Section 7 concludes this paper.

2. Problem Formulation and Preliminaries

2.1. USV Model

2.1.1. USV Model

To describe the motion of the USV, the North-East-Down coordinate system $\{XI\text{-}OI\text{-}YI\}$ and body coordinate system $\{YB\text{-}OB\text{-}YB\}$ are used in this paper, as shown in Figure 1.

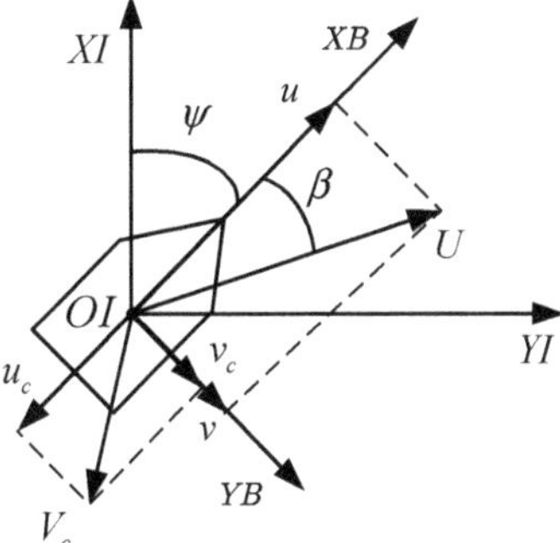

Figure 1. Definition coordinate systems and the vehicle states.

In Figure 1, u and v are surge and yaw velocity, U is the total speed, β is the sideslip angle, which is unknown, ψ is the heading angle of the USV, u_c and v_c are velocities of ocean currents, and V_c is the total speed.

The vehicle's model introduced in this subsection is similar to that given in [33]. This model can be used to describe an autonomous surface vehicle or an autonomous underwater vehicle moving in the plane. Under the coordinated system coordinate system $\{XI\text{-}OI\text{-}YI\}$ and $\{XB\text{-}OB\text{-}YB\}$, the dynamics of the USV can be described as follows:

$$\dot{\eta} = R(\psi)v$$
$$M\dot{v} = -C(v)v_r - Dv_r + H\tau + \tau_w \tag{1}$$

where $\eta = [x, y, \psi]^{\mathrm{T}}$ describes the position and the orientation of the vehicle with respect to the inertial frame $\{I\}$, $v_r = [u - u_c, v - v_c, r]^{\mathrm{T}}$ contains the surge, the sway, and the yaw velocities under current disturbance u_c, v_c, respectively (see Figure 1); $M > 0$ is the inertia matrix, $C(v)$ is the total Coriolis and centripetal acceleration matrix, D is the linear hydro-dynamic damping matrix, $\tau = [\tau_u, \tau_n]$ is the control input produced by two propellers, $\tau_w = [\tau_{w1}, \tau_{w2}, \tau_{w3}]$ is environmental disturbance, and the matrix $R(\psi), M, C(v), D, H$ is given by:

$$R(\psi) = \begin{bmatrix} \cos(\psi) & -\sin(\psi) & 0 \\ \sin(\psi) & \cos(\psi) & 0 \\ 0 & 0 & 1 \end{bmatrix}, \; M = \begin{bmatrix} m_{11} & 0 & 0 \\ 0 & m_{22} & m_{23} \\ 0 & m_{32} & m_{33} \end{bmatrix},$$

$$C(v) = \begin{bmatrix} 0 & 0 & c_{13} \\ 0 & 0 & c_{23} \\ -c_{13} & -c_{23} & 0 \end{bmatrix}, \; D = \begin{bmatrix} d_{11} & 0 & 0 \\ 0 & d_{22} & 0 \\ 0 & 0 & d_{33} \end{bmatrix}, \; H = \begin{bmatrix} 1 & 0 \\ 0 & 0 \\ 0 & 1 \end{bmatrix} \quad (2)$$

where $m_{11} = m - X_{\dot{u}}$, $m_{22} = m - Y_{\dot{v}}$, $m_{23} = mx_g - Y_{\dot{r}}$, $m_{32} = mx_g - N_{\dot{v}}$, $m_{33} = m - N_{\dot{r}}$, $c_{13} = -m_{22}v - m_{23}r$, $c_{23} = m_{11}u$. Here, m is total mass of the USV, $X_{\dot{u}}, Y_{\dot{v}}, Y_{\dot{r}}, N_{\dot{v}}, N_{\dot{r}}$ are the added masses due to hydrodynamics, and x_g is the XB-coordinate of the vehicle center of gravity. d_{11}, d_{22}, d_{33} denote the damping terms. It is worth noting that the linear damping matrix in (2) is reasonable for low-speed motion. Furthermore, since $\tau \in \mathbb{R}^2$, namely, the vehicle is underactuated in the workspace $\mathbb{R}^3$.

The model uncertainties are expressed by ζ, i.e., $\zeta = -M^{-1}[C(v)v_r - Dv_r + \tau_w] = [\zeta_u, \zeta_v, \zeta_r]^T$, then, (1) can be rewritten as:

$$\dot{v} = \zeta + M^{-1}H\tau \quad (3)$$

Substituting (2) into (1), the dynamic model (1) can be also expressed by component style as follows:

$$\begin{aligned} \dot{x} &= u\cos(\psi) - v\sin(\psi) \\ \dot{y} &= u\sin(\psi) + v\cos(\psi) \\ \dot{\psi} &= r \\ \dot{u} &= \zeta_u + \frac{\tau_u}{m_{11}} \\ \dot{v} &= \zeta_v \\ \dot{r} &= \zeta_r + \frac{\tau_n}{I_z} \end{aligned} \quad (4)$$

where $I_z = \frac{m_{22}}{m_{22}m_{33} - m_{23}m_{32}}$.

To facilitate the design and analysis of the control system, the following assumptions are taken.

Assumption 1. *The ocean current in the inertial frame $V_c = [u_c, v_c]^T$ is constant, irrotational, and bounded, and there exists a constant $V_{max} > 0$ such that $\sqrt{u_C^2 + v_c^2} \leq V_{max}$ [34]. The environmental disturbance τ_w and its derivative $\dot{\tau}_w$ are bounded.*

Remark 1. *Assumption 1 assures that the vehicle has enough thrust to resist the negative effect from ocean currents and environmental disturbances.*

Assumption 2. *For the unknown term ζ_i, $i = u, v, r$ are bounded and differentiable, i.e., there exists a positive constant ζ_i^* satisfying $|\zeta_i| \leq \zeta_i^*$, $\left|\dot{\zeta}_i\right| \leq \dot{\zeta}_i^*$.*

Remark 2. *Note that $\zeta = -M^{-1}[C(v)v_r - Dv_r + \tau_w]$, the velocity v, the derivative of v, and control input τ are all bounded due to the physical limits, and, according to Assumption 1, v_r and τ_w are also bounded, so ζ and $\dot{\zeta}$ are also bounded.*

2.1.2. Propeller Model

A catamaran USV is driven by two propellers with shaft speed n_1 and n_2 in rad/s. The thrusters generate surge force τ_u and yaw moment τ_n given as follows [30].

$$\begin{bmatrix} \tau_u \\ \tau_n \end{bmatrix} = \begin{bmatrix} 1 & 1 \\ l_1 & -l_1 \end{bmatrix} \begin{bmatrix} T_1 \\ T_2 \end{bmatrix} \quad (5)$$

where l_1 is the transverse distance from the center line of the USV to the center line of each propeller and T_1 and T_2 are the thrust force produced by left and right propeller, respectively. The relation between T_i and n_i (i = 1, 2) can be written as follows

$$\begin{aligned} T_i &= k_{pos} n_i |n_i|, \text{ if } n_i \geq 0 \\ T_i &= k_{neg} n_i |n_i|, \text{ if } n_i < 0 \end{aligned} \tag{6}$$

where k_{pos} and k_{neg} represent the scale coefficient of thrust; n_i ($n_{min} \leq n_i \leq n_{max}$, $n_{min} < 0, n_{max} > 0$) is the thrust shaft propeller revolving speed (rad/s).

2.2. Path-Following Problem

The curved path-following problem for an underactuated USV can be decomposed into two parts as follows [35].

(1) Geometric Task: Force the vehicle position to converge to a desired path by appropriate LOS guidance and path parameter update law. The LOS guidance law produces a desired heading angle.
(2) Dynamic Task: Force the vehicle to track the desired heading angle and desired forward speed by appropriate dynamic controller.

The above problems will be solved in Sections 3 and 4 separately. The following part establishes the error dynamics of path following. The schematic diagram of curved path following is shown in Figure 2.

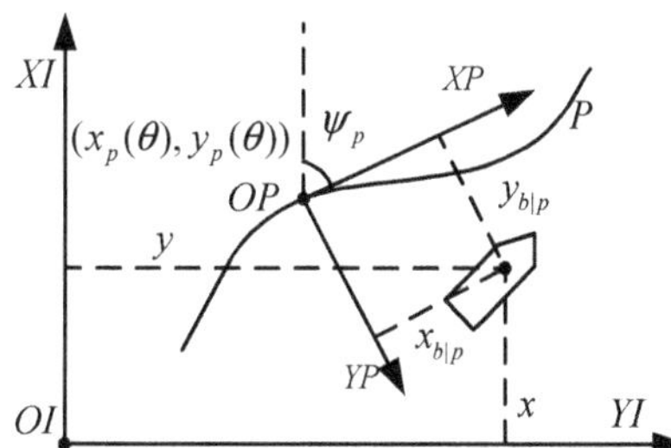

Figure 2. Schematic diagram of curved path following.

As illustrated in Figure 2, the path P is parameterized with a path variable θ. Moreover, for each virtual point on the given path, $(x_p(\theta), y_p(\theta)) \in P$, a path tangential frame {*XP-OP-YP*} is constructed to describe the position of the vessel, as illustrated in Figure 2. Hence, the path-following errors can be expressed by the coordinates of the USV in the frame {*XP-OP-YP*} denoted by $\boldsymbol{P}_{b|p} = \left[x_{b|p}, y_{b|p}\right]^T$, which are calculated by:

$$\begin{bmatrix} x_{b|p} \\ y_{b|p} \end{bmatrix} = \begin{bmatrix} \cos(\psi_p) & \sin(\psi_p) \\ -\sin(\psi_p) & \cos(\psi_p) \end{bmatrix} \begin{bmatrix} x - x_p(\theta) \\ y - y_p(\theta) \end{bmatrix} \tag{7}$$

where ψ_p is the angle of the path at the point $(x_p(\theta), y_p(\theta)) \in P$ with respect to the inertial XI-axis, $\psi_p = \arctan\left(y'_p(\theta)/x'_p(\theta)\right)$. The error dynamic can be calculated by substituting (4) in the derivative of (7), which is given by:

$$\begin{aligned} \dot{x}_{b|p} &= u\cos(\psi_e) - v\sin(\psi_e) + \left(k_c y_{b|p} - 1\right)u_p \\ \dot{y}_{b|p} &= u\sin(\psi_e) + v\cos(\psi_e) - k_c u_p x_{b|p} \end{aligned} \tag{8}$$

To facilitate the guidance law and controller design, Equation (8) can be rewritten as follows.

$$\dot{x}_{b|p} = U\cos(\psi_e)\cos(\beta) - U\sin(\psi_e)\sin(\beta) + \left(k_c y_{b|p} - 1\right)u_p$$
$$\dot{y}_{b|p} = U\sin(\psi_e)\cos(\beta) + U\cos(\psi_e)\sin(\beta) - k_c u_p x_{b|p} \tag{9}$$

where $\beta = \arctan(v/u)$ is sideslip angle; $U = \sqrt{u^2 + v^2}$ is course speed; k_c is curvature of the path at point $(x_p(\theta), y_p(\theta))$; $\psi_e = \psi - \psi_p$; u_p is the speed of the virtual point on the desired path, which is calculated as:

$$u_p = \dot{\theta}\sqrt{\left(x_p'(\theta)\right)^2 + \left(y_p'(\theta)\right)^2} \tag{10}$$

2.3. Preliminaries

Lemma 1 ([36]). *For* $(x,y) \in R^2$*, the following Young's inequality holds:*

$$xy \leq \frac{\varepsilon^p}{p}|x|^p + \frac{1}{p\varepsilon^q}|y|^q \tag{11}$$

where ε is a positive constant. The constants p and q should satisfy the conditions as $p > 1$, $q > 1$, and $(p-1)(q-1) = 1$.

Lemma 2 ([37]). *For* $\forall x_i \in \mathbb{R}$, $i = 1, 2, \ldots, n$ *and* $0 < q \leq 1$*, then*

$$\left(\sum_{i=1}^{n}|x_i|\right)^q \leq \sum_{i=1}^{n}|x_i|^q \leq n^{1-q}\left(\sum_{i=1}^{n}|x_i|\right)^q \tag{12}$$

Lemma 3 ([38]). *Consider the system of differential equations* $\dot{x} = f(x)$, $f(0) = 0$, $x \in \mathbb{R}^n$, *where* $f(\cdot): \mathbb{R}^n \to \mathbb{R}^n$ *is a continuous function. Suppose that there exists a continuous function* $V(x): U \to \mathbb{R}$ *such that the following conditions hold:*
(i) V is positive definite.
(ii) There exist real numbers $c > 0$ and $\alpha \in (0,1)$, and an open neighborhood U_0 of the origin such that:

$$\dot{V}(x) \leq -c(V(x))^\alpha, \; x \in U_0\backslash\{0\} \tag{13}$$

Then, the origin is a finite-time-stable equilibrium, and T is the settling-time function, then

$$T(x) \leq \frac{1}{c(1-\alpha)}V(x)^\alpha \tag{14}$$

Lemma 4 ([39]). *Suppose that there is a positive definite continuous Lyapunov function $V(x,t)$ defined on $U_1 \subset \mathbb{R}^n$ of the origin, and*

$$V(x,t) \leq -c_1 V^\alpha(x,t) + c_2 V(x,t), \; \forall x \in U_1\{0\} \tag{15}$$

where $c_1, c_2 > 0$ and $0 < \alpha < 1$. Thus, the origin of system $\dot{x} = f(x)$ is locally finite-time stable. The set $U_2 = \{x | V^{1-\alpha}(x,t) \leq c_1/c_2\}$ is contained in the domain of attraction of the origin. The settling time satisfies

$$T(x) \leq \ln(1 - (\frac{c_2}{c_1})V^{1-\alpha}(x_0, t_0))/(c_2\alpha - c_2) \tag{16}$$

for the given initial condition $x(t_0) = \{U_1 \cap U_2\}$.

3. Guidance System Design

The path-following control system can be divided into two parts: guidance system and control system. In this section, the guidance system will be designed to produce the desired heading angle to lead the USV to follow the desired path. To show the relation of the guidance system and control system, the block diagram of the whole control system is given as Figure 3.

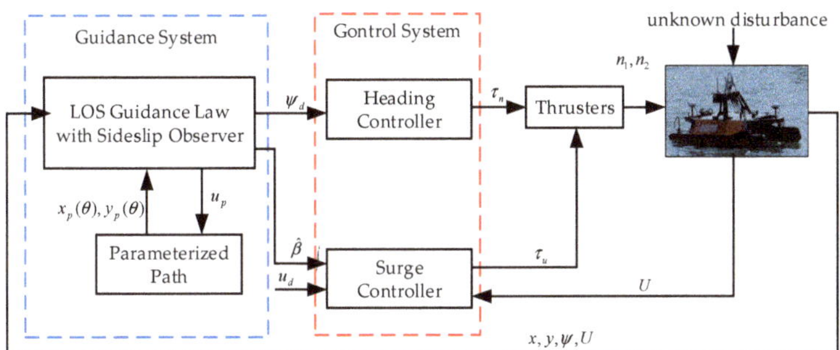

Figure 3. Block diagram of guidance and control system.

In Figure 3, the blue dashed box shows the guidance system and the red dashed box shows the control system. For the guidance system, the LOS guidance law with sideslip angle observer is designed to generate the desired heading angle for the USV and update the law for the path parameter. In addition, the unknown sideslip angle is estimated by the designed observer. For the control system, the heading controller and surge controller are designed for the USV to track the desired heading angle and desired surge speed based on SSMC and FTESO. The guidance system is designed in Section 3, and the control system will be designed in Section 4. The guidance system will be designed in two steps. For the first step, the sideslip angle observer is presented to estimate sideslip angle, and the finite-time stability of the observer is analyzed, which is given in Section 3.1. For the second step, the LOS guidance law is designed to produce the desired heading angle and the desired update law based on the estimated sideslip angle, which is given in Section 3.2.

3.1. Sideslip Angle Observer

In practice, the sideslip angle of the USV is not more than $20°$ for a USV with double thrusters. If the sideslip angle is small enough, then there are $\cos(\beta) \approx 1$ and $\sin(\beta) \approx \beta$. In addition, according to [16], the derivative of sideslip angle equals to zero, i.e., $\dot{\beta} = 0$. Although the sideslip angle is relatively small (typically less than $20°$), it largely affects the path-following properties of the vehicle, and if it is not properly compensated, this results in significant deviations from the desired path. Therefore, a finite-time observer is given in this subsection to estimate sideslip angle.

Due to the small sideslip angle, the tracking error dynamics (9) can be rewritten as follows:

$$\dot{x}_{b|p} = U\cos(\psi_e) - U\sin(\psi_e)\beta + \left(k_c y_{b|p} - 1\right)u_p$$
$$\dot{y}_{b|p} = U\sin(\psi_e) + U\cos(\psi_e)\beta - k_c u_p x_{b|p} \tag{17}$$

Then, the sideslip angle observer can be designed as:

$$\dot{\hat{x}}_{b|p} = U\cos(\psi_e) - U\sin(\psi_e)\hat{\beta} + \left(k_c \hat{y}_{b|p} - 1\right)u_p - k_{\tilde{x}} sig^\rho\left(\tilde{x}_{b|p}\right)$$
$$\dot{\hat{y}}_{b|p} = U\sin(\psi_e) + U\cos(\psi_e)\hat{\beta} - k_c u_p \hat{x}_{b|p} - k_{\tilde{y}} sig^\rho\left(\tilde{y}_{b|p}\right) \tag{18}$$
$$\dot{\hat{\beta}} = k_\beta\left(U\sin(\psi_e)\tilde{x}_{b|p} - U\cos(\psi_e)\tilde{y}_{b|p}\right)$$

where $\tilde{x}_{b|p} = \hat{x}_{b|p} - x_{b|p}$, $\tilde{y}_{b|p} = \hat{y}_{b|p} - y_{b|p}$, $\mathrm{sig}^\rho(*) = |*|^\rho \mathrm{sgn}(*)$, and $\mathrm{sgn}(*)$ is the sign function. Define the sideslip angle estimate error as $\tilde{\beta} = \hat{\beta} - \beta$, as mentioned above, there is $\dot{\tilde{\beta}} = \dot{\hat{\beta}}$, then the estimation error dynamics of (19) can be rewritten as:

$$
\begin{aligned}
\dot{\tilde{x}}_{b|p} &= -U\sin(\psi_e)\tilde{\beta} + \left(k_c y_{b|p} - 1\right)u_p - k_{\hat{x}}\mathrm{sig}^\rho\left(\tilde{x}_{b|p}\right) \\
\dot{\tilde{y}}_{b|p} &= U\cos(\psi_e)\tilde{\beta} - k_c u_p \tilde{x}_{b|p} - k_{\hat{y}}\mathrm{sig}^\rho\left(\tilde{y}_{b|p}\right) \\
\dot{\tilde{\beta}} &= k_{\hat{\beta}}\left(U\sin(\psi_e)\tilde{x}_{b|p} - U\cos(\psi_e)\tilde{y}_{b|p}\right)
\end{aligned}
\tag{19}
$$

where the variables have the constraints as $U_{min} \leq U \leq U_{max}$, $|\psi_e| \leq \pi$. U_{min} and U_{max} are the speed limits of the ship. u_p is a designed virtual point of the path, which is also bounded.

Theorem 1. *Consider the sideslip angle observer (18), the unknown term β can be identified very well, and the estimate error $\tilde{\beta}$, $\tilde{x}_{b|p}$, and $\tilde{y}_{b|p}$ in (19) asymptotically converge to zeros within finite time.*

Proof. Consider the positive definite and radially unbounded Lyapunov function candidate

$$
V_1 = \frac{1}{2}\tilde{x}^2_{\,b|p} + \frac{1}{2}\tilde{y}^2_{\,b|p} + \frac{1}{2k_{\hat{\beta}}}\tilde{\beta}^2
\tag{20}
$$

According to error dynamics (19), the time differentiation of V_1 can be obtained as follows:

$$
\dot{V}_1 = -k_{\hat{x}}\left|\tilde{x}_{b|p}\right|^{\rho+1} - k_{\hat{y}}\left|\tilde{y}_{b|p}\right|^{\rho+1}
\tag{21}
$$

The system described in Equation (19) is an autonomous system, so by applying the LaSalle theory, the set $\left\{\tilde{x}_{b|p}, \tilde{y}_{b|p}, \tilde{\beta} \middle| \dot{V}_1 = 0\right\}$ consists of the axis $\tilde{x}_{b|p} = 0, \tilde{y}_{b|p} = 0$, and only the invariant set inside $\tilde{x}_{b|p} = 0, \tilde{y}_{b|p} = 0$ is the origin $\tilde{x}_{b|p} = \tilde{y}_{b|p} = \tilde{\beta} = 0$. Thus, the asymptotic convergence of $\tilde{x}_{b|p}, \tilde{y}_{b|p}, \tilde{\beta}$ to zero is assured, i.e., there is $\left|\tilde{\beta}\right| \leq \epsilon_\beta$, $\epsilon_\beta > 0$. Then, (21) can be rewritten as follows:

$$
\begin{aligned}
\dot{V}_1 &= -k_{\hat{x}}\left|\tilde{x}_{b|p}\right|^{\rho_1+1} - k_{\hat{y}}\left|\tilde{y}_{b|p}\right|^{\rho_1+1} - \left(\frac{1}{k_{\hat{\beta}}}\right)^{\frac{\rho+1}{2}}\left|\tilde{\beta}\right|^{\rho+1} + \left(\frac{1}{k_{\hat{\beta}}}\right)^{\frac{\rho+1}{2}}\left|\tilde{\beta}\right|^{\rho+1} \\
&\leq -k_1\left(\left|\tilde{x}_{b|p}\right|^{\rho+1} + \left|\tilde{y}_{b|p}\right|^{\rho+1} + \left(\frac{1}{k_{\hat{\beta}}}\right)^{\frac{\rho+1}{2}}\left|\tilde{\beta}\right|^{\rho+1}\right) + \gamma(\tilde{\beta})
\end{aligned}
\tag{22}
$$

where $k_1 = \min(k_{\hat{x}}, k_{\hat{y}}, 1)$, $\gamma\left(\tilde{\beta}\right) = \left(\frac{1}{k_{\hat{\beta}}}\right)^{\frac{\rho+1}{2}}\left|\tilde{\beta}\right|^{\rho+1}$.

Using the inequality in Lemma 2, $\dot{V}_1$ can be calculated as:

$$
\begin{aligned}
\dot{V}_1 &\leq -2^{\frac{\rho+1}{2}}k_1\left(\frac{1}{2}\left|\tilde{x}_{b|p}\right|^2 + \frac{1}{2}\left|\tilde{y}_{b|p}\right|^2 + \frac{1}{2k_{\hat{\beta}}}\left|\tilde{\beta}\right|^2\right)^{\frac{\rho+1}{2}} + \gamma\left(\tilde{\beta}\right) \\
&\leq -k_1 V_1^{\frac{\rho+1}{2}} + \gamma\left(\tilde{\beta}\right)
\end{aligned}
\tag{23}
$$

According to Lemma 2, $\gamma\left(\tilde{\beta}\right)$ has the relations as follows:

$$
\begin{aligned}
\gamma\left(\tilde{\beta}\right) &\leq \left(\tfrac{1}{k_{\beta}}\right)^{\frac{\rho+1}{2}}\left|\tilde{\beta}\right|^{\rho+1} + \left|\tilde{x}_{b|p}\right|^{\rho+1} + \left|\tilde{y}_{b|p}\right|^{\rho+1} \\
&\leq 3^{\frac{1-\rho}{2}}\left(\left|\tilde{x}_{b|p}\right|^2 + \left|\tilde{y}_{b|p}\right|^2 + \tfrac{1}{k_{\beta}}\left|\tilde{\beta}\right|^2\right)^{\frac{\rho+1}{2}} \\
&= 2^{\frac{\rho+1}{2}}3^{\frac{1-\rho}{2}}\left(\begin{array}{c}\tfrac{1}{2}\left|\tilde{x}_{b|p}\right|^2 + \tfrac{1}{2}\left|\tilde{y}_{b|p}\right|^2 \\ + \tfrac{1}{2k_{\beta}}\left|\tilde{\beta}\right|^2\end{array}\right)^{\frac{\rho+1}{2}}
\end{aligned}
\tag{24}
$$

Combining (23) and (24), we can obtain

$$
\begin{aligned}
\dot{V}_1 &\leq -2^{\frac{\rho+1}{2}}\left(k_1 - 3^{\frac{1-\rho}{2}}\right)V_1^{\frac{\rho+1}{2}} \\
&= -c_1 V_1^{\alpha_1}
\end{aligned}
\tag{25}
$$

The term $\gamma\left(\tilde{\beta}\right)$ will approach to zero asymptotically. According to Lemma 3, we know the origin of observer error dynamics is locally finite-time stable. The settling time satisfies $T_1 \leq \frac{V_0^{1-\alpha_1}}{c(1-\alpha_1)}$, $V_0 = V\left(\tilde{x}_{b|p}(t_0), \tilde{y}_{b|p}(t_0), \tilde{\beta}(t_0), t_0\right)$.

The proof is concluded. $\square$

3.2. LOS Guidance Law

To stabilize the tracking error $x_{b|p}$, an update law is designed of the curved path parameter as an extra degree of freedom in the controller design. The update law is designed as follows:

$$
u_p = U\cos(\psi_e) - U\sin(\psi_e)\hat{\beta} - k_{x_{b|p}}x_{b|p}
\tag{26}
$$

where $k_{x_{b|p}} > 0$ is a positive constant.

To stabilize the tracking error $y_{b|p}$, the LOS law is chosen as the heading guidance law, and the form is as follows:

$$
\psi_d = \psi_p - \arctan\left(\frac{y_{b|p} + \Delta\hat{\beta}}{\Delta}\right)
\tag{27}
$$

where $\Delta > 0$.

Assumption 3. *The heading autopilot tracks the desired heading angle perfectly such that $\psi = \psi_d$.*

Theorem 2. *The tracking error dynamics (18) can be stabilized by adaptive LOS law (28) and the update law (27). Under Assumption 3, the origin $(x_e, y_e) = (0,0)$ is semi-globally asymptotically stable.*

Proof. Substitute guidance law (26) and (27) into (17), respectively, we can obtain

$$
\begin{aligned}
\dot{x}_{b|p} &= \Phi\left(t, y_{b|p}, U\right)\left(y_{b|p} + \Delta\hat{\beta}\right)\tilde{\beta} + \dot{\psi}_p - k_{x_{b|p}}x_{b|p} \\
\dot{y}_{b|p} &= -\Phi\left(t, y_{b|p}, U\right)y_{b|p} - \Phi\left(t, y_{b|p}, U\right)\Delta\tilde{\beta} - \dot{\psi}_p y_{b|p}
\end{aligned}
\tag{28}
$$

where $\Phi\left(t, y_{b|p}, U\right) = \dfrac{u}{\sqrt{\Delta^2 + \left(y_{b|p} + \Delta\hat{\beta}\right)^2}} > 0$.

Consider the positive definite and radially unbounded Lyapunov function candidate

$$V_2 = \frac{1}{2}x_{b|p}^2 + \frac{1}{2}y_{b|p}^2 \tag{29}$$

The derivative of V_2 is

$$
\begin{aligned}
\dot{V}_2 &= -k_{xe}x_{b|p}^2 - \Phi\left(t, y_{b|p}, U\right)y_{b|p}^2 - \Phi\left(t, y_{b|p}, U\right)\Delta\tilde{\beta}y_{b|p} - \Phi\left(t, y_{b|p}, U\right)\left(y_{b|p} + \Delta\hat{\beta}\right)\tilde{\beta}x_{b|p} \\
&\leq -k_{xe}x_{b|p}^2 - \Phi\left(t, y_{b|p}, U\right)y_{b|p}^2 + \Phi\left(t, y_{b|p}, U\right)\Delta\left|\tilde{\beta}\right|\left|y_{b|p}\right| + \Phi\left(t, y_{b|p}, U\right)\left(\left|y_{b|p}\right| + \Delta\hat{\beta}\right)\left|\tilde{\beta}\right|\left|x_{b|p}\right|
\end{aligned}
\tag{30}
$$

Using the inequality in Lemma 1, the following equation can be obtained:

$$\left|x_{b|p}\right|\left|y_{b|p}\right| \leq \frac{1}{2}\left|x_{b|p}\right|^2 + \frac{1}{2}\left|y_{b|p}\right|^2 \tag{31}$$

Substitute (31) into (30), it can be calculated as:

$$
\begin{aligned}
\dot{V}_2 &\leq -\left(k_{x_{b|p}} - \frac{1}{2}\Phi\left(t, y_{b|p}, U\right)\left|\tilde{\beta}\right|\right)x_{b|p}^2 - \Phi\left(t, y_{b|p}, U\right)\left(1 - \frac{1}{2}\left|\tilde{\beta}\right|\right)y_{b|p}^2 + \gamma\left(x_{b|p}, y_{b|p}, \left|\tilde{\beta}\right|\right) \\
&\leq -k_2 V_2 + \gamma\left(x_{b|p}, y_{b|p}, \left|\tilde{\beta}\right|\right)
\end{aligned}
\tag{32}
$$

where $\gamma\left(x_{b|p}, y_{b|p}, \left|\tilde{\beta}\right|\right) = \Phi\left(t, y_{b|p}, U\right)\Delta\left|\tilde{\beta}\right|\left(\hat{\beta}\left|x_{b|p}\right| + \left|y_{b|p}\right|\right)$, and $\gamma\left(x_{b|p}, y_{b|p}, 0\right) = 0$, k_2 satisfies that $k_2 = \frac{1}{2}\min\left\{k_{x_{b|p}} - \frac{1}{2}\Phi\left(t, y_{b|p}, U\right)\left|\tilde{\beta}\right|\right), \Phi\left(t, y_{b|p}, U\right)\left(1 - \frac{1}{2}\left|\tilde{\beta}\right|\right)y_{b|p}^2\right\}$.

It can be noted that $\left|\tilde{\beta}\right|$ will reach to zero in a finite time, therefore, $\gamma\left(x_{b|p}, y_{b|p}, \left|\tilde{\beta}\right|\right)$ will reach to zero in a finite time. According to the Lyapunov theorem, it is obvious that the equivalent point $\left(x_{b|p}, y_{b|p}\right) = (0, 0)$ is semi-globally exponential stable.

The proof is concluded. □

4. Control System Design

Thanks to the LOS guidance system, the whole path-following control system for the USV can be decoupled to two independent part to design: heading controller design and surge controller design. The heading controller is designed to generate the desired yaw moment to force the USV to track the desired heading angle given in Section 3.2. The surge controller is designed to generate the desired longitudinal thrust to force the USV to follow the desired surge speed, which is set manually. In the following, the heading controller and surge controller will be designed in Sections 4.1 and 4.2 based on the SSMC and FTESO techniques.

4.1. Heading Controller Design

The reference heading signal ψ_d is provided by the guidance system. The next step is to design the heading controller with the SSMC technique. Define $e_\psi = \psi - \psi_d$, then, the task of the SSMC controller includes the selection of sliding-mode surface and control strategy to stabilize the tracking error e_ψ.

Step 1: Select the sliding manifold in system state space according to the relative degree r.

Considering the USV dynamical model (4), let the second-order derivative of the plant output be proportional to the heading control input τ_r, $\ddot{\psi} \sim \tau_r$, so the relative degree is r = 2, then, the sliding manifold can be chosen as

$$\sigma = \dot{e}_\psi + c_1 e_\psi \tag{33}$$

where c_1 is a positive constant.

Step 2: Design a disturbance compensation observer. The derivative of σ can be calculated as:

$$\begin{aligned} \dot{\sigma} &= \zeta_r + \tfrac{\tau_n}{I_z} - \ddot{\psi}_d + c_1\left(r - \dot{\psi}_d\right) \\ &= \zeta_\sigma + \tfrac{\tau_n}{I_z} \end{aligned} \tag{34}$$

where $\zeta_\sigma = \zeta_r - \ddot{\psi}_d + c_1\left(r - \dot{\psi}_d\right)$ is the total disturbance.

Define the estimation error $\widetilde{\sigma} = \hat{\sigma} - \sigma$, $\widetilde{\zeta} = \hat{\zeta} - \zeta$. Then, the disturbance observer can be designed as follows:

$$\begin{aligned} \dot{\hat{\sigma}} &= \hat{\zeta} + \tfrac{\tau_n}{I_z} - k_{o1}\mathrm{sig}^{\rho_{o1}}\left(\widetilde{\sigma}\right) - L_{o1}\mathrm{sgn}\left(\widetilde{\sigma}\right) \\ \dot{\hat{\zeta}} &= -k_{o2}\mathrm{sig}^{\rho_{o2}}\left(\widetilde{\sigma}\right) - L_{o2}\mathrm{sgn}\left(\widetilde{\sigma}\right) \end{aligned} \tag{35}$$

Step 3: Design a heading controller to make the sliding surface hold that $\sigma = 0$. According to Lemma 5, the heading control law can be chosen as follows:

$$\begin{aligned} \tau_n &= I_z\left(-\hat{\zeta}_\sigma - k_{\sigma1}\mathrm{sig}^{\rho_{\sigma1}} + w_1 - L_\sigma\mathrm{sgn}\left(\widetilde{\sigma}\right)\right) \\ \dot{w}_1 &= -k_{\sigma2}|\sigma|^{\rho_{\sigma2}}\mathrm{sgn}(\sigma) \end{aligned} \tag{36}$$

where $0.5 < \rho_{\sigma1} < 1$; $\rho_{\sigma2} = 2\rho_{\sigma1} - 1$; $k_{\sigma1} > 0$ and $k_{\sigma2} > 0$ are constants.

Theorem 3. *For the sliding mode dynamics (34) with the total unknown dynamic ζ_σ, the FTESO is established in (35), which can estimate the unknown disturbance simultaneously, and the estimation error can converge to a bounded domain of zero.*

Proof. By combining with (34) and (35), the estimate error dynamics can be obtained as follows:

$$\begin{aligned} \dot{\widetilde{\sigma}} &= \widetilde{\zeta}_\sigma - k_{o1}\mathrm{sig}^{\rho_{o1}}\left(\widetilde{\sigma}\right) - L_{o1}\mathrm{sgn}\left(\widetilde{\sigma}\right) \\ \dot{\widetilde{\zeta}}_\sigma &= -k_{o2}\mathrm{sig}^{\rho_{o2}}\left(\widetilde{\sigma}\right) - L_{o2}\mathrm{sgn}\left(\widetilde{\sigma}\right) - \dot{\zeta}_\sigma \end{aligned} \tag{37}$$

Note that $\zeta_\sigma = \zeta_r - \ddot{\psi}_d + c_1\left(r - \dot{\psi}_d\right)$ and ψ_d is the expected heading angle, which is third-order derivative bounded, and according to Assumption 2, the $\dot{\zeta}_r$ is also bounded. Then, the total disturbance $\dot{\zeta}_\sigma$ is bounded, so there exists a positive constant l_1 satisfying $\left|\dot{\zeta}_\sigma\right| \le l_1$. First, the terms $-L_{o1}\mathrm{sgn}\left(\widetilde{\sigma}\right)$ and $-L_{o2}\mathrm{sgn}\left(\widetilde{\sigma}\right) - \dot{\zeta}_\sigma$ are removed in Equation (38) under the condition of $L_{o2} > l_1$, where l_1 is a positive constant satisfying $\left|\dot{\zeta}_\sigma\right| \le l_1$. Then it can be obtained:

$$\begin{aligned} \dot{\widetilde{\sigma}} &= \widetilde{\zeta} - k_{o1}\mathrm{sig}^{\rho_{o1}}\left(\widetilde{\sigma}\right) \\ \dot{\widetilde{\zeta}} &= -k_{o2}\mathrm{sig}^{\rho_{o2}}\left(\widetilde{\sigma}\right) \end{aligned} \tag{38}$$

Define f_a as the vector filed of system (38), and f_a is homogeneous of degree $\rho_{o1} - 1$ with respect to the dilation, $\Delta_k\left(\widetilde{\sigma}, \widetilde{\zeta}\right) = \left(k\widetilde{\sigma}, k^{\rho_{o1}}\widetilde{\zeta}\right)$, where $k > 0$. The Lyapunov function candidate is designed as follows:

$$V_3 = \frac{1}{2}Z^T Z \tag{39}$$

where $Z = \left[\mathrm{sig}^{\frac{1}{r}}\left(\widetilde{\sigma}\right), \mathrm{sig}^{\frac{1}{r\rho_{o1}}}\left(\widetilde{\zeta}\right)\right]^T$, $r = \rho_{o1}\rho_{o2}$, and we define $L_{f_{o1}}V_{3a}$ as the Lie derivative of V_{3a} along the f_a. Then, V_{3a} is homogeneous of degree $\frac{2}{r}$, with respect to

$\Delta_k\left(\tilde{\sigma},\tilde{\zeta}\right) = \left(k\tilde{\sigma},k^{\rho_{o1}}\tilde{\zeta}\right)$, and $L_{f_a}V_3 \leq -c_2 V_3^{\alpha_2}$, where $c_2 = -\max_{\{(\tilde{\sigma},\tilde{\zeta}):V_3=1\}} L_{f_a}V_3$,
$\alpha_2 = 1 + \frac{r\rho_{o1}}{2} - \frac{r}{2}$.

Then, take the time of derivative of (39) along (37), $\dot{V}_3$ can be computed as

$$
\dot{V}_3 = L_{f_a}V_3 + Z^T\left[\begin{array}{c} -\frac{1}{2}\mathrm{sig}^{\frac{1}{r}-1}\left(\tilde{\sigma}\right)L_{o1}\mathrm{sgn}\left(\tilde{\sigma}\right) \\ \frac{1}{r\rho_{o1}}\mathrm{sig}^{\frac{1}{r\rho_{o1}}-1}\left(\tilde{\zeta}\right)\left(-\dot{\zeta}-L_{o2}\mathrm{sgn}\left(\tilde{\sigma}\right)\right) \end{array}\right]
$$
$$
\leq -c_2 V_3^{\alpha_2} + \frac{L_{o1}}{r}\left|\tilde{\sigma}\right|^{\frac{2}{r}-1} + \frac{L_{o2}+l_1}{r\rho_{o1}}\left|\tilde{\zeta}\right|^{\frac{2}{r\rho_{o1}}-1}
$$
(40)

According to Lemma 2, the following relationship can be obtained as:

$$
\left|\tilde{\sigma}\right|^{\frac{2}{r}-1} \leq \left|\tilde{\sigma}\right|^{\frac{2}{r}\left(1-\frac{r}{2}\right)} + \left|\tilde{\zeta}\right|^{\frac{2}{r\rho_{o1}}\left(1-\frac{r}{2}\right)} \leq 2^{\frac{r}{2}}2^{1-\frac{r}{2}}V_3^{1-\frac{r}{2}}
$$
$$
\left|\tilde{\zeta}\right|^{\frac{2}{r\rho_{o1}}-1} \leq \left|\tilde{\sigma}\right|^{\frac{2}{r}\left(1-\frac{r\rho_{o1}}{2}\right)} + \left|\tilde{\zeta}\right|^{\frac{2}{r\rho_{o1}}\left(1-\frac{r\rho_{o1}}{2}\right)} \leq 2^{1-\frac{r\rho_{o1}}{2}}V_3^{1-\frac{r\rho_{o1}}{2}}
$$
(41)

Combining (40) and (41), the following inequality is given as:

$$
\dot{V}_3 \leq -c_2 V_3^{\alpha_2} + c_3 V_3^{\alpha_3} + c_4 V_3^{\alpha_4}
$$
(42)

where $c_3 = 2\frac{L_{o1}}{r}$, $c_4 = 2\frac{L_{o2}+l_1}{r\rho_{o1}}$, $\alpha_3 = 1 - \frac{r}{2}$, $\alpha_4 = 1 - \frac{r\rho_{o1}}{2}$, and there is $0 < \alpha_3 < \alpha_4 < \alpha_2 < 1$. The stability of V_3 can be divided into two parts [31,32]:

(1) If $V_3 \geq 1$, $\dot{V}_3 \leq -c_2 V_3^{\alpha_2} + c_5 V_3$, where $c_5 = c_3 + c_4$, according to Lemma 5, and V_3 will converges to 1 within finite time $t_1 \leq \ln\left[1 - \left(\frac{c_5}{c_2}\right)V_3(t_0)\right]/(c_4\alpha_2 - c_4)$.

(2) If $V_3 < 1$, $\dot{V}_3 \leq -c_2 V_3^{\alpha_2} + c_5 V_3^{\alpha_3}$. Select c_0, which satisfies $0 < c_0 < 1 - c_5/c_2$, then, $\dot{V}_3 \leq -c_2 c_0 V_3^{\alpha_2} - \left[c_2(1 - c_0)V_3^{\alpha_2-\alpha_3} - c_5\right]V_3^{\alpha_3}$. If $V_3^{\alpha_2-\alpha_3}$ satisfies $V_3^{\alpha_2-\alpha_3} > \frac{c_5}{c_2(1-c_0)}$, then there is $\dot{V}_3 \leq -c_2 c_0 V_3^{\alpha_2} \leq 0$. According to Lemma 4, V_3 will converge into $V_3^{\alpha_2-\alpha_3} < \frac{c_5}{c_2(1-c_0)}$ within finite time $t_2 \leq \frac{V_3^{1-\alpha_2}(t_1)}{c_2 c_0(1-\alpha_2)}$.

Finally, V_3 will converge into $V_3 \leq \left(\frac{c_5}{c_2(1-c_0)}\right)^{\frac{1}{\alpha_2-\alpha_3}}$ within finite time $T_2 \leq t_1 + t_2$. Then, the convergence domain of observer error can be obtained as follows:

$$
\left\|\left(\tilde{\sigma},\tilde{\zeta}\right)\right\| \leq \sqrt{2}\left(\frac{c_5}{c_2(1-c_0)}\right)^{\frac{1}{2(\alpha_2-\alpha_3)}}
$$
(43)

Finally, the estimation error can converge into a compact set $\Omega = \{\left(\tilde{\sigma},\tilde{\zeta}\right)\left|\left\|\left(\tilde{\sigma},\tilde{\zeta}\right)\right\|\right.$

$\leq \sqrt{2}\left(\frac{c_5}{c_2(1-c_0)}\right)^{\frac{1}{2(\alpha_2-\alpha_3)}}\}$.
The proof is concluded. $\square$

Theorem 4. *The control law (32) can make the sliding manifold σ in (38) approach zero within finite time.*

Proof. Substituting the control law (33) into (31), it can be calculated as:

$$
\dot{\sigma} = -\tilde{\zeta} - k_{\sigma 1}\mathrm{sig}^{\rho_{\sigma 1}}(\sigma) + w_1 - L_\sigma \mathrm{sgn}(\sigma)
$$
$$
\dot{w}_1 = -k_{\sigma 2}\mathrm{sig}^{\rho_{\sigma 2}}(\sigma)
$$
(44)

Consider the Lyapunov function candidate as follows:

$$V_4 = \frac{1}{2}w_1^2 + k_{\sigma 2}\int_0^\sigma sig^{\rho_{\sigma 2}}(z)dz \tag{45}$$

Combining Equation (45), the derivative of V_4 in (46) can be written as follows:

$$\begin{aligned}
\dot{V}_4 &= w_1\dot{w}_1 + k_{\sigma 2}sig^{\rho_{\sigma 2}}(\sigma)\left(\begin{array}{c}-\tilde{\zeta} - k_{\sigma 1}sig^{\rho_{\sigma 1}}(\sigma)\\ +w_1 - L_\sigma sgn(\sigma)\end{array}\right)\\
&\leq -k_{\sigma 1}k_{\sigma 2}|\sigma|^{\rho_{\sigma 1}+\rho_{\sigma 2}} - (L_\sigma - l_2)k_{\sigma 2}|\sigma|^{\rho_{\sigma 2}}\\
&\leq -k_{\sigma 1}k_{\sigma 2}|\sigma|^{\rho_{\sigma 1}+\rho_{\sigma 2}}
\end{aligned} \tag{46}$$

Apply the Lasalle theory. The set $\left\{\sigma, w_1\,\middle|\,\dot{V}_4 = 0\right\}$ consists of the axis $\sigma = 0$, and only the invariant set inside $\sigma = 0$ is the origin $\sigma = w_1 = 0$. Thus, the asymptotic convergence of σ and w_1 to zero is assured.

It can be seen from the above analysis that the term $-\tilde{\zeta} - L_\sigma sgn(\sigma)$ in (44) can be omitted reasonably. The $\dot{\sigma}$ and $\dot{w}_1$ can be obtained as:

$$\begin{aligned}
\dot{\sigma} &= -k_{\sigma 1}|\sigma|^{\rho_{\sigma 1}}sgn(\sigma) + w_1\\
\dot{w}_1 &= -k_{\sigma 2}|\sigma|^{\rho_{\sigma 2}}sgn(\sigma)
\end{aligned} \tag{47}$$

According to the analysis process of the stability of the system in (38), the dynamics described by (47) are finite-time stable.

The proof is concluded. $\square$

4.2. Surge Controller Design

In this subsection, the surge controller is designed for the USV to track the desired surge speed, which is set manually in advance.

To estimate the unknown term ζ_u, FTESO is introduced as follows:

$$\begin{aligned}
\dot{\hat{u}} &= \hat{\zeta}_u + \frac{\tau_u}{m_{11}} - k_{o3}sig^{\rho_{o3}}\left(\tilde{u}\right) - L_{o3}sgn\left(\tilde{u}\right)\\
\dot{\hat{\zeta}}_u &= -k_{o4}sig^{\rho_{o4}}\left(\tilde{u}\right) - L_{o4}sgn\left(\tilde{u}\right)
\end{aligned} \tag{48}$$

where k_{o3}, k_{o4} are positive constants.

Define estimation errors as $\tilde{u} = \hat{u} - u$, $\tilde{\zeta}_u = \hat{\zeta}_u - \zeta_u$. The derivatives of $\tilde{u}$ and $\tilde{\zeta}_u$ can be obtained as:

$$\begin{aligned}
\dot{\tilde{u}} &= \tilde{\zeta}_u - k_{o3}sig^{\rho_{o3}}\left(\tilde{u}\right) - L_{o3}sgn\left(\tilde{u}\right)\\
\dot{\tilde{\zeta}}_u &= -k_{o4}sig^{\rho_{o4}}\left(\tilde{u}\right) - L_{o4}sgn\left(\tilde{u}\right)
\end{aligned} \tag{49}$$

Define speed tracking error as $e_u = u - u_d$, where u_d is the desired surge speed, which is given manually in advance, then the surge control law can be given by:

$$\begin{aligned}
\tau_u &= m_{11}\left(-\hat{\zeta}_u - k_{u1}|e_u|^{\rho_{u1}}sgn(e_u) + w_2 - L_u sgn(e_u)\right)\\
\dot{w}_2 &= -k_{u2}|e_u|^{\rho_{u2}}sgn(e_u)
\end{aligned} \tag{50}$$

Substituting (50) into (4), $\dot{e}_u$ can be obtained as:

$$\begin{aligned}
\dot{e}_u &= -k_{u1}|e_u|^{\rho_{u1}}sgn(e_u) + w_2 - L_u sgn(e_u)\\
\dot{w}_2 &= -k_{u2}|e_u|^{\rho_{u2}}sgn(e_u)
\end{aligned} \tag{51}$$

Theorem 5 *For the surge motion dynamics (4) with the total unknown dynamic ζ_u, the FTESO established in (48) can be employed to obtain the value of the unknown disturbance simultaneously, and the estimation error in (50) is able to converge to a bounded domain of zero.*

Theorem 6. *The control law (50) can make the tracking error e_u converge to zero within finite time.*

Remark 3. *The proof processes of Theorems 5 and 6 are omitted because they are similar to the proof of Theorems 3 and 4.*

5. Stability Analysis

The stability of the guidance system and control system are given in Sections 3 and 4. In the following, the stability of the closed-loop system of the path-following control for the USV based on the proposed control method are explained.

Theorem 7. *The path-following errors $x_{b|p}$ and $y_{b|p}$ are uniformly ultimately bounded.*

Proof. According to where the variables have the constraints as $U_{min} \leq U \leq U_{max}$, $|\psi_e| \leq \pi$, U_{min} and U_{max} are the speed limits of the ship. u_p is a designed virtual point of the path, which is also bounded.

Theorems 1 and 3–6, the errors $\tilde{\beta}$, e_u, and sliding surface σ can converge to zero within finite time. Then, under the conditions of $\tilde{\beta} = 0$, $e_u = 0$, and $\sigma = 0$, the Lyapunov function candidate can be selected as $V = \frac{1}{2}x_{b|p}^2 + \frac{1}{2}y_{b|p}^2 + \frac{1}{2}e_\psi^2$, where $e_\psi = \psi - \psi_d$. Taking the time of derivative of V along (17), with the guidance law (26) and (27), it can be obtained:

$$
\begin{aligned}
\dot{V} &= -c_1 e_\psi^2 - k_{x_{b|p}} x_{b|p}^2 + U y_{b|p} \sin(e_\psi + \psi_d - \psi_p) + U y_{b|p} \cos(e_\psi + \psi_d - \psi_p)\beta \\
&= -c_1 e_\psi^2 - k_{x_{b|p}} x_{b|p}^2 + \Phi\left(t, y_{b|p}, U\right) \gamma\left(e_\psi, y_{b|p}\right) - \cos(e_\psi) \Phi\left(t, y_{b|p}, U\right) y_{b|p}^2 \\
&\leq -k_3 V + \Phi\left(t, y_{b|p}, U\right) \gamma\left(e_\psi, y_{b|p}\right)
\end{aligned} \tag{52}
$$

where $k_3 = 2\min\left(c_1, k_{x_{b|p}}, \cos(e_\psi) \Phi\left(t, y_{b|p}, U\right)\right)$, which satisfies $\gamma(0,0) = 0$. With a proper k_3, $\dot{V} \leq 0$. According to the Lyapunov stability theorem, the whole system is uniformly ultimately bounded.

The proof is concluded. □

6. Numerical Simulations

For the simulation, the mathematical model of the Otter USV is applied [40], with length $L = 2.0$ m, width 1.07 m, and weight 65 kg. The model is given in [41], and the parameters are $m_{11} = 70.5$ kg, $m_{22} = 147.5$ kg, $m_{23} = m_{32} = 11$ kg, $m_{33} = 43.2394$ kg, $l_1 = 0.395$ m, $k_{pos} = 0.0111$ kgm, $k_{neg} = 0.0064$ kgm. The desired path is a sinusoidal path described as $x_p(\theta) = 5\cos(0.8\theta)$, $y_p(\theta) = 5\theta$. The surge speed of the Otter USV is set as $u_d = 1.2$ m/s, and the constant ocean current speed and direction are set as $v_c = 0.2$ m/s and $\beta_c = 30°$, respectively. In addition, the time-varying exterior disturbances are selected as:

$$
\tau_w = \begin{bmatrix} 10 + 5\sin(0.8t)\cos(0.2t) \\ 1.5\sin(0.8t)\cos(0.2t) \\ 5 + 15\sin(0.8t)\cos(0.2t) \end{bmatrix} \tag{53}
$$

In order to verify the performance of SSMC for heading control, two other controllers are applied to the USV. The first one is the PID controller with reference feedback proposed as [20]:

$$
\tau_n = \tau_{ff} - K_p\left(e_\psi + \frac{1}{T_i}\int_0^t e_\psi d\tau + T_d \dot{e}_\psi\right) \tag{54}
$$

The reference feedback signal is defined as:

$$\tau_{ff} = \frac{T}{K}\dot{\psi}_d + \frac{1}{T}\psi_d \tag{55}$$

where K_p, T_i, T_d are proportional gain, integral time constant, and derivative constant, respectively. Nomoto gain and time constants K and T can be chosen according to the multiple maneuvering experiment.

The second one is the conventional SMC controller, which is defined as:

$$\tau_n = -(\beta_1 + d)\mathrm{sat}(a, \sigma_1) \tag{56}$$

where β_1 and d are designed positive constants; sliding mode is σ_1, which satisfies $\sigma_1 = \dot{e}_\psi + c_0 e_\psi$; $\mathrm{sat}(a, *)$ is the saturation function.

The control system and guidance system parameters are designed as Tables 1–3. It should be noted that the same surge controller is used to track the desired surge speed, and the same guidance system is chosen to produce expected heading angle, where the only difference is the heading controller. The control performances of PID, conventional SMC, and SSMC proposed by this paper are compared for USV curved path following.

Table 1. Heading controller parameters.

Control Strategies	Parameters
SSMC	$k_{o1} = 10, k_{o2} = 25, \rho_{o1} = 0.9, \rho_{o2} = 0.8,$ $L_{o1} = 0.01, L_{o2} = 0.1,$ $k_{\sigma1} = 2, k_{\sigma2} = 0.05, \rho_{\sigma1} = 0.8,$ $\rho_{\sigma2} = 0.6, L_\sigma = 0.001$
PID	$K_p = 93.15, T_i = 0.89, T_d = 6.67,$ $K = 0.0242, T = 1$
SMC	$\beta_1 = 1.2, d = 0.3, a = 0.05$

Table 2. Surge controller parameters.

Parameters
$k_{o1} = 10, k_{o2} = 25, \rho_{o1} = 0.9, \rho_{o2} = 0.8, L_{o1} = 0.01, L_{o2} = 0.1,$ $k_{\sigma1} = 2, k_{\sigma2} = 0.05, \rho_{\sigma1} = 0.8, \rho_{\sigma2} = 0.6, L_\sigma = 0.001,$ $K_p = 93.15, T_i = 0.89, T_d = 6.67, K = 0.0242, T = 1,$ $\beta_1 = 1.2, d = 0.3, a = 0.05$

Table 3. Guidance system parameters.

Parameters
$\Delta = 3$ m, $k_{x_{b|p}} = 2, k_{\hat{x}} = 5, k_{\hat{y}} = 5. \rho = 0.8, k_\beta = 10$

Figures 4–9. The desired path and the actual trajectories are shown in Figure 4. It is apparent that SSMC and SMC heading controllers have better performance than the PID controller under severe external disturbances. Around 20 s, the SSMC controller has higher tracking accuracy than the SMC controller. The sliding-mode dynamics of SMC and SSMC are shown in Figure 5. Both systems can reach the region of 0 bounded by ± 0.02 within 4 s. However, the convergence of SSMC is smoother than SMC. The path-following errors are plotted in Figure 6, and SSMC and SMC controllers have almost the same tracking performance. The heading tracking errors of SSMC and SMC converge to a small neighborhood of 0 from $20°$ within 8 s. The tracking errors of the PID controller are relatively large and choppy because of the influence of ocean current. It proves that SSMC and SMC controllers are more robust than the PID controller. The velocities of USVs are

shown in Figure 7. As shown in the third figure in Figure 7, the yaw velocities with SMC exhibit more severe oscillations than SSMC.

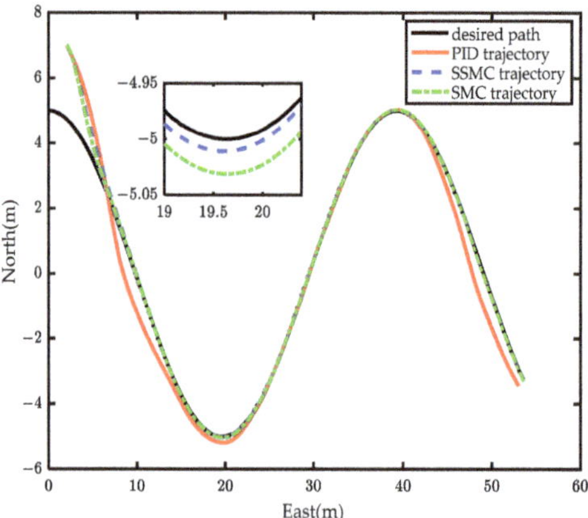

Figure 4. Desired path and actual trajectories of USVs with different control methods.

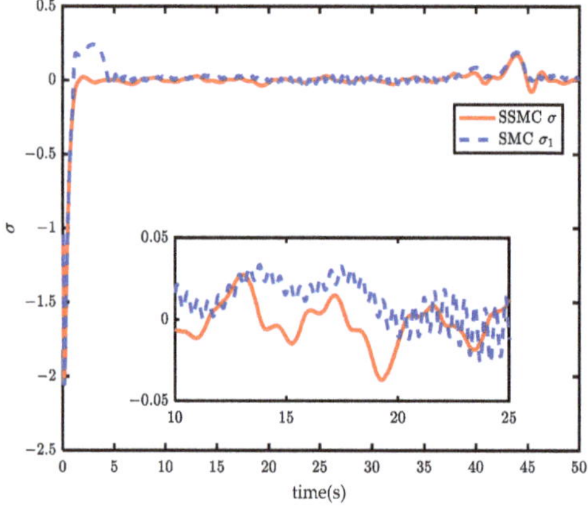

Figure 5. Sliding-mode dynamics σ and σ_1 of SSMC and SMC.

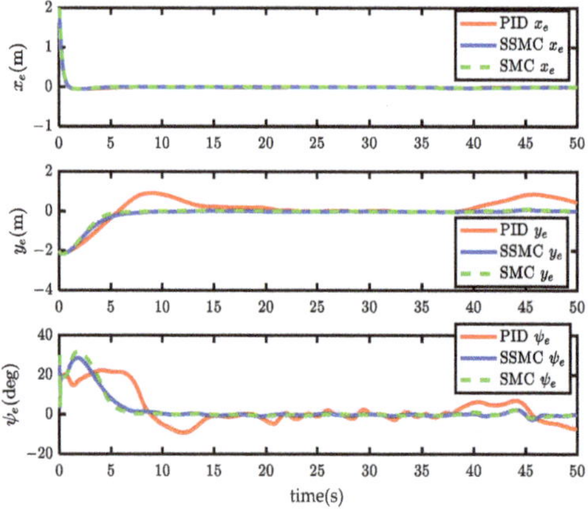

Figure 6. Path-following errors x_e, y_e, and ψ_e.

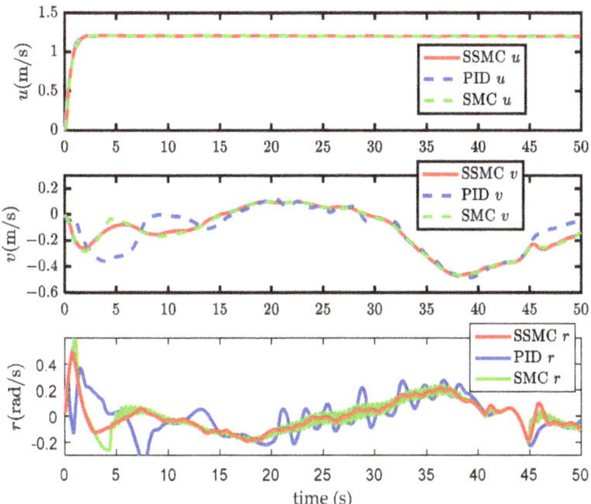

Figure 7. Velocities of USVs with different control methods.

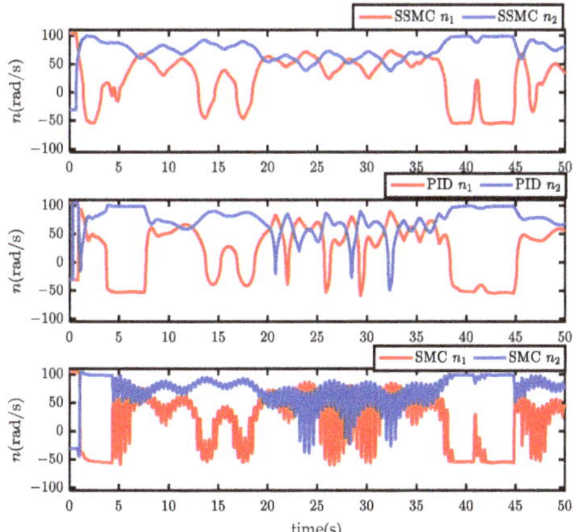

Figure 8. Propeller shaft speeds n_1 and n_2 with different control methods.

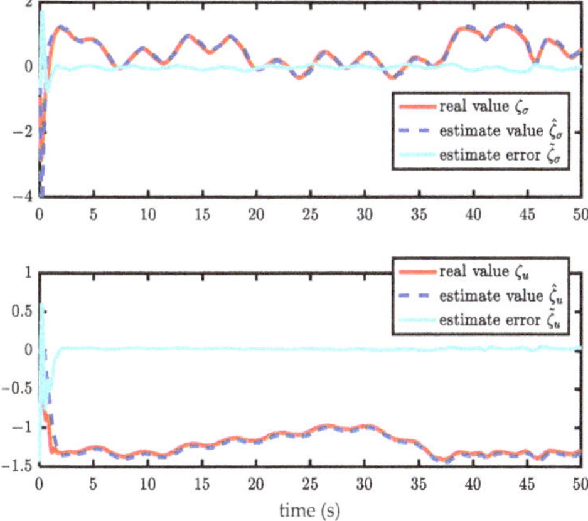

Figure 9. Effect of disturbances estimation ζ_σ and ζ_u with FTESO.

The propeller shaft speeds of the three controllers are shown in Figure 8. It is obvious that the very high frequency oscillation phenomenon occurs in the SMC controller. The changes of propeller shaft speeds of SSMC and PID controllers are comparatively gentle. It is shown that the high frequency auto-oscillation can be effectively avoided by the SSMC controller. The effect of disturbances estimation is shown in Figure 9. FTESO-based errors of $\tilde{\zeta}_\sigma$ reach the region of 0 bounded by ± 0.1 within 2.5 s, and the errors of $\tilde{\zeta}_u$ reach the region of 0 bounded by ± 0.01 within 2.1 s. It is illustrated that the estimation errors of FTESO are able to reach to a small bounded domain of zero within a finite time.

In summary, with the comparison analysis, the SSMC based on FTESO is more robust than the PID controller and smoother than the conventional SMC, thus the SSMC has the best path-following performance compared to the PID and SMC.

7. Conclusions

This paper develops a novel SSMC method based on FTESO for path following of the USV with unknown dynamics and exterior disturbances. The guidance and control scheme have simple and clear structures, and both the smoothness and robustness of the system are improved to visible levels, with the least possible modeling information and environmental disturbances. For the guidance system, the finite-time sideslip angle observer is incorporated into the LOS guidance law, which can make sideslip estimation error convergent to zero in finite time and speed up the convergence process. For the control system, FTESO is designed to estimate and compensate for the unknown disturbances on the sliding-mode surface instead of using a strong discontinuous control signal, which will cause a strong chattering phenomenon. At the same time, all errors of the dynamic system are proved to be bounded. The simulation results prove that the designed strategy has a better effectiveness than PID and conventional SMC, and the USV has satisfactory performance for path following under the unknown disturbances. It should be noted that the obstacle avoidance is not considered in the proposed control method. In the future, the artificial potential fields will be incorporated in the guidance system to realize collision avoidance.

Author Contributions: Conceptualization, Y.W.; methodology, Y.Q.; software, Y.Q. and R.C.; validation, Y.Q., R.C. and H.F.; writing—original draft preparation, Y.Q. and S.Z.; writing—review and editing, Y.W., Y.Q., S.Z., R.C. and H.F.; supervision, Y.W. and S.Z.; funding acquisition, Y.W. and H.F. All authors have read and agreed to the published version of the manuscript.

Funding: This research was funded by the National Natural Science Foundation of China, grant number 52271313, and the Innovative Research Foundation of Ship General Performance, grant number 21822216.

Institutional Review Board Statement: Not applicable.

Informed Consent Statement: Not applicable.

Data Availability Statement: Data are contained within the article.

Conflicts of Interest: The authors declare no conflict of interest.

References

1. Sonnenburg, C.R.; Woolsey, C.A. Modeling, Identification, and Control of an Unmanned Surface Vehicle. *J. Field Robot* **2013**, *30*, 371–398. [CrossRef]
2. Wang, N.; Sun, Z.; Yin, J.C. Fuzzy unknown observer-based robust adaptive path following control of underactuated surface vehicles subject to multiple unknowns. *Ocean Eng.* **2019**, *176*, 57–64. [CrossRef]
3. Bibuli, M.; Bruzzone, G.; Caccia, M. Path-Following Algorithms and Experiments for an Unmanned Surface Vehicle. *J. Field Robot* **2009**, *26*, 669–688. [CrossRef]
4. Fossen, T.I.; Breivik, M.; Skjetne, R. Line-of-sight path following of underactuated marine craft. *IFAC Proc. Vol.* **2003**, *36*, 211–216. [CrossRef]
5. Li, C.; Jiang, J.; Duan, F.; Liu, W. Modeling and Experimental Testing of an Unmanned Surface Vehicle with Rudderless Double Thrusters. *Sensors* **2019**, *19*, 2051. [CrossRef] [PubMed]

6. Hac, A.; Simpson, M.D. Estimation of vehicle side slip angle and yaw rate. In Proceedings of the SAE 2000 World Congress, Detroit, MI, USA, 6–9 March 2000.
7. Borhaug, E.; Pavlov, A.; Pettersen, K.Y. Integral LOS control for path following of underactuated marine surface vessels in the presence of constant ocean currents. In Proceedings of the 47th IEEE Conference on Decision and Control, Cancun, Mexico, 9–11 December 2008.
8. Fossen, T.I.; Pettersen, K.Y.; Galeazzi, R. Line-of-Sight Path Following for Dubins Paths with Adaptive Sideslip Compensation of Drift Forces. *IEEE Trans. Control Syst. Technol.* **2015**, *23*, 820–827. [CrossRef]
9. Caharija, W. Integral Line-of-Sight Guidance and Control of Underactuated Marine Vehicles: Theory, Simulations, and Experiments. *IEEE Trans. Control Syst. Technol.* **2016**, *24*, 1623–1642. [CrossRef]
10. Liu, T.; Dong, Z.P.; Du, H.W.; Song, L.F.; Mao, Y.S. Path Following Control of the Underactuated USV Based on the Improved Line-of-Sight Guidance Algorithm. *Pol. Marit. Res.* **2017**, *24*, 3–11. [CrossRef]
11. Mu, D.; Wang, G.; Fan, Y.; Sun, X.; Qiu, B. Adaptive LOS Path Following for a Podded Propulsion Unmanned Surface Vehicle with Uncertainty of Model and Actuator Saturation. *Appl. Sci.* **2017**, *7*, 1232. [CrossRef]
12. Su, Y.X.; Wan, L.L.; Zhang, D.H.; Huang, F.R. An improved adaptive integral line-of-sight guidance law for unmanned surface vehicles with uncertainties. *Appl. Ocean Res.* **2021**, *108*, 102488. [CrossRef]
13. Liu, L.; Wang, D.; Peng, Z.H.; Wang, H. Predictor-based LOS guidance law for path following of underactuated marine surface vehicles with sideslip compensation. *Ocean Eng.* **2016**, *124*, 340–348. [CrossRef]
14. Xia, G.Q.; Wang, X.W.; Zhao, B. LOS Guidance Law for Path Following of USV based on Sideslip Observer. In Proceedings of the 2019 Chinese Automation Congress, Hangzhou, China, 22–24 November 2019.
15. Wang, N.; Sun, Z.; Yin, J.; Su, S.F.; Sharma, S. Finite-Time Observer Based Guidance and Control of Underactuated Surface Vehicles with Unknown Sideslip Angles and Disturbances. *IEEE Access* **2018**, *6*, 14059–14070. [CrossRef]
16. Yu, Y.; Guo, C.; Yu, H. Finite-Time PLOS-Based Integral Sliding-Mode Adaptive Neural Path Following for Unmanned Surface Vessels with Unknown Dynamics and Disturbances. *IEEE Trans. Autom. Sci. Eng.* **2019**, *4*, 1500–1511. [CrossRef]
17. Qu, X.R.; Liang, X.; Hou, Y.H.; Li, Y.; Zhang, R.B. Finite-time sideslip observer-based synchronized path-following control of multiple unmanned underwater vehicles. *Ocean Eng.* **2020**, *217*, 107941. [CrossRef]
18. Yu, Y.; Guo, C.; Li, T. Finite-Time LOS Path Following of Unmanned Surface Vessels with Time-Varying Sideslip Angles and Input Saturation. *IEEE-ASME Trans. Mech.* **2022**, *27*, 463–474.
19. Wang, S.D.; Sun, M.W.; Liu, J. Fixed-Time Predictor-Based Path Following Control of Unmanned. *Surf. Veh. Control. Theory Appl.* **2022**, *39*, 1845–1853.
20. Fossen, S.; Fossen, T.I. Five-State Extended Kalman Filter for Estimation of Speed over Ground (SOG), Course over Ground (COG) and Course Rate of Unmanned Surface Vehicles (USVs): Experimental Results. *Sensors* **2021**, *21*, 7910. [CrossRef]
21. Fossen, T.I. Line-of-sight path-following control utilizing an extended Kalman filter for estimation of speed and course over ground from GNSS positions. *J. Mar. Sci. Technol.* **2022**, *27*, 806–813. [CrossRef]
22. Xia, G.; Chu, H.; Shao, Y.; Xia, B. DSC and LADRC Path Following Control for Dynamic Positioning Ships at High Speed. In Proceedings of the 2019 IEEE International Conference on Mechatronics and Automation (ICMA), Tianjin, China, 4–7 August 2019.
23. Chao, B.; Chang, X.; Liang, C.; Han, X. An Improved Model Predictive Control for Path-Following of USV Based on Global Course Constraint and Event-Triggered Mechanism. *IEEE Access* **2021**, *9*, 79725–79734.
24. Liu, Z.; Geng, C.; Zhang, J. Model predictive controller design with disturbance observer for path following of unmanned surface vessel. In Proceedings of the IEEE International Conference on Mechatronics and Automation, Takamatsu, Japan, 6–9 August 2017.
25. Woo, J.; Yu, C.; Kim, N. Deep reinforcement learning-based controller for path following of an unmanned surface vehicle. *Ocean Eng.* **2019**, *183*, 155–166. [CrossRef]
26. Gonzalez-Garcia, A.; Castañeda, H.; Garrido, L. USV Path-Following Control Based on Deep Reinforcement Learning and Adaptive Control. In Proceedings of the Global Oceans 2020: Singapore—U.S. Gulf Coast, Biloxi, MS, USA, 5–30 October 2020.
27. Gonzalez-Garcia, A.; Castañeda, H. Guidance and Control Based on Adaptive Sliding Mode Strategy for a USV Subject to Uncertainties. *IEEE J. Ocean. Eng.* **2021**, *46*, 1144–1154. [CrossRef]
28. Yu, Y.; Guo, C.; Shen, H.; Zhang, C. Sliding-mode Dynamic Surface Adaptive Path-following of Unmanned Vessels with Dynamic Uncertainties and Disturbances. In Proceedings of the 2018 13th World Congress on Intelligent Control and Automation (WCICA), Changsha, China, 4–8 July 2018.
29. Chen, Z.; Zhang, Y.; Nie, Y.; Tang, J.; Zhu, S. Disturbance-Observer-Based Sliding Mode Control Design for Nonlinear Unmanned Surface Vessel with Uncertainties. *IEEE Access* **2019**, *7*, 148522–148530. [CrossRef]
30. Šabanovic, A. Variable Structure Systems with Sliding Modes in Motion Control—A Survey. *IEEE Trans. Ind. Inform.* **2011**, *7*, 212–223. [CrossRef]
31. Nie, J.; Wang, H.X.; Lu, X.; Lin, X.G.; Shen, C.Y.; Zhang, Z.G.; Song, S.B. Finite-time output feedback path following control of underactuated MSV based on FTESO. *Ocean Eng.* **2021**, *224*, 108660. [CrossRef]
32. Chu, R.T.; Liu, Z.Q. Ship course sliding mode control system based on FTESO and sideslip angle compensation. *Chin. J. Ship Res.* **2022**, *17*, 71–79.

33. Fossen, T.I. *Handbook of Marine Craft Hydrodynamics and Motion Control*, 2nd ed.; John Wiley & Sons: Hoboken, NJ, USA, 2011; pp. 133–183.
34. Belleter, D.; Maghenem, M.A.; Paliotta, C.; Pettersen, K.Y. Observer based path following for underactuated marine vessels in the presence of ocean currents: A global approach. *Automatica* **2019**, *100*, 12. [CrossRef]
35. Skjetne, R.; Fossen, T.I.; Kokotović, P. Output Maneuvering for a Class of Nonlinear Systems. *IFAC Proc. Vol.* **2002**, *35*, 501–506. [CrossRef]
36. Hardy, G.H.; Littlewood, J.E.; Pólya, G. *Inequalities*; Cambridge University Press: Cambridge, UK, 1988.
37. Bhaty, S.P.; Bernstein, D.S. Finite-time stability of continuous autonomous systems. *SIAM J. Control Optim.* **2000**, *38*, 751–766. [CrossRef]
38. Bhat, S.P.; Bernstein, D.S. Finite-Time Stability of Homogeneous Systems. In Proceedings of the American Control Conference, Albuquerque, NM, USA, 6 June 1997.
39. Shen, Y.J.; Xia, X.H. Semi-global finite-time observers for nonlinear systems. *Automatica* **2008**, *44*, 3152–3156. [CrossRef]
40. Maritime Robotics AS. 2021. Available online: https://www.maritimerobotics.com (accessed on 25 October 2021).
41. Fossen, T.I.; Perez, T. Marine Systems Simulator (MSS). 2014. Available online: http://github.com/cybergalactic/MSS (accessed on 25 October 2021).

6.	Hac, A.; Simpson, M.D. Estimation of vehicle side slip angle and yaw rate. In Proceedings of the SAE 2000 World Congress, Detroit, MI, USA, 6–9 March 2000.

7.	Borhaug, E.; Pavlov, A.; Pettersen, K.Y. Integral LOS control for path following of underactuated marine surface vessels in the presence of constant ocean currents. In Proceedings of the 47th IEEE Conference on Decision and Control, Cancun, Mexico, 9–11 December 2008.

8.	Fossen, T.I.; Pettersen, K.Y.; Galeazzi, R. Line-of-Sight Path Following for Dubins Paths with Adaptive Sideslip Compensation of Drift Forces. *IEEE Trans. Control Syst. Technol.* **2015**, *23*, 820–827. [CrossRef]

9.	Caharija, W. Integral Line-of-Sight Guidance and Control of Underactuated Marine Vehicles: Theory, Simulations, and Experiments. *IEEE Trans. Control Syst. Technol.* **2016**, *24*, 1623–1642. [CrossRef]

10.	Liu, T.; Dong, Z.P.; Du, H.W.; Song, L.F.; Mao, Y.S. Path Following Control of the Underactuated USV Based on the Improved Line-of-Sight Guidance Algorithm. *Pol. Marit. Res.* **2017**, *24*, 3–11. [CrossRef]

11.	Mu, D.; Wang, G.; Fan, Y.; Sun, X.; Qiu, B. Adaptive LOS Path Following for a Podded Propulsion Unmanned Surface Vehicle with Uncertainty of Model and Actuator Saturation. *Appl. Sci.* **2017**, *7*, 1232. [CrossRef]

12.	Su, Y.X.; Wan, L.L.; Zhang, D.H.; Huang, F.R. An improved adaptive integral line-of-sight guidance law for unmanned surface vehicles with uncertainties. *Appl. Ocean Res.* **2021**, *108*, 102488. [CrossRef]

13.	Liu, L.; Wang, D.; Peng, Z.H.; Wang, H. Predictor-based LOS guidance law for path following of underactuated marine surface vehicles with sideslip compensation. *Ocean Eng.* **2016**, *124*, 340–348. [CrossRef]

14.	Xia, G.Q.; Wang, X.W.; Zhao, B. LOS Guidance Law for Path Following of USV based on Sideslip Observer. In Proceedings of the 2019 Chinese Automation Congress, Hangzhou, China, 22–24 November 2019.

15.	Wang, N.; Sun, Z.; Yin, J.; Su, S.F.; Sharma, S. Finite-Time Observer Based Guidance and Control of Underactuated Surface Vehicles with Unknown Sideslip Angles and Disturbances. *IEEE Access* **2018**, *6*, 14059–14070. [CrossRef]

16.	Yu, Y.; Guo, C.; Yu, H. Finite-Time PLOS-Based Integral Sliding-Mode Adaptive Neural Path Following for Unmanned Surface Vessels with Unknown Dynamics and Disturbances. *IEEE Trans. Autom. Sci. Eng.* **2019**, *4*, 1500–1511. [CrossRef]

17.	Qu, X.R.; Liang, X.; Hou, Y.H.; Li, Y.; Zhang, R.B. Finite-time sideslip observer-based synchronized path-following control of multiple unmanned underwater vehicles. *Ocean Eng.* **2020**, *217*, 107941. [CrossRef]

18.	Yu, Y.; Guo, C.; Li, T. Finite-Time LOS Path Following of Unmanned Surface Vessels with Time-Varying Sideslip Angles and Input Saturation. *IEEE-ASME Trans. Mech.* **2022**, *27*, 463–474.

19.	Wang, S.D.; Sun, M.W.; Liu, J. Fixed-Time Predictor-Based Path Following Control of Unmanned. *Surf. Veh. Control. Theory Appl.* **2022**, *39*, 1845–1853.

20.	Fossen, S.; Fossen, T.I. Five-State Extended Kalman Filter for Estimation of Speed over Ground (SOG), Course over Ground (COG) and Course Rate of Unmanned Surface Vehicles (USVs): Experimental Results. *Sensors* **2021**, *21*, 7910. [CrossRef]

21.	Fossen, T.I. Line-of-sight path-following control utilizing an extended Kalman filter for estimation of speed and course over ground from GNSS positions. *J. Mar. Sci. Technol.* **2022**, *27*, 806–813. [CrossRef]

22.	Xia, G.; Chu, H.; Shao, Y.; Xia, B. DSC and LADRC Path Following Control for Dynamic Positioning Ships at High Speed. In Proceedings of the 2019 IEEE International Conference on Mechatronics and Automation (ICMA), Tianjin, China, 4–7 August 2019.

23.	Chao, B.; Chang, X.; Liang, C.; Han, X. An Improved Model Predictive Control for Path-Following of USV Based on Global Course Constraint and Event-Triggered Mechanism. *IEEE Access* **2021**, *9*, 79725–79734.

24.	Liu, Z.; Geng, C.; Zhang, J. Model predictive controller design with disturbance observer for path following of unmanned surface vessel. In Proceedings of the IEEE International Conference on Mechatronics and Automation, Takamatsu, Japan, 6–9 August 2017.

25.	Woo, J.; Yu, C.; Kim, N. Deep reinforcement learning-based controller for path following of an unmanned surface vehicle. *Ocean Eng.* **2019**, *183*, 155–166. [CrossRef]

26.	Gonzalez-Garcia, A.; Castañeda, H.; Garrido, L. USV Path-Following Control Based on Deep Reinforcement Learning and Adaptive Control. In Proceedings of the Global Oceans 2020: Singapore—U.S. Gulf Coast, Biloxi, MS, USA, 5–30 October 2020.

27.	Gonzalez-Garcia, A.; Castañeda, H. Guidance and Control Based on Adaptive Sliding Mode Strategy for a USV Subject to Uncertainties. *IEEE J. Ocean. Eng.* **2021**, *46*, 1144–1154. [CrossRef]

28.	Yu, Y.; Guo, C.; Shen, H.; Zhang, C. Sliding-mode Dynamic Surface Adaptive Path-following of Unmanned Vessels with Dynamic Uncertainties and Disturbances. In Proceedings of the 2018 13th World Congress on Intelligent Control and Automation (WCICA), Changsha, China, 4–8 July 2018.

29.	Chen, Z.; Zhang, Y.; Nie, Y.; Tang, J.; Zhu, S. Disturbance-Observer-Based Sliding Mode Control Design for Nonlinear Unmanned Surface Vessel with Uncertainties. *IEEE Access* **2019**, *7*, 148522–148530. [CrossRef]

30.	Šabanovic, A. Variable Structure Systems with Sliding Modes in Motion Control—A Survey. *IEEE Trans. Ind. Inform.* **2011**, *7*, 212–223. [CrossRef]

31.	Nie, J.; Wang, H.X.; Lu, X.; Lin, X.G.; Shen, C.Y.; Zhang, Z.G.; Song, S.B. Finite-time output feedback path following control of underactuated MSV based on FTESO. *Ocean Eng.* **2021**, *224*, 108660. [CrossRef]

32.	Chu, R.T.; Liu, Z.Q. Ship course sliding mode control system based on FTESO and sideslip angle compensation. *Chin. J. Ship Res.* **2022**, *17*, 71–79.

33. Fossen, T.I. *Handbook of Marine Craft Hydrodynamics and Motion Control*, 2nd ed.; John Wiley & Sons: Hoboken, NJ, USA, 2011; pp. 133–183.
34. Belleter, D.; Maghenem, M.A.; Paliotta, C.; Pettersen, K.Y. Observer based path following for underactuated marine vessels in the presence of ocean currents: A global approach. *Automatica* **2019**, *100*, 12. [CrossRef]
35. Skjetne, R.; Fossen, T.I.; Kokotović, P. Output Maneuvering for a Class of Nonlinear Systems. *IFAC Proc. Vol.* **2002**, *35*, 501–506. [CrossRef]
36. Hardy, G.H.; Littlewood, J.E.; Pólya, G. *Inequalities*; Cambridge University Press: Cambridge, UK, 1988.
37. Bhaty, S.P.; Bernstein, D.S. Finite-time stability of continuous autonomous systems. *SIAM J. Control Optim.* **2000**, *38*, 751–766. [CrossRef]
38. Bhat, S.P.; Bernstein, D.S. Finite-Time Stability of Homogeneous Systems. In Proceedings of the American Control Conference, Albuquerque, NM, USA, 6 June 1997.
39. Shen, Y.J.; Xia, X.H. Semi-global finite-time observers for nonlinear systems. *Automatica* **2008**, *44*, 3152–3156. [CrossRef]
40. Maritime Robotics AS. 2021. Available online: https://www.maritimerobotics.com (accessed on 25 October 2021).
41. Fossen, T.I.; Perez, T. Marine Systems Simulator (MSS). 2014. Available online: http://github.com/cybergalactic/MSS (accessed on 25 October 2021).

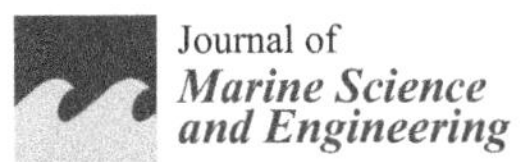

Journal of
Marine Science and Engineering

Article

Distributed Dual Closed-Loop Model Predictive Formation Control for Collision-Free Multi-AUV System Subject to Compound Disturbances

Mingyao Zhang *, Zheping Yan, Jiajia Zhou and Lidong Yue

College of Intelligent Systems Science and Engineering, Harbin Engineering University, Harbin 150001, China; yanzheping@hrbeu.edu.cn (Z.Y.); zhoujiajia@hrbeu.edu.cn (J.Z.); yuelidong@hrbeu.edu.cn (L.Y.)
* Correspondence: zhangmingyao@hrbeu.edu.cn

Abstract: This paper focuses on the collision-free formation tracking of autonomous underwater vehicles (AUVs) with compound disturbances in complex ocean environments. We propose a novel finite-time extended state observer (FTESO)-based distributed dual closed-loop model predictive control scheme. Initially, a fast FTESO is designed to accurately estimate both model uncertainties and external disturbances for each subsystem. Subsequently, the outer-loop and inner-loop formation controllers are developed by integrating disturbance compensation with distributed model predictive control (DMPC) theory. With full consideration of the input and state constraints, we resolve the local information-based DMPC optimization problem to obtain the control inputs for each AUV, thereby preventing actuator saturation and collisions among AUVs. Moreover, to mitigate the increased computation caused by the control structure, the Laguerre orthogonal function is applied to alleviate the computational burden in time intervals. We also demonstrate the stability of the closed-loop system by applying the terminal state constraint. Finally, based on a connected directed topology, comparative simulations are performed under various control schemes to verify the robustness and superior performance of the proposed scheme.

Keywords: multi-AUV system; formation tracking; finite-time extended state observer; distributed model predictive control; Laguerre function

Citation: Zhang, M.; Yan, Z.; Zhou, J.; Yue, L. Distributed Dual Closed-Loop Model Predictive Formation Control for Collision-Free Multi-AUV System Subject to Compound Disturbances. *J. Mar. Sci. Eng.* **2023**, *11*, 1897. https://doi.org/10.3390/jmse11101897

Academic Editor: Sergei Chernyi

Received: 12 September 2023
Revised: 25 September 2023
Accepted: 28 September 2023
Published: 29 September 2023

1. Introduction

Autonomous underwater vehicles (AUVs) have assumed indispensable roles in various underwater operations, such as ocean exploration and hydrologic surveys [1]. They can autonomously perform appropriate maneuvers to achieve predefined objectives. Compared with the operational capability of a single AUV, collaborative AUVs can respond more reliably and flexibly to complex missions and extended operational ranges, thereby improving the efficiency and robustness of undersea operations. Given this backdrop, numerous application cases about AUV coordinated formation have been triggered in both civilian and industrial fields for decades [2,3]. Irrespective of the specific collaborative missions undertaken by AUVs, the core challenge lies in ensuring motion stability of AUV formations within complex underwater environments and the constraints of their own models. To tackle this problem, several mainstream methodologies have been proposed by engineers and academics. Studies by Chen et al. [4] and Zhen et al. [5] proposed AUV formation control schemes combined with the virtual structure method. However, this approach suffers from limited flexibility and applicability. Wang et al. [6] utilized the leader–follower method to address the AUV formation tracking problem, but this approach relies on the state of the leader, reducing the robustness and fault tolerance of the formation. Conversely, leaderless formations have been proposed promisingly and have received more considerable attention [7]. Munir et al. [8] proposed a new arbitrary-order distributed control strategy based on the novel sliding surfaces of error dynamics, which addresses the cooperative tracking control of uncertain higher-order nonlinear systems. The strategies to mitigate the chattering issue caused by sliding surfaces are discussed in [9]. Despite the abundance of existing research, multiple-AUV formation tracking control remains a significantly challenging project.

One of the main challenges is the various disturbances resulting from the underwater environment and the motion model of the AUVs themselves [10]. On the one hand, unknown disturbances such as waves, tides, and currents, are inevitable in practical marine environments. On the other hand, AUVs exhibit highly nonlinear and coupled dynamics, leading to model uncertainties. These uncertainties are often induced by modeling errors and deviations in hydrodynamic coefficient measurement. According to the research by Cui et al. [11], these external disturbances and model uncertainties that degrade the system performance negatively are referred to as compound disturbances. In response to these challenges, researchers have developed diverse schemes, such as disturbance observers [12], fuzzy logic theory [13], and neural networks [14]. Among these, the extended state observer (ESO) initially proposed by Han [15] is an attractive option to estimate compound disturbances, as it does not rely on an accurate model. Lei et al. [16] designed a high-gain ESO to solve AUV horizontal trajectory tracking problems under the time-varying disturbances. Although many ESOs have been established for different platforms, most only guarantee asymptotic convergence of estimation errors, implying a potentially infinite convergence time. Some research works also lack a rigorous analysis of convergence. Considering the impact of severe underwater environments on estimation accuracy, the concept of finite-time ESO proves more beneficial for improving control performance [17]. Wang et al. [18] implemented a FTESO-based nonsingular terminal sliding mode controller to address unmanned surface vehicle (USV) trajectory tracking in disturbed environments. This approach ensured that the disturbance estimation errors converge within a finite time. However, there remains room for improvement and optimization of the design structure to further enhance observation performance.

AUV formation navigation also presents significant technical challenges due to various complex constraints. For instance, the AUV attitude has a certain desired range and navigation velocities are inherently limited. These intrinsic input and state constraints pose substantial challenges to control performance [19]. In practical applications, actuators often have input saturation constraints due to physical structure limitations. This results in a limitation of the actual active control force of the AUV. If a control signal exceeds this boundary, it may lead to system instability. However, most previous work assumes that the actuators can tolerate any level of control signals. To avoid actuator saturation, a nonlinear auxiliary system for filtering saturation errors was proposed [20]. Additionally, collisions between AUVs are undesirable during the formation configuration phase. Thus, the ability to avoid collisions is vital for AUV formation control. A wealth of solutions have been developed to this end, with Li and Wang [21] proposing a collision-free position consensus algorithm for AUVs based on potential function. Moreover, Xu et al. [22] presented an event-triggered algorithm based on deep reinforcement learning to avoid AUV collisions. However, the above studies disregard the physical constraints of AUVs. From the perspective of safe navigation, it is essential to integrate factors such as input, state restrictions, and collision avoidance into the design scheme.

Model predictive control (MPC) has garnered considerable attention due to its ability to simultaneously handle multiple composite constraints and offer superior dynamic performance. This is widely applied to MIMO systems affected by model distortions and complex constraints. Several MPC-based applications have been integrated into AUV control systems. Zhang et al. [23] proposed an MPC-based AUV trajectory tracking strategy under random disturbances. In [24], a robust model-predictive control scheme based on the active disturbance rejection control approach was developed for the AUV tracking task. The challenge of extending these systems to multi-AUV systems involves coordinating the control behavior of each subsystem and ensuring the closed-loop stability of the local MPC optimization problem under system constraints. This coordination aims to maximize the overall control performance. Hence, DMPC came into being. Zheng et al. [25] proposed a DMPC method based on local state information for MAS formation tracking. To the best of our knowledge, there are few studies that apply DMPC to multi-AUV formations. Wei et al. [26] developed a Lyapunov-based distributed predictive controller for AUV formation tracking, subject to current disturbances. The auxiliary controller was utilized to establish stability constraints to ensure the closed-loop stability of the system. However, this method only considers horizontal formations without uncertainties and state constraints. Furthermore, many works that design predictive controllers result in additional computational loads, which could impair the real-time execution capability of the controller. Shen and Shi [27] managed to reduce the MPC computational burden by decomposing the original AUV trajectory tracking optimization problem into smaller subproblems and then solving them in a distributed manner. Despite these efforts, there has been no research to address the heavy computation of DMPC applied to AUV formations. In order to improve the dynamic response and control accuracy of AUV formation tracking in three-dimensional (3-D) space, we adopt the Laguerre orthogonal function to reduce the computational load. In response to these discussions, it is imperative to develop a safe and efficient formation control scheme to solve the problems of disturbances, parameter uncertainties, and complicated constraints.

Motivated by the above observations, this paper investigates the collision-free formation tracking of multi-AUVs with compound disturbances under complicated constraints. A novel FTESO-based distributed dual closed-loop model predictive control scheme is proposed. This method satisfies the formation constraints and collision avoidance requirements while compensating for model uncertainties and external disturbances. We incorporate the Laguerre function to alleviate the computational burden of the DMPC optimization problem, also giving corresponding stability analysis. Based on the connected directed topology, comparative simulations under different schemes demonstrate the effectiveness and robustness of our proposed scheme. The main contributions of this paper are as follows:

1. Compared with the FTESO-based controllers presented in works [16,28], the proposed third-order fast FTESO can estimate the compound disturbances and their first derivatives, which effectively suppress the amplification and fluctuation of the generalized uncertainties. It has better estimation accuracy and convergence speed. Hence, the active disturbance rejection capability of AUV formation is enhanced;
2. Unlike the existing DMPC schemes depicted in works [29,30], a dual closed-loop structure is utilized to enhance the response speed of the DMPC system and the controllability of the AUV speed. The outer-loop controller sets the desired velocity and the inner-loop controller generates the driving force. By solving the constrained quadratic programming (QP) problems, the risks of actuator saturation and collision are reduced. The safety and robustness of formation tracking are improved;
3. In order to solve the issue of heavy computational burden in traditional predictive control, the Laguerre orthogonal function is incorporated to reconstruct the input matrices, which automatically trades off control performance and computational complexity, thus avoiding possible formation deviation due to slow computational speed. The stability of the closed-loop system is proved by exerting terminal state constraints.

The rest of this paper is organized as follows: Section 2 introduces some notations, lemmas, and graph theory, and describes the AUV model and control objective. Section 3 presents the methodology, including the design of the FTESO and dual closed-loop DMPC scheme, the application of the Laguerre function, and the corresponding stability analysis. Sections 4 and 5, respectively, provide simulation results and conclusions.

2. Preliminaries

2.1. Notations and Lemmas

Notation. $\mathbb{R}^n$ represents the n-dimensional Euclidean space, and $\mathbb{R}^{m \times n}$ denotes the set of $(m \times n)$ real matrix. I_n, 0_n, and $0_{p \times q}$ signify $(n \times n)$ identity matrix, $(n \times n)$, and $(p \times q)$

null matrices, respectively. $\|\cdot\|$ refers to the Euclidean vector norm and the induced matrix norm, while the infinity norm is denoted by $\|\cdot\|_{\infty}$. $\lambda_{\min}(\cdot)$ represents the minimum eigenvalue of the specified matrix $(\cdot)$, with its maximum eigenvalue denoted as $\lambda_{\max}(\cdot)$. For simplicity, some notations are defined as $sig^p(x) = \text{sign}(x)|x|^p$, $|x|^p = \left[|x_1|^p, |x_2|^p, \ldots, |x_n|^p\right]^T$, $x = [x_1, x_2, \ldots, x_n]^T$, $p \in \mathbb{R}$. $\text{sign}(\cdot)$ symbolizes the signum function with $\text{sign}(0) = 0$. Notably, $sig^0(x) = \text{sign}(x)$, $sig^0(x)|x|^p = sig^p(x)$.

Lemma 1 ([31]). Consider the system $\dot{x}(t) = f(x(t))$, $x(0) = x_0$, $f(0) = 0$, $x \in \mathbb{R}^n$, where $f : U \to \mathbb{R}^n$ is a continuous function. Suppose that this system has a unique solution in forward time for all initial conditions. If there exists a Lyapunov function $V(x)$, with $V(x_0)$ denoting its initial value, the following can be assumed: (1) The trajectory of this system is finite-time uniformly ultimately bounded stable within the region of $Q_1 = \left\{x \,\middle|\, V(x)^{\alpha_1 - \alpha_2} < \frac{\beta_2}{\gamma_1}\right\}$, if $\dot{V}(x) \leq -\beta_1 V(x)^{\alpha_1} + \beta_2 V(x)^{\alpha_2}$ for $\alpha_1 > \alpha_2$, $\beta_1 > 0$, $\beta_2 > 0$, $\gamma_1 \in (0, \beta_1)$. The settling time for the states reaching the stable residual set is subject to the constraint as $T_1 \leq \frac{V(x_0)^{1-\alpha_1}}{(\beta_1 - \gamma_1)(1 - \alpha_1)}$. (2) The trajectory of this system is fast finite-time uniformly ultimately bounded stable within $Q_2 = \left\{x \,\middle|\, \gamma_1 V(x)^{\alpha_1 - \alpha_2} + \gamma_2 V(x)^{1-\alpha_2} < \beta_3\right\}$, if $\dot{V}(x) \leq -\beta_1 V(x)^{\alpha_1} - \beta_2 V(x) + \beta_3 V(x)^{\alpha_2}$ for $\beta_3 > 0$, $\gamma_2 \in (0, \beta_2)$. The convergence time T_2 is bounded as $T_2 \leq \frac{\ln\left[(\beta_2 - \gamma_2) V(x_0)^{1-\alpha_1}/(\beta_1 - \gamma_1) + 1\right]}{(\beta_2 - \gamma_2)(1 - \alpha_1)}$.

2.2. Graph Theory

We introduce a directed topology graph $G = \{V, \varepsilon\}$ to describe the information interactions among the AUVs. Let the node set $V = \{V_1, V_2, \cdots, V_N\}$ to represent the N members in the formation, and an edge set $\varepsilon \subseteq V \times V$ to represent the communication from the node V_i to the node V_j. $A = [a_{ij}] \subset \mathbb{R}^{N \times N}$ is defined as an adjacency matrix, where a_{ij} represents the connection weight and $a_{ij} = 1$ if $(i, j) \in \varepsilon$, while $a_{ij} = 0$ if $(i, j) \notin \varepsilon$. It is assumed that the ith vehicle could receive information from the virtual leader and its neighbors $N_i = \{j \in V : (j, i) \in \varepsilon\}$. The graph is termed an undirected graph if bidirectional communication links exist among all members of the formation. Otherwise, it is referred to as a directed graph. A directed graph is considered strongly connected if a directed path can connect any point in the formation to any other.

2.3. AUV Model

As shown in Figure 1, it is convenient to describe the six-degree-of-freedom (DOF) AUVs with two reference frames: an earth-fixed frame $\{E\}$ and a body-fixed frame $\{B\}$. This paper employs a fully actuated torpedo-type AUV, referenced from [32], based on the control objectives. In addition, the AUV uses an ultra-short baseline acoustic positioning system for underwater localization. Since this AUV can be regarded as a highly metacentric stable vehicle with self-stable roll motion, the effect of roll is ignored (roll angle $\Phi_i = 0$, roll angular velocity $p_i = 0$). The kinematics and dynamics of the ith AUV are described as follows [33]:

$$\dot{\eta}_i = J(\eta_i)v_i \tag{1}$$

$$M_i \dot{v}_i + C_i(v_i)v_i + D_i(v_i)v_i + g_i(\eta_i) = \tau_i + \tau_{ic} \tag{2}$$

where $i = 1, 2, \ldots, N$, $\eta_i = [x_i, y_i, z_i, \theta_i, \psi_i]^T \in \mathbb{R}^5$, and $v_i = [u_i, v_i, w_i, q_i, r_i]^T \in \mathbb{R}^5$ denote the states of position, orientation, and velocity of the AUV, respectively. $J(\eta_i)$ is a rotation transformation matrix from the body-fixed frame to the earth-fixed frame, expressed as:

$$J(\eta_i) = \begin{bmatrix} \cos\psi_i \cos\theta_i & -\sin\psi_i & \cos\psi_i \sin\theta_i & 0 & 0 \\ \sin\psi_i \cos\theta_i & \cos\psi_i & \sin\psi_i \sin\theta_i & 0 & 0 \\ -\sin\theta_i & 0 & \cos\theta_i & 0 & 0 \\ 0 & 0 & 0 & 1 & 0 \\ 0 & 0 & 0 & 0 & 1/\cos\theta_i \end{bmatrix} \tag{3}$$

M_i represents the inertial matrix, which includes added mass. $C_i(v_i)$ and $D_i(v_i)$ denote the Coriolis and centripetal and hydrodynamic damping matrix, respectively, while $g_i(\eta_i)$ represents the restoring force and moment generated by gravity and buoyancy. $\tau_i = \left[\tau_{iu}, \tau_{iv}, \tau_{iw}, \tau_{iq}, \tau_{ir}\right]^T$ represents the control input, and τ_{ic} denotes the external disturbance. Detailed expressions of these matrices are available in [34].

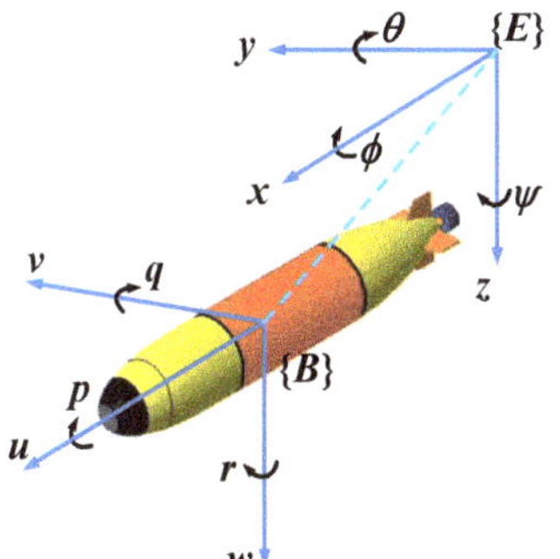

Figure 1. AUV coordinate system.

In practical engineering, we may not be able to obtain accurate hydrodynamic coefficients in the model, so the matrices in (2) are typically divided into two parts: the nominal value part and the uncertainty part caused by linear shifts, i.e., $M_i = M_i^* + \Delta M_i$, $C_i(v_i) = C_i^*(v_i) + \Delta C_i(v_i)$, $D_i(v_i) = D_i^*(v_i) + \Delta D_i(v_i)$, and $g_i(\eta_i) = g_i^*(\eta_i) + \Delta g_i(\eta_i)$, where $(\cdot)_i^*$ denotes the nominal value that can be obtained from the computational fluid dynamics (CFD) or experimental analysis. $\Delta(\cdot)_i$ symbolizes the difference between the real value and the nominal value.

Accordingly, the ith AUV dynamic model (2) can be reformulated as:

$$M_i^* \dot{v}_i = -C_i^*(v_i)v_i - D_i^*(v_i)v_i - g_i^*(\eta_i) + \tau_i + \tau_{id} \tag{4}$$

where $\tau_{id} = \tau_{ic} - \Delta M_i \dot{v}_i - \Delta C_i(v_i)v_i - \Delta D_i(v_i)v_i - \Delta g_i(\eta_i)$ is regarded as the compound disturbance, which includes uncertainties and unknown external disturbance. Typically, external disturbances are periodically varying and energy limited. The model uncertainties are related to the actual states and physical properties of the AUV. Based on the constraints of DMPC on the system state, in practice, we give the following reasonable assumption:

Assumption 1 ([11]). *The ocean current disturbance term τ_{ic} and the first time derivative $\dot{\tau}_{ic}$ are bounded, and the model uncertainties ΔM_i, ΔC_i, ΔD_i, and Δg_i are unknown and bounded. Hence, the compound disturbance τ_{id} of the ith AUV is bounded and satisfies $\|\tau_{id}\| \le \overline{\tau}_{id}$, where $\overline{\tau}_{id} \in \mathbb{R}^+$ represents the unknown upper bound.*

It should be noted that the above assumption is untenable if there are no system state constraints [35,36].

2.4. Control Objective

In this paper, the control objective is to develop a control scheme that enables AUV formation to track a reference trajectory while maintaining a predefined configuration. Initially, a FTESO is designed to compensate for external disturbances and model uncertainties of the AUV formation, so that the estimation errors converge to the origin. Subsequently, a dual closed-loop DMPC controller is designed. In this structure, the outer-loop controller enables the ith AUV to track the reference trajectory η_r by generating the desired velocity, resulting in the convergence of position tracking errors. The inner-loop controller is used to achieve the convergence of velocity tracking errors. The desired formation is implemented by setting the corresponding formation configuration vector r_{if} and the relative distance vector r_{ij}. The task must adhere to various constraints and ensure collision avoidance.

Because the navigation trajectory has a limited range and the speed is continuous without abrupt changes, we adopt the following reasonable assumptions to avoid singularities in the reference trajectory:

Assumption 2. The reference trajectory $\eta_r = [x_r, y_r, z_r, \theta_r, \psi_r]^T$ and its derivatives are smooth and bounded, i.e., $\|\eta_r\|_\infty \le \bar{\eta}_r$, $\|\dot{\eta}_r\|_\infty \le \bar{\eta}_{r1}$, and $\|\ddot{\eta}_r\|_\infty \le \bar{\eta}_{r2}$ with positive numbers $\bar{\eta}_r, \bar{\eta}_{r1}$, and $\bar{\eta}_{r2}$.

3. Methodology

This section develops the FTESO-based distributed dual closed-loop model predictive control scheme for the AUV formation to perform trajectory tracking. A novel FTESO is designed to compensate the compound disturbances. Based on the model information reconstructed by FTESO, the DMPC optimization problems are formulated for the outer and inner loops under constraints such as actuator saturation and collision avoidance, respectively. The Laguerre function is applied to alleviate the computational load. The block diagram of proposed control scheme is depicted in Figure 2.

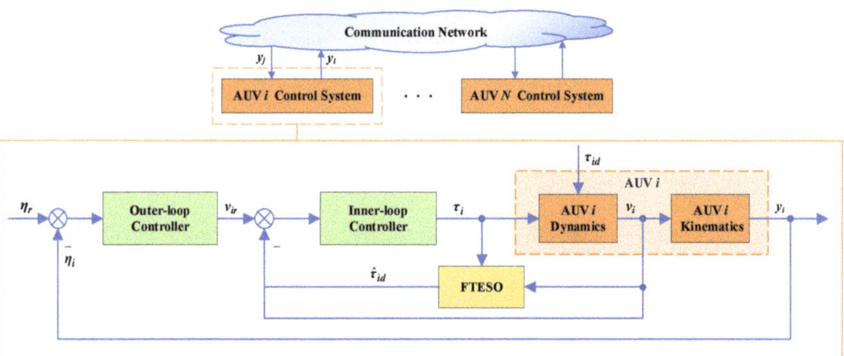

Figure 2. The FTESO-based DMPC dual closed-loop structure for the AUV formation.

3.1. FTESO Design and Convergence Analysis

The AUV model is fundamental to controller design, but obtaining an accurate model in practice is challenging. Considering the superiority and effectiveness of the ESO technique in estimating and compensating for synthetic uncertainty, a novel fast FTESO is designed to simultaneously reconstruct the external disturbance and model uncertainties of multiple AUVs.

First, define the auxiliary velocity variable as $\omega_i(v_i) = M_i^* v_i + \int v_i$, the derivative of $\omega_i(v_i)$ with respect to time can be obtained from (4)

$$\dot{\omega}_i(v_i) = v_i - C_i^*(v_i)v_i - D_i^*(v_i)v_i - g_i^*(\eta_i) + \tau_i + \tau_{id}. \tag{5}$$

For simplicity, denote $G_i(\eta_i, v_i) = v_i - C_i^*(v_i)v_i - D_i^*(v_i)v_i - g_i^*(\eta_i)$. Then, a new variable is defined as $z_{i1} = \omega_i(v_i)$, and the order of the system is extended by additional state variables, z_{i2} and z_{i3}, defined as $z_{i2} = \tau_{id}$ and $z_{i3} = \dot{z}_{i2}$ with $\dot{z}_{i3} = \sigma_i$. It should be noted that the compound disturbances z_{i2} are assumed to be bounded and continuously differentiable, and the components of its second derivative satisfies $|\sigma_{ip}| \le \bar{\sigma}_i, p = 1, 2, \ldots, 5$. where $\bar{\sigma}_i$ is an unknown positive constant. Afterward, the dynamic model of the ith AUV can be extended as follows:

$$\begin{cases} \dot{z}_{i1} = G_i(\eta_i, v_i) + \tau_i + z_{i2} \\ \dot{z}_{i2} = z_{i3} \\ \dot{z}_{i3} = \sigma_i. \end{cases} \tag{6}$$

Denote $\hat{z}_{i1}$, $\hat{z}_{i2}$, and $\hat{z}_{i3}$ as the observation values of states z_{i1}, z_{i2}, and z_{i3} in the above extended system, and $e_{i1} = \hat{z}_{i1} - z_{i1}$, $e_{i2} = \hat{z}_{i2} - z_{i2}$, and $e_{i3} = \hat{z}_{i3} - z_{i3}$ as the observation errors of the velocity, the compound disturbances, and its first derivatives, respectively. Then, a third-order fast FTESO is proposed as follows:

$$\begin{cases} \dot{\hat{z}}_{i1} = \hat{z}_{i2} - \beta_{i1}(sig^{\alpha_{i1}}(e_{i1}) + e_{i1}) + G_i(\eta_i, v_i) + \tau_i \\ \dot{\hat{z}}_{i2} = \hat{z}_{i3} - \beta_{i2}(sig^{\alpha_{i2}}(e_{i1}) + 2sig^{\alpha_{i1}}(e_{i1}) + e_{i1}) \\ \dot{\hat{z}}_{i3} = -\beta_{i3}(sig^{\alpha_{i3}}(e_{i1}) + 2sig^{\alpha_{i2}}(e_{i1}) + sig^{\alpha_{i1}}(e_{i1})) \end{cases} \tag{7}$$

where the observer gains satisfy $\beta_{ik} > 0$, $k = 1, 2, 3$, $\alpha_{i1} \in (2/3, 1)$ and $\alpha_{i2} = 2\alpha_{i1} - 1$, and $\alpha_{i3} = 3\alpha_{i1} - 2$. Although the actual value of z_{ik} is probably unavailable, its observed value $\hat{z}_{ik}$ can be obtained by the above FTESO. The analysis and proof that $\hat{z}_{ik}$ tracks the actual value are described below.

According to the extended system (6) and the proposed FTESO (7), we can obtain the observation error dynamics as follows:

$$\begin{cases} \dot{e}_{i1} = -\beta_{i1}(sig^{\alpha_{i1}}(e_{i1}) + e_{i1}) + e_{i2} \\ \dot{e}_{i2} = -\beta_{i2}(sig^{\alpha_{i2}}(e_{i1}) + 2sig^{\alpha_{i1}}(e_{i1}) + e_{i1}) + e_{i3} \\ \dot{e}_{i3} = -\beta_{i3}(sig^{\alpha_{i3}}(e_{i1}) + 2sig^{\alpha_{i2}}(e_{i1}) + sig^{\alpha_{i1}}(e_{i1})) - \sigma_i. \end{cases} \tag{8}$$

The stability and convergence of the proposed FTESO are stated in the following theorem:

Theorem 1. *Consider the AUV formation control system with the dynamic model (4) under Assumption 1. If the FTESO is proposed in the form of (9), with appropriate observer gains satisfying the prescribed constraints, then the observation errors $e_i = \left[e_{i1}^T, e_{i2}^T, e_{i3}^T\right]^T$ will converge to the small region Ω_i in finite time T_{if}. This implies that the error dynamics system (8) is finite-time uniformly ultimately bounded stable.*

Proof of Theorem 1. Consider a Lyapunov candidate function as $V_{i1}(e) = \varepsilon_i^T P_i \varepsilon_i$, where P_i is a positive definite symmetric matrix and $\varepsilon_i^T = \left[(sig^{\alpha_{i1}}(e_{i1}) + e_{i1})^T, e_{i2}^T, e_{i3}^T\right]$ is introduced as an auxiliary error variable. It should be noted that e_{i1}, e_{i2}, and e_{i3} will converge to origin in finite time, if the new state ε_i is finite-time stable. The time derivative of ε_i, invoking (8), yields:

$$\dot{\varepsilon}_i = \begin{bmatrix} \alpha_{i1}|e_{i1}|^{\alpha_{i1}-1}\dot{e}_{i1} + \dot{e}_{i1} \\ \dot{e}_{i2} \\ \dot{e}_{i3} \end{bmatrix} = \begin{bmatrix} \alpha_{i1}|e_{i1}|^{\alpha_{i1}-1}(e_{i2} - \beta_{i1}(sig^{\alpha_{i1}}(e_{i1}) + e_{i1})) \\ \frac{e_{i3}}{2} - \beta_{i2}(sig^{\alpha_{i2}}(e_{i1}) + sig^{\alpha_{i1}}(e_{i1})) \\ -\beta_{i3}(sig^{\alpha_{i3}}(e_{i1}) + sig^{\alpha_{i2}}(e_{i1})) \end{bmatrix}$$

$$+ \begin{bmatrix} e_{i2} - \beta_{i1}(sig^{\alpha_{i1}}(e_{i1}) + e_{i1}) \\ \frac{e_{i3}}{2} - \beta_{i2}(sig^{\alpha_{i1}}(e_{i1}) + e_{i1}) \\ -\beta_{i3}(sig^{\alpha_{i2}}(e_{i1}) + sig^{\alpha_{i1}}(e_{i1})) \end{bmatrix} + \begin{bmatrix} 0_5 \\ 0_5 \\ -\sigma_i \end{bmatrix} = diag\left(\left[|e_{i1}|^{\alpha_{i1}-1}, |e_{i1}|^{\alpha_{i1}-1}, |e_{i1}|^{\alpha_{i1}-1}\right]\right)A_{i1}\varepsilon_i + A_{i2}\varepsilon_i + \Phi_i \tag{9}$$

where $\Phi_i = \begin{bmatrix} 0_5 & 0_5 & -\sigma_i \end{bmatrix}^T$ and the coefficient matrices A_{i1} and A_{i2} are expressed as:

$$A_{i1} = \begin{pmatrix} -\alpha_{i1}\beta_{i1}I_5 & \alpha_{i1}I_5 & 0_5 \\ -\beta_{i2}I_5 & 0_5 & \bar{e}_i^{-1}I_5/2 \\ -\beta_{i3}\bar{e}_iI_5 & 0_5 & 0_5 \end{pmatrix}, \quad A_{2i} = \begin{pmatrix} -\beta_{i1}I_5 & I_5 & 0_5 \\ -\beta_{i2}I_5 & 0_5 & I_5/2 \\ -\beta_{i3}\bar{e}_iI_5 & 0_5 & 0_5 \end{pmatrix} \tag{10}$$

with $\bar{e}_i = |e_{i1}|^{\alpha_{i1}-1}$. From the characteristic polynomials of A_{i1} and A_{i2} that all their eigenvalues have negative real parts if the observer gains are set as $\beta_{ik} > 0$, indicating that

A_{i1} and A_{i2} are Hurwitz matrices. Thus, symmetric and positive definite matrices Q_{i1} and Q_{i2} exist that satisfy the following Lyapunov equations:

$$\begin{cases} A_{i1}^T P_i + P_i A_{i1} = -Q_{i1} \\ A_{i2}^T P_i + P_i A_{i2} = -Q_{i2}. \end{cases} \tag{11}$$

Differentiating $V_{i1}(e)$ with respect to time yields the following:

$$\begin{aligned} \dot{V}_{i1} &= \varepsilon_i^T \left[diag([\bar{e}_i, \bar{e}_i, \bar{e}_i]) \left(A_{i1}^T P_i + P_i A_{i1} \right) \right] \varepsilon_i + \varepsilon_i^T \left(A_{i2}^T P_i + P_i A_{i2} \right) \varepsilon_i + 2\varepsilon_i^T P_i \Phi_i \\ &= -\varepsilon_i^T \left[diag([\bar{e}_i, \bar{e}_i, \bar{e}_i]) Q_{i1} \right] \varepsilon_i - \varepsilon_i^T Q_{i2} \varepsilon_i + 2\varepsilon_i^T P_i \Phi_i \le -\bar{e}_i^{max} \varepsilon_i^T Q_{i1} \varepsilon_i - \varepsilon_i^T Q_{i2} \varepsilon_i + 2\|\varepsilon_i\| \|P_i\| \|\Phi_i\| \end{aligned} \tag{12}$$

where $\bar{e}_i^{max} = |e_{i1}|_{max}^{\alpha_{i1}-1}$ and $|e_{i1}|_{max} = \max\{|e_{i11}|, \ldots, |e_{i15}|\}$. Given the fact that $|e_{i1}|_{max} \le \|e_{i1}\| \le \|\varepsilon_i\|^{1/\alpha_{i1}}$ and $\alpha_{i1} \in \left(\frac{2}{3}, 1\right)$, we can obtain the following:

$$\begin{aligned} \dot{V}_{i1} &\le -\|\varepsilon_i\|^{\frac{\alpha_{i1}-1}{\alpha_{i1}}} \varepsilon_i^T Q_{i1} \varepsilon_i - \varepsilon_i^T Q_{i2} \varepsilon_i + 2\|\varepsilon_i\| \|P_i\| \|\Phi_i\| \\ &\le -\lambda_{min}(Q_{i1}) \|\varepsilon_i\|^{3-\frac{1}{\alpha_{i1}}} - \lambda_{min}(Q_{i2}) \|\varepsilon_i\|^2 + 2\|\varepsilon_i\| \|P_i\| \|\Phi_i\|. \end{aligned} \tag{13}$$

Since σ_i is assumed to be bounded reasonably by $|\sigma_{ip}| \le \bar{\sigma}_i$, we have $2\|\varepsilon_i\| \|P_i\| \|\Phi_i\| \le 2\sqrt{5}\bar{\sigma}_i \|\varepsilon_i\| \|P_i\| \le 2\sqrt{5}\bar{\sigma}_i \lambda_{min}(P_i)^{-\frac{1}{2}} V_{i1}^{\frac{1}{2}} \|P_i\|$, by using the inequality

$$\lambda_{min}(P_i) \|\varepsilon_i\|^2 \le V_{i1} \le \lambda_{max}(P_i) \|\varepsilon_i\|^2 \tag{14}$$

Then, inequality (13) becomes the following:

$$\begin{aligned} \dot{V}_{i1} &\le -\lambda_{min}(Q_{i1}) \lambda_{max}(P_i)^{\frac{1}{2\alpha_{i1}}-\frac{3}{2}} V_{1i}^{\frac{3}{2}-\frac{1}{2\alpha_{i1}}} - \lambda_{min}(Q_{i2}) \lambda_{max}(P_i)^{-1} V_{i1} + 2\sqrt{5}\bar{\sigma}_i \|P_i\| \lambda_{min}(P_i)^{-\frac{1}{2}} V_{i1}^{\frac{1}{2}} \\ &\le -\lambda_{i1} V_{i1}^{\frac{3}{2}-\frac{1}{2\alpha_{i1}}} - \lambda_{i2} V_{i1} + \lambda_{i3} V_{i1}^{\frac{1}{2}} \end{aligned} \tag{15}$$

where $\lambda_{i1} = -\lambda_{min}(Q_{i1}) \lambda_{max}(P_i)^{\frac{1}{2\alpha_{i1}}-\frac{3}{2}}$, $\lambda_{i2} = -\lambda_{min}(Q_{i2}) \lambda_{max}(P_i)^{-1}$, and $\lambda_{i3} = 2\sqrt{5}\bar{\sigma}_i \|P_i\| \lambda_{min}(P_i)^{-\frac{1}{2}}$.

It can be seen that (15) has the same form as the sufficient condition in Lemma 1 2. Thus, the error trajectories of the proposed FTESO (7) are fast finite-time uniformly ultimately bounded stable. The state observation errors e_i will converge to a small region Ω_i in the finite time T_{if}. Moreover, the settling time T_{if} is subject to the constraint:

$$T_{if} \le \frac{\ln\left((\lambda_{i2} - \bar{\lambda}_{i2}) V_{i1}(e_0)^{\frac{1}{2\alpha_{i1}}-\frac{1}{2}} / (\lambda_{i1} - \bar{\lambda}_{i1}) + 1 \right)}{(\lambda_{i2} - \bar{\lambda}_{i2})\left(\frac{1}{2\alpha_{i1}} - \frac{1}{2}\right)}. \tag{16}$$

And the stable region Ω_i is denoted as

$$\Omega_i = \left\{ e \mid \bar{\lambda}_{i1} V_{i1}(e)^{1-\frac{1}{2\alpha_{i1}}} + \bar{\lambda}_{i2} V_{i1}(e)^{\frac{1}{2}} < \lambda_{i3} \right\} \tag{17}$$

where $\bar{\lambda}_{i1}$ and $\bar{\lambda}_{i2}$ are arbitrary constants that meet the conditions $\bar{\lambda}_{i1} \in (0, \lambda_{i1})$ and $\bar{\lambda}_{i2} \in (0, \lambda_{i2})$. This completes the proof. $\square$

Remark 1. Contrasting our proposed FTESO (7) with the FTESO in [37], our approach factors in the dynamics of disturbances and uncertainties to achieve a higher degree of estimation accuracy. Our usage of fractional powers within the FTESO allows for a quick finite-time convergence. It can be noted that the size of the attraction region Ω_i hinges upon the selection of the observer gains β_{ik} and α_{i1}. By increasing β_{ik} or decreasing α_{i1}, the attraction region of the observation error system can be expanded and the convergence speed can be improved, but excessive tuning will lead to undesired overshoot and oscillation. As a result, a trade-off should be taken for β_{ik} and α_{i1}.

3.2. Outer-Loop Formation Prediction Control Law

In this subsection, we design a DMPC-based outer-loop formation controller. This controller, which draws on the information interaction with neighbors, facilitates the positional tracking of the ith AUV. The controller operates under composite constraints and ensures the avoidance of collisions. Then, we formulate a constrained QP problem in accordance with the control objective to obtain the optimal driving speed.

To facilitate the recursive model prediction and the implementation of the control law, the kinematic model (1) is discretized by using the Forward-Euler method with a sampling period T_s, resulting in following discrete model:

$$\eta_i(k+1) = \eta_i(k) + J_i(k)v_i(k)T_s. \tag{18}$$

To smoothen the speed change of the AUV, the velocity increment $\Delta u_{iv}(k) = v_i(k) - v_i(k-1)$ is taken as the control input. $x_{i\eta}(k) = \begin{bmatrix} \eta_i(k) & v_i(k-1) \end{bmatrix}^T$ is denoted as the state variable of the prediction model. The augmented state-space model of the outer-loop subsystem can be derived as:

$$x_{i\eta}(k+1) = A_{i\eta}x_{i\eta}(k) + B_{i\eta}\Delta u_{iv}(k) \tag{19}$$

$$y_{i\eta}(k) = C_{i\eta}x_{i\eta}(k) \tag{20}$$

where $A_{i\eta} = \begin{bmatrix} I_5 & J_i(k)T_s \\ 0_5 & I_5 \end{bmatrix} \in \mathbb{R}^{10\times10}$, $B_{i\eta} = \begin{bmatrix} J_i(k)T_s \\ I_5 \end{bmatrix} \in \mathbb{R}^{10\times5}$, and $C_{i\eta} = \begin{bmatrix} I_5 & 0_5 \end{bmatrix} \in \mathbb{R}^{5\times10}$.

According to the state prediction model (19) and (20), we can calculate the predicted state sequence of the system when given an input sequence. Let N_{p1} and N_{c1} denote the prediction and control horizon of the outer-loop controller, respectively. The predicted state sequence and the input incremental sequence are usually represented by compact vectors:

$$Y_{i\eta} = \begin{bmatrix} y_{i\eta}(k+1|k) \\ y_{i\eta}(k+2|k) \\ \vdots \\ y_{i\eta}(k+N_{p1}|k) \end{bmatrix} \in \mathbb{R}^{5N_{p1}}, \ x_{i\eta} = \begin{bmatrix} x_{i\eta}(k+1|k) \\ x_{i\eta}(k+2|k) \\ \vdots \\ x_{i\eta}(k+N_{p1}|k) \end{bmatrix} \in \mathbb{R}^{10N_{p1}} \tag{21}$$

$$\Delta U_{iv} = \begin{bmatrix} \Delta u_{iv}(k|k) \\ \Delta u_{iv}(k+1|k) \\ \vdots \\ \Delta u_{iv}(k+N_{c1}-1|k) \end{bmatrix} \in \mathbb{R}^{5N_{c1}} \tag{22}$$

where $y_{i\eta}(k+l|k)$ and $x_{i\eta}(k+l|k)$ are the output vector $y_{i\eta}(k+l)$ and state vector $x_{i\eta}(k+l)$ predicted at time k, respectively. $\Delta u_{iv}(k+j|k)$ denotes the input increment $\Delta u_{iv}(k+j)$ predicted at the same time k. Then, we characterize the relationship between the predicted output vector sequence and the control increment sequence through the following prediction equation based on the recurrence relations:

$$Y_{i\eta} = H^1_{ix}x_{i\eta}(k) + H^1_{iu}\Delta U_{iv} \tag{23}$$

where $x_{i\eta}(k)$ is the initial state, $H^1_{ix} = \begin{bmatrix} C_{i\eta}A_{i\eta}, C_{i\eta}A^2_{i\eta}, \dots, C_{i\eta}A^{N_{p1}}_{i\eta} \end{bmatrix}^T \in \mathbb{R}^{5N_{p1}\times10}$ and

$$H_{iu}^1 = \begin{bmatrix} C_{i\eta}B_{i\eta} & 0_5 & \cdots & 0_5 \\ C_{i\eta}A_{i\eta}B_{i\eta} & C_{i\eta}B_{i\eta} & \cdots & 0_5 \\ \vdots & \vdots & \ddots & \vdots \\ C_{i\eta}A_{i\eta}^{N_{p1}-1}B_{i\eta} & C_{i\eta}A_{i\eta}^{N_{p1}-2}B_{i\eta} & \cdots & C_{i\eta}A_{i\eta}^{N_{p1}-N_{c1}}B_{i\eta} \end{bmatrix} \in \mathbb{R}^{5N_{p1}\times 5N_{c1}}.$$

Considering the control objective, the constraints within the outer-loop subsystem are considered. First, we set upper and lower boundaries for the amplitude of the control input $u_{iv}(k)$ and the input increment $\Delta u_{iv}(k)$:

$$u_{iv}^{\min} \leq u_{iv}(k) \leq u_{iv}^{\max} \tag{24}$$

$$\Delta u_{iv}^{\min} \leq \Delta u_{iv}(k) \leq \Delta u_{iv}^{\max} \tag{25}$$

where $u_{iv}^{\min}$ and $\Delta u_{iv}^{\min}$ represent the predefined lower bounds, and $u_{iv}^{\max}$ and $\Delta u_{iv}^{\max}$ represent the predefined upper bounds.

Next, to assure safe navigation throughout the formation construction stage, we need to consider the collision avoidance constraints between AUVs. The primitive collision avoidance constraints of the ith AUV can be transformed into a convex constraint, as follows:

$$\left\| S\left(y_{i\eta}(k+l|k) - y_{j\eta}(k+l|k) \right) \right\| \geq r_s, \quad j \in \Xi_i \tag{26}$$

where $l = 1, 2, \ldots, N_{p1}$ and r_s is the preset minimum allowable distance between the ith AUV and the jth AUV. S denotes a scaling matrix. Let r_d be the radius of the safe detection zone for the ith AUV. Ξ_i is the set of those AUVs that contain within r_d. Let the nominal value $\bar{y}_{i\eta}$ represent an initial guess of the actual value $y_{i\eta}$ for convexifying the collision avoidance constraint. It follows from (26) that a sufficient condition for upholding the collision avoidance constraint is the following:

$$\bar{d}_{ij}^T(k+l|k)S^T S\left(y_{i\eta}(k+l|k) - \bar{y}_{j\eta}(k+l|k) \right) \geq r_s \left\| S\bar{d}_{ij}(k+l|k) \right\| \tag{27}$$

where $\bar{d}_{ij}(k+l|k) = \bar{y}_{i\eta}(k+l|k) - \bar{y}_{j\eta}(k+l|k)$. In order to express the constraints in a compact matrix form, define $\bar{R}_l = r_s \left\| S\bar{d}_{ij}(k+l|k) \right\| + \bar{d}_{ij}^T(k+l|k)S^T S\bar{y}_{j\eta}(k+l|k)$, $\bar{R}_{ij} = \left[\bar{R}_1, \bar{R}_2, \ldots, \bar{R}_{N_{p1}} \right]^T$ and $\bar{S}_{ij} = diag\left\{ \bar{S}_1, \bar{S}_2, \ldots, \bar{S}_{N_{p1}} \right\}$, and $\bar{S}_l = \bar{d}_{ij}^T(k+l|k)S^T S$. Then, (27) can be rewritten as $\bar{S}_{ij}Y_{i\eta} \geq \bar{R}_{ij}$. Substitute (23) to derive the collision avoidance constraint as follows:

$$\bar{S}_{ij}H_{iu}^1\Delta U_{iv} \geq \bar{R}_{ij} - \bar{S}_{ij}H_{ix}^1 x_{i\eta}(k). \tag{28}$$

The input amplitude constraint (24) can be converted to the input incremental constraint, associating (25) and (28), expressed in the compact linear constraint form as follows:

$$\Gamma_{i\eta}\Delta U_{iv} \leq \gamma_{i\eta} \tag{29}$$

where $\Gamma_{i\eta} = \begin{bmatrix} I_{5N_{c1}} \\ -I_{5N_{c1}} \\ I_{\eta1} \\ -I_{\eta1} \\ -\bar{S}_{ij}H_{iu}^1 \end{bmatrix}$, $\gamma_{i\eta} = \begin{bmatrix} \Delta U_{iv}^{\max} \\ -\Delta U_{iv}^{\min} \\ U_{iv}^{\max} - I_{\eta2}u_i(k-1) \\ -U_{iv}^{\min} + I_{\eta2}u_i(k-1) \\ \bar{S}_{ij}H_{ix}^1 x_{i\eta}(k) - \bar{R}_{ij} \end{bmatrix}$, $I_{\eta1} = \begin{bmatrix} I_5 & 0_5 & \cdots & 0_5 \\ I_5 & I_5 & \cdots & 0_5 \\ \vdots & \vdots & \ddots & \vdots \\ I_5 & I_5 & \cdots & I_5 \end{bmatrix} \in \mathbb{R}^{5N_{c1}\times 5N_{c1}}$, and $I_{\eta2} = [I_5, I_5, \ldots, I_5]^T \in \mathbb{R}^{5N_{c1}}$.

In order to achieve the control objective of formation positional tracking with low energy requirements, we define the local distributed cost function in the outer-loop subsystem of the ith AUV in a discretized form:

$$
\begin{aligned}
J_{i\eta}(k) = & \sum_{l=1}^{N_{p1}} \left\| \left(\boldsymbol{y}_{i\eta}(k+l|k) - \boldsymbol{y}_{if}(k+l) \right) \right\|_{Q_{if}}^2 + \sum_{l=0}^{N_{c1}-1} \|\Delta \boldsymbol{u}_{iv}(k+l|k)\|_{\boldsymbol{R}_{i1}}^2 \\
& + \sum_{l=1}^{N_{p1}} \sum_{j\in N_i} a_{ij} \left\| \left(\boldsymbol{y}_{i\eta}(k+l|k) - \boldsymbol{y}_{ij}(k+l) \right) \right\|_{Q_{ij}}^2
\end{aligned}
\tag{30}
$$

where $\boldsymbol{Q}_{if}$, $\boldsymbol{Q}_{ij}$, and $\boldsymbol{R}_{i1}$ are the weight matrices. $\boldsymbol{y}_{if}(k+l) = \boldsymbol{\eta}_r(k+l) + \boldsymbol{r}_{if}(k+l)$ with $\boldsymbol{r}_{if}(k+l)$ represents the formation configuration. $\boldsymbol{y}_{ij}(k+l) = \boldsymbol{y}_{j\eta}(k+l) + \boldsymbol{r}_{ij}(k+l)$ with $\boldsymbol{r}_{ij}(k+l)$ represents the predefined relative distance between the ith AUV and its neighbor jth AUV. N_{p1} indicates the degree of prediction of future tracking errors. The larger it is, the better the tracking accuracy and stability. The smaller N_{c1} is, the worse the dynamic response is, and conversely the more maneuverable the control is. $\boldsymbol{Q}_{if}$ is the position tracking matrix, the larger it is, the better the tracking accuracy and dynamic response. $\boldsymbol{Q}_{ij}$ is the relative position matrix, the larger it is, the better the ability of the formation to maintain the preset configuration. $\boldsymbol{R}_{i1}$ is the control increment weight matrix, mainly to limit the drastic change of $\Delta \boldsymbol{u}_{iv}$.

Based on the above derivations, we can formulate the optimization problem for the outer-loop subsystem of the ith AUV at the sampling instant k within the receding-horizon framework:

$$
\begin{aligned}
& \min_{\Delta U_{iv}} \; J_{i\eta}(k) \\
& s.t. \;\; \Gamma_{i\eta}\Delta U_{iv} \leq \gamma_{i\eta}.
\end{aligned}
\tag{31}
$$

To simplify the computation of (31), it can be transformed into a convex QP problem. This problem is solved over a finite receding horizon using a QP solver. The standard convex QP form of the DMPC problem (31) can be derived:

$$
\begin{aligned}
& \Delta U_{iv}^* = \underset{\Delta U_{iv}}{\operatorname{argmin}} \left(\tfrac{1}{2} \Delta U_{iv}^T W_{i\eta} \Delta U_{iv} + f_{i\eta}^T \Delta U_{iv} \right) \\
& s.t. \;\; \Gamma_{i\eta}\Delta U_{iv} \leq \gamma_{i\eta}
\end{aligned}
\tag{32}
$$

where $W_{i\eta} = \overline{\boldsymbol{R}}_{i1} + \boldsymbol{H}_{iu}^{1T}\overline{\boldsymbol{Q}}_{if}\boldsymbol{H}_{iu}^1 + \sum_{j\in N_i} a_{ij}\boldsymbol{H}_{iu}^{1T}\overline{\boldsymbol{Q}}_{ij}\boldsymbol{H}_{iu}^1$,

$$
f_{i\eta} = \boldsymbol{H}_{iu}^{1T}\overline{\boldsymbol{Q}}_{if}\left(\boldsymbol{H}_{ix}^1\boldsymbol{x}_{i\eta} - \boldsymbol{Y}_{if}\right) + \sum_{j\in N_i} a_{ij}\boldsymbol{H}_{iu}^{1T}\overline{\boldsymbol{Q}}_{ij}\left(\boldsymbol{H}_{ix}^1\boldsymbol{x}_{i\eta} - \boldsymbol{Y}_{ij}\right), \;\; \text{with} \;\; \boldsymbol{Y}_{if} =
$$

$\left[\boldsymbol{y}_{if}(k+1), \ldots, \boldsymbol{y}_{if}(k+N_{p1})\right]^T$, $\boldsymbol{Y}_{ij} = \left[\boldsymbol{y}_{ij}(k+1), \ldots, \boldsymbol{y}_{ij}(k+N_{p1})\right]^T$, $\overline{\boldsymbol{Q}}_{if} = diag\left\{\boldsymbol{Q}_{if}, \boldsymbol{Q}_{if}, \ldots, \boldsymbol{Q}_{if}\right\} \in \mathbb{R}^{5N_{p1}\times 5N_{p1}}$, $\overline{\boldsymbol{Q}}_{ij}$, and $\overline{\boldsymbol{R}}_{i1}$ are similar to $\overline{\boldsymbol{Q}}_{if}$, both corresponding compact matrices.

By solving the QP optimization problem in (32) online, we obtain the optimal control input increment sequence ΔU_{iv}^*. Of this sequence, we only utilize the first element $\Delta u_{iv}^*(k|k)$ for receding optimization. Once $\Delta u_{iv}^*(k)$ is determined, we obtain $v_i(k)$ which serves as the desired driving speed for the inner-loop controller of the ith AUV, i.e.,

$$
v_{ir}(k) = v_i(k) = v_i(k-1) + \Delta u_{iv}^*(k).
\tag{33}
$$

3.3. Inner-Loop Formation Prediction Control Law

In this subsection, with the aid of the proposed FTESO, we design a DMPC-based formation controller for the inner-loop subsystem to obtain the optimal driving force and moment for the ith AUV to track the desired speed.

The dynamic model (4) is discretized with a sampling period T_s, yielding the following discretized model:

$$
v_i(k+1) = \left(\boldsymbol{I} - \boldsymbol{M}_i^{*-1}T_s(\boldsymbol{C}_i^* + \boldsymbol{D}_i^*)\right)v_i(k) + \boldsymbol{M}_i^{*-1}T_s\boldsymbol{\tau}_i(k) + \boldsymbol{M}_i^{*-1}T_s\hat{\boldsymbol{\tau}}_{id}(k)
\tag{34}
$$

where $\hat{\tau}_{id}$ represents the compound disturbance compensated by FTESO (7), which is supposed to be invariant over a short period. It should be noted that we assume the center of gravity and buoyancy of the ith AUV to coincide, which allows $g_i(\eta_i)$ to approximate to zero. We select $x_{iv}(k) = \begin{bmatrix} v_i(k) & \tau_i(k-1) \end{bmatrix}^T$ as the state variable and take the increment $\Delta u_{i\tau}(k) = \tau_i(k) - \tau_i(k-1)$ as the control input. This allows us to reformulate the inner-loop predictive model as follows:

$$x_{iv}(k+1) = A_{iv}x_{iv}(k) + B_{iv}\Delta u_{i\tau}(k) + D_{iv} \tag{35}$$

$$y_{iv}(k) = C_{iv}x_{iv}(k) \tag{36}$$

where $A_{iv} = \begin{bmatrix} I_5 - M_i^{*-1}T_s(C_i^* + D_i^*) & M_i^{*-1}T_s \\ 0_5 & I_5 \end{bmatrix} \in \mathbb{R}^{10\times10}$, $B_{iv} = \begin{bmatrix} M_i^{*-1}T_s \\ I_5 \end{bmatrix} \in \mathbb{R}^{10\times5}$, $C_{iv} = \begin{bmatrix} I_5 & 0_5 \end{bmatrix} \in \mathbb{R}^{5\times10}$, and $D_{iv} = \begin{bmatrix} M_i^{*-1}T_s\hat{\tau}_{id} \\ 0_{5\times1} \end{bmatrix} \in \mathbb{R}^{10}$. Similar to our previous approach, we can characterize the relationship between the predicted output vector sequence and the control increment sequence using the following prediction equation:

$$Y_{iv} = H_{ix}^2 x_{iv}(k) + H_{iu}^2 \Delta U_{i\tau} + \overline{D}_{iv} \tag{37}$$

where $Y_{iv} = \begin{bmatrix} y_{iv}(k+1|k), y_{iv}(k+2|k), \ldots, y_{iv}(k+N_{p2}|k) \end{bmatrix}^T \in \mathbb{R}^{5N_{p2}}$,
$\Delta U_{i\tau} = \begin{bmatrix} \Delta u_{i\tau}(k|k), \Delta u_{i\tau}(k+1|k), \ldots, \Delta u_{i\tau}(k+N_{c2}-1|k) \end{bmatrix}^T \in \mathbb{R}^{5N_{c2}}$,
$H_{ix}^2 = \begin{bmatrix} C_{iv}A_{iv}, C_{iv}A_{iv}^2, \ldots, C_{iv}A_{iv}^{N_{p2}} \end{bmatrix}^T \in \mathbb{R}^{5N_{p2}\times10}$,

$$H_{iu}^2 = \begin{bmatrix} C_{iv}B_{iv} & 0_5 & \cdots & 0_5 \\ C_{iv}A_{iv}B_{iv} & C_{iv}B_{iv} & \cdots & 0_5 \\ \vdots & \vdots & \ddots & \vdots \\ C_{iv}A_{iv}^{N_{p2}-1}B_{iv} & C_{iv}A_{iv}^{N_{p2}-2}B_{iv} & \cdots & C_{iv}A_{iv}^{N_{p2}-N_{c2}}B_{iv} \end{bmatrix} \in \mathbb{R}^{5N_{p2}\times5N_{c2}}, \text{ and}$$

$\overline{D}_{iv} = \begin{bmatrix} C_{iv}D_{iv}, C_{iv}A_{iv}D_{iv} + C_{iv}D_{iv}, \ldots, C_{iv}\sum_{n=0}^{N_{p2}-1} A_{iv}^n D_{iv} \end{bmatrix}^T \in \mathbb{R}^{5N_{p2}}$. N_{p2} and N_{c2} denote the prediction and control horizon of the inner-loop controller, respectively.

According to the control objective, we assess the constraints on the control input increment and the actuator saturation in the inner-loop subsystem, as follows:

$$\Delta u_{i\tau}^{\min} \leq \Delta u_{i\tau}(k) \leq \Delta u_{i\tau}^{\max} \tag{38}$$

$$\tau_i^{\min} \leq \tau_i(k) \leq \tau_i^{\max} \tag{39}$$

where $\tau_i^{\min}$ and $\Delta u_{iv}^{\min}$ represent predefined lower bounds, while $\tau_i^{\max}$ and $\Delta u_{iv}^{\max}$ denote predefined upper bounds. The actuator saturation constraint (39) can be transformed into an input incremental constraint, and we can express the above constraints in a compact linear constraint form:

$$\Gamma_{iv}\Delta U_{i\tau} \leq \gamma_{iv} \tag{40}$$

where $\Gamma_{iv} = \begin{bmatrix} I_{5N_{c2}} \\ -I_{5N_{c2}} \\ I_{v1} \\ -I_{v1} \end{bmatrix}$, $\gamma_{i\eta} = \begin{bmatrix} \Delta U_{i\tau}^{\max} \\ -\Delta U_{i\tau}^{\min} \\ \overline{\tau}_i^{\max} - I_{v2}\tau_i(k-1) \\ -\overline{\tau}_{iv}^{\min} + I_{v2}\tau_i(k-1) \end{bmatrix}$, with $I_{v1} = \begin{bmatrix} I_5 & 0_5 & \cdots & 0_5 \\ I_5 & I_5 & \cdots & 0_5 \\ \vdots & \vdots & \ddots & \vdots \\ I_5 & I_5 & \cdots & I_5 \end{bmatrix} \in$

$\mathbb{R}^{5N_{c2}\times5N_{c2}}$ and $I_{v2} = \begin{bmatrix} I_5, I_5, \ldots, I_5 \end{bmatrix}^T \in \mathbb{R}^{5N_{c2}}$.

To achieve the convergence of the formation tracking velocity to the desired value, we define the local distributed cost function of the inner-loop subsystem as follows:

$$J_{iv}(k) = \sum_{l=1}^{N_{p2}} \|(y_{iv}(k+l|k) - v_{ir}(k+l))\|_{Q_{iv}}^2 + \sum_{l=0}^{N_{c2}-1} \|\Delta u_{i\tau}(k+l|k)\|_{R_{i2}}^2 \tag{41}$$

where $y_{iv}(k+l|k)$ and $\Delta u_{i\tau}(k+j|k)$ denote the predicted value of $y_{iv}(k+l)$ and $\Delta u_{i\tau}(k+j)$ at time k, respectively. Q_{iv} and R_{i2} are given weight matrices.

By substituting (37) into (41), we can formulate the DMPC optimization problem for the inner-loop subsystem of the ith AUV at sampling instant k as the following QP form:

$$\Delta U_{i\tau}^* = \underset{\Delta U_{i\tau}}{\mathrm{argmin}} \left(\tfrac{1}{2}\Delta U_{i\tau}^T W_{iv}\Delta U_{i\tau} + f_{iv}^T \Delta U_{i\tau} \right)$$
$$s.t. \ \Gamma_{iv}\Delta U_{i\tau} \le \gamma_{iv} \tag{42}$$

where $W_{iv} = \overline{R}_{i2} + H_{iu}^{2T}\overline{Q}_{iv}H_{iu}^2$ and $f_{iv} = H_{iu}^{2T}\overline{Q}_{iv}(H_{ix}^2 x_{iv} + \overline{D}_{iv} - V_{ir})$, with $v_{ir} = \left[v_{ir}(k+1),\dots,v_{ir}(k+N_{p2}) \right]^T \in \mathbb{R}^{5N_{p2}}$, $\overline{Q}_{iv} = diag\{Q_{iv}, Q_{iv}, \dots, Q_{iv}\} \in \mathbb{R}^{5N_{p2}\times 5N_{p2}}$, and $\overline{R}_{i2} = diag\{R_{i2}, R_{i2}, \dots, R_{i2}\} \in \mathbb{R}^{5N_{c2}\times 5N_{c2}}$.

The solution of the QP optimization problem (42) yields the optimal control input increment sequence $\Delta U_{i\tau}^*$ at time k. However, only the first element $\Delta u_{i\tau}^*(k|k)$ of the sequence is used for the ith AUV to obtain the optimal control force and moment $\tau_i^*(k) = \tau_i(k-1) + \Delta u_{i\tau}^*(k)$. The $\Delta u_{i\tau}^*(k)$ is recalculated at each sampling instant, the ith AUV repeatedly calculates and executes $\tau_i^*(k)$ to achieve receding optimization. The predicted state $x_{iv}(k+1)$ and the optimal input $\tau_i^*(k)$ are both determined solely by the current state $x_{iv}(k)$.

With the parallel optimization of N AUV subsystems, all local optimization problems are solved simultaneously at each sampling moment. One or more information interactions occur between local controllers to obtain the optimal input sequence for that moment. Thus, the proposed control law can compensate well for the compound disturbances, which consist of model uncertainties and external disturbances. This occurs throughout the iterative optimization process, while simultaneously ensuring collision avoidance and formation tracking control tasks under complex constraints.

3.4. Use of Laguerre Functions in the DMPC Design

This subsection introduces a strategy to handle the computational burden caused by a longer control horizon and dual closed-loop structure. This is the main difficulty in our theoretical analysis. The Laguerre orthogonal functions are leveraged in the DMPC design to decrease the order of the input incremental matrices. This approach permits a reduction in input variables during each control cycle, thereby reducing the computational burden in the time interval and improving real-time performance.

The Laguerre functions are a set of discrete orthogonal polynomial functions, let it be $l_1(k), l_2(k), \dots, l_M(k)$, the z-transfer of the mth Laguerre function is expressed as follows:

$$X_m(z) = \frac{\sqrt{1-a^2}}{z-a}\left[\frac{1-az}{z-a}\right]^{m-1} \tag{43}$$

where $0 \le a < 1$ denotes the pole of the Laguerre function, also known as the scaling factor. It can be verified that X_m satisfies the following orthogonality:

$$\begin{cases} \frac{1}{2\pi}\int_{-\pi}^{\pi} X_m(e^{j\omega})X_n(e^{j\omega})^* d\omega = 1 & m = n \\ \frac{1}{2\pi}\int_{-\pi}^{\pi} X_m(e^{j\omega})X_n(e^{j\omega})^* d\omega = 0 & m \ne n \end{cases} \tag{44}$$

where $(\cdot)^*$ denotes complex conjugate of $(\cdot)$.

The discrete Laguerre functions are defined by taking the inverse Z-transform of (43), i.e., $l_m(k) = Z^{-1}\{X_m(z)\}$. Given the network structure of $X_m(z)$ and the recurrence relation, the set of discrete Laguerre functions satisfies the following difference equation:

$$L(k+1) = \Xi L(k) \tag{45}$$

where

$$\Xi = \begin{bmatrix} a & 0 & 0 & \cdots & 0 \\ \beta & a & 0 & \cdots & 0 \\ -a\beta & \beta & a & \cdots & 0 \\ \vdots & \vdots & \vdots & \ddots & \vdots \\ (-a)^{M-2}\beta & (-a)^{M-3}\beta & (-a)^{M-4}\beta & \cdots & a \end{bmatrix} \quad \text{and} \quad L(k) =$$

$[l_1(k), l_2(k), \ldots, l_M(k)]^T$, with $\beta = 1 - a^2$ and initial condition $L(0) = \sqrt{\beta}\left[1, -a, a^2, -a^3, \ldots, (-a)^{M-1}\right]^T$. Note that at $a = 0$, the Laguerre functions are converted to impulse functions.

Assuming the current moment is k, the input increment of the single-input system at the next time l, represented by the Laguerre function, is:

$$\Delta u(k+l) = \sum_{m=1}^{M} \kappa_m l_m(l) = L(l)^T K \tag{46}$$

where $K = [\kappa_1, \kappa_2, \ldots, \kappa_M]^T$. When we extend this to the multi-AUV system, each AUV has five independent control inputs, and the input increment of the ith AUV is as follows:

$$\Delta u_i(k)^T = \left[L_i^1(k)^T K_i^1, L_i^2(k)^T K_i^2, \ldots, L_i^5(k)^T K_i^5\right] = \overline{L}_i(k)^T \overline{K}_i \tag{47}$$

where $L_i^p(k) = \left[l_{i1}^p(k), l_{i2}^p(k), \ldots, l_{iM}^p(k)\right]^T$ and $K_i^p = \left[\kappa_{i1}^p, \kappa_{i2}^p, \ldots, \kappa_{iM}^p\right]^T$, with $p = 1, 2, \ldots 5$. $\overline{L}_i(k) = diag\left\{L_i^1(k)^T, L_i^2(k)^T, \ldots, L_i^5(k)^T\right\}$, and $\overline{K}_i = \left[K_i^{1T}, K_i^{2T}, \ldots, K_i^{5T}\right]^T$. Note that within a multi-input structure, the scaling factor a_p and the number of polynomial terms M_p can be selected independently for each input signal.

For illustrative purposes, the inner-loop predictive controller of the ith AUV is taken as an example. If we partition the input matrix into $B_{iv} = \begin{bmatrix} B_{iv}^1 & B_{iv}^2 & \cdots & B_{iv}^5 \end{bmatrix}$, the prediction of the system output in the next l steps can be derived as follows:

$$\begin{aligned} y_{iv}(k+l|k) = & \sum_{j=0}^{l-1} C_{iv}A_{iv}^{l-j-1}\left[B_{iv}^1 L_i^1(j)^T K_{i\tau}^1 \quad B_{iv}^2 L_i^2(j)^T K_{i\tau}^2 \quad \cdots \quad B_{iv}^5 L_i^5(j)^T K_{i\tau}^5 \right] \\ & + C_{iv}A_{iv}^l x_{iv}(k) + \sum_{j=0}^{l-1} C_{iv}A_{iv}^{l-j-1}D_{iv}. \end{aligned} \tag{48}$$

For a compact notation, we denote (48) by the following:

$$y_{iv}(k+l|k) = C_{iv}A_{iv}^l x_{iv}(k) + \mu_i(l)^T \overline{K}_{i\tau} + D_{iv}^l \tag{49}$$

where $\mu_i(l)^T = \sum_{j=0}^{l-1} C_{iv}A_{iv}^{l-j-1}\left[B_{iv}^1 L_i^1(j)^T \quad B_{iv}^2 L_i^2(j)^T \quad \cdots \quad B_{iv}^5 L_i^5(j)^T\right]$ and $D_{iv}^l = \sum_{j=0}^{l-1} C_{iv}A_{iv}^{l-j-1}D_{iv}$. $\overline{K}_{i\tau}$ as the parameter vector that is to be optimized.

First, we employ the Laguerre function to optimize the constraint terms (38) and (39), leading to the following constraint form:

$$\Delta u_{i\tau}^{\min} \leq \overline{L}_{i\tau}^{\ T}\overline{K}_{i\tau} \leq \Delta u_{i\tau}^{\max} \tag{50}$$

$$\tau_i^{\min} \leq \widehat{L}_{i\tau}\overline{K}_{i\tau} + \tau_i(k-1) \leq \tau_i^{\max} \tag{51}$$

where $\overline{L}_{i\tau} = diag\left\{L_{i\tau}^1(l)^T, \ldots, L_{i\tau}^5(l)^T\right\}$ and $\widehat{L}_{i\tau} = diag\left\{\sum_{j=0}^{l-1} L_{i\tau}^1(j)^T, \sum_{j=0}^{l-1} L_{i\tau}^2(j)^T, \ldots, \sum_{j=0}^{l-1} L_{i\tau}^5(j)^T\right\}$.

Given that the Laguerre functions are orthonormal for a sufficiently large control horizon N_{c2}. Substituting (47) into (41) and using the orthogonality (44) of the Laguerre function (i.e., the inner product of different terms is 0 and the same term is 1), the following derivation can be performed to obtain the reconstructed form of the cost function (41):

$$
\begin{aligned}
J_{iv}(k) &= \sum_{l=1}^{N_{p2}} \|(y_{iv}(k+l|k) - v_{ir}(k+l))\|_{Q_{iv}}^2 + \sum_{l=0}^{N_{c2}-1} \Delta u_{i\tau}(k+l|k)^T R_{i2} \Delta u_{i\tau}(k+l|k) \\
&= \sum_{l=1}^{N_{p2}} \|(y_{iv}(k+l|k) - v_{ir}(k+l))\|_{Q_{iv}}^2 + \sum_{l=0}^{N_{c2}-1} \left(\overline{L}_{i\tau}(l)^T \overline{K}_{i\tau}\right) R_{i2} \left(\overline{L}_{i\tau}(l)^T \overline{K}_{i\tau}\right)^T \\
&= \sum_{l=1}^{N_{p2}} [y_{iv}(k+l|k) - v_{ir}(k+l)]^T Q_{iv} [y_{iv}(k+l|k) - v_{ir}(k+l)] \\
&\quad + \sum_{l=0}^{N_{c2}-1} \left(diag\{L_{i\tau}^1(l), L_{i\tau}^2(l), \ldots, L_{i\tau}^5(l)\}\overline{K}_{i\tau}\right) R_{i2} \left(diag\{L_{i\tau}^1(l), L_{i\tau}^2(l), \ldots, L_{i\tau}^5(l)\}\overline{K}_{i\tau}\right)^T \\
&= \sum_{l=1}^{N_{p2}} [y_{iv}(k+l|k) - v_{ir}(k+l)]^T Q_{iv} [y_{iv}(k+l|k) - v_{ir}(k+l)] + \overline{K}_{i\tau}^T R_{i2} \overline{K}_{i\tau}.
\end{aligned}
\tag{52}
$$

By substituting (49) into (52), we can rewrite the DMPC optimization problem (42) for the inner-loop subsystem of the *i*th AUV as:

$$
\begin{aligned}
\min_{\overline{K}_{i\tau}^*} J_{iv}(k) &= \min_{\overline{K}_{i\tau}^*} \left(\tfrac{1}{2}\overline{K}_{i\tau}^T W_{iL} \overline{K}_{i\tau} + f_{iL}^T \overline{K}_{i\tau}\right) \\
s.t. \quad &(50), \ (51)
\end{aligned}
\tag{53}
$$

where $W_{iL} = \sum_{l=1}^{N_{p2}} \mu_i(l) Q_{iv} \mu_i(l)^T + R_{i2}$ and $f_{iL} = \sum_{l=1}^{N_{p2}} \mu_i(l) Q_{iv} \left(C_{iv} A_{iv}^l x_{iv}(k) + D_{iv}^l - v_{ir}(k+l)\right)$.

The QP optimization Equation (53), with constraints, can be solved to obtain the optimal parameter vector $\overline{K}_{i\tau}^*$. This vector replaces the conventional DMPC method calculation of $\Delta u_{i\tau}^*$. Thus, the optimal input increment of the inner-loop subsystem is indirectly obtained by the rolling optimized control law, $\Delta u_{i\tau}(k)^T = \overline{L}_{i\tau}(0)^T \overline{K}_{i\tau}$, until the control variables at the next moment are calculated. This iterative process ensures the achievement of receding horizon optimization. The use of the Laguerre function in the design of the outer-loop predictive controller is not included here, as its analysis parallels that of the inner-loop controller described above.

Remark 2. By parameterizing the input increment sequence using the Laguerre function, the input matrix order in the prediction horizon can be lowered, thereby reducing the computational load online. This property enables its application in large-scale and real-time AUV control systems. With the employment of the Laguerre function, the coefficients a_p and M_p can also be served as tuning parameters, in addition to the control and prediction horizon and weighting matrices. Larger a_p and M_p lead to faster closed-loop responses [38].

3.5. Stability Analysis

A notable attribute of the MPC is the potential for establishing the stability of a closed-loop system under certain conditions. Extending this to cases using Laguerre polynomials, a terminal state constraint is utilized to analyze the stability of the closed-loop system. Specifically, for the inner-loop subsystem, an additional constraint is attached to the final state of the receding optimization problem: $x_{iv}(k + N_{p2}) = 0$, where $x_{iv}(k + N_{p2})$ is the terminal state produced under the effect of the control sequence, $\Delta u_{i\tau}(k+l)^T = \overline{L}_{i\tau}(l)^T \overline{K}_{i\tau}$.

Theorem 2. *Consider the inner-loop subsystem (35) and (36) of the ith AUV in the formation control system, which has a local cost function (41) with constraints (38) and (39). The inner-loop predictive control subsystem is asymptotically stable if for each sampling instant k, there exists a solution $\overline{K}_{i\tau}$ such that the performance index J_{iv} is minimized subject to the terminal state constraint.*

Proof of Theorem 2. Constructing an appropriate Lyapunov function is key to ensuring the stability of the DMPC system. Select the cost function $J_{iv}(k)$ as the Lyapunov function $V_{i2}(y(k),k)$:

$$V_{i2}(y(k),k) = \sum_{l=1}^{N_{p2}} \tilde{y}_{iv}(k+l|k)^T Q_{iv}\tilde{y}_{iv}(k+l|k) + \sum_{l=0}^{N_{c2}-1} \Delta u_{i\tau}(k+l)^T R_{i2}\Delta u_{i\tau}(k+l) \quad (54)$$

where $\tilde{y}_{iv}(k+l|k) = y_{iv}(k+l|k) - v_{ir}(k+l)$ and $y_{iv}(k+l|k) = C_{iv}A_{iv}^l x_{iv}(k) + \mu_i(l)^T \overline{K}_{i\tau}^k + D_{iv}^l$, $\overline{K}_{i\tau}^k$ is the parameter vector solution of the cost function (41) under the original and terminal constraints at moment k, and input increment $\Delta u_{i\tau}(k+l)^T = \overline{L}_{i\tau}(l)^T \overline{K}_{i\tau}^k$. It is clear that $V_{i2}(y(k),k)$ is positive definite and tends to infinity as $y_{iv}(k)$ tends to infinity. Similarly, the Lyapunov function at moment $k+1$ can be derived as:

$$\begin{aligned} V_{i2}(y(k+1),k+1) =\ & \sum_{l=1}^{N_{p2}} \tilde{y}_{iv}(k+1+l|k+1)^T Q_{iv}\tilde{y}_{iv}(k+1+l|k+1) \\ & + \sum_{l=0}^{N_{c2}-1} \Delta u_{i\tau}(k+1+l)^T R_{i2}\Delta u_{i\tau}(k+1+l) \end{aligned} \quad (55)$$

where $y_{iv}(k+1+l|k+1) = C_{iv}A_{iv}^l x_{iv}(k+1) + \mu_i(l)^T \overline{K}_{i\tau}^{k+1} + D_{iv}^l$, $\overline{K}_{i\tau}^{k+1}$ is the parameter vector solution at time $k+1$, and $\Delta u_{i\tau}(k+1+l)^T = \overline{L}_{i\tau}(l)^T \overline{K}_{i\tau}^{k+1}$. Given that $y_{iv}(k+1)$ is the response one step ahead of $y_{iv}(k)$ and $y_{iv}(k+1) = C_{iv}A_{iv}x_{iv}(k) + C_{iv}B_{iv}\Delta u_{i\tau}(k) + C_{iv}D_{iv}$, the feasible solution of $\overline{K}_{i\tau}^{k+1}$ corresponding to the initial output $y_{iv}(k+1)$ in the receding horizon is $\overline{K}_{i\tau}^k$. Therefore, the feasible solution sequence at moment $k+1$ is to move the elements in $\overline{L}_{i\tau}(0)^T\overline{K}_{i\tau}^k$, $\overline{L}_{i\tau}(1)^T\overline{K}_{i\tau}^k$, ..., $\overline{L}_{i\tau}(N_{c2}-1)^T\overline{K}_{i\tau}^k$ one step forward and substitute the last element with 0, i.e., $\overline{L}_{i\tau}(1)^T\overline{K}_{i\tau}^k$, $\overline{L}_{i\tau}(2)^T\overline{K}_{i\tau}^k$, ..., $\overline{L}_{i\tau}(N_{c2}-1)^T\overline{K}_{i\tau}^k$, $0_{5\times1}$. Due to the optimality of the solution $\overline{K}_{i\tau}^{k+1}$ at $k+1$, it follows that

$$V_{i2}(y(k+1),k+1) \leq \hat{V}_{i2}(y(k+1),k+1) \quad (56)$$

where $\hat{V}_{i2}(y(k+1),k+1)$ is identical to (55) except that the parameter vector solution $\overline{K}_{i\tau}^{k+1}$ in the control sequence is replaced by the feasible solution $\overline{K}_{i\tau}^k$. The difference between $V_{i2}(y(k),k)$ and $V_{i2}(y(k+1),k+1)$ is then bounded by the following:

$$V_{i2}(y(k+1),k+1) - V_{i2}(y(k),k) \leq \hat{V}_{i2}(y(k+1),k+1) - V_{i2}(y(k),k). \quad (57)$$

Eliminate the same terms in the control sequence and output sequence of $\hat{V}_{i2}(y(k+1),k+1)$ and $V_{i2}(y(k),k)$ at moments $k+1$, $k+2$,..., $k+N_{p2}-1$, and we can derive the following equation:

$$\begin{aligned} \hat{V}_{i2}(y(k+1),k+1) - V_{i2}(y(k),k) =\ & \tilde{y}_{iv}(k+N_{p2}|k)^T Q_{iv}\tilde{y}_{iv}(k+N_{p2}|k) \\ & -\tilde{y}_{iv}(k+1|k)^T Q_{iv}\tilde{y}_{iv}(k+1|k) - \Delta u_{i\tau}(k)^T R_{i2}\Delta u_{i\tau}(k). \end{aligned} \quad (58)$$

Given the terminal constraint $x_{iv}(k+N_{p2}) = 0$ is applied, equivalent to $y_{iv}(k+N_{p2}) = 0$, we have the following:

$$\begin{aligned} \hat{V}_{i2}(y(k+1),k+1) - V_{i2}(y(k),k) =\ & -v_{ir}(k+N_{p2})^T Q_{iv}v_{ir}(k+N_{p2}) \\ & -\tilde{y}_{iv}(k+1|k)^T Q_{iv}\tilde{y}_{iv}(k+1|k) - \Delta u_{i\tau}(k)^T R_{i2}\Delta u_{i\tau}(k). \end{aligned} \quad (59)$$

This allows inequality (57) to be converted into:

$$\begin{aligned} V_{i2}(y(k+1),k+1) - V_{i2}(y(k),k) \leq\ & -v_{ir}(k+N_{p2})^T Q_{iv}v_{ir}(k+N_{p2}) \\ & -\tilde{y}_{iv}(k+1|k)^T Q_{iv}\tilde{y}_{iv}(k+1|k) - \Delta u_{i\tau}(k)^T R_{i2}\Delta u_{i\tau}(k) < 0. \end{aligned} \quad (60)$$

Namely, $V_{i2}(y(k+1), k+1) < V_{i2}(y(k), k)$; the Lyapunov function is monotonically decreasing. This proves that the inner loop subsystem is asymptotically stable. $\square$

Next, we analyze the stability of the entire closed-loop system. Analogous to the proof of Theorem 2, we select $J_{i\eta}(k)$ as the Lyapunov function $V_{i3}(y(k), k)$ of the outer-loop subsystem:

$$
\begin{aligned}
V_{i3}(y(k), k) = & \sum_{l=1}^{N_{p1}} \left\| \left(y_{i\eta}(k+l|k) - y_{if}(k+l) \right) \right\|_{Q_{if}}^2 + \sum_{l=0}^{N_{c1}-1} \| \Delta u_{iv}(k+l|k) \|_{R_{i1}}^2 \\
& + \sum_{l=1}^{N_{p1}} \sum_{j \in N_i} a_{ij} \left\| \left(y_{i\eta}(k+l|k) - y_{ij}(k+l) \right) \right\|_{Q_{ij}}^2 .
\end{aligned}
\tag{61}
$$

According to the idea of the proof of Theorem 2, the following inequality can be obtained:

$$
\begin{aligned}
V_{i3}(y(k+1), k+1) - V_{i3}(y(k), k) \leq\ & -y_{if}(k+N_{p1})^T Q_{if} y_{if}(k+N_{p1}) - \tilde{y}_{i\eta}(k+1|k)^T Q_{if} \tilde{y}_{i\eta}(k+1|k) \\
& - N(N-1) r_{ij}(k+N_{p1})^T Q_{ij} r_{ij}(k+N_{p1}) - \Delta u_{iv}(k)^T R_{i1} \Delta u_{iv}(k) < 0.
\end{aligned}
\tag{62}
$$

Next, we set the Lyapunov function of the entire closed-loop system as follows:

$$
V_{i4}(y(k), k) = V_{i2}(y(k), k) + V_{i3}(y(k), k).
\tag{63}
$$

From inequalities (60) and (62), we have the following:

$$
\begin{aligned}
V_{i4}(y(k+1), k+1) - V_{i4}(y(k), k)\ &= V_{i2}(y(k+1), k+1) - V_{i2}(y(k), k) \\
&+ V_{i3}(y(k+1), k+1) - V_{i3}(y(k), k) < 0
\end{aligned}
\tag{64}
$$

As a result, the entire closed-loop system is asymptotically stable.

4. Simulation

In this section, some simulation analyses are conducted to verify the effectiveness and robustness of the proposed control scheme. A formation system consisting of four AUVs ($N = 4$, $i = 1, 2, 3, 4$) with a virtual leader (AUV0) is considered. The directed communication topology for the simulation is depicted in Figure 3, the meaning of the arrows is the direction of the communication or information flow between the nodes in the formation network. Initial values for x_i, y_i, and z_i are randomly distributed within the intervals $[10, 40]$, $[0, 30]$, and $[-10, 0]$, respectively, while the attitude angles θ_i and ψ_i lie within the intervals $[-\pi/18, \pi/18]$ and $[0, \pi]$, respectively. The parameters related to the AUVs are based on previous research [39]. A diamond formation was predefined to facilitate omnidirectional exploration, with the desired formation configuration preset to $r_{1f} = [0, 0, 6.5, 0, 0]^T$, $r_{2f} = [0, -7.5, 0, 0, 0]^T$, $r_{3f} = [0, 0, -6.5, 0, 0]^T$, and $r_{4f} = [0, 7.5, 0, 0, 0]^T$. $r_{12} = -r_{21} = [0, 7.5, 6.5, 0, 0]^T$, $r_{13} = -r_{31} = [0, 0, 13, 0, 0]^T$, $r_{14} = -r_{41} = [0, -7.5, 6.5, 0, 0]^T$, $r_{23} = -r_{32} = [0, -7.5, 6.5, 0, 0]^T$, $r_{24} = -r_{42} = [0, -15, 0, 0, 0]^T$ and $r_{34} = -r_{43} = [0, -7.5, -6.5, 0, 0]^T$. The safety distance during the formation construction stage is set as $r_s = 3\,m$, while the detection distance measured using sonar is set as $r_d = 6\,m$. To reflect model uncertainties, 20% of the nominal values were taken as model errors, meaning that the parameters for the AUVs in the simulation represented only 80% of the nominal system dynamics. External disturbances were applied to each AUV to evaluate the formation robustness, modeled as follows [34]:

$$
\left\{
\begin{aligned}
\tau_{icu} &= 0.1\mathrm{sign}(u_i) + 0.2\sin(0.1t)\ N \\
\tau_{icv} &= 0.1\mathrm{sign}(v_i) + 0.3\sin(0.3t)\ N \\
\tau_{icw} &= 0.08\mathrm{sign}(w_i) + 0.2\sin(0.5t)\ N \\
\tau_{icq} &= 0.02\mathrm{sign}(q_i) + 0.1\sin(0.3t)\ N \cdot m \\
\tau_{icr} &= 0.05\mathrm{sign}(r_i) + 0.1\sin(0.3t)\ N \cdot m
\end{aligned}
\right.
\tag{65}
$$

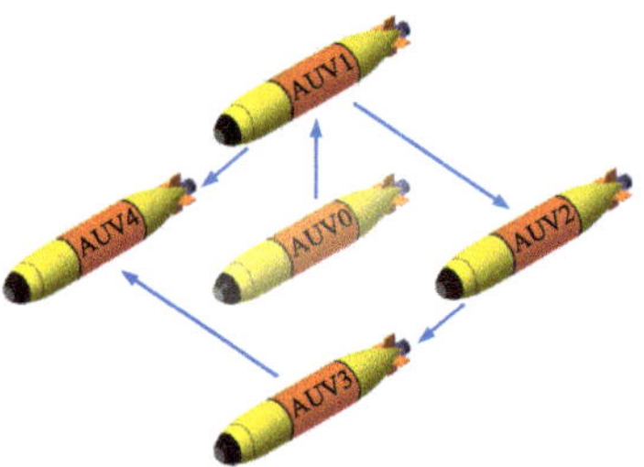

Figure 3. Structure of communication topology.

Each control parameter has its settings guidelines: Given the low driving speed of the AUV in this paper, smaller N_{p1} and N_{p2} are intended to be used. During debugging, reduce it if the rapidity is not enough, and increase it if the stability is not good; the selection of N_{c1} and N_{c2} is based on a trade-off between performance and computation [40]; since we value the position tracking performance more than the velocity tracking performance, $\boldsymbol{Q}_{if}$ is set slightly larger than $\boldsymbol{Q}_{iv}$; to weaken the interaction of angles between AUVs, the orientation weight in $\boldsymbol{Q}_{ij}$ is set slightly smaller; when tuning $\boldsymbol{R}_{i1}$ and $\boldsymbol{R}_{i2}$, it can be set very small first, and then increase it slightly if the system is stable and the control variable does not change too drastically [41]. By solving the Lyapunov Equation (11), the relationship between the observer gains β_{ik} and α_{i1}, such that $\boldsymbol{A}_{i1}$ and $\boldsymbol{A}_{i2}$ are Hurwitz matrices, can be obtained, and tuned to select the appropriate values [42]; the Laguerre parameter a_p is adjusted within the constraint interval, and a smaller M_p is selected to coordinate the number of constraints in the optimization problem, and to make a trade-off between response speed and control complexity [38]. Following the above guidelines, we dealt with the main difficulties in the simulation and selected the parameters that produced the optimal simulation results and listed them in Table 1.

Table 1. Control parameters of the proposed scheme.

Parameter	Value	Parameter	Value
$\boldsymbol{Q}_{if}$	$\mathrm{diag}(10^2, 10^2, 10^2, 10^2, 10^2)$	T_s	0.5 s
$\boldsymbol{Q}_{ij}$	$\mathrm{diag}(10^2, 10^2, 10^2, 10, 10)$	N_{p1}	20
$\boldsymbol{Q}_{iv}$	$\mathrm{diag}(10, 10, 10, 10, 10)$	N_{c1}	8
$\boldsymbol{R}_{i1}$	$\mathrm{diag}(10^{-2}, 10^{-2}, 10^{-2}, 10^{-2}, 10^{-2})$	N_{p2}	10
$\boldsymbol{R}_{i2}$	$\mathrm{diag}(10^{-1}, 10^{-1}, 10^{-1}, 10^{-1}, 10^{-1})$	N_{c2}	5
β_{i1}	$\mathrm{diag}(20, 20, 20, 10, 10)$	α_{i1}	0.75
β_{i2}	$\mathrm{diag}(160, 160, 160, 80, 80)$	a_p	0.7
β_{i3}	$\mathrm{diag}(160, 160, 160, 80, 80)$	M_p	3

Moreover, based on the actual speed limit of the thruster, we provide the state and input constraints as follows: $\Delta \boldsymbol{U}_{iv}^{\max} = -\Delta \boldsymbol{U}_{iv}^{\min} = [0.2, 0.1, 0.2, 0.05, 0.05]^T$, $\boldsymbol{U}_{iv}^{\max} = -\boldsymbol{U}_{iv}^{\min} = [1.5, 1, 1, 0.05, 0.2]^T$, and $\Delta \boldsymbol{U}_{i\tau}^{\max} = -\Delta \boldsymbol{U}_{i\tau}^{\min} = [50, 50, 100, 5, 5]^T$. To avoid actuator saturation for each AUV, the bounds of force and moment are set as $\tau_i^{\min} = [-200, -500, -500, -7, -10]^T$ and $\tau_i^{\max} = [300, 500, 500, 7, 10]^T$. The reference trajectory generated by the virtual leader is a 3-D spiral curve, defined as follows:

$$\begin{cases} x_r(t) = 30\cos(0.005\pi t) \\ y_r(t) = 30\sin(0.005\pi t) \\ z_r(t) = -0.05t - 3 \end{cases} \tag{66}$$

To verify the disturbance compensation performance of the proposed FTESO (7), we conducted comparative simulations with the ESO (67) from [43] and the FTESO (68) from [18]. Figure 4 shows the norms of the compound disturbance estimation errors $\|e_{i2}\| = \|\hat{\tau}_{id} - \tau_{id}\|$ for the four AUVs under the three observers, characterizing insights into transient and steady-state responses. We calculated the settling time of the designed FTESO in the simulation and highlighted it on the plots. It is clear from Figure 4 that our proposed third-order fast FTESO can achieve finite-time stabilization, with the estimation errors converging to a small neighborhood of the origin. And the dynamic convergence speed and estimation accuracy of the proposed FTESO are better than ESO (67) and FTESO (68) with less chattering. This shows the advantages of our approach. Thus, each AUV can accurately compensate for model uncertainties and external disturbances of its corresponding subsystem in finite time.

$$\begin{cases} \dot{\hat{z}}_{i1} = \hat{z}_{i2} - \beta_{i1}a_{i1}e_{i1} + G_i(\eta_i, v_i) + \tau_i \\ \dot{\hat{z}}_{i2} = -\beta_{i1}^2 a_{i2}e_{i1} \end{cases}. \tag{67}$$

$$\begin{cases} \dot{\hat{z}}_{i1} = \hat{z}_{i2} - \beta_{i1}sig^{3/4}(e_{i1}) + G_i(\eta_i, v_i) + \tau_i \\ \dot{\hat{z}}_{i2} = -\beta_{i2}sig^{1/2}(e_{i1}) \end{cases}. \tag{68}$$

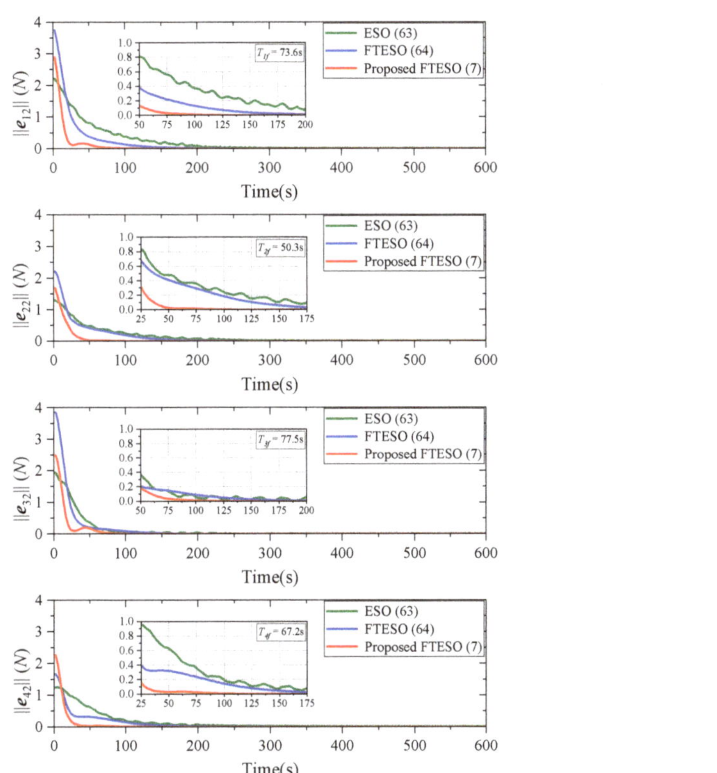

Figure 4. Compound disturbance estimation errors e_{i2} of the ith AUV.

The collision avoidance performance of the AUV formation was tested via a set of comparison experiments with and without collision avoidance constraints based on our proposed scheme. Since the initial positions of the four AUVs are randomly distributed, the risk of collision is increased. The formation trajectory without collision avoidance constraints is shown in Figure 5 (top). Here, the four AUVs track the reference trajectory while keeping the preset shape, but AUV3 and AUV4 collide at 10 s, followed by a collision between AUV1 and AUV2 at 20 s. Specifically, as presented in Figure 6 (top), the relative distance between AUV1 and AUV2 during the formation configuration stage exceeds the safe distance, resulting in a collision. The same situation occurs with AUV3 and

AUV4. However, when collision avoidance constraints are considered, the formation trajectory (shown in Figure 5 (**bottom**)) indicates that the four AUVs can perform the formation tracking task while avoiding collision during the configuration stage. The collision avoidance performance is visualized in Figure 6 (**bottom**), where the distances among AUVs within the detection zone are always greater than the safe distance, indicating that inter-vehicle collision avoidance can be achieved. Therefore, the proposed control scheme can provide real-time collision avoidance capability for AUV formation maneuvers.

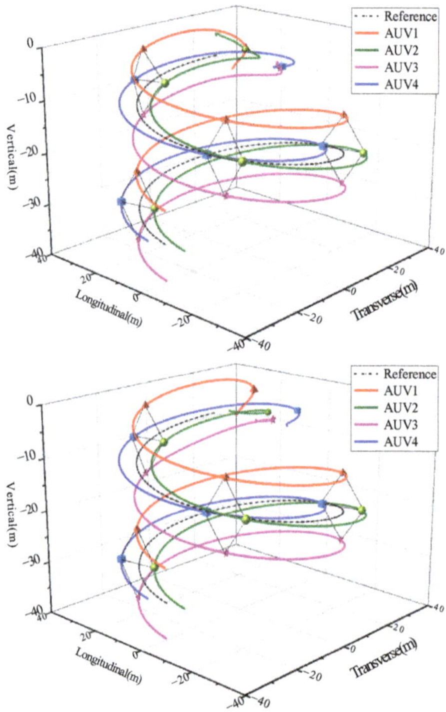

Figure 5. 3-D formation trajectories without (**top**) and with (**bottom**) collision avoidance constraints.

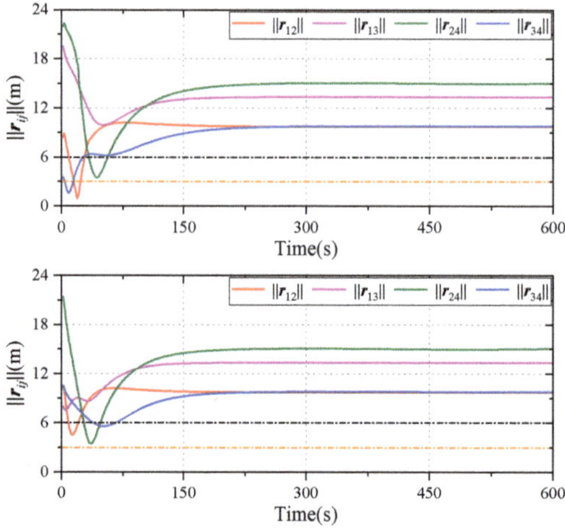

Figure 6. Relative distances among AUVs without (**top**) and with (**bottom**) collision avoidance constraints.

In order to assess the feasibility and superiority of the proposed scheme, we conducted three sets of comparative simulations with the same parameters, constraints, and disturbance settings: (a) the proposed FTESO-based dual closed-loop DMPC with Laguerre function; (b) a FTESO-based dual closed-loop DMPC without Laguerre function; (c) a standard DMPC without FTESO. Figures 7–16 plot the tracking performance curves of AUV positional and velocity states under the three schemes. It can be easily observed that, in all scenarios, the four AUVs are able to successfully track the desired state despite the differing tracking errors. In scheme (a), full-state stable tracking is achieved within 200 s. Meanwhile, in scheme (b), the process takes about 300 s, which suggests that the use of the Laguerre function improves both the response speed and control accuracy. Although the standard DMPC scheme (c) can also achieve formation tracking, the settling time of the state variables is longer and accompanied by oscillations due to the uncompensated compound disturbance effects. Compared with the other schemes, our proposed method delivers superior formation tracking control performance.

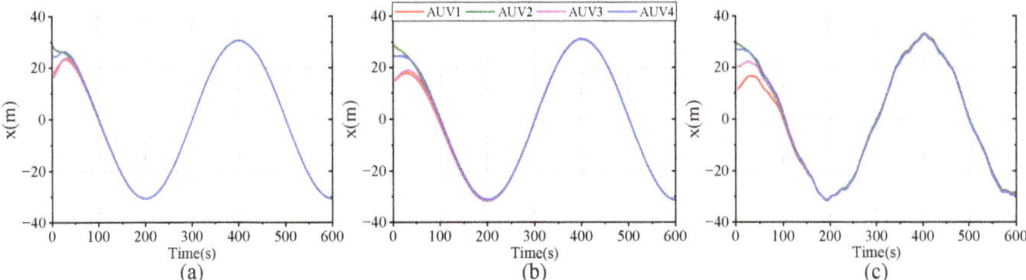

Figure 7. State x_i of the AUV formation system. (**a**) The proposed FTESO-based dual closed-loop DMPC with Laguerre function. (**b**) The FTESO-based dual closed-loop DMPC without Laguerre function. (**c**) The standard DMPC without FTESO.

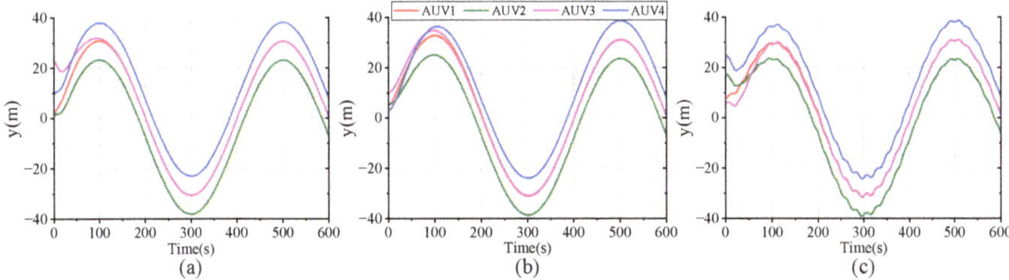

Figure 8. State y_i of the AUV formation system. (**a**) The proposed FTESO-based dual closed-loop DMPC with Laguerre function. (**b**) The FTESO-based dual closed-loop DMPC without Laguerre function. (**c**) The standard DMPC without FTESO.

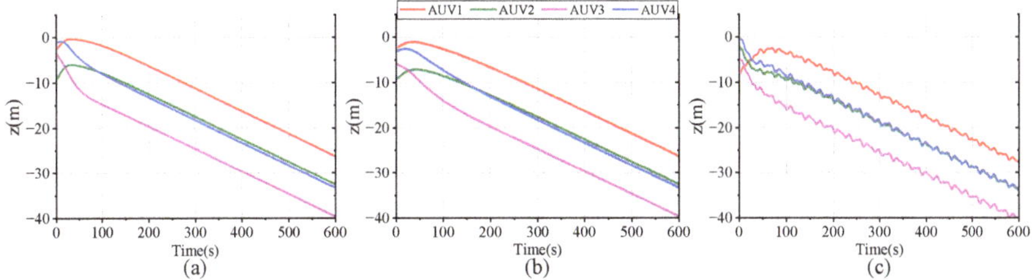

Figure 9. State z_i of the AUV formation system. (**a**) The proposed FTESO-based dual closed-loop DMPC with Laguerre function. (**b**) The FTESO-based dual closed-loop DMPC without Laguerre function. (**c**) The standard DMPC without FTESO.

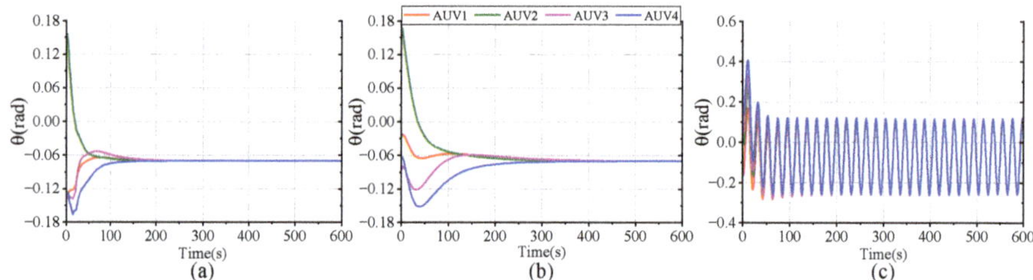

Figure 10. State θ_i of the AUV formation system. (**a**) The proposed FTESO-based dual closed-loop DMPC with Laguerre function. (**b**) The FTESO-based dual closed-loop DMPC without Laguerre function. (**c**) The standard DMPC without FTESO.

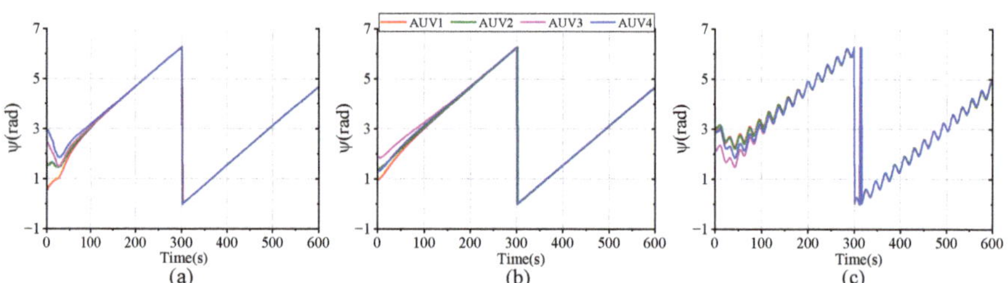

Figure 11. State ψ_i of the AUV formation system. (**a**) The proposed FTESO-based dual closed-loop DMPC with Laguerre function. (**b**) The FTESO-based dual closed-loop DMPC without Laguerre function. (**c**) The standard DMPC without FTESO.

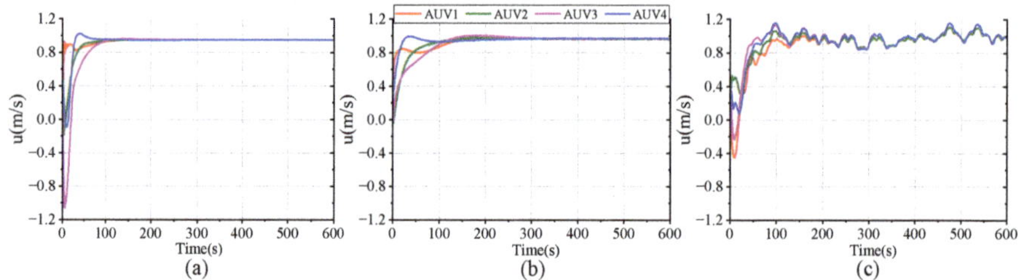

Figure 12. State u_i of the AUV formation system. (**a**) The proposed FTESO-based dual closed-loop DMPC with Laguerre function. (**b**) The FTESO-based dual closed-loop DMPC without Laguerre function. (**c**) The standard DMPC without FTESO.

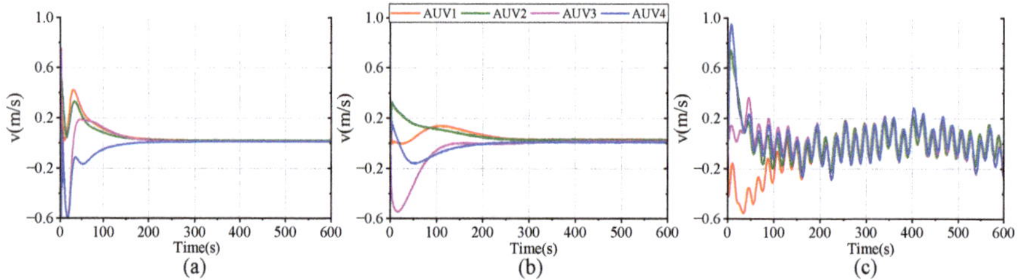

Figure 13. State v_i of the AUV formation system. (**a**) The proposed FTESO-based dual closed-loop DMPC with Laguerre function. (**b**) The FTESO-based dual closed-loop DMPC without Laguerre function. (**c**) The standard DMPC without FTESO.

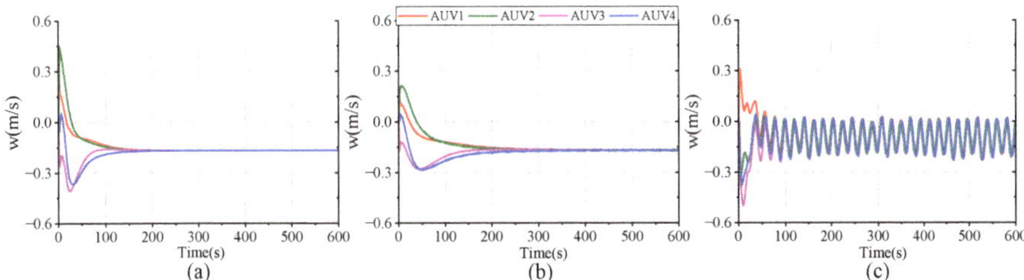

Figure 14. State w_i of the AUV formation system. (**a**) The proposed FTESO-based dual closed-loop DMPC with Laguerre function. (**b**) The FTESO-based dual closed-loop DMPC without Laguerre function. (**c**) The standard DMPC without FTESO.

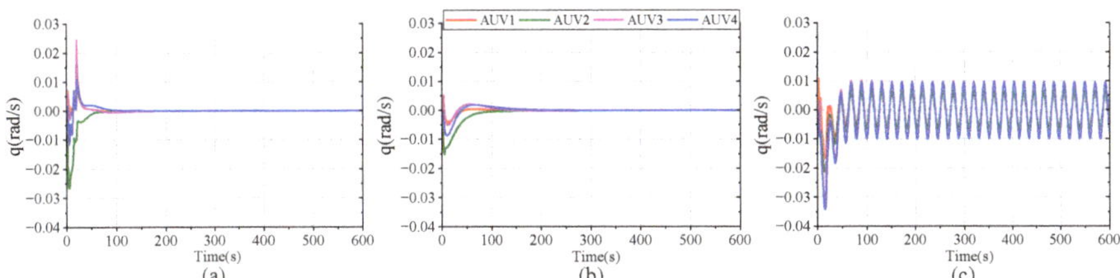

Figure 15. State q_i of the AUV formation system. (**a**) The proposed FTESO-based dual closed-loop DMPC with Laguerre function. (**b**) The FTESO-based dual closed-loop DMPC without Laguerre function. (**c**) The standard DMPC without FTESO.

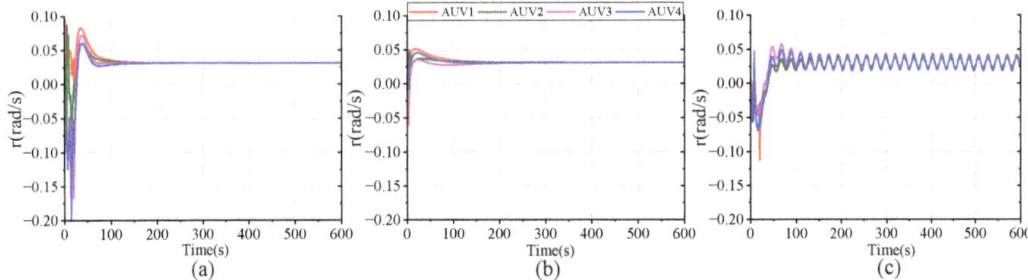

Figure 16. State r_i of the AUV formation system. (**a**) The proposed FTESO-based dual closed-loop DMPC with Laguerre function. (**b**) The FTESO-based dual closed-loop DMPC without Laguerre function. (**c**) The standard DMPC without FTESO.

Figure 17 intuitively presents a 3-D formation trajectory tracking. Combined with Figures 7–16, it implies that all three schemes can successfully accomplish formation spiral tracking under the specified input and state constraints. However, when the formation faces harsh compound disturbances, the tracking performance of the controller without disturbance compensation performs poorly, demonstrating a tracking error significantly larger than that of the FTESO-based controller. This is because the compound disturbances cause the AUV to deviate from the desired trajectory. By comparing the results of (a,b), it can be further observed that the proposed control scheme with Laguerre function allows the AUV to form the preset formation more quickly and converge to the desired trajectory more smoothly. This implies a faster response at the onset of the task. Thus, the dual closed-loop structure and Laguerre function enable the AUV formation to track the reference trajectory with better speed and accuracy.

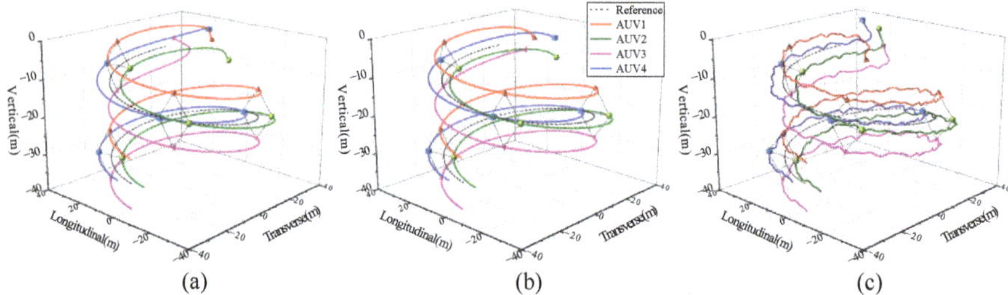

Figure 17. 3-D trajectories of the AUV formation under different schemes. (**a**) The proposed FTESO-based dual closed-loop DMPC with Laguerre function. (**b**) The FTESO-based dual closed-loop DMPC without Laguerre function. (**c**) The standard DMPC without FTESO.

Without loss of generality, Figure 18 shows the actual control forces and moments versus time for AUV1 under the three schemes. Without the benefit of FTESO to compensate for compound disturbances, the fluctuations of the control force and moment are relatively drastic and unstable (Figure 18(c1,c2)). This is attributed to the need for the AUV to significantly rectify the driving forces and moments to more rapidly approach the deviated reference trajectory. Under the proposed scheme, as shown in Figure 18(a1,a2), the AUV forces and moments vary relatively smoothly, which makes the AUV track the trajectory steadily when disturbed. Comparing Figure 18(a1,a2) and Figure 18(b1,b2), the Laguerre-based controller has the fastest control signal response with the smallest amplitude when the disturbances are accurately compensated. This confirms that our proposed scheme (a) provides superior control performance. It is worth noting that the variation of control forces and moments always remains within the prescribed limits. This reflects the ability of the DMPC to handle the actuator saturation effectively, ensuring that the control input for each DOF does not exceed the maximum force provided by the actuator, thus reducing actuator losses.

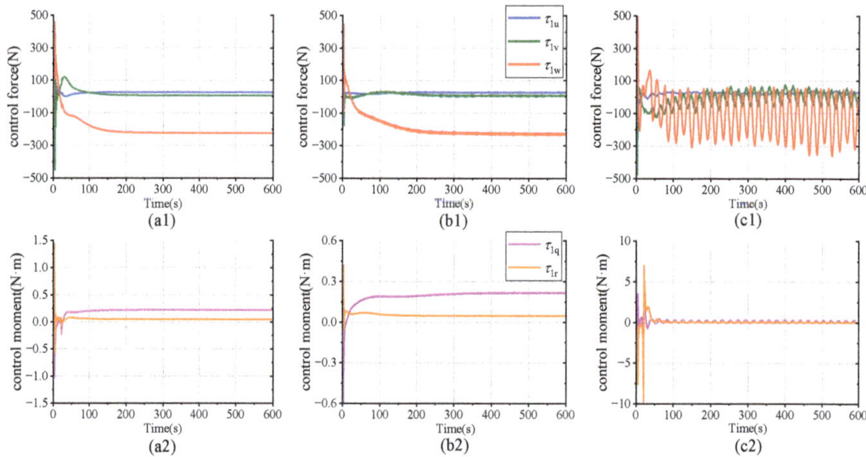

Figure 18. Actual control force and moment for AUV1 under different schemes. (**a1**) Control force of scheme (a). (**a2**) Control moments of scheme (a). (**b1**) Control force of scheme (b). (**b2**) Control moments of scheme (b). (**c1**) Control force of scheme (c). (**c2**) Control moments of scheme (c).

To differentiate between the computational demands among the three schemes, we recorded the emulator execution times under the same configurations. The detailed simulation times corresponding to Figure 17 are given in Table 2. It can be observed that the actual running time of the standard DMPC system is approximately 43.62 s. In contrast, the system with a dual closed-loop DMPC requires 57.91 s, which is about 32.8% longer than the standard DMPC. This increase is due to the greater complexity of the dual closed-loop structure as opposed to the simpler DMPC structure. Although there is improvement in control efficacy, the execution of the dual closed-loop structure is sacrificed to some extent. However, the proposed system, which employs a Laguerre-based dual closed-loop DMPC, the computation time only requires 11.75 s. This suggests that, despite the inclusion of both the dual closed-loop structure and FTESO, the use of the Laguerre function makes the system solution faster. Thus, the proposed scheme can simultaneously improve the computational speed and control performance.

Table 2. Comparison of controller execution time.

Control Scheme	Simulation Time	Run Time
Scheme (a)	600 s	11.75 s
Scheme (b)	600 s	57.91 s
Scheme (c)	600 s	43.62 s

5. Conclusions

In conclusion, this study presents a FTESO-based distributed dual closed-loop model predictive control scheme for the AUV formation subject to compound disturbances. The designed FTESO can compensate for model uncertainties and external disturbances of each AUV faster and more accurately. Control inputs are determined by solving a constrained DMPC optimization problem based on local information, while avoiding both collisions among AUVs and actuator saturation. The Laguerre orthogonal function is applied to alleviate the heavy computational burden, and the corresponding stability proof is provided. Finally, based on a connected directed topology, simulation results of different schemes are investigated under the same compound disturbances and system constraints. It is confirmed that our proposed scheme provides the best tracking effect and superior active disturbance rejection capability. Control signals show smaller oscillations and enhanced stability. In addition, the computation time of our proposed formation control system, which utilizes the Laguerre function, is reduced by 73.1% and 79.7% compared to the standard DMPC system and the dual closed-loop DMPC system, respectively. This verifies that our proposed scheme can respond quickly to minimize control costs and improve real-time execution and dynamic performance of the system.

The proposed method does not consider the impact of communication burden on AUV formation. Therefore, in future work, we will focus on the control scheme based on the event-triggered mechanisms. Considering the limitations of the optimization accuracy of discrete predictive control, we want to carry out research on continuous predictive control with faster control response. In addition, it is essential to conduct formation obstacle avoidance research and real AUV experiments.

Author Contributions: Conceptualization, M.Z. and Z.Y.; Data curation, J.Z. and L.Y.; Funding acquisition, Z.Y.; Investigation, J.Z. and L.Y.; Methodology, M.Z.; Resources, Z.Y.; Software, L.Y.; Validation, M.Z.; Visualization, J.Z.; Writing—original draft, M.Z.; Writing—review & editing, M.Z. All authors have read and agreed to the published version of the manuscript.

Funding: This work was supported by National Natural Science Foundation of China under grant No. 52071102, and in part by National Natural Science Foundation of China under grant No. 51679057.

Institutional Review Board Statement: Not applicable.

Informed Consent Statement: Not applicable.

Data Availability Statement: Not applicable.

Conflicts of Interest: The authors declare no conflict of interest.

References

1. Shi, Y.; Shen, C.; Fang, H.; Li, H. Advanced control in marine mechatronic systems: A survey. *IEEE ASME Trans. Mechatron.* **2017**, *22*, 1121–1131. [CrossRef]
2. Liu, G.; Chen, L.; Liu, K.; Luo, Y. A swarm of unmanned vehicles in the shallow ocean: A survey. *Neurocomputing* **2023**, *531*, 74–86. [CrossRef]
3. Yu, H.; Zeng, Z.; Guo, C. Coordinated formation control of discrete-time autonomous underwater vehicles under alterable communication topology with time-varying delay. *J. Mar. Sci. Eng.* **2022**, *10*, 712. [CrossRef]
4. Chen, Y.L.; Ma, X.W.; Bai, G.Q.; Sha, Y.; Liu, J. Multi-autonomous underwater vehicle formation control and cluster search using a fusion control strategy at complex underwater environment. *Ocean Eng.* **2020**, *216*, 108048. [CrossRef]
5. Zhen, Q.; Wan, L.; Li, Y.; Jiang, D. Formation control of a multi-AUVs system based on virtual structure and artificial potential field on SE(3). *Ocean Eng.* **2022**, *253*, 111148. [CrossRef]
6. Wang, J.; Wang, C.; Wei, Y.; Zhang, C. Sliding mode based neural adaptive formation control of underactuated AUVs with leader-follower strategy. *Appl. Ocean Res.* **2020**, *94*, 101971. [CrossRef]
7. He, X.; Geng, Z. Globally convergent leaderless formation control for unicycle-type mobile robots. *IET Contr. Theory Appl.* **2020**, *14*, 2651–2662. [CrossRef]
8. Munir, M.; Khan, Q.; Ullah, S.; Syeda, T.M.; Algethami, A.A. Control Design for Uncertain Higher-Order Networked Nonlinear Systems via an Arbitrary Order Finite-Time Sliding Mode Control Law. *Sensors* **2022**, *22*, 2748. [CrossRef]
9. Ullah, S.; Khan, Q.; Mehmood, A.; Kirmani, S.A.M.; Mechali, O. Neuro-adaptive fast integral terminal sliding mode control design with variable gain robust exact differentiator for under-actuated quadcopter UAV. *ISA Trans.* **2022**, *120*, 293–304. [CrossRef]
10. Zhang, W.; Wu, W.; Li, Z.; Du, X.; Yan, Z. Three-Dimensional Trajectory Tracking of AUV Based on Nonsingular Terminal Sliding Mode and Active Disturbance Rejection Decoupling Control. *J. Mar. Sci. Eng.* **2023**, *11*, 959. [CrossRef]
11. Cui, R.; Chen, L.; Yang, C.; Chen, M. Extended state observer-based integral sliding mode control for an underwater robot with unknown disturbances and uncertain nonlinearities. *IEEE Trans. Ind. Electron.* **2017**, *64*, 6785–6795. [CrossRef]
12. Ding, S.; Chen, W.H.; Mei, K.; Murray-Smith, D.J. Disturbance observer design for nonlinear systems represented by input-output models. *IEEE Trans. Ind. Electron.* **2019**, *67*, 1222–1232. [CrossRef]
13. Liang, X.; Qu, X.; Wan, L.; Ma, Q. Three-dimensional path following of an underactuated AUV based on fuzzy backstepping sliding mode control. *Int. J. Fuzzy Syst.* **2018**, *20*, 640–649. [CrossRef]
14. Zhang, G.; Yin, S.; Huang, C.; Zhang, W. Intervehicle Security-Based Robust Neural Formation Control for Multiple USVs via APS Guidance. *J. Mar. Sci. Eng.* **2023**, *11*, 1020. [CrossRef]
15. Han, J. From PID to active disturbance rejection control. *IEEE Trans. Ind. Electron.* **2009**, *56*, 900–906. [CrossRef]
16. Lei, M.; Li, Y.; Pang, S. Extended state observer-based composite-system control for trajectory tracking of underactuated AUVs. *Appl. Ocean Res.* **2021**, *112*, 102694. [CrossRef]
17. Nie, J.; Wang, H.; Lu, X.; Lin, X.; Sheng, C.; Zhang, Z.; Song, S. Finite-time output feedback path following control of underactuated MSV based on FTESO. *Ocean Eng.* **2021**, *224*, 108660. [CrossRef]
18. Wang, N.; Zhu, Z.; Qin, H.; Deng, Z.; Sun, Y. Finite-time extended state observer-based exact tracking control of an unmanned surface vehicle. *Int. J. Robust Nonlinear Control.* **2021**, *31*, 1704–1719. [CrossRef]
19. Sankaranarayanan, V.N.; Yadav, R.D.; Swayampakula, R.K.; Ganguly, S.; Roy, S. Robustifying payload carrying operations for quadrotors under time-varying state constraints and uncertainty. *IEEE Robot. Autom. Lett.* **2022**, *7*, 4885–4892. [CrossRef]
20. Chu, Z.; Xiang, X.; Zhu, D.; Luo, C.; Xie, D. Adaptive fuzzy sliding mode diving control for autonomous underwater vehicle with input constraint. *Int. J. Fuzzy Syst.* **2018**, *20*, 1460–1469. [CrossRef]
21. Li, S.; Wang, X. Finite-time consensus and collision avoidance control algorithms for multiple AUVs. *Automatica* **2013**, *49*, 3359–3367. [CrossRef]
22. Xu, J.; Huang, F.; Wu, D.; Cui, Y.; Yan, Z.; Du, X. A learning method for AUV collision avoidance through deep reinforcement learning. *Ocean Eng.* **2022**, *260*, 112038. [CrossRef]
23. Zhang, Y.; Liu, X.; Luo, M.; Yang, C. MPC-based 3-D trajectory tracking for an autonomous underwater vehicle with constraints in complex ocean environments. *Ocean Eng.* **2019**, *189*, 106309. [CrossRef]
24. Arcos-Legarda, J.; Gutiérrez, Á. Robust Model Predictive Control Based on Active Disturbance Rejection Control for a Robotic Autonomous Underwater Vehicle. *J. Mar. Sci. Eng.* **2023**, *11*, 929. [CrossRef]
25. Zheng, Y.; Li, S.E.; Li, K.; Borrelli, F.; Hedrick, J.K. Distributed model predictive control for heterogeneous vehicle platoons under unidirectional topologies. *IEEE Trans. Control Syst. Technol.* **2016**, *25*, 899–910. [CrossRef]
26. Wei, H.; Shen, C.; Shi, Y. Distributed Lyapunov-based model predictive formation tracking control for autonomous underwater vehicles subject to disturbances. *IEEE Trans. Syst. Man Cybern. Syst.* **2019**, *51*, 5198–5208. [CrossRef]
27. Shen, C.; Shi, Y. Distributed implementation of nonlinear model predictive control for AUV trajectory tracking. *Automatica* **2020**, *115*, 108863. [CrossRef]
28. Li, B.; Hu, Q.; Yu, Y.; Ma, G. Observer-based fault-tolerant attitude control for rigid spacecraft. *IEEE Trans. Aerosp. Electron. Syst.* **2017**, *53*, 2572–2582. [CrossRef]
29. Wei, H.; Sun, Q.; Chen, J.; Shi, Y. Robust distributed model predictive platooning control for heterogeneous autonomous surface vehicles. *Control Eng. Pract.* **2021**, *107*, 104655. [CrossRef]

30. Zhao, R.; Miao, M.; Lu, J.; Wang, Y.; Li, D. Formation control of multiple underwater robots based on ADMM distributed model predictive control. *Ocean Eng.* **2022**, *257*, 111585. [CrossRef]

31. Hu, Q.; Jiang, B. Continuous finite-time attitude control for rigid spacecraft based on angular velocity observer. *IEEE Trans. Aerosp. Electron. Syst.* **2018**, *54*, 1082–1092. [CrossRef]

32. Yan, Z.; Gong, P.; Zhang, W.; Li, Z.; Teng, Y. Autonomous underwater vehicle vision guided docking experiments based on L-shaped light array. *IEEE Access* **2019**, *7*, 72567–72576. [CrossRef]

33. Fossen, T.I. *Handbook of Marine Craft Hydrodynamics and Motion Control*; John Wiley & Sons: Hoboken, NJ, USA, 2011.

34. Xu, J.; Cui, Y.; Xing, W.; Huang, F.; Yan, Z.; Wu, D.; Chen, T. Anti-disturbance fault-tolerant formation containment control for multiple autonomous underwater vehicles with actuator faults. *Ocean Eng.* **2022**, *266*, 112924. [CrossRef]

35. Tao, T.; Roy, S.; De Schutter, B.; Baldi, S. Distributed Adaptive Synchronization in Euler Lagrange Networks with Uncertain Interconnections. *IEEE Trans. Autom. Control*, 2013; *online ahead of print.*

36. Roy, S.; Baldi, S.; Fridman, L.M. On adaptive sliding mode control without a priori bounded uncertainty. *Automatica* **2020**, *111*, 108650. [CrossRef]

37. Kong, S.; Sun, J.; Wang, J.; Zhou, Z.; Shao, J.; Yu, J. Piecewise Compensation Model Predictive Governor Combined with Conditional Disturbance Negation for Underactuated AUV Tracking Control. *IEEE Trans. Ind. Electron.* **2022**, *70*, 6191–6200. [CrossRef]

38. Wang, L. Continuous time model predictive control design using orthonormal functions. *Int. J. Control.* **2001**, *74*, 1588–1600. [CrossRef]

39. Yan, Z.; Wang, M.; Xu, J. Integrated guidance and control strategy for homing of unmanned underwater vehicles. *J. Frankl. Inst.-Eng. Appl. Math.* **2019**, *356*, 3831–3848. [CrossRef]

40. Hosen, M.A.; Hussain, M.A.; Mjalli, F.S. Control of polystyrene batch reactors using neural network based model predictive control (NNMPC): An experimental investigation. *Control Eng. Pract.* **2011**, *19*, 454–467. [CrossRef]

41. Cortes, P.; Kouro, S.; La Rocca, B.; Vargas, R.; Rodriguez, J.; Leon, J.I.; Vazquez, S.; Franquelo, L.G. Guidelines for weighting factors design in Model Predictive Control of power converters and drives. In Proceedings of the IEEE International Conference on Industrial Technology, Churchill, VIC, Australia, 10–13 February 2009; pp. 1–7.

42. Zhang, C.; Zhang, G.; Dong, Q. Multi-variable finite-time observer-based adaptive-gain sliding mode control for fixed-wing UAV. *IET Contr. Theory Appl.* **2021**, *15*, 223–247. [CrossRef]

43. Ma, C.; Tang, Y.; Lei, M.; Jiang, D.; Luo, W. Trajectory tracking control for autonomous underwater vehicle with disturbances and input saturation based on contraction theory. *Ocean Eng.* **2022**, *266*, 112731. [CrossRef]

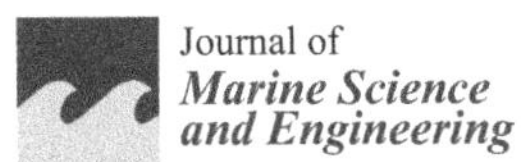

Journal of
*Marine Science
and Engineering*

Article

Path Planning of Deep-Sea Landing Vehicle Based on the Safety Energy-Dynamic Window Approach Algorithm

Zuodong Pan [1,2], Wei Guo [1,3,*], Hongming Sun [1,3], Yue Zhou [2,*] and Yanjun Lan [1]

[1] Institute of Deep-Sea Science and Engineering, Chinese Academy of Sciences, Sanya 572000, China; zuodongpan@163.com (Z.P.); sunhm@idsse.ac.cn (H.S.); lanyj@idsse.ac.cn (Y.L.)
[2] School of Engineering, Shanghai Ocean University, Shanghai 201306, China
[3] University of Chinese Academy of Sciences, Beijing 100049, China
* Correspondence: guow@idsse.ac.cn (W.G.); y-zhou@shou.edu.cn (Y.Z.)

Abstract: To ensure the safety and energy efficiency of autonomous sampling operations for a deep-sea landing vehicle (DSLV), the Safety Energy-Dynamic Window Approach (SE-DWA) algorithm was proposed. The safety assessment sub-function formed from the warning obstacle zone and safety factor addresses the safety issue arising from the excessive range measurement error of forward-looking sonar. The trajectory comparison evaluation sub-function with the effect of reducing energy consumption achieves a reduction in path length by causing the predicted trajectory to deviate from the historical trajectory when encountering "U"-shaped obstacles. The pseudo-power evaluation sub-function with further energy consumption reduction ensures optimal linear and angular velocities by minimizing variables when encountering unknown obstacles. The simulation results demonstrate that compared with the Minimum Energy Consumption-DWA algorithm, the SE-DWA algorithm improves the minimum distance to an actual obstacle zone by 68% while reducing energy consumption by 11%. Both the SE-DWA algorithm and the Maximum Safety-DWA (MS-DWA) algorithm ensure operational safety with minimal distance to the actual obstacle zone, yet the SE-DWA algorithm achieves a 24% decrease in energy consumption. In conclusion, the path planned by the SE-DWA algorithm ensures not only safety but also energy consumption reduction during autonomous sampling operations by a DSLV in the deep sea.

Keywords: deep-sea landing vehicle; the Safety Energy-Dynamic Window Approach algorithm; safe; low energy consumption; autonomous sampling operations

Citation: Pan, Z.; Guo, W.; Sun, H.; Zhou, Y.; Lan, Y. Path Planning of Deep-Sea Landing Vehicle Based on the Safety Energy-Dynamic Window Approach Algorithm. *J. Mar. Sci. Eng.* **2023**, *11*, 1892. https://doi.org/10.3390/jmse11101892

Academic Editor: Kamal Djidjeli

Received: 24 August 2023
Revised: 24 September 2023
Accepted: 26 September 2023
Published: 28 September 2023

1. Introduction

The abundance of marine resources has attracted global attention as resources, energy, and space available on land are gradually diminishing [1,2]. The Deep-Sea Landing Vehicle (DSLV) is an autonomous deep-sea equipment capable of long-term, large-scale, and short-distance refinement operations in the deep-sea environment [3,4]. There is a huge challenge for DSLV to complete autonomous sampling operations because the forward-looking sonar has a large ranging error caused by-deep sea noise and the battery power of the DSLV is limited. Therefore, planning a safe and energy-efficient path holds significant importance for DSLV in autonomous sampling operations.

Path planning is categorized into global path planning and local path planning. Global path planning relies on accurate global maps to efficiently achieve collision-free and shortest path planning [5,6]. Local path planning is real-time path planning in unknown environments based on data collected by relevant sensors such as LiDAR and forward-looking sonar, with higher requirements for path safety, smoothness, and traceability [7–9]. The Dynamic Window Approach (DWA) algorithm, one of the classical algorithms for two-dimensional local path planning, generates trajectories and assigns scores through feasible velocity combinations to directly determine the optimal velocity [10,11]. The algorithm considers the robot's physical constraints, environmental factors, and current state comprehensively without requiring path following.

Xin et al. [12] proposed an Enhanced DWA algorithm that takes the distance function as the weight of the target-oriented coefficient, which effectively optimizes the stability of the mobile robot during operation with lower angular velocity dispersion and less energy consumption. However, this algorithm has a longer runtime compared with the DWA algorithm. While using this algorithm for autonomous sampling operations can reduce energy consumption in the propulsion system, it may increase energy consumption in other systems such as positioning and communication. Additionally, it does not account for safety concerns in the deep-sea environment.

Wang et al. [13] proposed an ACO-DWA algorithm that addresses the issue of poor obstacle avoidance performance for robots in high-density obstacle environments and unknown obstacle static environments. This algorithm effectively addresses safety concerns for DSLVs in the deep-sea environment. However, it lacks an energy-related evaluation sub-function, leading to an inability to reduce the DSLV's energy consumption during autonomous sampling operations.

Masato Kobayashi et al. [14] proposed the DWV (Dynamic Window Approach with Virtual Manipulators) algorithm, which enhances the robot's success rate in reaching the target point by generating more predictive trajectories in narrow or dynamic environments. While this algorithm can ensure a certain degree of energy reduction by reducing the path length, the safety of DSLV is not guaranteed when applying this algorithm for autonomous sampling operations.

Figure 1 illustrates the deep-sea autonomous sampling operations for DSLV. The application of the DWA algorithm for autonomous sampling operations in the deep sea has two deficiencies. (1) The safety of the DSLV is considerably compromised when encountering ground obstacles such as underwater ridges in unknown deep-sea environments. This compromise is mainly due to the degradation of the accuracy of forward sonar ranging caused by deep-sea noise. The detected obstacle volume may be smaller than the actual volume due to the decrease in accuracy [15,16]. (2) The limited energy of DSLV will be depleted more quickly due to the absence of energy consumption-related evaluation sub-functions in the DWA algorithm [3].

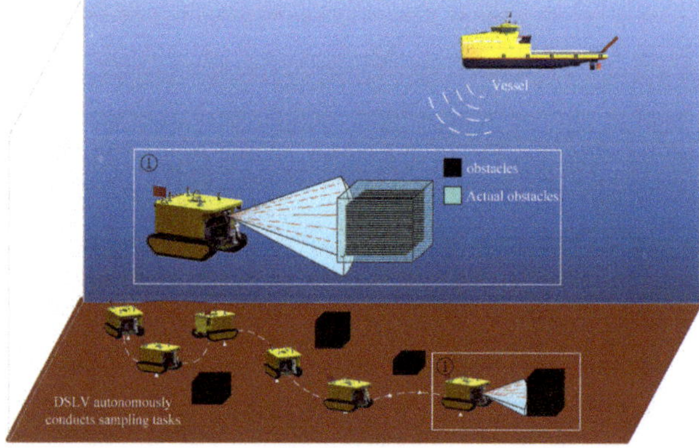

Figure 1. DSLV deep-sea autonomous sampling operations. ① The safety of the DSLV.

To address the above issues, improvements and additional evaluation sub-functions have been introduced to ensure the safety and reduce energy consumption during autonomous sampling operations. The safety assessment sub-function, formed of warning obstacle zones and safety factors, addresses safety concerns arising from the decreased accuracy of forward-looking sonar due to underwater noise and other factors. The trajectory comparison evaluation sub-function reduces energy consumption by decreasing the path length when encountering "U"-shaped obstacles. The pseudo-power evaluation sub-function reduces energy consumption by optimizing linear and angular velocities when encountering unknown deep-sea obstacles.

The subsequent sections of this paper are outlined as follows: Section 2 presents the kinematic modelling of the DSLV. Section 3 introduces the fundamental principles of the SE-DWA algorithm and the flows of applying it to the DSLV for deep-sea autonomous sampling operations. Section 4 determines the coefficients of the SE-DWA algorithm's evaluation functions and compares them with the DWA algorithm in terms of safety and energy consumption under a deep-sea environmental map. Section 5 concludes the findings of this article.

2. Kinematic Modelling of DSLV

Flat abyssal plain areas are commonly chosen to enhance the success rate of autonomous sampling operations for DSLV [17]. The problem is simplified to a two-dimensional plane for algorithmic research, as the terrain changes in these areas are minor and negligible.

The DSLV developed in this paper utilizes a dual-motor rear-wheel drive system, controlled differentially to manage its forward, backward and steering movements. The schematic diagram of bilateral motor drive of DSLV is shown in Figure 2.

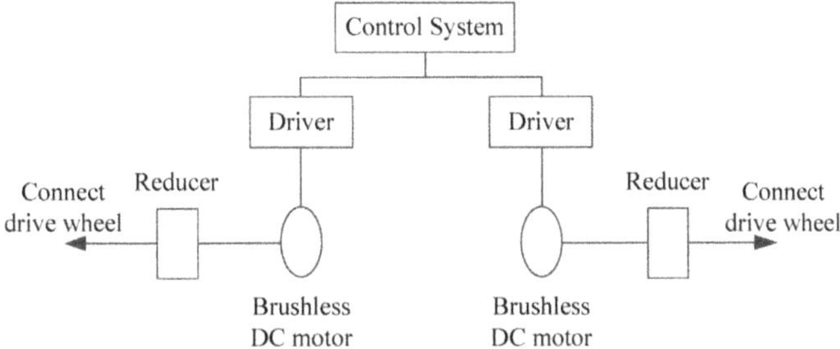

Figure 2. The schematic diagram of bilateral motor drive of DSLV.

The DSLV kinematic modelling diagram is shown in Figure 3. In the Cartesian coordinate system O-XY, let $x_{(t)}$ and $y_{(t)}$ represent the horizontal and vertical coordinates of the geometric center of the DSLV, respectively, and θ denotes the heading angle. The project team reinforced the track structure's tensioner brackets to withstand the impact of landing on the seabed during the design phase of the DSLV. Furthermore, the DSLV maintains a slow crawling speed when operating on a flat sandy bottom. For analysis purposes, it can be assumed that the DSLV's center of mass aligns with the geometric center. Consequently, the linear and angular velocities of its center of mass are as follows:

$$\begin{cases} v = \omega R = \frac{v_o + v_i}{2} \\ \omega = \frac{v_o - v_i}{B} \end{cases} \tag{1}$$

where the expressions for v_o and v_i are as follows:

$$\begin{cases} v_o = (1 - s_1) \cdot r_1 \cdot \omega_1 \\ v_i = (1 - s_2) \cdot r_1 \cdot \omega_2 \end{cases} \tag{2}$$

where R represents the steering radius; v_o and v_i denote the linear velocities of the left and right tracks, respectively; B indicates the center distance between the left and right tracks; s_1 and s_2 refer to the slip rates of the left and right tracks, respectively; r_1 stands for the radius of the left and right track drive wheels; and ω_1 and ω_2 represent the angular velocities of the left and right track drive wheels, respectively.

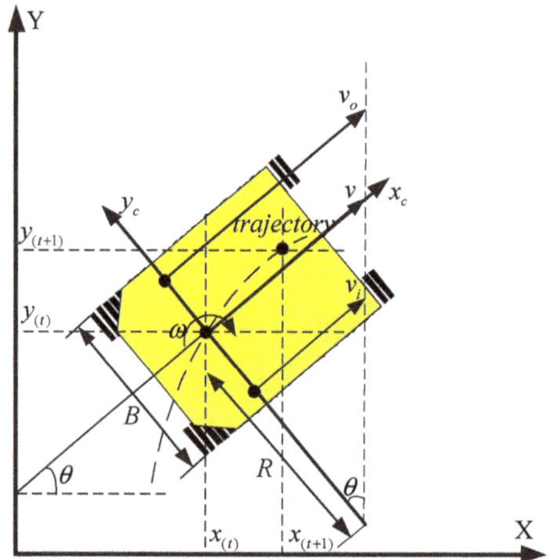

Figure 3. DSLV kinematic modelling diagram.

By combining Equation (1) and Figure 3, the position coordinates of the DSLV at moment t can be obtained as $(x(t), y(t), \varphi(t))$, as shown in Equation (3), and the position coordinates corresponding to moment $t + 1$ can be obtained as $(x(t+1), y(t+1), \varphi(t+1))$, as shown in Equation (4).

$$\begin{cases} x(t) = v \cdot \cos\theta \cdot t \\ y(t) = v \cdot \sin\theta \cdot t \\ \varphi(t) = \omega \cdot t \end{cases} \tag{3}$$

$$\begin{cases} x(t+1) = x(t) + \int_t^{t+1} (v \cdot \cos\theta \cdot t)dt \\ y(t+1) = y(t) + \int_t^{t+1} (v \cdot \sin\theta \cdot t)dt \\ \varphi(t+1) = \varphi(t) + \int_t^{t+1} (\omega)dt \end{cases} \tag{4}$$

The DSLV autonomous sampling control module model is shown in Figure 4.

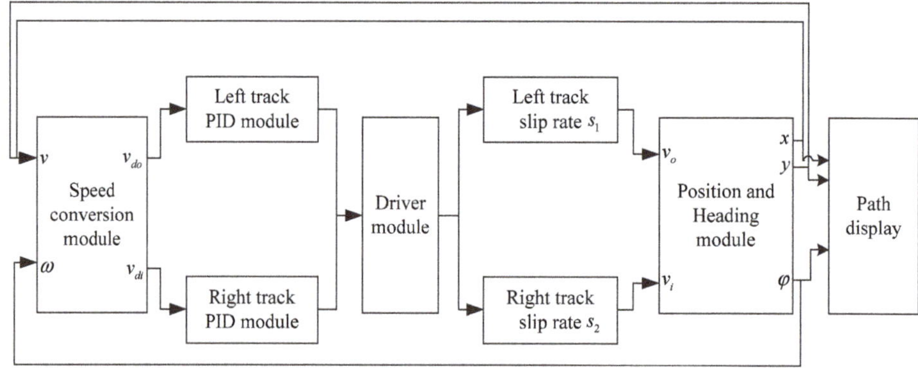

Figure 4. The DSLV control module model.

3. The SE-DWA Algorithm

3.1. Velocity Constraints

The speed window of the SE-DWA algorithm is illustrated in Figure 5. It is constrained by speed limit (E_v), safety limits (E_o), and the acceleration and deceleration limits of the motor drive (E_a) [18]. The expression (E_{all}) for the velocity window of the SE-DWA algorithm is as follows:

$$E_{all} = E_v \cap E_o \cap E_a \tag{5}$$

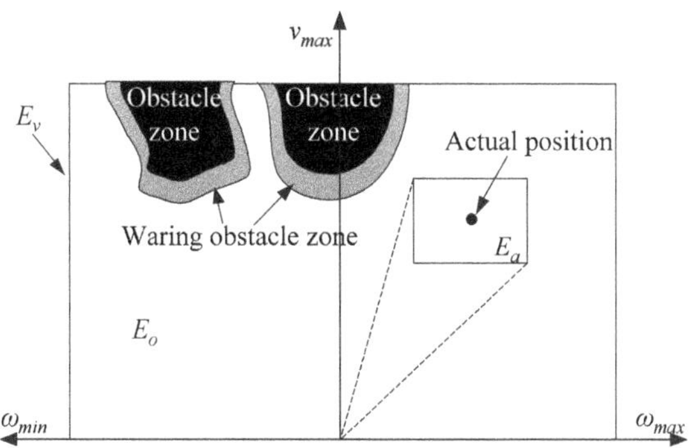

Figure 5. The speed window of the SE-DWA algorithm.

(1)   Speed limit

The maximum and minimum speed limits for DSLV are as follows:

$$E_v = \{(v, \omega)|v \in [0, v_{max}], \omega \in [\omega_{min}, \omega_{max}]\} \tag{6}$$

where v_{max} represent the maximum linear velocities, and ω_{min} and ω_{max} represent the minimum and maximum angular velocities.

(2)   Security restriction

The DSLV limits the speed window to avoid collisions with obstacles, and the limited speed window is as follows:

$$E_o = \begin{cases} \left\{(v, \omega)|v \leq \sqrt{2 \cdot dist_{obs}(v, \omega) \cdot \dot{v}} \cap \omega \leq \sqrt{2 \cdot dist_{obs}(v, \omega) \cdot \dot{\omega}}\right\} \ L_{dist} > 2\sigma_{max}^{sonar} \\ \left\{(v, \omega)|v \leq \sqrt{2 \cdot dist_{warn}(v, \omega) \cdot \dot{v}} \cap \omega \leq \sqrt{2 \cdot dist_{warn}(v, \omega) \cdot \dot{\omega}}\right\} \ 0 < L_{dist} \leq 2\sigma_{max}^{sonar} \end{cases} \tag{7}$$

where $dist_{obs}(v, \omega)$ represents the distance on the trajectory corresponding to the velocity combination to the nearest obstacle zone; $dist_{warn}(v, \omega)$ represents the distance on the trajectory corresponding to the velocity combination to the nearest warning obstacle zone; L_{dist} represents the distance between the edges of the two nearest obstacle zone; and σ_{max}^{sonar} represents the maximum error value of the forward sonar ranging distance.

(3)   Reachable speed limit

The range of DSLV speed combinations in a sampling period is as follows:

$$E_a = \{(v, \omega)|v \in [v_0 - \dot{v} \cdot \Delta t, v_0 + \dot{v} \cdot \Delta t], \omega \in [\omega_0 - \dot{\omega} \cdot \Delta t, \omega_0 + \dot{\omega} \cdot \Delta t]\} \tag{8}$$

where v_0 and ω_0 represent the current linear and angular velocities of the DSLV; $\dot{v}$ and $\dot{\omega}$ represent the maximum linear accelerations and angular accelerations of the DSLV; and Δt represents the sampling period.

3.2. Evaluation Function

3.2.1. Safety Evaluation Sub-Function

The DWA algorithm requires real-time acquisition of obstacle positions as a component of its local path-planning approach. However, the precision of forward sonar ranging data in the deep-sea environment is lower and the errors are larger compared with LiDAR ranging data on land [16,19]. The inability to guarantee the safety of autonomous sampling operations on the seafloor emphasizes the need to address this issue. In this section, the obstacle avoidance evaluation sub-function within the DWA algorithm will be enhanced and will be named the safety evaluation sub-function.

The safety evaluation sub-function calculates its cost based on the distance between the predicted trajectory and obstacles, multiplied by the safety coefficient corresponding to the warning obstacle zone. Figure 6 illustrates the method for computing this cost, using the example of cost calculation for the i-th predicted trajectory and j-th obstacle. To prevent a sharp increase in cost due to a high number of obstacles in the local map, a threshold radius R_{safe} is established, with the geometric center of the DSLV as the point of origin. When the j-th obstacle located within this circle, the cost associated with the i-th predicted trajectory encountering the j-th obstacle is as follows:

$$cost_{ij}^{safe} = \eta L_{ij}^{safe} \tag{9}$$

where η represents the safety coefficient and L_{ij}^{safe} represents the distance from the endpoint of predicted trajectory i to the center of the j-th obstacle. Their respective values are as follows:

$$\eta = \begin{cases} 0.8 & 0 < L_{ij}^{obsmin} \leq \sigma_{max}^{sonar}/2 \\ 0.9 & \sigma_{max}^{sonar}/2 < L_{ij}^{obsmin} \leq \sigma_{max}^{sonar} \\ 1.0 & \sigma_{max}^{sonar} < L_{ij}^{obsmin} \end{cases} \tag{10}$$

$$L_{ij}^{safe} = \left(\sqrt{(x_j - x_i)^2} + \sqrt{(y_j - y_i)^2}\right) \tag{11}$$

where L_{ij}^{obsmin} represents the closest distance from the endpoint of predicted trajectory i to the edge of the j-th obstacle; x_i and y_i represent the x and y coordinates of the endpoint of predicted trajectory i, respectively; and x_j and y_j represent the x and y coordinates of the center of the j-th obstacle, respectively.

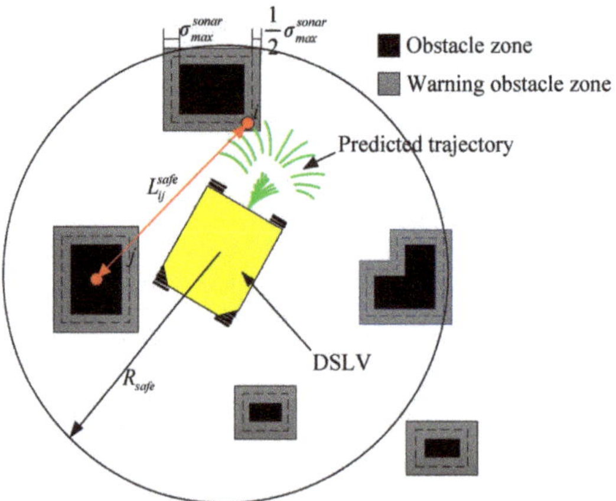

Figure 6. Computation method of safety sub-function cost values.

The expression for the safety evaluation sub-function of predicted trajectory i within the current sampling cycle is as follows:

$$safe(v,\omega)_i = \frac{1}{jmax} \sum_{j=1}^{jmax} cost_{ij}^{safe} \tag{12}$$

where $jmax$ represents the number of obstacles within a radius of R_{safe}.

The construction of the obstacle zone, actual obstacle zone, and warning obstacle zone is illustrated in Figure 7. In the figure, $\lambda = rand[-\sigma_{max}^{sonar}, \sigma_{max}^{sonar}]$. Both the obstacle zone and the actual obstacle zone are impassable areas. The obstacle zone is a known area, while the actual obstacle zone is unknown. Both the DWA algorithm and the SE-DWA algorithm plan paths based on the obstacle zone. However, a certain distance from the actual obstacle zone is required to ensure the safety of DSLV's autonomous sampling operations. In trajectory one, obstacles are enlarged by the DSLV's width, resulting in a shorter path length but lower safety. The algorithm corresponding to trajectory one will be referred to as the Minimum Energy Consumption-DWA (MEC-DWA) algorithm. In trajectory two, obstacles are expanded by the DSLV's width and σ_{max}^{sonar}, resulting in a longer path length but higher safety. The algorithm corresponding to trajectory two will be referred to as the Maximum Safety-DWA (MS-DWA) algorithm.

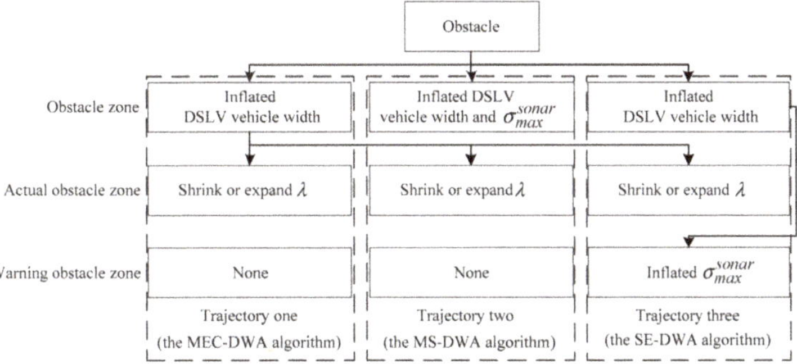

Figure 7. Region construction.

Figure 8 illustrates the mechanism of the safety evaluation sub-function. In trajectory one, obstacles are expanded only by the DSLV's width. The volume of the actual obstacle zone can easily exceed the obstacle zone, leading to safety concerns. However, trajectory two involves obstacles that are expanded by both the DSLV's width and σ_{max}^{sonar}. Although the volume of the actual obstacle zone cannot exceed the obstacle zone, the excessive inflation requires DSLV to crawl a longer distance to reach the target point. In trajectory three, despite expanding the obstacles only by the DSLV's width, the presence of the warning obstacle zone and the safety evaluation sub-function lead to a different choice. Under the same linear and angular velocities, when compared with the DWA algorithm, it selects position 2 instead of position 1. This decision keeps it away from the actual obstacle zone, ensuring the safety of the DSLV.

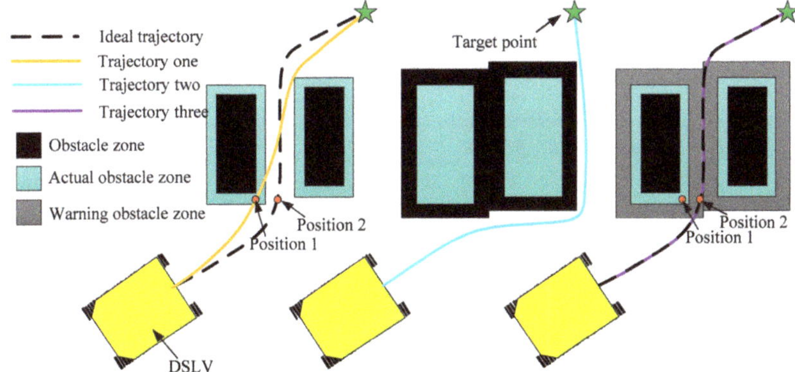

Figure 8. The mechanism of the safety evaluation sub-function.

3.2.2. Trajectory Comparison Evaluation Sub-Function

The seafloor is filled with irregular obstacles, making it prone to the formation of a "U"-shaped obstacle environment [20]. The DWA algorithm, due to the scoring mechanism of its evaluation functions, may end up crawling a longer distance while attempting to find an exit from the "U"-shaped obstacle environment. In some cases, it might even become stuck without finding a way out. To address this issue, this section introduces a new trajectory comparison evaluation sub-function.

The trajectory comparison evaluation sub-function entails evaluating the proximity of predicted trajectories to the grid cells occupied by historical trajectories, which contributes to the calculation of the cost value used for scoring. Figure 9 illustrates the method for computing the cost value of the trajectory comparison evaluation sub-function. The shaded area represents grid cells within a radius of R_{locus} around the historical trajectory. Taking the calculation of predicted trajectory m and historical trajectory n as an example, a threshold radius of R_{locus} is established with the DSLV's geometric center as the origin. When historical trajectory n falls within the circle, the corresponding cost value for predicted trajectory m in relation to historical trajectory n is as shown below:

$$cost_{mn}^{locus} = \left(R_{locus} - L_{mn}^{locus} \right) \cdot \left|\left| \frac{v - v_0}{v_{max}} \right|\right| \tag{13}$$

where L_{mn}^{locus} represents the distance from the endpoint of the predicted trajectory m to the grid cell where the historical trajectory n is located, and v and v_{max}, respectively, denote the current linear velocity at the endpoint of the predicted trajectory m and the maximum linear velocity achievable within the current velocity window. (The addition of velocity terms aims to prevent a sharp increase in the cost values for the surrounding grid cells when the vehicle travels at low speeds.)

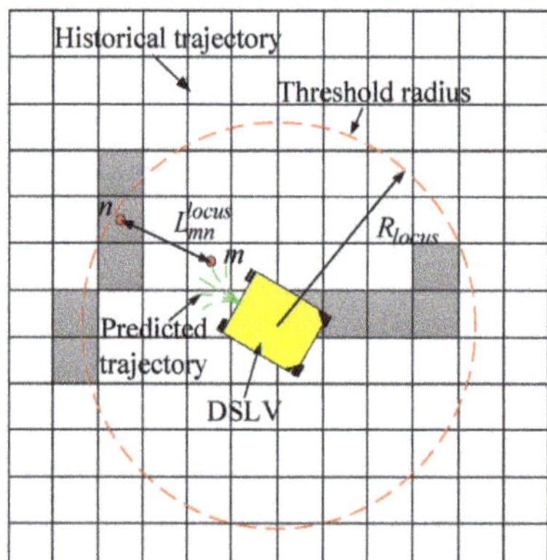

Figure 9. Computation method of trajectory contrast sub-function cost values.

The expression for the trajectory comparison evaluation sub-function of the predicted trajectory m is as follows:

$$locus(v, \omega) = \frac{1}{n_{max}} \sum_{n=1}^{n_{max}} cost_{mn}^{locus} \tag{14}$$

where n_{max} represents the number of grid cells within a radius of R_{locus} that contain historical trajectories.

Figure 10 illustrates the mechanism of the trajectory comparison evaluation sub-function. Trajectories four and five are generated by the DWA algorithm and the SE-DWA algorithm, respectively. When faced with a 'U'-shaped obstacle environment, trajectory five is more favorable than trajectory four because of its ability to escape local optima. When scoring is conducted using the evaluation function of the DWA algorithm, satisfying both conditions $\theta_1 < \theta_2$ and $d_1 < d_2$, the two trajectories will receive similar scores, making it challenging to determine their relative superiority. However, with the inclusion of the trajectory comparison evaluation sub-function, the cost of grid cells within the purple circle for trajectory four is lower compared with that for trajectory five. As a result, trajectory five receives a higher score, allowing the DSLV to identify the exit of the 'U'-shaped obstacle environment.

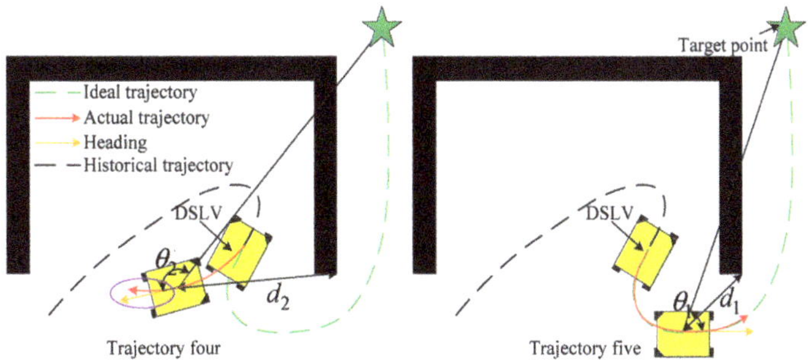

Figure 10. The mechanism of the trajectory contrast sub-function.

3.2.3. Pseudo-Power Evaluation Sub-Function

The trajectory comparison evaluation sub-function aims to decrease energy consumption by minimizing the distance crawled by the DSLV. According to the reference literature, the DSLV maintains optimal velocity during crawling to minimize variations in the linear and angular velocities, which can reduce energy consumption to a certain extent [21,22]. Therefore, the constructed pseudo-power evaluation sub-function is as follows:

$$energy(v,\omega) = 1 - \frac{v \cdot \dot{v} + \omega \cdot \dot{\omega}}{\dot{v}_{max} \cdot \dot{v}_{max} + \dot{\omega}_{max} \cdot \dot{\omega}_{max}} \tag{15}$$

where $\dot{v}$ and $\dot{\omega}$ represent linear and angular accelerations, respectively, and $\dot{v}_{max}$ and $\dot{\omega}_{max}$ represent the maximum linear velocity and the maximum angular acceleration, respectively.

3.3. Evaluation Function of the SE-DWA Algorithm

To satisfy the dynamic constraint conditions of the DSLV, the evaluation sub-functions need to undergo a smoothing process, specifically normalization. The calculation formula is shown as follows:

$$\begin{cases} head(v,\omega)_k^{smooth} = \dfrac{head(v,\omega)_k}{\sum\limits_{k=1}^{K} head(v,\omega)_k} \\[2ex] safe(v,\omega)_k^{smooth} = \dfrac{safe(v,\omega)_k}{\sum\limits_{k=1}^{K} sec(v,\omega)_k} \\[2ex] vel(v,\omega)_k^{smooth} = \dfrac{vel(v,\omega)_k}{\sum\limits_{k=1}^{K} vel(v,\omega)_k} \\[2ex] locus(v,\omega)_k^{smooth} = \dfrac{locus(v,\omega)_k}{\sum\limits_{k=1}^{K} locus(v,\omega)_k} \\[2ex] energy(v,\omega)_k^{smooth} = \dfrac{energy(v,\omega)_k}{\sum\limits_{k=1}^{K} energy(v,\omega)_k} \end{cases} \tag{16}$$

where K represents the total number of predicted trajectories; k represents the current predicted trajectory under evaluation; $head(v,\omega)$ represents the navigation evaluation sub-function, which indicates the complement angle between the velocity direction and the target point; $safe(v,\omega)$ represents the safety evaluation sub-function; $vel(v,\omega)$ represents the velocity evaluation sub-function, which indicates the speed magnitude in the trajectory; $locus(v, \omega)$ represents the trajectory comparison evaluation sub-function; and $energy(v, \omega)$ represents the pseudo-power evaluation sub-function.

The final evaluation function of the SE-DWA algorithm is as follows:

$$G(v, \omega) = \quad \gamma_1 head(v, \omega)_k^{smooth} + \gamma_2 safe(v, \omega)_k^{smooth} + \gamma_3 vel(v, \omega)_k^{smooth} \\ + \gamma_4 locus(v, \omega)_k^{smooth} + \gamma_5 energy(v, \omega)_k^{smooth} \tag{17}$$

where γ_1, γ_2, γ_3, γ_4, and γ_5 represent the weights of the five evaluation sub-functions, respectively.

3.4. Application of the SE-DWA Algorithm in DSLV

The process diagram of DSLV applying the SE-DWA algorithm is depicted in Figure 11, with the following main steps. (1) Initialization: the deep-sea environment data acquired using sensors such as forward sonar are used to construct a map using the grid-based method. Subsequently, the destination point is established. (2) The SE-DWA algorithm: after acquiring a DSLV's state parameters, the SE-DWA algorithm's evaluation function is employed to determine the optimal predicted trajectory for the sampling period. (3) Real-time environmental detection: sensors continuously monitor the surrounding environment. When a potential collision with an actual obstacle zone or a local optimum is detected, adjustments in heading angle are executed. (4) Destination detection: at the end of each sampling period, a check is performed to verify whether the DSLV has reached the target point. If the target point has been reached, the path planning for this cycle is concluded. Otherwise, steps (2) and (3) are reiterated until the target point is attained.

Figure 11. Flowchart of the SE-DWA algorithm for DSLV applications.

4. Simulation Experiment

4.1. Determination of Evaluation Function Coefficients

The SE-DWA algorithm comprises five evaluation sub-functions associated with four weights. These weights are designated as the navigation weight, safety weight, velocity weight, and energy consumption weight. The navigation weight corresponds to the navigation evaluation sub-function, which is used to control the DSLV's motion direction. The safety weight corresponds to the safety evaluation sub-function, which is employed to prevent collisions between the DSLV and obstacles. The velocity weight pertains to the velocity evaluation sub-function, governing the DSLV's maximum linear velocity. The energy consumption weight is associated with both the trajectory comparison evaluation sub-function and the pseudo-power sub-function. It serves the purpose of reducing path length and ensuring optimal velocity, respectively, thus lowering energy consumption.

The coefficients for the navigation evaluation sub-function, safety evaluation sub-function, and velocity evaluation sub-function in the SE-DWA algorithm are referenced from the coefficients set in the DWA algorithm [18]. To determine the coefficients for the trajectory comparison evaluation sub-function and the pseudo-power evaluation sub-function, simulation experiments are conducted in the simulated deep-sea environmental map (80 m × 80 m) shown in Figure 12. In the figure, S(2,2) and G(78,78) represent the starting point and the goal point, respectively. Assuming that DSLV behaves as a point mass, obstacles were constructed as shown in Figure 7 to form obstacle zones, actual obstacle zones, and warning obstacle zones. The algorithm was validated using Matlab 2018a. The software was run on a 64-bit operating system with an Intel(R) Core(TM) i5-5200 CPU processor.

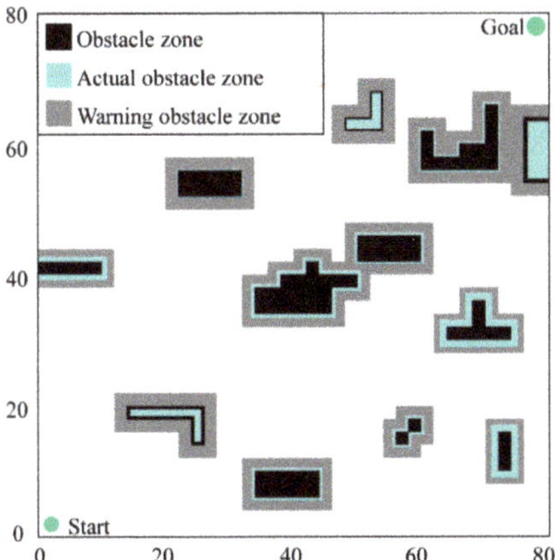

Figure 12. Map of the deep-sea environment.

To determine the ratio between the coefficients of the safety evaluation sub-function, trajectory comparison evaluation sub-function, and pseudo-power evaluation sub-function, a coefficient ratio ($ratio = \gamma_2/(\gamma_4 + \gamma_5)$) was set from 0 to 2 in increments of 0.1 for ten path-planning experiments. During these experiments, the closest distance to an actual obstacle zone and the average energy consumption when moving to the target point were recorded. Finally, the relationship between the coefficient ratio and the closest distance to obstacles and energy consumption is illustrated in Figure 13.

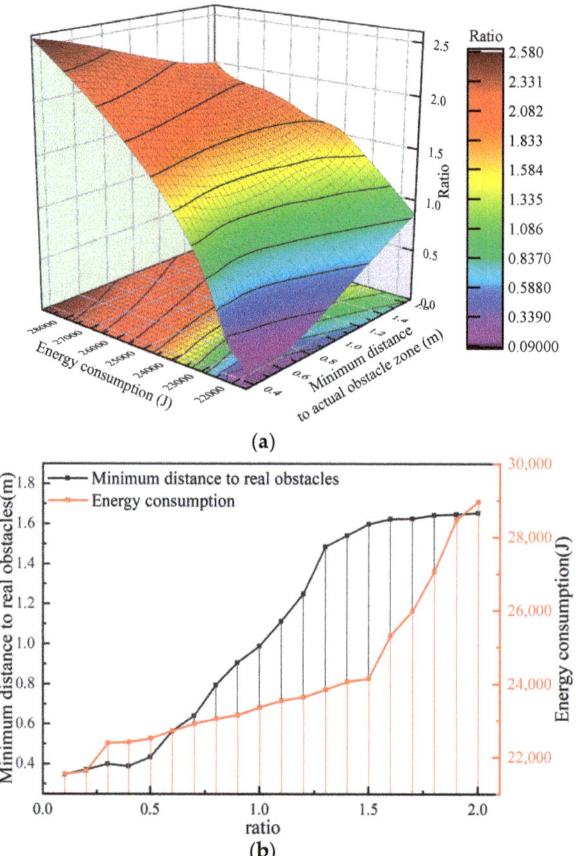

(a)

(b)

Figure 13. The relationship between average energy consumption and the average distance to actual obstacle zone. (**a**) The average energy consumption and the average distance to actual obstacle zone of the three-dimensional graph; (**b**) the average energy consumption and the average distance to actual obstacle zone of the two-dimensional graph.

From Figure 13, it can be observed that the closest distance to the actual obstacle zone increased from 0.5 m to 3 m when the coefficient ratio ranged from 0.5 to 1.5, indicating a significant improvement in safety distance. However, in the range of 1.5 to 2.0, the value remained relatively stable at around 3 m, indicating little change in safety distance. As for energy consumption, it gradually increased within the coefficient ratio of 0.6 to 1.5, while a noticeable upward trend was observed in the range of 1.5 to 2.0. To strike a balance between energy consumption and safety considerations, the coefficient ratio of 1.5 was selected, corresponding to a ratio of 3:2 for the coefficients. As a result, the final coefficient ratios for the safety evaluation sub-function, trajectory comparison evaluation sub-function, and pseudo-power evaluation sub-function were determined to be in the ratio of 3:1:1.

4.2. Simulated Experiments in the Deep-Sea Environment

To validate the effectiveness of applying the SE-DWA algorithm to autonomous sampling operations in the deep sea, a simulation-based verification is conducted in the underwater environment depicted in Figure 12. The simulation parameters for the DSLV are outlined in Table 1, where the DSLV slip rate (s_1 and s_2) is referenced from the literature [23]. The simulation parameters for both the DWA algorithm and the SE-DWA algorithm are detailed in Tables 2 and 3, respectively. In these tables, t_{pre} refers to prediction time, R_1 represents the inflation of the DSLV's body width, R_2 denotes the inflation maximum error, R_3 signifies the safety distance, and σ_{max}^{sonar} is referenced from the literature [16,24].

Table 1. Simulation parameters for the DSLV.

v_{max}	ω_{max}	ω_{min}	$\dot{v}_{max}$	$\dot{\omega}_{max}$	s_1,s_2
0.2 m/s	0.4 rad/s	−0.4 rad/s	0.1 m/s^2	0.15	0.85~1.0

Table 2. Simulation parameters for the DWA algorithm.

α_1	α_2	α_3	t_{pre}	R_1	R_2
0.1	0.3	0.2	5 s	1 m	2 m

Table 3. Simulation parameters for the SE-DWA algorithm.

γ_1	γ_2	γ_3	γ_4	γ_5	t_{pre}	R_1	R_3	σ_{max}^{sonar}	R_{safe}	R_{locus}
0.1	0.3	0.2	0.1	0.1	5 s	1 m	0.5 m	2 m	30 m	25 m

Constructing the environmental map in Figure 10 based on the scheme in Figure 7, three approaches—namely the MEC-DWA algorithm, the MS-DWA algorithm, and the SE-DWA algorithm—are employed for 50 simulation experiments. The distinction between the MEC-DWA algorithm and the MS-DWA algorithm resides solely in their size-to-obstacle expansion. They exemplify two extremes concerning safety and energy consumption. The paths produced by the MEC-DWA algorithm demonstrate the lowest attainable energy consumption within the realm of DWA algorithms, albeit at the expense of the safety performance. Conversely, the paths generated by the MS-DWA algorithm attain the highest level of safety among DWA algorithms but do so at the cost of the highest energy consumption.

The comparison of distances to actual obstacle zones and energy consumption for these 50 simulations is illustrated in Figures 14 and 15. The statistical results, including the average values for the simulation data, are presented in Table 4.

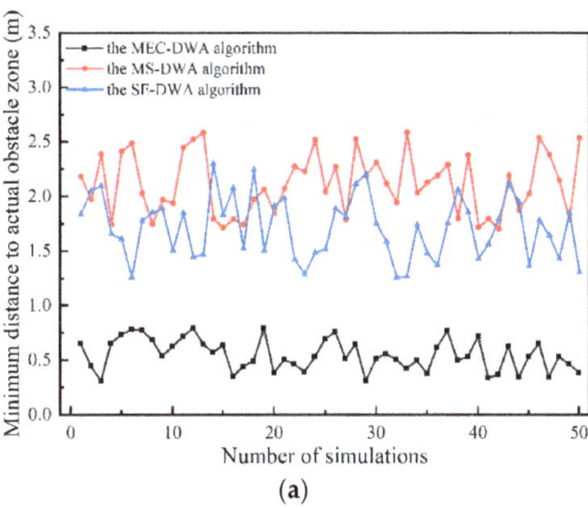

Figure 14. *Cont.*

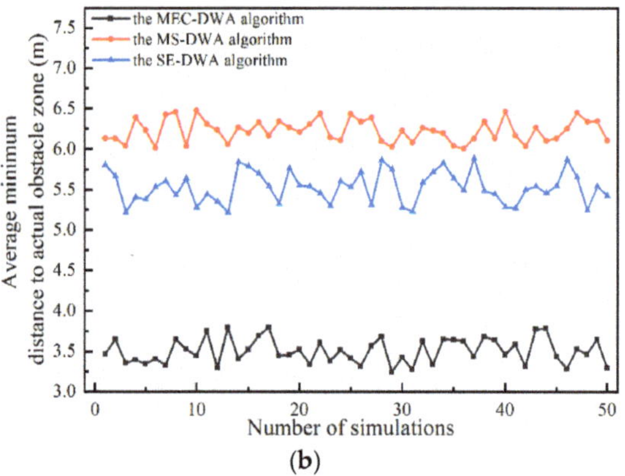

Figure 14. Comparison analysis of distance to actual obstacle zone. (**a**) Closest distance; (**b**) average closest distance.

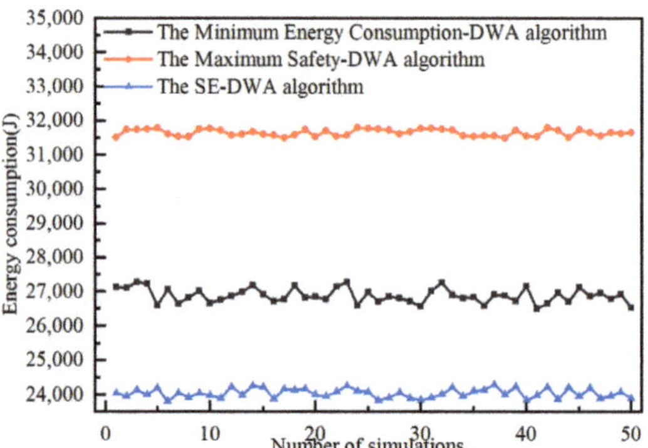

Figure 15. Comparison of energy consumption data.

Table 4. Average of 50 simulation data.

Algorithm	Minimum Distance to Actual Obstacle Zone	Average Minimum Distance to Actual Obstacle Zone	Energy Consumption	Path Length
The MEC-DWA algorithm	0.547 m	3.505 m	26,880.3 J	134.752 m
The MS-DWA algorithm	2.113 m	6.231 m	31,651.7 J	160.051 m
The SE-DWA algorithm	1.715 m	5.534 m	24,049.4 J	124.981 m

Based on Figure 14 and Table 4, it is evident that both the SE-DWA and MS-DWA algorithms have planned paths with a minimum distance from actual obstacle zones that exceeds 1.5 times the width of the DSLV (1 m). This distance provides a substantial safety margin to accommodate imperfect sensing systems, control errors, or other sources of uncertainty. In contrast, the paths generated by the MEC-DWA algorithm result in a minimum distance from actual obstacle zones that is less than the width of the vehicle. In such a scenario, the risk of collision between the DSLV and obstacles significantly increases due to factors such as sensor inaccuracies.

To validate the safety of the planned paths, besides checking if the minimum distance from actual obstacle zones meets the requirements, a safety assessment can also be employed. Safety assessment (s_d) is represented as the ratio of the minimum distance from actual obstacle zones to the average minimum distance from actual obstacle zones. For the paths generated by the SE-DWA algorithm and the MS-DWA algorithm, s_d is 0.339 and 0.310, respectively. This indicates that the DSLV is relatively distant from obstacles, resulting in a lower collision risk. Conversely, for the paths generated by the MEC-DWA algorithm, s_d is 0.156. This indicates that the DSLV is relatively closer to obstacles, resulting in a higher collision risk. The safety assessment for paths generated by the SE-DWA algorithm and the MS-DWA algorithm shows little difference, which confirms the validity of the SE-DWA algorithm's safety evaluation sub-function.

According to Table 4, it is evident that the path length for the SE-DWA algorithm is the shortest, showing a 21.9% reduction compared with the path length of the MS-DWA algorithm. When compared with the energy-efficient MEC-DWA algorithm, it also demonstrates a 7.2% reduction. At the same velocity, less time and energy are consumed when crawling to the target point. The trajectory comparison evaluation sub-function of the SE-DWA algorithm has been validated.

Based on Figure 15 and Table 4, it is evident that the SE-DWA algorithm plans paths with the lowest energy consumption. The unit energy consumption is the ratio of energy consumption to path length, indicating shorter travel times and lower energy consumption for the same path length. The unit energy consumption corresponding to the MEC-DWA algorithm, MS-DWA algorithm, and SE-DWA algorithm is 199.47, 197.76, and 192.42, respectively. The pseudo-power evaluation sub-function of the SE-DWA algorithm has been validated.

For the specific analysis, the path-planning results of the three approaches from one of the fifty trials were selected for examination. The path-planning results for the DWA algorithm and the SE-DWA algorithm are depicted in Figures 16 and 17, respectively. The variations in linear and angular velocities for the DWA algorithm and the SE-DWA algorithm are displayed in Figures 18 and 19, respectively. The energy consumption profiles for all three approaches are illustrated in Figure 20.

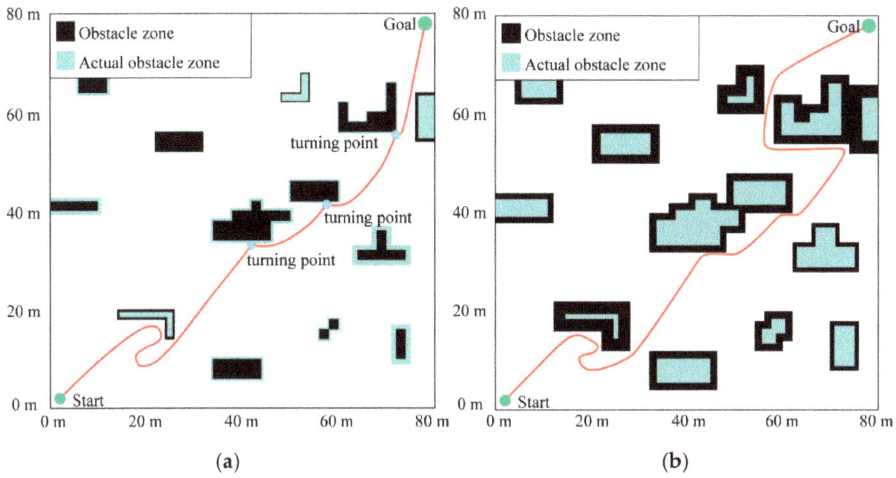

Figure 16. Path-planning results of the DWA algorithm. (**a**) The MEC-DWA algorithm; (**b**) The MS-DWA algorithm.

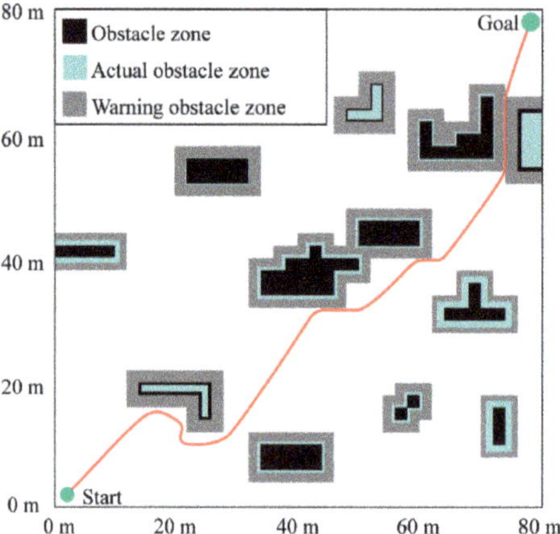

Figure 17. Path-planning result of the SE-DWA algorithm.

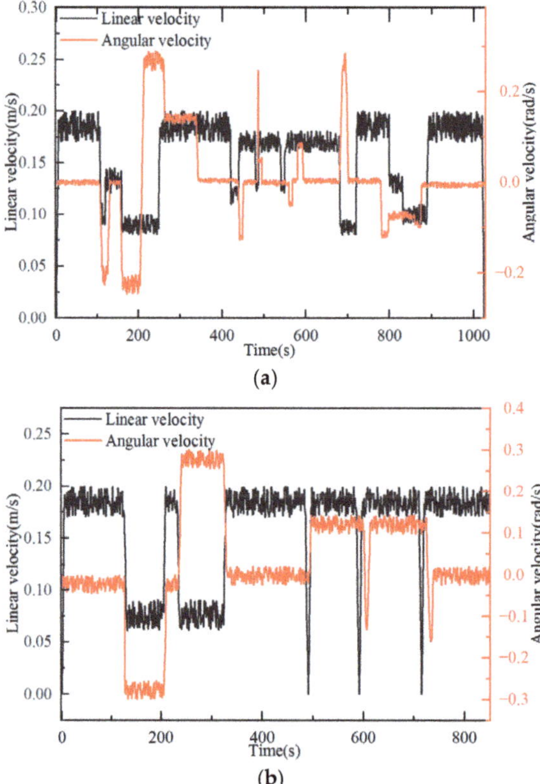

Figure 18. Linear velocity–angular velocity variation in the DWA algorithm. (**a**) The MEC-DWA algorithm; (**b**) the MS-DWA algorithm.

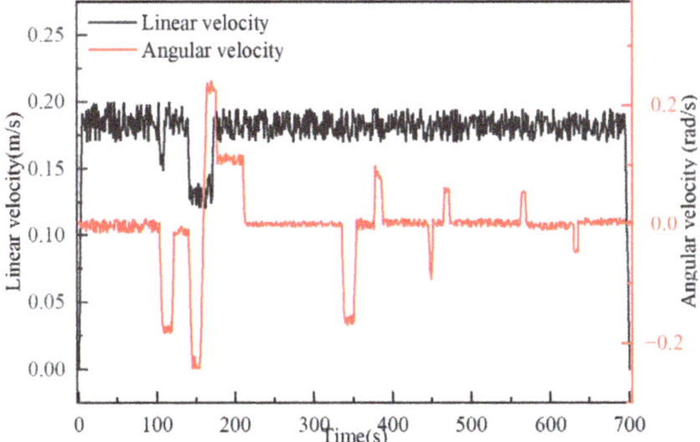

Figure 19. Linear velocity–angular velocity variation in the SE-DWA algorithm.

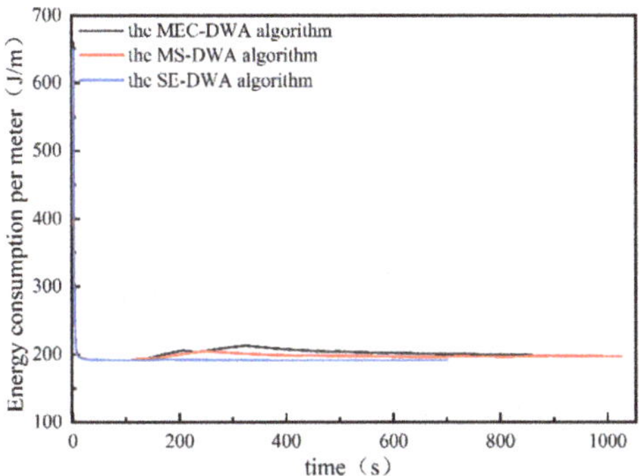

Figure 20. Unit energy consumption variation graph.

Based on the results from Figure 16 and Table 4, it is evident that the path generated by the MEC-DWA algorithm contains three turning points. These turning points indicate instances where the DSLV detects an impending collision with the actual obstacle zone, triggering emergency braking and adjustments to the heading angle. The minimum distance from the generated path to the actual obstacle zone is 0.547 m, which is smaller than the DSLV's vehicle width. This result does not ensure the autonomous sampling safety of the DSLV.

The path planned by the MS-DWA algorithm maintains a minimum distance of 2.113 m from the actual obstacle zones. This distance is significantly greater than the 0.547 m maintained by the MEC-DWA algorithm. Due to the larger inflation, the DSLV is required to travel a greater distance to reach the target point when encountering narrow areas. As a result, the path length increases by 18.7% compared with that of the MEC-DWA algorithm.

Based on the outcomes presented in Figure 17 and Table 4, it is evident that the SE-DWA algorithm generates a path with a minimum distance of 1.715 m from the actual obstacle zone, which is greater than the width of a DSLV's width. This ensures the safety of a DSLV's autonomous sampling operations. It validates the effectiveness of the added safety evaluation sub-function and the warning obstacle zone. The path length is reduced by 7.8% compared with that of the MEC-DWA algorithm and further reduced by 21.9% compared with that of the MS-DWA algorithm. This confirms the effectiveness of the trajectory comparison evaluation sub-function.

As evident from Figures 18 and 19, it is clear that when comparing the SE-DWA algorithm with the DWA algorithm, there are smaller variations in both the linear and angular velocities. This ensures that the DSLV remains at an optimal speed, contributing to a reduction in energy consumption to some extent. This finding validates the effectiveness of the pseudo-power evaluation sub-function. Simultaneously, Figures 17 and 19 reveal that the path-planning times for the DWA algorithm are 1010 s and 830 s, respectively. In contrast, the SE-DWA algorithm completes path planning in only 700 s. This efficiency is a result of shorter path lengths and optimal velocities, further confirming the effectiveness of the trajectory comparison evaluation sub-function and the pseudo-power evaluation sub-function.

From Figure 20 and Table 4, it is evident that the SE-DWA algorithm has the lowest energy consumption per meter. Compared with the MEC-DWA algorithm, the energy consumption is reduced by 10.5%. When compared with the MS-DWA algorithm, the energy consumption is reduced by 24%.

5. Conclusions

We have proposed the SE-DWA algorithm to address safety and energy consumption challenges faced by DSLVs during autonomous sampling operations in the deep-sea environment. Based on the dynamic window derived from the analysis of a DSLV's kinematics, we have devised three crucial sub-functions: safety evaluation sub-function, trajectory comparison evaluation sub-function, and pseudo-power evaluation sub-function. The safety evaluation sub-function tackles safety issues arising from reduced accuracy of forward-looking sonar due to underwater noise and other factors. The trajectory comparison evaluation sub-function reduces energy consumption by decreasing the path length when encountering "U"-shaped obstacles. The pseudo-power evaluation sub-function optimizes linear velocity and angular velocity to reduce energy consumption when encountering unknown deep-sea obstacles. Eventually, we conducted simulation experiments using deep-sea environmental maps. The simulation results demonstrate that the path planned by the SE-DWA algorithm compared with that from the DWA algorithm not only ensures enhanced safety performance but also results in at least an 11% reduction in energy consumption. The SE-DWA algorithm aligns better with the requirements of DSLVs for autonomous sampling in the deep-sea environment.

In the safety evaluation sub-function of the SE-DWA algorithm, the value used to construct the warning obstacle zone is currently set as the maximum forward-looking sonar error, which is a fixed value. When this value can be adaptively adjusted in the future, path security will be further enhanced and energy consumption will be further reduced. In the future, the focus will be on applying the SE-DWA algorithm to DSLVs' autonomous sampling operations at a depth of 4500 m in the deep sea to validate its feasibility and effectiveness in real-world environments.

Author Contributions: Conceptualization, Z.P. and Y.Z.; methodology, Z.P.; software, Z.P.; validation, Z.P., H.S., and W.G.; formal analysis, Z.P. and H.S.; investigation, Z.P., W.G., H.S., and Y.Z.; resources, H.S.; data curation: Z.P. and Y.L.; writing—original draft preparation, Z.P.; writing—review and editing, Z.P., H.S., and W.G.; visualization, Z.P., H.S., and Y.L.; supervision, W.G. and Y.Z.; project administration, Y.Z. and Y.L.; funding acquisition, W.G. All authors have read and agreed to the published version of the manuscript.

Funding: This research was funded by The Major Scientific and Technological Projects of Hainan Province, grant number ZDKJ202016.

Institutional Review Board Statement: Not applicable.

Informed Consent Statement: Not applicable.

Data Availability Statement: The data presented in this study are available in this article (tables and figures).

Acknowledgments: The authors thank the other members of the research group for their contributions to the research on the SE-DWA algorithm. And, the authors would like to thank the reviewers for their careful work.

Conflicts of Interest: The authors declare no conflict of interest.

References

1. Sun, H.; Gao, S.; Liu, J.; Liu, W. Research on comprehensive benefits and reasonable selection of marine resources development types. *Open Geosci.* **2022**, *14*, 141–150. [CrossRef]
2. David, M.B.; Charlotte, R.H. Sustainable use of ocean resources. *Mar. Policy* **2023**, *154*, 105672. [CrossRef]
3. Sun, H.; Guo, W.; Lan, Y.; Wei, Z.; Gao, S.; Sun, Y.; Fu, Y. Black-Box Modelling and Prediction of Deep-Sea Landing Vehicles Based on Optimised Support Vector Regression. *J. Mar. Sci. Eng.* **2022**, *10*, 575. [CrossRef]
4. Sun, H.; Guo, W.; Zhou, Y.; Sun, P.; Zhang, Y. Full-sea deep landing vehicle mechanism design and analysis of submersible and snorkeling performance. *Robot* **2020**, *42*, 207–214.
5. Sen, H.; Lei, W.; Yiting, W.; Huacheng, H. An efficient motion planning based on grid map: Predicted Trajectory Approach with global path guiding. *Ocean Eng.* **2021**, *238*, 109696. [CrossRef]
6. Huixia, Z.; Yadong, T.; Wenliang, Z. Global Path Planning of Unmanned Surface Vehicle Based on Improved A-Star Algorithm. *Sensors* **2023**, *23*, 6647. [CrossRef]
7. Daoud, M.A.; Mehrez, M.W.; Rayside, D.; Melek, W.W. Simultaneous Feasible Local Planning and Path-Following Control for Autonomous Driving. *IEEE Trans. Intell. Transp. Syst.* **2022**, *23*, 16358–16370. [CrossRef]
8. Szczepanski, R.; Tarczewski, T.; Erwinski, K. Energy Efficient Local Path Planning Algorithm Based on Predictive Artificial Potential Field. *IEEE Access* **2022**, *10*, 39729–39742. [CrossRef]
9. Chinonso, E.O.; Mohd Murtadha, M.; Nur Haliza Abdul, W.; Olakunle, E.; Abdulaziz, A.-N.; Zaleha, H.S. An Overview of Machine Learning Techniques in Local Path Planning for Autonomous Underwater Vehicles. *IEEE Access* **2023**, *11*, 24894–24907. [CrossRef]
10. Mingpeng, S.; Xiangyu, Z.; Yi, Z. Improvement of Dynamic Window Approach in Dynamic Obstacle Environment. *J. Phys. Conf. Ser.* **2023**, *2477*, 012059. [CrossRef]
11. Lee, D.H.; Lee, S.S.; Ahn, C.K.; Shi, P.; Lim, C.-C. Finite Distribution Estimation-Based Dynamic Window Approach to Reliable Obstacle Avoidance of Mobile Robot. *IEEE Trans. Ind. Electron.* **2020**, *68*, 9998–10006. [CrossRef]
12. Lai, X.; Wu, D.; Wu, D.; Li, J.H.; Yu, H. Enhanced DWA algorithm for local path planning of mobile robot. *Ind. Robot Int. J. Robot. Res. Appl.* **2023**, *50*, 186–194. [CrossRef]
13. Wang, Y.; Tian, Y.; Li, X.; Li, L. Adaptive dynamic window method for crossing dense obstacles. *Control Decis. -Mak.* **2019**, *34*, 927–936. [CrossRef]
14. Masato, K.; Naoki, M. Local Path Planning: Dynamic Window Approach with Virtual Manipulators Considering Dynamic Obstacles. *IEEE Access* **2022**, *10*, 17018–17029. [CrossRef]
15. Sun, Y.; Luo, X.; Ran, X.; Zhang, G. A 2D Optimal Path Planning Algorithm for Autonomous Underwater Vehicle Driving in Unknown Underwater Canyons. *J. Mar. Sci. Eng.* **2021**, *9*, 252. [CrossRef]
16. Zhang, Y.; Zhang, H.; Liu, J.; Zhang, S.; Liu, Z.; Lyu, E.; Chen, W. Submarine pipeline tracking technology based on AUVs with forward looking sonar. *Appl. Ocean Res.* **2022**, *122*, 103128. [CrossRef]
17. van Haren, H. Abyssal plain hills and internal wave turbulence. *Biogeosciences* **2018**, *15*, 4387–4403. [CrossRef]
18. Fox, D.; Burgard, W.; Thrun, S. The dynamic window approach to collision avoidance. *IEEE Robot. Autom. Mag. A Publ. IEEE Robot. Autom. Soc.* **1997**, *4*, 23–33. [CrossRef]
19. Yun, P.; Xiang, H.; ShuangGao, L. A measurement point planning method based on lidar automatic measurement technology. *Rev. Sci. Instrum.* **2023**, *94*, 015104. [CrossRef]
20. Li, Y.; Ma, T.; Chen, P.; Jiang, Y.; Wang, R.; Zhang, Q. Autonomous underwater vehicle optimal path planning method for seabed terrain matching navigation. *Ocean Eng.* **2017**, *133*, 107–115. [CrossRef]
21. Riazi, S.; Wigstrom, O.; Bengtsson, K.; Lennartson, B. Energy and Peak Power Optimization of Time-Bounded Robot Trajectories. *IEEE Trans. Autom. Sci. Eng.* **2017**, *14*, 646–657. [CrossRef]
22. Liu, Y.; Huang, H.; Fan, Q.; Zhu, Y.; Chen, X.; Han, Z. Mobile Robot Path Planning Based on Improved A*_DWA Algorithm. *Comput. Integr. Manuf. Syst.* **2022**, 1–20. Available online: https://kns.cnki.net/kcms/detail/11.5946.TP.20221125.1957.004.html (accessed on 23 August 2023).
23. Li, L.; Zou, X. Undersea robot automatically tracks and controls scheduled mining path. *J. Mech. Eng.* **2007**, *43*, 152–157. [CrossRef]
24. Yang, B.; Wang, W.; Liu, Y.; Li, X.; Liang, T.; Guo, W.; Liao, J.; Pan, F. Forward-looking scanning sonar imaging radial error analysis and compensation. *J. Electron. Inf.* **2021**, *43*, 796–802.

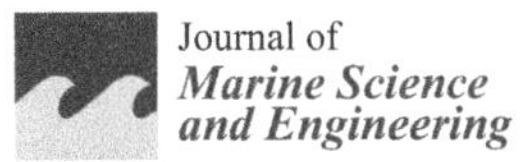

Journal of
*Marine Science
and Engineering*

Article

Trajectory Tracking Predictive Control for Unmanned Surface Vehicles with Improved Nonlinear Disturbance Observer

Huixuan Fu [1], Wenjing Yao [1], Ricardo Cajo [2] and Shiquan Zhao [1,*]

[1] College of Intelligent Systems Science and Engineering, Harbin Engineering University, Harbin 150001, China; fuhuixuan@hrbeu.edu.cn (H.F.); yaowenjing@hrbeu.edu.cn (W.Y.)
[2] Facultad de Ingeniería en Electricidad y Computación, Escuela Superior Politécnica del Litoral, ESPOL, Campus Gustavo Galindo Km 30.5 Vía Perimetral, P.O. Box 09-01-5863, Guayaquil 090150, Ecuador; rcajo@espol.edu.ec
* Correspondence: zhaoshiquan@hrbeu.edu.cn

Abstract: The motion of unmanned surface vehicles (USVs) is frequently disturbed by ocean wind, waves, and currents. A poorly designed controller will cause failures and safety problems during actual navigation. To obtain a satisfactory motion control performance for the USVs, a model predictive control (MPC) method based on an improved Nonlinear Disturbance Observer (NDO) is proposed. First, the USV model is approximately linearized and MPC is designed for the multivariable system with constraints. To compensate for the influence of disturbances, an improved NDO is designed where the calculation time for MPC is reduced. Finally, comparison simulations are conducted between MPC with the original NDO and MPC with an improved NDO, and the results show that they have similar performances to the USVs. However, the proposed method has fewer parameters that need to be tuned and is much more time-saving compared to MPC with a traditional NDO.

Keywords: unmanned surface vehicle; trajectory tracking; nonlinear disturbance observer; model predictive control

Citation: Fu, H.; Yao, W.; Cajo, R.; Zhao, S. Trajectory Tracking Predictive Control for Unmanned Surface Vehicles with Improved Nonlinear Disturbance Observer. *J. Mar. Sci. Eng.* **2023**, *11*, 1874. https://doi.org/10.3390/jmse11101874

Academic Editor: Alessandro Ridolfi

Received: 2 September 2023
Revised: 22 September 2023
Accepted: 23 September 2023
Published: 26 September 2023

1. Introduction

As technology has developed, USVs have been extensively used in various fields. However, due to disturbances from the sea wind, waves, and current, trajectory tracking control is of widespread concern. There have many studies undertaken on the control of the USV trajectory tracking technologies, including the PID controller [1–4], sliding mode control [5–8], backstepping control [9–12], MPC [13–16], adaptive control [17–20] and intelligent control [21–24].

MPC has been developed as an advanced optimization control algorithm based on the superiorities of feedback correction and rolling optimization. MPC has low requirements for the model, and it can solve constrained and multivariable problems. Although the computational load increases with fractional order, presently the development of microprocessors has made it possible for such computation. Predictive compensator-based event-triggered MPC with an NDO strategy has been proposed, and the trajectory tracking control problem of USV subject to input constraints, external disturbances, and cyber-attacks has been addressed [14]. However, the output constraints are considered in this paper, and the NDO has been discretized with an approximate discretization method. The adaptive line-of-sight algorithm was developed to obtain an expected heading angle. In addition, MPC was applied to reduce the lateral error, where the sideslip angle compensation was considered [25]. In addition, to obtain accurate state variables in real time, a linear extended-state observer was designed to overcome the influence of environmental disturbances and the nonlinearity of the model. However, the linearization still caused certain deviations in model disturbance estimation. To adjust the controller parameters, the MPC controller has been used to carry out both control allocation and trajectory tracking in real time [26]. Furthermore, it has concurrently optimized closed-loop performance with reinforcement learning-based and system-identification methods. To convert chance constraints into deterministic convex constraints, a convex conditional value of risk approximation has been introduced [27]. The converted constraints were further transformed into second-order cone constraints. Then, to account for the external disturbances and fulfill physical constraints, a stochastic model predictive control (SMPC) scheme was used to design the controller. The authors in [28] designed the path planning and control of the USVs simultaneously to overcome the disadvantage of the "first-planning-then-tracking" structure, and the artificial potential field and MPC were combined to solve the planning and tracking problem. The authors in [29] proposed the finite control set model predictive control (FCS-MPC). The more practical control commands formed a limited set of control, namely the thruster propulsion angle and speed of the USV. In addition, a fast and safe collision-avoidance system was designed according to the basics of FCS-MPC, which was applicable to varying environments. The authors in [30], to guarantee precise and stable trajectory tracking performance for AUVs, proposed a novel control architecture based on model-free control principles. The combination of intelligent PID and PD feedforward control had good performance for trajectory tracking accuracy, disturbance rejections, and initial tracking error compensations. The authors in [31], to solve the surge-motion tracking control problem of an autonomous undersea vehicle (AUV) with system constraints, proposed adaptive backstepping control scheme. Both a state feedback control scheme and an output feedback control scheme were developed for AUVs with deferred output constraints. The authors in [32], to pay more attention to the characteristics of flexibility and accessibility, proposed a fusion framework of field theoretical planning and an MPC algorithm. In addition, the trajectory smoothness and collision-avoidance constraints under a complex environment were considered. The authors in [33], to solve the fault-tolerant trajectory tracking control problem of twin-propeller non-rudder USVs subject to propeller faults, proposed an adaptive fault-tolerant trajectory tracking control scheme by utilizing the excellent nonlinearity approximation performance of neural networks. The authors in [34], to achieve autonomous cooperative formation control of underactuated USVs in a complex ocean environment, adopted a dual MPC approach based on a virtual trajectory. The authors in [35] showed that time-varying external disturbances affected the accuracy of trajectory tracking. To ensure trajectory tracking accuracy, a reduced-order extended-state observer and the super-twisting second-order sliding mode controller were proposed. The authors in [36], to solve the lumped uncertainties caused by input quantization, actuator faults, and dead zones, proposed an adaptive sliding mode tracking controller for USVs with predefined time performance. In the control design, adaptive control gains were established based on barrier functions. In addition, a predefined time-adaptive SMC scheme was adopted by introducing an auxiliary function. The authors in [37], to set the velocity of the USV converge to zero at the berth, adopted an interpolation approach to densify the waypoints at the end of the berthing trajectory. In addition, to improve computational performance during the USV berthing, an event-triggered adaptive horizon MPC approach was adopted.

Controllers are usually poorly tuned to USV motion systems, and the disturbance-rejection performance of controllers is not satisfactory due to disturbances from ocean waves, wind, and currents. Therefore, an improved NDO-based MPC method for trajectory tracking control of USVs is proposed in this paper. First, the MPC is designed for USV trajectory tracking. Then, an NDO is designed to estimate the disturbance of ocean wind, waves, and currents, which has fewer parameters. In addition, Lyapunov stability is analyzed for the overall system. Finally, the proposed method is verified by simulation experiments. The main contributions of this work can be summarized as follows: the output constraints are considered in this paper, and the NDO has been discretized with an approximate discretization method; the disturbances are compensated for by an improved NDO, and better trajectory tracking performances are obtained for USVs; also, with the improved NDO, calculation time is saved compared with the traditional NDO.

The rest of the paper is structured as follows: Section 2 introduces the USV kinematics model and dynamic model. In Section 3, the improved NDO-based MPC is designed for the USV, and Lyapunov stability is analyzed for the overall system. Comparison simulations and discussion of the results is performed in Section 4. Section 5 presents the conclusions.

2. State-Space Model of Unmanned Surface Vehicles

In general, the kinematics modeling of USVs is represented by the North–East–Down coordinate system, while the dynamics model is built in the ship coordinate system. The North–East–Down coordinate system is also called the geodetic coordinate system or the inertial coordinate system. It is usually used as the reference system, and any point on the sea can be used as the origin of the system. The ship coordinate system changes with the motion of the ship and can describe the force, moment, linear velocity, and angular velocity of a USV in various degrees of freedom.

To simplify the model, a USV model is utilized with three degrees of freedom for trajectory tracking control. The motions of yaw, surge, and sway are the most important for the trajectory tracking of the USVs, so the roll, pitch, and heave of the USVs are ignored. Thus, the USV model can be represented in the ship coordinate system and the geodetic coordinate system. This is shown in Figure 1.

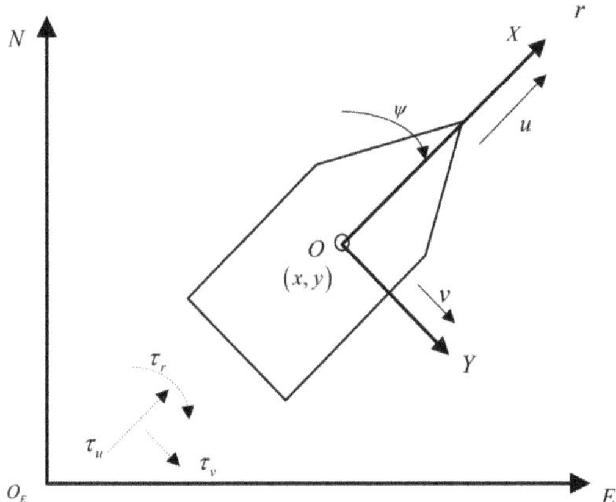

Figure 1. Three degrees of freedom of motion model for USVs.

In Figure 1, it can be seen the North–East–Down coordinate system is represented by NO_EE, and the coordinate system for the ship is described by XOY. The kinematics and dynamics model of the ship can be represented as:

$$\dot{\eta} = \mathbf{R}(\psi)\upsilon \tag{1}$$

$$\mathbf{M}\dot{\upsilon} + \mathbf{C}(\upsilon)\upsilon + \mathbf{D}(\upsilon)\upsilon = \tau + \tau_\mathbf{d} \tag{2}$$

where $\eta = [x, y, \psi]^T$ represents the x, y position and heading angle vector of the USV in the inertial coordinate system; x and y represent the position of the ship with regard to the North–East–Down coordinate system, ψ represents the yaw angle information of the USV; $\upsilon = [u, v, r]^T$ is the vector of the velocity and angular velocity for the USV in the ship coordinate system; u, v, and r are the surge velocity, sway velocity, and yaw angular velocity of USV, respectively; τ_u, τ_v, and τ_r represent the surge thrust, sway thrust, and yaw moment, which are the control inputs of the system; $\tau_\mathbf{d} = [\tau_{ud}, \tau_{vd}, \tau_{rd}]^T$ is the corresponding thrust and moment caused by the time-varying external disturbances. $\mathbf{R}(\psi)$ is the rotation matrix with the relationship of $\mathbf{R}^{-1}(\psi) = \mathbf{R}^T(\psi)$; $\mathbf{M}$ represents the inertial matrix of USV, where $\mathbf{M} = \mathbf{M}^T > 0$; $\mathbf{C}(\upsilon)$ represents the Coriolis centripetal force matrix, and $\mathbf{C}(\upsilon) = -\mathbf{C}(\upsilon)^T$; and $\mathbf{D}(\upsilon)$ is the nonlinear hydrodynamic damping matrix. The detailed information for the matrixes is shown as follows:

$$\mathbf{R}(\psi) = \begin{bmatrix} \cos\psi & -\sin\psi & 0 \\ \sin\psi & \cos\psi & 0 \\ 0 & 0 & 1 \end{bmatrix}, \ \mathbf{M} = \begin{bmatrix} m_{11} & 0 & 0 \\ 0 & m_{22} & m_{23} \\ 0 & m_{32} & m_{33} \end{bmatrix} \tag{3}$$

$$\mathbf{C}(\upsilon) = \begin{bmatrix} 0 & 0 & -m_{22}v \\ 0 & 0 & m_{11}u \\ m_{22}v & -m_{11}u & 0 \end{bmatrix}, \ \mathbf{D}(\upsilon) = -\begin{bmatrix} d_{11} & 0 & 0 \\ 0 & d_{22} & d_{23} \\ 0 & d_{32} & d_{33} \end{bmatrix} \tag{4}$$

According to Equations (3) and (4), the reduced kinematics and dynamics equations can be represented as:

$$\begin{cases} \dot{x} = u\cos\psi - v\sin\psi \\ \dot{y} = u\sin\psi + v\cos\psi \\ \dot{\psi} = r \end{cases} \tag{5}$$

$$\begin{cases} m_{11}\dot{u} - m_{22}vr + d_{11}u = \tau_u + \tau_{ud} \\ m_{22}\dot{v} + m_{23}\dot{r} + m_{11}ur + d_{22}v + d_{23}r = \tau_v + \tau_{vd} \\ m_{32}\dot{v} + m_{33}\dot{r} + (m_{22} - m_{11})uv + d_{32}v + d_{33}r = \tau_r + \tau_{rd} \end{cases} \tag{6}$$

For the USVs, there are constraints for the actuators and the outputs. In addition, they are described as follows.

The increment constraints for inputs can be represented as:

$$\Delta\mathbf{u}_{\min}(t+k) \leq \Delta\mathbf{u}(t+k) \leq \Delta\mathbf{u}_{\max}(t+k)$$
$$k = 0, 1, \cdots, N_c - 1 \tag{7}$$

where N_c denotes the control horizon.

The upper and lower-limit constraints for inputs can be represented as:

$$\mathbf{u}_{\min}(t+k) \leq \mathbf{u}(t+k) \leq \mathbf{u}_{\max}(t+k)$$
$$k = 0, 1, \cdots, N_c - 1 \tag{8}$$

The outputs of speed increment constraints can be represented as:

$$\Delta\upsilon_{\min}(t+k) \leq \Delta\upsilon(t+k) \leq \Delta\upsilon_{\max}(t+k)$$
$$k = 0, 1, \cdots, N_c - 1 \tag{9}$$

The outputs of speed upper and lower-limit constraints can be represented as:

$$\upsilon_{\min}(t+k) \leq \upsilon(t+k) \leq \upsilon_{\max}(t+k)$$
$$k = 0, 1, \cdots, N_c - 1 \tag{10}$$

The terminal equality constraint can be represented as:

$$\|\boldsymbol{\eta}(k+N|t) - \boldsymbol{\eta}_{\mathbf{r}}(k+N|t)\|^2 = 0 \tag{11}$$

where $\boldsymbol{\eta}$ denotes the vector to be controlled, and $\boldsymbol{\eta}_{\mathbf{r}}$ denotes the reference trajectory.

3. Nonlinear Disturbance Observer-Based Model Predictive Control

In this section, NDO-based MPC is designed for the three-degrees-of-freedom kinematics and dynamics of a USV with the state-space model. The design schematic diagram of NDO-based MPC is shown in Figure 2.

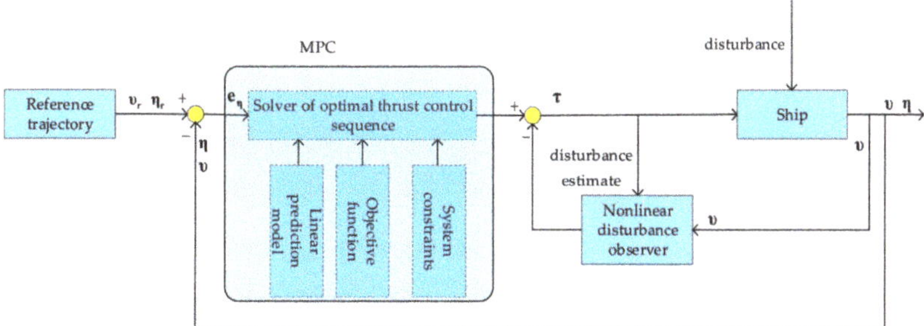

Figure 2. Schematic diagram of MPC based on a nonlinear observer.

According to Equations (1) and (2), the state-space equation for the ship can be rewritten as:

$$\begin{cases} \dot{\eta} = \mathbf{R}(\varphi)\upsilon \\ \dot{\upsilon} = \mathbf{M}^{-1}(\tau + \tau_{\mathbf{d}} - \mathbf{C}(\upsilon)\upsilon - \mathbf{D}(\upsilon)\upsilon) \end{cases} \tag{12}$$

From Equation (12), the matrixes **R**, **C**, and **D** are nonlinear variables, which gives the model strong nonlinearity.

3.1. Model Predictive Control Design of an Unmanned Surface Vehicle

3.1.1. Discrete Linearization of an Unmanned Surface Vehicle Model

USV is a complex system with large inertia and time-delay characteristics. During navigation, it is subjected to various forces such as thrust, hydrodynamic force, hydrostatic force, and external disturbances. Therefore, the USV model has obvious nonlinear characteristics. In this paper, a simplified model of the USV is applied, and the uncertainty is dealt with by the NDO together with the disturbances from the ocean environment. In addition, the model is linearized.

In this paper, the NDO is designed to compensate for the disturbance. Therefore, a linear model of the USV is sufficient for MPC design, where the uncertainty from the linearization of the USV model can be solved with an NDO. Therefore, the model is linearized first and then discretized. Finally, the optimal control sequence is obtained according to the linear model predictive control.

The linearization of a nonlinear system can be divided into approximate linearization and exact linearization. Among them, the approximate linearization method is relatively simple with high applicability but low accuracy. The precision of the accurate linearization method is high, but it should be noted that a special case analysis of a single system is difficult and has poor universality.

The reference trajectory is represented in Equation (13) with environmental disturbances. The first-order Taylor expansion can be obtained at any point $(\mathbf{x_r}, \mathbf{u_r})$; then approximate linearization is achieved by leaving the higher-order terms. This can be seen in Equation (14).

$$\dot{\mathbf{x}}_{\mathbf{r}} = f(\mathbf{x_r}, \mathbf{u_r}) \tag{13}$$

$$\dot{\mathbf{x}} = f(\mathbf{xr}, \mathbf{ur}) + \left.\frac{\partial f}{\partial \mathbf{x}}\right|_{\substack{\mathbf{x}=\mathbf{xr}\\\mathbf{u}=\mathbf{ur}}}(\mathbf{x} - \mathbf{xr}) + \left.\frac{\partial f}{\partial \mathbf{u}}\right|_{\substack{\mathbf{x}=\mathbf{xr}\\\mathbf{u}=\mathbf{ur}}}(\mathbf{u} - \mathbf{ur}) \tag{14}$$

Subtract Equation (13) from (14), and the following equation can be achieved.

$$\dot{\tilde{\mathbf{x}}} = \mathbf{A}\tilde{\mathbf{x}} + \mathbf{B}\tilde{\mathbf{u}} \tag{15}$$

and $\tilde{\mathbf{x}} = \mathbf{x} - \mathbf{x_r}$, $\tilde{\mathbf{u}} = \mathbf{u} - \mathbf{u_r}$, $\mathbf{A} = \left.\frac{\partial f}{\partial \mathbf{x}}\right|_{\substack{\mathbf{x}=\mathbf{xr}\\\mathbf{u}=\mathbf{ur}}}$, $\mathbf{B} = \left.\frac{\partial f}{\partial \mathbf{u}}\right|_{\substack{\mathbf{x}=\mathbf{xr}\\\mathbf{u}=\mathbf{ur}}}$.

The discrete form for Equation (15) is shown as follows:

$$\begin{cases} \tilde{\mathbf{x}}(k+1) = \mathbf{A_d}\tilde{\mathbf{x}}(k) + \mathbf{B_d}\tilde{\mathbf{u}}(k) \\ \tilde{\mathbf{y}}(k) = \mathbf{C}\tilde{\mathbf{x}}(k) \end{cases} \tag{16}$$

where $\mathbf{A_d} = T\mathbf{A} + \mathbf{I}$, $\mathbf{B_d} = T\mathbf{B}$, $\tilde{\mathbf{y}}(k) = \mathbf{y}(k) - \mathbf{y_r}(k)$, T is the discretization step.

3.1.2. Objective Function Design

To ensure that the USV can track the trajectory smoothly and quickly, the cost function shown in Equation (17) is applied, which concerns the increments of the control variables and the errors of the system states.

$$\min \quad J = \min\left\{ \sum_{i=1}^{N_p} \|\boldsymbol{\eta}(k+i) - \boldsymbol{\eta_r}(k+i)\|_{\mathbf{Q}}^2 + \sum_{i=1}^{N_c-1} \|\Delta\mathbf{u}(k+i)\|_{\mathbf{R}}^2 \right\} \tag{17}$$

where $\mathbf{Q}$ and $\mathbf{R}$ are the weight matrixes for tracking errors and increments of the control variables, respectively; and N_p is the prediction horizon. $\Delta\mathbf{u}(k+i)$ is the variable of the increments of the control variables, so it can be obtained as follows:

$$\xi(k) = \begin{bmatrix} \tilde{\mathbf{x}}(k) \\ \tilde{\mathbf{u}}(k-1) \end{bmatrix} \tag{18}$$

The new state-space equation is represented as:

$$\begin{cases} \xi(k+1) = \tilde{\mathbf{A}}\xi(k) + \tilde{\mathbf{B}}\Delta\mathbf{u}(k) \\ \tilde{\mathbf{y}}(k) = \tilde{\mathbf{C}}\xi(k) \end{cases} \tag{19}$$

where $\tilde{\mathbf{A}} = \begin{bmatrix} \mathbf{A_d} & \mathbf{B_d} \\ \mathbf{0} & \mathbf{I_{Nu}} \end{bmatrix}$, $\tilde{\mathbf{B}} = \begin{bmatrix} \mathbf{B_d} \\ \mathbf{I_{Nu}} \end{bmatrix}$, $\tilde{\mathbf{C}} = [\mathbf{I_{Nx}} \quad \mathbf{0}]$, N_u denotes the number of control variables, N_x denotes the number of state variables.

Therefore, the system prediction outputs can be calculated as follows:

$$\begin{aligned} \mathbf{Y} &= \boldsymbol{\Psi}\xi(k) + \mathbf{H}\tilde{\mathbf{u}}(k) \\ J &= \tfrac{1}{2}\Delta\mathbf{u}(k)^T\mathbf{H}_J\Delta\mathbf{u}(k) + \mathbf{f}_J\Delta\mathbf{u}(k) \end{aligned} \tag{20}$$

where $\mathbf{H}_J = 2(\mathbf{H}^T\mathbf{Q_c}\mathbf{H} + \mathbf{R_c})$, $\mathbf{f}_J = 2\boldsymbol{\Psi}\xi(k)\mathbf{Q_c}\mathbf{H}$, $\mathbf{Q_c} = \begin{bmatrix} \mathbf{Q} & & \\ & \ddots & \\ & & \mathbf{Q} \end{bmatrix}_{N_p \times N_p}$,

$\mathbf{R_c} = \begin{bmatrix} \mathbf{R} & & \\ & \ddots & \\ & & \mathbf{R} \end{bmatrix}_{N_c \times N_c}$.

3.2. Nonlinear Disturbance Observer Design of Unmanned Surface Vehicles

To make MPC for a USV more applicable, a nonlinear disturbance observer is designed, which can estimate and compensate for the external environmental disturbance received by the USV. Therefore, the stability and anti-disturbance performance is improved for the USV, while capsizing and unnecessary navigation accidents are avoided for the USV.

According to the mathematical model of the USV, the state equation is designed as follows:

$$\dot{\hat{\boldsymbol{\tau}}}_\mathbf{d} = \mathbf{K_0}(\boldsymbol{\tau_d} - \hat{\boldsymbol{\tau}}_\mathbf{d}) = -\mathbf{K_0}\hat{\boldsymbol{\tau}}_\mathbf{d} + \mathbf{K_0}(\mathbf{M}\dot{\boldsymbol{\upsilon}} + \mathbf{C}(\boldsymbol{\upsilon})\boldsymbol{\upsilon} + \mathbf{D}(\boldsymbol{\upsilon})\boldsymbol{\upsilon} - \boldsymbol{\tau}) \tag{21}$$

where $\mathbf{K_0}$ is a three-dimensional positive definite matrix; the estimated disturbance values $\hat{\boldsymbol{\tau}}_\mathbf{d}$ can be specifically written as $\hat{\boldsymbol{\tau}}_\mathbf{d} = [\hat{\tau}_{ud}, \hat{\tau}_{vd}, \hat{\tau}_{rd}]^T$, which are the estimated values of surge disturbance, sway disturbance, and yaw direction disturbance.

It can be seen from Equation (12) that υ of a USV can be obtained directly, but the derivative term of the speed state variable $\dot{\upsilon}$ cannot be obtained directly. Therefore, it is necessary to improve the disturbance observer. The variable β can be selected as the intermediate assignment variable of the observer, which is expressed as follows:

$$\beta = \hat{\boldsymbol{\tau}}_\mathbf{d} - \mathbf{K_0 M}\upsilon \tag{22}$$

Then,

$$\begin{aligned}
\dot{\beta} &= \dot{\hat{\boldsymbol{\tau}}}_\mathbf{d} - \mathbf{K_0 M}\dot{\upsilon} \\
&= \mathbf{K_0}(\mathbf{M}\dot{\upsilon} + \mathbf{C}(\upsilon)\upsilon + \mathbf{D}(\upsilon)\upsilon - \boldsymbol{\tau}) - \mathbf{K_0}\hat{\boldsymbol{\tau}}_\mathbf{d} - \mathbf{K_0 M}\dot{\upsilon} \\
&= -\mathbf{K_0}(\beta + \mathbf{K_0 M}\upsilon) + \mathbf{K_0}(\mathbf{C}(\upsilon)\upsilon + \mathbf{D}(\upsilon)\upsilon - \boldsymbol{\tau}) \\
&= -\mathbf{K_0}\beta - \mathbf{K_0}(\mathbf{K_0 M}\upsilon - \mathbf{C}(\upsilon)\upsilon - \mathbf{D}(\upsilon)\upsilon + \boldsymbol{\tau})
\end{aligned} \tag{23}$$

Therefore, the new equation of the improved NDO is:

$$\begin{aligned}
\hat{\boldsymbol{\tau}}_\mathbf{d} &= \beta + \mathbf{K_0 M}\upsilon \\
\dot{\beta} &= -\mathbf{K_0}\beta - \mathbf{K_0}(\mathbf{K_0 M}\upsilon - \mathbf{C}(\upsilon)\upsilon - \mathbf{D}(\upsilon)\upsilon + \boldsymbol{\tau})
\end{aligned} \tag{24}$$

The improved equation avoids the calculation of $\dot{\upsilon}$ and simplifies the calculation process. Therefore, it can improve calculation efficiency and save calculation time.

Equation (24) is in a continuous form, and cannot be directly used for MPC design. Therefore, it is discretized with the approximate discretization method:

$$\begin{aligned}
\boldsymbol{\tau_\mathbf{d}}(k+1) &= \beta_\mathbf{d} + \mathbf{K_{d1} M}\upsilon + \boldsymbol{\tau_\mathbf{d}}(k) \\
\beta(k+1) &= -\mathbf{K_{d2}}\beta(k) - \mathbf{K_{d1}}(\mathbf{K_0 M}\upsilon - \mathbf{C}(\upsilon)\upsilon - \mathbf{D}(\upsilon)\upsilon + \boldsymbol{\tau})
\end{aligned} \tag{25}$$

and $\beta_\mathbf{d} = T\beta$, $\mathbf{K_{d1}} = T\mathbf{K_0}$, and $\mathbf{K_{d2}} = T\mathbf{K_0} - \mathbf{I}$.

3.3. Stability Analysis of Unmanned Surface Vehicles

3.3.1. Stability Analysis of the Model Predictive Control of Unmanned Surface Vehicles

To verify the stability of the USV control system under MPC, the Lyapunov function defined as $V^0(k)$ is selected:

$$V^0(k) = \min_{\Delta v} \sum_{i=1}^{N_p} \|\eta(k+i|t) - \eta_\mathbf{d}(k+i|t)\|_\mathbf{Q}^2 + \sum_{i=1}^{N_c-1} \|\Delta\mathbf{u}(k+i|t)\|_\mathbf{R}^2 \tag{26}$$

If the control horizon N_c is defined to be equal to the prediction horizon N_p, the above equation can be simplified as follows:

$$V^0(k) = \min_{\Delta v} \sum_{i=1}^{N} \|\eta(k+i|t) - \eta_\mathbf{d}(k+i|t)\|_\mathbf{Q}^2 + \|\Delta\mathbf{u}(k+i|t)\|_\mathbf{R}^2 \tag{27}$$

The quadratic function is always greater than 0, so its positive definiteness is proven:

$$V^0(k) \geq 0 \tag{28}$$

Then, we only need to prove that $V^0(k)$ is decreasing, then its stability is proven.

$$
\begin{aligned}
V^0(k+1) \quad &= \min_{\Delta v}\left\{\sum_{i=1}^{N}\|\,\eta(k+i+1|t)-\eta_{\mathbf{d}}(k+i+1|t)\,\|_{\mathbf{Q}}^2 + \|\,\Delta\mathbf{u}(k+i+1|t)\,\|_{\mathbf{R}}^2\right\} \\[4pt]
&= \min_{\Delta v}\left\{\begin{aligned} &\sum_{i=1}^{N}\Big(\|\,\eta(k+i|t)-\eta_{\mathbf{d}}(k+i|t)\,\|_{\mathbf{Q}}^2 + \|\,\Delta\mathbf{u}(k+i|t)\,\|_{\mathbf{R}}^2\Big)- \\ &\|\,\eta(k+1|t)-\eta_{\mathbf{d}}(k+1|t)\,\|_{\mathbf{Q}}^2 - \|\,\Delta\mathbf{u}(k+1|t)\,\|_{\mathbf{R}}^2+ \\ &\|\,\eta(k+1+N|t)-\eta_{\mathbf{d}}(k+1+N|t)\,\|_{\mathbf{Q}}^2 + \|\,\Delta\mathbf{u}(k+1+N|t)\,\|_{\mathbf{R}}^2 \end{aligned}\right\} \\[4pt]
&\stackrel{\Delta v}{=} -\|\,\eta(k+1|t)-\eta_{\mathbf{d}}(k+1|t)\,\|_{\mathbf{Q}}^2 - \|\,\Delta\mathbf{u}(k+1|t)\,\|_{\mathbf{R}}^2+ \\[4pt]
&\quad \min_{\Delta v}\left\{\begin{aligned} &\sum_{i=1}^{N}\Big(\|\,\eta(k+i|t)-\eta_{\mathbf{d}}(k+i|t)\,\|_{\mathbf{Q}}^2 + \|\,\Delta\mathbf{u}(k+i|t)\,\|_{\mathbf{R}}^2\Big)+ \\ &\|\,\eta(k+1+N|t)-\eta_{\mathbf{d}}(k+1+N|t)\,\|_{\mathbf{Q}}^2 + \|\,\Delta\mathbf{u}(k+1+N|t)\,\|_{\mathbf{R}}^2 \end{aligned}\right\} \\[4pt]
&\leq -\|\,\eta(k+1|t)-\eta_{\mathbf{d}}(k+1|t)\,\|_{\mathbf{Q}}^2 - \|\,\Delta\mathbf{u}(k+1|t)\,\|_{\mathbf{R}}^2 + V^0(k)+ \\[4pt]
&\quad \min_{\Delta v}\left\{\|\,\eta(k+1+N|t)-\eta_{\mathbf{d}}(k+1+N|t)\,\|_{\mathbf{Q}}^2 + \|\,\Delta\mathbf{u}(k+1+N|t)\,\|_{\mathbf{R}}^2\right\}
\end{aligned}
\tag{29}
$$

The terminal equation constraint is:

$$
\|\eta(k+1+N|t)-\eta_{\mathbf{r}}(k+1+N|t)\|_{\mathbf{Q}}^2 = 0
\tag{30}
$$

Furthermore,

$$
\min_{\Delta v}\left\{\|\eta(k+1+N|t)-\eta_{\mathbf{d}}(k+1+N|t)\|_{\mathbf{Q}}^2 + \|\Delta\mathbf{u}(k+1+N|t)\|_{\mathbf{R}}^2\right\} = 0
\tag{31}
$$

$$
\|\eta(k+i|t)-\eta_{\mathbf{r}}(k+i|t)\|_{\mathbf{Q}}^2 + \|\Delta\mathbf{u}(k+i|t)\|_{\mathbf{R}}^2 = 0
\tag{32}
$$

Therefore, $V^0(k+1) \leq V^0(k)$, and the stability of MPC is proven.

3.3.2. Stability Analysis of the Nonlinear Disturbance Observer of Unmanned Surface Vehicles

To verify the stability of the NDO and ensure that it can be applied to the trajectory tracking control system, it is necessary first to define the variable of the difference between the observer's estimated value and the actual value of the external disturbance to the USV:

$$
\tilde{\tau}_{\mathbf{d}} = \tau_{\mathbf{d}} - \hat{\tau}_{\mathbf{d}}
\tag{33}
$$

Considering the kinematic Equations (2), (24) and (33) of the USV, then by calculating the derivative of time on both sides of Equation (27), the formula can be represented as follows:

$$
\begin{aligned}
\dot{\hat{\tau}}_{\mathbf{d}} &= \dot{\beta} + \mathbf{K}_0\mathbf{M}\dot{\upsilon} \\
&= -\mathbf{K}_0\beta - \mathbf{K}_0(\mathbf{K}_0\mathbf{M}\upsilon - \mathbf{C}(\upsilon)\upsilon - \mathbf{D}(\upsilon)\upsilon + \tau) + \mathbf{K}_0(-\mathbf{C}(\upsilon)\upsilon - \mathbf{D}(\upsilon)\upsilon + \tau + \tau_{\mathbf{d}}) \\
&= \mathbf{K}_0(\tau_{\mathbf{d}} - (\beta + \mathbf{K}_0\mathbf{M}\upsilon)) \\
&= \mathbf{K}_0\tilde{\tau}_{\mathbf{d}}
\end{aligned}
\tag{34}
$$

From Equation (34), one can calculate the derivative of both sides of Equation (33) with respect to time, and simplify it to obtain:

$$
\dot{\tilde{\tau}}_{\mathbf{d}} = \dot{\tau}_{\mathbf{d}} - \dot{\hat{\tau}}_{\mathbf{d}} = \dot{\tau}_{\mathbf{d}} - \mathbf{K}_0\tilde{\tau}_{\mathbf{d}}
\tag{35}
$$

The Lyapunov method is used to verify the stability of the disturbance observer, and the appropriate Lyapunov function is selected as follows:

$$
V_d = \frac{1}{2}\tilde{\tau}_{\mathbf{d}}^{\mathrm{T}}\tilde{\tau}_{\mathbf{d}}
\tag{36}
$$

According to Equation (35), the derivative of both sides of Equation (36) with respect to time can be obtained as follows:

$$\dot{V}_d = \tilde{\boldsymbol{\tau}}_{\mathbf{d}}^{\mathrm{T}}\dot{\tilde{\boldsymbol{\tau}}}_{\mathbf{d}} = -\tilde{\boldsymbol{\tau}}^{\mathrm{T}}\mathbf{K}_0\tilde{\boldsymbol{\tau}} + \tilde{\boldsymbol{\tau}}_{\mathbf{d}}^{\mathrm{T}}\dot{\boldsymbol{\tau}}_{\mathbf{d}} \tag{37}$$

According to Young's inequality theory,

$$\tilde{\boldsymbol{\tau}}_{\mathbf{d}}^{\mathrm{T}}\dot{\boldsymbol{\tau}}_{\mathbf{d}} \leq a_1\tilde{\boldsymbol{\tau}}_{\mathbf{d}}^{\mathrm{T}}\tilde{\boldsymbol{\tau}}_{\mathbf{d}} + \frac{C_d^2}{4a_1} \tag{38}$$

and $a_1 > 0$, C_d is the limit of the disturbance change rate.

From Equations (37) and (38), this can be written as inequality:

$$\dot{V}_d \leq -\tilde{\boldsymbol{\tau}}^{\mathrm{T}}\mathbf{K}_0\tilde{\boldsymbol{\tau}} + a_1\tilde{\boldsymbol{\tau}}_{\mathbf{d}}^{\mathrm{T}}\tilde{\boldsymbol{\tau}}_{\mathbf{d}} + \frac{C_d^2}{4a_1} \leq -2(\lambda_{\min}(\mathbf{K}_0) - a_1)V_d + \frac{C_d^2}{4a_1} \tag{39}$$

Take

$$\begin{cases} \mu_0 = 2(\lambda_{\min}(\mathbf{K}_0) - a_1) > 0 \\ C_0 = \frac{C_d^2}{4a_1} > 0 \end{cases} \tag{40}$$

and $\lambda_{\min}(\mathbf{K}_0) > a_1$. Equation (39) can be abbreviated as follows:

$$\dot{V}_d \leq -\mu_0 V_d + C_0 \tag{41}$$

According to Equation (36), V_d is always greater than 0, and from (41), the result can be obtained as follows:

$$0 \leq V_d \leq \frac{C_0}{\mu_0} + (V_d(0) - \frac{C_0}{\mu_0})e^{-\mu_0 t} \tag{42}$$

According to Equation (42), it shows that the Lyapunov function V_d stays in a closed ball of some radius whose origin is the center of the sphere. In addition, it is uniformly ultimately bounded. In addition, the radius of the sphere is $R_{V_d} = \frac{C_0}{\mu_0} = \frac{C_d^2}{8a_1(\lambda_{\min}(\mathbf{K}_0)-a_1)}$. According to Equation (36), it can be found that the disturbance estimation error variable $\tilde{\boldsymbol{\tau}}_{\mathbf{d}}$ also converges to the sphere radius $R_{\tilde{\boldsymbol{\tau}}_{\mathbf{d}}} = \sqrt{2\frac{C_0}{\mu_0}} = \frac{C_d}{\sqrt{4a_1(\lambda_{\min}(\mathbf{K}_0)-a_1)}}$ with the origin as the center of the sphere. At the same time, it can also be known that if the external environmental disturbance value $\boldsymbol{\tau}_{\mathbf{d}}$ of USV is an arbitrary unknown constant value, then the boundary of the disturbance charge rate is $C_d = 0$. According to Equation (39), the observer estimation error value $\tilde{\boldsymbol{\tau}}_{\mathbf{d}}$ can converge to the origin.

According to Equation (40), as long as the appropriate observer parameters a_1 and $\mathbf{K}_0$ can be selected, an arbitrarily small error convergence radius $R_{\tilde{\boldsymbol{\tau}}_d}$ can be obtained. In other words, NDO can estimate the external environmental disturbance suffered by the USV according to an arbitrarily small error, and the estimation accuracy depends on the selected parameters.

4. Results and Discussions

To verify the influence of the improved NDO-based MPC, it is applied for trajectory tracking control of a USV called CyberShip II. The tracking errors and performance of the USV are shown in this section. In addition, the computational efficiency of the improved NDO is verified.

4.1. Model Parameters of Unmanned Surface Vehicle

In Equations (3) and (4), $\mathbf{M} = \begin{bmatrix} 25.8 & 0 & 0 \\ 0 & 33.8 & 1 \\ 0 & 1 & 2.8 \end{bmatrix}$, $\mathbf{D} = \begin{bmatrix} 0.72 & 0 & 0 \\ 0 & 0.86 & -0.11 \\ 0 & -0.11 & 1.90 \end{bmatrix}$. In (16),

$$\mathbf{C} = \begin{bmatrix} 1 & 0 & 0 & 0 & 0 & 0 \\ 0 & 1 & 0 & 0 & 0 & 0 \\ 0 & 0 & 1 & 0 & 0 & 0 \\ 0 & 0 & 0 & 1 & 0 & 0 \\ 0 & 0 & 0 & 0 & 1 & 0 \\ 0 & 0 & 0 & 0 & 0 & 1 \end{bmatrix}.$$ In (17), $\mathbf{Q} = \begin{bmatrix} 10 & 0 & 0 & 0 & 0 & 0 \\ 0 & 10 & 0 & 0 & 0 & 0 \\ 0 & 0 & 10 & 0 & 0 & 0 \\ 0 & 0 & 0 & 10 & 0 & 0 \\ 0 & 0 & 0 & 0 & 10 & 0 \\ 0 & 0 & 0 & 0 & 0 & 10 \end{bmatrix}$,

$$\mathbf{R} = \begin{bmatrix} 0.01 & 0 & 0 \\ 0 & 0.01 & 0 \\ 0 & 0 & 0.01 \end{bmatrix}.$$

In the simulation, $\boldsymbol{\eta} = [0\ 0\ 0°]^{\mathrm{T}}$, which is set as the initial state of USV; $\boldsymbol{\upsilon} = [0\ 0\ 0]^{\mathrm{T}}$, which is set as the initial speed of USV. In addition, the reference trajectories of USV are shown as follows:

$$\begin{cases} x_d = 2\sin(0.02t), & 0 \leq t \leq 500 \\ y_d = 2 - 2\cos(0.02t), & 0 \leq t \leq 500 \\ \psi_d = \arctan(\dot{x}_d/\dot{y}_d), & 0 \leq t \leq 500 \end{cases} \tag{43}$$

$$\begin{cases} x_d = 8\sin(0.02t), & 0 \leq t \leq 500 \\ y_d = t, & 0 \leq t \leq 500 \\ \psi_d = \arctan(\dot{x}_d/\dot{y}_d), & 0 \leq t \leq 500 \end{cases} \tag{44}$$

$T = 0.1$, which is set as the simulation sampling time; $N_x = 6$, which is set as the number of states; $N_u = 3$, which is set as the number of control variables; $N_p = 20$, which is set as the prediction horizon; $N_c = 10$, which is set as the control horizon.

4.2. Simulation Results and Analysis

In MATLAB 2021a, NDO-based MPC for trajectory tracking control of a USV is compared with MPC without an observer. Simulations with different disturbances are performed to verify the anti-disturbance and robustness performances. The performances of the improved NDO are discussed.

The composite model of external disturbances meets the requirements of Level 3 sea conditions. The specific external wind, wave, and current disturbances settings are as follows:

$$\begin{cases} \tau_{du} = m_{11}h(s)w_u(s) \\ \tau_{dv} = m_{22}h(s)w_v(s) \\ \tau_{dr} = m_{33}h(s)w_r(s) \end{cases} \tag{45}$$

where wave transfer function $h(s) = \frac{K_\omega s}{s^2 + 2\zeta\omega_0 s + \omega_0^2}$, $K_\omega = 2\zeta\omega_0\sigma_0$, ω_0, ζ, and σ_0 represent wave frequency, wave strength gain, and damping constant, respectively; $K_\omega = 0.255$, $\omega_0 = 0.808$. $w_u(s)$, $w_v(s)$, and $w_r(s)$ represent random white-noise disturbances, then the noise power is set to 0.01, 0.005, and 0.1, respectively.

The simulation results of improved NDO-based MPC for trajectory tracking control and traditional MPC with disturbances, when the reference trajectory is a circle, are shown in Figure 3. The comparisons of improved NDO-based MPC for trajectory tracking control and traditional MPC with disturbances, when the reference trajectory is sinusoidal, are shown in Figure 4. The calculation times between the two NDO methods with disturbances are shown in Figure 5. The trajectory tracking errors of the three trajectory tracking methods with different parameters are shown in Figure 6.

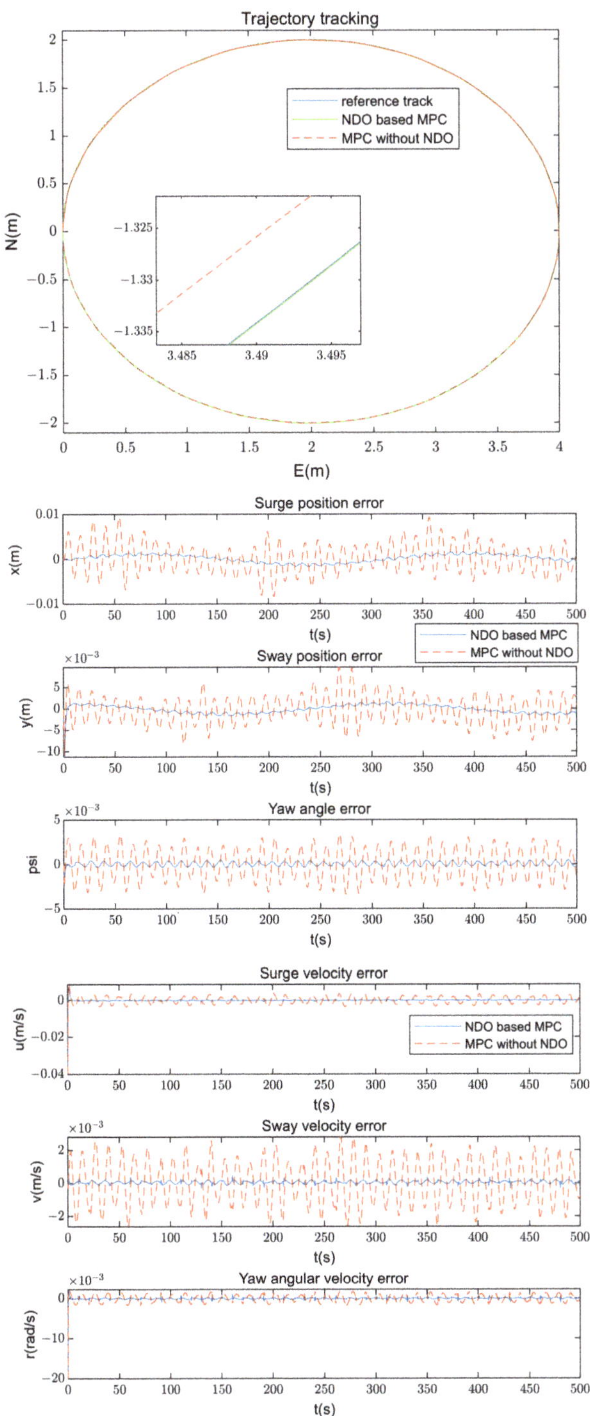

Figure 3. Comparisons of improved NDO-based MPC for trajectory tracking control and traditional MPC with disturbances when the reference trajectory is a circle.

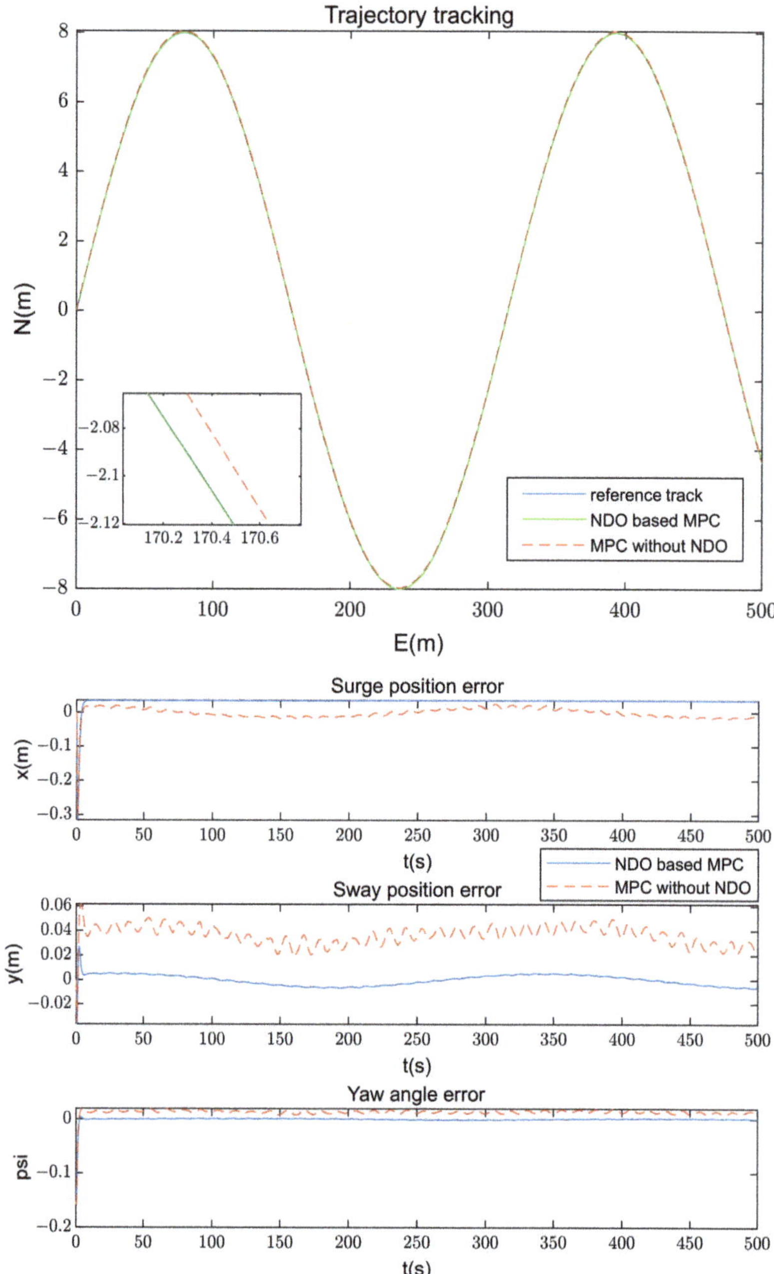

Figure 4. *Cont.*

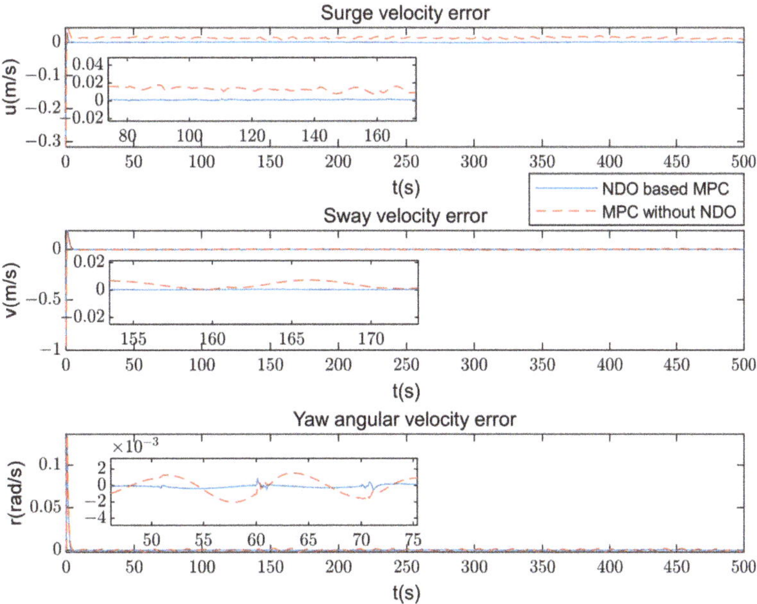

Figure 4. Comparisons of improved NDO-based MPC for trajectory tracking control and traditional MPC with disturbances when the reference trajectory is sinusoidal.

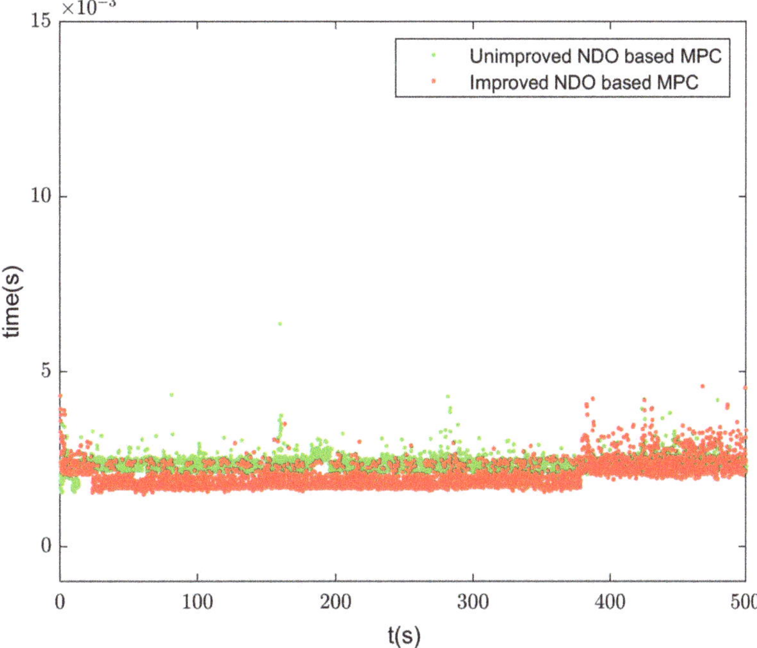

Figure 5. Comparison of the calculation time between the two NDO methods with disturbances.

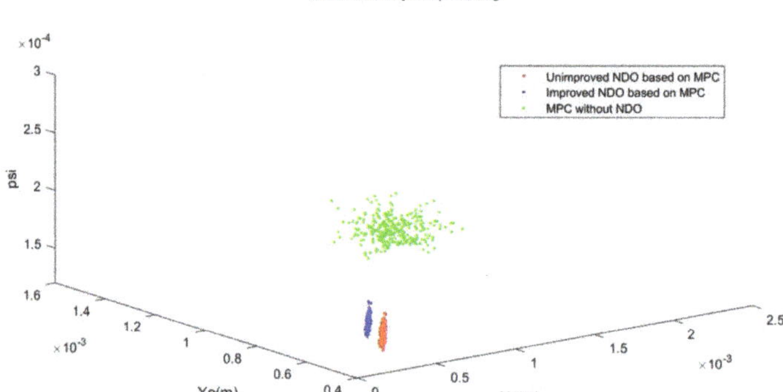

Figure 6. Trajectory tracking errors of the three trajectory tracking methods with different parameters.

Figure 3 shows that the improved NDO-based MPC has better disturbance-rejection performance than MPC without an observer when the reference trajectory is a circle. It also can be seen from the simulations that both methods can track the reference trajectory. However, the former can track the reference trajectory smoothly, while the latter fluctuates a lot. The surge position error range of MPC without an observer for trajectory tracking control is −0.1 to 0.1 m; the sway position error range is −0.1 to 0.1 m; and the yaw angle error range is −0.05 to 0.05. Although the surge position error range of the NDO-based MPC is −0.005 to 0.005 m, the sway position error range is −0.005 to 0.005 m and the yaw angle error range is −0.005 to 0.005. Figure 4 shows the comparisons of the two trajectory tracking methods with disturbances when the reference trajectory is sinusoidal. In addition, the improved NDO-based MPC has lower tracking errors than MPC without an observer. The surge position error range of MPC without an observer for trajectory tracking control is −0.05 to 0.05 m; the sway position error range is −0.02 to 0.02 m; and the yaw angle error range is 0.00 to 0.05. Although the surge position error range of NDO-based MPC is 0.05 to 0.05 m, the sway position error range is 0.02 to 0.06 m, and the yaw angle error range is 0.000 to 0.005. Figure 5 shows that the improved NDO has better performance than the unimproved NDO in terms of the calculation time with disturbances.

We do not have a real ship for this experiment. To overcome this shortage, the ship parameters were changed when we did some comparisons between different methods to ensure the robustness of the proposed method. The USV model has uncertainty, so three methods were used for trajectory tracking for the USV with different model parameters. The three methods are unimproved NDO-based MPC, improved NDO-based MPC, and MPC without NDO. Figure 6 shows that the improved NDO and unimproved NDO effectively reduce the roughness caused by model uncertainty. In addition, the unimproved NDO has similar performance of tracking errors compared to the improved NDO-based MPC.

In addition, a comparison of the calculation time of two NDOs with disturbances is shown in Table 1. It shows the average calculation time and maximum single calculation time of the two NDOs. $IAE = \int_0^t |e(\zeta)| d\zeta$ and $RMSE = \left(\frac{1}{t} \int_0^t e^2(\zeta) d\zeta\right)^{1/2}$ are used to evaluate the tracking effect and steady-state performance. The smaller the values of IAE and RMSE, the better the control performance of the scheme applied. In addition, the comprehensive performance comparisons of the position and speed tracking errors of the two methods are shown in Table 2. Table 2 shows the IAE and $RMSE$ of the two methods.

From Table 1, the calculation time of the improved NDO is much lower than that of the traditional NDO. The average calculation time of the improved NDO is 0.0020, and it is 0.0024 for the traditional NDO. The maximum single calculation time of the improved NDO is 0.0046, and it is 0.0064 for the traditional NDO. The results show that, compared with the traditional NDO, the average calculation time of the improved NDO is decreased by 16.67%, and the maximum individual calculation time is decreased by 28.13%.

Table 1. Comparison of the calculation time of the two methods.

	Improved NDO Based MPC	Unimproved NDO Based MPC
Average calculation time(s)	0.0020	0.0024
Maximum single calculation time(s)	0.0046	0.0064

Table 2. Comparison of position and velocity errors of the two methods.

Tracking Error	Computing Method	Improved NDO Based MPC	Non-Observer
y_e	IAE	9.3209	141.6562
	$RMSE$	0.0072	0.1104
x_e	IAE	9.2273	135.6914
	$RMSE$	0.0071	0.1061
ψ_e	IAE	10.7869	79.6175
	$RMSE$	0.0077	0.0574
$\|u_e\|$	IAE	6.3167	55.5531
	$RMSE$	0.0055	0.0435
$\|v_e\|$	IAE	3.9132	58.2295
	$RMSE$	0.0030	0.0425
$\|r_e\|$	IAE	6.1470	40.2591
	$RMSE$	0.0050	0.0290

From Table 2, the IAE and $RMSE$ of NDO-based MPC are lower than the MPC without an observer. NDO-based MPC effectively enhances the anti-disturbance performance of the system. MPC without observer trajectory tracking control has the characteristics of the predictive model, rolling optimization, and feedback correction, which can resist external disturbances and model mismatches to some extent.

With the data shown in Tables 1 and 2, NDO-based MPC is superior to MPC without an observer for trajectory tracking control in terms of position and speed tracking errors.

The NDO is designed to estimate the external disturbances suffered, to improve the anti-disturbance performance of the USV. Therefore, the comparison of estimated values of the NDO and actual disturbances is shown in Figure 7.

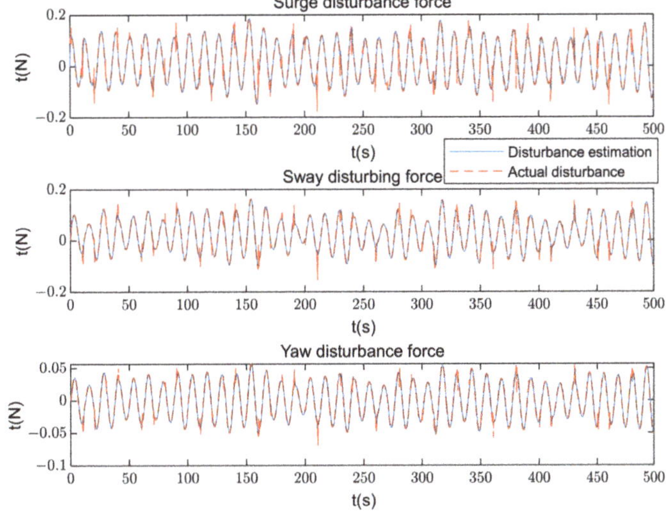

Figure 7. Comparison of estimated values of the NDO and actual disturbances.

Figure 7 shows the relationship between the estimated values of the NDO and the actual values of the disturbance. The NDO has good estimation performance in terms of the disturbances, including surge disturbance force, sway disturbance force, and yaw disturbance force.

5. Conclusions

In this paper, an improved NDO-based MPC for trajectory tracking is proposed to guarantee the stable motion of a USV, which suffers various disturbances from the ocean wind, waves, and currents. MPC is used to optimize the system torque based on the measured position and speed state variables. Then, the NDO is designed to estimate the disturbances, and the estimated torque is compensated for in the controller. Estimation errors can converge to zero in a finite time. The simulation results show that NDO-based MPC can effectively compensate for external disturbances and obtain good tracking and disturbance-rejection performance. The proposed method has a similar tracking performance to the USVs with the MPC based on unimproved NDO, but the improved NDO-based MPC is far quicker. However, due to the linearization of the model of the USV, the method only shows good performance in a near-neighbor area around the operation point. For large-difference operation points, the parameters need to be retuned.

Author Contributions: Conceptualization, H.F.; methodology, H.F. and S.Z.; software, W.Y.; validation, W.Y., R.C. and S.Z.; writing—original draft preparation, W.Y. and S.Z.; writing—review and editing, H.F., R.C. and S.Z. All authors have read and agreed to the published version of the manuscript.

Funding: This research was funded by the National Natural Science Foundation of China, grant number 52271313; Innovative Research Foundation of Ship General Performance, grant number 21822216; Fundamental Research Funds for the Central Universities, grant number XK2040021004025.

Institutional Review Board Statement: Not applicable.

Informed Consent Statement: Not applicable.

Data Availability Statement: Not applicable.

Conflicts of Interest: The authors declare no conflict of interest. The funders had no role in the design of the study.

References

1. Miao, R.; Dong, Z.; Wan, L.; Zeng, J. Heading control system design for a micro-USV based on an adaptive expert S-PID algorithm. *Pol. Marit. Res.* **2018**, *25*, 6–13. [CrossRef]
2. Li, W.; Ge, Y.; Guan, Z.; Ye, G. Synchronized Motion-Based UAV–USV Cooperative Autonomous Landing. *J. Mar. Sci. Eng.* **2022**, *10*, 1214. [CrossRef]
3. Liao, Y.; Jia, Z.; Zhang, W.; Jia, Q.; Li, Y. Layered berthing method and experiment of unmanned surface vehicle based on multiple constraints analysis. *Appl. Ocean Res.* **2019**, *86*, 47–60. [CrossRef]
4. Jin, J.; Liu, D.; Wang, D.; Ma, Y. A Practical Trajectory Tracking Scheme for a Twin-Propeller Twin-Hull Unmanned Surface Vehicle. *J. Mar. Sci. Eng.* **2021**, *9*, 1070. [CrossRef]
5. Jiang, X.; Xia, G.; Feng, Z.; Wu, Z.G. Nonfragile Formation Seeking of Unmanned Surface Vehicles: A Sliding Mode Control Approach. *IEEE Trans. Netw. Sci. Eng.* **2021**, *9*, 431–444. [CrossRef]
6. Jiang, X.; Xia, G. Sliding mode formation control of leaderless unmanned surface vehicles with environmental disturbances. *Ocean Eng.* **2022**, *244*, 110301. [CrossRef]
7. Velueta, M.J.; Rullan, J.L.; Ruz-Hernandez, J.A.; Alazki, H. A Strategy of Robust Control for the Dynamics of an Unmanned Surface Vehicle under Marine Waves and Currents. *Math. Probl. Eng.* **2019**, *2019*, 4704567. [CrossRef]
8. Wu, Y.; Yang, S.; Li, W.; Liu, D.; Hou, K. Dynamic Analysis and Motion Control of an Underactuated Unmanned Surface Vehicle (WL-II). *Mar. Technol. Soc. J.* **2017**, *51*, 10–20. [CrossRef]
9. Zhou, W.; Wang, Y.; Ahn, C.K.; Cheng, J.; Chen, C. Adaptive fuzzy backstepping-based formation control of unmanned surface vehicles with unknown model nonlinearity and actuator saturation. *IEEE Trans. Veh. Technol.* **2020**, *69*, 14749–14764. [CrossRef]
10. Wang, C.; Xie, S.; Chen, H.; Peng, Y.; Zhang, D. A decoupling controller by hierarchical backstepping method for straight-line tracking of unmanned surface vehicle. *Syst. Sci. Control Eng.* **2019**, *7*, 379–388. [CrossRef]
11. Jin, J.; Zhang, J.; Liu, D. Design and verification of heading and velocity coupled nonlinear controller for unmanned surface vehicle. *Sensors* **2018**, *18*, 3427. [CrossRef] [PubMed]

12. Weng, Y.; Wang, N. Data-driven robust backstepping control of unmanned surface vehicles. *Int. J. Robust Nonlinear Control* **2020**, *30*, 3624–3638. [CrossRef]
13. Zhao, S.; Wang, S.; Cajo, R.; Ren, W.; Li, B. Power Tracking Control of Marine Boiler-Turbine System Based on Fractional Order Model Predictive Control Algorithm. *J. Mar. Sci. Eng.* **2022**, *10*, 1307. [CrossRef]
14. Feng, N.; Wu, D.; Yu, H.; Yamashita, A.S.; Huang, Y. Predictive compensator based event-triggered model predictive control withnonlinear disturbance observer for unmanned surface vehicle under cyber-attacks. *Ocean Eng.* **2022**, *259*, 111868. [CrossRef]
15. Zhao, S.; Cajo, R.; De Keyser, R.; Ionescu, C.M. The potential of fractional order distributed MPC applied to steam/water loop in large scale ships. *Processes* **2020**, *8*, 451. [CrossRef]
16. Han, X.; Zhang, X. Tracking control of ship at sea based on MPC with virtual ship bunch under Frenet frame. *Ocean. Eng.* **2022**, *247*, 110737. [CrossRef]
17. Liu, Z.; Zhang, Y.; Yuan, C.; Luo, J. Adaptive path following control of unmanned surface vehicles considering environmental disturbances and system constraints. *IEEE Trans. Syst. Man Cybern. Syst.* **2018**, *51*, 339–353. [CrossRef]
18. Liao, Y.; Du, T.; Jiang, Q. Model-free adaptive control method with variable forgetting factor for unmanned surface vehicle control. *Appl. Ocean Res.* **2019**, *93*, 101945. [CrossRef]
19. Liao, Y.; Jiang, Q.; Du, T.; Jiang, W. Redefined output model-free adaptive control method and unmanned surface vehicle heading control. *IEEE J. Ocean. Eng.* **2019**, *45*, 714–723. [CrossRef]
20. Li, Y.; Wang, L.; Liao, Y.; Jiang, Q.; Pan, K. Heading MFA control for unmanned surface vehicle with angular velocity guidance. *Appl. Ocean Res.* **2018**, *80*, 57–65. [CrossRef]
21. Huang, Z.; Liu, X.; Wen, J.; Zhang, G.; Liu, Y. Adaptive navigating control based on the parallel action-network ADHDP metho for unmanned surface vessel. *Adv. Mater. Sci. Eng.* **2019**, *2019*, 7697143. [CrossRef]
22. Wen, Y.; Tao, W.; Zhu, M.; Zhou, J.; Xiao, C. Characteristic model-based path following controller design for the unmanned surface vessel. *Appl. Ocean Res.* **2020**, *101*, 102293. [CrossRef]
23. Wang, R.; Li, D.; Miao, K. Optimized radial basis function neural network based intelligent control algorithm of unmanned surface vehicles. *J. Mar. Sci. Eng.* **2020**, *8*, 210. [CrossRef]
24. Zhou, Y.; Wu, N.; Yuan, H.; Pan, F.; Shan, Z.; Wu, C. PDE Formation and Iterative Docking Control of USVs for the Straight-Line-Shaped Mission. *J. Mar. Sci. Eng.* **2022**, *10*, 478. [CrossRef]
25. Liu, Z.; Song, S.; Yuan, S.; Ma, Y.; Yao, Z. ALOS-Based USV Path-Following Control with Obstacle Avoidance Strategy. *J. Mar. Sci. Eng.* **2022**, *10*, 1203. [CrossRef]
26. Martinsen, A.B.; Lekkas, A.M.; Gros, S. Reinforcement learning-based NMPC for tracking control of ASVs: Theory and experiments. *Control Eng. Pract.* **2022**, *120*, 105024. [CrossRef]
27. Tan, Y.; Cai, G.; Li, B.; Teo, K.L.; Wang, S. Stochastic model predictive control for the set point tracking of unmanned surface vehicles. *IEEE Access* **2019**, *8*, 579–588. [CrossRef]
28. Wang, X.; Liu, J.; Peng, H.; Qie, X.; Zhao, X.; Lu, C. A Simultaneous Planning and Control Method Integrating APF and MPC to Solve Autonomous Navigation for USVs in Unknown Environments. *J. Intell. Robot. Syst.* **2022**, *105*, 36. [CrossRef]
29. Sun, X.; Wang, G.; Fan, Y.; Mu, D.; Qiu, B. Collision avoidance using finite control set model predictive control for unmanned surface vehicle. *Appl. Sci.* **2018**, *8*, 926. [CrossRef]
30. Bingul, Z.; Gul, K. Intelligent-PID with PD Feedforward Trajectory Tracking Control of an Autonomous Underwater Vehicle. *Machines* **2023**, *11*, 300. [CrossRef]
31. Zhang, Y.; Xu, O. Adaptive Backstepping Axial Position Tracking Control of Autonomous Undersea Vehicles with Deferred Output Constraint. *Appl. Sci.* **2023**, *13*, 2219. [CrossRef]
32. Peng, Y.; Li, Y. Autonomous Trajectory Tracking Integrated Control of Unmanned Surface Vessel. *J. Mar. Sci. Eng.* **2023**, *11*, 568. [CrossRef]
33. Liu, Z.Q.; Wang, Y.L.; Han, Q.L. Adaptive fault-tolerant trajectory tracking control of twin-propeller non-rudder unmanned surface vehicles. *Ocean Eng.* **2023**, *285*, 115294. [CrossRef]
34. Dong, Z.P.; Zhang, Z.Q.; Qi, S.J.; Zhang, H.S.; Li, J.K.; Liu, Y.C. Autonomous cooperative formation control of underactuated USVs based on improved MPC in complex ocean environment. *Ocean Eng.* **2023**, *270*, 113633. [CrossRef]
35. Liu, W.; Ye, H.; Yang, X. Super-Twisting Sliding Mode Control for the Trajectory Tracking of Underactuated USVs with Disturbances. *J. Mar. Sci. Eng.* **2023**, *11*, 636. [CrossRef]
36. Jiang, T.; Yan, Y.; Yu, S.-H. Adaptive Sliding Mode Control for Unmanned Surface Vehicles with Predefined-Time Tracking Performances. *J. Mar. Sci. Eng.* **2023**, *11*, 1244. [CrossRef]
37. Yuan, S.Z.; Liu, Z.L.; Sun, Y.X.; Wang, Z.X.; Zheng, L.H. An event-triggered trajectory planning and tracking scheme for automatic berthing of unmanned surface vessel. *Ocean Eng.* **2023**, *273*, 113964. [CrossRef]

Journal of
Marine Science and Engineering

Article

Underwater Acoustically Guided Docking Method Based on Multi-Stage Planning

Hongli Xu *, Hongxu Yang , Zhongyu Bai and Xiangyue Zhang

Faculty of Robot Science and Engineering, Northeastern University, Shenyang 110819, China; yanghongxu@stumail.neu.edu.cn (H.Y.); zhongyubai@stumail.neu.edu.cn (Z.B.); zhangxiangyue@mail.neu.edu.cn (X.Z.)
* Correspondence: xuhongli@mail.neu.edu.cn

Abstract: Autonomous underwater vehicles (AUVs) are important in areas such as underwater scientific research and underwater resource collection. However, AUVs suffer from data portability and energy portability problems due to their physical size limitation. In this work, an acoustic guidance method for underwater docking is proposed to solve the problem of persistent underwater operation. A funnel docking station and an autonomous remotely operated vehicle (ARV) are used as the platform for designing the guidance algorithms. First, the underwater docking guidance is divided into three stages: a long-range approach stage, a mid-range adjustment stage and a short-range docking stage. Second, the relevant guidance strategy is designed for each stage to improve the docking performance. Third, a correction method based on an ultra-short baseline (USBL) system is proposed for the ARV's estimate of the depth, relative position and orientation angle of the docking station. To verify the feasibility of the docking guidance method, in this work, tests were performed on a lake and in a shallow sea. The success rate of autonomous navigation docking on the lake was 4 out of 7. The success rate of acoustic guidance docking on the lake and in the shallow sea were 11 out of 14 and 6 out of 8, respectively. The experimental results show the effectiveness of the docking guidance method in lakes and shallow seas.

Keywords: underwater docking; acoustic guidance; autonomous navigation; autonomous underwater vehicle; autonomous remotely operated vehicle; docking station

Citation: Xu, H.; Yang, H.; Bai, Z.; Zhang, X. Underwater Acoustically Guided Docking Method Based on Multi-Stage Planning. *J. Mar. Sci. Eng.* **2023**, *11*, 1629. https://doi.org/10.3390/jmse11081629

Academic Editor: Sergei Chernyi

Received: 19 July 2023
Revised: 11 August 2023
Accepted: 16 August 2023
Published: 21 August 2023

1. Introduction

With the continuous deepening of ocean exploration in various countries, autonomous underwater vehicles (AUVs) have become an important tool for exploring the marine environment [1–4]. AUVs have important applications in areas such as underwater rescue, military reconnaissance, resource exploration and marine scientific research [5,6]. The small size of the AUV makes it highly concealable during operation, however, the small size of the AUV also limits its range of motion, resulting in an inability to operate underwater for extended periods of time [7–10]. In addition, the low transmission rate of underwater acoustic data prevents the AUV from uploading the data in a timely manner [11]. Therefore, the AUV requires multiple manual recoveries and deployments, which greatly reduces efficiency, increases costs and is less concealable [12,13].

To enable the AUV to stay underwater for long periods, a docking station for data and power transmission has been deployed underwater [14]. After the mission, the AUV can navigate autonomously to the underwater docking station and perform operations such as energy replenishment and data upload [15,16]. Underwater docking technology greatly improves the concealment, continuity and mobility of AUV underwater operations and is a key research direction in the future [17–19].

Underwater docking guidance methods can be divided into three methods depending on the used sensor: acoustic guidance [20], optical guidance [21] and electromagnetic guidance [22]. Underwater acoustic guidance refers to the application of acoustic equipment such as long baseline (LBL) systems, short baseline (SBL) systems and ultra-short baseline (USBL) systems for guidance. The advantage of acoustic guidance is that it works over long distances, up to 3 km, and can be omnidirectional [23,24]. The disadvantage is that the positioning accuracy is low, the real-time performance is poor and it is easily exposed [25,26]. The following are typical results of docking using acoustic guidance:

Stokey et al. proposed a funnel docking scheme based on the REMUS AUV [20]. They designed an acoustic localisation method based on a USBL, a clamp and contact motors for locking, charging and communication circuits. Tests were performed in the summer of 1997. Allen et al. proposed a second generation of the REMUS docking system as an improvement on the original docking scheme [23]. In their work, they adopted a low-profile, bottom-mounted docking station and upgraded the sensors of the docking station and the docking vehicle. They also successfully performed docking tests in 2005.

Singh et al. proposed a pole docking scheme based on the Odyssey IIB AUV [27]. They used acoustic guidance and divided the docking into five stages: homing, docking, core alignment, power transfer and undocking. Tests were conducted at Cape Hatteras in May 1997 to demonstrate the reliability of the docking system.

Kawasaki et al. proposed a platform-based underwater docking solution based on the Marine Bird AUV [28,29]. They used a super short base line (SSBL) for acoustic guidance and divided the docking process into six stages: approaching the transponder service area, approaching the base, holding the guide, connecting to the base, recharging the batteries and undocking. In 2001–2003, trials were carried out at a dock and at sea to demonstrate the performance of the docking system.

McEwen et al. proposed a docking method for the Bluefin-21 AUV and a fixed cone docking station [30,31]. In their work, the docking process is as follows: use pure pursuit guidance while homing to within USBL range of the beacon, then use the USBL for positioning and sail along the centreline of the cone, when the AUV is close to the dock, slow down and complete the final alignment and then latch. Docking tests were performed in 2005–2006 in a seawater test tank and in Monterey Bay.

Hayato et al. proposed a docking method using acoustic guidance and optical guidance [32]. They used passive acoustic guidance to guide the vehicle into a docking cage from a long distance and used optical guidance at short distance to provide high positioning accuracy during terminal guidance. Due to the limitations of the experimental conditions, they tested the optical guidance system to demonstrate the validity of the proposed method.

Maki et al. proposed a docking method for a hovering Tri-TON AUV at a seafloor station [33]. In their work, an acoustic localisation and communication device (ALOC) was used to estimate the rough position of the AUV at long distance, while, at short distance, image processing was used to measure the precise relative position of the AUV. Tests were performed in a tank, and the docking success rate was 50%. Sato et al. conducted acoustic-optical docking tests with a Tri-TON 2 AUV [34,35]. Compared to the Tri-TON AUV, the maximum depth of the Tri-TON 2 was extended from 800 m to 2000 m. The performance of the docking method was verified through a series of tank and sea trials in 2015.

Vallicrosa et al. proposed a combined acoustic–optical-based guidance method using the Sparus II AUV and conducted tests in a water tank and in a shallow sea [36–40]. In their work, the docking procedure is as follows: (1) Estimate the position and orientation of the docking station and guide the AUV to a location 40 m in front of the docking station. (2) If the range is available, use an acoustic localisation algorithm to estimate the position of the docking station. (3) If the docking station is detected, navigate the AUV to a location 10 m in front of the docking station. (4) Follow the track towards the docking station until light beacons are detected. Otherwise, return to (3). (5) Navigate the AUV to the entrance of the docking station.

Existing acoustic guidance docking methods mostly rely on pre-positioning the AUV within the range of acoustic equipment and then using the acoustic equipment to obtain the precise position of the docking station and then homing. However, these studies focus on the design of the docking structure and the selection of the docking sensors, and the docking guidance process is not studied enough. During the docking process, as the AUV approaches the docking station, the docking accuracy requirement also increases. Therefore, the docking process needs to be divided into several stages according to the distance from the AUV to the docking station, and guidance strategies need to be designed for each stage in order to improve the docking success rate. To address these issues, this paper proposes a docking method based on acoustic guidance for autonomous remotely operated vehicles (ARV). First, the docking process is divided into three stages according to the distance from the ARV to the docking station. Second, the corresponding guidance algorithm for each docking stage is designed. Third, an acoustic-based method is proposed to update the ARV's perception of the docking station, including the depth, relative position and orientation angle of the docking station.

The paper is organised as follows: Section 2 describes the parameters, structure and functions of the ARV for docking and the docking station. Section 3 presents the docking procedure, describes the specific methods for each stage of docking and proposes a method for correcting the ARV's perception of the docking station. Section 4 presents lake and shallow sea tests and analyses the results of the tests. Section 5 outlines the conclusions and future work.

2. Docking Platform

2.1. Underwater Docking ARV

An ARV is a composite underwater robot that combines the features of a remotely operated vehicle (ROV) and an AUV, which can operate either remotely or autonomously for underwater tasks [41]. The ARV used for underwater docking is shown in Figure 1. The dimensions of the ARV are 2.980 (L) × 0.697 (W) × 0.779 (H) (m), the weight is 350 kg and it is equipped with 6 underwater cameras, 6 lights, a USBL beacon terminal, a doppler velocity log (DVL), a wi-fi module, an altimeter, a depth gauge, an inertial navigation system (INS), a global positioning system (GPS), an attitude sensor and other modules. The ARV can realise motion states such as forward and backward, transverse, in situ rotation, surfacing and diving, its speed can reach 3 knots, the radius of its turning circle is 5 m and the working water depth is 0–4500 m. The ARV can be divided into two types of path planning mode: waypoint path mode and heading path mode [42].

Figure 1. Underwater docking ARV.

The structure of the ARV autonomous docking module is shown in Figure 2. The docking module is an NVIDIA Jetson Nano, which is used to receive status information and optical guidance video information from the ARV main controller and store them into a shared data area. The acoustic guidance module and the optical guidance module generate ARV control commands by processing the shared data and send them to the ARV to realise ARV docking control.

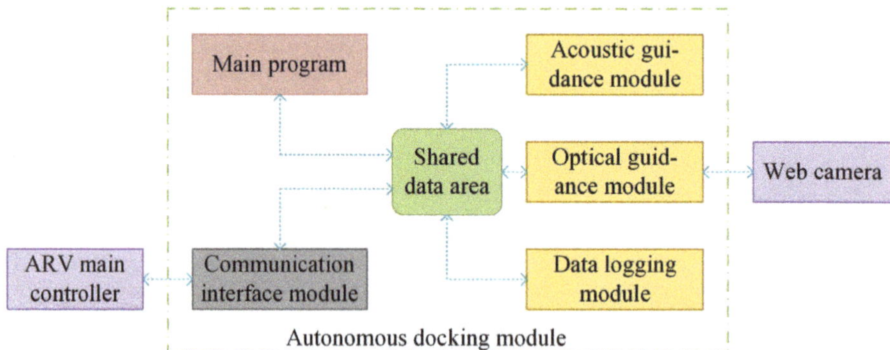

Figure 2. Structure diagram of the docking control module.

2.2. Underwater Docking Station

The docking station used for underwater docking is shown in Figure 3. The overall outline of the docking station consists of two main parts: the base and the dock, where the dock can be divided into two parts: the frame and the guide funnel. The overall dimensions of the dock are 2.067 (L) × 2.065 (W) × 2.065 (H) (m) and the weight is 210 kg, where the guide funnel has a length of 1.17 m, a maximum inner diameter of 1.82 m, a minimum inner diameter of 0.86 m and a weight of 66 kg. The dock is installed on top of the base so that it can be suspended in the water. The internal structure of the docking station includes 8 guide lights, a USBL host terminal, an underwater camera, an altimeter, a depth gauge, an electronic compass and other equipment.

Figure 3. Underwater docking station.

3. Docking Method

The docking flowchart is shown in Figure 4. Due to the limitation of the entrance nozzle angle of the docking station, it is necessary to reduce the error of the docking attitude angle and the position of the ARV. In addition, during the docking process, the actual depth, orientation angle and position may deviate due to ocean currents or other reasons. Therefore, it needs to be continuously corrected during the docking process. To overcome the above problems, the docking guidance process is designed to be divided into three stages: a long-range approach stage, a mid-range adjustment stage and a short-range docking stage. In the long-range approach stage, the ARV approaches a position on the centreline of the docking station, which is far from its entrance nozzle, to make adjustments with sufficient distance; in the mid-range adjustment stage, the ARV gradually approaches the docking station at a slow speed along its centreline to reduce the lateral deviation during navigation; in the short-range docking stage, the ARV uses inertial navigation to sail directly to the docking station or uses optics for more precise guidance.

Figure 4. Docking flowchart.

3.1. Long-Range Approach Stage

When the ARV receives the docking command, it begins the remote docking procedure. The purpose of the long-range approach stage is to allow the ARV to approach the centreline of the docking station, at a heading angle approximately towards its entrance nozzle, to adjust to a more suitable position and heading angle and then enter the mid-range adjustment stage.

In the long-range approach stage, the ARV will descend to the preset depth of the docking station and perform fixed-depth navigation mode so that the ARV and the docking station are at the same level. It selects several points between 80 m and 300 m away from the docking station along the centreline of the docking station (the last point is the 80 m point), and performs a waypoint path mode from far to near. Once the ARV has reached the target position at 80 m, it enters the mid-range adjustment stage.

The method for controlling the ARV to reach the target position during waypoint path mode is shown in Figure 5. The ARV starts from the start position p_s and sails towards the target position p_e. The formula for calculating the coordinates of p_e is as follows:

$$\begin{cases} x_e = x_d + l_{de} \times \sin \varphi_d \\ y_e = y_d + l_{de} \times \cos \varphi_d \end{cases}, \tag{1}$$

where x_e and y_e are the along-axis and lateral coordinates of p_e, x_d and y_d are the along-axis and lateral coordinates of the docking station p_d, l_{de} is the distance between p_d and p_e and φ_d is the orientation angle of p_d in the geodetic coordinate system.

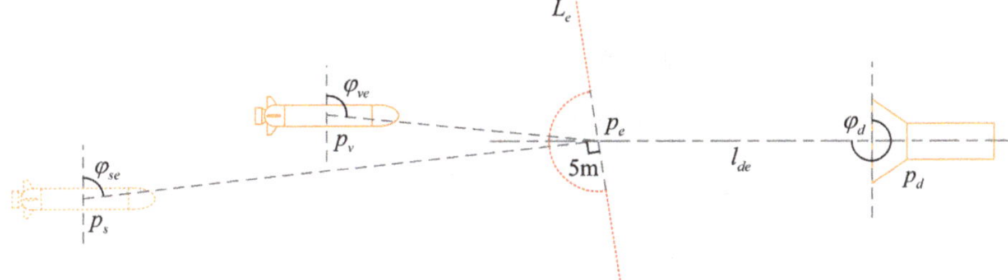

Figure 5. Judgment of reaching the target position.

If the distance between p_e and the current ARV position p_v is less than or equal to 5 m, or if the ARV sails past L_e, which is a vertical plane perpendicular to $p_s p_e$ that passes through p_e (as shown in the shaded part of Figure 5), it can be determined that the ARV has reached the target position. The judgement formula is as follows:

$$\begin{cases} \sqrt{(x_v - x_e)^2 + (y_v - y_e)^2} \leq 5 \\ |\varphi_{se} - \varphi_{ve}| \geq 90 \end{cases}, \tag{2}$$

where x_v and y_v are the horizontal and vertical coordinates of p_v and φ_{se} and φ_{ve} are the azimuth angles of p_e relative to p_s and p_v in the geodetic coordinate system.

To prevent the ARV from being unable to reach the target position due to obstacles and other reasons during waypoint path mode, set the maximum time limit for ARV waypoint path mode:

$$t_{\max} = 2 \times \frac{\sqrt{(x_s - x_e)^2 + (y_s - y_e)^2}}{v}, \tag{3}$$

where $t_{\max}$ is the maximum time limit of waypoint path mode, x_s and y_s are the horizontal and vertical coordinates of p_s and v is the ARV forward speed in waypoint path mode.

3.2. Mid-Range Adjustment Stage

The purpose of the mid-range adjustment stage is to adjust the position and heading angle of the ARV. Compared to [20], to enter the short-range docking stage with a better attitude, a lateral deviation adjustment algorithm is added to stabilise the ARV near the centreline of the docking station with a heading angle towards the entrance nozzle of the docking station.

In the mid-range adjustment stage, the start position and the target position are respectively 80 m and 5 m in front of the entrance nozzle of the docking station, and heading path mode is used. Once the ARV has reached the target position, it enters the short-range docking stage.

The path tracking method in the mid-range adjustment stage is shown in Figure 6. When the distance l_3 between ARV current position p_v and the docking station centreline L is less than or equal to 1 m, the ARV will dock with the heading angle φ_{t1} of the target position p_1. Otherwise, the ARV will dock with the heading angle φ_{t2} of the target position p_2, which is the midpoint between p_1 and the projection point p_3 of p_v on L. The judgment formula is as follows:

$$l_3 \begin{cases} \leq 1, & \text{target heading is } \varphi_{t1} \\ > 1, & \text{target heading is } \varphi_{t2} \end{cases}, \tag{4}$$

where the distance l_3 is between p_v and L, the target heading φ_{t1} and φ_{t2} can be calculated by the following formula:

$$\begin{cases} l_3 = l_1 \times |\sin \varphi_1| \\ \varphi_{t1} = \operatorname{atan2}(x_1 - x_v, y_1 - y_v) \\ \varphi_{t2} = \operatorname{atan2}(x_2 - x_v, y_2 - y_v) \end{cases}, \tag{5}$$

where x_1 and y_1 are the horizontal and vertical coordinates of the target point p_1:

$$\begin{cases} x_1 = x_d + 5 \times \sin \varphi_d \\ y_1 = y_d + 5 \times \cos \varphi_d \end{cases}, \tag{6}$$

l_1 is the distance between p_v and p_1:

$$l_1 = \sqrt{(x_v - x_1)^2 + (y_v - y_1)^2}, \tag{7}$$

φ_1 is the angle between $p_v p_1$ and L:

$$\varphi_1 = \varphi_d - \operatorname{atan2}(x_v - x_1, y_v - y_1), \tag{8}$$

x_2 and y_2 are the horizontal and vertical coordinates of p_2, which is the midpoint of p_1 and p_3:

$$\begin{cases} x_2 = x_1 + \dfrac{l_1 \times |\cos \varphi_1| \times \sin \varphi_d}{2} \\ y_2 = y_1 + \dfrac{l_1 \times |\cos \varphi_1| \times \cos \varphi_d}{2} \end{cases}. \tag{9}$$

If the distance between p_v and p_1 is less than or equal to 2 m, or if the ARV sails past L_1, which is the vertical plane perpendicular to L that passes through p_1, it can be determined that the ARV has reached the p_1 point and enters the short-range docking stage.

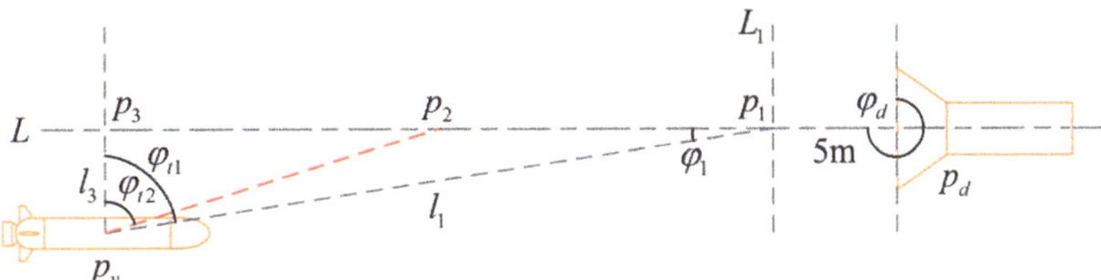

Figure 6. Path tracking method in the mid-range adjustment stage.

3.3. Short-Range Docking Stage

The purpose of the short-range docking stage is to control the ARV to navigate accurately to the docking station. The short-range docking stage can be appoached using an optical guidance method, or an autonomous navigation guidance method. In this paper, the autonomous navigation guidance method is used.

In the short-range docking stage, a heading path mode is used, the ARV sails from 5 m in front of the entrance nozzle of the docking station and the direction of the entrance nozzle of the docking station is used as the heading angle for docking. As the ARV approaches the docking station, the influence of the docking station positioning error on the heading angle increases. To avoid the above problem, the ARV will use a fixed-heading mode if the distance between the ARV and the docking station is less than 1.5 m.

As the docking system does not have a capture mechanism, it is impossible to judge whether the docking is in place by hardware. To stop the ARV after it has been guided to the docking station, it is necessary to calculate the sailing time through actual tests.

3.4. Correction of Docking Station Status Information

During the docking process, as the ARV approaches the docking station, it receives status information from the docking station measured by the USBL and the depth gauge through a hydroacoustic communicator. The docking station status information includes its depth, relative position and orientation angle, and the ARV uses this information to correct its docking strategy in real time.

The docking station depth correction method is as follows: if the difference between the received docking station depth and the docking station depth before update is less than 5 m, update the depth information.

The correction of the docking station relative position and orientation angle is obtained by the positioning information measured by the USBL. The method is shown in Figure 7:

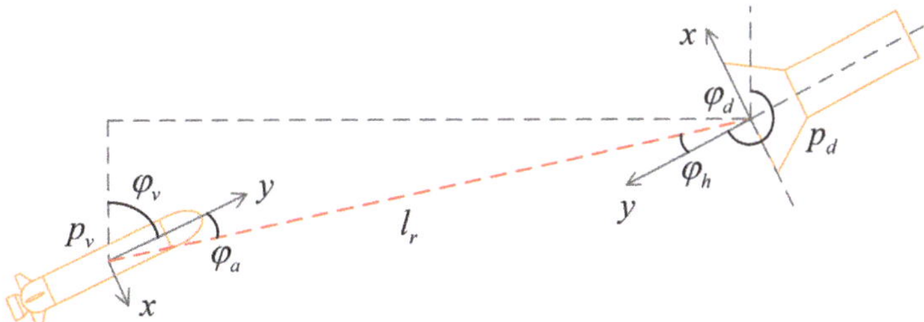

Figure 7. Correction of the relative position and orientation angle of the docking station.

If the difference between the depth of the ARV and the preset depth of the docking station is less than 0.5 m, the ARV and the docking station are considered to be in the same horizontal plane. The formula for calculating the relative position of the docking station is as follows:

$$\begin{cases} x_d = x_v + l_r \times \sin(\varphi_v + \varphi_a) \\ y_d = y_v + l_r \times \cos(\varphi_v + \varphi_a) \end{cases}, \tag{10}$$

where l_r is the distance between the ARV and the docking station measured by USBL, φ_v is the heading angle of the ARV and φ_a is the azimuth angle of the docking station in the ARV coordinate system measured by USBL. If the distance between the new docking station position and the docking station position before update is less than 15 m, update the position information.

The formula for calculating the orientation angle of the docking station is as follows:

$$\varphi_d = \text{atan2}(x_v - x_d, y_v - y_d) - \varphi_h, \tag{11}$$

where φ_h is the azimuth angle of the ARV in the docking station coordinate system measured by the USBL. If the difference between the new docking station orientation angle and the docking station orientation angle before update is less than 15 degrees, update the orientation angle information.

The method for updating the position and orientation angle is shown in Figure 8. DP... and DO... are the currently stored docking station positions and orientation angles, 5 groups are stored in total and the initial values are the preset position and orientation angle of the docking station. DP and DO are the new status information measured by USBL, DP_{mean} and DO_{mean} are the average docking station position and orientation angle. If DP and DO satisfy the conditions for updating the position and orientation angle of the docking station, they are pushed into the array and DP_0 and DO_0 are pushed out of the array. The ARV uses the updated position and orientation angle of the docking station DP_{mean} and DO_{mean}.

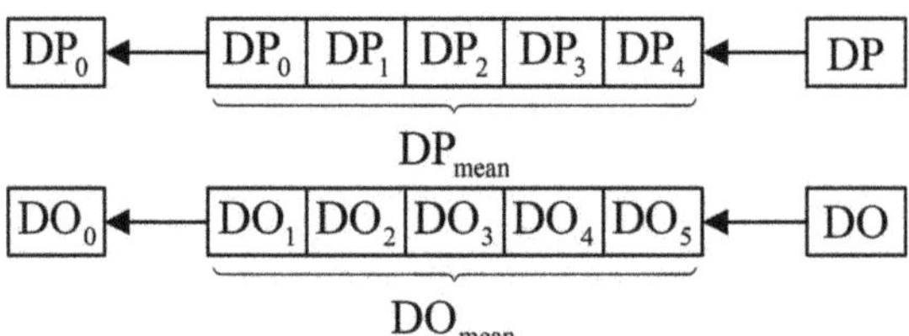

Figure 8. Method of updating the position and orientation angle.

4. Test Verification and Analysis

To verify the feasibility of the guidance method, tests on lakes and shallow seas are performed from July 2022 to August 2022. The tests include autonomous navigation docking tests and acoustic guidance docking tests.

4.1. Autonomous Navigation Docking Tests on the Lake

Autonomous navigation docking uses only position information provided by inertial navigation for guidance without using USBL to correct the docking station depth, position and orientation angle. The ARV starts the docking task at a random position between 20 m and 50 m away from the docking station and a random initial heading angle. The long-guidance waypoints are set at 100 m and 80 m in front of the docking station, and the sailing speed is set at 2 knots for the first waypoint path mode and 0.3 m/s for the remaining stages.

The docking station is deployed 1.5 m underwater, and the position and the orientation angle of the docking station are measured and input to the ARV autonomous docking module as the preset docking station position and the preset orientation angle.

Real images and video screenshots of the autonomous docking tests on the lake are shown in Figure 9. A total of seven autonomous navigation docking tests were performed, of which four docking tests were successful. The records of the docking tests are shown in Table 1, where the lateral control error is the distance between the ARV position when docking is complete and the vector, which is in the direction of the docking station orientation angle through the docking station position. A positive value means that the ARV is to the right side of the vector, and a negative value means that the ARV is to the left side of the vector.

Figure 9. Autonomous docking tests on the lake.

Table 1 shows that the lateral deviation of the docking control using the position information provided by the inertial navigation has reached the accuracy required for docking. However, the large positioning error of inertial navigation causes the actual arrival position of the AUV to deviate, resulting in docking failure.

Table 1. Records of autonomous navigation docking tests on the lake.

Index	ARV Initial Heading Angle (°)	Docking Station Initial Orientation Angle (°)	Initial Distance between ARV and Docking Station (m)	Lateral Control Error (m)	Success or Not
1	161.237	196	45.271	0.106	No
2	184.230	196	47.307	0.023	No
3	233.973	196	39.273	0.106	No
4	249.888	196	45.216	−0.091	Yes
5	278.954	196	41.150	−0.091	Yes
6	174.634	196	42.331	−0.061	Yes
7	175.202	196	49.301	−0.061	Yes

The track and the deviation curve of the sixth autonomous navigation docking test are shown in Figure 10. It can be seen that as the ARV sails from the 100 m waypoint to the 80 m waypoint, there is some fluctuation in the lateral deviation at this stage due to the change in target heading. As the ARV approaches the 80 m waypoint, the lateral deviation has stabilised around 0, that is, the ARV has been adjusted to be close to the centreline of the docking station. As the ARV sails from the 80 m waypoint to the docking station, the track is always stable on the centreline of the docking station. When entering the docking station, the lateral deviation is −0.061 m.

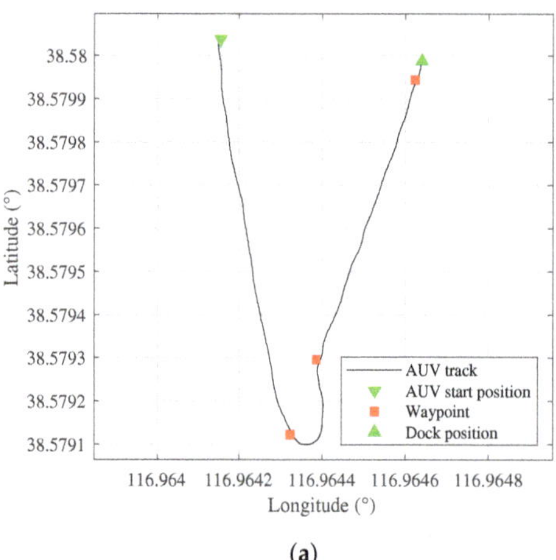 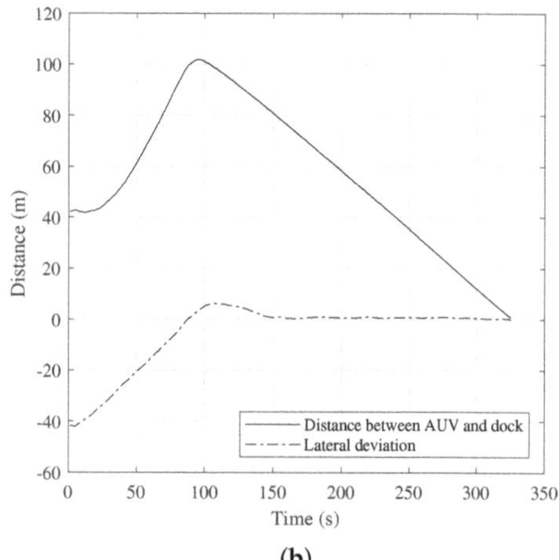

(a)   (b)

Figure 10. (**a**) Track of autonomous navigation docking test on the lake. (**b**) Deviation curve of autonomous navigation docking test on the lake.

4.2. Acoustic Guidance Docking Tests on the Lake

Acoustic guidance docking uses inertial navigation position for guidance and uses underwater acoustic communication to correct the docking station depth, position and orientation angle in real time. The ARV starts the docking task at a random position between 20 m and 50 m away from the docking station and starts the docking task with a random initial heading angle. The long-guidance waypoints are set at 120 m and 80 m in front of the docking station, and the navigation speed at each stage is the same as that of the autonomous navigation docking tests.

The docking station is deployed 1.5 m underwater, and the position and the orientation angle of the docking station are measured and input to the ARV autonomous docking module as preset values.

The acoustic guidance docking tests are divided into three stages: acoustic guidance performance tests, docking station orientation angle adjustment tests and docking station position adjustment tests.

4.2.1. Acoustic Guidance Performance Tests

In this stage, the USBL functioned only to adjust the position of the docking station, the method for updating the docking station orientation angle mentioned in Section 3.4 is not used, and the orientation angle of the docking station is kept at the preset value.

In the first stage of acoustic guidance docking, a total of 10 tests were performed, of which 8 docking tests were successful. The records of the docking tests are shown in Table 2.

The track and the deviation curve of the ninth acoustic guidance docking test is shown in Figure 11. The apparent position of the docking station after several updates has some deviation from the preset position because the inertial navigation of the ARV has a cumulative error during its navigation, resulting in some offset in its self-estimated position. As the ARV navigates from the 80 m waypoint to the docking station, its track is gradually corrected as the position of the docking station is continuously updated. When the docking is complete, the lateral deviation of the ARV relative to the updated docking station position is −0.909 m. It can be seen that the acoustic guidance docking algorithm has sufficient adjustment capability in the case of drift in autonomous navigation.

Table 2. Records of acoustic guidance docking tests on the lake (stage 1).

Index	ARV Initial Heading Angle (°)	Docking Station Initial Orientation Angle (°)	Initial Distance between ARV and Docking Station (m)	Update Times of Docking Station Status Information	Lateral Control Error (m)	Success or Not
1	268.108	196	40.246	20	−0.589	Yes
2	269.674	196	36.265	28	−0.702	Yes
3	220.765	185	37.316	12	0.200	Yes
4	242.820	185	21.649	30	0.556	No
5	152.327	185	21.937	14	0.431	Yes
6	269.948	185	20.919	36	−1.256	No
7	139.144	185	21.771	45	−0.238	Yes
8	242.691	185	23.836	16	−0.242	Yes
9	179.415	185	22.938	38	−0.909	Yes
10	242.952	185	21.615	24	−0.536	Yes

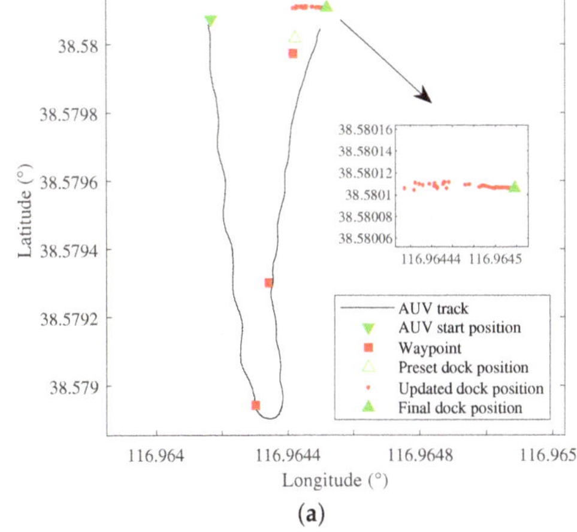

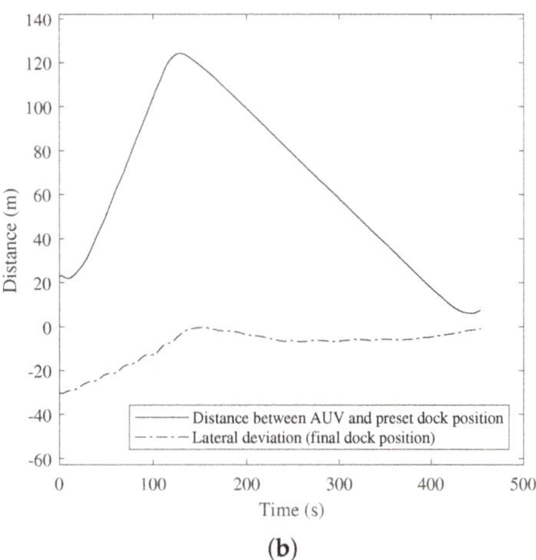

Figure 11. (**a**) Track of acoustic guidance docking test on the lake (stage 1). (**b**) Deviation curve of acoustic guidance docking test on the lake (stage 1).

Figure 11 shows that when the ARV is successfully docked, its position has not reached the updated docking station position that it has estimated using data provided by the USBL. After analysis, due to reasons such as data delay caused by ARV movement and the USBL operating mechanism, the estimated docking station position is directly behind the actual docking station position after multiple adjustments. This error affects the lateral deviation and the success rate of acoustic guidance docking.

4.2.2. Docking Station Orientation Angle Adjustment Tests

At this stage, the orientation angle of the docking station received by the USBL is used to update the docking strategy in real time and control the docking of the ARV.

In the second stage of acoustic guidance docking, a total of two tests were performed, both of which were successful. The records of the docking tests are shown in Table 3.

The track and heading angle update curve of the first acoustic guidance docking test are shown in Figure 12. The orientation angle of the docking station will fluctuate to some extent due to factors such as water flow. The ARV successfully used the updated docking station orientation angle to modify the docking path in real time, and the docking can still be successful. The lateral deviation of the ARV when it reaches the docking station was

0.023 m, which meets the accuracy requirements for docking. This proves that the ARV docking strategy can overcome fluctuations in the orientation angle of the docking station.

Table 3. Records of acoustic guidance docking tests on the lake (stage 2).

Index	ARV Initial Heading Angle (°)	Docking Station Initial Orientation Angle (°)	Initial Distance between ARV and Docking Station (m)	Update Times of Docking Station Status Information	Updated Docking Station Orientation Angle (°)	Lateral Control Error (m)	Success or Not
1	159.068	210	22.747	67	209.898	0.023	Yes
2	286.052	210	22.348	45	213.571	−0.726	Yes

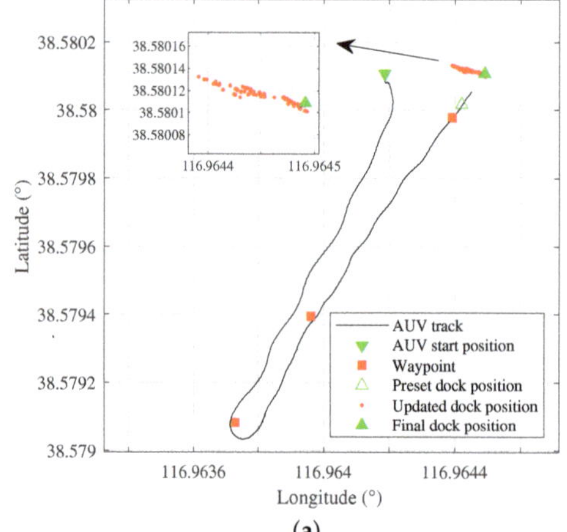

(a)

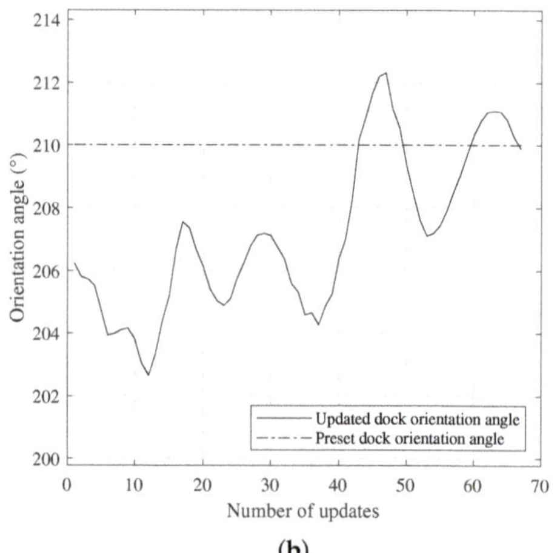

(b)

Figure 12. (**a**) Track of acoustic guidance docking test on the lake (stage 2). (**b**) Orientation angle update curve of acoustic guidance docking test on the lake (stage 2).

4.2.3. Docking Station Position Adjustment Tests

At this stage, the preset position of the docking station was altered so that there was a certain deviation from the actual measured position, forcing the ARV to correct its estimate of the position of the docking station in real time according to the information from the USBL.

In the third stage of acoustic guidance docking, a total of two tests were performed, of which one docking test was successful. The records of the docking tests are shown in Table 4.

Table 4. Records of acoustic guidance docking tests on the lake (stage 3).

Index	ARV Initial Heading Angle (°)	Docking Station Initial Orientation Angle (°)	Initial Distance between ARV and Docking Station (m)	Update Times of Docking Station Status Information	Lateral Control Error (m)	Success or Not
1	290.271	200	24.521	33	0.956	Yes
2	269.674	200	25.127	38	1.479	No

The track of the first acoustic guidance docking test is shown in Figure 13. The preset docking station position is 2.608 m east of the measured docking station position, so the ARV adjusts the relative position and the orientation angle of the docking station using the positioning information from the USBL and finally successfully completes the docking. The lateral deviation is 0.956 m when the docking is complete.

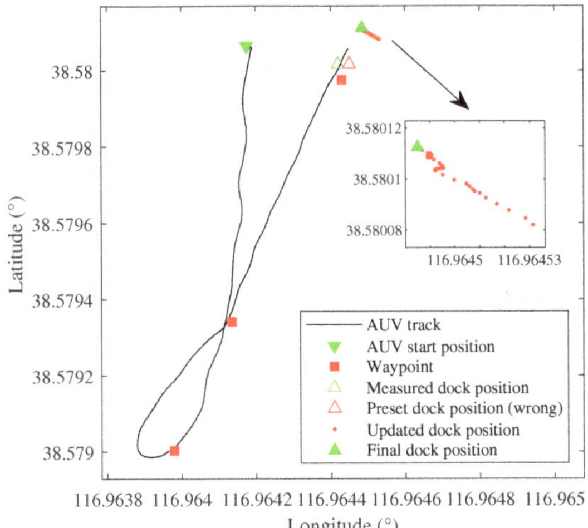

Figure 13. Track of acoustic guidance docking test on the lake (stage 3).

4.3. Acoustic Guidance Docking Tests in the Shallow Sea

In the shallow sea acoustic guidance docking tests, the ARV started the docking task with a random initial heading angle at a random position between 20 m and 500 m from the docking station. The long-guidance waypoints were set at 220 m, 180 m, 120 m and 80 m ahead of the docking station, and the sailing speed was set at 2 knots for the first waypoint path mode and 1 knot for the remaining stages.

The docking station was deployed 1.5 m to 2.5 m underwater, and the position and the orientation angle of the docking station were measured and input to the ARV autonomous docking module as preset values.

Real images and video screenshots of the autonomous docking tests in the shallow sea are shown in Figure 14. A total of eight acoustic guidance docking tests were performed, of which six docking tests were successful. The records of the docking tests are shown in Table 5.

Table 5. Records of acoustic guidance docking tests in the shallow sea.

Index	ARV Initial Heading Angle (°)	Docking Station Initial Orientation Angle (°)	Initial Distance between ARV and Docking Station (m)	Update Times of Docking Station Status Information	Lateral Control Error (m)	Success or Not
1	174.546	225	72.906	12	0.472	Yes
2	145.463	225	23.004	19	0.448	Yes
3	352.805	225	460.635	6	−1.424	No
4	339.969	205	295.552	19	−0.347	Yes
5	23.115	215	332.006	44	0.635	No
6	330.825	215	359.362	36	0.085	Yes
7	349.211	215	356.190	30	−0.972	Yes
8	316.172	215	355.217	21	0.163	Yes

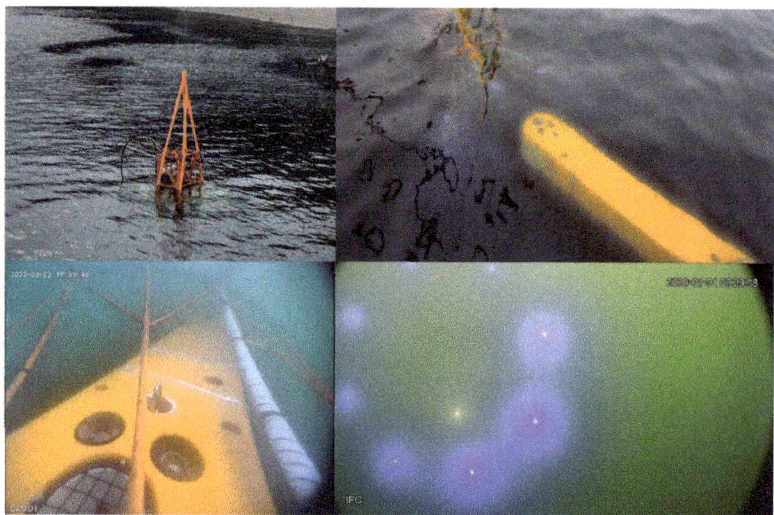

Figure 14. Autonomous docking tests in the shallow sea.

The track and the deviation curve of the seventh acoustic guidance docking test are shown in Figure 15. To overcome the influence of the complex underwater environment on ARV navigation and USBL reception, more waypoints were set in the long-range approach stage, adjustments were made from a longer distance to reduce lateral deviation during docking and more USBL positioning information was received to locate the relative position of the docking station more accurately. It can be seen that the ARV has tracked to the centreline of the docking station when it sailed to 180 m ahead of the docking station. During the approach, as the position and orientation angle of the docking station are updated, the centreline of the docking station will also change, resulting in an increase in the lateral deviation of the ARV. Through continuous adjustment, the ARV can track the centreline of the docking station in real time and complete the docking successfully.

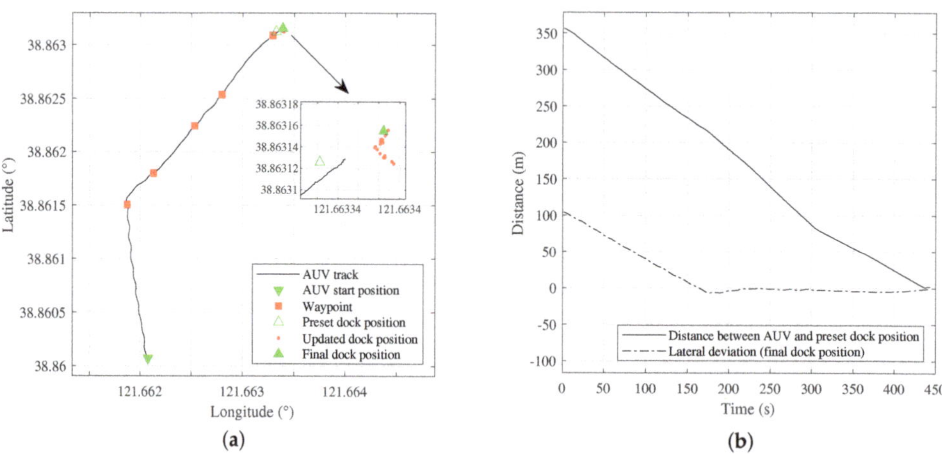

Figure 15. (**a**) Track of acoustic guidance docking test in the shallow sea. (**b**) Deviation curve of acoustic guidance docking test in the shallow sea.

We have compared the success rates of some classical docking tests in the sea, as shown in Table 6. It can be seen that the docking algorithm in this paper has improved the docking success rate, demonstrating the reliability of the docking algorithm. The heading adjustment algorithm in the mid-range adjustment stage effectively reduced the lateral deviation during docking. The real-time adjustment of the depth, relative position and orientation of the docking station using USBL effectively overcame the depth and orientation changes of the docking station in the water as well as the cumulative errors generated by the ARV inertial navigation. All these were helpful for the improvement of docking success rate.

Table 6. Comparison of success rates of sea docking tests.

Docking System	Success Rate
Tri-Ton 2 AUV	2 out of 3 (66.7%)
REMUS AUV	17 out of 29 (58.6%)
Odyssey IIB AUV (electromagnetic guidance)	5 out of 8 (62.5%)
This paper	6 out of 8 (75.0%)

After analysis, the following reasons are summarised for the failures of the acoustic guidance docking tests in the shallow sea:

1. The ocean environment is complex and changing; sea currents and waves will also affect the navigation of the ARV, resulting in docking failure.
2. There is a drift in autonomous navigation; if the ARV has a long continuous underwater navigation distance, the inertial navigation will accumulate significant error; if the error is large, it will be difficult for the ARV to adjust to the centreline of the docking station after it begins to receive information from the USBL, resulting in docking failure.
3. If the ARV does not receive a sufficient amount of docking station status information, it may not be able to locate the correct position of the docking station, resulting in docking failure.

5. Conclusions

Aiming at persistent operation of autonomous underwater vehicles, this paper proposes an underwater docking method based on acoustic guidance. According to the structural properties of the funnel docking station, the underwater docking is divided into three stages. To solve the problem of autonomous navigation deviation in underwater docking, a USBL-based correction method for the docking station depth, relative position and orientation angle is proposed. Autonomous navigation docking tests on a lake were first performed, with a docking success rate of 4 out of 7. Next, acoustic guidance docking tests on the lake were performed, and the docking success rate was 11 out of 14. Finally, through shallow sea tests, acoustic guidance docking tests were performed, with a docking success rate of 6 out of 8. The following conclusions can be drawn from the analysis of the test data: the average lateral error of the autonomous navigation docking was 0.077 m, which met the required accuracy for docking, however, due to the cumulative error of inertial navigation, the docking success rate was low; acoustic guidance docking tests on the lake improved the docking success rate by 17.9% by correcting the ARV's estimate of the depth, relative position and orientation angle of the docking station, verifying the reliability of the acoustic guidance algorithm on the lake; finally, in acoustic guidance docking tests in a shallow sea, due to the complexity of the ocean environment, the success rate of the docking was reduced. Therefore, improving the docking success rate of acoustic guidance in the sea and increasing the stability of the docking guidance algorithm are the focus of future research.

Author Contributions: Conceptualisation, H.X.; methodology, H.X., Z.B., X.Z. and H.Y.; software, Z.B. and H.Y.; validation, H.X., Z.B., X.Z. and H.Y.; formal analysis, H.X. and H.Y.; data curation, H.Y.; writing—original draft preparation, H.Y.; writing—review and editing, Z.B.; project administration, H.X.; funding acquisition, H.X. All authors have read and agreed to the published version of the manuscript.

Funding: This research was supported by the National Defense Preliminary Research Project under grant No. 50911020604, by the Fundamental Research Funds for the Central Universities under grant N2126006, by the Fundamental Research Funds for the Central Universities under grant N2326004 and by the National Natural Science Foundation of China under grant 62203099.

Institutional Review Board Statement: Not applicable.

Informed Consent Statement: Not applicable.

Data Availability Statement: Not applicable.

Abbreviations

The following abbreviations are used in this manuscript:

ALOC	Acoustic localisation and communication device
ARV	Autonomous remotely operated vehicle
AUV	Autonomous underwater vehicle
DVL	Doppler velocity log
GPS	Global positioning system
INS	Inertial navigation system
LBL	Long baseline
ROV	Remotely operated vehicle
SBL	Short baseline
SSBL	Super short baseline
USBL	Ultra-short baseline

References

1. Floreano, D.; Wood, R.J. Science, technology and the future of small autonomous drones. *Nature* **2015**, *521*, 460–466. [CrossRef]
2. Li, Z.; Zhao, S.; Duan, J.; Su, C.; Yang, C.; Zhao, X. Human cooperative wheelchair with brain–machine interaction based on shared control strategy. *IEEE/ASME Trans. Mechatron.* **2016**, *22*, 185–195. [CrossRef]
3. Chin, C.; Lau, M. Modeling and testing of hydrodynamic damping model for a complex-shaped remotely-operated vehicle for control. *J. Mar. Sci. Appl.* **2012**, *11*, 150–163. [CrossRef]
4. Cui, R.; Yang, C.; Li, Y.; Sharma, S. Adaptive neural network control of AUVs with control input nonlinearities using reinforcement learning. *IEEE Trans. Syst. Man Cybern. Syst.* **2017**, *47*, 1019–1029. [CrossRef]
5. Cheng, C.; Fallahi, K.; Leung, H.; Chi, K.T. A genetic algorithm-inspired UUV path planner based on dynamic programming. *IEEE Trans. Syst. Man Cybern. Part C (Appl. Rev.)* **2012**, *42*, 1128–1134. [CrossRef]
6. Shi, Y.; Shen, C.; Fang, H.; Li, H. Advanced control in marine mechatronic systems: A survey. *IEEE/ASME Trans. Mechatron.* **2017**, *22*, 1121–1131. [CrossRef]
7. Teeneti, C.R.; Truscott, T.T.; Beal, D.N.; Pantic, Z. Review of wireless charging systems for autonomous underwater vehicles. *IEEE J. Ocean. Eng.* **2019**, *46*, 68–87. [CrossRef]
8. Yoo, C.; Fitch, R.; Sukkarieh, S. Online task planning and control for aerial robots with fuel constraints in winds. In Proceedings of the Algorithmic Foundations of Robotics XI: Selected Contributions of the Eleventh International Workshop on the Algorithmic Foundations of Robotics, Istanbul, Turkey, 3–5 August 2014; pp. 711–727.
9. Jaffe, J.S.; Franks, P.J.; Roberts, P.L.; Mirza, D.; Schurgers, C.; Kastner, R.; Boch, A. A swarm of autonomous miniature underwater robot drifters for exploring submesoscale ocean dynamics. *Nat. Commun.* **2017**, *8*, 14189. [CrossRef]
10. Yang, C.; Jiang, Y.; Li, Z.; He, W.; Su, C. Neural control of bimanual robots with guaranteed global stability and motion precision. *IEEE Trans. Ind. Inform.* **2016**, *13*, 1162–1171. [CrossRef]
11. Villagra, J.; Milanés, V.; Pérez, J.; Godoy, J. Smooth path and speed planning for an automated public transport vehicle. *Robot. Auton. Syst.* **2012**, *60*, 252–265. [CrossRef]
12. Kimball, P.W.; Clark, E.B.; Scully, M.; Richmond, K.; Flesher, C.; Lindzey, L.E.; Harman, J.; Huffstutler, K.; Lawrence, J.; Lelievre, S. The ARTEMIS under-ice AUV docking system. *J. Field Robot.* **2018**, *35*, 299–308. [CrossRef]
13. Tian, Q.; Wang, T.; Song, Y.; Wang, Y.; Liu, B. Autonomous Underwater Vehicle Path Tracking Based on the Optimal Fuzzy Controller with Multiple Performance Indexes. *J. Mar. Sci. Eng.* **2023**, *11*, 463. [CrossRef]

14. Stokey, R.; Allen, B.; Austin, T.; Goldsborough, R.; Forrester, N.; Purcell, M.; Von Alt, C. Enabling technologies for REMUS docking: An integral component of an autonomous ocean-sampling network. *IEEE J. Ocean. Eng.* **2001**, *26*, 487–497. [CrossRef]
15. Singh, H.; Bellingham, J.G.; Hover, F.; Lemer, S.; Moran, B.A.; von der Heydt, K.; Yoerger, D. Docking for an autonomous ocean sampling network. *IEEE J. Ocean. Eng.* **2001**, *26*, 498–514. [CrossRef]
16. Zhang, W.; Wu, W.; Li, Z.; Du, X.; Yan, Z. Three-Dimensional Trajectory Tracking of AUV Based on Nonsingular Terminal Sliding Mode and Active Disturbance Rejection Decoupling Control. *J. Mar. Sci. Eng.* **2023**, *11*, 959. [CrossRef]
17. Meng, L.; Lin, Y.; Gu, H.; Su, T. Study on dynamic docking process and collision problems of captured-rod docking method. *Ocean Eng.* **2019**, *193*, 106624. [CrossRef]
18. Wadhams, P. The use of autonomous underwater vehicles to map the variability of under-ice topography. *Ocean Dyn.* **2012**, *62*, 439–447. [CrossRef]
19. Wynn, R.B.; Huvenne, V.A.I.; Le Bas, T.P.; Murton, B.J.; Connelly, D.P.; Bett, B.J.; Ruhl, H.A.; Morris, K.J.; Peakall, J.; Parsons, D.R. Autonomous Underwater Vehicles (AUVs): Their past, present and future contributions to the advancement of marine geoscience. *Mar. Geol.* **2014**, *352*, 451–468. [CrossRef]
20. Stokey, R.; Purcell, M.; Forrester, N.; Austin, T.; Goldsborough, R.; Allen, B.; von Alt, C. A docking system for REMUS, an autonomous underwater vehicle. In Proceedings of the Oceans' 97—MTS/IEEE Conference Proceedings, Halifax, NS, Canada, 6–9 October 1997; Volume 2, pp. 1132–1136.
21. Park, J.; Jun, B.; Lee, P.; Lee, F.; Oh, J. Experiment on Underwater Docking of an Autonomous Underwater Vehicle 'ISiMI' using Optical Terminal Guidance. In Proceedings of the OCEANS 2007—Europe, Aberdeen, UK, 18–21 June 2007; pp. 1–6.
22. Feezor, M.D.; Blankinship, P.R.; Bellingham, J.G.; Sorrell, F.Y. Autonomous underwater vehicle homing/docking via electromagnetic guidance. In Proceedings of the Oceans '97—MTS/IEEE Conference Proceedings, Halifax, NS, Canada, 6–9 October 1997; Volume 2, pp. 1137–1142.
23. Allen, B.; Austin, T.; Forrester, N.; Goldsborough, R.; Kukulya, A.; Packard, G.; Purcell, M.; Stokey, R. Autonomous Docking Demonstrations with Enhanced REMUS Technology. In Proceedings of the OCEANS 2006, Boston, MA, USA, 18–21 September 2006; pp. 1–6.
24. Liang, J.; Liu, L. Optimal Path Planning Method for Unmanned Surface Vehicles Based on Improved Shark-Inspired Algorithm. *J. Mar. Sci. Eng.* **2023**, *11*, 1386. [CrossRef]
25. Morgado, M.; Oliveira, P.; Silvestre, C. Design and experimental evaluation of an integrated USBL/INS system for AUVs. In Proceedings of the 2010 IEEE International Conference on Robotics and Automation, Anchorage, AK, USA, 3–7 May 2010; pp. 4264–4269.
26. Cho, G.R.; Kang, H.; Kim, M.-G.; Lee, M.-J.; Li, J.-H.; Kim, H.; Lee, H.; Lee, G. An Experimental Study on Trajectory Tracking Control of Torpedo-like AUVs Using Coupled Error Dynamics. *J. Mar. Sci. Eng.* **2023**, *11*, 1334. [CrossRef]
27. Singh, H.; Bowen, M.; Hover, F.; LeBas, P.; Yoerger, D. Intelligent docking for an autonomous ocean sampling network. In Proceedings of the Oceans '97—MTS/IEEE Conference Proceedings, Halifax, NS, Canada, 6–9 October 1997; Volume 2, pp. 1126–1131.
28. Kawasaki, T.; Fukasawa, T.; Noguchi, T.; Baino, M. Development of AUV "Marine Bird" with underwater docking and recharging system. In Proceedings of the 2003 International Conference Physics and Control, Proceedings (Cat. No.03EX708), Tokyo, Japan, 25–27 June 2003; pp. 166–170.
29. Fukasawa, T.; Noguchi, T.; Kawasaki, T.; Baino, M. "MARINE BIRD", a new experimental AUV with underwater docking and recharging system. In Proceedings of the Oceans 2003—Celebrating the Past ... Teaming toward the Future (IEEE Cat. No. 03CH37492), San Diego, CA, USA, 22–26 September 2003; Volume 4, pp. 2195–2200.
30. McEwen, R.S.; Hobson, B.W.; McBride, L.; Bellingham, J.G. Docking control system for a 54-cm-diameter (21-in) AUV. *IEEE J. Ocean. Eng.* **2008**, *33*, 550–562. [CrossRef]
31. Hobson, B.W.; McEwen, R.S.; Erickson, J.; Hoover, T.; McBride, L.; Shane, F.; Bellingham, J.G. The Development and Ocean Testing of an AUV Docking Station for a 21" AUV. In Proceedings of the OCEANS 2007, Vancouver, BC, Canada, 29 September–4 October 2007; pp. 1–6.
32. Kondo, H.; Okayama, K.; Choi, J.; Hotta, T.; Kondo, M.; Okazaki, T.; Singh, H.; Chao, Z.; Nitadori, K.; Igarashi, M.; et al. Passive acoustic and optical guidance for underwater vehicles. In Proceedings of the 2012 Oceans—Yeosu, Yeosu, Republic of Korea, 21–24 May 2012; pp. 1–6.
33. Maki, T.; Shiroku, R.; Sato, Y.; Matsuda, T.; Sakamaki, T.; Ura, T. Docking method for hovering type AUVs by acoustic and visual positioning. In Proceedings of the 2013 IEEE International Underwater Technology Symposium (UT), Tokyo, Japan, 5–8 March 2013; pp. 1–6.
34. Sato, Y.; Maki, T.; Masuda, K.; Matsuda, T.; Sakamaki, T. Autonomous docking of hovering type AUV to seafloor charging station based on acoustic and visual sensing. In Proceedings of the 2017 IEEE Underwater Technology (UT), Busan, Republic of Korea, 21–24 February 2017; pp. 1–6.
35. Sato, Y.; Maki, T.; Matsuda, T.; Sakamaki, T. Detailed 3D seafloor imaging of Kagoshima Bay by AUV Tri-TON2. In Proceedings of the 2015 IEEE Underwater Technology (UT), Chennai, India, 23–25 February 2015; pp. 1–6.
36. Vallicrosa, G.; Ridao, P.; Ribas, D.; Palomer, A. Active Range-Only beacon localization for AUV homing. In Proceedings of the 2014 IEEE/RSJ International Conference on Intelligent Robots and Systems, Chicago, IL, USA, 14–18 September 2014; pp. 2286–2291.

37. Vallicrosa, G.; Bosch, J.; Palomeras, N.; Ridao, P.; Carreras, M.; Gracias, N. Autonomous homing and docking for AUVs using range-only localization and light beacons. *IFAC-PapersOnLine* **2016**, *49*, 54–60. [CrossRef]
38. Palomeras, N.; Vallicrosa, G.; Mallios, A.; Bosch, J.; Vidal, E.; Hurtos, N.; Carreras, M.; Ridao, P. AUV homing and docking for remote operations. *Ocean Eng.* **2018**, *154*, 106–120. [CrossRef]
39. Vallicrosa, G.; Ridao, P.; Ribas, D. AUV single beacon range-only SLAM with a SOG filter. *IFAC-PapersOnLine* **2015**, *48*, 26–31. [CrossRef]
40. Vallicrosa, G.; Ridao, P. Sum of gaussian single beacon range-only localization for AUV homing. *Annu. Rev. Control* **2016**, *42*, 177–187. [CrossRef]
41. Murashima, T.; Aoki, T.; Tsukioka, S.; Hyakudome, T.; Yoshida, H.; Nakajoh, H.; Ishibashi, S.; Sasamoto, R. Thin cable system for ROV and AUV in JAMSTEC. In Proceedings of the Oceans 2003—Celebrating the Past . . . Teaming toward the Future (IEEE Cat. No. 03CH37492), San Diego, CA, USA, 22–26 September 2003; Volume 5, pp. 2695–2700.
42. Wu, B.; Li, S.; Zeng, J.; Li, Y.; Wang, X. ARV navigation and control system at Arctic research. In Proceedings of the OCEANS 2009, Biloxi, MS, USA, 26–29 October 2009; pp. 1–6.

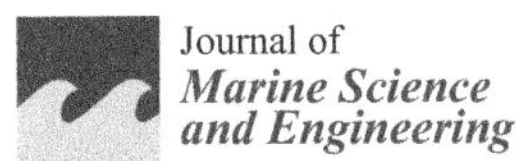

Journal of
*Marine Science
and Engineering*

Article

Optimal Path Planning Method for Unmanned Surface Vehicles Based on Improved Shark-Inspired Algorithm

Jingrun Liang [1,2,]* and Lisang Liu [1,2,]*

1 School of Electronic, Electrical Engineering and Physics, Fujian University of Technology, Fuzhou 350118, China
2 Fujian Province Industrial Integrated Automation Industry Technology Development Base, Fuzhou 350118, China
* Correspondence: liangjingrun@smail.fjut.edu.cn (J.L.); liulisang@fjut.edu.cn (L.L.)

Abstract: As crucial technology in the auto-navigation of unmanned surface vehicles (USVs), path-planning methods have attracted scholars' attention. Given the limitations of White Shark Optimizer (WSO), such as convergence deceleration, time consumption, and nonstandard dynamic action, an improved WSO combined with the dynamic window approach (DWA) is proposed in this paper, named IWSO-DWA. First, circle chaotic mapping, adaptive weight factor and the simplex method are used to improve the initial solution and spatial search efficiency and accelerate the convergence of the algorithm. Second, optimal path information planned by the improved WSO is put into the DWA to enhance the USV's navigation performance. Finally, the COLREGs rules are added to the global dynamic optimal path planning method to ensure the USV's safe navigation. Compared with the WSO, the experimental simulation results demonstrate that the path length cost, steering cost and time cost of the proposed method are decreased by 13.66%, 18.78% and 79.08%, respectively, and the improvement in path smoothness cost amounts to 19.85%. Not only can the proposed IWSO-DWA plan an optimal global navigation path in an intricate marine environment, but it can also help a USV avoid other ships dynamically in real time and meets the COLREGs rules.

Keywords: circle chaotic mapping; simplex method; White Shark Optimizer; dynamic window approach; COLREGs rules; path planning method

Citation: Liang, J.; Liu, L. Optimal Path Planning Method for Unmanned Surface Vehicles Based on Improved Shark-Inspired Algorithm. *J. Mar. Sci. Eng.* **2023**, *11*, 1386. https://doi.org/10.3390/jmse11071386

Academic Editor: Fausto Pedro García Márquez

Received: 11 June 2023
Revised: 24 June 2023
Accepted: 27 June 2023
Published: 7 July 2023

1. Introduction

As small surface vehicles with autonomous navigation capability, USVs are widely used in maritime patrolling, resource exploration and marine rescue [1]. Comprehensively considering conditions such as reefs, water depth and no-sail areas, planning safe and efficient routes for USVs in a complex marine environment has gradually become a hot topic for scholars [2]. To address USV path planning issues in real marine environments, Sing et al. [3] improved the Dijkstra method and planned a global navigation path in a workspace with dynamic and static obstacles. However, the algorithm has high computational complexity and slow path search efficiency. Rui et al. [4] presented an enhanced A-star algorithm for USV path planning. The algorithm has three path smoothers, which are capable of generating a smooth and continuous path in a marine environment, but are not capable of avoiding moving obstacles in real time. Recently, some nature-inspired meta-heuristic algorithms have gradually been adopted in USV path planning. Guo et al. [5] presented an enhanced particle swarm optimization (PSO) algorithm to plan a global USV path that could avoid collisions. Cui et al. [6] enhanced the ant colony algorithm (ACO) and implemented it in USV path planning. Ma Y. et al. [7] presented a dynamic enhanced PSO, which constrained USV path planning in terms of three aspects: collision avoidance, boundary movement and speed. Sahoo et al. [8] combined the advantages of the grey wolf algorithm (GWO) and the genetic algorithm (GA) and proposed a hybrid grey wolf algorithm (HGWO) for path planning and obstacle avoidance of autonomous underwater vehicles (AUVs). Gu et al. [9] proposed an improved RRT algorithm for ship path planning, which clustered the data from an automatic identification system (AIS) and then improved the sampling strategy to accelerate convergence. The improved Douglas–Peucker (DP) and RRT algorithms are combined to optimize paths. Ma D. et al. [10] presented an enhanced Gaussian pseudospectral method (RGPM) for continuous optimal control of USVs, which can obtain an optimal smooth path. Han et al. [11] introduced a mixed approach to path planning based on enhanced Theta* and the DWA. Theta* was utilized to globally plan an optimal path and then the improved DWA was used to enhance the vehicle's dynamic collision avoidance ability. Wang et al. [12] improved the velocity obstacle method (VO) and integrated it into the set-based guidance (SBG) framework to establish a dynamic collision avoidance (DCA) model known as USV-DCA. In order to respond to the dynamic ocean environment, Hu et al. [13] applied the A-Star algorithm and DWA method to safe USV navigation, and the real-time collision avoidance behavior of USVs conforms to COLREGs rules. Zhao et al. [14] put forward an adaptive elite GA with fuzzy inference (AEGAfi), which can control the USVs to optimize its global trajectory, and its dynamic behavior conforms to COLREGs. Li et al. [15] combined the artificial potential field (APF) with the ACO and proposed an improved APF-ACO algorithm, which overcame the local optimum shortcomings in the APF method, and achieved the path planning and collision avoidance of ships. Hao et al. [16] proposed a dynamic fast Q-learning algorithm (DFQL) to plan global USV paths in known marine environments. The algorithm initializes the Q table in combination with the APF method and provides static and dynamic rewards to motivate the USV to move toward the target point. Guo S. et al. [17] proposed a model based on deep reinforcement learning, which combined the Depth Deterministic Strategy Gradient (DDPG) algorithm with the APF method for autonomous path planning of USVs. Sang et al. [18] proposed an improved APF method and combined it with the A-Star algorithm for the formation control and path planning of the USVs.

These above researchers have carried out commendable work, however, there are still some existing problems such as falling into local optimum, time consumption and lack of smoothness in the planned path. White Shark Optimizer (WSO) is an innovative intelligent method that was developed in 2022 to imitate the foraging behavior of white sharks. Compared to other nature-inspired methods such as Butterfly Optimization Algorithm (BOA) [19], Grey Wolf Optimizer (GWO) [20], Manta Ray Foraging Optimizer (MRFO) [21], Whale Optimization Algorithm (WOA) [22] and Sparrow Search Algorithm (SSA) [23], the WSO algorithm offers the benefits of simplicity, high flexibility, strong robustness and rapid convergence [24]. However, since the population of WSO is not rich enough in its initial stage, it will decelerate the convergence in later iterations, and the risk of being caught in a local optimum should be considered. Therefore, the traditional WSO algorithm needs to be further improved. In line with the above research, this study focuses on innovative improvements and applications of WSO. When the previous literature is reviewed, the research on combining the improved WSO and DWA to solve the optimal USV path planning problem has not been found. To overcome the limitations of the traditional WSO algorithm, guide the USV to plan its global optimal path and avoid the obstacles in time, this paper proposes an enhanced WSO algorithm named IWSO-DWA, which combines the advanced techniques of circle chaotic mapping, adaptive weight factor, the simplex method and the DWA. First of all, the population of white sharks is initialized by using circle chaotic mapping to increase its diversity and speed up the algorithm's convergence. Secondly, the adaptive weight factor method is used to update the location of the best white shark, which maintains a balance between both exploration and exploitation to promote the algorithm's capacity. Then, the simplex method is adopted to refresh the location of white sharks as they move toward the best white shark, so as to enhance the ability of escaping the local optimal value. Finally, the enhanced DWA is utilized for avoiding obstacles dynamically. Furthermore, the azimuth evaluation function of the DWA is improved to incorporate the COLREGs rules for dynamic obstacle avoidance. By combining the improved WSO algorithm with the improved DWA method, the USV can not only sail along the optimal path, but also avoid other obstacle ships regularly in real time, and its dynamic behavior conforms to the COLREGs rule.

The following are the primary contributions of this paper:

1. Aiming at insufficient search ability caused by uneven population distribution of the WSO, the white shark population is initialized by using circle chaotic mapping to enrich the diversity and enhance the initial solution quality of the algorithm.
2. In the proposed IWSO, the adaptive weight factor is utilized to refresh the best white shark's location to balance the exploration and exploitation capacity.
3. To address the issue that the WSO slips into the regional optimum easily in the later iteration, the simplex method is used to update the other white sharks' movement position toward the best white shark, which increases the probability of breaking out the local optimum.
4. An innovative fusion method known as the IWSO-DWA algorithm is created by combining the improved WSO with the enhanced DWA. The proposed IWSO-DWA can not only plan a global optimal path of navigation in an intricate marine environment, but also can help USV avoid the other ships dynamically in real time and meet the COLREGs.

The rest of this paper is arranged as follows: Section 2 introduces the WSO algorithm and its improvement with multi-strategies innovatively. Then, in Section 3, both the standard DWA and its enhancement are presented. Section 4 introduces a novel global optimal path planning method called IWSO-DWA. The experimental simulation results of the IWSO-DWA are presented in Section 5, which also includes a comparison of its performance advantages with those of conventional algorithms. Section 6 concludes the research and outlines future work.

2. White Shark Optimizer and Its Improvement

2.1. Traditional White Shark Optimizer (WSO)

The White Shark Optimizer, a novel nature-inspired algorithm introduced in 2022, mimics the foraging behavior of white sharks. The WSO has the superiorities of simplicity, high flexibility and strong robustness. However, it also has certain drawbacks, including population diversity deficiency, limited search range and a tendency to slip into the regional optimum.

Assume that the set matrix of the white shark population is:

$$W = [w_1, w_2, w_3, \ldots, w_n]^T, w_i = [w_{i,1}, w_{i,2}, w_{i,3}, \ldots, w_{i,d}] \tag{1}$$

Let n denote the number of white sharks, with $i = (1,2,\ldots,n)$. The dimension of the problem definition is represented by the variable d.

According to Equation (1), the fitness function of white sharks is expressed as follows:

$$F(w) = [f(w_1), f(w_2), \ldots, f(w_n)]^T \tag{2}$$

where $f(w_i) = [f(w_{i,1}), f(w_{i,2}), \ldots, f(w_{i,d})]$. The i-th white shark's fitness value is represented by $f(w_i)$. The white shark population's fitness value is denoted by $F(w)$.

In the traditional WSO algorithm, white sharks search for prey extensively through their sensitive hearing, smell and sight. While prey is moving in the sea, it will produce hesitation of the waves and special smells. Once a white shark perceives the prey's position, it approaches the prey in a wave motion. The white sharks' motion speed can be expressed as follows:

$$v_{k+1}^i = \mu[v_k^i + p_1(\omega_{gbest_k} - \omega_k^i) \times c_1 + p_2(\omega_{best}^{v_k^i} - \omega_k^i) \times c_2] \tag{3}$$

where k denotes the current iterations. v_{k+1}^i is the i-th white shark's new velocity vector in $(k+1)$-th iteration. v_k^i is the i-th white shark's current velocity vector in k-th iteration.

ω_{gbest_k} denotes the white shark's global optimal position. In the k-th step, the i-th white shark's current position is denoted by ω_k^i. The i-th optimal known position in the white shark population is denoted by $\omega_{best}^{v_k^i}$. c_1 and c_2 are selected from [0,1] randomly. v_k^i denotes the white sharks' i-th index vector when they have reached their optimal location.

Great white sharks usually hunt for food in the ocean's depths randomly. What is more, great white sharks approach the optimal prey's location. The location of white sharks near the optimal prey is updated as follows:

$$\omega_{k+1}^i = \begin{cases} \omega_k^i \cdot \neg \oplus \omega_0 + u \cdot a + l \cdot b; & rand < mv \\ \omega_k^i + v_k^i / f; & rand \geq mv \end{cases} \tag{4}$$

where ω_{k+1}^i represents the new position of i-th white shark. $\neg$ is a negation operator. The search space bound is indicated by l and u. mv represents the increasing movement force of the white shark as it approaches its prey. a,b and ω_0 represent the vector in one dimension.

The best white shark is closely situated to the optimal prey. By using fish school behavior, all white sharks will migrate towards the best white shark, and the position is updated as follows:

$$\omega_{k+1}'^i = \omega_{gbest_k} + r_1 \vec{D}_\omega \mathrm{sgn}(r_2 - 0.5), \quad r_3 < s_s \tag{5}$$

where $\omega_{k+1}'^i$ denotes the i-th white shark's position relative to the prey. $\mathrm{sgn}(r_2 - 0.5)$ stands for the search direction of the white shark. $\vec{D}_\omega$ stands for the distance between the white shark and its prey. s_s stands for the strength of white sharks' sense organs. R_1, r_2 and r_3 are selected from [0,1] randomly.

The fish school behavior of the traditional WSO algorithm can be expressed as follows:

$$\omega_{k+1}^i = \frac{\omega_k^i + \omega_{k+1}'^i}{2 \times rand} \tag{6}$$

where the related variables have been explained in Equations (3) and (5), so they will not be described here again.

The flow chart of the traditional WSO algorithm is depicted in Figure 1.

To sum up, the traditional WSO algorithm exhibits the following limitations:

1. Since the white shark's initialization population is created randomly, it is prone to problems such as uneven population distribution, poor diversity and low quality of the initial solution, which will not only decelerate the convergence, but also may fall into local optimum.
2. Since white sharks hunt for prey amid the ocean's depths randomly, they may not be close enough to the optimal prey, which will lead to an imbalance in both exploration and exploitation capacity.
3. In the fish school behavior, other white sharks' position may not be optimal to the best white shark, which will slip into the regional optimum easily.

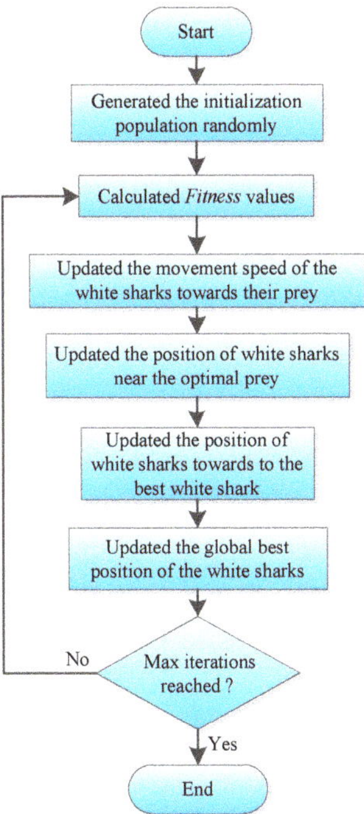

Figure 1. Flow chart of traditional WSO algorithm.

2.2. Improved White Shark Optimizer (IWSO)

Due to the traditional WSO algorithm's drawbacks, the following aspects will be improved in this paper:

1. Considering the shortcomings of the WSO, such as uneven distribution and insufficient diversity of the population, a circle chaotic mapping algorithm is used to initialize the white shark population, thus further improving the quality of the initial solution.
2. To address the imbalance of exploration and exploitation capacity, the adaptive weight factor method is utilized to update the best white shark's position so that it can strengthen the balance between the exploration and exploitation capability.
3. To deal with the issue that the WSO slips into the regional optimum easily, the simplex method is utilized to update the other great white sharks' position near the best white shark to increase the possibility of the algorithm escaping the local region.

2.2.1. Circle Chaotic Mapping

Circle chaotic mapping has gained considerable attention owing to its simple structure and strong uniformity. It exhibits complex, unpredictable and random behaviors, and is often employed to enhance the diversity of the population. Circle chaotic mapping outperforms other kinds of chaotic mappings like logistic chaotic mapping and tent chaotic mapping in terms of ergodic uniformity, randomness and diversity.

In the traditional WSO algorithm, the initialization population of white sharks is generated randomly, which may lead to the disadvantages of uneven population distribution, poor diversity and slipping into the regional optimum easily in the later iteration. For this, the circle chaotic mapping method is employed to generate the initial circle population of white sharks, which is then combined with a random population. The resulting group is evaluated, and the best sharks are selected to form the optimal white shark population of the next generation. The optimized white shark individuals are more similar to the initial optimal solution than the sharks in the random population and initial circle population, which evens out white shark population distribution, broadens the algorithm's search range and improves its efficacy.

Let w_i denote the individuals within the white shark population, and then the initialization formula for the white shark population using the circle chaotic mapping method can be expressed as follows [25]:

$$w_{i+1} = mod(w_i + 0.2 - \frac{0.5}{2\pi} \cdot sin(2\pi \cdot w_i), 1) \tag{7}$$

where *mod* indicates remainder.

2.2.2. Adaptive Weight Factor

When white sharks hunt for prey amid the ocean's depths randomly, they may not be close enough to the optimal prey, which may lead to an imbalance between exploration and exploitation. So, the adaptive weight factor is used to update the best white shark's position in this paper, which makes the algorithm have outstanding exploration ability in the earlier iteration and excellent exploitation ability in the later iteration. By introducing the adaptive weight factor in the process of white shark hunting prey, it is beneficial to balance the capacity for both exploration and exploitation. The adaptive weight factor proposed in this paper is expressed as follows:

$$\alpha = 0.2 + \frac{1}{0.6 + e^{(-f(w_i)/\mu')^k}} \tag{8}$$

where $f(w_i)$ represents the *i*-th white shark's fitness value. μ' represents the white shark's best fitness value in the first iteration. α is the dynamic nonlinear factor, which is used to update the best white shark's position. The refined formula is presented below:

$$\omega^i_{k+1} = \begin{cases} \alpha \cdot \omega^i_k \cdot \neg \oplus \omega_0 + u \cdot a + l \cdot b; & rand < mv \\ (1 - \alpha) \cdot \omega^i_k + v^i_k / f; & rand \geq mv \end{cases} \tag{9}$$

As can be seen from the formula, the best white shark's position is adjusted by the adaptive weight factor adaptively, so that the capacity for both exploration and exploitation can be balanced.

2.2.3. Simplex Method

The simplex method is a direct search algorithm for optimizing multi-dimensional unconstrained problems proposed by Nelder et al. [26] in 1965. The algorithm takes $d + 1$ points in d-dimensional space to form a simplex and then calculates the function value of its vertices. The sub-optimal points are obtained by internal compression, external compression, reflection and expansion of the worst point of the simplex. Then, the worst point of the simplex is replaced by the sub-optimal point, and the simplex is reconstructed to approach the global optimum continuously [27]. Since the simplex method is not affected by the continuity and derivability of the objective function, it has an excellent optimization ability, thus improving its capacity to break out the regional optimum. For this, the simplex method is utilized to update the other white sharks' location in the fish school behavior, so as to urge them to approach the best white shark continuously and make their positions close to the global optimum, which accelerates the convergence to the optimal solution and enables it to overcome regional optimum. The diagram of the optimization process for the simplex method is shown in Figure 2.

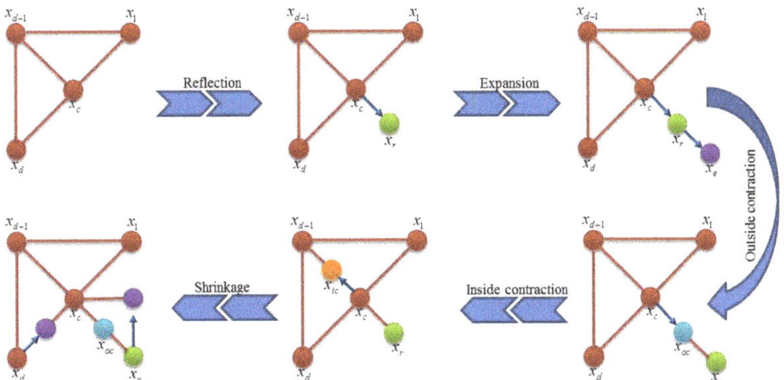

Figure 2. Diagram of optimization process for the simplex method.

As is shown in the figure, the optimization steps of the simplex method can be summarized as follows:

Step 1: Ranking and evaluating. All individuals of the white shark population in the d-dimensional space were ranked and evaluated for fitness values, and the current best white shark x_g, the second best white shark x_b and the abandoned white shark x_s will be selected, which is expressed as follows:

$$f(x_{d+1}) \geq \cdots \geq f(x_b) \geq f(x_c) = f(\frac{x_g + x_b}{2}) \geq f(x_g) \geq f(x_1) \tag{10}$$

Step 2: Reflection. Performing a reflection operation to obtain a reflection point x_r. The formula of reflection operation is:

$$x_r = (1 + \delta) \cdot x_c - \delta \cdot x_s \tag{11}$$

Here, x_r refers to the reflective point obtained from x_s. δ is the reflective factor, typically set at 1.

Step 3: Expansion. If $f(x_g) > f(x_r)$, the expansion process is carried out to obtain the expansion point x_e. The basic formula of expansion operation is as follows:

$$x_e = (1 - \chi) \cdot x_c + \chi \cdot x_r \tag{12}$$

where χ is the expansion factor. If $f(x_e) > f(x_r)$, substitutes x_r for x_s.

Step 4: Outside Contraction. When $f(x_r) < f(x_s)$, the outside contraction point x_{oc} can be obtained by the outside contraction operation, which can be expressed as follows:

$$x_{oc} = x_c + \phi \cdot (x_r - x_c) \tag{13}$$

where ϕ is the contraction coefficient. If $f(x_r) > f(x_{oc})$, substitutes x_{oc} for x_s.

Step 5: Inside Contraction. If $f(x_s) < f(x_r)$, the inside contraction operation can be expressed as follows:

$$x_{ic} = x_c - \phi \cdot (x_r - x_c) \tag{14}$$

where x_{ic} is the inside contraction point. If $f(x_{ic}) < f(x_s)$, substitutes x_{ic} for x_s.

Step 6: Shrinkage. For vertices x_i in d-dimensional space, the shrinkage operation is expressed as follows:

$$x_i = x_1 + \xi \cdot (x_i - x_1) \tag{15}$$

where ξ is the shrinkage coefficient.

After optimization by the simplex method, the position of the white shark individuals is closer to global optimum, which helps improve the probability of the algorithm breaking out the regional optimum.

2.2.4. Performance of IWSO on IEEE CEC-2005

In this paper, a series of advanced strategies such as circle chaotic mapping, adaptive weight factor method and simplex method is used to enhance the WSO. Moreover, the CEC-2005 test suite is used to verify the improved effect of the presented IWSO and its outstanding performance. CEC-2005 is a test suite containing several challenging benchmark functions, which has a large number of local optimal solutions, so it could be used to simulate the complexity of real search space and further verify the IWSO's capability and reliability. The presented IWSO is compared with WSO and other five highly respected meta-heuristic algorithms such as BOA, GWO, MRFO, WOA and SSA over 25 independent runs in some benchmark functions of CEC-2005. The population size of the IWSO and other algorithms is set to 300, and the search agent is set to 50. The experimental simulation results are displayed in Table 1.

Table 1. Optimization results of IWSO and other algorithms (BOA,GWO,MRFO,WOA,SSA,WSO) running on the CEC-2005 test function.

Function		BOA	GWO	MRFO	WOA	SSA	WSO	IWSO
F1	Best	9.16×10^{-6}	7.19×10^{-20}	2.40×10^{-269}	2.20×10^{-55}	0.00×10^{0}	7.21×10^{1}	0.00×10^{0}
	Worst	2.99×10^{-5}	7.00×10^{-18}	2.73×10^{-255}	8.42×10^{-49}	2.26×10^{-175}	3.75×10^{2}	0.00×10^{0}
	Mean	2.00×10^{-5}	1.80×10^{-18}	1.20×10^{-256}	3.44×10^{-50}	9.03×10^{-177}	2.25×10^{2}	0.00×10^{0}
	Std	5.34×10^{-6}	1.85×10^{-18}	0.00×10^{0}	1.68×10^{-49}	0.00×10^{0}	7.07×10^{1}	0.00×10^{0}
F2	Best	1.85×10^{-8}	8.05×10^{-12}	1.14×10^{-136}	1.40×10^{-35}	0.00×10^{0}	2.78×10^{0}	0.00×10^{0}
	Worst	5.21×10^{-1}	6.99×10^{-11}	5.96×10^{-127}	1.04×10^{-31}	2.31×10^{-61}	7.02×10^{0}	0.00×10^{0}
	Mean	5.52×10^{-2}	1.75×10^{-11}	2.71×10^{-128}	1.28×10^{-32}	9.24×10^{-63}	4.94×10^{0}	0.00×10^{0}
	Std	1.31×10^{-1}	1.27×10^{-11}	1.19×10^{-127}	2.66×10^{-32}	4.62×10^{-62}	1.17×10^{0}	0.00×10^{0}
F3	Best	8.23×10^{-6}	1.17×10^{-5}	4.60×10^{-261}	2.10×10^{4}	0.00×10^{0}	3.76×10^{2}	0.00×10^{0}
	Worst	2.74×10^{-5}	2.66×10^{-2}	7.44×10^{-245}	6.91×10^{4}	5.76×10^{-121}	1.42×10^{3}	0.00×10^{0}
	Mean	1.81×10^{-5}	2.73×10^{-3}	6.66×10^{-246}	4.73×10^{4}	2.31×10^{-122}	8.60×10^{2}	0.00×10^{0}
	Std	4.48×10^{-6}	5.69×10^{-3}	0.00×10^{0}	1.27×10^{4}	1.15×10^{-121}	2.75×10^{2}	0.00×10^{0}
F4	Best	1.14×10^{-3}	1.54×10^{-5}	1.10×10^{-136}	3.57×10^{-2}	0.00×10^{0}	6.73×10^{0}	0.00×10^{0}
	Worst	2.66×10^{-3}	2.81×10^{-4}	7.69×10^{-122}	8.50×10^{1}	8.76×10^{-97}	1.17×10^{1}	0.00×10^{0}
	Mean	1.81×10^{-3}	8.91×10^{-5}	3.25×10^{-123}	4.22×10^{1}	3.51×10^{-98}	9.04×10^{0}	0.00×10^{0}
	Std	3.89×10^{-4}	5.82×10^{-5}	1.54×10^{-122}	2.65×10^{1}	1.75×10^{-97}	1.35×10^{0}	0.00×10^{0}
F5	Best	2.88×10^{1}	2.57×10^{1}	2.21×10^{1}	2.76×10^{1}	2.35×10^{-9}	8.02×10^{2}	2.87×10^{-7}
	Worst	2.89×10^{1}	2.87×10^{1}	2.51×10^{1}	2.87×10^{1}	5.60×10^{-4}	2.27×10^{4}	2.24×10^{-6}
	Mean	2.89×10^{1}	2.67×10^{1}	2.35×10^{1}	2.80×10^{1}	8.45×10^{-5}	7.75×10^{3}	1.03×10^{-6}
	Std	2.47×10^{-2}	7.96×10^{-1}	5.87×10^{-1}	3.10×10^{-1}	1.50×10^{-4}	5.41×10^{3}	1.06×10^{-6}
F6	Best	3.48×10^{0}	9.60×10^{-5}	3.32×10^{-9}	7.01×10^{-2}	1.99×10^{-10}	5.18×10^{1}	9.98×10^{-9}
	Worst	6.28×10^{0}	1.36×10^{0}	1.71×10^{-7}	9.51×10^{-1}	2.22×10^{-6}	4.60×10^{2}	2.35×10^{-8}
	Mean	4.98×10^{0}	6.08×10^{-1}	2.70×10^{-8}	3.29×10^{-1}	4.36×10^{-7}	2.50×10^{2}	1.88×10^{-8}
	Std	6.48×10^{-1}	3.91×10^{-1}	3.38×10^{-8}	2.07×10^{-1}	5.68×10^{-7}	1.05×10^{2}	7.65×10^{-9}
F7	Best	8.60×10^{-4}	6.36×10^{-4}	2.06×10^{-5}	1.49×10^{-4}	5.74×10^{-6}	1.55×10^{-2}	1.29×10^{-5}
	Worst	6.09×10^{-3}	5.87×10^{-3}	3.18×10^{-4}	1.68×10^{-2}	9.01×10^{-4}	1.15×10^{-1}	1.00×10^{-4}
	Mean	3.13×10^{-3}	2.41×10^{-3}	1.57×10^{-4}	3.62×10^{-3}	2.88×10^{-4}	4.02×10^{-2}	4.56×10^{-5}
	Std	1.28×10^{-3}	1.24×10^{-3}	8.72×10^{-5}	3.36×10^{-3}	2.25×10^{-4}	2.27×10^{-2}	4.74×10^{-5}
F8	Best	-1.71×10^{7}	-7.42×10^{3}	-9.41×10^{3}	-1.26×10^{4}	-1.26×10^{4}	-4.11×10^{3}	-1.23×10^{4}
	Worst	-6.46×10^{4}	-4.57×10^{3}	-7.32×10^{3}	-7.05×10^{3}	-5.63×10^{3}	-2.94×10^{3}	-1.22×10^{4}
	Mean	-1.48×10^{6}	-6.18×10^{3}	-8.51×10^{3}	-9.84×10^{3}	-8.76×10^{3}	-3.40×10^{3}	-1.23×10^{4}
	Std	3.37×10^{6}	8.07×10^{2}	5.36×10^{2}	1.87×10^{3}	2.36×10^{3}	2.90×10^{2}	7.79×10^{1}

Table 1. *Cont.*

Function		BOA	GWO	MRFO	WOA	SSA	WSO	IWSO
F9	Best	3.42×10^{-8}	1.65×10^{-12}	0.00×10^{0}	0.00×10^{0}	0.00×10^{0}	3.12×10^{1}	0.00×10^{0}
	Worst	3.58×10^{-4}	2.03×10^{1}	0.00×10^{0}	5.68×10^{-14}	0.00×10^{0}	1.73×10^{2}	0.00×10^{0}
	Mean	2.50×10^{-5}	7.42×10^{0}	0.00×10^{0}	2.27×10^{-15}	0.00×10^{0}	1.15×10^{2}	0.00×10^{0}
	Std	7.67×10^{-5}	5.76×10^{0}	0.00×10^{0}	1.14×10^{-14}	0.00×10^{0}	3.90×10^{1}	0.00×10^{0}
F10	Best	6.63×10^{-4}	6.33×10^{-11}	8.88×10^{-16}	8.88×10^{-16}	8.88×10^{-16}	3.80×10^{0}	8.88×10^{-16}
	Worst	1.87×10^{-3}	6.28×10^{-10}	8.88×10^{-16}	1.51×10^{-14}	8.88×10^{-16}	6.75×10^{0}	8.88×10^{-16}
	Mean	1.35×10^{-3}	2.28×10^{-10}	8.88×10^{-16}	5.58×10^{-15}	8.88×10^{-16}	5.25×10^{0}	8.88×10^{-16}
	Std	2.77×10^{-4}	1.38×10^{-10}	0.00×10^{0}	2.85×10^{-15}	0.00×10^{0}	7.50×10^{-1}	0.00×10^{0}

The table presents the objective function values for the IWSO and other algorithms in terms of their best, worst, average and standard deviation. Based on the results, it can be concluded that the IWSO effectively identifies the global optimum solution for the majority of the CEC-2005 test functions, and its standard deviation is smaller than that of the WSO algorithm and the other five meta-heuristic algorithms, which shows that IWSO algorithm is effective in improving WSO algorithm. Therefore, when solving complex optimization problems, the proposed IWSO algorithm is robust and reliable.

3. Dynamic Window Approach and Its Improvement

3.1. USV Modeling

Since there are many parameters in the actual USV motion model, it may be difficult to directly model it. Therefore, the following assumptions are used to simplify the USV motion model [28].

1. USV is considered a rigid body with uniform mass distribution and no geometric deformation.
2. The roll, pitch and heave motions of USV can be ignored.
3. The xz-plane of USV is symmetrical, and the center of mass lies in the geometric symmetry plane.
4. During the voyage of USV, the temporal and spatial variability of ocean currents and wind in the selected area is considered to be quasi-static.

For the DWA method, it is important to construct the motion model of USV first. According to Ref. [29], the USV has restricted mobility and its motion trajectory can be regarded as consisting of each small arc. If Δt is very small, the motion of the USV may be modeled as a uniform linear motion. In Ref. [30], USV's motion planning problem can be simplified to the motion of a rigid body with three freedom degrees (surge, sway and yaw) in plane space. Based on the above analysis, ignoring the influence of wind and ocean currents, USV's model can be formulated as:

$$\begin{cases} x_{t+1} - x_t = v_t \cdot \Delta t \cdot \cos \theta_t \\ y_{t+1} - y_t = v_t \cdot \Delta t \cdot \sin \theta_t \\ \theta_{t+1} - \theta_t = \omega_t \cdot \Delta t \\ \dot{p} = R_\psi(\psi) \cdot v \\ \dot{\psi} = r \end{cases} \tag{16}$$

where at time t, (x_t, y_t) and θ_t are the USV's location and orientation, respectively. Similarly, (x_{t+1}, y_{t+1}) and θ_{t+1} represent the USV's location and orientation at time $t + 1$, respectively. v_t represents the USV's linear velocity at time t. ω_t represents the USV's angular velocity at time t. $p = [x, y]^T$ stands for the USV's spatial vector. $v = [u, v]^T$ represents the USV's velocity vector. $R_\psi(\psi)$ is a rotation matrix, which is expressed by the following formula:

$$R_\psi(\psi) = \begin{bmatrix} \cos \psi & -\sin \psi \\ \sin \psi & \cos \psi \end{bmatrix} \tag{17}$$

The USV's motion model is displayed in Figure 3.

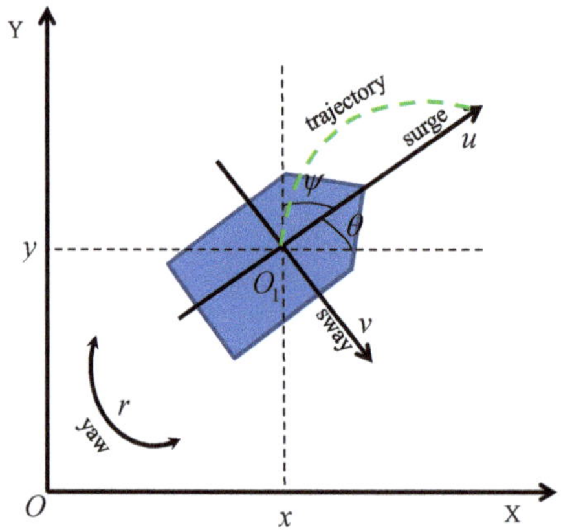

Figure 3. Diagram of motion model for USV.

3.2. Velocity Sampling

Due to countless groups (v, ω) in the domain of motion vectors, sampling these velocities based on the real USV restrictions is required to obtain a workable velocity range [31].

1. Speed constraint: limited by the USV's maximal and minimal velocity:

$$V_m = \{(v,\omega)|v \in [v_{\min}, v_{\max}], \omega \in [\omega_{\min}, \omega_{\max}]\} \tag{18}$$

where the minimal and maximal linear velocities are represented by $v_{\min}$ and $v_{\max}$, respectively. The minimal and maximal angular velocities are represented by $\omega_{\min}$ and $\omega_{\max}$, respectively.

2. Dynamic constraint: influenced by the motor acceleration and deceleration performance of USV, which is expressed as follows:

$$V_d = \left\{(v,\omega)\middle| v \in [v_g - v_f' \cdot \Delta t, v_g + v_e' \cdot \Delta t] \wedge \omega \in [\omega_g - \omega_f' \cdot \Delta t, \omega_g + \omega_e' \cdot \Delta t]\right\} \tag{19}$$

where v_g, ω_g represents the USV's current linear and angular velocity, respectively. v_e', ω_e' represent the USV's maximal linear acceleration and maximal angular acceleration, respectively. v_f', ω_f' represent the USV's maximal linear deceleration and maximal angular deceleration, respectively.

3. Braking distance constraint: To prevent the USV from colliding with other ships or obstacles, the USV will be constrained by the braking distance, and the speed will be reduced to zero within the braking distance according to its maximum deceleration. The braking distance constraint is presented in the following formula:

$$V_a = \left\{(v,\omega)\middle| v \leq \sqrt{2 \cdot dist(v,\omega) \cdot v_f'} \wedge \omega \leq \sqrt{2 \cdot dist(v,\omega) \cdot \omega_f'}\right\} \tag{20}$$

where $dist(v, \omega)$ stands for the distance between the nearest obstacle to the USV and the end of the deduced trajectory.

3.3. Evaluation Function and Its Improvement

After sampling the velocity of USV, the DWA method will deduce the trajectory based on the sampled velocity, and the scoring mechanism is used to sort these trajectories, and the greatest score trajectory will be selected as the final trajectory of USV. Among them, the scoring mechanism of the DWA method is composed of three functions, speed function, azimuth evaluation function and obstacle distance function [32], which are expressed as follows:

$$F(v, \omega) = \alpha \cdot vel(v, \omega) + \beta \cdot head(v, \omega) + \gamma \cdot dist(v, \omega) \tag{21}$$

where α, β and γ are the weight factor of the three functions. However, owing to the absence of knowledge on the global path, the DWA method is prone to slip into the regional optimum when encountering a complex marine environment. Therefore, global path information planned by the IWSO will be incorporated into the enhanced DWA method, so that the USV can break out the regional optimum.

Aiming at the shortcomings of the traditional DWA, some strategies are used to improve it in this paper. Firstly, $head(v, \omega)$ is changed to the tangent angle between the global optimal navigation path and the USV. Then, the current azimuth angle from the USV to the nearest sub-target point can be expressed by the following formula:

$$\theta_c = tan(\frac{y_2 - y_1}{x_2 - x_1}) \cdot \frac{180^\circ}{\pi} \tag{22}$$

where θ_c represents the current azimuth of the USV. (x_1, y_1) represents the current position coordinates of USV. (x_2, y_2) represents the position coordinates of the sub-target point nearest the USV. What is more, the improved azimuth cost function is expressed as follows:

$$head'(v, \omega) = |\theta_c - \theta_{st}| \tag{23}$$

where θ_{st} represents the azimuth between the predicted trajectory and the target point. The improved azimuth evaluation function can guide USV along the global optimal path planned by the IWSO while avoiding other obstacle ships or dynamic obstacles.

Secondly, in order to hasten USV arrival at the target point, the distance cost function within the USV's present location and the sub-target point is constructed, which is expressed as follows:

$$path(v, \omega) = \sqrt{(x_E - x_{ST})^2 + (y_E - y_{ST})^2} \tag{24}$$

where (x_E, y_E) represents the sub-target point coordinates. (x_{ST}, y_{ST}) represents the current predicted trajectory coordinates.

To sum up, the assessment function for the enhanced DWA is:

$$F'(v, \omega) = \sigma(\alpha \cdot vel(v, \omega) + \beta \cdot head'(v, \omega) + \gamma \cdot dist(v, \omega) + \eta \cdot path(v, \omega)) \tag{25}$$

where σ represents the smoothing factor. η represents path cost weight coefficient.

4. The Proposed Fusion Algorithm IWSO-DWA

To further smooth the USV's navigation path and endow it with the ability of real-time dynamic collision avoidance, this paper combines the proposed IWSO algorithm with the improved DWA method and proposes a novel global dynamic optimal path planning method, which is named IWSO-DWA. Due to the complex maritime navigation environment and many ships coming and going, the COLREGs is introduced in this paper to construct the collision avoidance model of USV, so that it can avoid other obstacle ships reasonably while navigating along the optimal path globally. The pseudo-code of the proposed IWSO-DWA is illustrated in Algorithm 1.

Algorithm 1: IWSO-DWA

Input:
The set of population size: P;
The map information: G;
The maximum number of iterations: K.
Output: Optimal navigation path.
1. Initializing population by circle chaotic mapping;
2. **While** $k < K$ **do**
3. Updating the parameters of WSO;
4. Identifying the current optimal solution;
5. **for** $i = 1$ **to** P **do**
6. Updating the motion velocity of white sharks;
7. **end for**
8. **for** $i = 1$ **to** P **do**
9. Refreshing the best white shark's location by the adaptive weight factor;
10. **end for**
11. **for** $i = 1$ **to** P **do**
12. **If** rand $\leq s_s$ **then**
13. $\vec{D}_\omega = \left| rand \times (\omega_{gbest_k} - \omega^i{}_k) \right|$;
14. **If** $i == 1$ **then**
15. $\omega^i{}_{k+1} = \omega_{gbest_k} + r_1 \vec{D}_\omega \mathrm{sgn}(r_2 - 0.5)$;
16. **else**
17. $\omega'^i{}_{k+1} = \omega_{gbest_k} + r_1 \vec{D}_\omega \mathrm{sgn}(r_2 - 0.5)$;
18. $\omega^i{}_{k+1} = \dfrac{\omega^i{}_k + \omega'^i{}_{k+1}}{2 \times rand}$;
19. **end if**
20. **end if**
21. Using the simplex method to update the white sharks' position;
22. **end for**
23. Modifying the position of any white shark that exceeds the boundary;
24. Assessing and revising the updated positions;
25. $k = k + 1$;
26. **end while**
27. Obtaining the optimal path globally and incorporating it into DWA;
28. Considering the COLREGs rules;
29. **return** optimal navigation path.

As can be seen from the pseudo-code of the proposed IWSO-DWA, the IWSO is responsible for planning the global optimal path under a given environment model. The global optimal path information is obtained to choose the present route's start point and sub-target point, and fed into the local path planner subsequently. Under the action of the IWSO-DWA algorithm, USV can travel along the global optimal path planned by the IWSO, and the other obstacle ships will be detected in real time as the process proceeds. When other obstacle ships approach, USV will avoid it in real time, and its dynamic behavior meets the COLREGs. After successful collision avoidance, USV will continue to move along the global optimal path. Finally, refresh the current path's status until the USV reaches the final target point. The flow chart of IWSO-DWA for global dynamic optimal path planning is displayed in Figure 4.

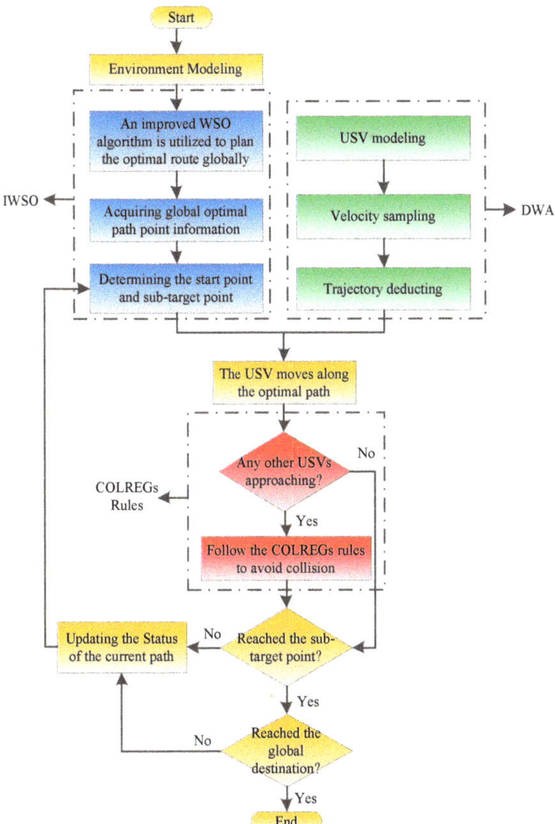

Figure 4. Flow chart of global dynamic optimal path planning method (IWSO-DWA).

4.1. COLREGs Rules

The International Regulations for Preventing Collisions at Sea (COLREGs) is a kind of sea traffic regulation that aims to avoid collisions between ships navigating the open seas.

According to Ref. [33], there are four representative rules of COLREGs: overtaking, head-on, port side crossing and starboard crossing. The four representative rules of COLREGs are shown in Figure 5.

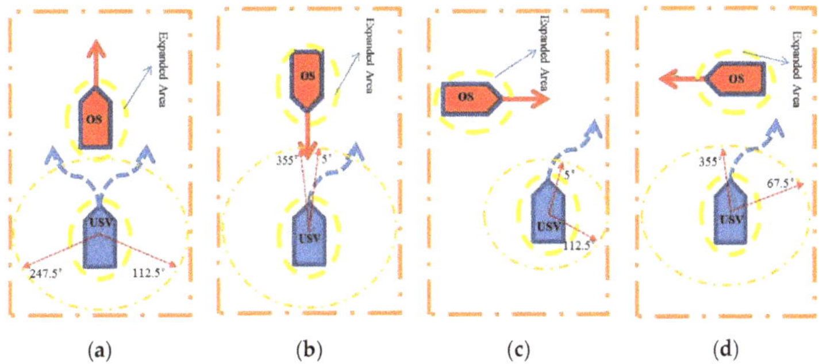

Figure 5. Four representative rules of COLREGs: (**a**) overtaking situation; (**b**) head-on situation; (**c**) port side crossing situation; (**d**) starboard crossing situation.

The blue pentagon stands for USV, the red pentagon stands for obstacle ship, and the yellow oval dotted line stands for the shape range of the ship. In the overtaking situation, both USV and obstacle ship move from bottom to top. When the USV is behind the obstacle ship and on the same route, USV can overtake the obstacle ship from the port side or starboard. In the head-on situation, the obstacle ship moves from top to bottom, while the USV sails from bottom to top and meets the front of the obstacle ship, and USV can avoid the obstacle ship through starboard. In the port side crossing situation, the obstacle ship moves from left to right, while the USV sails from bottom to top and meets the obstacle ship. At this time, the obstacle ship has the obligation to avoid a collision. However, if the red dynamic obstacle ship (OS) fails to take relevant collision avoidance actions, USV should adjust the starboard in time to avoid collision with it. When the red dynamic obstacle is far away, the USV continues to move to the target point. In the starboard crossing situation, when the obstacle ship moves from right to left and the USV moves from bottom to top and meets the obstacle ship, the obstacle ship has no obligation to avoid a collision at this time. The USV needs to adjust the starboard and quickly cross the obstacle ship to avoid collision with it.

4.2. Complexity Analysis

An algorithm's time complexity might be determined by the magnitude of the input problem (d), the population size (n), the algorithm's iterations (K) and the cost function evaluation (c). In this paper, the total time complexity of the IWSO-DWA algorithm can be expressed as:

$$
\begin{aligned}
O(\text{IWSO-DWA}) \ &= O(optimal\ path\ problem) + O(initialization) \\
&+ O(cost\ function\ evaluation) + O(Solution\ update) \\
&= O(1 + d \cdot n + n \cdot c \cdot K + d \cdot n \cdot K) \\
&\cong O(n \cdot c \cdot K + d \cdot n \cdot K)
\end{aligned}
\tag{26}
$$

5. Experimental Results and Analysis

5.1. Environment Modeling

To replicate the intricate navigational marine conditions for USV simulation purposes, two map environment models of USV are established, both of which are 500 m × 500 m. The light blue arrow represents the direction of ocean currents, the black block represents the islands, the heavy blue triangle represents the starting point of USV with the coordinate (10,10), and the red star represents the target point of USV with the coordinate (490,490), as shown in Figure 6.

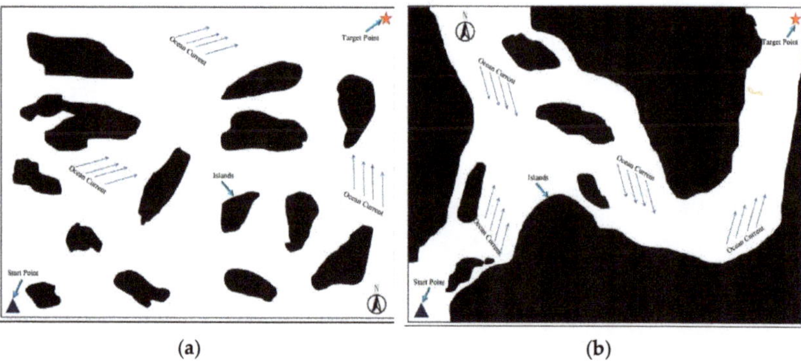

(a)   (b)

Figure 6. Environmental models for USV: (**a**) environmental model 1 (ENV.1); (**b**) environmental model 2 (ENV.2).

Additionally, the experiment was simulated on a laptop with an Intel(R) Core(TM) i7-5500 processor clocked at 2.40 GHz, 8 GB memory and Windows 7 64-bit operating system with MATLAB R2017b software.

5.2. Static Path Planning Simulation Experiment

There are two sets of static path planning simulation experiments to validate the proposed IWSO-DWA's advantages in the USV's path planning problems. The static path planning simulation experiment of USV is carried out by using the proposed IWSO-DWA, IWSO, WSO and five other common meta-heuristic algorithms (BOA, GWO, MRFO, WOA and SSA) in the same map. The parameters of the DWA part of the proposed IWSO-DWA are set as follows: The maximal linear velocity v_{max} is 5 m/s and the maximal angular velocity ω_{max} is 60 rad/s. The minimal linear velocity v_{min} is 1 m/s and the minimal angular velocity ω_{min} is 10 rad/s. The maximal linear acceleration v'_e is 0.7 m/s^2 and the maximal angular acceleration ω'_e is 75 rad/s^2. The maximal linear deceleration v'_f is 0.8 m/s^2 and the maximal angular deceleration ω'_f is 80 rad/s^2. The weight α, β and γ of the evaluation function are set to 0.3, 0.06 and 0.4, respectively. The population size of white sharks and five other meta-heuristic algorithms are all set to 50, and the maximal iteration is set to 300. The mentioned algorithms (BOA, GWO, MRFO, WOA, SSA, WSO, IWSO and IWSO-DWA) are used for the static path planning of USV in the ENV.1, and the results obtained from the simulation experiment are displayed in Figure 7.

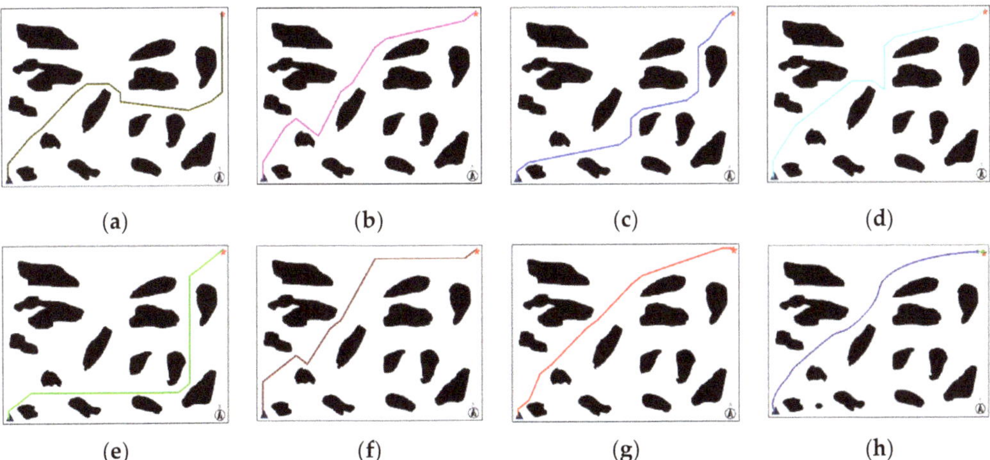

Figure 7. Static path planned by the algorithms (BOA, GWO, MRFO, WOA, SSA, WSO, IWSO, IWSO-DWA) in ENV.1: (**a**) planned by BOA; (**b**) planned by GWO; (**c**) planned by MRFO; (**d**) planned by WOA; (**e**) planned by SSA; (**f**) planned by WSO; (**g**) planned by IWSO; (**h**) planned by IWSO-DWA. The dark blue triangle represents the start point and the red star represents the target point.

When taking into account the metrics of path length, steering times, path smoothness and time cost systematically, the static path planning performance of IWSO-DWA proposed in this study surpasses that of IWSO, WSO and other meta-heuristic algorithms (BOA, GWO, MRFO, WOA, SSA). Compared with the WSO, the path length, steering times and time cost planned by the IWSO decreased by 16.12%, 28.57% and 76.97%, respectively. And the path smoothness planned by the IWSO is improved by 30.22%.

Since the proposed IWSO-DWA algorithm is based on the IWSO algorithm to increase its dynamic characteristics, the proposed IWSO-DWA algorithm and the IWSO algorithm have equivalent effects on convergence performance when solely focusing on their static characteristics. Thus, when analyzing the algorithm's convergence, it suffices to only evaluate the IWSO. The convergence curves of the mentioned algorithms (BOA, GWO, MRFO, WOA, SSA, WSO and IWSO) in ENV.1 are displayed in Figure 8.

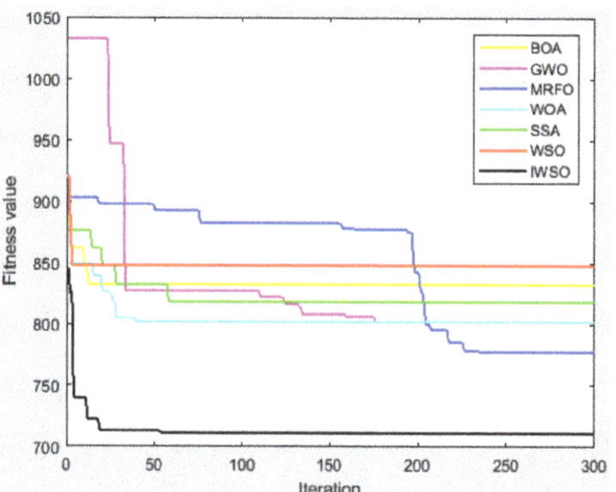

Figure 8. Convergence curve of mentioned algorithms (BOA, GWO, MRFO, WOA, SSA, WSO, IWSO) in ENV.1.

In the convergence curves, the horizontal axis label represents the iteration of the algorithms, and the vertical axis represents the fitness value of the algorithms. When compared with the WSO and five other meta-heuristic algorithms, the proposed IWSO algorithm has demonstrated the fastest convergence speed and highest accuracy of final convergence accuracy.

Based on its demonstrated performance, the second group of static path planning simulation experiments is executed to further validate the advancements of the proposed IWSO-DWA. The mentioned algorithms (BOA, GWO, MRFO, WOA, SSA, WSO, IWSO and IWSO-DWA) are utilized in the static path planning simulation experiment in ENV.2, and the results are illustrated in Figure 9.

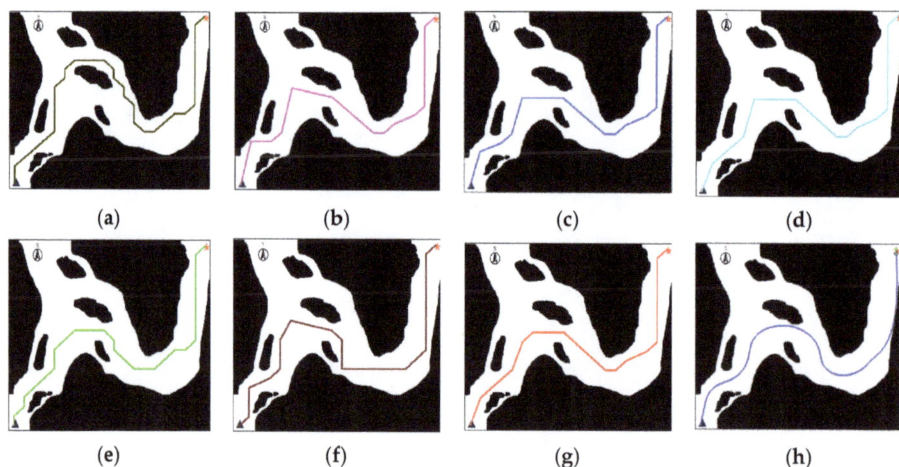

Figure 9. Static path planned by the algorithms (BOA, GWO, MRFO, WOA, SSA, WSO, IWSO, IWSO-DWA) in ENV.2: (**a**) planned by BOA; (**b**) planned by GWO; (**c**) planned by MRFO; (**d**) planned by WOA; (**e**) planned by SSA; (**f**) planned by WSO; (**g**) planned by IWSO; (**h**) planned by IWSO-DWA. The dark blue triangle represents the start point and the red star represents the target point.

Similarly, when synthetically considering measurement criteria such as path length, steering times, path smoothness, and time cost, the proposed IWSO-DWA algorithm exhibited superior performance in the static path planning simulation experiments compared to the IWSO, the WSO and the five other meta-heuristic algorithms (BOA, GWO, MRFO, WOA and SSA). Compared with the WSO, the path length, steering times and time cost planned by the IWSO are decreased by 11.2%, 9% and 81.19%, respectively. Meanwhile, the path smoothness planned by the IWSO is improved by 9.49%. The convergence curves of the mentioned algorithms (BOA, GWO, MRFO, WOA, SSA, WSO, IWSO) in ENV.2 are shown in Figure 10.

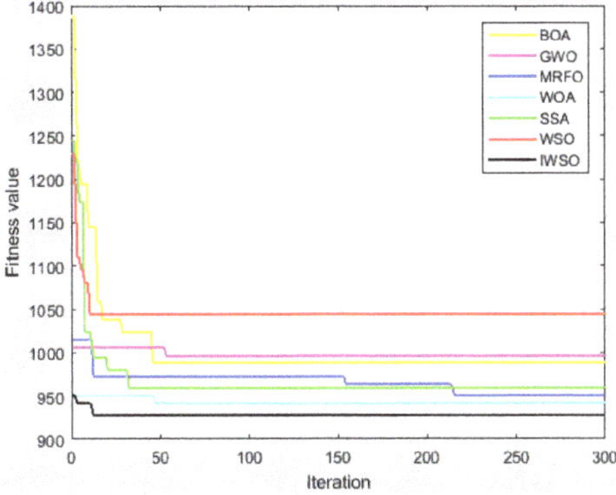

Figure 10. Convergence curve of mentioned algorithms (BOA, GWO, MRFO, WOA, SSA, WSO, IWSO) in ENV.2.

In the convergence curves, the horizontal axis label represents the iteration of the algorithms and the vertical axis represents the fitness value of the algorithms. When compared with the WSO and five other meta-heuristic algorithms, the proposed IWSO algorithm reaches stability in about 15 iterations, which excelled in both convergence speed and accuracy.

After the completion of two static path planning simulation experiment sets, it is necessary to summarize the performance of the BOA, GWO, MRFO, WOA, SSA, WSO and IWSO algorithms in numerical format. Supposed the planned path of the algorithms can be represented by a group of points set L, and $L = \{L_1, L_2, \ldots, L_\lambda\}$. λ is the number of path points in the set L. Then, the continuous steering angle between the path point L_i and the subsequent path point L_{i+1} is denoted by θ_i. To better assess the smoothness of the path planned by the mentioned algorithms, a path smoothness cost metric denominated *mot* has been established, which is defined as follows:

$$mot = \sum_{i=1}^{\lambda-1} \frac{\varepsilon \cdot \pi \cdot \theta_i}{180 \cdot (1 + \theta_{i+1})^{\frac{3}{2}}} \tag{27}$$

where $i = 1, 2, \ldots, \lambda - 1$. ε is the number of turns of the L. θ_{i+1} is the next rotation angle of the continuous rotation angle θ_i. The smaller the value of *mot*, the smoother the path.

In addition, to better evaluate the optimal path planning performance of the proposed algorithm, some metrics such as the steering cost, planning time cost and shortest path length cost are considered as the measure criteria of the mentioned algorithms. The steering cost indicates the total number of turns of the path planned by the algorithms. The planning time cost refers to the time it takes for an algorithm to plan its path in a static obstacle environment. The shortest path length cost means that the algorithm plans the shortest safe and collision-free path from the starting point to the target point, which can be defined as follows:

$$C_L = C_{safe} \cdot \sum_{i=2}^{\lambda} \sqrt{(x_{L_i} - x_{L_{i-1}})^2 + (y_{L_i} - y_{L_{i-1}})^2} \tag{28}$$

where C_L represents the shortest path length cost, which is the sum of the Euclidean distances between the points L_{i-1} and L_i in the set L of path points planned by the algorithms. C_{safe} represents the safety path cost. In this paper, all the paths provided by the algorithms must be safe and collision-free, so here, $C_{safe} = 1$.

In summary, the simulation experiments of the static path planning demonstrate that the proposed IWSO-DWA can effectively plan an optimal path globally that is both secure and smooth in the established environmental models, irrespective of any changes to the distribution of obstacles. As the proposed IWSO-DWA algorithm enhances its dynamic qualities based on the IWSO algorithm, it can be deemed equivalent to the IWSO algorithm when solely considering the static characteristics of path planning. Thus, when summarizing the simulation comparison experiments of the static path planning in digital form, it only needs to compare the performance of the proposed IWSO with WSO and five other algorithms (BOA, GWO, MRFO, WOA and SSA). The algorithms' performance in the simulation comparison experiments of the static path planning is summarized in Table 2.

Table 2. Comparison performance of the mentioned algorithms (BOA, GWO, MRFO, WOA, SSA, WSO, IWSO).

ENV. Model	Algorithm	Shortest Path Length Cost (m)	Steering Cost	Smoothness Cost (*mot*)	Time Cost (s)
	BOA	914.530	8	0.591	2.056
	GWO	802.370	9	0.456	6.564
ENV.1	MRFO	777.267	11	0.794	1.920
	WOA	801.650	8	0.455	7.037
	SSA	862.133	5	0.628	3.472
	WSO	847.487	7	0.309	5.416
	IWSO	710.873	5	0.215	1.247
	BOA	1144.975	19	3.097	7.846
	GWO	996.773	11	0.947	2.228
ENV.2	MRFO	950.803	11	0.930	2.256
	WOA	941.590	11	0.977	3.355
	SSA	959.620	15	2.220	6.317
	WSO	1044.975	11	1.024	5.490
	IWSO	927.925	10	0.926	1.033

5.3. Dynamic Avoidance Simulation Experiment

Two sets of dynamic collision avoidance simulation experiments were conducted to validate whether the proposed IWSO-DWA conforms to COLREGs rules and effectively avoids collisions in dynamic scenarios. Four situations are established in environmental model 1 and environmental model 2 of the COLREGs respectively: overtaking situation, head-on situation, port side crossing situation and starboard crossing situation. In the figures, the blue boat indicates the USV and the red boat indicates the obstacle ship. In ENV.1, the overtaking situation between the USV and the red obstacle ship is shown in Figure 11.

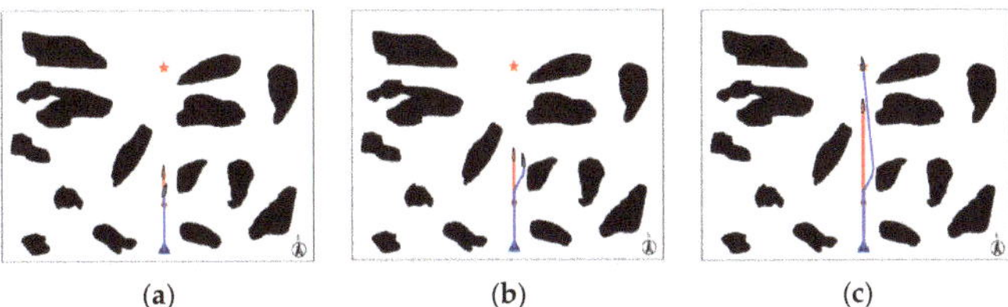

(a)   (b)   (c)

Figure 11. Overtaking situation in ENV.1: (**a**) the preparatory state; (**b**) the meeting state; (**c**) the completion state. The dark blue triangle and red star represent the start point and target point of the USV (in blue color), respectively, and the purple circle represents the start point of the obstacle ship (in red color).

The starting coordinate of the red dynamic obstacle ship is (260,120), and it moves in a straight line from bottom to top at a velocity of 2 m/s. The coordinate of the starting point of USV is (260,30), the target point of USV is (260,380), and it moves in a straight line from bottom to top at a velocity of 4 m/s. In the overtaking situation, when encountering the red dynamic obstacle ship, the USV initiates collision avoidance by veering toward the upper right direction at an angle of approximately 65 degrees. It expertly navigates past the red dynamic obstacle ship from its starboard side and continues towards the target point, following a previous path, thereby successfully avoiding a rear-end collision, and the dynamic collision avoidance behavior of the USV conforms to the COLREGs. The x, y position and yaw angle of the USV for its dynamic collision avoidance behavior are depicted in Figure 12.

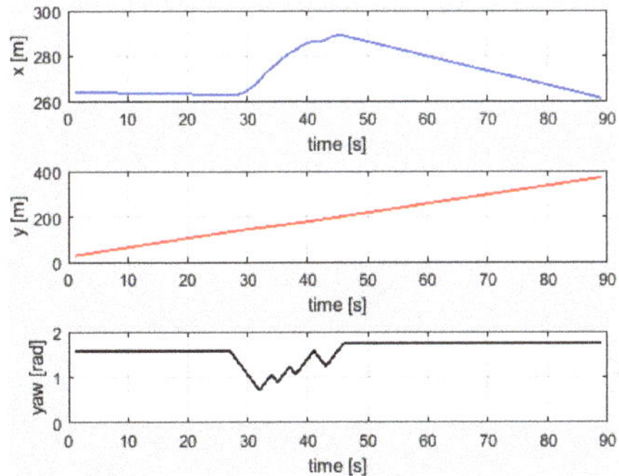

Figure 12. The motion states of the USV for overtaking situation in ENV.1.

After completing the overtaking situation of USV in ENV.1, the head-on situation experiment of USV is carried out, and the results are displayed in Figure 13.

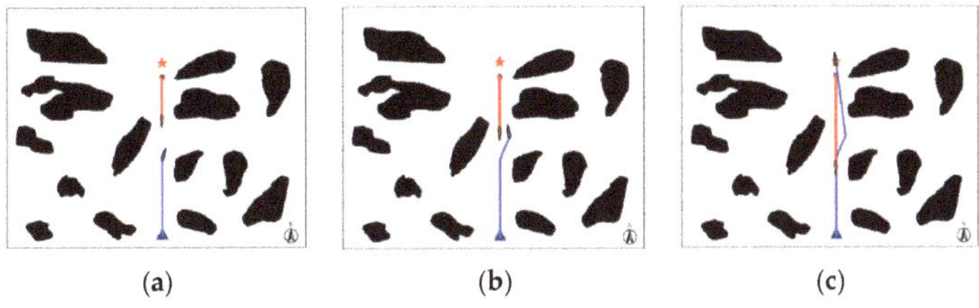

(a)   (b)   (c)

Figure 13. The head-on situation in ENV.1: (**a**) the preparatory state; (**b**) the meeting state; (**c**) the completion state. The dark blue triangle and red star represent the start point and target point of the USV (in blue color), respectively, and the purple circle represents the start point of the obstacle ship (in red color).

The starting point coordinate of the red dynamic obstacle ship is (260,350), and it moves in a straight line from top to bottom with a moving speed of 3 m/s. The coordinate of the starting point of USV is (260,30), the target point of USV is (260,380), and it moves in a straight line from bottom to top with a moving speed of 4 m/s. In the head-on situation, when encountering the red dynamic obstacle ship, the USV initiates adjusting starboard of the ship in a direction approximately 70 degrees towards the upper right direction, then skillfully navigates past the red dynamic obstacle ship's upper region from the USV's starboard side. After successfully avoiding the head-on collision with the red dynamic obstacle ship, the USV then progresses toward the target point, and the dynamic collision avoidance behavior of the USV conforms to the COLREGs. The x, y position and yaw angle of the USV for its dynamic collision avoidance behavior are depicted in Figure 14.

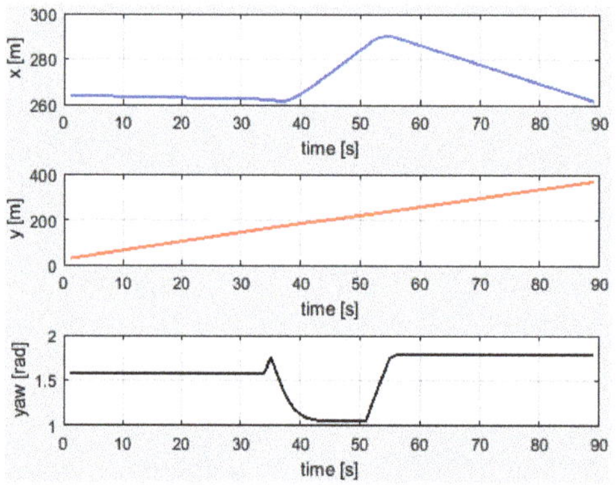

Figure 14. The motion states of the USV for head-on situation in ENV.1.

After completing the head-on situation of USV in ENV.1, the port side crossing situation experiment of USV is carried out, and the results are displayed in Figure 15.

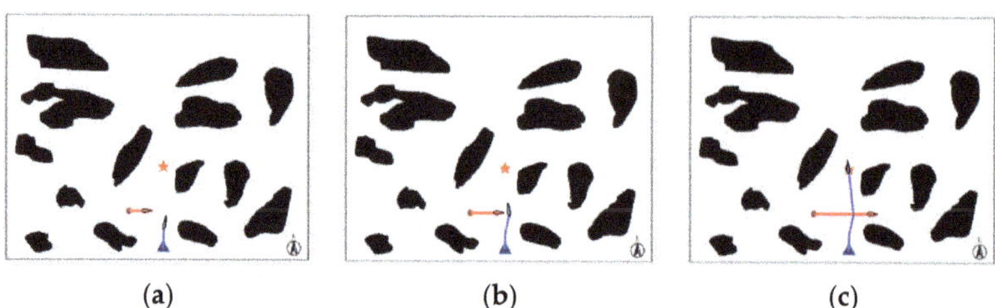

(a)   (b)   (c)

Figure 15. Port side crossing situation in ENV.1: (a) the preparatory state; (b) the meeting state; (c) the completion state. The dark blue triangle and red star represent the start point and target point of the USV (in blue color), respectively, and the purple circle represents the start point of the obstacle ship (in red color).

The starting point coordinate of the red dynamic obstacle ship is (180,110), and it moves in a straight line from left to right with a moving speed of 4 m/s. The coordinate of the starting point of USV is (260,30), the target point of USV is (260,180), and it moves in a straight line from bottom to top with a moving speed of 4 m/s. In the port side crossing situation, since the red dynamic obstacle ship is a giving way vessel, it should stop to let the USV pass when it encounters the USV. However, if the red dynamic obstacle ship did not stop, the USV must take evasive action to prevent a collision. When the red obstacle ship enters the evasive range, the USV actively adjusts the starboard and keeps a safe distance from the red obstacle ship. After collision avoidance, the USV continued to move along the original path to the target point, thus finishing the port side crossing situation, and the dynamic collision avoidance behavior of the USV conforms to the COLREGs. The x, y position and yaw angle of the USV for its dynamic collision avoidance behavior are depicted in Figure 16.

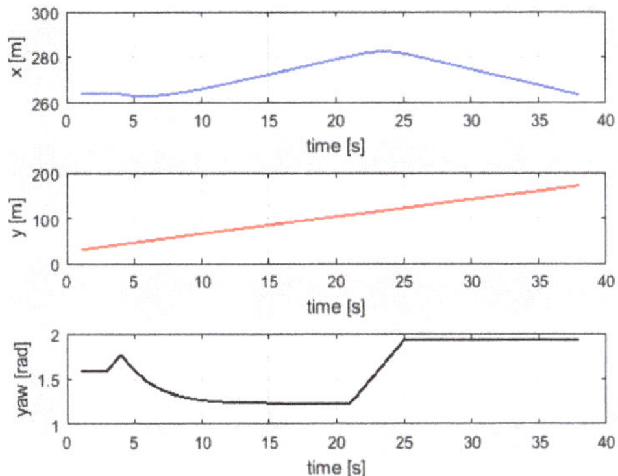

Figure 16. The motion states of the USV for port side crossing situation in ENV.1.

After completing the port side crossing situation of USV in ENV.1, the starboard crossing situation experiment of USV is carried out, and the results are displayed in Figure 17.

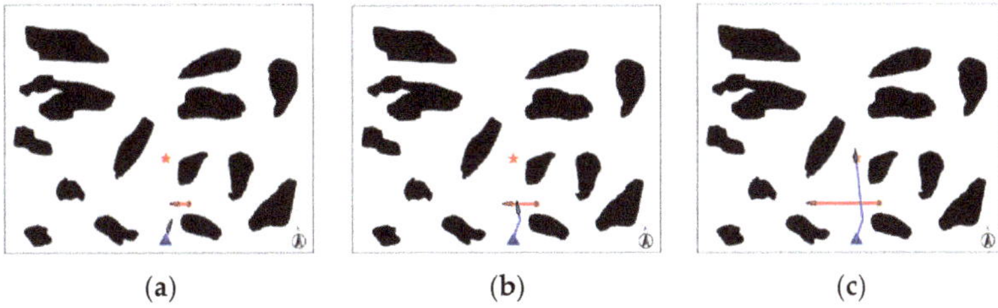

(a)   (b)   (c)

Figure 17. Starboard crossing situation in ENV.1: (**a**) the preparatory state; (**b**) the meeting state; (**c**) the completion state. The dark blue triangle and red star represent the start point and target point of the USV (in blue color), respectively, and the purple circle represents the start point of the obstacle ship (in red color).

The starting coordinate of the red dynamic obstacle ship is (330,110), and it moves in a straight line from right to left with a moving speed of 4 m/s. The coordinate of the starting point of USV is (260,30), the target point of USV is (260,180), and it moves in a straight line from bottom to top with a moving speed of 4 m/s. In the starboard crossing situation, since the USV is a giving way vessel, it should stop to let the red dynamic obstacle ship pass when it encounters the red dynamic obstacle ship. When the red obstacle ship enters the evasive range, the USV adjusts its starboard side at approximately 47 degrees to avoid the red dynamic obstacle vessel and stops to wait for it to move away from the evasive range. Once the red obstacle vessel is out of the evasive range, the USV continues to move to the upper left to the target point, thus finishing the starboard crossing situation of the USV, and the dynamic collision avoidance behavior of USV conforms to the COLREGs. The x, y position and yaw angle of the USV for its dynamic collision avoidance behavior are depicted in Figure 18.

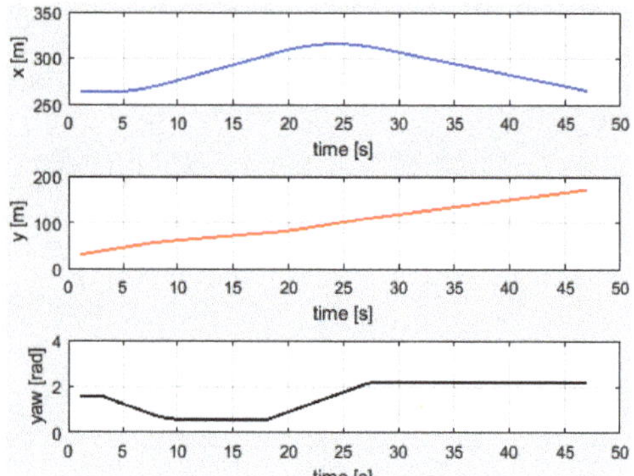

Figure 18. The motion states of the USV for starboard crossing situation in ENV.1.

Similarly, in ENV.2, four dynamic avoidance simulation experiments were conducted to validate the effectiveness of the proposed IWSO-DWA in line with the COLREGs. The overtaking situation between the USV and the red dynamic obstacle ship is shown in Figure 19.

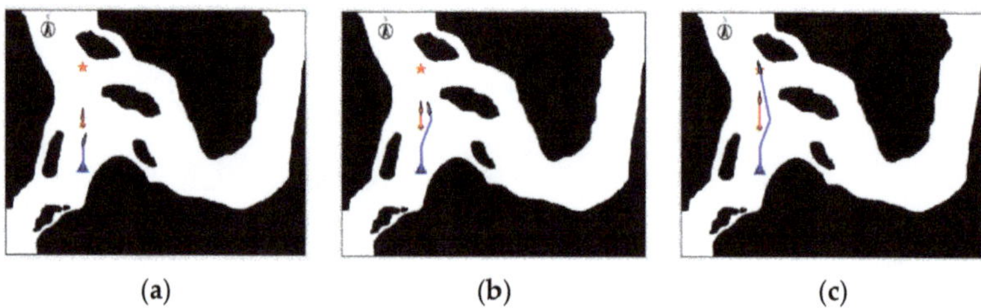

(a)   (b)   (c)

Figure 19. Overtaking situation in ENV.2: (**a**) the preparatory state; (**b**) the meeting state; (**c**) the completion state. The dark blue triangle and red star represent the start point and target point of the USV (in blue color), respectively, and the purple circle represents the start point of the obstacle ship (in red color).

The starting point coordinate of the red dynamic target obstacle is (130,260), and it moves in a straight line from bottom to top with a moving speed of 2.5 m/s. The coordinate of the starting point of USV is (130,180), the target point of USV is (130,380), and it moves vertically upwards at a velocity of 4 m/s. In the overtaking situation, when the USV enters the evasive range, it avoids collision by steering approximately 68 degrees to the upper right and proactively sails across the upper section of the red dynamic obstacle ship from the USV's port side. Once it safely overtakes the red dynamic obstacle ship, the USV proceeds to advance toward the target point in the upper left direction, thus finishing the overtaking situation between the USV and the red dynamic obstacle ship, and the dynamic collision avoidance behavior of the USV conforms to the COLREGs. The x, y position and yaw angle of the USV for its dynamic collision avoidance behavior are depicted in Figure 20.

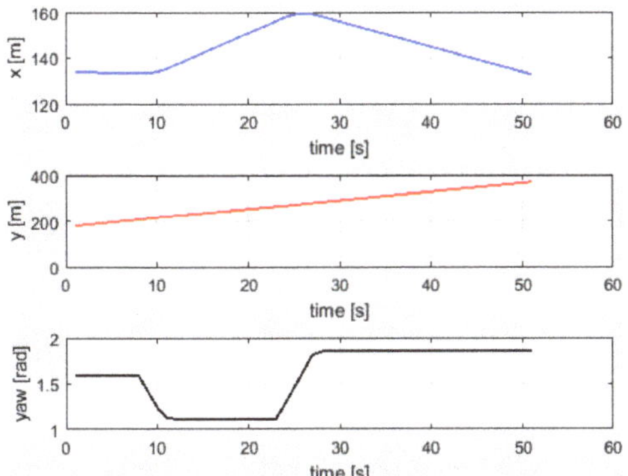

Figure 20. The motion states of the USV for overtaking situation in ENV.2.

After completing the overtaking situation of USV in ENV.2, the head-on situation experiment of USV is carried out, and the results are displayed in Figure 21.

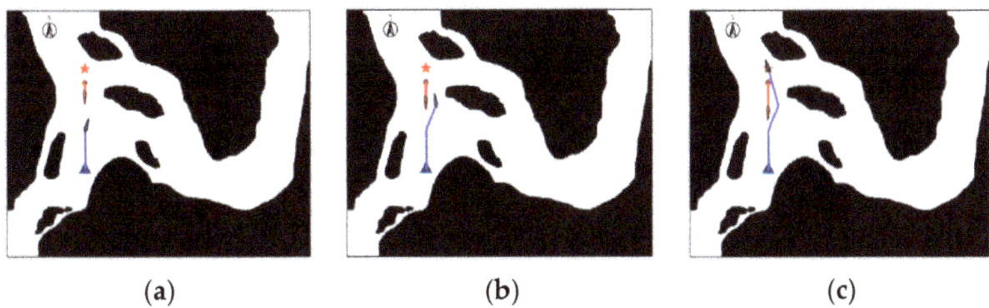

(a) (b) (c)

Figure 21. The head-on situation in ENV.2: (**a**) the preparatory state; (**b**) the meeting state; (**c**) the completion state. The dark blue triangle and red star represent the start point and target point of the USV (in blue color), respectively, and the purple circle represents the start point of the obstacle ship (in red color).

The starting point coordinate of the red dynamic obstacle ship is (130,350), and it travels vertically downwards at a velocity of 2 m/s. The coordinate of the starting point of USV is (130,180), the target point of USV is (130,380), and it moves vertically upwards at a speed of 4 m/s. In the head-on situation, when the USV encounters the red dynamic obstacle ship, the USV avoids the collision by turning approximately 63 degrees to the starboard and proceeding to cross the upper section of the red dynamic obstacle ship. Once it is far away from the red dynamic obstacle ship, the USV moves to the target point in the direction of around 18 degrees to the upper left, thus finishing the head-on situation between the USV and the red dynamic obstacle ship, and the dynamic collision avoidance behavior of the USV conforms to the COLREGs. The x, y position and yaw angle of the USV for its dynamic collision avoidance behavior are depicted in Figure 22.

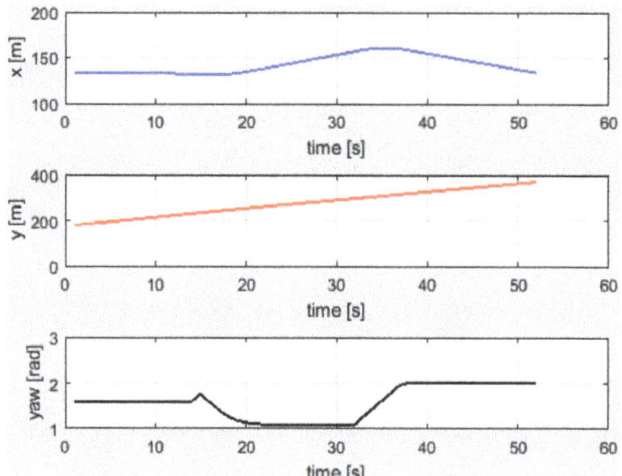

Figure 22. The motion states of the USV for head-on situation in ENV.2.

After completing the head-on situation of USV in ENV.2, the port side situation experiment of USV is carried out, and results are displayed in Figure 23.

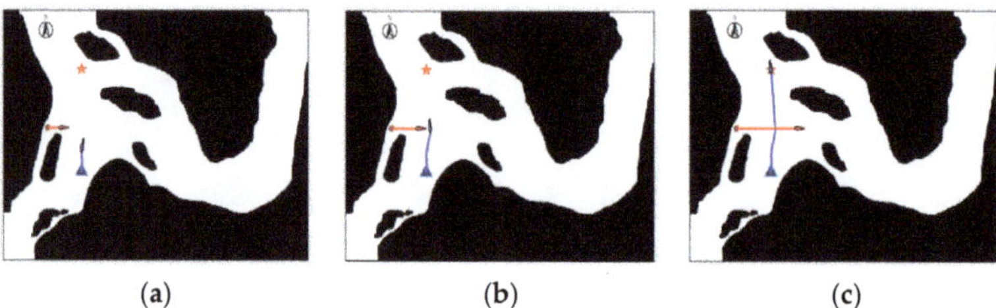

(a) (b) (c)

Figure 23. Port side situation in ENV.2: (**a**) the preparatory state; (**b**) the meeting state; (**c**) the completion state. The dark blue triangle and red star represent the start point and target point of the USV (in blue color), respectively, and the purple circle represents the start point of the obstacle ship (in red color).

The starting point coordinate of the red dynamic obstacle ship is (75,260), it moves in a straight line from left to right with a moving speed of 3 m/s. The coordinate of the starting point of USV is (130,180), the target point of USV is (130,380), and it moves in a straight line from bottom to top with a moving speed of 4 m/s. In the port side situation, since the red obstacle ship is a giving way vessel when encountering the USV, it should stop to let the USV pass. However, if the red dynamic obstacle ship did not stop, the USV should take evasive action to prevent a collision. When the red obstacle ship enters the evasive range, the USV adjusts its starboard at roughly 80 degrees to avoid the red obstacle ship. Once the USV moves away from the red obstacle ship, it continues to move to the upper left towards the target point, thus finishing the port side situation between the USV and the red dynamic obstacle ship, and the dynamic collision avoidance behavior of the USV conforms to the COLREGs. The x, y position and yaw angle of the USV for its dynamic collision avoidance behavior are depicted in Figure 24.

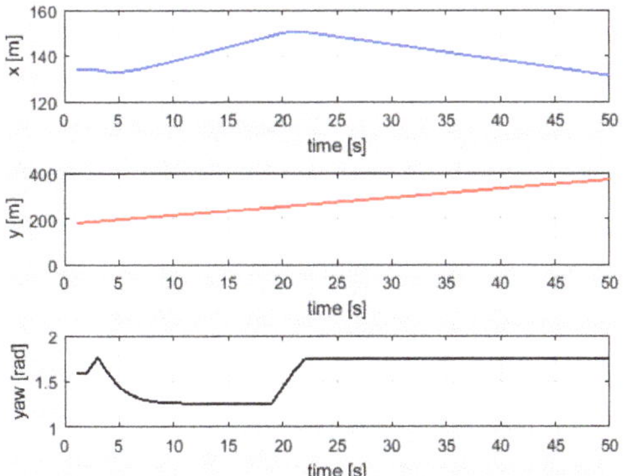

Figure 24. The motion states of the USV for portside crossing situation in ENV.2.

After completing the port side situation of USV in ENV.2, the starboard situation experiment of USV is carried out, and the results are displayed in Figure 25.

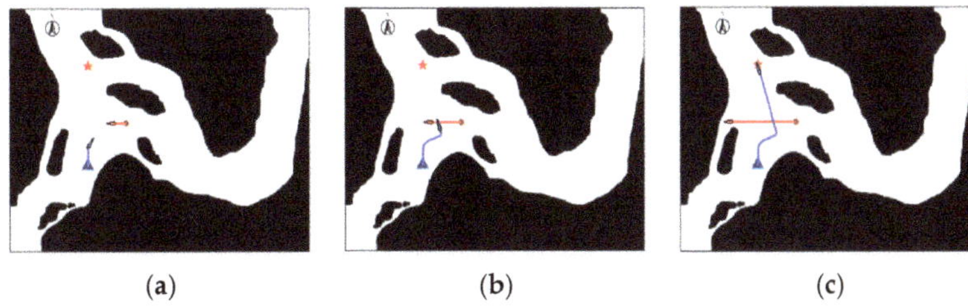

(a)   (b)   (c)

Figure 25. Starboard situation in ENV.2: (**a**) the preparatory state; (**b**) the meeting state; (c) the completion state. The dark blue triangle and red star represent the start point and target point of the USV (in blue color), respectively, and the purple circle represents the start point of the obstacle ship (in red color).

The starting point coordinate of the red dynamic obstacle ship is (200,260), and it moves in a straight line from right to left with a moving speed of 3 m/s. The coordinate of the starting point of USV is (130,180), the target point of USV is (130,380), and it moves in a straight line from bottom to top at a velocity of 4 m/s. In the starboard situation, since the USV is a giving way vessel when encountering the red obstacle ship, it should stop to let the red dynamic obstacle ship pass. When the red obstacle ship enters the evasive range, the USV steers away from collision by adjusting its starboard and maintaining a safe distance from the red obstacle ship. Once the red obstacle ship is far away, the USV crosses the upper section of the red dynamic obstacle ship and moves to the target point in the upper left direction, thus finishing the starboard situation between the USV and the red dynamic obstacle ship, and the dynamic collision avoidance behavior of the USV conforms to the COLREGs. The x, y position and yaw angle of the USV for its dynamic collision avoidance behavior are depicted in Figure 26.

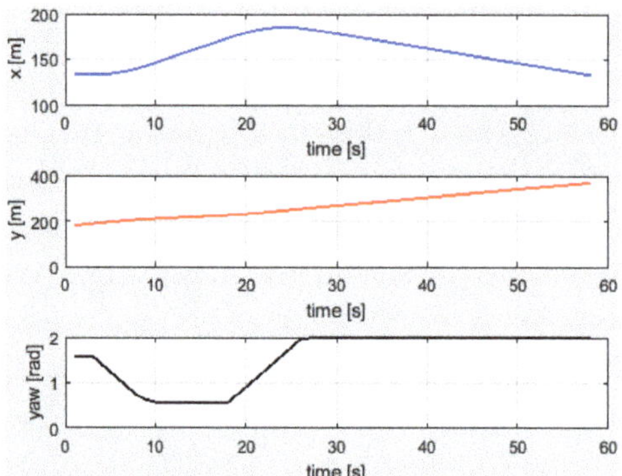

Figure 26. The motion states of the USV for starboard crossing situation in ENV.2.

6. Conclusions and Future Work

This research proposes a new IWSO-DWA algorithm to address the optimal path planning issue for USV. First of all, aiming at the disadvantages of uneven distribution and insufficient diversity of the white shark population, a circle chaotic mapping algorithm is employed to improve the initial solution's quality. Then, the adaptive weight factor technique is used to update the best white shark's position, ensuring a balance between global exploration and local exploitation. Furthermore, the simplex method is used to update the other white sharks' position near the best white shark, enhancing the algorithm's ability to escape the local optimum solution. Finally, a novel global dynamic optimal path planning method called the IWSO-DWA algorithm is developed by combining the improved WSO and the enhanced DWA. The performance of the IWSO-DWA algorithm is tested through two sets of static path planning simulation comparison experiments and two sets of dynamic avoidance simulation experiments. The study found that the IWSO-DWA algorithm outperformed traditional WSO algorithms and five other heuristic algorithms (BOA, GWO, MRFO, WOA and SSA) in the simulation experiments. Thus, the proposed IWSO-DWA algorithm not only addresses the issues encountered in the traditional WSO algorithm, but also guides USV to plan a global optimal path in challenging marine environments and possesses path smoothing capability and dynamic collision avoidance ability, and its collision avoidance behavior conforms to the COLREGs. However, the proposed IWSO-DWA has only been evaluated through simulations, and future research is required to focus on assessing its effectiveness in practical engineering optimization problems in real USV.

Author Contributions: Conceptualization, L.L.; methodology, J.L.; software, J.L.; validation, L.L. and J.L.; writing—original draft preparation, J.L.; writing—review and editing, L.L.; visualization, J.L.; project administration L.L. and J.L.; funding acquisition, L.L. All authors have read and agreed to the published version of the manuscript.

Funding: This research was funded by the Natural Science Foundation of Fujian Province, grant number 2022H6005.

Institutional Review Board Statement: Not applicable.

Informed Consent Statement: Not applicable.

Data Availability Statement: Not applicable.

Acknowledgments: The authors would like to thank the editors and anonymous reviewers for their constructive comments.

Conflicts of Interest: The authors declare no conflict of interest.

References

1. Bai, X.; Li, B.; Xu, X.; Xiao, Y. A Review of Current Research and Advances in Unmanned Surface Vehicles. *J. Mar. Sci. Appl.* **2022**, *21*, 47–58. [CrossRef]
2. Zhou, C.; Gu, S.; Wen, Y.; Du, Z.; Xiao, C.; Huang, L.; Zhu, M. The review unmanned surface vehicle path planning: Based on multi-modality constraint. *Ocean Eng.* **2020**, *200*, 107043. [CrossRef]
3. Singh, Y.; Sharma, S.; Sutton, R.; Hatton, D. Towards use of Dijkstra Algorithm for Optimal Navigation of an Unmanned Surface Vehicle in a Real-Time Marine Environment with results from Artificial Potential Field. *TransNav Int. J. Mar. Navig. Saf. Sea Transp.* **2018**, *12*, 125–131. [CrossRef]
4. Song, R.; Liu, Y.; Bucknall, R. Smoothed A* algorithm for practical unmanned surface vehicle path planning. *Appl. Ocean Res.* **2019**, *83*, 9–20. [CrossRef]
5. Guo, X.; Ji, M.; Zhao, Z.; Wen, D.; Zhang, W. Global path planning and multi-objective path control for unmanned surface vehicle based on modified particle swarm optimization (PSO) algorithm. *Ocean Eng.* **2020**, *216*, 107693. [CrossRef]
6. Cui, Y.; Ren, J.; Zhang, Y. Path Planning Algorithm for Unmanned Surface Vehicle Based on Optimized Ant Colony Algorithm. *IEEJ Trans. Electr. Electron. Eng.* **2022**, *17*, 1027–1037. [CrossRef]
7. Ma, Y.; Hu, M.; Yan, X. Multi-objective path planning for unmanned surface vehicle with currents effects. *ISA Trans.* **2018**, *75*, 137–156. [CrossRef]
8. Sahoo, S.P.; Das, B.; Pati, B.B.; Garcia Marquez, F.P.; Segovia Ramirez, I. Hybrid Path Planning Using a Bionic-Inspired Optimization Algorithm for Autonomous Underwater Vehicles. *J. Mar. Sci. Eng.* **2023**, *11*, 761. [CrossRef]
9. Qiyong, G.; Rong, Z.; Jialun, L.; Chen, L. An improved RRT algorithm based on prior AIS information and DP compression for ship path planning. *Ocean Eng.* **2023**, *279*, 114595. [CrossRef]
10. Ma, D.; Hao, S.; Ma, W.; Zheng, H.; Xu, X. An optimal control-based path planning method for unmanned surface vehicles in complex environments. *Ocean Eng.* **2022**, *245*, 110532. [CrossRef]
11. Han, S.; Wang, L.; Wang, Y.; He, H. A dynamically hybrid path planning for unmanned surface vehicles based on non-uniform Theta* and improved dynamic windows approach. *Ocean Eng.* **2022**, *257*, 111655. [CrossRef]
12. Wang, W.; Du, J.; Tao, Y. A dynamic collision avoidance solution scheme of unmanned surface vessels based on proactive velocity obstacle and set-based guidance. *Ocean Eng.* **2022**, *248*, 110794.
13. Hu, S.; Tian, S.; Zhao, J.; Shen, R. Path Planning of an Unmanned Surface Vessel Based on the Improved A-Star and Dynamic Window Method. *J. Mar. Sci. Eng.* **2023**, *11*, 1060. [CrossRef]
14. Zhao, L.; Bai, Y.; Paik, J.K. Global-local hierarchical path planning scheme for unmanned surface vehicles under dynamically unforeseen environments. *Ocean Eng.* **2023**, *280*, 114750. [CrossRef]
15. Li, M.; Li, B.; Qi, Z.; Li, J.; Wu, J. Optimized APF-ACO Algorithm for Ship Collision Avoidance and Path Planning. *J. Mar. Sci. Eng.* **2023**, *11*, 1177. [CrossRef]
16. Bing, H.; He, D.; Zheping, Y. A path planning approach for unmanned surface vehicles based on dynamic and fast Q-learning. *Ocean Eng.* **2023**, *270*, 113632. [CrossRef]
17. Guo, S.; Zhang, X.; Zheng, Y.; Du, Y. An Autonomous Path Planning Model for Unmanned Ships Based on Deep Reinforcement Learning. *Sensors* **2020**, *20*, 426. [CrossRef]
18. Sang, H.; You, Y.; Sun, X.; Zhou, Y.; Liu, F. The hybrid path planning algorithm based on improved A* and artificial potential field for unmanned surface vehicle formations. *Ocean Eng.* **2021**, *223*, 108709. [CrossRef]
19. Arora, S. Butterfly optimization algorithm: A novel approach for global optimization. *Soft Comput.* **2019**, *23*, 715–734. [CrossRef]
20. Mirjalili, S.; Mirjalili, S.M.; Lewis, A. Grey Wolf Optimizer. *Adv. Eng. Softw.* **2014**, *69*, 46–61. [CrossRef]
21. Zhao, W.; Zhang, Z.; Wang, L. Manta ray foraging optimization: An effective bio-inspired optimizer for engineering applications. *Eng. Appl. Artif. Intell.* **2020**, *87*, 103300. [CrossRef]
22. Mirjalili, S.; Lewis, A. The Whale Optimization Algorithm. *Adv. Eng. Softw.* **2016**, *95*, 51–67. [CrossRef]
23. Xue, J.; Shen, B. A novel swarm intelligence optimization approach: Sparrow search algorithm. *Syst. Sci. Control Eng.* **2020**, *8*, 22–34. [CrossRef]
24. Braik, M.; Hammouri, A.; Atwan, J.; Al-Betar, M.A.; Awadallah, M.A. White Shark Optimizer: A novel bio-inspired meta-heuristic algorithm for global optimization problems. *Knowl.-Based Syst.* **2022**, *243*, 108457. [CrossRef]
25. Hu, S.; Liu, H.; Feng, Y.; Cui, C.; Ma, Y.; Zhang, G.; Huang, X. Tool Wear Prediction in Glass Fiber Reinforced Polymer Small-Hole Drilling Based on an Improved Circle Chaotic Mapping Grey Wolf Algorithm for BP Neural Network. *Appl. Sci.* **2023**, *13*, 2811. [CrossRef]
26. Nelder, J.A.; Mead, R. A Simplex Method for Function Minimization. *Comput. J.* **1965**, *4*, 308–313. [CrossRef]
27. Zhou, L.; Zhou, X.; Yi, C. A Hybrid STA Based on Nelder–Mead Simplex Search and Quadratic Interpolation. *Electronics* **2023**, *12*, 994. [CrossRef]
28. Vu, M.T.; Van, M.; Bui, D.H.P.; Do, Q.T.; Huynh, T.-T.; Lee, S.-D.; Choi, H.-S. Study on Dynamic Behavior of Unmanned Surface Vehicle-Linked Unmanned Underwater Vehicle System for Underwater Exploration. *Sensors* **2020**, *20*, 1329. [CrossRef]
29. Liu, L.; Yao, J.; He, D.; Chen, J.; Guo, J. Global Dynamic Path Planning Fusion Algorithm Combining Jump-A* Algorithm and Dynamic Window Approach. *IEEE Access* **2021**, *9*, 19632–19638. [CrossRef]
30. Gu, N.; Wang, D.; Peng, Z.; Wang, J.; Han, Q.-L. Advances in Line-of-Sight Guidance for Path Following of Autonomous Marine Vehicles: An Overview. *IEEE Trans. Syst. Man Cybern. Syst.* **2023**, *53*, 12–28. [CrossRef]

31. Liu, L.; Liang, J.; Guo, K.; Ke, C.; He, D.; Chen, J. Dynamic Path Planning of Mobile Robot Based on Improved Sparrow Search Algorithm. *Biomimetics* **2023**, *8*, 182. [CrossRef] [PubMed]
32. Sun, X.; Wang, G.; Fan, Y.; Mu, D. Collision avoidance control for unmanned surface vehicle with COLREGs compliance. *Ocean Eng.* **2023**, *267*, 113263. [CrossRef]
33. Woo, J.; Kim, N. Collision avoidance for an unmanned surface vehicle using deep reinforcement learning. *Ocean Eng.* **2020**, *199*, 107001. [CrossRef]

Journal of
*Marine Science
and Engineering*

Article

An Experimental Study on Trajectory Tracking Control of Torpedo-like AUVs Using Coupled Error Dynamics

Gun Rae Cho [1,*], Hyungjoo Kang [1], Min-Gyu Kim [1], Mun-Jik Lee [1], Ji-Hong Li [1], Hosung Kim [2], Hansol Lee [2] and Gwonsoo Lee [3]

[1] Korea Institute of Robotics and Technology Convergence, Pohang 37666, Republic of Korea ; hjkang@kiro.re.kr (H.K.); zxdwa0817@kiro.re.kr (M.-G.K.); mcklee@kiro.re.kr (M.-J.L.); jhli5@kiro.re.kr (J.-H.L.)
[2] Hanwha Systems, Gumi 39376, Republic of Korea; hosung0608.kim@hanwha.com (H.K.); hansol.lee@hanwha.com (H.L.)
[3] Department of Mechatronics Engineering, Chungnam National University, Daejeon 34134, Republic of Korea; kali55@o.cnu.ac.kr
* Correspondence: sandman@kiro.re.kr; Tel.: +82-54-279-0459

Abstract: In this paper, we propose a trajectory tracking controller with experimental verification for torpedo-like autonomous underwater vehicles (AUVs) with underactuation characteristics. The proposed controller overcomes the underactuation problem by designing the desired error dynamics in a coupled form using state variables in body-fixed and world coordinates. Unlike the back-stepping control requiring high-order derivatives of state variables, the proposed controller only requires the first derivatives of the states, which can alleviate noise magnification issues due to differentiation. We adopt time delay estimation to estimate the dynamics indirectly using control inputs and vehicle outputs, making the proposed controller relatively easy to apply without requiring the all of the vehicle dynamics. We also address some practical issues that commonly arise in experimental environments: handling measurement noises and actuation limits. To mitigate the effects of noise on the controller, a filtering technique using a moving window average is employed. Additionally, to account for the actuation limits, we design an anti-windup structure that takes into consideration the nonlinearity between the thrusting force and rotating speed of the thruster. We verify the tracking performance of the proposed controller through experimentation using an AUV. The experimental results show that the 3D motion control of the proposed controller exhibits an RMS error of 0.3216 m and demonstrate that the proposed controller achieves accurate tracking performance, making it suitable for survey missions that require tracking errors of less than one meter.

Keywords: autonomous underwater vehicles; robust trajectory tracking; coupled desired error dynamics; time delay estimation

Citation: Cho, G.R.; Kang, H.; Kim, M.-G.; Lee, M.-J.; Li, J.-H.; Kim, H.; Lee, H.; Lee, G. An Experimental Study on Trajectory Tracking Control of Torpedo-like AUVs Using Coupled Error Dynamics. *J. Mar. Sci. Eng.* **2023**, *11*, 1334. https://doi.org/10.3390/jmse11071334

Academic Editors: Bing Li and Bowen Xing

Received: 3 May 2023
Revised: 26 June 2023
Accepted: 28 June 2023
Published: 30 June 2023

1. Introduction

Designing a trajectory tracking controller for torpedo-like autonomous underwater vehicles (AUVs) is challenging due to their underactuation characteristics. These vehicles have only three control inputs—surge force, pitch, and yaw moment—to control their 3D motion in space. The lack of control inputs leads to dissatisfaction of the matching condition, where certain uncertain terms in the state equation cannot be directly compensated for by the control inputs [1,2]. Additionally, the vehicles have nonlinear dynamics involving both rigid body dynamics and hydrodynamics [2]. In the development of AUVs and their control systems, it is crucial to verify their performance in experimental environments. During such verification, numerous issues can affect the system's performance, including sensor measurement noise, modeling errors in system dynamics, disturbances, and imperfections in control systems, such as jitter in the sampling time. As a result, developing a trajectory tracking controller for AUVs is challenging in two aspects: the design of the control algorithm and the experimental verification process.

Regarding the controller design manner, there have been several research works to propose trajectory tracking control scheme for AUVs. To overcome the matching condition issues of the vehicles, several research works have proposed control schemes based on back-stepping control (BC). BC using the vehicle dynamic model was proposed in [3–6]. BC with time delay estimation (BCTDE), an indirect estimator of the vehicle dynamics, has also been studied [2,7,8]. BC provides an excellent and systematic method to handle matching condition issues. However, it requires high-order derivatives of the state variables, which may result in instability in experimental environments due to the magnification of noise effects. There have been several studies on controlling the vehicles using other schemes. Hierarchical design of the controllers has been researched to address the underactuation characteristics [9,10]. Sliding mode control has been applied to gain robustness against model errors and disturbances [11–14]. Adaptive schemes have been employed to resolve model uncertainty [15,16]. Neural networks have been used for robust path following [17,18].

Regarding the experimental verification manner, however, it is hard to find previous research works proposing trajectory tracking algorithms with experimental results. For example, the aforementioned previous research works primarily demonstrate control performance with simulation results and lack experimental verification. The reasons for the absence of experiments in previous works are not explicitly mentioned, but this could be attributed to the need for further investigation to address practical issues such as sensor noises or modeling errors in AUV dynamics, including disturbances. For example, the authors have attempted to verify the performance of the BCTDE [2,7,8] and found it difficult to determine stable gains for the BCTDE. Figure 1 illustrates the experimental results for depth control using the BCTDE, which indicate unstable responses. This instability arises because the BCTDE requires high-order differentiation of the state variables to handle unmatched dynamics and disturbances, thereby amplifying the effects of noise in the state measurements. As another reason for the lack of experimentation, the difficulty of obtaining a suitable experimental platform can be considered, as AUVs are costly plaforms. In contrast to the trajectory tracking problem, however, there have been several research works that proposed via point tracking control of underactuated AUVs with experimental verification [19–21]. These research works utilize traditional approaches employing PID-type controllers for the forward velocity, pitch angle (or depth), and heading angle, combined with a desired heading angle planner such as the line of sight (LOS) [22–24]. Designing via point tracking controllers is relatively straightforward since each controller focuses only on stabilizing states with dynamics matched to the control inputs. However, they mainly focus on waypoint tracking and are unable to handle time-varying trajectory tracking problems. There have also been research works suggesting trajectory tracking controllers for underwater vehicles with full degree-of-freedom (DOF) actuation [25–29]. In such cases, the vehicle can generate the control actuations for every controlled state, eliminating issues arising from a lack of satisfaction of the matching condition.

In this paper, we propose a robust trajectory tracking controller for the 3D motion of underwater vehicles, along with experimental verification of its control performance. In terms of controller design, the proposed controller incorporates an appropriate design of the desired error dynamics and time delay estimation (TDE). To address the underactuation issues of the vehicle, the desired error dynamics is formulated in a coupled form between the state variables in body-fixed and world coordinates. By utilizing the TDE [2,30], the controller effectively compensates for the nonlinear dynamics and disturbances of the vehicle while maintaining a simple structure. The proposed controller only requires the state variables and their first derivatives, mitigating the issues of noise amplification due to differentiation. An initial version of the proposed controller was presented in [31], and in this paper, we extend the controller for motion control in a 3D space. In terms of experimental implementation, practical issues related to measurement noise and actuation limitations are addressed. A moving average filter is employed to mitigate the effects of sensor noise on control performance, and actuation limitations combined with nonlinear dynamics are also considered. We verify the tracking performance of the proposed controller through experiments conducted on an AUV.

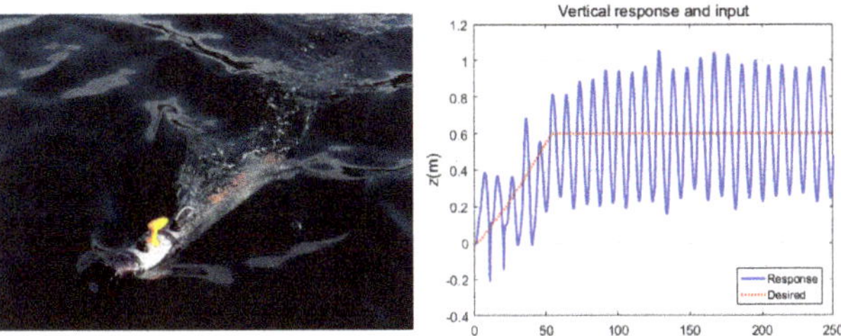

Figure 1. Experimental results for depth control using the BCTDE. The results indicate an almost unstable response due to the high−order differentiation of states required by the controller, which amplifies the noise effect in the state measurements.

2. Controller Design for the AUV

2.1. AUV Systems and Motion-Governing Equations

Figure 2 illustrates the AUV platform utilized in this research, which was developed by Hanwha Systems [32]. The AUV serves as a testbed for underwater docking tasks [33]. The linear velocities of the vehicle are measured using a Doppler velocity log (DVL), and the angular velocities are measured using an inertial measurement unit (IMU). The position of the vehicle in the world coordinate is estimated using an extended Kalman filter (EKF) that utilizes navigation sensor data from an IMU, a DVL, a depth sensor, a digital compass, and a global navigation satellite system (GNSS) [34–36]. The vehicle is equipped with a thruster for forward propulsion, as well as rudder fins and stern fins for lateral and vertical moments, respectively. The actuators are controlled by an ARM-based embedded system with a control sampling frequency of 10 Hz.

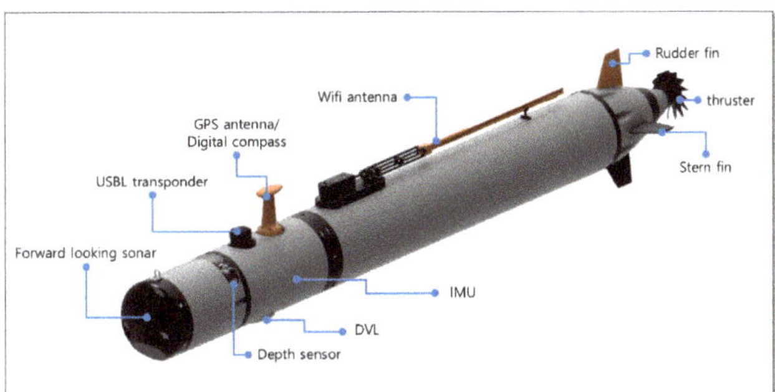

Figure 2. AUV used in the experiment. The vehicle was developed by Hanwha Systems [32]. The AUV serves as a testbed for underwater docking tasks [33].

To formulate the motion-governing equation, let us consider the control problem of the vehicle in a 3D space as shown in Figure 3. Assuming that the roll motion of the vehicle can be neglected, the governing equations for motion are given as follows [8,37]:

$$\dot{\eta} = \mathbf{R}v$$
$$\dot{\theta} = q,$$
$$\dot{\psi} = r/c\theta, \quad and, \tag{1}$$

$$m_{11}\dot{u} - m_{22}vr + m_{33}wq + f_u(u)u + \tau_{eu} = \tau_u,$$
$$m_{22}\dot{v} + m_{11}ur + f_v(v)v + m_{22}\tau_{ev} = 0,$$
$$m_{33}\dot{w} - m_{11}uq + f_w(w)w - d_1 + \tau_{ew} = 0, \qquad (2)$$
$$m_{55}\dot{q} - (m_{33} - m_{11})uw + f_q(q)q + (d_2 + \tau_{eq}) = \tau_q,$$
$$m_{66}\dot{r} - (m_{11} - m_{22})uv + f_r(r)r + \tau_{er} = \tau_r,$$

where $c\bullet$ and $s\bullet$ denote $cos(\bullet)$ and $sin(\bullet)$, respectively; $\eta = [x, y, z]^T$; $v = [u, v, w]^T$; x, y, z, θ, and ψ are the positions and orientations of the vehicle in the world coordinate; u, v, w, q, and r are the translational velocities and angular velocities; τ_u, τ_q, and τ_r are the control inputs; τ_{eu}, τ_{ev}, τ_{ew}, τ_{eq}, τ_{eu}, and τ_{er} are the bounded external disturbances, such as ocean currents and waves; $m_{ii}(i = 1,2,3,5,6)$ represent the terms for the combined mass and intertia parameters; $d_1 = (W - B)c\theta$; $d_2 = (z_g W - z_b B)s\theta$; W is the gravity, B is the buoyancy of the vehicle; $f_k(k)(k = u, v, w, q, r)$ represents the hydrodynamic damping and friction terms; and $\mathbf{R} \equiv \mathbf{R}_z(\psi)\mathbf{R}_y(\theta)$ is the rotation matrix of $\{B\}$ with respect to $\{W\}$. From Equation (2), the dynamics of the controllable states are rearranged as follows:

$$\overline{m}_u\dot{u} + h_u = \tau_u,$$
$$\overline{m}_q\dot{q} + h_q = \tau_q, \qquad (3)$$
$$\overline{m}_r\dot{r} + h_r = \tau_r,$$

where $\overline{m}_u$, $\overline{m}_q$, and $\overline{m}_r$ are the positive constants which represent the known ranges of inertia and h_u, h_q, and h_r are nonlinear terms, defined as

$$h_u \equiv (m_{11} - \overline{m}_u)\dot{u} - m_{22}vr + m_{33}wq + f_u(u)u + \tau_{eu},$$
$$h_q \equiv (m_{55} - \overline{m}_q)\dot{q} - (m_{33} - m_{11})uw + f_q(q)q + (d_2 + \tau_{eq}), \qquad (4)$$
$$h_r \equiv (m_{66} - \overline{m}_r)\dot{r} - (m_{11} - m_{22})uv + f_r(r)r + \tau_{er}.$$

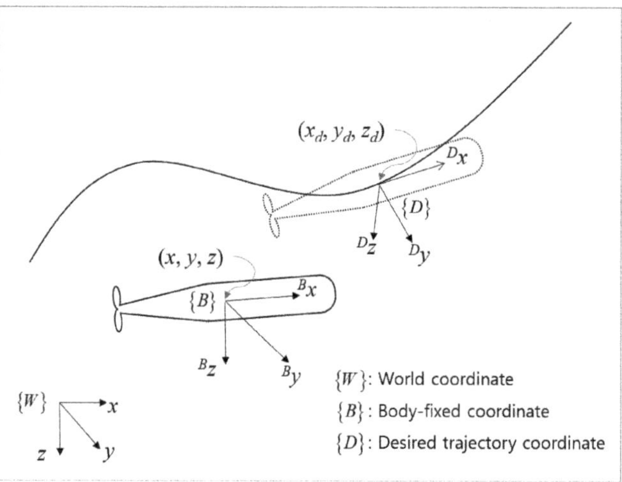

Figure 3. Definition of coordinates.

2.2. Desired Trajectory

Due to the underactuated nature, only three trajectory variables can be designed independently. Note that in Equation (2), there are only three independent control inputs to control five DOFs in the world coordinate. We can set the independent trajectory variables as x, y, and z in $\{W\}$. The trajectories can be represented by the following continuous time functions [8]:

$$\boldsymbol{\eta}_d = [x_d(t), y_d(t), z_d(t)]^T. \tag{5}$$

Note that, when taking into account Equation (1), the angular and velocity trajectories according to Equation (5) satisfy the following relationship [8]:

$$
\begin{aligned}
\theta_d &= -sin^{-1}\left(\dot{z}_d / \sqrt{\dot{x}_d^2 + \dot{y}_d^2 + \dot{z}_d^2}\right), \\
\psi_d &= atan2(\dot{y}_d, \dot{x}_d), \quad and, \\
u_d &= \sqrt{\dot{x}_d^2 + \dot{y}_d^2 + \dot{z}_d^2}, \\
q_d &= \dot{\theta}_d, \\
r_d &= \dot{\psi}_d c\theta_d.
\end{aligned}
\tag{6}
$$

The goal of this paper is to design a controller for an AUV system represented by Equations (1) and (3) to track the desired trajectories defined in Equations (5) and (6). The controller is specifically designed to reduce the tracking error along $\boldsymbol{\eta}_d$ by using control inputs.

2.3. Dynamics of Tracking Error

In this subsection, we arrange the tracking error dynamics of the vehicle by describing them in the desired trajectory coordinate $\{D\}$. This approach helps to minimize changes in the relationship between variables in different coordinates [38,39]. When there is no tracking error, u affects ^Dx for every attitude of the vehicle, and the transformation can be easily achieved as follows [40]:

$$^D\boldsymbol{\eta} = \mathbf{R}_d^T\boldsymbol{\eta} - \mathbf{R}_d^T\boldsymbol{\eta}_d, \tag{7}$$

where $^D\boldsymbol{\eta} = [^Dx, {}^Dy, {}^Dz]^T$ denotes the translational position of the vehicle with respect to $\{D\}$ and $\mathbf{R}_d \equiv \mathbf{R}_z(\psi_d)\mathbf{R}_y(\theta_d)$. Using Equation (7), the desired trajectories in Equation (5) can be transformed into those in $\{D\}$ as follows:

$$^D\boldsymbol{\eta}_d = \mathbf{0}. \tag{8}$$

Subtracting Equation (7) from Equaiton (8) yields the following relationship:

$$^D\boldsymbol{\eta}_e = \mathbf{R}_d^T\boldsymbol{\eta}_e, \tag{9}$$

where $\bullet_e \equiv \bullet_d - \bullet$. Note that from Equation (9), the tracking problem in $\{D\}$ is identical to that in $\{W\}$. Convergence of the tracking error in $\{D\}$ guarantees convergence in $\{W\}$. This is because $\mathbf{R}_d$ is a rotation matrix and cannot be singular. By taking the derivatives of Equation (9) with Equations (1) and (5) and introducing positive constants $\bar{\alpha}_u$, $\bar{\alpha}_\psi$, and $\bar{\alpha}_\theta$, the error dynamics of the state variable in $\{D\}$ can be rearranged as follows [8]:

$$
\begin{bmatrix} ^D\dot{x}_e \\ ^D\dot{y}_e \\ ^D\dot{z}_e \end{bmatrix} = \begin{bmatrix} \bar{\alpha}_u u_e \\ \bar{\alpha}_\psi \psi_e \\ -\bar{\alpha}_\theta \theta_e \end{bmatrix} + \begin{bmatrix} \lambda_x \\ \lambda_y \\ \lambda_z \end{bmatrix}.
\tag{10}
$$

Refer to [8] for a detailed derivation of Equation (10). The last term in the above equation represents the nonlinear terms defined as follows:

$$
\begin{bmatrix} \lambda_x \\ \lambda_y \\ \lambda_z \end{bmatrix} = -\begin{bmatrix} \bar{\alpha}_u u_e \\ \bar{\alpha}_\psi \psi_e \\ -\bar{\alpha}_\theta \theta_e \end{bmatrix} + v_d - \mathbf{R}_d^T\mathbf{R}v - \omega_d^{\times\, D}\boldsymbol{\eta}_e,
\tag{11}
$$

where $v_d = [u_d, 0, 0]^T$ and $\omega_d^\times$ is a skew-symmetric matrix of $\omega_d \equiv [0, q_d, r_d]^T$ [8]. Aside from that, from Equations (1) and (6), the attitude error is obtained as follows:

$$\begin{aligned} \dot{\theta}_e &= q_e, \\ \dot{\psi}_e &= \bar{\alpha}_r r_e + \lambda_\psi, \end{aligned} \tag{12}$$

where $\bar{\alpha}_r$ is a positive constant and

$$\lambda_\psi = -\bar{\alpha}_r r_e + r_d / c\theta_d - r / c\theta. \tag{13}$$

As a result, the tracking error dynamics are expressed in Equations (10) and (12). The objective of this paper is to design a control input in Equation (3) to stabilize the error dynamics, particularly ${}^D x_e$, ${}^D y_e$, and ${}^D z_e$.

2.4. Controller Design Using the Coupled Error Dynamics and Time Delay Estimation

In this paper, our goal is to design the desired error dynamics for u_e, r_e, and q_e that can asymptotically stabilize the tracking errors of ${}^D x_e$, ${}^D y_e$, and ${}^D z_e$, respectively. Note that from Equation (3), one can directly control the variables in $\{B\}$, namely u, q, and r, by designing appropriate control inputs τ_u, τ_q, and τ_r, respectively. From Equations (10) and (12), u adjusts ${}^D x_e$, q affects θ_e and consequently ${}^D z_e$, and r determines ψ_e and therefore ${}^D y_e$. Thus, coupled error dynamics between the variables in $\{B\}$ and $\{D\}$ can be designed to stabilize the variables in $\{D\}$. The desired error dynamics of u_e, r_e, and q_e are designed as follows:

$$\begin{aligned} \dot{u}_e + K_u u_e + K_x {}^D x_e &= 0, \\ \dot{q}_e + K_q q_e + K_\theta \theta_e - K_z {}^D z_e &= 0, \\ \dot{r}_e + K_r r_e + K_\psi \psi_e + K_y {}^D y_e &= 0, \end{aligned} \tag{14}$$

where $K_\bullet > 0$ represents the control gains.

In the manner of the computed torque control, the controller for Equation (3) can be designed as follows:

$$\begin{aligned} \tau_u &= \widehat{h}_u + \overline{m}_u \mu_u, \\ \tau_v &= \widehat{h}_q + \overline{m}_q \mu_q, \\ \tau_r &= \widehat{h}_r + \overline{m}_r \mu_r, \end{aligned} \tag{15}$$

where $\widehat{\bullet}$ denotes the estimate of $\bullet$ and μ_u, μ_q, and μ_r are the command inputs to insert the desired dynamics for u, q, and r, respectively. When $\widehat{\bullet} = \bullet$, the controlled dynamics is obtained from Equations (3) and (15) as follows:

$$\begin{aligned} \mu_u - \dot{u} &= 0, \\ \mu_q - \dot{q} &= 0, \\ \mu_r - \dot{r} &= 0. \end{aligned} \tag{16}$$

To induce the desired error dynamics in Equations (14)–(16), the command inputs are designed as follows:

$$\begin{aligned} \mu_u &= \dot{u}_d + K_u u_e + K_{xp} {}^D x_e, \\ \mu_q &= \dot{q}_d + K_q q_e - K_\theta \theta_e - K_{zp} {}^D z_e, \\ \mu_r &= \dot{r}_d + K_r r_e + K_\psi \psi_e + K_{yp} {}^D y_e. \end{aligned} \tag{17}$$

To implement the controller in Equation (15), it is necessary to obtain $\widehat{h}_u$, $\widehat{h}_q$, and $\widehat{h}_r$, the estimates of Equation (4). However, obtaining an exact dynamic model of Equation (4) is difficult and time-consuming. To address this, we employ the TDE [2,30,41,42] for robust and efficient estimation. The key idea behind TDE is that if the system dynamics are given as a continuous or piece-wise continuous function, then the variation in the dynamics during a very short time can be negligible. Thus, the value of the dynamics at the current time can be estimated by using the value of the dynamics at a short time before. Based on this idea, the system dynamics can be estimated indirectly by utilizing previous information on the system input and output. From Equation (3), the dynamics can be estimated as follows:

$$\begin{aligned}
\widehat{h}_{u(t)} &= h_{u(t-L)} = \tau_{u(t-L)} - \overline{m}_u \dot{u}_{(t-L)}, \\
\widehat{h}_{q(t)} &= h_{q(t-L)} = \tau_{q(t-L)} - \overline{m}_q \dot{q}_{(t-L)}, \\
\widehat{h}_{r(t)} &= h_{r(t-L)} = \tau_{r(t-L)} - \overline{m}_r \dot{r}_{(t-L)},
\end{aligned} \tag{18}$$

where L denotes a short time delay which is commonly set as the sampling time of the control system. As a result, the final form of the proposed controller is Equation (15) with Equations (17) and (18).

It is noteworthy that the proposed controller in Equation (15) with Equations (17) and (18) only requires the first derivative of the state variables and the states themselves. The linear and angular velocities can be measured using IMU and DVL, while the vehicle's position can be obtained through a navigation algorithm such as a Kalman filter, utilizing sensors such as IMU, DVL, the depth sensor, and the digital compass [34]. The advantage of not requiring high-order differentiation of the states is that it helps stabilize the controller in experimental environments. This is because differentiating the states amplifies the noise effect present in the state measurements. In comparison, the BCTDE [2,7,8] necessitates third-order differentiation of the states. Therefore, one can expect that the proposed controller is relatively easier to stabilize in experimental environments. In addition, note that the TDE method does not require the entire vehicle dynamics model. One can design the proposed controller by only selecting the inertial gains, such as $\overline{m}_u, \overline{m}_q, \overline{m}_r, \overline{\alpha}_u, \overline{\alpha}_\psi, \overline{\alpha}_\theta,$ and $\overline{\alpha}_r$, as well as the feedback gains, such as $K_x, K_u, K_y, K_\psi, K_r, K_z, K_\theta,$ and K_q.

2.5. Error Dynamics of the Proposed Controller

The TDE provides an efficient way to estimate the nonlinear dynamics of the vehicle, but it cannot estimate the dynamics variation exactly during a sampling time L. Thus, an estimation error of the dynamics remains. Taking into account the estimation error of the TDE, one can rearrange the error dynamics in Equaiton (14) by utilizing Equations (3) and (15) with Equations (17) and (18) as follows:

$$\begin{aligned}
\dot{u}_e + K_u u_e + K_x{}^D x_e &= \epsilon_u, \\
\dot{q}_e + K_q q_e + K_\theta \theta_e - K_z{}^D z_e &= \epsilon_q, \\
\dot{r}_e + K_r r_e + K_\psi \psi_e + K_y{}^D y_e &= \epsilon_r,
\end{aligned} \tag{19}$$

where $\epsilon_\bullet$ denotes the TDE errors, which are defined as follows:

$$\begin{aligned}
\epsilon_u &\equiv \overline{m}_u^{-1}\left(h_{u(t)} - h_{u(t-L)}\right), \\
\epsilon_q &\equiv \overline{m}_q^{-1}\left(h_{q(t)} - h_{q(t-L)}\right), \\
\epsilon_r &\equiv \overline{m}_r^{-1}\left(h_{r(t)} - h_{r(t-L)}\right).
\end{aligned} \tag{20}$$

From Equations (10), (12), and (19), the error dynamics of the controlled system for the forward, lateral, and vertical directions, respectively, are as follows:

$$\begin{bmatrix} {}^{D}\dot{x}_e \\ \dot{u}_e \end{bmatrix} = \underbrace{\begin{bmatrix} 0 & \overline{\alpha}_u \\ -K_x & -K_u \end{bmatrix}}_{\mathbf{A}_x} \begin{bmatrix} {}^{D}x_e \\ u_e \end{bmatrix} + \underbrace{\begin{bmatrix} \lambda_x \\ \epsilon_u \end{bmatrix}}_{\mathbf{B}_x},$$

$$\qquad(21\text{a})$$

$$\begin{bmatrix} {}^{D}\dot{y}_e \\ \dot{\psi}_e \\ \dot{r}_e \end{bmatrix} = \underbrace{\begin{bmatrix} 0 & \overline{\alpha}_\psi & 0 \\ 0 & 0 & \overline{\alpha}_r \\ -K_y & -K_\psi & -K_r \end{bmatrix}}_{\mathbf{A}_y} \begin{bmatrix} {}^{D}y_e \\ \psi_e \\ r_e \end{bmatrix} + \underbrace{\begin{bmatrix} \lambda_y \\ \lambda_\psi \\ \epsilon_r \end{bmatrix}}_{\mathbf{B}_y},$$

$$\qquad(21\text{b})$$

$$\begin{bmatrix} {}^{D}\dot{z}_e \\ \dot{\theta}_e \\ \dot{q}_e \end{bmatrix} = \underbrace{\begin{bmatrix} 0 & -\overline{\alpha}_\theta & 0 \\ 0 & 0 & 1 \\ K_z & -K_\theta & -K_q \end{bmatrix}}_{\mathbf{A}_z} \begin{bmatrix} {}^{D}z_e \\ \theta_e \\ q_e \end{bmatrix} + \underbrace{\begin{bmatrix} \lambda_z \\ 0 \\ \epsilon_q \end{bmatrix}}_{\mathbf{B}_z}.$$

$$\qquad(21\text{c})$$

By taking the Laplace transform of Equations (21a)–(21c), one can examine the stability of the error dynamics and the influence of the forcing functions $\mathbf{B}_x$, $\mathbf{B}_y$, and $\mathbf{B}_z$. The Laplace-transformed error dynamics can be obtained as follows:

$$\begin{bmatrix} {}^{D}x_e \\ u_e \end{bmatrix} = (s\mathbf{I} - \mathbf{A}_x)^{-1}\mathbf{B}_x = \frac{1}{|s\mathbf{I} - \mathbf{A}_x|} \begin{bmatrix} s + K_u & \overline{\alpha}_u \\ -K_x & s \end{bmatrix} \begin{bmatrix} \lambda_x \\ \epsilon_u \end{bmatrix},$$

$$\qquad(22\text{a})$$

$$\begin{bmatrix} {}^{D}y_e \\ \psi_e \\ r_e \end{bmatrix} = \frac{1}{|s\mathbf{I} - \mathbf{A}_y|} \begin{bmatrix} s^2 + K_r s + \overline{\alpha}_r K_\psi & \overline{\alpha}_\psi s + \overline{\alpha}_\psi K_r & \overline{\alpha}_\psi \overline{\alpha}_r \\ -\overline{\alpha}_r K_y & s^2 + K_r s & \overline{\alpha}_r s \\ K_y s & -K_\psi s - \overline{\alpha}_\psi K_y & s^2 \end{bmatrix} \begin{bmatrix} \lambda_y \\ \lambda_\psi \\ \epsilon_r \end{bmatrix},$$

$$\qquad(22\text{b})$$

$$\begin{bmatrix} {}^{D}z_e \\ \theta_e \\ q_e \end{bmatrix} = \frac{1}{|s\mathbf{I} - \mathbf{A}_z|} \begin{bmatrix} s^2 + K_q s + K_\theta & -\overline{\alpha}_\theta \\ K_z & s \\ K_z s & s^2 \end{bmatrix} \begin{bmatrix} \lambda_z \\ \epsilon_q \end{bmatrix},$$

$$\qquad(22\text{c})$$

where s denotes the Laplace operator and

$$|s\mathbf{I} - \mathbf{A}_x| = s^2 + K_u s + \overline{\alpha}_u K_x,$$

$$\qquad(23\text{a})$$

$$|s\mathbf{I} - \mathbf{A}_y| = s^3 + K_r s^2 + \overline{\alpha}_r K_\psi s + \overline{\alpha}_\psi \overline{\alpha}_r K_y,$$

$$\qquad(23\text{b})$$

$$|s\mathbf{I} - \mathbf{A}_z| = s^3 + K_q s^2 + K_\theta s + \overline{\alpha}_\theta K_y.$$

$$\qquad(23\text{c})$$

From Equations (22a)–(22c), the influence of each term of the forcing functions $\mathbf{B}_x$, $\mathbf{B}_y$, and $\mathbf{B}_z$ can be estimated. The tracking errors in a steady state can be analyzed by using the final value theorem of the Laplace transform: $\lim_{t\to\infty} x(t) = \lim_{s\to 0} sx(s)$. For example, from Equations (22a) and (23a), ${}^{D}x_e(s)/\lambda_x(s) = (s + K_u)/(s^2 + K_u s + \overline{\alpha}_u K_x)$. Assume that $\lambda_x(s)$ is given as a step function: $\lambda_x(s) = l/s$ with a constant l. Then, ${}^{D}x_e(t)|_{t\to\infty} = K_u l/(\overline{\alpha}_u K_x)$, and one can estimate that in the forward direction error, ${}^{D}x_e(t)$, there will be a steady state error dependent on the amount of disturbed dynamics l and the control gains $\overline{\alpha}_u$, K_u, and K_x. The controller cannot perfectly compensate for the influence of the forcing function, but it can attenuate the influence by selecting appropriate control gains.

The characteristic equations in Equations (23a)–(23c) are useful for selecting appropriate control gains. In order to ensure stable error dynamics, the characteristic equations must satisfy the Hurwitz condition, and the gains $K_\bullet$ must be chosen accordingly. Additionally, the inertial gains $\overline{m}_u$, $\overline{m}_q$, $\overline{m}_r$, $\overline{\alpha}_u$, $\overline{\alpha}_\psi$, $\overline{\alpha}_\theta$, and $\overline{\alpha}_r$ can be obtained through tuning. Previous research works utilizing the TDE have suggested selecting inertial gains within a known range of the vehicle's inertial terms. If the vehicle model is unknown, however, then the inertial gains can be obtained through tuning [2,30].

3. Practical Issues for the Experiments

When setting up the controller for the experiment, practical issues such as the noise effects of the measurements and actuator limitations have to be considered. These issues will be discussed in the following subsections.

3.1. Handling the Noise Effect in the TDE

The TDE provides an effective and efficient method for estimating the nonlinear dynamics of vehicles. However, the use of state derivatives in the TDE amplifies the noise effect in the measurement of the state variables. In the case of underwater vehicles, the vehicle's position is usually estimated by Kalman filtering of the sensor data, such as acceleration from the IMU and velocity from the DVL, which have considerable measurement noise. The amplification of noise can undermine the stability conditions. One way to handle this is to use the low-pass filtering effects of the inertial gains of the TDE [30]. Decreasing the inertial gains shows a similar effect to low-pass filtering. However, we found experimentally that adjusting the inertial gains was not enough. Therefore, we devised a method to attenuate the noise effect of the TDE. The idea is quite simple: cut down the direct TDE value and supplement the remaining part with the averaged value of the TDE. Figure 4 shows the noise attenuation method in the TDE. Note that the use of an average filter in Figure 4 can attenuate the noise effect because the filter also averages the noise. However, the filter may slightly degrade the performance of the TDE because the filtered value of the TDE cannot estimate exactly any quick changes in vehicle dynamics. In the case of underwater vehicles, dynamic changes occur due to the vehicle dynamics and disturbance changes such as sea currents, which depend on the mission (or desired trajectory) and the environment. In the case of AUVs, dynamic changes may not be fast because they are commonly used for surveys of large areas, and disturbances such as sea currents change slowly according to tide variation.

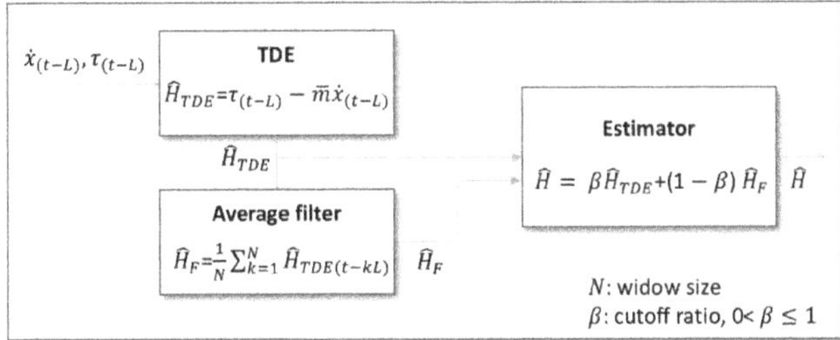

Figure 4. Filtering the TDE to reduce the noise effect. The algorithm reduces the direct TDE value and supplements the remaining part with the averaged value of the TDE. This mitigates the noise present in the TDE thanks to the averaging effect.

3.2. Handling Nonlinearity and the Limits of the Actuators

When designing a controller, it is important to consider the actuator characteristics, such as nonlinearity and the actuation limits. For instance, the thruster of a vehicle exhibits nonlinear dynamics between the propulsion force and rotation speed of the thruster. Moreover, the thruster has limits on rotational velocity and acceleration because the thruster is a mechanical system. In this subsection, we address compensation methods for the actuator characteristics, focusing on the thrusters in particular and briefly touching on the rudder fins and stern fins. The thruster dynamics are as follows [23,43]:

$$\tau_u = T_{|n|n}|n|n + T_{|n|u}|n|u, \tag{24}$$

where n denotes the rotation velocity of the thruster and $T_{|n|n}$ and $T_{|n|u}$ represent the actuator coefficients corresponding to the rotational speed of the actuator and fluid speed around the actuator, respectively. By ignoring $T_{|n|u}|n|u$ and $T_{|n|n}$ from Equation (24), we adopted a simple thruster model, which is as follows:

$$\tau_u' = |n|n. \tag{25}$$

This is because the effect of $T_{|n|u}|n|u$, the term included in the RHS of Equation (24), can be compensated for by the feedback loop of the controller, and the scale coefficient $T_{|n|n}$ in Equation (24) can be adjusted by tuning the controller gain. Note that when substituting Equation (24) into the first equation of Equation (3), $T_{|n|u}|n|u$ can be treated as part of the nonlinear term h_u (i.e., $h'_u = h_u - T_{|n|u}|n|u$). The coefficient $T_{|n|n}$ simply scales the value of $\overline{m}_u$ (i.e., $\overline{m}'_u = \overline{m}_u / T_{|n|n}$).

Regarding the actuation limit of the thrusting force, we considered the limits on the rotational velocity and acceleration of the thrusting propeller as follows:

$$\underline{n} = \begin{cases} n, & when \; |n| < n_{max}, \; |\dot{n}| < \dot{n}_{max}, \; n \geq 0, \\ f(n), & when \; any \; of \; above \; conditions \; is \; not \; satisfied, \end{cases} \tag{26}$$

where $\underline{n}$ represents n with the actuation limit and $f(n)$ is a limiting function that considers the limit of n and $\dot{n}$ as well as the sign of n. Note that in Equation (26), the sign condition $n \geq 0$ is included because only the forward thrusting force is available. In the case of the controllers using the TDE, handling the actuation limit to prevent the wind-up phenomenon is as straightforward as incorporating a limit block to ensure that the calculated control input for the TDE matches the actual value of the actuation applied to the vehicle [44]. Figure 5 shows the limiter block for handling the actuation limit of the thrusting force, while Figure 6 shows the structure of the limiter block, which is explained by Equation (26). In the case of the rudder fins and stern fins, one can handle the actuation limit in similar ways to the case of the thrusting force. The only differences are that (1) the actuation forces are linear with the fin motions [23] and (2) both positive and negative forces are available. By simply removing the condition $n \geq 0$ in Equation (26) and eliminating the blocks of sqrt() and square() in Figures 5 and 6, the same limiting algorithm can be applied to the rudder fins and the stern fins.

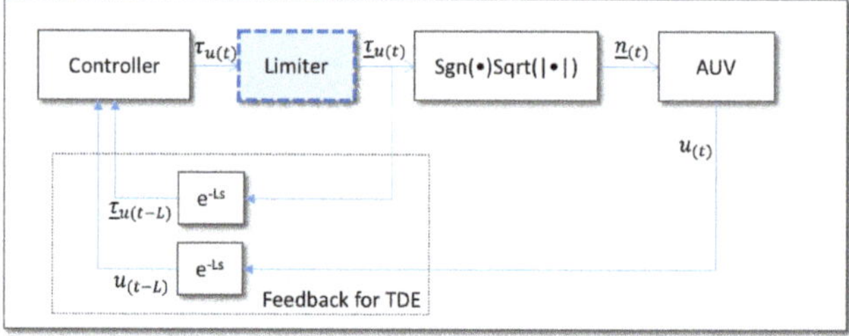

Figure 5. TDE feedback block for h_u of the controller. $Sgn(\bullet)$ denotes a signum function, and $Sqrt(\bullet) \equiv \sqrt{\bullet}$. The 'Limiter' block is included to prevent the wind-up phenomenon due to the actuation limit [44]. The '$Sgn(\bullet)Sqrt(|\bullet|)$' block is for compensating for the actuation dynamics in Equation (25).

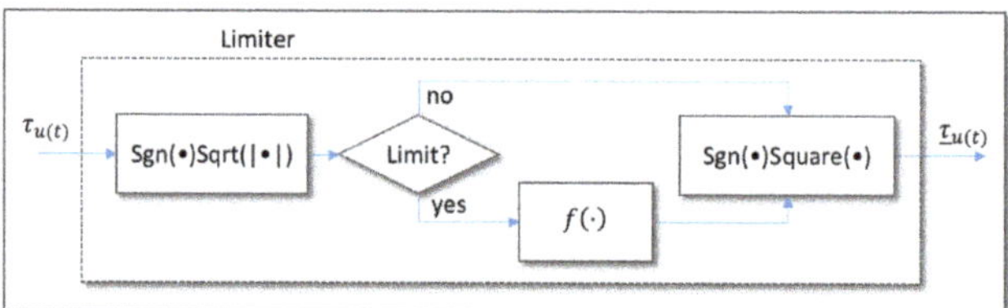

Figure 6. Limiter block of the thrusting force τ_u: The thruster has actuation limits on the rotation speed and acceleration. Therefore, before checking the limitation, the '$Sgn(\bullet)Sqrt(|\bullet|)$' block is included to convert the control force τ_u into the rotation speed according to Equation (25).

4. Experimental Study

The tracking performance of the proposed controller was experimentally verified using the AUV platform depicted in Figure 2. The experiments were performed in the seawater at a port located in the South Sea of Korea. For the experiment, the control gains were set as follows. The inertial gains were set to $\overline{m}_u = 3000$, $\overline{m}_q = 0.7$, $\overline{m}_r = 1.0$, $\overline{\alpha}_u = 1.0$, $\overline{\alpha}_\psi = 1.5$, $\overline{\alpha}_\theta = 1.5$, and $\overline{\alpha}_r = 1.5$ by tuning, and the feedback gains $K_x = 1.0$, $K_u = 2.0$, $K_y = 0.216$, $K_\psi = 1.08$, $K_r = 1.8$, $K_z = 0.125$, $K_\theta = 0.75$, and $K_q = 1.5$ were selected for the desired error dynamics having poles at $p_{dx} = -1.0$ (double poles), $p_{dy} = -0.6$ (triple poles), and $p_{dz} = -0.5$ (triple poles). In this case, the characteristic equations in Equations (23a)–(23c) had poles at $p_{cx} = -1.0$ (double poles), $p_{cy} = -0.502, -0.649 \pm 0.740i$, and $p_{cz} = -0.897, -0.302 \pm 0.344i$, which were placed in the LHP. Regarding the noise-handling algorithm in Figure 4, the window size for the average filter was set at $N = 128$, and the β values for $\widehat{h}_u$, $\widehat{h}_q$, and $\widehat{h}_r$ in Equation (18) were $\beta_u = 0.7$, $\beta_q = 0.5$, and $\beta_r = 0.9$, respectively.

4.1. Experimental Verification of the Noise-Handling Issue

In this subsection, the noise handling method described in Section 3.1 is experimentally verified. The experiments were conducted on the surface of the sea, with actuations in the XY plane. The thrusters and rudder fins were activated, while the stern fins were deactivated. The tracking performances were compared between the case where the noise-handling algorithm was not applied and the case where the algorithm was applied. A trajectory involving linear motion was used, as shown in Figure 7. The experimental results are presented in Figures 7–12. Figures 7–9 show the responses for the case without the noise-handling algorithm, while Figures 10–12 show the responses when the algorithm was applied. When comparing Figures 7 and 10, it may be difficult to recognize a significant difference in performance between the two cases. Aside from the initial errors in Figure 10, the error bounds appear to be similar in both cases. However, when comparing Figures 9 and 12, it is evident that the noise-handling algorithm was effective. In the case without the algorithm (Figure 9), the control input (rudder angle) switched frequently, resulting in erratic responses in r, ψ, and $^D y$. This behavior was due to the TDE in Equation (18), which included the first-order derivative of the velocity state and amplified the noise present in the state measurement. In contrast, when the noise-handling algorithm was applied (Figure 12), the control input (rudder angle) reacted smoothly to tracking errors, leading to smooth convergence in the velocity and position responses. From Figures 8 and 11, it can be observed that the chatterings in the forward direction responses were alleviated when the noise-handling algorithm was adopted.

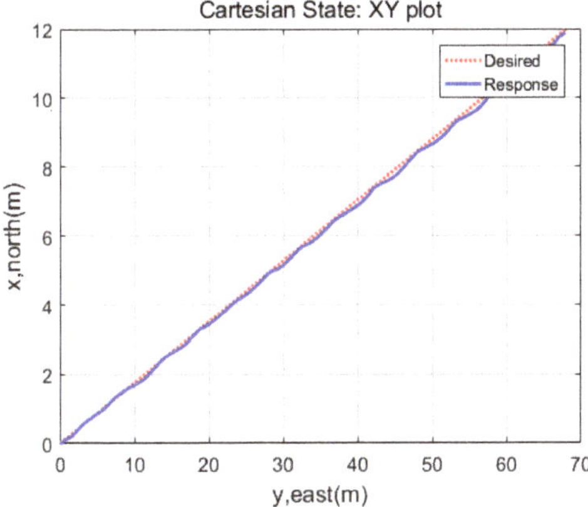

Figure 7. XY plot of the responses when noise-handling algorithm was not applied. The responses exhibited chattering behaviors due to the influence of measurement noise on the TDE.

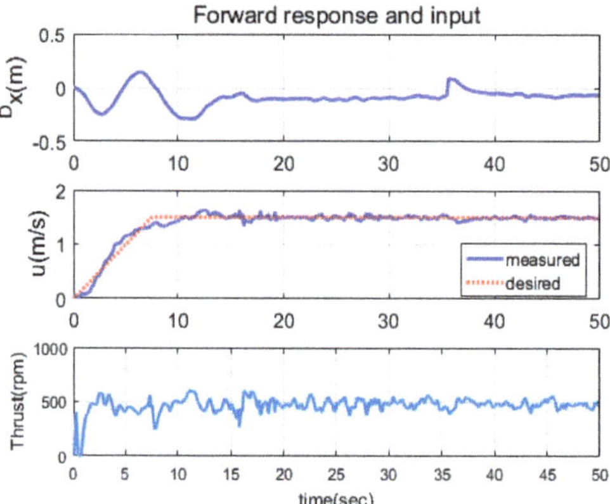

Figure 8. Forward direction responses when noise-handling algorithm was not applied. The responses were stable; however, there was some minor chattering in the control input.

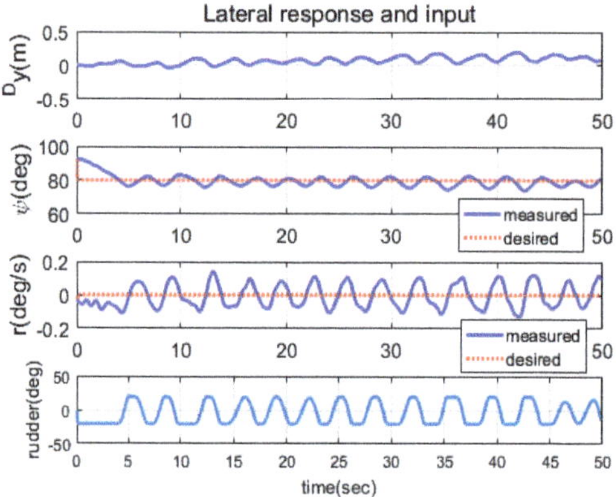

Figure 9. Lateral direction responses when noise-handling algorithm was not applied. The control input switched frequently, leading to chattering responses in r, ψ, and ^Dy.

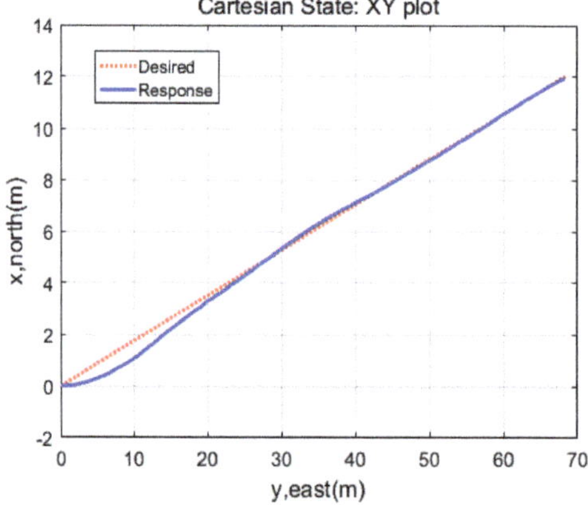

Figure 10. XY plot of the responses when noise-handling algorithm in Figure 4 was applied. The responses smoothly converged to the desired trajectories even in the presence of an initial angular error.

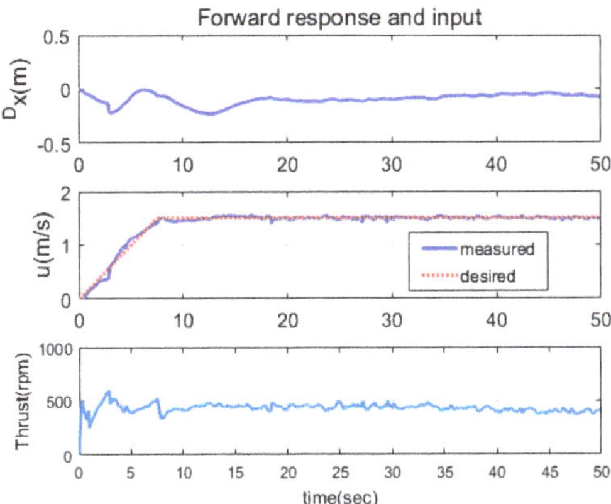

Figure 11. Forward direction responses when noise-handling algorithm in Figure 4 was applied. The responses exhibited smoother convergence compared with the responses shown in Figure 8.

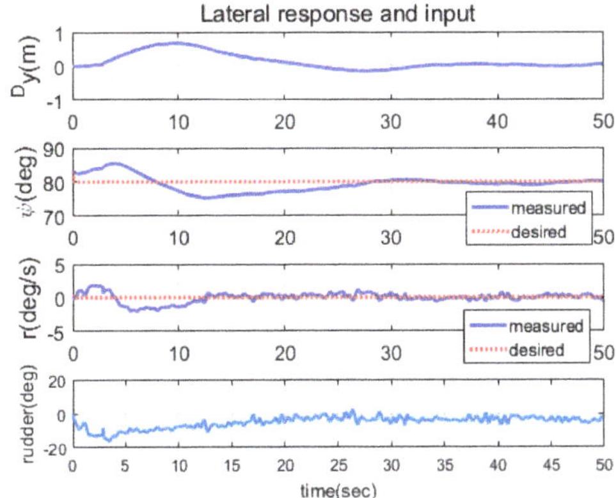

Figure 12. Lateral direction responses when noise-handling algorithm in Figure 4 was applied. The responses exhibited smoother convergence compared with the responses shown in Figure 9.

4.2. Experimental Verification of Tracking Performance in 3D Space Motion

The trajectory tracking performance of the proposed controller in 3D space motion was verified experimentally. As shown in Figure 13, the desired trajectory, drawing the shape of the number eight, was applied. The experimental results are presented in Figures 13–17. Figures 13 and 14 show the tracking responses in the world coordinate $\{W\}$, demonstrating that the controlled system had stable responses and followed the desired trajectory with bounded errors. The root mean square (RMS) errors in Table 1 indicate that the tracking performances were accurate enough to perform surveying missions requiring tracking errors of less than one meter. It can be observed from Table 1 that the RMS error in the vertical direction was slightly larger than those in the other directions due to buoyancy acting as a disturbance in the vertical direction. Figures 15–17 show the responses and control inputs for each direction. Note that in Figure 17, the response of θ exhibited a large tracking error induced by the controller to reduce the tracking error of z. In Figure 15, it can be observed that the tracking error of ^{D}x bounced around at times $t = 66$ s, 87 s, and 141 s, which was caused by irregular DVL signals resulting from stiff changes in the seabed.

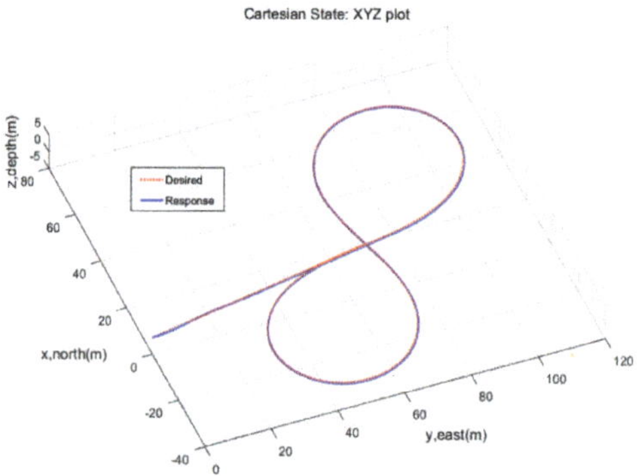

Figure 13. XYZ plot of the tracking responses.

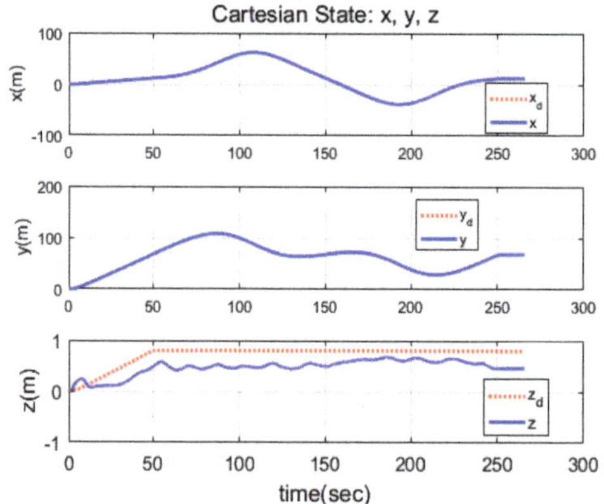

Figure 14. The time responses of the positions in a Cartesian space.

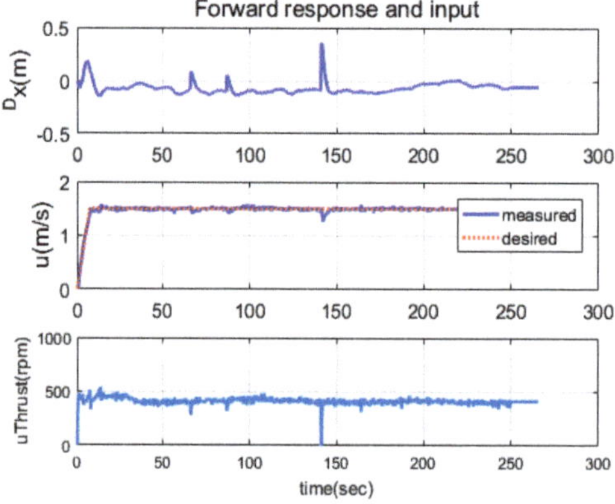

Figure 15. The responses and actuation in the forward direction. ^Dx converged to zero because, as explained in Equation (8), the desired trajectory expressed in the desired trajectory coordinate $\{D\}$ was zero. The responses of ^Dx and u followed their desired trajectories smoothly, demonstrating the effectiveness of the noise filtering depicted in Figure 4. That aside, the tracking response of ^Dx bounced around at times $t = 66$ s, 87 s, and 141 s, which was caused by irregular DVL signals resulting from stiff changes in the seabed.

Table 1. The root mean square (RMS) errors in $\{D\}$.

Direction	Forward	Lateral	Vertical	Total
RMS error (m)	0.0838	0.1595	0.2663	0.3216

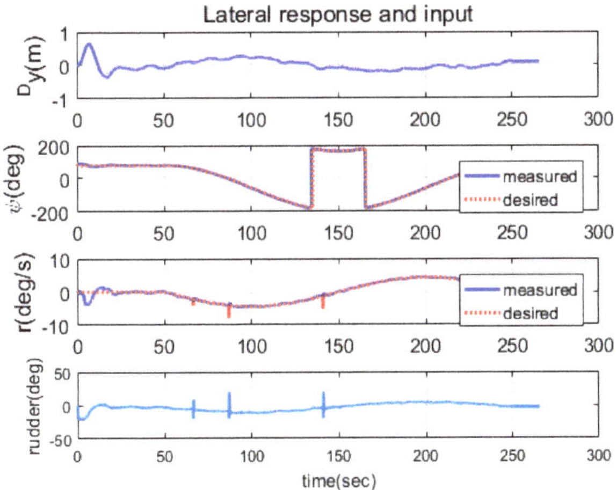

Figure 16. The responses and actuation in the lateral direction. The responses of ^{D}y, ψ, and r followed their desired trajectories smoothly, demonstrating the effectiveness of the noise filtering depicted in Figure 4. The responses did not exhibit any significant chattering, in contrast to the responses in Figure 9, which represents the case without noise filtering.

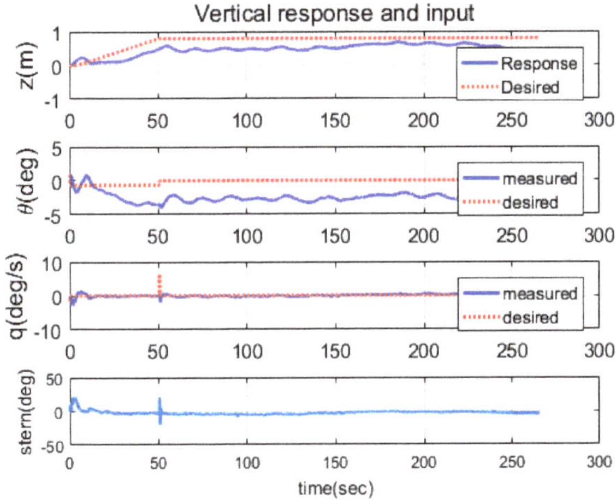

Figure 17. The responses and actuation in the vertical direction. The responses of z, θ, and q followed their desired trajectories smoothly. The response of z showed a steady state error because the buoyancy acted as disturbances in the vertical direction. The error of θ induced a reduction in the error of z, corresponding to the desired error dynamics in (14). The responses having non-zero errors were matched with the error dynamics analysis in Equations (22a–22c), which explains that the proposed controller may not converge to zero if there are non-zero disturbances on the vehicle dynamics.

5. Conclusions

In this paper, we proposed a trajectory tracking controller for AUVs in 3D space motion, along with experimental verification on the sea. The main concept of the proposed controller is to design the desired error dynamics that combine the state variables in the body-fixed coordinate and the world coordinate (i.e., the desired trajectory coordinate) to address the underactuated nature of the vehicle. The TDE, an indirect estimation method utilizing control inputs and vehicle outputs, was employed to estimate the nonlinear dynamics and disturbances of the vehicle. Consequently, the proposed controller is relatively easy to implement as it does not require the entire dynamic model of the vehicle. In terms of experimental implementation, the controller is relatively easy to stabilize since it only requires the first derivatives of the states and the states themselves, thereby potentially mitigating noise amplification arising from differentiation. Practical issues related to implementation in experimental environments were also addressed. A noise-filtering algorithm for the TDE was developed, and compensation methods for the mechanical limitations and nonlinear dynamics of the actuators were devised. The performance of the proposed controller was validated through experiments in seawater. The experimental results demonstrate the effectiveness of the noise-filtering algorithm in stabilizing the control performance. Through trajectory tracking control experiments in 3D space motion, it was verified that the proposed controller achieves accurate tracking performance, rendering it suitable for survey missions requiring precise tracking performance with errors of less than one meter.

Author Contributions: Conceptualization, G.R.C. and J.-H.L.; methodology, G.R.C.; software, G.R.C. and H.K. (Hyungjoo Kang); validation, G.R.C., H.K. (Hyungjoo Kang), M.-G.K., M.-J.L., H.K. (Hosung Kim), H.L. and G.L.; formal analysis, G.R.C.; investigation, G.R.C.; resources, J.-H.L.; data curation, G.R.C.; writing—original draft preparation, G.R.C.; writing—review and editing, G.R.C.; visualization, G.R.C.; supervision, G.R.C.; project administration, G.R.C.; funding acquisition, G.R.C. All authors have read and agreed to the published version of the manuscript.

Funding: This research was supported by the Korea Institute of Marine Science & Technology Promotion (KIMST) and funded by the Ministry of Oceans and Fisheries of Korea (20220567, Development of standard manufacturing technology for marine leisure vessels and safety support robots for underwater leisure activities).

Institutional Review Board Statement: Not applicable.

Informed Consent Statement: Not applicable.

Data Availability Statement: Data are contained within the article.

Conflicts of Interest: The authors declare no conflict of interest.

Abbreviations

The following abbreviations are used in this manuscript:

AUV	Autonomous underwater vehicle
BC	Back-stepping control
BCTDE	Back-stepping control with time delay estimation
TDE	Time delay estimation
DOF	Degree of freedom
IMU	Inertial measurement unit
DVL	Doppler velocity log
GNSS	Global navigation satellite system
RMS	Root mean square

References

1. Khalil, H.K. *Nonlinear Systems*, 3rd ed.; Prentice Hall: Hoboken, NJ, USA, 2002.
2. Cho, G.R.; Li, J.H.; Park, D.; Jung, J.H. Robust trajectory tracking of autonomous underwater vehicles using back-stepping control and time delay estimation. *Ocean Eng.* **2020**, *201*, 107131. [CrossRef]
3. Li, J.H.; Lee, P.M. Path tracking in dive plane for a class of torpedo-type underactuated AUVs. In Proceedings of the 7th Asian Control Conference, ASCC 2009, Hong Kong, China, 27–29 August 2009; pp. 360–365.
4. Repoulias, F.; Papadopoulos, E. Planar trajectory planning and tracking control design for underactuated AUVs. *Ocean Eng.* **2007**, *34*, 1650–1667. [CrossRef]
5. Liang, X.; Qu, X.; Hou, Y.; Zhang, J. Three-dimensional path following control of underactuated autonomous underwater vehicle based on damping backstepping. *Int. J. Adv. Robot. Syst.* **2017**, *14*, 1729881417724179. [CrossRef]
6. Juan, L.; Zhang, Q.; Cheng, X.; Mohammed, N.F. Path following backstepping control of underactuated unmanned underwater vehicle. In Proceedings of the 2015 IEEE International Conference on Mechatronics and Automation (ICMA), Beijing, China, 2–5 August 2015; pp. 2267–2272.
7. Cho, G.R.; Park, D.G.; Kang, H.; Lee, M.J.; Li, J.H. Horizontal Trajectory Tracking of Underactuated AUV using Backstepping Approach. *IFAC-PapersOnLine* **2019**, *52*, 174–179. [CrossRef]
8. Cho, G.R.; Kang, H.; Lee, M.J.; Kim, M.G.; Li, J.H. 3D Space Trajectory Tracking of Underactuated AUVs using Back-Stepping Control and Time Delay Estimation. *IFAC-PapersOnLine* **2021**, *54*, 238–244. [CrossRef]
9. Elmokadem, T.; Zribi, M.; Youcef-Toumi, K. Terminal sliding mode control for the trajectory tracking of underactuated Autonomous Underwater Vehicles. *Ocean Eng.* **2017**, *129*, 613–625. [CrossRef]
10. Yu, C.; Xiang, X.; Lapierre, L.; Zhang, Q. Nonlinear guidance and fuzzy control for three-dimensional path following of an underactuated autonomous underwater vehicle. *Ocean Eng.* **2017**, *146*, 457–467. [CrossRef]
11. Karkoub, M.; Wu, H.M.; Hwang, C.L. Nonlinear trajectory-tracking control of an autonomous underwater vehicle. *Ocean Eng.* **2017**, *145*, 188–198. [CrossRef]
12. Yang, X.; Yan, J.; Hua, C.; Guan, X. Trajectory Tracking Control of Autonomous Underwater Vehicle with Unknown Parameters and External Disturbances. *IEEE Trans. Syst. Man Cybern. Syst.* **2019**, *51*, 1054–1063. [CrossRef]
13. Elmokadem, T.; Zribi, M.; Youcef-Toumi, K. Control for dynamic positioning and way-point tracking of underactuated autonomous underwater vehicles using sliding mode control. *J. Intell. Robot. Syst.* **2019**, *95*, 1113–1132. [CrossRef]
14. Yan, Z.; Wang, M.; Xu, J. Robust adaptive sliding mode control of underactuated autonomous underwater vehicles with uncertain dynamics. *Ocean Eng.* **2019**, *173*, 802–809. [CrossRef]
15. Tabataba'i-Nasab, F.S.; Keymasi Khalaji, A.; Moosavian, S.A.A. Adaptive nonlinear control of an autonomous underwater vehicle. *Trans. Inst. Meas. Control* **2019**, *41*, 3121–3131. [CrossRef]
16. Li, J.; Du, J.; Sun, Y.; Lewis, F.L. Robust adaptive trajectory tracking control of underactuated autonomous underwater vehicles with prescribed performance. *Int. J. Robust Nonlinear Control* **2019**, *29*, 4629–4643. [CrossRef]
17. Wang, J.; Wang, C.; Wei, Y.; Zhang, C. Three-Dimensional Path Following of an Underactuated AUV Based on Neuro-Adaptive Command Filtered Backstepping Control. *IEEE Access* **2018**, *6*, 74355–74365. [CrossRef]
18. Li, J.H.; Lee, M.J.; Kang, H.; Kim, M.G.; Cho, G.R. Neural-net based robust adaptive control for 3D path following of torpedo-type AUVs. In Proceedings of the 2020 59th IEEE Conference on Decision and Control (CDC), Jeju Island, Republic of Korea, 14–18 December 2020; pp. 5261–5266.
19. Jun, B.H.; Park, J.Y.; Lee, F.Y.; Lee, P.M.; Lee, C.M.; Kim, K.; Lim, Y.K.; Oh, J.H. Development of the AUV 'ISiMI'and a free running test in an Ocean Engineering Basin. *Ocean Eng.* **2009**, *36*, 2–14. [CrossRef]
20. Rout, R.; Subudhi, B. NARMAX self-tuning controller for line-of-sight-based waypoint tracking for an autonomous underwater vehicle. *IEEE Trans. Control Syst. Technol.* **2016**, *25*, 1529–1536. [CrossRef]
21. Refsnes, J.E.; Sorensen, A.J.; Pettersen, K.Y. Model-based output feedback control of slender-body underactuated AUVs: Theory and experiments. *IEEE Trans. Control Syst. Technol.* **2008**, *16*, 930–946. [CrossRef]
22. Healey, A.J.; Lienard, D. Multivariable sliding mode control for autonomous diving and steering of unmanned underwater vehicles. *IEEE J. Ocean. Eng.* **1993**, *18*, 327–339. [CrossRef]
23. Fossen, T.I. *Guidance and Control of Ocean Vehicles*; John Wiley & Sons Inc.: Hoboken, NJ, USA, 1994.
24. Park, D.; Li, J.H.; Ki, H.; Kang, H.; Kim, M.G.; Suh, J.H. Selective AUV guidance scheme for structured environment navigation. In Proceedings of the OCEANS 2019—Marseille, Marseille, France, 17–20 June 2019; pp. 1–5.
25. González-García, J.; Gómez-Espinosa, A.; García-Valdovinos, L.G.; Salgado-Jiménez, T.; Cuan-Urquizo, E.; Escobedo Cabello, J.A. Experimental Validation of a Model-Free High-Order Sliding Mode Controller with Finite-Time Convergence for Trajectory Tracking of Autonomous Underwater Vehicles. *Sensors* **2022**, *22*, 488. [CrossRef]
26. Dai, X.; Xu, H.; Ma, H.; Ding, J.; Lai, Q. Dual closed loop AUV trajectory tracking control based on finite time and state observer. *Math. Biosci. Eng.* **2022**, *19*, 11086–11113. [CrossRef] [PubMed]
27. Borlaug, I.L.G.; Pettersen, K.Y.; Gravdahl, J.T. Comparison of two second-order sliding mode control algorithms for an articulated intervention AUV: Theory and experimental results. *Ocean Eng.* **2021**, *222*, 108480. [CrossRef]
28. Guerrero, J.; Torres, J.; Creuze, V.; Chemori, A. Trajectory tracking for autonomous underwater vehicle: An adaptive approach. *Ocean Eng.* **2019**, *172*, 511–522. [CrossRef]

29. Kim, J.; Joe, H.; Yu, S.c.; Lee, J.S.; Kim, M. Time-delay controller design for position control of autonomous underwater vehicle under disturbances. *IEEE Trans. Ind. Electron.* **2015**, *63*, 1052–1061. [CrossRef]

30. Cho, G.R.; Chang, P.H.; Park, S.H.; Jin, M. Robust tracking under nonlinear friction using time-delay control with internal model. *IEEE Trans. Control Syst. Technol.* **2009**, *17*, 1406–1414.

31. Cho, G.R.; Park, D.G.; Li, J.H.; Kang, H.; Kim, M.G. Path Tracking Control of AUV using Nonholonomic Error Dynamics. In Proceedings of the 18th International Conference on Control, Automation and Systems (ICCAS 2018), PyeongChang, Republic of Korea, 17–20 October 2018; pp. 271–275.

32. Hanwha Systems. Unmanned Maritime System. Available online: https://www.hanwhasystems.com/en/business/defense/naval/marine_index.do (accessed on 21 June 2023).

33. Kang, H.; Cho, G.R.; Kim, M.G.; Lee, M.J.; Li, J.H.; Kim, H.S.; Lee, H.; Lee, G. Mission Management Technique for Multi-sensor-based AUV Docking. *J. Ocean Eng. Technol.* **2022**, *36*, 181–193. [CrossRef]

34. Lee, G.; Lee, P.Y.; Kim, H.S.; Lee, H.; Lee, J. Learning-based Localization of AUV with Outlier Sensor Data. In Proceedings of the 2021 21st International Conference on Control, Automation and Systems (ICCAS), Jeju, Republic of Korea, 12–15 October 2021; pp. 2257–2260.

35. Kim, G.; Lee, J.; Lee, P.; Kim, H.; Lee, H. A study on docking guidance navigation algorithm of AUV by combining inertial navigation sensor and docking guidance sensor. *J. Inst. Control. Robot. Syst.* **2019**, *25*, 647–656. [CrossRef]

36. Choi, K.; Lee, G.; Lee, P.Y.; Kim, H.S.; Lee, H.; Kang, H.; Lee, J. Localization algorithm of multiple-AUVs utilizing relative 3D observations. *J. Korea Robot. Soc.* **2022**, *17*, 110–117. [CrossRef]

37. Wang, J.; Wang, C.; Wei, Y.; Zhang, C. On the fuzzy-adaptive command filtered backstepping control of an underactuated autonomous underwater vehicle in the three-dimensional space. *J. Mech. Sci. Technol.* **2019**, *33*, 2903–2914. [CrossRef]

38. Xia, Y.; Xu, K.; Li, Y.; Xu, G.; Xiang, X. Improved line-of-sight trajectory tracking control of under-actuated AUV subjects to ocean currents and input saturation. *Ocean Eng.* **2019**, *174*, 14–30. [CrossRef]

39. Peng, Z.; Wang, J.; Han, Q.L. Path-following control of autonomous underwater vehicles subject to velocity and input constraints via neurodynamic optimization. *IEEE Trans. Ind. Electron.* **2018**, *66*, 8724–8732. [CrossRef]

40. Craig, J.J. *Introduction to Robotics: Mechanics and Control*; Addision Wesley: Boston, MA, USA, 1989.

41. Youcef-Toumi, K.; Ito, O. A time delay controller for systems with unknown dynamics. *J. Dyn. Syst. Meas. Control* **1990**, *112*, 133–142. [CrossRef]

42. Hsia, T.; Gao, L. Robot manipulator control using decentralized linear time-invariant time-delayed joint controllers. In Proceedings of the IEEE International Conference on Robotics and Automation, Cincinnati, OH, USA, 13–18 May 1990; pp. 2070–2075.

43. Cho, G.R.; Kang, H.; Lee, M.J.; Li, J.H. Heading Control of URI-T, an Underwater Cable Burying ROV: Theory and Sea Trial Verification. *J. Ocean Eng. Technol.* **2019**, *33*, 178–188. [CrossRef]

44. Chang, P.H.; Park, S.H. The development of anti-windup scheme and stick-slip compensator for time delay control. In Proceedings of the 1998 American Control Conference. ACC (IEEE Cat. No.98CH36207), Philadelphia, PA, USA, 26 June 1998; Volume 6, pp. 3629–3633.

Journal of
Marine Science and Engineering

Article

Long-Term Trajectory Prediction for Oil Tankers via Grid-Based Clustering

Xuhang Xu [1], Chunshan Liu [1,*], Jianghui Li [2,*], Yongchun Miao [3] and Lou Zhao [1]

[1] School of Communication Engineering, Hangzhou Dianzi University, Hangzhou 310018, China; xuxuhang@hdu.edu.cn (X.X.); lou.zhao@hdu.edu.cn (L.Z.)
[2] State Key Laboratory of Marine Environmental Science, College of Ocean and Earth Sciences, Xiamen University, Xiamen 361005, China
[3] School of Electronic and Information Engineering, Anhui University, Hefei 230039, China; ycmiao@ahu.edu.cn
[*] Correspondence: chunshan.liu@hdu.edu.cn (C.L.); jli@xmu.edu.cn (J.L.)

Abstract: Vessel trajectory prediction is an important step in route planning, which could help improve the efficiency of maritime transportation. In this article, a high-accuracy long-term trajectory prediction algorithm is proposed for oil tankers. The proposed algorithm extracts a set of waymark points that are representative of the key traveling patterns in an area of interest by applying DBSCAN clustering to historical AIS data. A novel path-finding algorithm is then developed to sequentially identify a subset of waymark points, from which the predicted trajectory to a fixed destination is produced. The proposed algorithm is tested using real data offered by the Danish Maritime Authority. Numerical results demonstrate that the proposed algorithm outperforms state-of-the-art vessel trajectory prediction algorithms and is able to make high-accuracy long-term trajectory predictions.

Keywords: trajectory prediction; AIS data; clustering

1. Introduction

About 90% of global trade is carried by maritime transportation [1]. With the continuing growth of international trade, modern maritime transportation calls for more intelligent methods for transportation management to achieve larger capacity, faster traveling speed and higher safety levels. To achieve these goals, accurate predictions of vessels' future movement is important and can be used in many maritime applications, such as port management, anomaly detection and collision avoidance [2].

Despite the importance of vessel trajectory prediction, it remains a challenging task due to the diverse navigation environments and the stochastic nature of vessel movements. While most of the existing works on vessel trajectory prediction have focused on making short-term predictions [3–7], being particularly useful for collision warning and avoidance, this work investigates the problem of long-term vessel trajectory prediction. Long-term trajectory predictions can be used to guide the captains to operate the ship in a more fuel-efficient way and hence reduce the carbon dioxide emissions, with assistance also from additional information, including weather and current forecasts on the route. Long-term vessel trajectory predictions can also be used by the agencies of port management to obtain more accurate estimates of the remaining navigation distance and subsequently obtain more accurate predictions for vessel arrivals.

The prevalence of maritime transportation data makes it possible to take a data-driven approach to develop high-accuracy vessel trajectory algorithms [2,8]. For instance, the Automatic Identification System (AIS), a global autonomous tracking system that has been made compulsory for ships exceeding 300 tons, provides abundant and near real-time information about ships. Apart from static information such as MMSI, ship name and ship type, the AIS messages also contain the location information of the ship (longitude (Lon) and latitude (Lat)) and information about its traveling (e.g., speed over ground (SOG),

Citation: Xu, X.; Liu, C.; Li, J.; Miao, Y.; Zhao, L. Long-Term Trajectory Prediction for Oil Tankers via Grid-Based Clustering. *J. Mar. Sci. Eng.* **2023**, *11*, 1211. https://doi.org/10.3390/jmse11061211

Academic Editor: Sergei Chernyi

Received: 19 May 2023
Revised: 7 June 2023
Accepted: 8 June 2023
Published: 11 June 2023

course over ground (COG) and heading) [9]. This information, when gathered in large scale, can be used to characterize the voyage patterns in a corresponding area and can further be exploited to predict vessel trajectory.

This paper presents an AIS data based long-term vessel trajectory prediction algorithm, aiming to predict the trajectory of oil tankers from any location to a pre-defined destination, e.g., a port. The proposed algorithm takes a data-driven approach. First, the traveling patterns of tankers in an area of interest are extracted from the historical AIS data via key point clustering. The output of clustering is a set of *waymark points*, with each corresponding to a cluster. Each waymark point is characterized by the average Lon, Lat, COG and the number of key points in the cluster that it represents. In real-time trajectory prediction, the algorithm uses the Lon and Lat to filter the waymark points and uses the information of COG and the number of points to calculate a weighted distance between the filtered waymark points and the current reference point. A segment of the predicted trajectory is generated by connecting the reference point to the waymark point that has the smallest weighted distance. The complete long-term trajectory prediction is made by repeating this process until the reference point reaches the destination. For the proposed algorithm, the key point clustering only needs to be perform once. Once the set of waymark points is obtained, all subsequent oil tankers arriving at the same destination can use the obtained set of waymark points to achieve trajectory prediction. We verify the effectiveness of the proposed algorithm using real AIS data provided by the Danish Maritime Authority (DMA) [10].

The rest of this paper is organized as follows. Section 2 provides a brief review of related work in the area of trajectory prediction. Section 3 describes the proposed trajectory prediction algorithm. In Section 4, the proposed algorithm is applied to real AIS data and is compared to state-of-the-art algorithms. Finally, Section 5 draws conclusions and discusses possible future research directions. For ease of exposition, Table 1 lists the key notation used in this paper.

Table 1. List of key notations.

Key Notation	
λ_k	Longitude
ϕ_k	Latitude
θ_i	COG $\in [0°, 360°)$
$\tilde{p}_k = (\lambda_k, \phi_k, \theta_k)$	The k-th key point
$p_k = (\lambda_k, \phi_k)$	The location of the k-th key point
$q_m = (\bar{\lambda}_m, \bar{\phi}_m, \bar{\theta}_m, \bar{\rho}_m)$	The m-th waymark point, $\bar{\lambda}_m, \bar{\phi}_m, \bar{\theta}_m, \bar{\rho}_m$ denotes the Lon, Lat, COG and the point density
p_{DES}	Destination location
$dire(p_m, p_n)$	The heading direction from position p_m to position p_n
$dist(p_m, p_n)$	The spherical distance between position p_m and p_n
$\vec{a}$	Direction $\vec{a}$, a unit vector of the origin of a two-dimensional coordinate system
$LDM\{\vec{a_1}, \vec{a_2}, ..., \vec{a_m}\}$	The linear directional mean of direction $\vec{a_1}, \vec{a_2}, ..., \vec{a_m}$, defined in Equation (2)
$\Delta_{\vec{a}, \vec{b}}$	The angular change from direction $\vec{a}$ to direction $\vec{b}$, defined in Equation (3)
$\mathcal{Q}_n$	Candidate set identified in the n-th iteration, defined in Equation (4)
$\mathcal{Q}'_n$	Refined candidate set at the n-th iteration, defined in Equation (8), $\mathcal{Q}'_n \subseteq \mathcal{Q}_n$
$\vec{d}_{cur_n}$	The heading trend at the beginning of the n-th iteration, defined in Equation (5)
$\vec{d}_{\mathcal{Q}'_n}$	The mean COGs of the waymark points in $\mathcal{Q}'_n$

Table 1. *Cont.*

Key Notation	
$\vec{d}_n$	The heading direction from position p_{n-1} to position p_n
$\vec{d}_{DES_n}$	The direction from position p_n to the destination
$\delta'_m = \Delta_{\vec{d}_{p_n,q_m},\vec{d}_{DES_n}}$	Directional difference between $\vec{d}_{p_n,q_m}$ and $\vec{d}_{DES_n}$, defined in Equation (6)
$\delta_m = \Delta_{\theta_n,\theta_m}$	Directional difference between the COG of p_n and the COG of q_m, defined in Equation (7)

2. Related Work

With the advances in high-precision positioning technologies such as the Global Positioning Systems and real-time radars, it is possible to acquire accurate information about positions for vehicles, aircraft, ships and pedestrians. The availability of high-accuracy positioning data, together with other information such as speed and acceleration, makes it possible to predict the trajectories of the targets of interest automatically, which could find applications in many areas, including terrestrial navigation [11,12], autonomous driving [13,14], and maritime traffic management [2,8,15,16]. This paper focuses on the trajectory prediction problem for vessels, for which the existing works can be classified into two categories, i.e., short-term trajectory prediction [3–7] and long-term trajectory prediction [17,18].

For short-term vessel trajectory prediction, the time horizon over which the predictions are made is usually from a few seconds to a few tens of minutes. Such vessel prediction algorithms are often developed for collision avoidance to ensure navigation safety. Hence, apart from the prediction accuracy, timeliness is also important [19]. In this category, classical algorithms follow from mathematical models of mobility and statistical techniques for accurate trajectory predictions. Examples of such methods include the Kalman filter-based algorithm [20], the Gaussian process-based algorithm [3], and the Markov process-based algorithm [21]. However, due to the complicated nature of vessel trajectories, these methods may not be able to capture the characteristics of the trajectories to predict and thus fail to provide accurate predictions.

The recent advances in machine learning techniques, along with the assistance of abundant AIS data, have stimulated a burst of research on machine-learning-based vessel trajectory algorithms [5–7]. For instance, Zhang et al. [5] used a hybrid method based on LSTM and KNN for short-term trajectory prediction. The algorithm switches between LSTM and KNN based on the trajectory densities of different areas. That is, in dense areas, KNN is used for trajectory prediction, while in sparse areas, LSTM is adopted. Using publicly available AIS data collected near Xiamen Port, Fujian Province from 2018 to 2019 [5], the results show that this method has better prediction accuracy compared with classical prediction algorithms developed based on Kalman filtering. You et al. [6] proposed a sequence-to-sequence model based on GRU to predict vessel trajectory at a time horizon of 5 min. The model works by first encoding the trajectory as a context vector to maintain the temporal relationship of the trajectory position, and then using GRU as a decoder to output the future trajectory. Numerical results based on the data collected near Chongqing and Wuhan of Yangtze River Channel demonstrate good short-term trajectory predictions. Murray et al. [7] proposed a data-driven trajectory prediction method which first applies Principal Component Analysis (PCA) to each trajectory to generate feature vectors and then uses the extracted feature vectors as inputs for the Gaussian Mixer model to predict multiple possible trajectories at the same time. The time horizon of prediction is about 30 min. All of the above work uses deep learning or data-driven methods and achieves better results.

Compared to the problem of short-term trajectory prediction, the progress on the long-term trajectory prediction problem is much less reported, partly due to the more challenging nature of the long-term prediction problem. Existing short-term prediction algorithms may be used to produce long-term predictions through recursions; however, the accuracies are expected to decrease as the number of prediction steps increases [6]. One relevant work that tackles the problem of long-term trajectory can be found in [22], where DBSCAN was used to cluster historical trajectories, and the trajectory predictions are produced by pre-trained deep learning models. This work was tested using the publicly available AIS data from MarineCadastre in Zones 15 and 16. While [22] also adopted DBSCAN, the overall methodology is completely different from our proposed approach. For instance, in [22], DBSCAN was used to cluster trajectories based on trajectory statistics such as the trimmed mean of the longitude (latitude). In contrast, our work adopts DBSCAN to cluster the key points from many historical trajectories, where key points sharing similar characteristics are grouped into the same cluster. Two other relevant works on long-term vessel trajectory predictions can be found in [17,18]. Both these two works attempt to predict the remaining path of ships to a destination in order to better estimate the remaining traveling distance and to achieve a more accurate estimate of the arrival times. Ogura et al. [17] evaluated the difference between the predicted weather and the weather of historical trajectories, and then adopts the historical trajectory with the smallest difference as the prediction. The AIS data of four cargos traveling to and from Japan was used to test the weather-based algorithm [17]. Alessandrini et al. [18] adopted a graph-based approach by calculating the raster of density and directionality of all the historical data and applying a path-finding algorithm to identify a path from the current location to the destination. This method was tested using AIS data recorded in October 2015 near the port of Trieste, an Italian city on the Northeastern Adriatic Sea, with the assistance of historical LRIT data from 2009 to 2014 [18].

3. Methodology

In this section, we describe the proposed long-term vessel trajectory prediction algorithm. Figure 1 presents a flow chart of the proposed algorithm. As shown in the figure, the algorithm has the following two modules: the key point clustering module and the trajectory finding module. In the remaining part of this section, we present the details of the two modules.

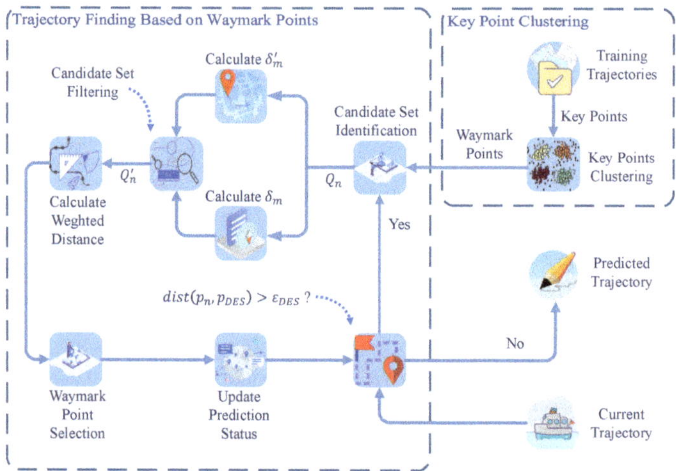

Figure 1. Flow chart of the trajectory prediction algorithm; (1) pre-processed original trajectory is used in key point clustering, (2) filter the candidate set and calculate weighted distance, (3) select the point with the smallest weighted distance as the next location point, (4) calculate the distance from the current location to the destination. If it is less than the threshold value, continue to execute the key point connection step; otherwise, output the predicted trajectory to end the prediction process.

3.1. Key Point Clustering

The key point clustering module aims to extract from pre-processed training trajectories a set of waymark points that are representative of the traveling patterns of the oil tankers in an area of interest. Instead of keeping all the raw AIS data points in the training trajectories, each trajectory is pre-processed to remove redundant data and to reduce the number of points retained in the trajectory. (A detail description of data pre-processing will be presented in Section 4.1 along with the experiments.) The points retained in the training trajectories are called the *key* points. The key point clustering module treats all key points (from all different training trajectories) as the input and produces a set of *waymark* points. As a result of clustering, each waymark point not only reflects the Lon and Lat of tankers passing through the vicinity of the corresponding cluster, but also contains information such as the COG to better capture the motion of tankers in the corresponding area.

Each predicted trajectory consists of a number of waymark points. Hence, to achieve accurate predictions, there should be a sufficient number of waymark points that can comprehensively capture all possible routes in the area of interest. Additionally, waymark points that are close in space should have distinguishing features related to the motion of the vessels to make clear selections among the waymark points. To fulfill these purposes, we adopt a COG-based clustering process to generate the waymark points [23].

To facilitate the description of the clustering process, we use λ_k, ϕ_k to represent the Lon and Lat of the key point $\tilde{p}_k$, respectively, and θ_k the COG of $\tilde{p}_k$. Hence, each key point is represented by $\tilde{p}_k = (\lambda_k, \phi_k, \theta_k)$. Let $\mathcal{D} = \{\tilde{p}_1, \tilde{p}_2, \tilde{p}_3, \ldots, \tilde{p}_K\}$ be the set of key points in the training trajectories. The core idea of the clustering algorithm is first to discretize the entire area of interest into a set of grids and then run DBSCAN for each grid to cluster the key points falling into the corresponding area. Specifically, DBSCAN is used to cluster the key points within the grid area based only on the COG to extract representative traveling directions. Each cluster identified by DBSCAN forms a waymark point q_m, which contains four-dimensional information, including the average Lon $\bar{\lambda}_m$ and average Lat $\bar{\phi}_m$ of all the key points in the corresponding cluster, the median COG of the key points, i.e., $\bar{\theta}_m$, and the number of key points $\bar{\rho}_m$ in the cluster, i.e., $q_m = (\bar{\lambda}_m, \bar{\phi}_m, \bar{\theta}_m, \bar{\rho}_m)$.

Each of the grid areas can be represented as:

$$G_{i,j} = \{(\lambda, \phi) | \lambda \in (\bar{\lambda}_i - \delta_\lambda, \bar{\lambda}_i + \delta_\lambda), \phi \in (\bar{\phi}_j - \delta_\phi, \bar{\phi}_j + \delta_\phi)\}, \tag{1}$$

where $\bar{\lambda}_i$ and $\bar{\phi}_j$ are the Lon and Lat of the center of the grid, and δ_λ and δ_ϕ are the widths in the longitude and latitude direction. The difference of the center Lon (Lat) between two adjacent grid areas can be made smaller than $2\delta_\phi$ $(2\delta_\lambda)$, e.g., $|\bar{\lambda}_i - \bar{\lambda}_{i+1}| < 2\delta_\phi$. In such cases, two adjacent grid areas may partially overlap with each other, and the key points falling into the overlapped areas are used multiple times during clustering. Such a setup makes it easier to obtain a representative set of waymark points from non-uniformly distributed key points by using a relatively large grid width.

It is noted that a customized distance metric is adopted when applying DBSCAN for COG clustering. Specifically, suppose $\theta_i \in [0°, 360°)$ and $\theta_j \in [0°, 360°)$ are two COGs. Then, the corresponding DBSCAN distance metric is calculated as $|\theta_i - \theta_j|$ if $|\theta_i - \theta_j| < 180°$ and $360 - |\theta_i - \theta_j|$ otherwise. This customized metric is to avoid the errors caused by COGs close to the north, i.e., COG values close to $0°$ or $360°$. Additionally, it is noted that the Lon and Lat are not used in the clustering process. This is due to the fact that the COG clustering is performed for each grid area, where the key points have relatively close values of Lon and Lat.

As an illustrative example, Figure 2 presents the key points obtained from the trajectories of the oil tankers passing through the shown area between 2017 to 2019. It can be seen that there are areas where the key points are sparse, while there are also areas where the key points are much denser. It is known that DBSCAN requires one to specify a value on the minimum number of points in a cluster so the algorithm can operate. If there are less key points than this value, then all points in the grid area will be classified as noise, and no cluster can be formed. Therefore, in a sparse area, the width of the grid area has to be set sufficiently large. However, if overlapping is not permitted, then a large width will lead to a reduced number of waymark points, which will make the representation of the traveling patterns less accurate.

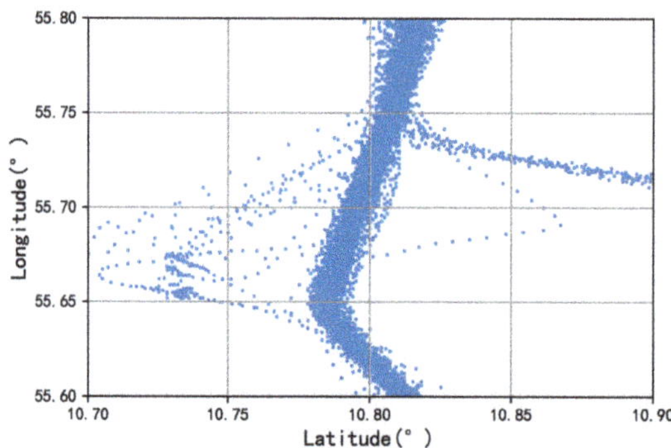

Figure 2. Key points obtained from oil tankers' trajectories between 2017 to 2019.

3.2. Trajectory Finding Based on Waymark Points

Without loss of generality, suppose a prediction of the future trajectory is made at time instant 0, where the historical trajectory consisted of the set of previous location points $\{p_{-I} \ldots, p_{-2}, p_{-1}, p_0\}$, and $p_n = (\lambda_n, \phi_n)$ contains the Lat and Lon of the n-th position. Denote $Q = \{q_1, q_2, \ldots, q_M\}$ as the set of all waymark points obtained from the key point clustering. The task of the trajectory prediction is to select from Q a set of N points to produce the predicted trajectory $T = \{p_1, \ldots, p_n, \ldots, p_N, p_{DES}\}$, where p_{DES} is the destination of interest. This process is implemented in an iterative way by sequentially identifying the N waymark points. Note that the number of point N is determined by the algorithm automatically and is not a preset value.

In the n-th iteration, the algorithm first identifies a candidate waymark point set $Q_n \subseteq Q$ based on $T_n = \{p_{-I} \ldots, p_0, \ldots, p_n\}$. The candidate set Q_n is then filtered to generate a refined candidate set Q'_n, from which the predicted point p_{n+1} is chosen based on a weighted distance to be explained later in this section. In this process, the Lat, Lon and the information of direction and point density are all utilized.

For ease of exposition, we introduce the following notation:

- $dire(p_m, p_n)$: the heading direction from position p_m to position p_n;
- $dist(p_m, p_n)$: the spherical distance between p_m to position p_n, calculated using the Python package Geopy;
- $LDM\{\vec{a}_1, \vec{a}_2, \ldots, \vec{a}_m\}$: the linear directional mean of direction $\vec{a}_1, \vec{a}_2, \ldots, \vec{a}_m$, calculated as:

$$LDM\{\vec{a}_1, \vec{a}_2, \ldots, \vec{a}_m\} = \angle\left(\frac{\vec{a}_1 + \vec{a}_2 + \ldots + \vec{a}_m}{m}\right), \tag{2}$$

  where $\vec{a}_i, i = 1, \ldots, m$ is a unit vector from the origin.

- $\Delta_{\vec{a},\vec{b}}$: the angular change from direction $\vec{a}$ to direction $\vec{b}$, measured in degrees. $\Delta_{\vec{a},\vec{b}}$ is calculated as :

$$\Delta_{\vec{a},\vec{b}} = \begin{cases} 360° - \Delta & \Delta \geq 180° \\ \Delta & \Delta \leq 180° \end{cases} \tag{3}$$

where $\Delta = |\angle\vec{a} - \angle\vec{b}|$. Here, angle $\angle\vec{a}$, $\angle\vec{b}$ both take value in $[0°, 360°)$, where $0°$ corresponds to the north.

At the n-th iteration, the distance from p_n to p_{DES} is compared against a pre-determined threshold ε_{DES}. If

$$dist(p_n, p_{DES}) < \varepsilon_{DES},$$

i.e, the distance to the destination is sufficiently small, then the iterative process terminates, and p_{DES} is appended to T_n to form the completed trajectory prediction. Otherwise, the algorithm continues to identify p_{n+1}. The threshold ε_{DES} needs to be chosen with care, as setting it too large can lead to premature terminations of the trajectory finding, while setting it too small may make the prediction susceptible to interference from other routes in the areas of high concentrations of multiple routes towards different directions, e.g., the areas near a port. We recommend to set ε_{DES} to be the distance from the destination from which various routes begin to diverge.

3.2.1. Candidate Set $\mathcal{Q}_n$ Identification and Navigation Status Update

At iteration $n \geq 1$, the algorithm first identifies the candidate set $\mathcal{Q}_n$, which is made up of all the waymark points within a rectangle centered at the latest point in the trajectory, i.e., around point $p_n = (\bar{\lambda}_n, \bar{\phi}_n)$:

$$\mathcal{Q}_n = \{(\bar{\lambda}_m, \bar{\phi}_m, \bar{\theta}_m, \bar{\rho}_m) | \lambda_m \in (\bar{\lambda}_n - \varepsilon_\lambda, \ \bar{\lambda}_n + \varepsilon_\lambda), \phi_m \in (\bar{\phi}_n - \varepsilon_\phi, \ \bar{\phi}_n + \varepsilon_\phi)\} \tag{4}$$

where $\varepsilon_\lambda > 0$ and $\varepsilon_\phi > 0$ are used to define the size of the rectangular area.

To ensure that the waymark point p_{n+1} locates in the direction that is consistent with the recent trajectory trend, we update the current heading trend $\vec{d}_{cur_n}$ as:

$$\vec{d}_{cur_n} = \begin{cases} \vec{d}_n & n = 1 \\ LDM\left(\vec{d}_{cur_{n-1}}, \vec{d}_n, \vec{d}_{\mathcal{Q}'_{n-1}}\right) & n > 1 \end{cases} \tag{5}$$

where $\vec{d}_n = dire(p_{n-1}, p_n)$ and $\vec{d}_{\mathcal{Q}'_{n-1}}$ is the mean of the COGs of the waymark points in set $\mathcal{Q}'_{n-1}$, and $\mathcal{Q}'_{n-1}$ is the filtered candidate set in the $(n-1)$-th iteration (see Section 3.2.2 for details). From the above equation, it can be seen that when $n > 1$, $\vec{d}_{cur_n}$ is the linear directional mean of the (predicted) traveling direction of iteration $n - 1$, the heading direction $\vec{d}_n$ from point p_{n-1} to p_n (the current position), and the mean COG of the key points in $\mathcal{Q}'_{n-1}$.

3.2.2. Candidate Set Filtering

In this step, the waymark points in $\mathcal{Q}_n$ that deviate significantly from the expected traveling direction are filtered out from further consideration in the current prediction iteration. To do so, we first denote $\vec{d}_{DES_n} \doteq dire(p_n, p_{DES})$ as the direction from the current position to the destination and then introduce the following angle-dependent parameters:

$$\delta'_m = \Delta_{\vec{d}_{p_n, q_m}, \vec{d}_{DES_n}} \tag{6}$$

$$\delta_m = \Delta_{\theta_n, \bar{\theta}_m} \tag{7}$$

where $\vec{d}_{p_n, q_m} = dire(p_n, q_m)$ is the heading direction from p_n to q_m. Parameter δ'_m measures the degree of difference between $\vec{d}_{p_n, q_m}$ and the direction of the current position towards the destination, while δ_m measures the degree of difference between the COG of p_n and the COG of q_m. An illustration of the above notation is given in Figure 3.

Ideally, if a tanker does not make a sharp turn, δ_m is expected to take small values. Additionally, as the tanker is traveling towards the destination, it is expected that the heading from p_n to q_m does not deviate too drastically from $\vec{d}_{DES_n}$. Therefore, it is expected that δ'_m does not take extremely large values. Therefore, we identify the refined candidate set by applying filtering based on δ_m and δ'_m as follows:

$$\mathcal{Q}'_n = \{q_m | q_m \in Q_n, \delta_m \leq \epsilon_m, \delta'_m \leq \epsilon'_m\}, \tag{8}$$

where ϵ_m and ϵ'_m are two thresholds.

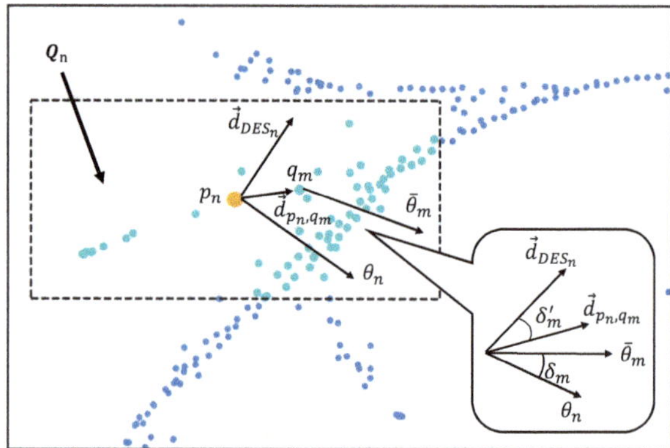

Figure 3. An illustration of the mathematical notation for candidate set filtering.

3.2.3. Waymark Point Selection

Once the filtered candidate set of waymark points, $\mathcal{Q}'_n$, has been identified, the algorithm can proceed to make a selection for the next waymark point. As each waymark point $q_m = (\bar{\lambda}_m, \bar{\phi}_m, \bar{\theta}_m, \bar{\rho}_m)$ represents a cluster of key points, a higher value of $\bar{\rho}_m$ means that the corresponding grid area has been visited more frequently. Hence, it is reasonable to allocate more weight to such waymark points in the selection. Additionally, the expected trajectory trend, i.e., $\vec{d}_{cur_n}$, is expected to be close to the heading from p_n to q_m. Based on these considerations, we propose a weighted distance to account for both the frequency that a waymark point is visited and the angular change from p_n to q_m. Specifically, the weighted distance between the current point p_n and a candidate waymark point $q_m \in \mathcal{Q}'_n$ can be calculated as:

$$Dis(p_n, q_m) = \kappa\sqrt{\rho_{max}/\bar{\rho}_m} + (1 - \kappa)\Delta_{\vec{d}_{cur_n}, \vec{d}_{p_n, q_m}} \tag{9}$$

where $\Delta_{\vec{d}_{cur_n}, \vec{d}_{p_n, q_m}}$ is the angular change defined in (3), and $\kappa \in (0, 1)$ is the weight. Usually, $\Delta_{\vec{d}_{cur_n}, \vec{d}_{p_n, q_m}}$ is much larger than $\sqrt{\rho_{max}/\bar{\rho}_m}$; thus, κ can be set close to 1 to better balance the two terms. Parameter $\rho_{max} = \max\{\bar{\rho}_m | q_m \in \mathcal{Q}'_n\}$ is the maximum density of all the waymark points in the refined candidate set $\mathcal{Q}'_n$. Based on the weighted distance, the next waymark point is then selected as the one with the smallest weighted distance, i.e.,

$$p_{n+1} = \arg\min_{q_m \in \mathcal{Q}'_n} Dis(p_n, q_m). \tag{10}$$

The weighted distance defined in (9) does not consider the geo-distance between two waymark points. This is due to the fact that the refined candidate set $\mathcal{Q}'_n \subseteq \mathcal{Q}_n$ contains waymark points that are sufficiently close to the current reference point, as shown in (4). In this case, it is sufficient to focus on the angular and density differences to identify the next waymark point. The proposed trajectory prediction algorithm is finally summarized in Algorithm 1:

Algorithm 1 Trajectory prediction algorithm

Require: p_{DES}, $T_0 = \{\dots, p_{-1}, p_0\}$, $\mathcal{Q} = \{q_1, q_2, \dots, q_M\}$
Parameters: ε_{DES}, ε_λ, ε_ϕ, ε_{δ_k}, $\varepsilon_{\delta'_k}$, κ, ϵ_m, ϵ'_m
Initialization: $T = \{p_0\}$
    while $dist(p_n, p_{DES}) \geq \varepsilon_{DES}$ **do**
        $\mathcal{Q}_n = \{(\bar{\lambda}_m, \bar{\phi}_m, \bar{\theta}_m, \bar{\rho}_m) | \bar{\lambda}_m \in (\bar{\lambda}_n - \varepsilon_\lambda, \bar{\lambda}_n + \varepsilon_\lambda),$
            $\bar{\phi}_m \in (\bar{\phi}_n - \varepsilon_\phi, \bar{\phi}_n + \varepsilon_\phi)\}$
        $\mathcal{Q}'_n = \{q_m | q_m \in \mathcal{Q}_n, \delta_m \leq \epsilon_m, \delta'_m \leq \epsilon'_m\}$
        for all q_m *in* $\mathcal{Q}'_n$ **do**
            $Dis(p_n, q_m) = \kappa \sqrt{\rho_{max}/\bar{\rho}_m} + (1 - \kappa)\Delta_{\vec{d}_{cur_n}, \vec{d}_{p_n, q_m}}$
        end for
        $p_{n+1} = \arg\min_{q_m \in \mathcal{Q}'_n} Dis(p_n, q_m)$
        Append p_{n+1} to T
    end while
    return T

---

4. Experimental Results

In this section, we present the experimental results that evaluate the performance of the proposed trajectory prediction method. We start this section by introducing the dataset and the pre-processing techniques used to process the raw AIS data. The results obtained from key point clustering are then presented to illustrate the operation of the proposed method, followed by the evaluations of trajectory prediction accuracies.

4.1. Dataset and Pre-Processing

In the experiment, we evaluate the proposed trajectory prediction method using the historical AIS data provided by the Danish Maritime Authority (DMA) [10]. The raw data contain AIS messages from various types of vessels traveling around Danish waters. The AIS data have 26 fields, including latitude, longitude, COG, destination, etc. In these experiments, we will mainly use latitude, longitude and COG fields for trajectory prediction. A comprehensive list and description of other fields can be found in Appendix A of [23]. The experiment uses the AIS data from oil tankers traveling to Port Skagen during the period between 2017 and 2019. This dataset contains the AIS data for 1278 vessels and has over 40 million AIS records. In this dataset, the data between 1 January 2017 and 30 June 2019 are used to extract the training trajectories and produce the waymark points. The AIS data from oil tankers traveling to the same port from 1 July 2019 to 31 December 2019 are used to test the proposed trajectory prediction algorithm.

The raw data are pre-processed before being used to produce the waymark points. Key steps of the pre-processing include trajectory extraction and integrity check and downsampling. We refer to Appendix B of [23] for a detailed description of the pre-processing steps. In the experiments, a total of 1046 training trajectories were obtained after the pre-processing, and there are 232 trajectories in the test period. Each trajectory undergoes a route to the destination, i.e., Port Skagen.

4.2. Results for Grid-Based Key Point Clustering

In the training dataset, there are over 550,000 key points. During the key point clustering, the difference in latitude/longitude between the center of the adjacent grid areas is set to 0.05 degrees, while the width of the grid area is 0.05 degrees for both the latitude and the longitude. The parameters for DBSCAN are listed in Table 2.

Table 2. Key point clustering: DBSCAN parameter settings.

		Minimum Number of Points		ϵ
		$100 \leq n$	40	
COG clustering	key points n	$10 \leq n < 100$	5	1
		$n < 10$	$\lceil n/3 \rceil$	

Figure 4 plots all the key points from the training trajectories, from which the results show that there are some relatively clear routes in the eastern part of the area. In contrast, the routes are much less clear in the western part of the area. After DBSCAN clustering, a total of 6127 waymark points are obtained, which are plotted in Figure 5. Comparing Figure 4 to Figure 5, the results show that although the number of waymark points is only about 1.1% of the key points, they capture all the main route patterns in both the western and the eastern part of the area.

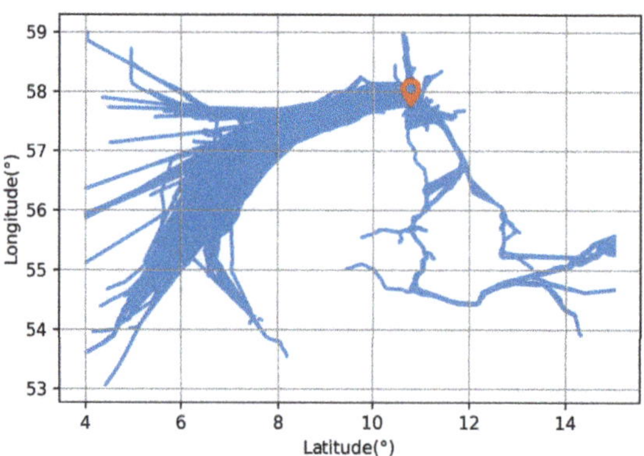

Figure 4. All key points in the training trajectories.

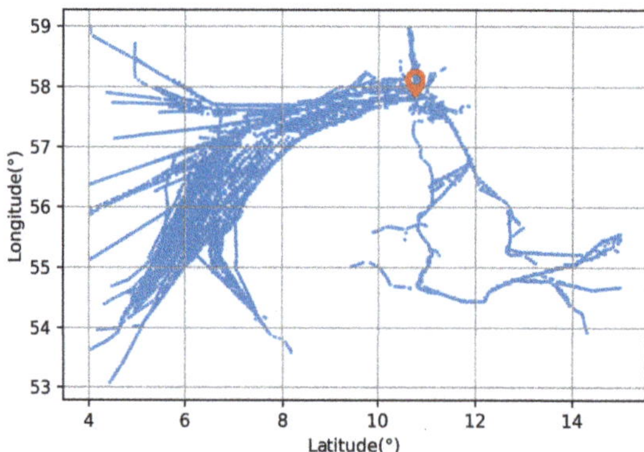

Figure 5. All waymark points obtained after clustering of the key points.

Figure 6 shows the empirical cumulative distribution (CDF) of the COG standard deviation (STD) before and after clustering. Before applying the clustering algorithm, each COG STD is estimated based on the key points within the same grid area. After the clustering, the COG STD is for the key points within the same cluster. Therefore, as expected and shown in Figure 6, the clustering significantly reduces the STD of the COG. For example, before applying the COG clustering, the 95% STD value is about 67.8°. This means that for about 5% of the clusters (before applying the COG clustering), the COG spread measured by the STD is $\geq 67.8°$. This result suggests that although the grid areas are relatively small, the traveling directions of the tankers within the same small area can differ significantly. In contrast, after applying the clustering, the 95% STD of the COG reduces to 2.2°. In other words, after clustering, the traveling direction of the cluster becomes more consistent, which is more indicative for the prediction of future positions of the vessel of interest.

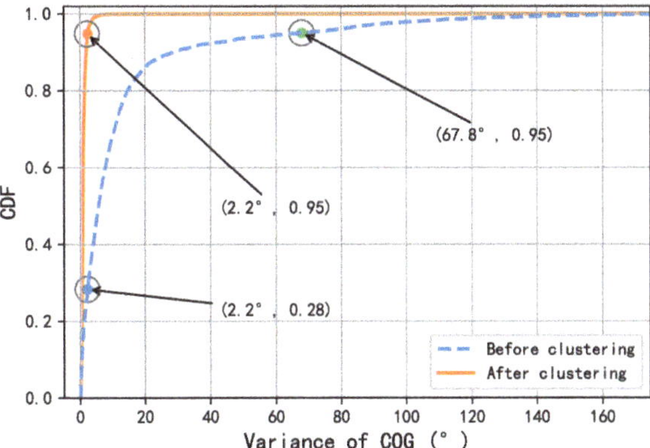

Figure 6. The empirical distribution of the COG variance before and after grid clustering, showing that the variance of the COG is greatly reduced after clustering.

4.3. Results for Trajectory Prediction

We now evaluate the performance of the proposed trajectory prediction. The parameters for waymark point connection are set as follows. The distance threshold $\varepsilon_{DES} = 20$ km, the length and the width of the rectangular region used to identify $\mathcal{Q}_n$ were set to $\varepsilon_\lambda = 0.2°$, $\varepsilon_\varphi = 0.2°$, and the weight coefficient of the weighted distance is set to $\kappa = 10/11$. In the filtering waypoint step, we have chosen two more lenient values to filter out impossible points, which are $\epsilon'_m = 180°$ and $\epsilon_m = 120°$.

Figure 7 shows the number of test samples versus the prediction horizon measured by the number of hours. It can be seen that the test samples contain a mixed of short-term and long-term trajectory prediction tests. About 53.5% of the tests correspond to trajectory predictions in 10 h or more.

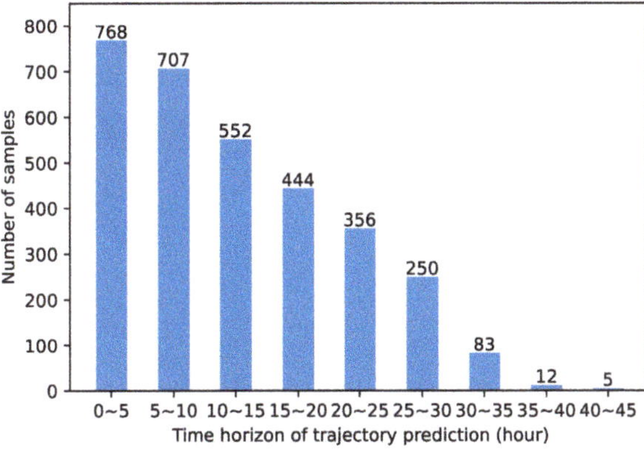

Figure 7. Number of test samples vs. time horizon of trajectory prediction.

To test the proposed algorithm, we implemented a trajectory prediction algorithm based on graph theory and the Dijkstra algorithm [18] and use it as the baseline. Additionally, we also implemented a straightforward trajectory prediction algorithm that simply selects the training trajectory that most matches the historical segment of the trajectory under prediction. This baseline is termed *best-matched*. Directed Hausdorff distance is adopted to measure the similarity. Figure 8 presents four examples of the trajectory prediction results of the proposed algorithm (the red curves), the graph-based baseline from [18] (the greed curves), and the baseline of the best matched (the orange curves). The ground truth trajectory is also plotted as a reference in each example (the grey solid curves). In the figures, the grey dotted curves are the historical segment of the trajectory prior to the prediction moment. The results show that while the baseline algorithms can provide trajectory predictions fairly close to the ground truth, the smoothness of the predicted trajectories is not good enough. As a comparison, the proposed algorithm can offer more accurate predictions in the sense that the predicted trajectories are visually closer to the true ones and retain the smoothness of realistic trajectories.

From a quantitative perspective, we further evaluate the similarities between the predicated and the ground truth trajectories for the proposed algorithm and the baseline algorithms. To measure the similarity, we consider the DTW distance [24–27] and the Hausdorff distance [28], which are commonly used to measure the similarities of two curves/sequences of data. A smaller value of this metric suggests a higher degree of similarity and thus a more accurate trajectory prediction.

As the area of interest contains different routes to Port Skagen from different directions, to better demonstrate the details of the performance evaluation, we classified the trajectories into eight trajectory clusters, as shown in Figure 9, for the tested data. Table 3 shows the number of test instances for each cluster in the experiment and the corresponding average DTW and HD. Each test instance refers to a time instance of a trajectory in the test dataset. Since each track lasts more than 25 h on average, the number of test instances is significantly larger than the number of trajectories.

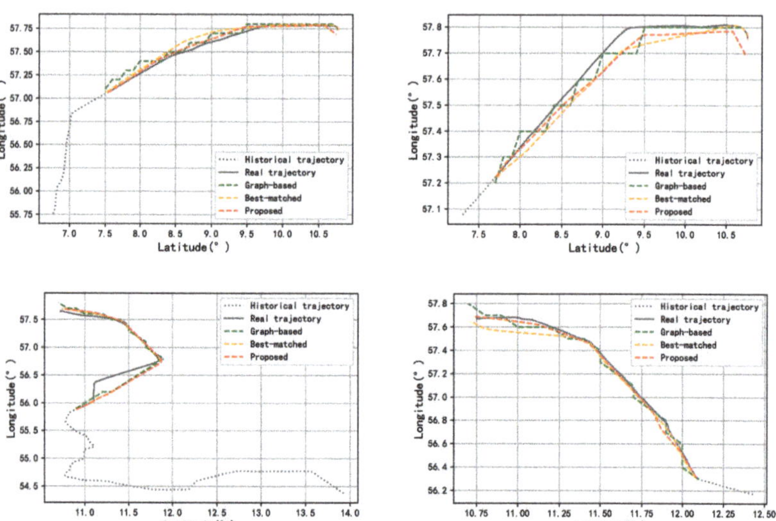

Figure 8. Examples of trajectory prediction results.

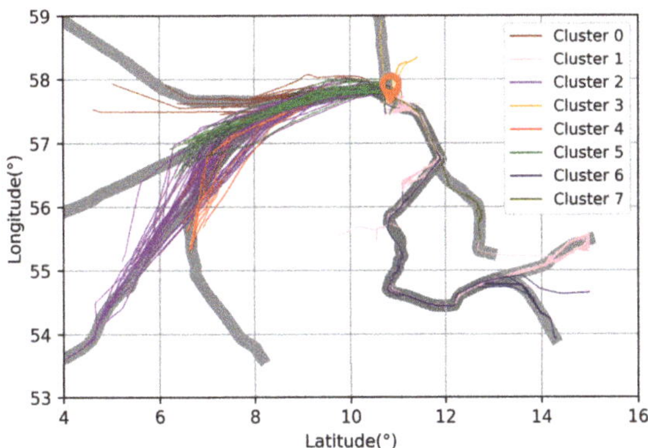

Figure 9. An illustration of the trajectories of the 8 clusters in the test dataset.

From Table 3, it can be seen that the proposed algorithm outperforms the baseline algorithms on both the performance metrics and for all the eight trajectory classes. Specifically, there is a 12.08% and 25.18% improvement in the DTW distance compared to the best-history and the graph-based baselines, respectively. For Hausdorff distance, the improvement is 11.05% and 42.04% over the two baselines. It can also be seen that for trajectory clusters 1, 6 and 7, which locate on the eastern side of the sea, the advantages of the proposed algorithm over the graph-based approach are even more significant due to the complexity of the shipping lanes and the extremely narrow passable channels.

To further show the stability of the performance of the algorithms, box plots of the performance metrics of the three algorithms are presented in Figures 10 and 11, where the orange line indicates the median, and the green triangle is the location of the mean. It can be seen that for both metrics, the proposed algorithm has a lower median and 75% quantile of the mean than the baseline algorithm, demonstrating higher robustness.

Table 3. Performance comparison between different trajectory prediction algorithms based on DTW and Hausdorff distance.

Cluster	Sample Size	DTW Distance (Unit: km)			Hausdorff Distance (Unit: km)		
		Proposed	Best-Matched	Graph-Based	Proposed	Best-Matched	Graph-Based
0	137	555.68	676.44	772.76	8.93	9.96	11.68
1	1734	759.10	908.50	1181.15	11.68	16.04	19.06
2	903	1872.65	2086.40	2221.03	13.46	15.80	15.16
3	13	201.43	218.75	421.26	5.10	8.18	9.86
4	105	1475.34	1552.61	1668.01	11.80	10.42	12.79
5	162	1075.40	1050.57	1204.27	13.61	14.70	15.07
6	115	522.76	548.78	943.52	7.83	8.85	14.81
7	8	167.82	408.43	359.49	3.28	8.01	14.45
Average	3177	1094.31	1244.72	1462.59	11.98	15.14	17.02
Percentage			−12.08%	−25.18%		−11.05%	−42.04%

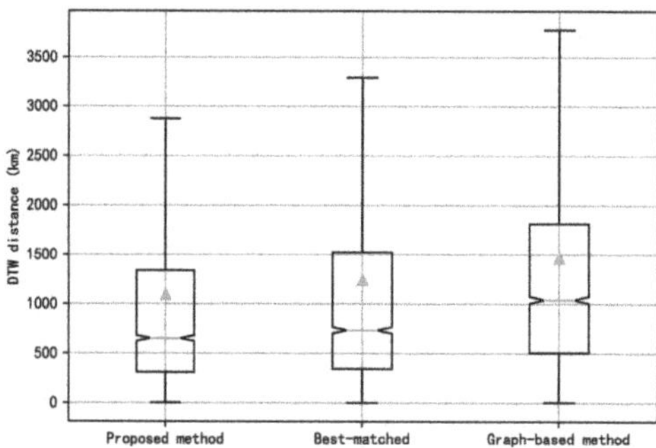

Figure 10. Boxplot of the DTW distance obtained for the proposed algorithm and the two baselines.

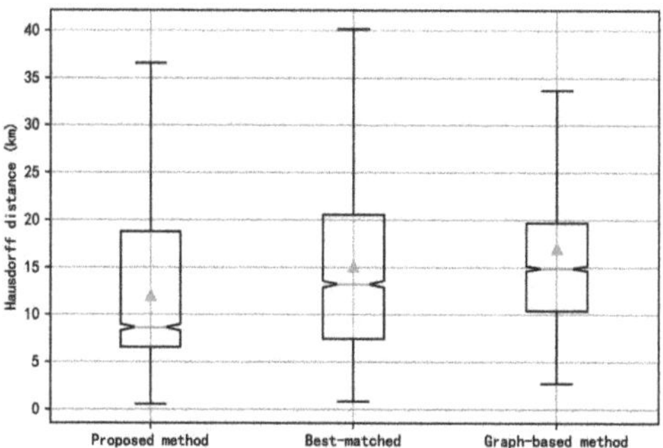

Figure 11. Boxplot of the Hausdorff distance obtained for the proposed algorithm and the two baselines.

5. Conclusions and Discussion

In this paper, a long-term trajectory prediction algorithm has been developed for oil tankers traveling to a known destination port. The proposed algorithm utilizes key points from historical training trajectories that are extracted from historical AIS data in an area of interest. The algorithm then applies the DBSCAN clustering algorithm to generate a set of waymark points that are much fewer than the key points but still retain all the traveling patterns of the oil tankers. Based on the waymark points, a novel path-finding algorithm has been developed that sequentially identifies a set of waymark points to form the predicted trajectory. The proposed algorithm was tested on real AIS data for oil tankers in Danish waters with a fixed destination of port Skagen. Experimental results show that compared to existing trajectory prediction algorithms, including the graph-based and the best-matched approach, the proposed method can achieve more accurate trajectory predictions with higher trajectory smoothness. Specifically, measured by DTW distance and Hausdorff distance, the proposed method offers a reduction of 25.18% and 42.04% over the graph-based baseline and 12.08% and 11.05% over the best-matched baseline, respectively.

While the proposed algorithm has been tested using real AIS data, the tested scenario is relatively limited. Further testings with AIS data from different areas, on different types of vessels, and to different destinations are needed to obtain a more comprehensive evaluation of the performance of the algorithm. Additionally, the focus of the current work was only on the prediction of the trajectory path. Further effort is needed to match the positions of the predicted trajectory to timestamps. In this direction, a combination of the current algorithm with machine learning techniques may be needed to better capture the motion dynamics of vessels in different areas.

Author Contributions: X.X.: onceptualization, methodology, formal analysis, software, visualization, writing—original draft. C.L.: conceptualization, methodology, formal analysis, visualization, supervision, writing—original draft, writing—review and editing, funding acquisition. J.L.: conceptualization, methodology, visualization, validation, supervision, writing—review and editing, funding acquisition. Y.M.: conceptualization, methodology, visualization, supervision, writing—review and editing. L.Z.: conceptualization, methodology, visualization, supervision, writing—review and editing. All authors have read and agreed to the published version of the manuscript.

Funding: This work was jointly supported by the Zhejiang Provincial Natural Science Foundation of China (Grant No. LZ22F010001) and the Fundamental Research Funds for the Provincial Universities of Zhejiang (Grant No. GK229909299001-013).

Institutional Review Board Statement: Not applicable.

Informed Consent Statement: Not applicable.

Data Availability Statement: The raw AIS data are available at "https://dma.dk/safety-at-sea/navigational-information/ais-data" (accessed on 6 June 2023). The processed data that support the findings of this study are available on request from the corresponding authors.

Conflicts of Interest: The authors declare no conflict of interest.

References

1. Zhang, X.; Fu, X.; Xiao, Z.; Xu, H.; Qin, Z. Vessel Trajectory Prediction in Maritime Transportation: Current Approaches and Beyond. *IEEE Trans. Intell. Transp. Syst.* **2022**, *23*, 19980–19998. [CrossRef]
2. Tu, E.; Zhang, G.; Mao, S.; Rachmawati, L.; Huang, G. Modeling Historical AIS Data For Vessel Path Prediction: A Comprehensive Treatment. *arXiv* **2020**, arXiv:2001.01592.
3. Rong, H.; Teixeira, A.P.; Soares, C.G. Ship trajectory uncertainty prediction based on a Gaussian Process model. *Ocean Eng.* **2019**, *182*, 499–511. [CrossRef]
4. Altché, F.; de La Fortelle, A. An LSTM Network for Highway Trajectory Prediction. In Proceedings of the 2017 IEEE 20th International Conference on Intelligent Transportation Systems (ITSC), Yokohama, Japan, 16–19 October 2017; IEEE: Piscataway, NJ, USA, 2017; pp. 353–359.
5. Zhang, L.; Zhu, Y.; Su, J.; Lu, W.; Li, J.; Yao, Y. A Hybrid Prediction Model Based on KNN-LSTM for Vessel Trajectory. *Mathematics* **2022**, *10*, 4493. [CrossRef]
6. You, L.; Xiao, S.; Peng, Q.; Claramunt, C.; Zhang, J. ST-Seq2Seq: A Spatio-Temporal Feature-Optimized Seq2Seq Model for Short-Term Vessel Trajectory Prediction. *IEEE Access* **2020**, *8*, 218565–218574. [CrossRef]
7. Murray, B.; Perera, L.P. An AIS-based multiple trajectory prediction approach for collision avoidance in future vessels. In Proceedings of the International Conference on Offshore Mechanics and Arctic Engineering, Glasgow, UK, 9–14 June 2019; American Society of Mechanical Engineers: New York, NY, USA, 2019; Volume 58851.
8. Kim, E.; Cho, H.; Kim, N.; Kim, E.; Ryu, J.; Park, H. Sensitive Resource and Traffic Density Risk Analysis of Marine Spill Accidents Using Automated Identification System Big Data. *J. Mar. Sci. Appl.* **2020**, *19*, 173–181. [CrossRef]
9. Artana, K.; Pitana, T.; Dinariyana, D.; Ariana, M.; Kristianto, D.; Pratiwi, E. Real-time monitoring of subsea gas pipelines, offshore platforms, and ship inspection scores using an Automatic Identification System. *J. Mar. Sci. Appl.* **2018**, *17*, 101–111. [CrossRef]
10. DMA. AIS Data. 2022. Available online: https://dma.dk/safety-at-sea/navigational-information/ais-data (accessed on 6 June 2023).
11. Hacinecipoglu, A.; Konukseven, E.I.; Koku, A.B. Multiple Human Trajectory Prediction and Cooperative Navigation Modeling in Crowded Scenes. *Intell. Serv. Robot.* **2020**, *13*, 479–493. [CrossRef]
12. Shang, J.; Liu, C.; Shi, H.; Cheng, T.; Yue, K. A New Algorithm for Navigation Trajectory Prediction of Land Vehicles Based on a Generalised Extended Extrapolation Model. *J. Navig.* **2019**, *72*, 1217–1232. [CrossRef]
13. Ni, Y.; Zhao, X. 3DTRIP: A General Framework for 3D Trajectory Recovery Integrated With Prediction. *IEEE Robot. Autom. Lett.* **2023**, *8*, 512–519. [CrossRef]
14. Greer, R.; Deo, N.; Trivedi, M. Trajectory Prediction in Autonomous Driving With a Lane Heading Auxiliary Loss. *IEEE Robot. Autom. Lett.* **2021**, *6*, 4907–4914. [CrossRef]
15. Kim, K.I.; Lee, K.M. Deep Learning-Based Caution Area Traffic Prediction with Automatic Identification System Sensor Data. *Sensors* **2018**, *18*, 3172. [CrossRef]
16. Xu, X.; Bai, X.; Xiao, Y.; He, J.; Xu, Y.; Ren, H. A Port Ship Flow Prediction Model Based on the Automatic Identification System and Gated Recurrent Units. *J. Mar. Sci. Appl.* **2021**, *20*, 572–580. [CrossRef]
17. Ogura, T.; Inoue, T.; Uchihira, N. Prediction of Arrival Time of Vessels Considering Future Weather Conditions. *Appl. Sci.* **2021**, *11*, 4410. [CrossRef]
18. Alessandrini, A.; Mazzarella, F.; Vespe, M. Estimated Time of Arrival Using Historical Vessel Tracking Data. *IEEE Trans. Intell. Transp. Syst.* **2019**, *20*, 7–15. [CrossRef]

19. Petrou, P.; Tampakis, P.; Georgiou, H.V.; Pelekis, N.; Theodoridis, Y. Online Long-Term Trajectory Prediction Based on Mined Route Patterns. In Proceedings of the Multiple-Aspect Analysis of Semantic Trajectories: First International Workshop, MASTER 2019, Wurzburg, Germany, 16 September 2020; pp. 34–49.

20. Perera, L.P.; Soares, C.G. Ocean vessel trajectory estimation and prediction based on extended Kalman filter. In Proceedings of the Second International Conference on Adaptive and Self-Adaptive Systems and Applications, Lisbon, Portugal, 21–26 November 2010; Citeseer: Princeton, NJ, USA, 2010; pp. 14–20.

21. Vasquez, D.; Fraichard, T.; Laugier, C. Growing Hidden Markov Models: An Incremental Tool for Learning and Predicting Human and Vehicle Motion. *Int. J. Robot. Res.* **2009**, *28*, 1486–1506. [CrossRef]

22. Bautista-Sánchez, R.; Barbosa-Santillan, L.I.; Sánchez-Escobar, J.J. Method for select best AIS data in prediction vessel movements and route estimation. *Appl. Sci.* **2021**, *11*, 2429. [CrossRef]

23. Xu, X.; Liu, C.; Li, J.; Miao, Y. Trajectory clustering for SVR-based Time of Arrival estimation. *Ocean Eng.* **2022**, *259*, 111930. [CrossRef]

24. Li, H.; Liu, J.; Liu, R.W.; Xiong, N.; Wu, K.; Kim, T.H. A dimensionality reduction-based multi-step clustering method for robust vessel trajectory analysis. *Sensors* **2017**, *17*, 1792. [CrossRef]

25. Yuan, G.; Sun, P.; Zhao, J.; Li, D.; Wang, C. A review of moving object trajectory clustering algorithms. *Artif. Intell. Rev.* **2017**, *47*, 123–144. [CrossRef]

26. Li, H.; Liu, J.; Yang, Z.; Liu, R.W.; Wu, K.; Wan, Y. Adaptively constrained dynamic time warping for time series classification and clustering. *Inf. Sci.* **2020**, *534*, 97–116. [CrossRef]

27. Zhao, L.; Shi, G. A trajectory clustering method based on Douglas-Peucker compression and density for marine traffic pattern recognition. *Ocean Eng.* **2019**, *172*, 456–467. [CrossRef]

28. Wang, L.; Chen, P.; Chen, L.; Mou, J. Ship AIS Trajectory Clustering: An HDBSCAN-Based Approach. *J. Mar. Sci. Eng.* **2021**, *9*, 566. [CrossRef]

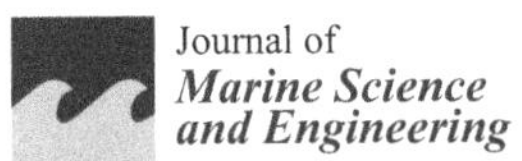

Journal of
*Marine Science
and Engineering*

Article

Optimized APF-ACO Algorithm for Ship Collision Avoidance and Path Planning

Mingze Li, Bing Li, Zhigang Qi *, Jiashuai Li and Jiawei Wu

College of Intelligent Science and Engineering, Harbin Engineering University, Harbin 150001, China;
limingze123@hrbeu.edu.cn (M.L.); libing265@hrbeu.edu.cn (B.L.); lijiashuai@hrbeu.edu.cn (J.L.);
wujiawei@hrbeu.edu.cn (J.W.)
* Correspondence: qizhigang@hrbeu.edu.cn

Abstract: The primary objective of this study is to investigate maritime collision avoidance and trajectory planning in the presence of dynamic and static obstacles during navigation. Adhering to safety regulations is crucial when executing ship collision avoidance tasks. To address this issue, we propose an optimized APF-ACO algorithm for collision avoidance and path planning. First, a ship collision avoidance constraint model is constructed based on COLREGs to enhance the safety and applicability of the algorithm. Then, by introducing factors such as velocity, position, and shape parameters, the traditional APF method is optimized, creating a dynamic APF gradient for collision avoidance decision making in the face of dynamic obstacles. Furthermore, the optimized APF method is integrated with the ant colony optimization algorithm, the latter modified to overcome the inherent local optimality issues in the APF method. Ultimately, validations are conducted in three areas: static avoidance and planning in restricted sea areas, avoidance under conditions of mixed static and dynamic obstacles, and avoidance in situations of multiple ship encounters. These serve to illustrate the feasibility and efficacy of the proposed algorithm in achieving dynamic ship collision avoidance while simultaneously completing path-planning tasks.

Keywords: path planning; collision avoidance; the optimal APF; ACO

Citation: Li, M.; Li, B.; Qi, Z.; Li, J.; Wu, J. Optimized APF-ACO Algorithm for Ship Collision Avoidance and Path Planning. *J. Mar. Sci. Eng.* **2023**, *11*, 1177. https://doi.org/10.3390/jmse11061177

Academic Editor: Mihalis Golias

Received: 11 May 2023
Revised: 29 May 2023
Accepted: 2 June 2023
Published: 4 June 2023

1. Introduction

In the wake of sustained growth within the maritime and shipbuilding industries, the significance of oceanic transport has become increasingly conspicuous. According to estimations by the International Maritime Organization (IMO), approximately 90% of global commerce is conducted via sea transportation [1]. Nevertheless, ship collision incidents are becoming progressively prevalent. The prevention of inter-ship collisions and the avoidance of surface obstacles are integral components of successful navigation. Statistical data regarding ship collisions reveal that human factors are the primary cause, with approximately 96% of accidents attributed to human error [2]. In an effort to mitigate such errors, autonomous vessels supporting intelligent navigation have garnered significant interest, with automatic collision avoidance and path planning constituting a pivotal challenge in the development of intelligent ships [3]. Path-planning technology directly influences the degree of vessel intelligence. In accordance with the International Regulations for Preventing Collisions at Sea (COLREGs) [4], vessels must continually adapt their collision avoidance and obstacle evasion plans during navigation, adjusting the obstacle trajectory accordingly [5]. Consequently, it is imperative to investigate a swift and effective path-planning method for implementation in autonomous ship control systems [6].

Currently, prevalent ship path-planning and collision avoidance algorithms include APF, genetic algorithm (GA) [7], ant colony optimization (ACO) [8], particle swarm optimization (PSO), model predictive control, and deep reinforcement learning, among others [9]. However, these methods have certain limitations in practical applications. For instance, while the APF method is simple and easy to implement, it is prone to local optima [10]; the GA is suitable for handling complex mathematical models and nonlinear constraint problems but has higher computational complexity [11]. Consequently, researchers often improve or combine multiple methods to overcome their respective drawbacks and enhance path planning and collision avoidance performance.

Numerous scholars have achieved significant advancements in the fusion and optimization of various path-planning algorithms. Xia Chen et al. [12] addressed the shortcomings of the rapidly exploring random tree star (RRT*) algorithm, such as slow convergence speed and large randomness in the search range, by proposing a UAV trajectory planning method based on a goal-biased APF-RRT* algorithm. This method employs a goal-biased strategy to guide the generation of random sampling points, accelerating the convergence speed of the algorithm, and introduces an improved APF method into the random search tree to significantly reduce the number of iterations, generate higher quality new nodes, and decrease path length. M. F. D. Santos et al. [13] addressed the issue of robust control for surface vessels, employing a robust optimal control approach based on continuous loop closure and optimal control theory. They adjusted the PID controller to handle uncertainties arising from different parameter variations. Y. Su et al. [14] conducted asymptotic dynamic positioning of ships in the presence of constraints on the actuators and partial failures of the actuators. The authors put forth a nonlinear PD fault-tolerant controller. Yangsheng Liu et al. [15] presented an improved simulated annealing-APF (SA-APF) algorithm to address path-planning problems in 3D space. This method uses the simulated annealing (SA) algorithm to optimize distance costs and combines it with the APF, realizing large-scale multi-objective 3D space path planning. Zhe Zhang et al. [16] designed a fusion algorithm combining an improved APF method and rolling window method for the local path-planning problem of unmanned underwater vehicles. The rolling window method models the local environment detected by UUV sensors, and the position factor is introduced into the repulsive force equation of the traditional APF method, making the approach effective and real-time. Hang Zhang et al. [17] developed an adaptive particle swarm optimization APF method. By using the APSO algorithm to preliminarily obtain the global virtual path, the method improves the APF approach and solves the local minimum value problem in traditional APF. Sarada Prasanna Sahoo et al. [18] proposed a hybrid algorithm combining grey wolf optimization (GWO) and GA and extended the hybrid path-planning to the cooperative path-planning application for autonomous underwater vehicles.

Moreover, numerous scholars have conducted extensive research on dynamic collision avoidance decision algorithms. Yanshuang Du et al. [19] addressed flight safety issues in dynamic airspace for unmanned aerial vehicles by proposing a real-time reactive collision-free path generation method, capable of adjusting the safe distance based on the relative motion of surrounding obstacles. Wenjun Zhang et al. [20] examined the relationship between a ship's encounter state and the COLREGs in the context of collision avoidance for ship path planning, which can help reduce the cost of planned routes. Hongguang Lyu et al. [21] introduced an autonomous trajectory planning algorithm based on a modified APF, aimed at ensuring that ships or unmanned surface vessels effectively address collision avoidance problems in dynamic environments while adhering to the COLREGs. Yufei Zhuang et al. [22] proposed a waypoint generation algorithm that considers USV turning radius constraints by establishing an obstacle detection mechanism, reducing redundant turning angles and path lengths when USVs traverse obstacles. They devised a multi-dimensional path evaluation function for reasonably assessing planned routes, enabling improved USV path planning in complex environments. Zhongxian Zhu et al. [23] proposed a precise environmental potential field model based on electronic navigational chart (ENC) surface objects and an improved APF, which can obtain collision-free paths under different weather conditions and in narrow waters. Sung-Wook Ohn et al. [24] considered open-sea, restricted-water interactions between two or multiple vessels and the COLREGs while constructing optimal local path-planning routes for ships.

In practical scenarios, the schematics of ship trajectory planning and collision evasion necessitate a holistic consideration of an array of determinants encompassing vessel dynamics, marine environments, and established navigational safety norms [1]. Hence, the cruciality of selecting and fine-tuning apt methodologies to circumnavigate real-world dilemmas becomes paramount. The APF algorithm exhibits structural clarity and computational swiftness, making it an indispensable asset in local ship route planning and dynamic obstacle avoidance. Despite the algorithm's inherent drawbacks in terms of global planning and intricate movements, its theoretical framework is nonetheless amenable to enhancements and pragmatic applications. Further, the synergistic application of the APF and ACO algorithms, particularly if the ACO algorithm can bolster its computational speed and global planning within a broad scale, could offer a more precise resolution to the extensive obstacle path-planning problem. Throughout the research journey, simulation experiments and practical trials can serve as credible tools to authenticate the effectiveness and applicability of the chosen methods, providing invaluable insights for the ongoing study and advancement of intelligent ship navigation.

2. Problem Description

In light of the current research landscape and inspiration, this paper proposes an optimized artificial potential field method for ship collision avoidance and path planning, adhering to COLREG requirements. This approach achieves superior global and local path planning for vessels while facilitating avoidance decision making for both static and dynamic obstacles. The research primarily addresses three issues.

Given the current state of research and inspiration, this study proposes an optimized artificial potential field method for ship collision avoidance and path planning. In comparison to prior methodologies, this novel approach fully considers ship dynamics, marine environment, and COLREG requirements. Through the innovative integration of APF and ACO algorithms, it addresses path planning among large-scale obstacles and local optimum issues, achieving global and local path planning for ships and providing support for collision avoidance decision making against static and dynamic obstacles. This research primarily addresses the following three issues:

(1) Ship collision avoidance rule modeling: By considering ship dynamics and COLREGs in collision avoidance decision making and trajectory planning, this paper constructs a rule constraint model for vessel navigation through the study of the COLREGs, thereby enhancing practical applicability;

(2) Construction of an optimized artificial potential field: To achieve collision avoidance planning for static and dynamic obstacles, this paper, inspired by prior work [25], introduces endpoint and dynamic obstacle information, such as position, speed, and type parameters, to optimize the attractive and repulsive potential field models, fulfilling dynamic ship collision avoidance requirements;

(3) Incorporating ant colony optimization to address local optimality issues: To resolve the local optimality problem prevalent in traditional artificial potential field methods for global planning, this paper combines the artificial potential field method with ant colony optimization and introduces improvements to the latter, achieving dynamic obstacle avoidance and global path planning for vessels.

3. Ship Collision Avoidance Rule Modeling

In open-water navigation, examining ship collision avoidance decision-making and path-planning algorithms requires the integration of the COLREGs as constraints. Key aspects include the following: (1) Feasibility of real-time environmental data collection during navigation and concurrent hazard assessment; (2) Determination of ship collision avoidance decisions, timing, and evasive actions; (3) Consideration of ship dynamics and COLREGs within decision making and trajectory planning.

In compliance with the COLREGs, vessels must ensure that other ships remain outside the safe encounter distance throughout their journey, necessitating navigation beyond the safe encounter distance as a performance criterion.

When evaluating collision risks with obstacles, the vessel safety field and safety distance are commonly used as vital factors. Unlike static obstacles, dynamic obstacles call for the consideration of both the current safety field and safety distance, as well as the prediction of safety field and safety distance trends for future temporal dynamic obstacles. As illustrated in Figure 1, assume our vessel is located at point O, and the obstructing vessel moves from point A to point B, decreasing the distance between the two vessels from OA to OB and altering the azimuth angle $\Delta\delta$ between them.

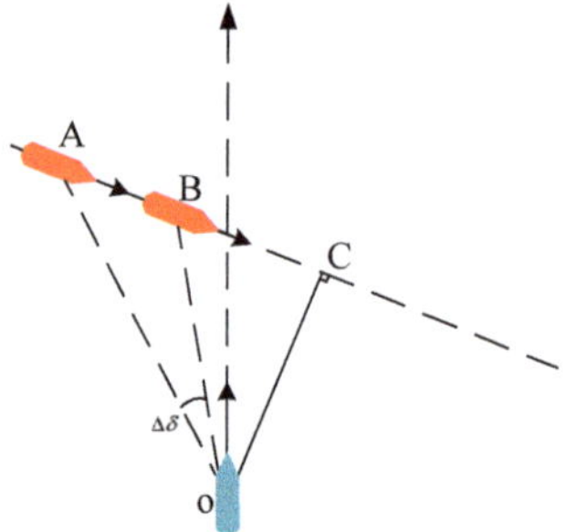

Figure 1. Relationship between distance and bearing among vessels.

The ship safety field is an area centered on the vessel's center of gravity. Although various safety field models with different shapes and judgments are employed in ship safety research, the safety distance standard remains widely utilized in the navigation process. The selection of a ship safety field must consider numerous factors. Existing research that overemphasizes the form of the ship safety field may not be conducive to the practical application of domain models.

For computational convenience, this paper treats the safety field of ships and other obstacle ships as circles, as illustrated in Figure 2. P_0 represents the center of gravity of our ship's safety field circle, while P_1 denotes the center of gravity of other ships. L_{pp} signifies the total length of ships contained within the circumcircle, s_0 and s_1 are the adequate safety margins for both ships. When selecting the safety margin, the position data of the own ship and other ships should be considered, as well as the increase in field range due to the proportional effect of the ship. The product of L_{pp} and the set factor μ indicates that the radii of the expanded circles for the ship and other obstacle ships are R_0 and R_1, respectively.

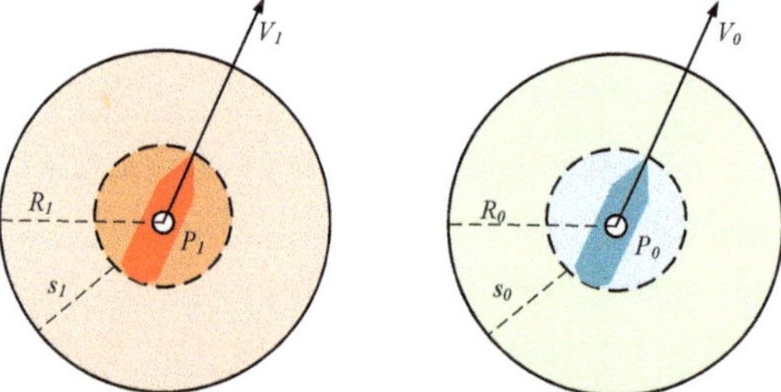

Figure 2. Diagram Depicting Ships Safety Distance.

By treating the safety field of ships and other obstacle ships as circles, the proposed method simplifies calculations while still providing a reasonable safety margin for both ships. This approach enables efficient and effective collision avoidance, considering the position data of both the own ship and other ships in the navigation process.

Building upon this, the distance between the expanded boundaries of the ship and other obstacle ships is denoted as D_s. This distance considers the maneuverability parameters of the ship and other ships, their respective speeds, relative speeds, hydrological conditions, and other factors. During the voyage, the distance is also related to the respective headings of the own ship and other ships. Different encounter situations necessitate varying D_m values. The radius of the ship safety field established in this paper is the safety distance between the ship and other ships, expressed as follows:

$$D_m = R_0 + D_s + R_1. \tag{1}$$

The illustration of ship safety field is shown in Figure 3.

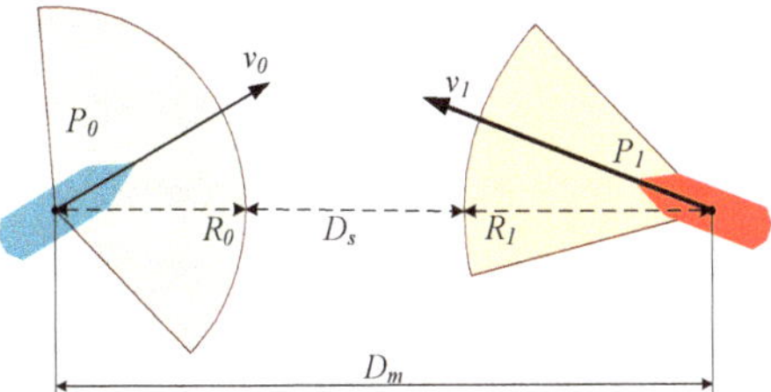

Figure 3. Illustration of ship safety field.

It is worth noting that there is a distinct difference between the ship safety field calculated based on the own ship and the ship safety field calculated according to other obstacle ships. The D_s of the other ship differs from ours. Factors contributing to this difference include vessel size and maneuverability.

The expression for the ship safety distance D_s is as follows:

$$D_s = \int |v| dt + \mu_e D_e + \mu_w D_w, \tag{2}$$

where dt is the sampling period; $|v|$ is the absolute value of the speed at which the two ships approach each other; D_n is the safe distance in the current navigation environment; D_e is the safe distance in the current environment; D_w is the safe distance in the current waters; μ_e and μ_w are the adjustment coefficients of D_e and D_w, respectively. In practical computations, according to the type of vessel and its suitable navigational waters, the values of D_e and D_w typically range from 0.3 to 1.2 nautical miles. Moreover, the values of μ_e and μ_w are adjusted based on the current environment and waters.

In order to effectively avoid collisions during a ship's navigation, it is essential to have a well-designed collision avoidance decision algorithm that considers various factors and stages of the collision development process. The stages include long distance without danger, collision risk, emergency situation, urgent danger, and final collision. These stages represent the process of the distance between the own ship and the obstacle ship developing from near to far.

The collision avoidance decision algorithm should be capable of taking the most appropriate action to best avoid a collision when other ships are not maneuvering in accordance with the COLREGs or if the two ships are in close proximity due to some other reason. The algorithm should also be able to determine when to initiate collision avoidance maneuvers, as initiating them too early or too late could negatively impact the ship's navigation.

When analyzing risk situations, the influence of other factors must be considered, particularly the distance between our ship and other obstacle ships as a constraint. Figure 4 elucidates that in compliance with the COLREGs, the most crucial aspect is the closest point of approach D_A between the obstacle ship and the own ship and the collision risk detection distance D_c, which represents the distance allowing the vessel to undertake collision avoidance measures upon identifying a collision risk situation.

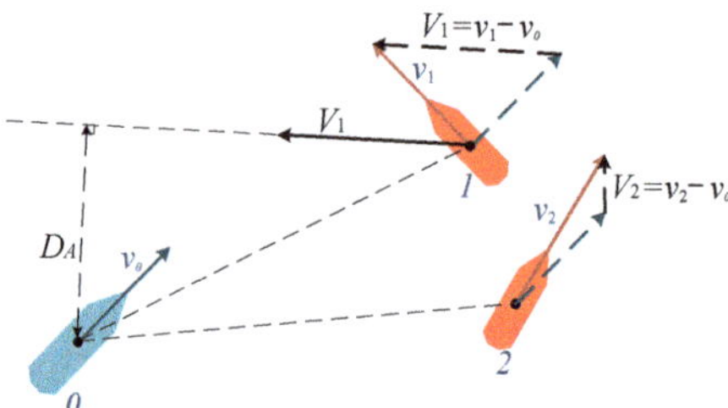

Figure 4. Distance closest point of approach.

In this paper, D_c is defined as the sum of the safety field radius D_m and the potential field influence range radius α_0. The value of α_0 is determined by the maximum value of the product of f_k times the obstacle's bulking circle radius and α_{omin}, with α_{omin} set at 0.4. This definition allows for a comprehensive assessment of collision risk, considering both the safety and the potential fields' influence on the ship's navigation.

$$D_c = D_m + \alpha_o \tag{3}$$

In this context, $\alpha_o = max\{D_f \cdot R_1, \alpha_{omin}\}$, D_f is a proportional coefficient referred to as the obstacle distance influence factor, which adjusts the potential field's influence range depending on the distance between the ship and the obstacle.

4. Path Planning Based on Optimized APF

During the navigation process, the ship can use the APF method to take appropriate collision avoidance actions according to the COLREGs. Once the risk of collision is eliminated, the ship can resume sailing on the planned route and continue towards its destination. The APF method can adapt to changing environmental conditions and the movements of other ships, making it a suitable choice for intelligent ship navigation systems.

4.1. Construction and Calculation of Attractive Potential Field

At present, in the artificial potential field algorithm, the potential energy equation of the destination's attractive potential field is usually transformed into a standard parabolic equation; $P_a(p,v)$ is the attractive potential field; $D(p_0, p_s)$ is the distance between the ship and the goal; $D(v_0, v_s)$ is the equation of the resultant velocity of our ship and the goal.

$$\begin{cases} P_a(p,v) = P_a(p) + P_a(v) \\ P_a(p) = \dfrac{1}{2}\sigma_p D^2(p_0, p_s) \ , \\ P_a(v) = \dfrac{1}{2}\sigma_v D^2(v_0, v_s) \end{cases} \tag{4}$$

where σ_p is scale weight factor for location, σ_v is speed factor, p_0 is the current position of the ship, p_s is the current position of the goal, v_0 is the current speed of the ship, v_s is the current speed of the goal. When the ship sails in the two-dimensional space on the sea surface, the attractive potential field that the ship experiences is non-negative. Only when the relative position $D(p_0, p_s)$ and relative velocity $D(v_0, v_s)$ of the ship and the target point are all 0, is the attractive potential field 0.

According to the attractive potential field equation $P_a(p,v)$, the attractive forces of equation $F_a(p,v)$ can be obtained:

$$F_a(p,v) = \frac{\partial P_a(p,v)}{\partial p} + \frac{\partial P_a(p,v)}{\partial v} \tag{5}$$

The attractive forces equation $F_a(p,v)$ can be rewritten as

$$\begin{cases} F_a(p,v) = F_a(p) + F_a(v) \\ F_a(p) = \varepsilon_p D(p_0, p_s)\mu_p \\ F_a(v) = \varepsilon_v D(v_0, v_s)\mu_v \end{cases} \tag{6}$$

where the attractive force $F_a(p)$ represents the magnitude of the location-based influence, while μ_p denotes the position vector. Additionally, $F_a(v)$ signifies the magnitude of the velocity-based attractive force, ensuring that the ship's speed aligns with the target endpoint's speed in both magnitude and direction upon arrival. $F_a(v)$ exhibits a positive correlation with $D(p_0, p_s)$, with the direction coinciding with the target point's movement relative to the ship and μ_v representing the unit vector in the direction of velocity.

4.2. Construction and Calculation of Repulsive Potential Field

When the relative position between the vessel and other ships falls within the collision risk detection threshold, our vessel should adhere to the COLREGs and implement collision avoidance measures to achieve dynamic and static obstacle avoidance. The equation for the repulsive potential field of dynamic obstacles P_{rd} in complex waters is

$$\begin{cases} P_{rd} = \varepsilon_d R_1 \omega^* \left(D_d^*\right)^2 d_s^2 \\ \omega^* = \left|e^{\omega_0 - \omega} - 1\right| \\ D_d^* = \dfrac{1}{D - D_m} - \dfrac{1}{\rho_0} \end{cases} \tag{7}$$

where ε_d represents the proportional coefficient of the repulsive potential field generated by the dynamic obstacle, ω_0 denotes the maximum relative position line angle, and d_s is the distance between the vessel and the goal.

The equation for the repulsive potential field of a static obstacle P_{rs} in open water is

$$\begin{cases} P_{rs} = \varepsilon_s R_1 \left(D_d^*\right)^2 d_s^2 \\ D_s^* = \dfrac{1}{D - D_L} - \dfrac{1}{\rho_0} \end{cases} \tag{8}$$

where ε_s represents the proportional coefficient of the repulsive potential field generated by the obstacle.

When ship encounters a static obstacle, the repulsive potential field it experiences depends on the distance between the ship and the static obstacle. Unlike the traditional repulsive potential field model, this model introduces a smaller constant: D_L. In this paper, D_L is chosen as the product of γ times the L_{PP} of the ship as the radius.

$$D_L = \gamma \cdot L_{PP}, \tag{9}$$

where γ is an adjustment factor with a value range from 1 to 3. Its value is determined by considering factors such as the maneuverability of our ship, the ship's length, the accuracy of sensor data (particularly position and distance), and an appropriate safety margin. The circle with the ship's position as its center and D_L as its radius represents an area where no obstacles are allowed to enter. If any obstacle enters this area, it is considered a collision accident, so this area is called the forbidden zone. Within this range, the repulsive potential field is large enough and bounded to prevent our ship from colliding with other obstacles.

The total repulsive potential field equation is

$$P_r = P_{rd} + P_{rs}. \tag{10}$$

By calculating the negative gradient of the repulsive potential field function $P_r(p,v)$ with respect to position and velocity, the corresponding repulsive force function expression $F_r(p,v)$ can be derived:

$$F_r(p,v) = -\frac{\partial P_r(p,v)}{\partial p} - \frac{\partial P_r(p,v)}{\partial v} \tag{11}$$

The equation of the ship under the resultant force is

$$F(p, v) = F_a(p, v) + F_r(p, v) \tag{12}$$

Considering the presence of multiple static obstacles and dynamic obstacle ships, the resulting expression for the repulsive force acting on our vessel is as follows:

$$F_r = \sum_{i=1}^{n} F_{r_i}, \tag{13}$$

where F_{r_i} represents the repulsive force exerted by the i-th ship or obstacle on the vessel and n refers to the total number of obstacles and other ships encountered by the vessel.

4.3. Ant Colony Optimization Algorithm

In path planning, the quality of a route often depends on multiple factors, such as distance, safety, and time consumption, thus forming a multi-objective optimization problem. The ant colony algorithm is a nature-inspired heuristic algorithm, drawing inspiration from the path optimization behavior of ants in search of food and avoidance of obstacles [26]. Utilizing the global search capability and parallelism of the ant colony algorithm, satisfactory solutions can be effectively found in multi-objective optimization problems.

Specifically regarding local optima, the search may stall or cycle infinitely among several singular points when a local optimum or singularity is encountered. However, the ant colony algorithm selects the next node to visit based on the roulette wheel selection method, rather than choosing directly according to the size of the probability [27]. This method expands the search range, enabling the search to possibly escape local optima and find the global optimum. The application of the ant colony algorithm in a global optimization strategy not only effectively enhances search efficiency but also avoids becoming trapped in local optima.

At time t, ant k selects the direction of travel from point i to point j based on the calculation of the state transition probability. By calculating the transition probability $P_{ij}^k(t)$ of the available paths adjacent to point i, the transition probability of moving to grid point j is determined using the roulette wheel selection rule:

$$P_{ij}^k(t) = \begin{cases} \dfrac{\left[ph_{ij}(t)\right]^{\gamma}\left[hf_{ij}(t)\right]^{\sigma}}{\sum_{j \in m}\left[ph_{ij}(t)\right]^{\gamma}\left[hf_{ij}(t)\right]^{\sigma}} & ,j \in m, \\ 0 & ,j \notin m \end{cases} \tag{14}$$

where m is the set of available nodes for the ant's next step, $P_{ij}^k(t)$ is the node transition probability, $ph_{ij}(t)$ is the pheromone concentration on the path, $hf_{ij}(t)$ is the heuristic function, γ is the pheromone concentration coefficient, and σ is the heuristic function coefficient. The definition of the heuristic function $hf_{ij}(t)$ is

$$hf_{ij}(t) = \frac{1}{D_{ij}} \tag{15}$$

In the formula, D_{ij} represents the Euclidean distance between node i and node j. The definition of D_{ij} is

$$D_{ij} = \sqrt{\left(x_i - x_j\right)^2 \left(y_i - y_j\right)^2} \tag{16}$$

To prevent an abundance of pheromones from leading subsequent ants astray in their path selection, the information along the path will be updated using the following method after each generation of ants completes their search:

$$ph_{ij}(t+1) = ph_{ij}(t) - \beta ph_{ij}(t) + \Delta ph_{ij}(t, t+1) \tag{17}$$

In the formula, β represents the pheromone evaporation coefficient, and $\Delta ph_{ij}(t, t+1)$ denotes the increment of pheromones between points i and j during the current iteration.

$$\Delta ph_{ij} = \sum_{a=1}^{N} \Delta ph_{ij}^{a} \tag{18}$$

where N represents the predefined number of ants in the colony, and Δph_{ij}^{a} refers to the pheromones released by ant a as it traverses the path. The definition of Δph_{ij}^{a} is

$$\Delta ph_{ij}^{a} = P/D_a \tag{19}$$

P represents the initial pheromone value; D_a denotes the total length of the path traversed by ant a.

In traditional ant colony algorithms, the initial distribution of pheromones in the environment is set to the same value, leading to a certain degree of blindness in the initial path search by ants. This affects the search efficiency of the algorithm and increases the search time. In this paper, an improved method for initial pheromone distribution is proposed. This method calculates the initial pheromone $ph_{ij}(0)$ amount based on the number of obstacles surrounding a node. The specific implementation is as follows:

$$ph_{ij}(0) = \vartheta \cdot f(j) \tag{20}$$

$$f(j) = \frac{1 - C_U(m)}{8} \tag{21}$$

In the formula, ϑ is an enhancement coefficient that can be determined based on the actual situation; $f(j)$ is an obstacle avoidance coefficient; C_U represents the complement symbol; and U is the set of adjacent pixels for the current pixel position. The $f(j)$ function can be used to calculate the proportion of free pixels among the eight adjacent pixels of the current pixel. When there are fewer obstacle pixels in the neighborhood of the pixel, the initial pheromone of the pixel is larger; otherwise, the initial pheromone is smaller. This guides ants to advance and avoid searching in areas with too many obstacles, thus accelerating the convergence speed of the algorithm.

4.4. Collision Avoidance and Path Planning Algorithm Design

The method proposed in this paper first extracts key information from electronic chart data, such as ships, obstacles, and target points. The electronic chart is then converted into a rasterized image, where each grid cell represents either a navigable area or an obstacle. The system employs an ant colony algorithm for global path planning, initializing the ant population, heuristic information, and improved pheromone allocation, allowing ants to search and select the next grid cell on the rasterized image. Upon reaching the target point, the pheromone concentration is updated, the path length is recorded, and iterations are repeated, with the shortest path chosen as the global plan.

Local planning is achieved through an optimized artificial potential field method, calculating the attractive potential field (pointing towards the target) and repulsive potential field (moving away from obstacles) for static and dynamic obstacles. The composite potential field is obtained by combining these fields, which is used to calculate the ship's moving direction. Dynamic obstacles are detected, and repulsive potential fields are applied, realizing local static and dynamic collision avoidance. The ship's position is updated, and steps are repeated until the target is reached.

The global planning path of the ant colony algorithm and the local planning path of the optimized artificial potential field method are integrated to form the final ship path. The visualized final path is output for further analysis and use. Through this process, the combination of the optimized artificial potential field method and the ant colony algorithm is achieved, realizing ship path planning and dynamic collision avoidance in complex environments.

The traditional APF algorithm often encounters multiple local optima during the computation process. When it reaches these points, the search may stagnate or loop infinitely among several singular points, rendering the program unable to stop. Therefore, the search is likely to become trapped in these local optima, preventing it from finding the global optimum. To address this situation, this study proposes a fusion algorithm that incorporates an optimized ant colony algorithm to implement a global optimization strategy. The algorithm employs a roulette wheel selection method to choose the next node to visit, rather than making a direct choice based on the size of the probability. This approach broadens the search range, thereby locating the global optimum and avoiding becoming mired in local optima. For the optimization of the local planning APF method, a maximum number of iterations is set in this study. If the search exceeds this number without finding a satisfactory solution, the search is forcibly halted. The algorithm process proposed in this paper is illustrated in Figure 5.

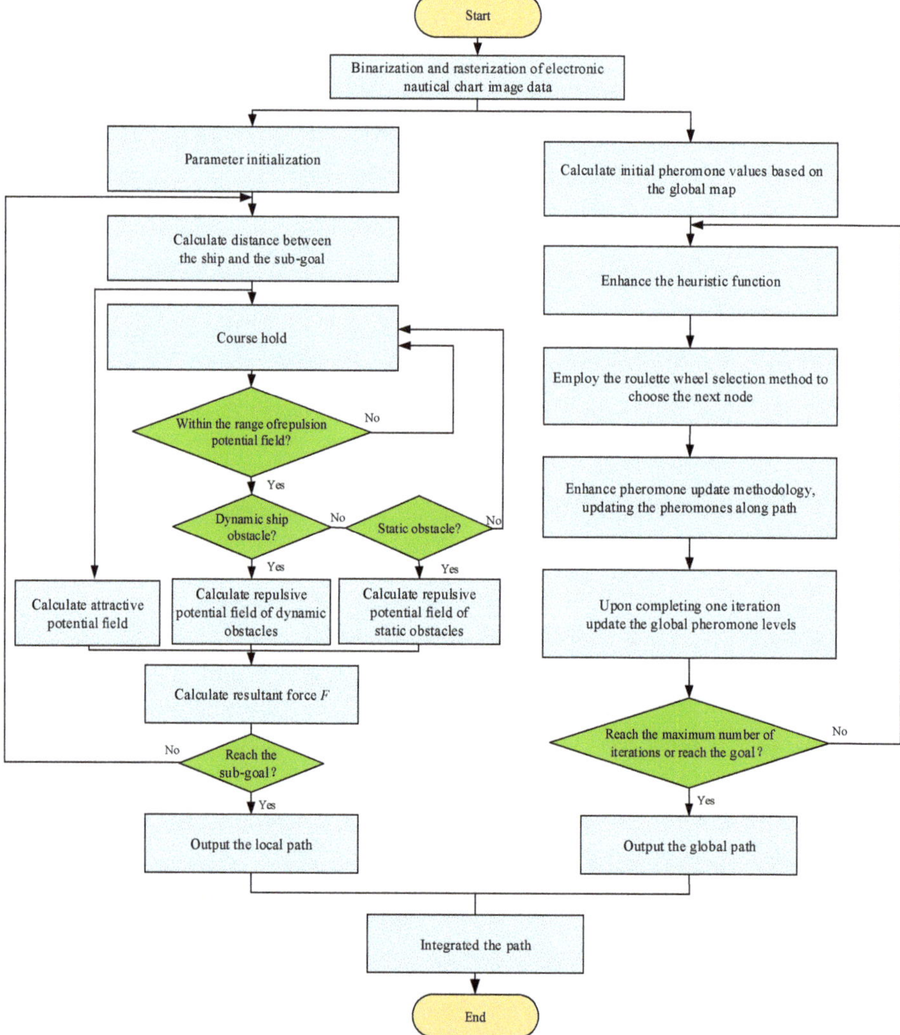

Figure 5. Algorithm flow chart.

5. Results and Analysis

In this study, the traditional artificial potential field algorithm was optimized by considering ship types and parameters in the waters, as well as adhering to the requirements and specifications of the COLREGs. This enabled the establishment of a collision avoidance decision system for static and dynamic obstacles at sea and the realization of the ship's autonomous obstacle avoidance decision and path-planning algorithm. To further validate the effectiveness, reliability, and real-time performance of the ship obstacle avoidance decision-making and path-planning algorithms based on the artificial potential field method, a simulation test of autonomous collision avoidance and navigation planning was conducted using the aforementioned ship autonomous collision avoidance system.

To meet real-world navigational task requirements, multiple simulation validations have been conducted in this study. Initially, tests for static avoidance and planning in three types of water areas, such as coastlines, islands, and reefs, were carried out to validate the performance of the proposed algorithm under varying static obstacles. Subsequently, tests were conducted under mixed dynamic and static obstacles to verify the algorithm's stability and robustness under composite obstacle interference. Last, based on the COLREGs' classification of vessel encounters, multiple encounter simulation tests were performed under different meeting patterns, simulating navigational situations in real waters. An in-depth discussion and analysis of the results followed the simulations.

In the simulation experiment, the positive direction of the vertical axis in the coordinate system represents north, while the positive direction of the horizontal axis represents east. The planned path of our ship is depicted by a blue trajectory, dynamic obstacle ships are represented by red dotted lines, and static obstacles are indicated by black color blocks.

5.1. Preprocessing of Electronic Nautical Chart Image Data

Prior to implementing the proposed algorithm, electronic nautical chart image data must undergo preprocessing. As shown in Figure 6, this preprocessing includes image binarization and rasterization, which simplifies the data for subsequent simulation and validation. First, the electronic nautical chart image is binarized, converting elements, such as ships, obstacles, and target points, into black and white pixels to streamline the image information. Next, rasterization divides the electronic nautical chart into regular smaller areas, with each raster cell representing a navigable area or an obstacle. The raster cell size can be adjusted based on specific requirements and computational resources. In this study, the raster window value is set to five to reduce computation time without sacrificing accuracy.

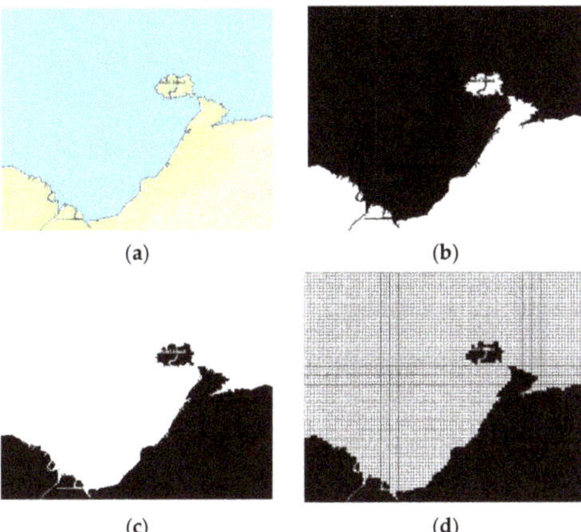

Figure 6. Image preprocessing: (**a**) Original electronic nautical chart; (**b**) Binarization; (**c**) Binarization inversion; (**d**) Rasterization.

This preprocessing method helps to simplify the geographical environment, providing clear and easily processed input data for the subsequent simulation and validation of the ant colony algorithm and the optimized artificial potential field method, thereby enhancing the efficiency and accuracy of path planning.

5.2. Static Obstacle Avoidance Simulation

In order to avoid collisions in restricted water areas, vessels need to consider different types of static obstacles. In this simulation experiment, an environment map corresponding to the electronic chart is established to accomplish autonomous ship obstacle avoidance decision-making and route-planning experiments. Analyses are conducted for coastlines, islands, and reefs, based on their specific characteristics.

(1) Coastline: The coastline, where land meets the sea, usually exhibits complex topography, possibly comprising beaches, cliffs, and bay currents. Coastal areas present a challenge for ship route planning due to the potential for shallow waters and complex terrain. Ships must maintain a certain distance to avoid grounding and collisions. Moreover, maritime currents and tidal factors in coastal areas must be factored into route planning, thus the repulsive potential field has a large range of influence;

(2) Islands: Islands are pieces of land in the sea, varying in size. The presence of islands may necessitate detours, especially in areas with many small islands, such as archipelagos or coral reefs. Concealed hazards, such as reefs and sandbars, may be present around some islands, posing higher demands for route planning;

(3) Reefs: Reefs are rocks or stones in the sea that can appear anywhere, including coastlines, around islands, and even in the open sea. Reefs pose a significant risk for ship route planning as they often lie below the water's surface and are difficult to observe directly. If a ship strikes a reef, it may sustain severe damage or even sink. Therefore, safety and local optimization issues must be considered in ship route planning.

For ships navigating in restricted waters to avoid collisions, different types of static obstacles must be considered individually. In this section's simulation experiment, obstacles are categorized into three types: coastlines, islands, and reefs, which are analyzed separately. The basic parameters of the subject vessel are presented in Table 1.

Table 1. Basic parameters of target ship.

Parameter	Target Ship
L_{pp}	82 m
Width	21 m
Weight	5430 t
Safety field radii R	246 m

Based on the electronic chart, a corresponding water environment map is created to carry out the ship's autonomous collision avoidance decision-making and path-planning experiments.

Simulation results are shown in Figure 7, where black represents the projection of obstacles on the horizontal plane, and white areas indicate navigable regions for the ship. The simulation statistics are shown in Table 2.

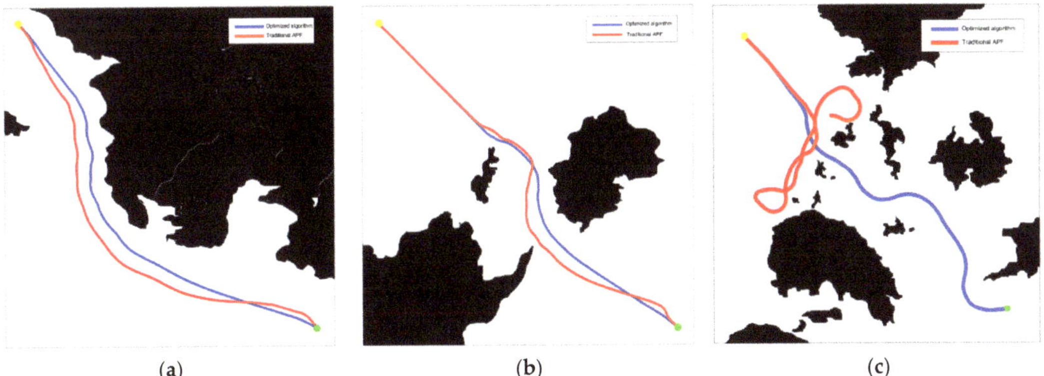

Figure 7. Trajectory schematics amidst static obstacles: (**a**) Coastlines; (**b**) Islands; (**c**) Reefs.

Table 2. Statistical analysis table for static obstacles.

Obstacle Types	Traditional APF			Optimized Algorithm		
	Reach the Goal	Simulation Duration	Path Length	Reach the Goal	Simulation Duration	Path Length
Coastlines	Yes	18.75 s	17.57 n mile	Yes	46.35 s	14.80 n mile
Islands	Yes	22.42 s	21.24 n mile	Yes	53.76 s	16.92 n mile
Reefs	No	-	-	Yes	64.25 s	7.95 n mile

The simulation results presented in Figure 7 and Table 2 allow us to compare the trajectory planning of the traditional APF method and the optimized algorithm under three common marine environment obstacles. As can be seen clearly from Figure 7a,b, the path of the optimized algorithm is shorter than that of the traditional APF method. By repeating this experiment 10 times, each result exhibits minor differences, demonstrating some randomness. The average path length is reduced by 17%, and the simulated curve is smoother.

Moreover, as shown in Figure 7c, the traditional APF algorithm, under complex environmental conditions, stagnates due to the influence of multiple repulsive potential fields or becomes trapped in an infinite loop between several singular points, causing the program to become non-responsive. The path planning becomes stuck in these local optimal solutions, unable to find the global optimum. However, the optimized ant colony algorithm in this paper implements a global optimization strategy. It uses a roulette wheel selection method to choose the next node to be visited, instead of directly selecting according to the size of the probability. This strategy broadens the search range to find the global optimum and avoids falling into local optima, effectively resolving the local optimum problem present in the traditional APF method. However, due to the complexity of the optimized algorithm, the simulation computation time is longer than that of the traditional APF method, indicating a certain limitation.

5.3. Dynamic Obstacle Avoidance Simulation

Dynamic obstacle collision avoidance simulation refers to target ship navigating through and encountering obstacle ships, the parameters for the obstacle ship is delineated as depicted in Table 3.

Table 3. Basic parameters of obstacle ship.

Parameter	Obstacle Ship
L_{pp}	54 m
Width	8 m
Weight	1042 t
Safety field radii R	135 m

When the obstacle ship maintains its sailing state unchanged, our ship makes an emergency collision avoidance decision outside the safety domain. The starting position of my ship is designed, with an initial heading angle of $-45°$ and a maximum speed of 16 knots. The obstacle ship has a sailing direction angle of $45°$ and maintains a speed of 10 knots. Under these circumstances, the simulation results of the proposed optimized algorithm are compared with the traditional dynamic APF method. The simulation results are illustrated in Figure 8.

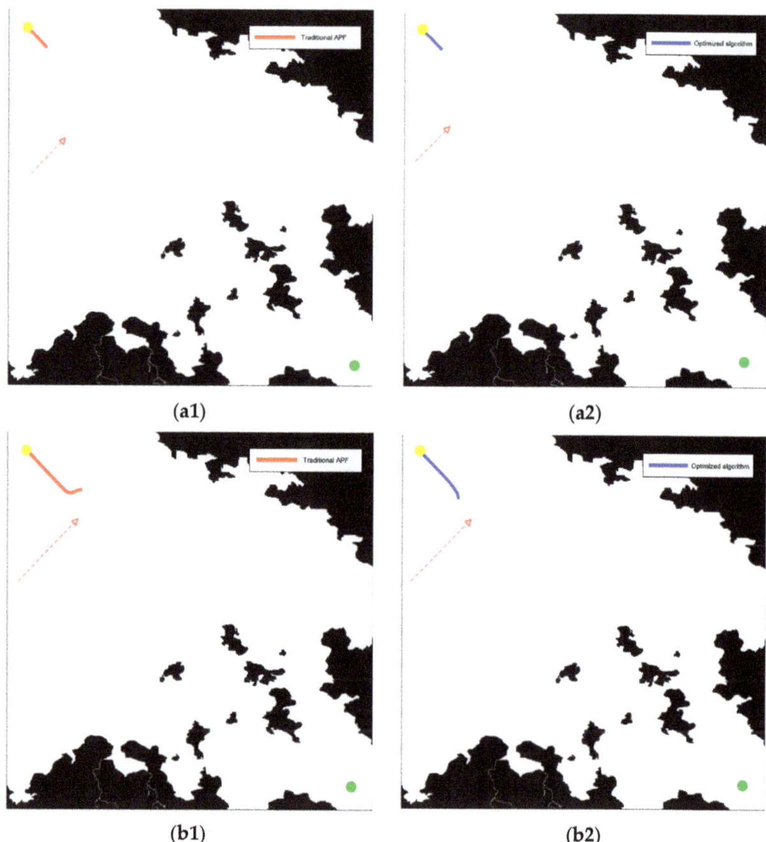

(a1)

(a2)

(b1)

(b2)

Figure 8. *Cont.*

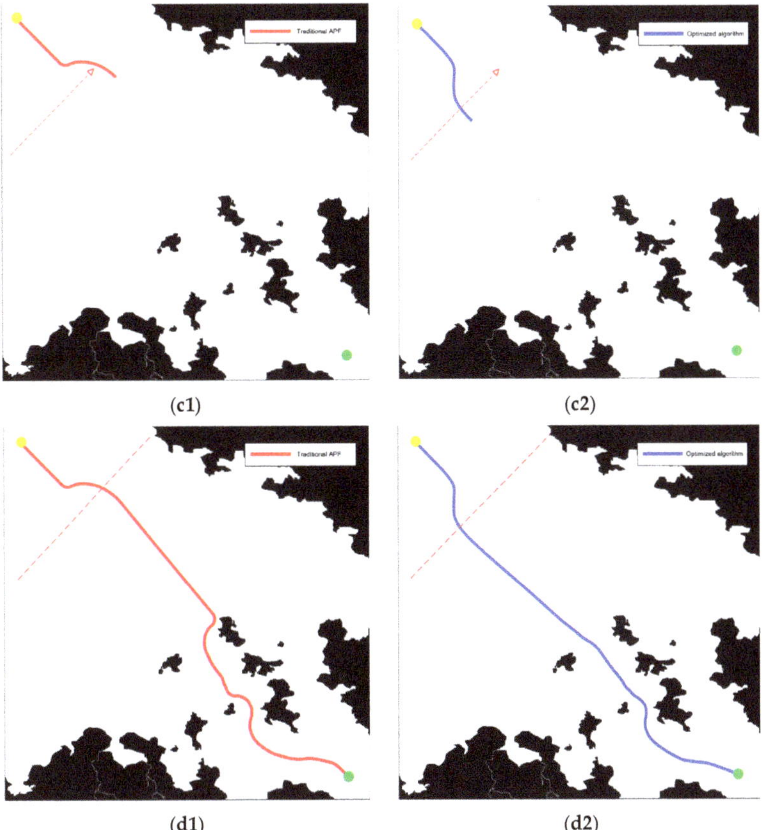

Figure 8. Path-planning trajectory diagram under dynamic obstacles: (**a1**) The ship and the obstacle ship form a cross encounter situation; (**a2**) Ships form a cross encounter situation; (**b1**) The ship is directed to turn left to avoid; (**b2**) The ship is directed to turn right to avoid (**c1**) The ship ends collision avoidance operation and resumes sailing; (**c2**) The ship end avoidance and resume navigation; (**d1**) The ship arrives at the finish line after avoiding static obstacles; (**d2**) The ship reaches its terminus subsequent to circumvent stationary impediments.

The simulation statistics are shown in Table 4.

Table 4. Statistical analysis table for dynamic obstacles.

Algorithm Categories	Simulation Duration	Path Length	Minimum Distance between Ships	Maximum Turning Angle
Traditional APF	59.25 s	18.72 n mile	0.24 n mile	$-71°$
Optimized algorithm	143.40 s	15.81 n mile	1.16 n mile	$49°$

Analyzing the simulation results using the traditional dynamic artificial potential field method for collision avoidance simulations, the target vessel performs a 71° left turn when encountering a crossing situation with the obstacle vessel due to the lack of constraint from the vessel collision avoidance rule model, as shown in Figure 8(b1). Furthermore, the distance between the target vessel and the obstacle ship is too small in Figure 8(c1), and the vessel does not resume its course after avoiding the collision. These actions do not comply with COLREG requirements for vessel handling and safe navigation.

In contrast, as shown in Figure 8(a2) and based on the optimized algorithm proposed in this paper, the target vessel detects an impending obstacle ship outside the safe zone. In Figure 8(b2), it promptly takes evasive action and performs a collision avoidance maneuver by turning right 49°. This complies with COLREG requirements for vessel handling, adhering to the principle of turning right at an appropriate angle to yield and effectively clears the path in Figure 8(c2). After completing the collision avoidance maneuver, the vessel resumes its course and eventually reaches its destination. During the path planning and collision avoidance process, the minimum distance between the target vessel and the obstacle ship is significantly greater than $R_0 + R_1$, fulfilling the safety requirements for collision avoidance. After conducting this segment of the experiment five times, it is evident that the results of each trial vary slightly due to the stochastic roulette wheel structure of the ant colony algorithm. Despite the inherent randomness, the stability of the system remains unaffected. The average navigational distance is 18.84 n miles, with the greatest recorded distance being 19.30 n miles.

5.4. Validation of Multi-Ship and Multi-State Collision Avoidance

As per the COLREG ship encounter categorization, the encounters between the target ship and the obstacle ships are divided into three types: overtaking, head-on meeting, and crossing. This section will carry out multiple encounter simulation tests under different encounter modes, simulating the navigation situation in real waters. The parameters of the obstacle ships in the three typical meeting situations are shown in Table 5.

Table 5. Parameters of the obstacle ships in three typical encounter scenarios.

Parameter	Obstacle Ship 1 (OS1)	Obstacle Ship 1 (OS2)	Obstacle Ship 1 (OS3)
L_{PP}	68 m	104 m	78 m
Width	9 m	11 m	9 m
Weight	1566 t	4650 t	2080 t
Safety field radii R	186 m	306 m	244 m
Encounter scenario	Overtaking	Head-on	Crossing

The experiment is set in an open-water scenario, where three distinct obstructive vessels intersect the path of the target ship in different encounter scenarios. Figure 9 demonstrates the result of the multi-vessel, multi-state collision avoidance test.

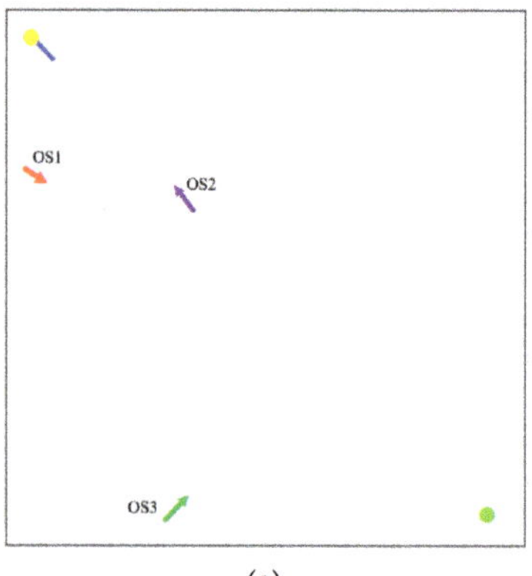

(a)

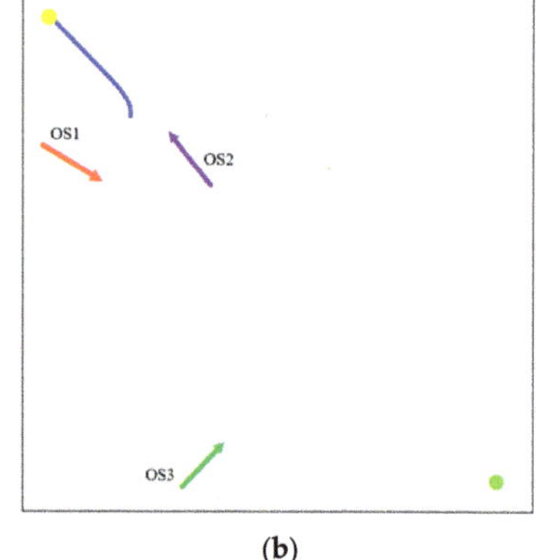

(b)

Figure 9. *Cont.*

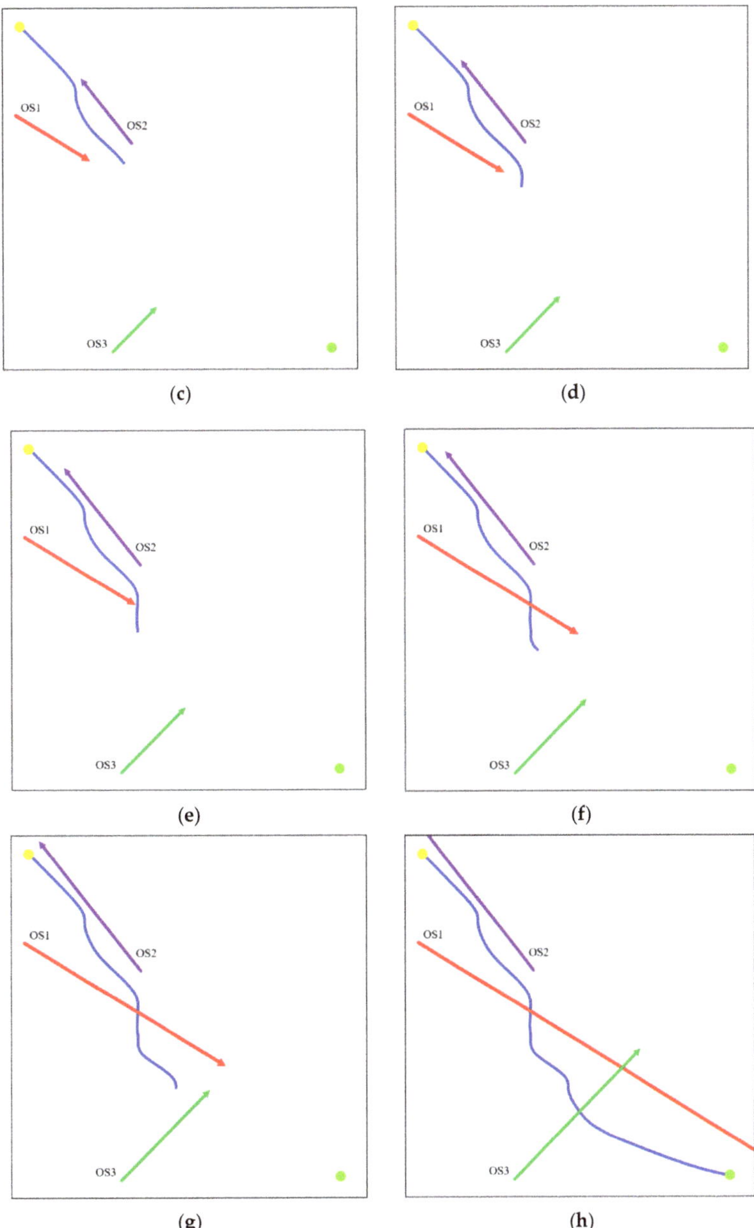

Figure 9. Results of multi-vessel, multi-state collision avoidance validation: (**a**) Numerous ships navigate concurrently within the same maritime expanse; (**b**) The ship and OS1 constitute a head-on encounter scenario; (**c**) The ship turns right to avoid and continues to sail; (**d**) The ship and OS2 constitute a overtaking encounter scenario; (**e**) The ship turning right at a wide angle to avoid; (**f**) The ship resume navigation after completing a right turn to avoid; (**g**) The ship and OS3 constitute a crossing encounter scenario; (**h**) The ship arrives at the end of the line after completing avoidance.

Based on the validation curves presented in Figure 9, the statistical data garnered from simulation results, encompassing multiple vessels and diverse circumstantial collision avoidance, are articulated as exhibited in Table 6.

Table 6. Statistics of the simulation data results for multi-vessel and multi-condition collision avoidance validation.

Encounter Scenario	Minimum Distance between Ships	Maximum Turning Angle
Overtaking	1.05 n mile	46°
Head-on	1.20 n mile	42°
Crossing	1.42 n mile	53°

According to the validation results in Figure 9a, the target vessel has a course of $-45°$ and is heading straight to the destination along the pre-planned route. Despite the presence of several obstructing vessels in this open sea, the distances between them are considerable and do not yet invade the safety distance criterion. Consequently, the target vessel is not within the influence range of the repulsive force field of the obstructing vessels and therefore does not take evasive actions.

Figure 9b shows that over time, both the target vessel and the obstructing vessel OS2 have traveled a certain distance, forming a head-on meeting situation. The target vessel enters the influence range of OS2's repulsive field and makes a significant 42° right turn. The closest point of approach is 1.20 nautical miles, which meets the COLREGs and safety requirements, fully demonstrating the robustness and stability of the algorithm.

Figure 9c–e illustrate that after successfully avoiding OS2 with a right turn, the target vessel encounters a situation of overtaking OS1. Upon entering OS2's repulsive field, the vessel makes a substantial 46° right turn following the algorithm's constraints on vessel operations. The closest point of approach is 1.05 nautical miles. After evasion, the vessel resumes its course to the endpoint, meeting COLREG requirements.

Figure 9f–h demonstrate that after resuming its course, the target vessel intersects with the obstructing vessel OS3. The target vessel executes a substantial 53° right turn and passes from the rear side of OS3. The closest point of approach is 1.42 nautical miles. After completing the evasion, the vessel resumes its course and eventually reaches the endpoint.

Through the multi-vessel and multi-condition collision avoidance validation simulation, the algorithm's capability to handle multiple meeting scenarios under COLREGs and multiple encounters in real waters is proven. The algorithm shows stability and robustness, and the evasive actions meet the normative requirements of the COLREGs.

In summary, the three sets of experiments validate the algorithm's performance in static evasion and planning in restricted waters, collision avoidance under mixed static and dynamic obstacles, and collision avoidance in multiple encounters. The vessel's evasive actions comply with regulations, and the planned route meets safety requirements. This indicates that the algorithm can prevent collisions in emergencies and that the planned route is a smooth curve, demonstrating its practicality.

6. Conclusions

This paper primarily investigates collision avoidance and path-planning problems for ships encountering dynamic and static obstacles during navigation. To address these challenges, a collision avoidance and path-planning algorithm based on an optimized APF-ACO algorithm is proposed. First, a ship collision avoidance constraint model is established according to the COLREGs, enhancing the algorithm's safety and applicability. Subsequently, by introducing factors such as velocity, position, and shape parameters, the traditional APF method is optimized, creating a dynamic APF gradient to enable collision avoidance decision making for dynamic obstacles. Furthermore, this research combines the optimized APF method with an ant colony optimization algorithm, improving the latter to resolve inherent local optimality issues in the APF method and achieve collision avoidance decision-making and trajectory-planning methods for static/dynamic obstacles in complex waters based on the APF method.

Comparative simulation experiments demonstrate the feasibility and effectiveness of the proposed algorithm in accomplishing dynamic ship collision avoidance and path-planning tasks. This research holds significant practical implications for the safe navigation of ships in complex environments and offers valuable insights for the development and study of intelligent ship navigation systems. Future research may focus on further optimization of the algorithm to enhance its performance in handling more complex scenarios and diverse obstacle types, laying the foundation for the realization of intelligent and automated ship navigation.

Author Contributions: Conceptualization, B.L. and M.L.; methodology, J.L. and M.L.; software, M.L. and Z.Q.; validation, B.L. and M.L.; formal analysis, Z.Q. and J.L.; writing—original draft preparation, M.L., B.L. and J.W.; project administration, J.L. and J.W.; funding acquisition, J.L. and J.W.; writing—review and editing, M.L., B.L. and Z.Q. All authors have read and agreed to the published version of the manuscript.

Funding: This research was funded by the Natural Science Foundation of Heilongjiang Province grant number KY10400210217, Fundamental Strengthening Program Technical Field Fund grant number 2021-JCJQ-JJ-0026, Natural Science Foundation of Heilongjiang Province (LH2019E035), the Project of Education Science Planning in Heilongjiang Province (GJB1320064), and the Harbin Engineering University Education and Teaching Program (JG2021B06).

Institutional Review Board Statement: Not applicable.

Informed Consent Statement: Not applicable.

Data Availability Statement: The data used to support the findings of this study are included within the article and are also available from the corresponding authors upon request.

Conflicts of Interest: The authors declare no conflict of interest.

References

1. Tu, E.; Zhang, G.; Rachmawati, L.; Rajabally, E.; Huang, G.-B. Exploiting AIS Data for Intelligent Maritime Navigation: A Comprehensive Survey from Data to Methodology. *IEEE Trans. Intell. Transp. Syst.* **2018**, *19*, 1559–1582. [CrossRef]
2. Chiang, H.-T.L.; Tapia, L. COLREG-RRT: An RRT-Based COLREGS-Compliant Motion Planner for Surface Vehicle Navigation. *IEEE Robot. Autom. Lett.* **2018**, *3*, 2024–2031. [CrossRef]
3. González, D.; Pérez, J.; Milanés, V.; Nashashibi, F. A Review of Motion Planning Techniques for Automated Vehicles. *IEEE Trans. Intell. Transp. Syst.* **2016**, *17*, 1135–1145. [CrossRef]
4. Zaccone, R.; Martelli, M.; Figari, M. A COLREG-Compliant Ship Collision Avoidance Algorithm. In Proceedings of the 2019 18th European Control Conference (ECC), Naples, Italy, 25–28 June 2019; pp. 2530–2535. [CrossRef]
5. Mannarini, G.; Subramani, D.N.; Lermusiaux, P.F.J.; Pinardi, N. Graph-Search and Differential Equations for Time-Optimal Vessel Route Planning in Dynamic Ocean Waves. *IEEE Trans. Intell. Transp. Syst.* **2020**, *21*, 3581–3593. [CrossRef]
6. Lyu, H.; Hao, Z.; Li, J.; Li, G.; Sun, X.; Zhang, G.; Yin, Y.; Zhao, Y.; Zhang, L. Ship Autonomous Collision-Avoidance Strategies—A Comprehensive Review. *J. Mar. Sci. Eng.* **2023**, *11*, 830. [CrossRef]
7. Tsai, C.-C.; Huang, H.-C.; Chan, C.-K. Parallel Elite Genetic Algorithm and Its Application to Global Path Planning for Autonomous Robot Navigation. *IEEE Trans. Ind. Electron.* **2011**, *58*, 4813–4821. [CrossRef]
8. Nian, R.; Shen, Z.; Ding, W. Research on Global Path Planning of Unmanned Sailboat Based on Improved Ant Colony Optimization. In Proceedings of the 2021 6th International Conference on Automation, Control and Robotics Engineering (CACRE), Dalian, China, 15–17 July 2021; pp. 428–432. [CrossRef]
9. Xing, B.; Wang, X.; Yang, L.; Liu, Z.; Wu, Q. An Algorithm of Complete Coverage Path Planning for Unmanned Surface Vehicle Based on Reinforcement Learning. *J. Mar. Sci. Eng.* **2023**, *11*, 645. [CrossRef]
10. Malone, N.; Chiang, H.-T.; Lesser, K.; Oishi, M.; Tapia, L. Hybrid Dynamic Moving Obstacle Avoidance Using a Stochastic Reachable Set-Based Potential Field. *IEEE Trans. Robot.* **2017**, *33*, 1124–1138. [CrossRef]
11. Zhang, Q. A Hierarchical Global Path Planning Approach for AUV Based on Genetic Algorithm. In Proceedings of the 2006 International Conference on Mechatronics and Automation, Luoyang, China, 25–28 June 2006; pp. 1745–1750. [CrossRef]
12. Chen, X.; Fan, J. UAV trajectory planning based on APF-RRT* algorithm with goal-biased strategy. In Proceedings of the 2022 34th Chinese Control and Decision Conference (CCDC), Hefei, China, 15–17 August 2022; pp. 3253–3258. [CrossRef]
13. Dos Santos, M.F.; Neto, A.F.D.S.; Honorio, L.D.M.; Da Silva, M.F.; Mercorelli, P. Robust and Optimal Control Designed for Autonomous Surface Vessel Prototypes. *IEEE Access* **2023**, *11*, 9597–9612. [CrossRef]

14. Su, Y.; Zheng, C.; Mercorelli, P. Nonlinear PD Fault-Tolerant Control for Dynamic Positioning of Ships with Actuator Constraints. *IEEE/ASME Trans. Mechatron.* **2017**, *22*, 1132–1142. [CrossRef]

15. Liu, Y.; Qi, J.; Wang, M.; Wu, C.; Sun, H. Path Planning for Large-scale UAV Formation Based on Improved SA-APF Algorithm. In Proceedings of the 2022 41st Chinese Control Conference (CCC), Hefei, China, 25–27 July 2022; pp. 4472–4478. [CrossRef]

16. Zhang, Z.; Chen, S.; Li, Y.; Wang, L.; Ren, R.; Xu, L.; Wang, J.; Zhang, X. Local Path Planning of Unmanned Underwater Vehicle Based on Improved APF and Rolling Window Method. In Proceedings of the 2022 International Conference on Cyber-Physical Social Intelligence (ICCSI), Nanjing, China, 18–21 November 2022; pp. 542–549. [CrossRef]

17. Zhang, H.; Luo, F. An improved UAV path planning method based on APSOvnp-APF algorithm. In Proceedings of the 2022 34th Chinese Control and Decision Conference (CCDC), Hefei, China, 15–17 August 2022; pp. 5458–5463. [CrossRef]

18. Sahoo, S.P.; Das, B.; Pati, B.B.; Marquez, F.P.G.; Ramirez, I.S. Hybrid Path Planning Using a Bionic-Inspired Optimization Algorithm for Autonomous Underwater Vehicles. *J. Mar. Sci. Eng.* **2023**, *11*, 761. [CrossRef]

19. Du, Y.; Zhang, X.; Nie, Z. A Real-Time Collision Avoidance Strategy in Dynamic Airspace Based on Dynamic Artificial Potential Field Algorithm. *IEEE Access* **2019**, *7*, 169469–169479. [CrossRef]

20. Zhang, W.; Yan, C.; Lyu, H.; Wang, P.; Xue, Z.; Li, Z.; Xiao, B. COLREGS-based Path Planning for Ships at Sea Using Velocity Obstacles. *IEEE Access* **2021**, *9*, 32613–32626. [CrossRef]

21. Lyu, H.; Yin, Y. Ships trajectory planning for collision avoidance at sea based on modified artificial potential field. In Proceedings of the 2017 2nd International Conference on Robotics and Automation Engineering (ICRAE), Shanghai, China, 29–31 December 2017; pp. 351–357. [CrossRef]

22. Zhuang, Y.; Dong, H.; Huang, H.; Kao, Y. Dynamic Path Planning of USV Based on Improved Artificial Potential Field Method in Harsh Environment. In Proceedings of the 2022 China Automation Congress (CAC), Xiamen, China, 25–27 November 2022; pp. 5391–5396. [CrossRef]

23. Zhu, Z.; Lyu, H.; Zhang, J.; Yin, Y. Environment Potential Field Modeling for Ship Automatic Collision Avoidance in Restricted Waters. *IEEE Access* **2022**, *10*, 59290–59307. [CrossRef]

24. Ohn, S.-W.; Namgung, H. Requirements for Optimal Local Route Planning of Autonomous Ships. *J. Mar. Sci. Eng.* **2023**, *11*, 17. [CrossRef]

25. Lyu, H.; Yin, Y. COLREGS-Constramed Real-time Path Planning for Autonomous Ships Using Modified Artificial Potential Fields. *J. Navig.* **2019**, *72*, 588–608. [CrossRef]

26. Chen, L.; Su, Y.; Zhang, D.; Leng, Z.; Qi, Y.; Jiang, K. Research on path planning for mobile robots based on improved ACO. In Proceedings of the Youth Academic Annual Conference of Chinese Association of Automation IEEE, Nanchang, China, 28–30 May 2021.

27. Li, B.; Qi, X.; Yu, B.; Liu, L. Trajectory Planning for UAV Based on Improved ACO Algorithm. *IEEE Access* **2019**, *8*, 2995–3006. [CrossRef]

Journal of
Marine Science and Engineering

Article

A Novel Algorithm for Ship Route Planning Considering Motion Characteristics and ENC Vector Maps

Qinghua He [1], Zhenyu Hou [2] and Xiaoxiao Zhu [1,*]

[1] College of Mechanical and Transportation Engineering, China University of Petroleum, Beijing 100100, China; qinghuaharry1204@gmail.com
[2] College of Safety Science and Engineering, Civil Aviation University of China, Tianjin 300399, China
* Correspondence: x.zhu@cup.edu.cn

Abstract: Global route planning is a pivotal function of unmanned surface vehicles (USVs). For ships, the safety of navigation is the priority. This paper presents the VK-RRT* algorithm as a way of designing the planned route automatically. Different from other algorithms or studies, this study employs electronic navigation chart (ENC) vector data instead of grid maps as the basis of the search, which reduces data error when converting the vector map into the grid map. In addition, Delaunay triangulation is employed to organize vector data, in which the depth value is taken as a factor to ensure the safety of the planning route. Furthermore, the initial planned route is not suitable for ship tracking as it does not consider the ship motion characteristics. Therefore, the planned route needs to be further optimized. In the final part, we also conducted experiments to verify the effectiveness and advantages of the proposed algorithm. The results show that the proposed algorithm could reduce the lengths of paths by about 23% on average and save planning time; these are largely dependent on the environment.

Keywords: route planning; ENC map; sample-based algorithm

1. Introduction

Recently, the number of studies on unmanned surface vehicles (USV) has attracted much attention in the marine technology field. In fact, one of the premises of USV is generating the ship planning route (SPR) automatically. Then, the ship can track the route and arrive at the destination safely.

The ship route planning problem is a sub-field of route planning, where various algorithms and strategies were proposed and applied.

At first, there was a large number of studies on algorithms based on graphs. Reference [1] proposed the Dijkstra algorithm to address two problems in the route planning field: (1) Refer to a data structure that consists of paths connecting any two points; (2) Find the optimal path with a minimum total length between two given nodes. Then, the Dijkstra algorithm was employed and improved to solve practice issues. To solve the route planning problem involving the length of paths, a novel algorithm was created based on the Dijkstra algorithm, which referred to fuzzy theory [2]. Except the length of paths, time costs were also considered in [3,4], where traditional Dijkstra was improved. Furthermore, there were also various studies on the ship route planning problem employing the Dijkstra algorithm. There were also researchers who created a three-dimensional Dijkstra algorithm that supported the ship to plan the motion in which the speeds and courses were determined at each second [5]. Generating global optimum solutions for ship routes was expected. Moreover, the weather was also taken as an index of the Dijkstra algorithm when planning the minimum time cost for a route [6].

Then, based on the Dijkstra algorithm, a novel method called A* was also proposed and improved. In the original A* algorithm, the path was connected by the vertices of

Citation: He, Q.; Hou, Z.; Zhu, X. A Novel Algorithm for Ship Route Planning Considering Motion Characteristics and ENC Vector Maps. *J. Mar. Sci. Eng.* **2023**, *11*, 1102. https://doi.org/10.3390/jmse11061102

Academic Editor: Alessandro Ridolfi

Received: 2 April 2023
Revised: 15 May 2023
Accepted: 18 May 2023
Published: 23 May 2023

the grid map used, which means that the length of the route was not optimal. Therefore, Theta* was applied to [7] to make up for the drawback. Moreover, to save searching time, time-efficient A* was employed in [8] and simulated successfully. To address route planning problems, researchers considered multiple factors when applying the A* algorithm. For example, collision avoidance, rules on the sea, and the motion characteristics of ships were taken as examples [9]. Moreover, some researchers even considered the space characteristics of the ship [10].

Moreover, sampling theory was also extended and applied to route planning problems. RRT (rapid-exploring random tree) is a common algorithm used in the field of path planning. The RRT algorithm is completed in probability if reaches 100% if enough time is given for exploration. The main idea is to construct the searching tree and find line segments connecting the start point to the destination. Compared to other sample-based algorithms, not only does the RRT* have their advantages but it also has a higher degree of freedom [11]. Moreover, there were several algorithms that were generated based on the RRT algorithm. For instance, the RRT* is a kind of sample-based algorithm, which optimized the length of the route given by the RRT algorithm. In reference [12], the RRT* was compared with the A* via simulations to illustrate their advantages and disadvantages. In fact, the RRT* had been improved in various ways, taking into account different optimization goals or environmental conditions. To solve the field programmable gate arrays problem, the RRT* was improved to consider the terrain, planning speed, etc., [13]. Furthermore, a novel series of improved RRT* algorithms, named Quick-RRT*, PQ-RRT*, and P-RRT* were proposed by researchers [14,15]. All of them focused on fast speed converges to improve the characteristic of the route planning module of the ship. To do so, they combined the RRT* and potential function to give a better solution with a fast speed of convergence. Except for the above algorithms, the Connect-RRT algorithm was also developed to plan routes with fast speed [16]. In fact, the Connect RRT algorithm maintains two trees: one of them started from the departure point and another began from the destination. When the two trees encountered each other during the searching process, the route planning task could be ended. Based on that, various studies imitating the principle of Connect-RRT emerged, such as RRT*-connect [17], bidirectional potential guided RRT* [18], informed RRT*-connect [19], smooth RRT-connect [20], and so on.

RRT and its extended methods in the marine technology field (ship route planning) was applied to generate planned routes in a canal [21]. In such area, the current had to be considered, as is reasonable, as it would affect the navigation safety of ships. Moreover, the rule of local governments regarding canal navigation shall be also considered to avoid collisions between surface vehicles. To make up for the disadvantages of the RRT algorithm, reference [22] gave a hybrid step size and target attractive force RRT algorithm, which mainly improved the accuracy of the planned route in narrow waters. Moreover, the RRT algorithm was developed to have abilities of enhanced adaptability in [23]. Furthermore, it was necessary to include regulations when considering the route planning algorithm. Based on that, reference [24] gave a novel algorithm that combined the rule on the sea. Moreover, reference [25] gave the Bi-RRT algorithm, which can make decisions when faced with obstacles to prevent collisions.

Recently, with the development of artificial intelligence, many researchers have started using the reinforce leaning theory to address route planning problems [26–28]. For instance, the algorithm called deep reinforcement learning was provided in [29] to solve collision avoidance problem and optimize the length of the route at the same time. Moreover, the reinforcement learning algorithm was also engaged in unsupervised learning [30].

Different from robots and other vehicles, the navigation safety of ships involves two factors: water depth and obstacles. It means that the SPR could not cross land, reefs, and shallow water areas. However, few of the above studies consider both factors. Instead, they paid more attention to avoiding obstacles. Moreover, it is well-known that SPR is a polyline. However, the ship could not follow polylines accurately because of kinetic constraints [31]. Therefore, it is essential to consider ship motion characteristics in waypoint areas in optimization. In addition, most of the studies reviewed employed grid maps as configuration spaces. It means that converting from ENC or other sources of charts to raster maps should be conducted in advance. Errors occurred during such processes.

To make up for the above drawbacks, this study proposes a novel approach called the VK-RRT* algorithm. The main contributions of it are as follows:

(1) Unlike in most research, ENC vector data are utilized instead of simulated environmental data or raster map data. The use of vector data could reduce errors emerging from the raster map data process and could also accelerate the application of the path planning algorithm in navigational practice;

(2) We propose a novel strategy for the implementation of the path planning algorithm to account for ship kinetic constraints. Under constant speed, turning trajectories in the form of arcs are predicted and checked for safety in the path planning process. The strategy could largely decrease the pressure of controllers to track the planed path in the turning area;

(3) Compared to RRT and RRT* algorithms, the VK-RRT* method could give solutions faster.

The remainder of this study is arranged as follows: Section 2 presents the preliminary knowledge involved in this study. Section 3 gives RRT, RRT*, and the proposed VK-RRT* algorithm descriptions in detail for readers to compare and understand. Section 4 presents two case studies to illustrate the advantages and effectiveness of the proposed algorithm. Section 5 gives some conclusions and drawbacks, which need to be addressed in near future.

2. Preliminary Knowledge

In this section, we give the definition of the route planning of ships, motion constrains considered, and the Delaunay triangulation explanations used in our algorithms.

2.1. Problem Definition

This section presents the ship route planning problem that is investigated in this study. Let $\chi \subseteq R^d$ be the configuration space where ships navigate where $d \geq 2$. In addition, there are two sub-spaces χ_{obs} and χ_{free} in χ, which represent the obstacle space and the free space of navigation, respectively. p_{init} represents the initial position, whereas p_{goal} is the position of the destination that is to be reached. Path σ is a continuous function $[0,1] \mapsto \chi$ and it has bounded variation. Moreover, σ is free of collision when $\sigma(\tau) \in \chi_{free}$, $\tau \in [0,1]$.

Ship route planning problem: It is designed to safely find a proper or optimal σ guiding ship from the departure position to the destination.

Additionally, the three sub-problems are defined as follows:

Sub-problem 1. Considering a triplet $\left\{ p_{init}, p_{goal}, \chi_{obstacle} \right\}$, generate a feasible route if there is one. Otherwise, report failure.

For ship navigation, the shorter route means less cost. Therefore, this study also considers route optimization in the process of planning. Let Σ be the set of all routes and $\Sigma_{feasible}$ be the set of all feasible routes.

Sub-problem 2. Considering a triplet $\left\{ p_{init}, p_{goal}, \chi_{obstacle} \right\}$, define a cost function c and find a feasible route σ_f, such as $c(\sigma_f) = \min(c(\sigma) : \sigma \in \Sigma_{feasible})$. Report failure if there is no σ_f.

Sub-problem 3. Considering a triplet $\left\{ p_{init}, p_{goal}, \chi_{obstacle} \right\}$, define a cost function c and find a feasible route σ_k, such as $c(\sigma_k) = \min(c(\sigma) : \sigma \in \Sigma_{feasible})$, and find out if it is feasible to follow σ_k, considering ship motion characteristics. Report failure if there is no σ_k.

2.2. ENC Vector Data

To address the above problems, electronic nautical chart (ENC) vector data are employed to generate configuration space χ without rasterization. This type of chart could be used in the marine field, such as in navigation, fishing, etc.

According to the International Hydrographic Organization (IHO) S57 standard [32], ENC data consist of a certain amount of features that play a vital role in navigation. However, in this study, sound/depth in χ_{free}, land, reef or other static obstacles instead of all the features in ENC are employed in χ_{obs} to generate a planning route. In short, this study only considers the effects of depth and obstacle positions on the route planning problem.

Before implementing the ENC vector data, it is important to clarify the types of data. In ENC, an object is defined as an identifiable set of information. An object may have attributes and may be related to other objects. Feature objects have descriptive attributes but no geometry (i.e., information about the shape and position of a real-world entity). Spatial objects may have descriptive attributes and must have geometry. To be specific, there are three elements that represent geometry: point, line, and area.

In the proposed algorithm, ENC data are employed to form the configuration space χ. Obstacle areas make up χ_{obs} while depth points compose χ_{free}.

2.3. Ship Motion Characteristics

The motion characteristics of a vessel refer to the physical phenomena that describe the vessel's movements and behaviors on a waterway. These characteristics are influenced by several factors, including the physical characteristics of a vessel, such as its size, shape, weight, speed, and the hydrodynamic conditions of the water through which it moves. In particular, the motion of a vessel can be characterized by its heave (vertical motion), pitch (rotational motion around the transverse axis), roll (rotational motion around the longitudinal axis), and yaw (rotational motion around the vertical axis). These motions are indicative of the vessel's dynamic behavior and are influenced by external forces, such as wind, waves, and currents. A thorough understanding of the motion characteristics of a vessel is essential for ensuring the safe and efficient operation of a marine craft.

For route planning control problems, the turning constraint are usually taken into consideration. To be specific, when a ship alters its course, its trajectory is an arc. While most routes designed by researchers are presented as broken lines, straight line segments are connected in sequence. It is possible that ships ground in waypoint areas. Therefore, we have to take ship motion into account for safety purposes.

In order to simulate the ship motions, the MMG model is applied in this paper and Table 1 gives explanation of the variables used in Equations (1) and (2).

$$\begin{cases} (m + m_x)\dot{u} - (m + m_y)vr = X_H + X_P + X_R + X_{wind} + X_{wave} \\ (m + m_y)\dot{v} + (m + m_x)ur = Y_H + Y_P + Y_R + Y_{wind} + Y_{wave} \\ (I_{xx} + J_{xx})\dot{p} = K_H + K_P + K_R + K_{wind} + K_{wave} \\ (I_{zz} + J_{zz})\dot{r} = N_H + N_P + N_R + N_{wind} + N_{wave} \end{cases}, \qquad (1)$$

$$\begin{cases} \dot{x} = u\cos(\psi) - v\cos(\varphi)\sin(\psi) \\ \dot{y} = u\sin(\psi) + v\cos(\varphi)\cos(\psi) \\ \dot{\varphi} = p \\ \dot{\psi} = r\cos(\varphi) \end{cases}, \qquad (2)$$

Table 1. Explanation of variables.

Variables	Explanation	Variables	Explanation
m	mass	I_{xx}	Roll moment of inertia
m_x, m_y	additional mass	I_{zz}	yaw moment
x, y	position	J_{xx}	additional roll moment
ψ, φ	course and roll angle	J_{zz}	additional yaw moment
X, Y, K, N with H, P, R, wind, wave subscripts		Forces and moments	

Reference [33] gives more information about this model.

2.4. Delaunay Triangulation

The Delaunay triangulation is a geometric algorithm that constructs a triangulation using a set of points in a plane, where each triangle in the triangulation satisfies the Delaunay criterion; no point in the point set is inside the circumcircle of any triangle. This feature ensures that the triangulation avoids forming skinny triangles, which is advantageous for various applications, such as mesh generation and image processing.

In practice, there are many situations that we need in order to design routes using discrete points. The Delaunay triangulation (DT) method is such a method that can be used to solve the above problem. In electronic nautical charts, the depth points are unordered and discrete. The DT method could reconstruct them as the initial map for route planning.

The principle of DT is simple: for a given set P, any two points could construct an edge of DT map when there is a circle above two points and there is no other point in the circle [34].

It should be made clear that different ways of generating maps largely affect the planed route. In this paper, we only discuss the effectiveness of the proposed algorithm.

3. Route Planning Algorithm

In this section, we propose the VK-RRT* algorithm for the route planning problem of USV. Based on traditional RRT and considering motion constraints, the VK-RRT* could be applied using data from the ECDIS platform, which has great potential for commercial realisation.

3.1. Rapid-Exploring Random Tree

Rapid-exploring random tree (RRT) is a conventional algorithm in the field of path planning. It gives feasible solutions faster than other methods, such as A*, genetic algorithm (GA). Therefore, in this study, we improve RRT to VK-RRT* to produce a planned route quickly. In fact, RRT is a data structure. It will be sampled in the configuration space χ, and connect adjacent points constructing tree Γ.

Figure 1 shows the working principles of the RRT algorithm. The function *sample()* will randomly find a point from space χ; nearest() is responsible for selecting the nearest node from X_{rand} in tree Γ; steer() gives the X_{new} according to X_{rand} and $X_{nearest}$.

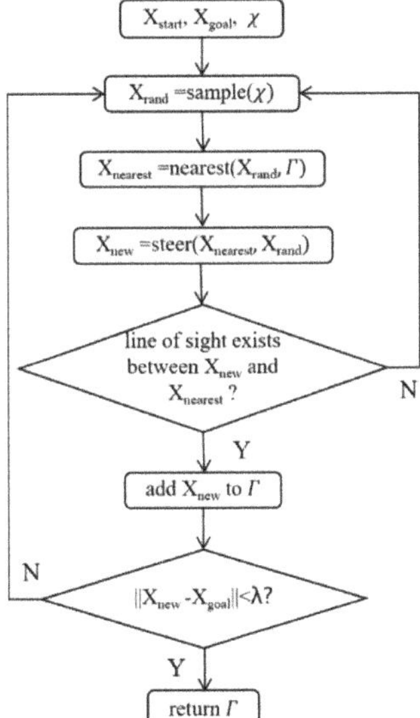

Figure 1. Flow chart of rapid-exploring random tree.

Theoretically, the RRT algorithm could find feasible paths connecting departure points and destinations if time cost is not considered, which means that the algorithm reaches 100%. However, the RRT algorithm cannot be used in practical applications because the given path is in the form of a polyline, making it impossible to track ships.

Therefore, the RRT* algorithm has been proposed by previous researchers to improve it. Figure 2 describes the RRT* algorithm. The difference between RRT and RRT* is that there is a rewire function in the RRT* algorithm. It optimizes the relationships of nodes in Γ to reduce the account of nodes.

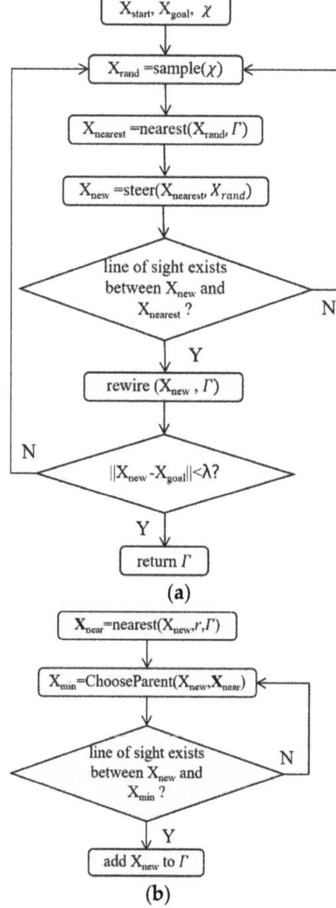

Figure 2. Flow chart of RRT* algorithm. (**a**) Outlines of RRT* algorithm, (**b**) Rewire function of RRT* algorithm.

3.2. VK-RRT Algorithm*

Taking ship motion characteristics into account, we propose the VK-RRT* algorithm.

Before introducing the algorithm, the usage of the Delaunay triangulation is illustrated. The depth points $E_i \in \mathbf{E}$ and obstacle positions are extracted from the electronic nautical chart. There are five attributes of E_i: (φ, λ) represents the latitude and longitude; d denotes the depth value; η is minimum depth of around E_i; ε means the minimum distance from adjacent objects. Similarly, to enlarge the waypoints data set, we also created $C_i \in \mathbf{C}$ to represent the nodes in the center of every triangular (Figures 3 and 4).

By obtaining data sets of depth points and obstacles, Delaunay triangulation is applied for generating the DT map, which is represented as χ. It is noted that there are two areas in space χ: free space χ_{free} and obstacle space χ_{obs}. **E** and **C** make up the configuration space χ_{free}.

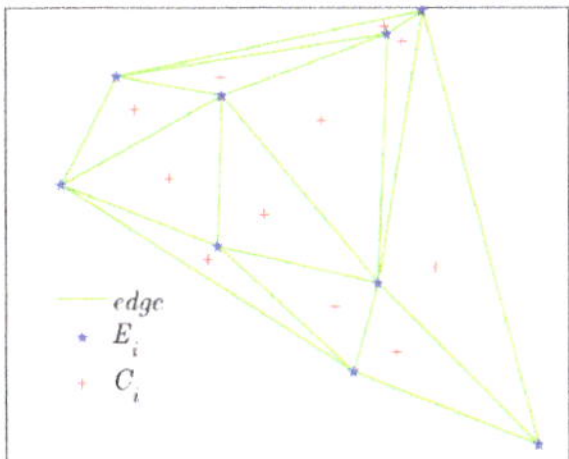

Figure 3. Sketch map of Delaunay triangle constructing **E**.

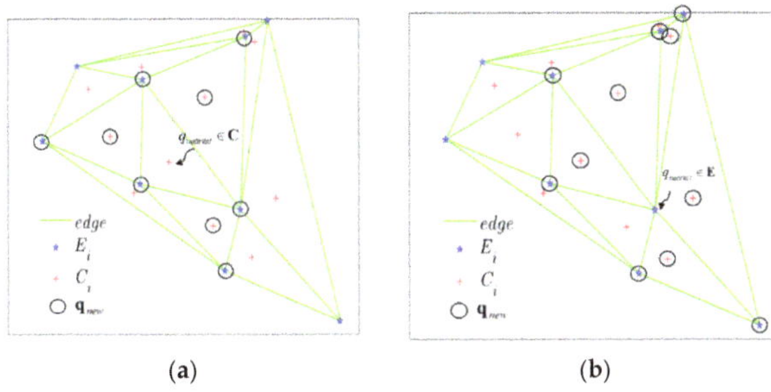

(a)   (b)

Figure 4. Methods to generate the new node $\mathbf{q}_{new}$. (**a**) Method when the nearest node is in the triangular; (**b**) Method when the nearest node is on the vertex of a triangular.

Figure 5 describes the working principle of VK-RRT*. Except for the data process, the rewire module is different from traditional RRT*, as it considers ship motion characteristics.

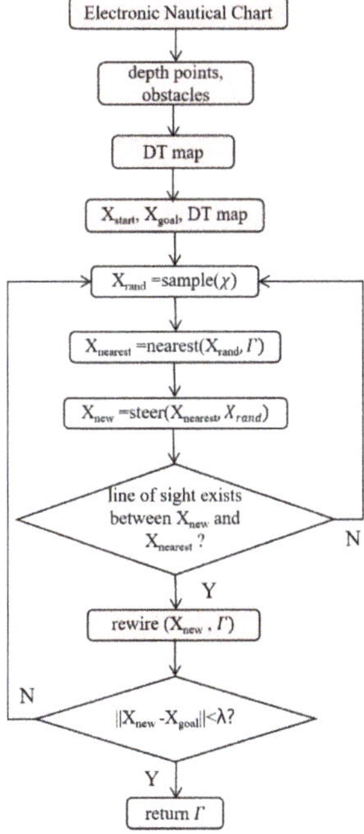

Figure 5. Flow chart of VK-RRT*.

As shown in Figure 6, the dark blue line is the route given by RRT*. However, it is impossible for surface ships to track the polyline as the large inertia.

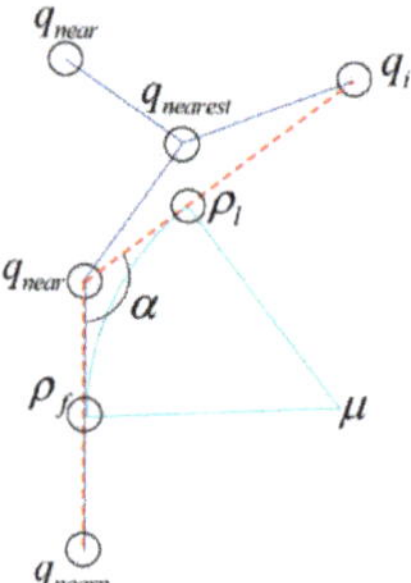

Figure 6. Route optimization considering ship motions.

Moreover, ships may encounter dangers in waypoint areas, as such in areas that have not been checked for safety in the route planning stage.

In fact, when the ship changes its course, the trajectory will be an arc line. Therefore, arc line segments in waypoint areas are designed. Navigating along the arc line, the ship only needs to give a fixed rudder angle to achieve tracking.

For example, Figure 7 shows the sketch map. The route $q_{nearq} \rightarrow q_{near} \rightarrow q_{nearest}$ is not suitable for a ship to track. Instead, we used $q_{nearq} \rightarrow \rho_f \rightarrow \rho_l \rightarrow q_i$ to replace it. The new route is determined by using Equations (1) and (2).

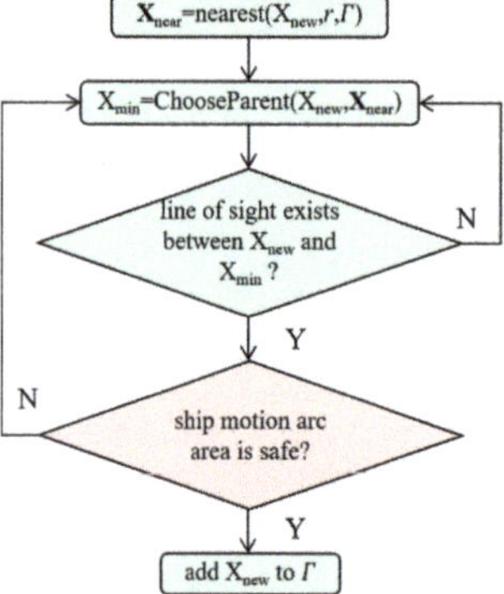

Figure 7. Rewire function of VK-RRT*.

Specifically, the main task is to determine the distance of $\rho_f \rightarrow q_{near}$: v. We suggest the following equations

$$v = \frac{R}{\tan((DN(q_{nearp} \rightarrow q_{near}) - DN(q_{near} \rightarrow q_i))/2)},\tag{3}$$

where DN is the function used to measure the direction of a vector from the north; R is the turning radius determined by Equations (1) and (2).

Then, we can obtain the latitude and longitude of ρ_f and ρ_l.

$$\begin{cases} P_{\rho_f}(x,y) = & q_{nearp}(x,y) + (1 - \frac{v}{\|q_{near}(x,y) - q_{nearp}(x,y)\|}) \\ & \cdot (q_{near}(x,y) - q_{nearp}(x,y)) \\ P_{\rho_l}(x,y) = & q_{near}(x,y) + (1 - \frac{v}{\|q_{near}(x,y) - q_i(x,y)\|}) \\ & \cdot (q_i(x,y) - q_{near}(x,y)) \end{cases},\tag{4}$$

In detail, VK-RRT* could plan the route that satisfies the requirement of ship motion constraints.

4. Simulation and Analysis

In order to verify the effectiveness and advantages of VK-RRT*, we carried out two simulations. Moreover, in our tests, we used the ships described in [33]. In the first scenario, we aimed to examine the ability of the proposed algorithm to plan route passing through narrow channels. In the second scenario, we tried to examine it in an open area. In each experiment, we compared the proposed algorithm with the RRT and RRT* algorithms to present the advantages of our algorithm. Both experiments have been conducted on ECDIS platform, which means that the frame we propose could be applied after professional packaging.

Moreover, in the process of choosing new points, we could add goal orientation to accelerate the search. This kind of strategy has been extensively studied in [35–40].

$$q_{rand} \begin{cases} q_{goal}, & \omega > \omega_d \\ x \in \chi_{free}, & \omega \leq \omega_d \end{cases}, \tag{5}$$

where ω is a random number, ω_d is set parameter of the adjustment.

4.1. Simulation in Islands

In this simulation, we use the electronic nautical chart numbered "US5M16IM.000", shown in Figure 8.

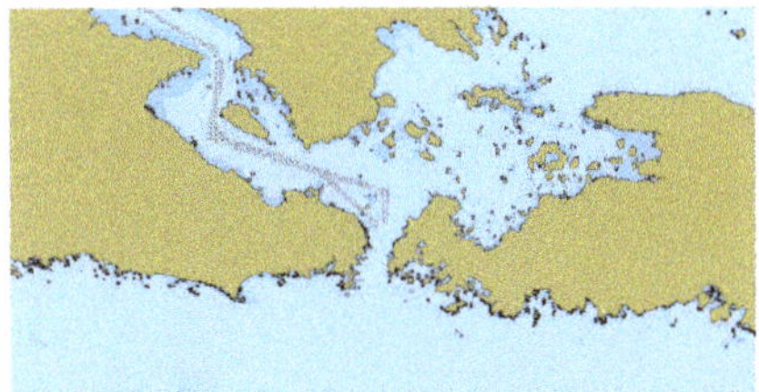

Figure 8. Electronic nautical chart numbered "US5M16IM.000".

First, the positions of the departure point and destination are (89.909° W, 45.924° N) and (83.853° W, 46.924° N) separately. Moreover, we assume that the ship will steer in a rudder angle of $\delta = 15°$ altering its course. In this experiment, the ability to search the route passing through narrow channels is mostly validated.

The results are illustrated in Figures 9–13, in which the red line is the planned route given by algorithms with different strategies; the blue lines construct the rapid-explore random tree, and the black lines represent the search process.

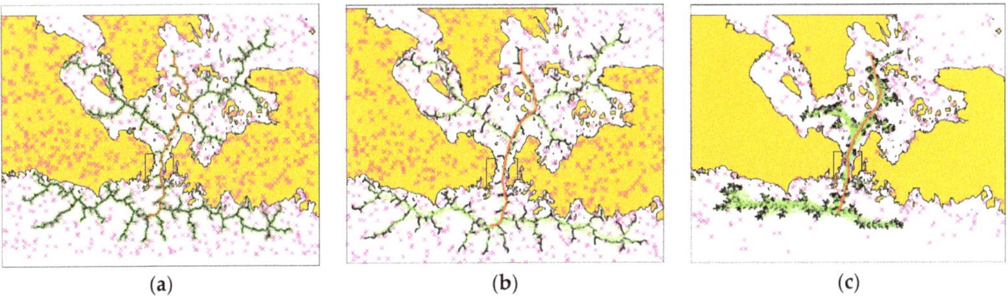

(a)   (b)   (c)

Figure 9. Planned routes without goal orientation in islands areas: (**a**) Given by the RRT algorithm; (**b**) Given by the RRT* algorithm; (**c**) Given by the proposed algorithm.

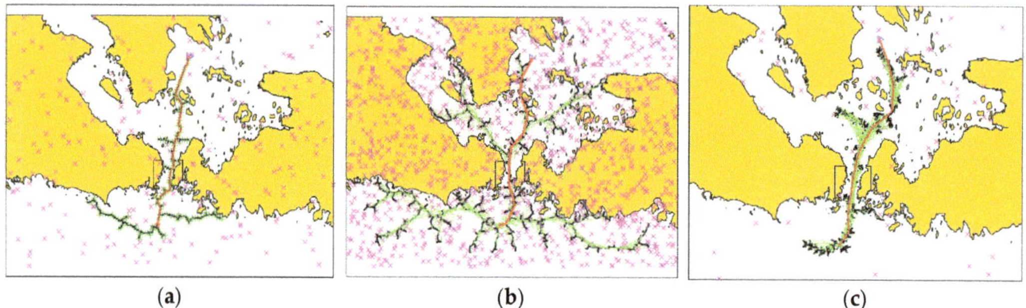

Figure 10. Planned routes with goal orientation in island areas: (**a**) Given by the RRT algorithm;
(**b**) Given by the RRT* algorithm; (**c**) Given by the proposed algorithm.

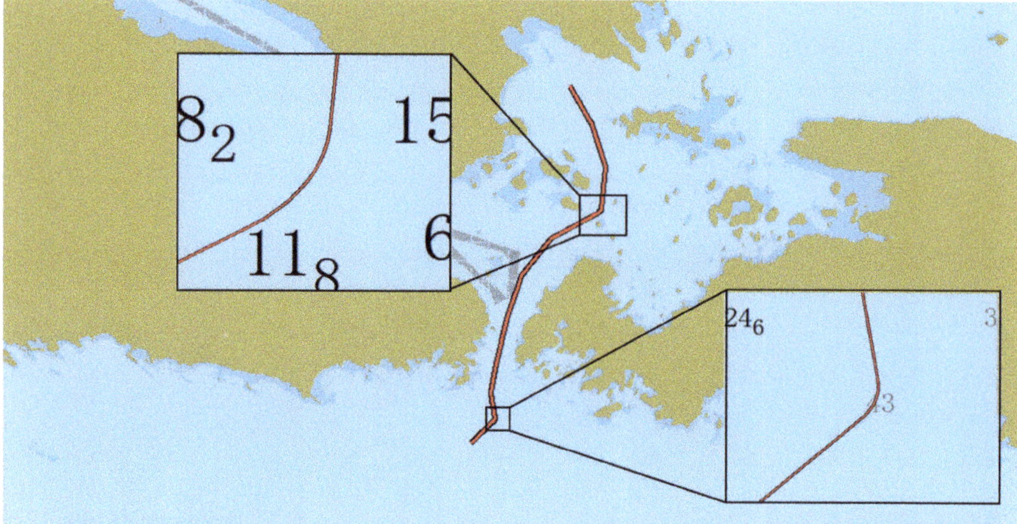

Figure 11. Overview of the planned route designed by the proposed algorithm on electric nautical chart.

Figure 12. Electronic nautical chart of Dalian port.

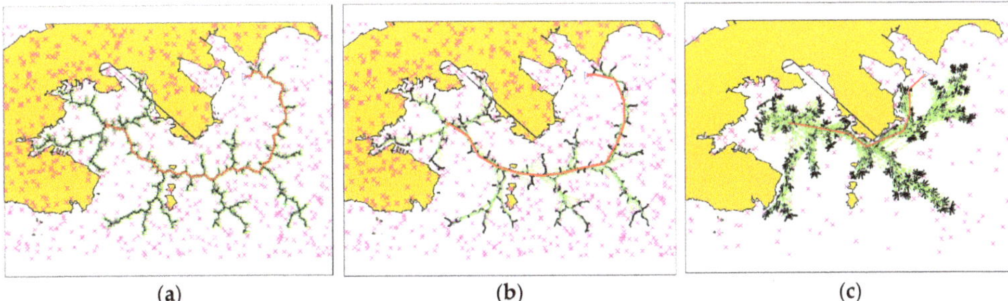

(a)   (b)   (c)

Figure 13. Planned routes without goal orientation in the port: (**a**) Given by the RRT algorithm; (**b**) Given by the RRT* algorithm; (**c**) Given by the proposed algorithm.

Specifically, in Figure 9a, the RRT algorithm gives the planning route in 170 s. It sampled the entire area and constructed bigger trees compared to other algorithms. The length of the route is 22.62 nautical miles, with 84 waypoints. Then, in Figure 9b, the RRT* spends 167.24 s to plan the route. Because of its advance, the length of the route is 19.65, which is shorter than that given by the RRT algorithm. The result of the proposed algorithm is shown in Figure 9c. The time cost has been reduced significantly and it only took 129.41 s to provide the planned route. The length of that route is also the shortest: 18.72 nautical miles. We can also observe that in the proposed algorithm, most of nodes searched were lump together, which improves search effectiveness.

Figure 10 illustrates the results of comparisons of the RRT, RRT*, and proposed algorithms considering goal orientations. The time consumed using the above three algorithms is reduced significantly compared to the algorithms without goal orientation. Their time costs are 41.47 s, 161.72 s, and 39.81 s, respectively. The lengths of the routes are 19.55 nm, 18.91 nm, and 18.71 nm.

Moreover, we should also focus on the characteristics of the waypoint areas. In such places, ships will alter their courses, which could not be tracked by the route planned by the RRT and RRT* algorithms, as both of them create a polyline-based route.

Figure 11 shows the planned route on an electronic nautical chart. It is noticeable that the planned route consists of straight-line segments and arc-line segments. In waypoint areas, the route guides ship navigation from a straight line segment to another line segment in an arc-shaped trajectory. This kind of route takes ship motion characteristics into account. The arc line segments are checked for safety in the search process to could ensure safety.

To be specific, in navigation practice, the planned route was not considered by route optimization in waypoints areas. In previous studies, many researchers have used different kinds of technologies to smooth out the route, making it smoother to track. However, for ship control engineers, it is difficult to design a controller that could track curves well, especially considering the large inertia of ships. Instead, our strategy could let ship navigate at a fix rudder angle to track arc lines as they are designed in reference to ship turning trajectories.

4.2. Simulation in the Port

Here, we provide another simulation in the port area to further validate the feasibility of the proposed algorithm. The area is shown as Figure 12: Dalian Port, China.

In this scenario, the departure point and destination are set as (121.727° E, 38.964° N) and (121.944° E, 39.02° N), respectively. The other parameters of the simulation are the same as those of the first simulation.

Figures 13 and 14 demonstrate the search process of the RRT, RRT*, and proposed algorithms with or without goal orientation strategy. Without considering goal orientation (shown in Figure 12), the time cost of RRT, RRT*, and proposed algorithms is 97.08 s, 87.59 s, and 34.01 s, respectively. It is clear that the proposed algorithm has advantages in the time cost index. Moreover, the lengths of the routes of these algorithms are 32.79 nm, 27.45 nm, and 17.57 nm, respectively.

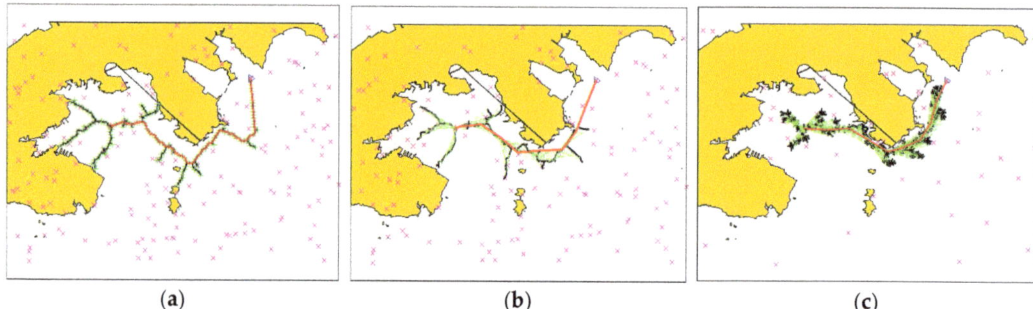

Figure 14. Planned routes with goal orientation in island areas: (**a**) Given by the RRT algorithm; (**b**) Given by the RRT* algorithm; (**c**) Given by the proposed algorithm.

Figure 14 illustrates the planned route given by three algorithms with goal orientation. Their time costs are 26.02 s, 19.94 s, and 9.49 s, respectively. The lengths of the routes given by the RRT, RRT*, and the proposed algorithms are 22.98 nm, 17.73 nm, and 17.22 nm. For the proposed algorithm, the goal orientation strategy only affects the time cost but has little impact on the length of the route.

Similarly, we also show the planned route on the electronic nautical chart for the convenience of analyzing waypoint areas.

We can see that in waypoint areas in Figure 15, the planned route is optimized to adapt ship motion characteristics.

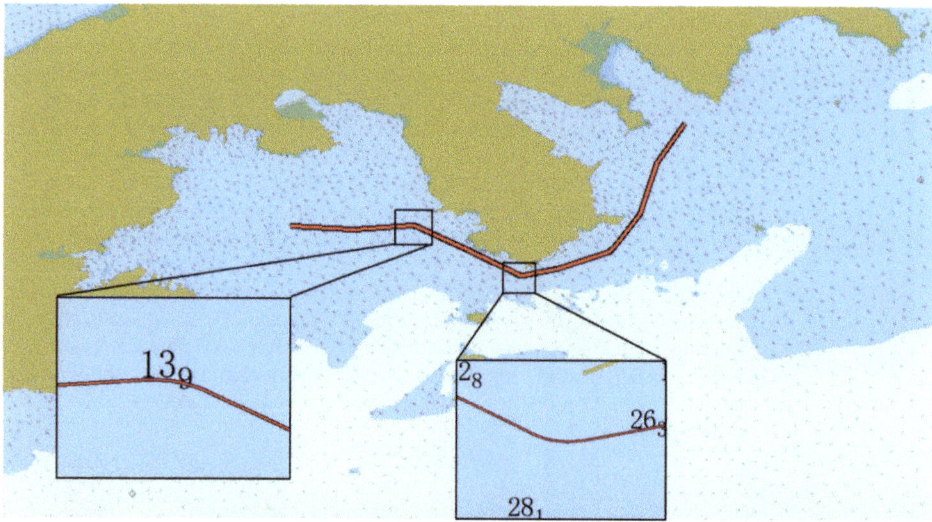

Figure 15. Overview of the planned route designed by the proposed algorithm on electric nautical charts.

In this section, we carry out two simulations validating the feasibility and advantages of the proposed algorithm. The result shows that the proposed algorithm meets the requirements of navigation and ensures safety.

5. Conclusions

This study introduces VK-RRT*, a novel algorithm for addressing SPR problems using ENC data. The algorithm uses Delaunay triangulation to organize ENC vector data, and searches for routes within it. Two cases are examined to demonstrate the effectiveness of VK-RRT*.

However, the study has some limitations that need to be addressed in the future. Firstly, this research only considers depth and land/reef areas; other traffic schemes, such as traffic separation, should also be considered. Secondly, local path planning, or collision avoidance, is not considered, which is also an important part for intelligent ships.

Moving forward, future research should focus on addressing these limitations to improve the effectiveness of the algorithm in practical applications. In addition, it is worth developing a route planning model that could be applied on the ECDIS platform. Finally, there should be a focus on local path planning considering dynamic vessels or obstacles if a better global route is to be obtained and smart navigation is to be achieved.

Author Contributions: Conceptualization, Q.H. and X.Z.; methodology, Q.H.; software, Z.H.; validation, Q.H. and Z.H., writing—original draft preparation, Q.H., writing—review and editing, X.Z.; visualization, Z.H.; supervision, X.Z. All authors have read and agreed to the published version of the manuscript.

Funding: This research received no external funding.

Conflicts of Interest: The authors declare no conflict of interest.

References

1. Dijkstra, E.W. A note on two problems in connexion with graphs. *Numer. Math.* **1959**, *1*, 269–271. [CrossRef]
2. Deng, Y.; Chen, Y.X.; Zhang, Y.J.; Mahadevan, S. Fuzzy Dijkstra algorithm for shortest path problem under uncertain environment. *Appl. Soft Comput.* **2012**, *12*, 1231–1237. [CrossRef]
3. Noto, M.; Sato, H. A method for the shortest path search by extended Dijkstra algorithm. In Proceedings of the 2000 IEEE International Conference on Systems, Man and Cybernetics, Nashville, TN, USA, 8–11 October 2000; pp. 2316–2320.
4. Broumi, S.; Bakal, A.; Talea, M.; Smarandache, F.; Vladareanu, L. Applying Dijkstra algorithm for solving neutrosophic shortest path problem. In Proceedings of the 2016 International Conference on Advanced Mechatronic Systems (ICAMechS), Melbourne, VIC, Australia, 30 November–3 December 2016; pp. 412–416.
5. Wang, H.L.; Mao, W.G.; Eriksson, L. A Three-Dimensional Dijkstra's algorithm for multi-objective ship voyage optimization. *Ocean. Eng.* **2019**, *186*, 106131. [CrossRef]
6. Zhu, X.; Wang, H.; Shen, Z.; Lv, H. Ship weather routing based on modified Dijkstra algorithm. In Proceedings of the 2016 6th International Conference on Machinery, Materials, Environment, Biotechnology and Computer, Tianjin, China, 11–16 June 2016; Atlantis Press: Amsterdam, The Netherlands, 2016.
7. Daniel, K.; Nash, A.; Koenig, S. Theta*: Angle angle path finding on grids. *J. Artif. Intell. Res.* **2010**, *39*, 533–579. [CrossRef]
8. Guruji, A.K.; Agarwal, H.; Parsediya, D.K. Time-efficient A* Algorithm for Robot Path Planning. *Procedia Technol.* **2016**, *23*, 144–149. [CrossRef]
9. Liu, C.; Mao, Q.; Chu, X.; Xie, S. An improved A-star algorithm considering water current, traffic separation and berthing for vessel path planning. *Appl. Sci.* **2019**, *9*, 1057. [CrossRef]
10. Xie, W.; Fang, X.; Wu, S. 2.5 D Navigation Graph and Improved A-Star Algorithm for Path Planning in Ship inside Virtual Environment. In Proceedings of the 2020 Prognostics and Health Management Conference (PHM-Besançon), Besancon, France, 4–7 May 2020; pp. 295–299.
11. LaValle, S.M. *Rapidly-Exploring Random Trees: A New Tool for Path Planning*; University of Illinois Urbana Champaign: Champaign, IL, USA, 1998.
12. Braun, J.; Brito, T.; Lima, J.; Costa, P.; Costa, P.; Nakano, A. A Comparison of A* and RRT* algorithms with dynamic and real time constraint scenarios for mobile robots. In Proceedings of the 9th International Conference on Simulation and Modeling Methodologies, Technologies and Applications, Prague, Czech Republic, 29–31 July 2019; pp. 398–405.
13. Xiao, S.; Bergmann, N.; Postula, A. Parallel RRT∗ architecture design for motion planning. In Proceedings of the 2017 27th International Conference on Field Programmable Logic and Applications (FPL), Ghent, Belgium, 4–8 September 2017; pp. 1–4.
14. Jeong, I.-B.; Lee, S.-J.; Kim, J.-H. Quick-RRT*: Triangular inequality-based implementation of RRT* with improved initial solution and convergence rate. *Expert Syst. Appl.* **2019**, *123*, 82–90. [CrossRef]
15. Li, Y.; Wei, W.; Gao, Y.; Wang, D.; Fan, Z. PQ-RRT*: An improved path planning algorithm for mobile robots. *Expert Syst. Appl.* **2020**, *152*, 113425. [CrossRef]

16. Kuffner, J.J.; LaValle, S.M. RRT-connect: An efficient approach to single-query path planning. In Proceedings of the 2000 ICRA. Millennium Conference. IEEE International Conference on Robotics and Automation. Symposia Proceedings (Cat. No. 00CH37065), San Francisco, CA, USA, 24–28 April 2000; pp. 995–1001.
17. Klemm, S.; Oberländer, J.; Hermann, A.; Roennau, A.; Schamm, T.; Zollner, J.M.; Dillmann, R. RRT*-Connect: Faster, asymptotically optimal motion planning. In Proceedings of the 2015 IEEE International Conference on Robotics and Biomimetics (ROBIO), Zhuhai, China, 6–9 December 2015; pp. 1670–1677.
18. Xinyu, W.; Xiaojuan, L.; Yong, G.; Jiadong, S.; Rui, W. Bidirectional potential guided RRT* for motion planning. *IEEE Access* **2019**, *7*, 95046–95057. [CrossRef]
19. Mashayekhi, R.; Idris, M.Y.I.; Anisi, M.H.; Ahmedy, I.; Ali, I. Informed RRT*-connect: An asymptotically optimal single-query path planning method. *IEEE Access* **2020**, *8*, 19842–19852. [CrossRef]
20. Lau, C.; Byl, K. Smooth RRT-connect: An extension of RRT-connect for practical use in robots. In Proceedings of the 2015 IEEE International Conference on Technologies for Practical Robot Applications (TePRA), Woburn, MA, USA, 11–12 May 2015; pp. 1–7.
21. Enevoldsen, T.T.; Reinartz, C.; Galeazzi, R. COLREGs-Informed RRT* for Collision Avoidance of Marine Crafts. *arXiv* **2021**, arXiv:2103.14426.
22. Zhang, Z.; Wu, D.; Gu, J.; Li, F. A path-planning strategy for unmanned surface vehicles based on an adaptive hybrid dynamic stepsize and target attractive force-RRT algorithm. *J. Mar. Sci. Eng.* **2019**, *7*, 132. [CrossRef]
23. Xiong, C.; Zhou, H.; Lu, D.; Zeng, Z.; Lian, L.; Yu, C. Rapidly-Exploring Adaptive Sampling Tree*: A Sample-Based Path-Planning Algorithm for Unmanned Marine Vehicles Information Gathering in Variable Ocean Environments. *Sensors* **2020**, *20*, 2515. [CrossRef] [PubMed]
24. Chiang, H.-T.L.; Tapia, L. COLREG-RRT: An RRT-based COLREGS-compliant motion planner for surface vehicle navigation. *IEEE Robot. Autom. Lett.* **2018**, *3*, 2024–2031. [CrossRef]
25. Zilu, O.; Hongdong, W.; Jianyao, W.; Hong, Y. Automatic collision avoidance algorithm for unmanned surface vessel based on improved Bi-RRT algorithm. *Chin. J. Ship Res.* **2019**, *14*, 9.
26. Orozco-Rosas, U.; Picos, K.; Montiel, O. Hybrid path planning algorithm based on membrane pseudo-bacterial potential field for autonomous mobile robots. *IEEE Access.* **2019**, *7*, 156787–156803. [CrossRef]
27. Orozco-Rosas, U.; Montiel, O.; Sepúlveda, R. Mobile robot path planning using membrane evolutionary artificial potential field. *Appl. Soft Comput.* **2019**, *77*, 236–251. [CrossRef]
28. Orozco-Rosas, U.; Picos, K.; Pantrigo, J.J.; Montemayor, A.S.; Cuesta-Infante, A. Mobile robot path planning using a QAPF learning algorithm for known and unknown environments. *IEEE Access* **2022**, *10*, 84648–84663. [CrossRef]
29. Shen, H.; Hashimoto, H.; Matsuda, A.; Taniguchi, Y.; Terada, D.; Guo, C. Automatic collision avoidance of multiple ships based on deep Q-learning. *Appl. Ocean. Res.* **2019**, *86*, 268–288. [CrossRef]
30. Chen, C.; Chen, X.-Q.; Ma, F.; Zeng, X.-J.; Wang, J. A knowledge-free path planning approach for smart ships based on reinforcement learning. *Ocean. Eng.* **2019**, *189*, 106299. [CrossRef]
31. Zhao, B.; Zhang, X.; Liang, C. A novel path-following control algorithm for surface vessels based on global course constraint and nonlinear feedback technology. *Appl. Ocean. Res.* **2021**, *111*, 102635. [CrossRef]
32. IHO. *IHO Transfer Standard for Digital Hydrographc Data*, 3.1 ed.; Special Publication No. 57; IHO: Monaco, France, 2000.
33. Zhang, Q.; Zhang, X.-K.; Im, N.-K. Ship nonlinear-feedback course keeping algorithm based on MMG model driven by bipolar sigmoid function for berthing. *Int. J. Nav. Archit. Ocean. Eng.* **2017**, *9*, 525–536. [CrossRef]
34. Nguyen, C.M.; Rhodes, P.J. Delaunay triangulation of large-scale datasets using two-level parallelism. *Parallel Comput.* **2020**, *98*, 102672. [CrossRef]
35. Karaman, S.; Frazzoli, E. Sampling-based algorithms for optimal motion planning. *Int. J. Robot. Res.* **2011**, *30*, 846–894. [CrossRef]
36. Liu, H.; Zhang, X.; Wen, J.; Wang, R.; Chen, X. Goal-biased bidirectional RRT based on curve-smoothing. *IFAC-Pap.* **2019**, *52*, 255–260. [CrossRef]
37. Urmson, C.; Simmons, R. Approaches for heuristically biasing RRT growth. In Proceedings of the 2003 IEEE/RSJ International Conference on Intelligent Robots and Systems (IROS 2003), Las Vegas, NV, USA, 27–31 October 2003; Volume 2, pp. 1178–1183.
38. Jin, X.; Yan, Z.; Yang, H.; Wang, Q.; Yin, G. A goal-biased RRT path planning approach for autonomous ground vehicle. In Proceedings of the 2020 4th CAA International Conference on Vehicular Control and Intelligence (CVCI), Hangzhou, China, 18–20 December 2020; pp. 743–746.
39. Ayawli, B.B.; Mei, X.; Shen, M.; Appiah, A.Y.; Kyeremeh, F. Optimized RRT-A* path planning method for mobile robots in partially known environment. *Inf. Technol. Control* **2019**, *48*, 179–194. [CrossRef]
40. Liang, C.; Zhang, X.; Han, X. Route planning and track keeping control for ships based on the leader-vertex ant colony and nonlinear feedback algorithms. *Appl. Ocean. Res.* **2020**, *101*, 102239. [CrossRef]

Journal of
*Marine Science
and Engineering*

Article

A Novel Robust IMM Filtering Method for Surface-Maneuvering Target Tracking with Random Measurement Delay

Chen Chen [1], Weidong Zhou [1,*] and Lina Gao [2]

[1] Department of Intelligent Systems Science and Engineering, Harbin Engineering University, Harbin 150001, China; chenchen2019@hrbeu.edu.cn
[2] Department of Measurement and Control Engineering, School of Electronics and Information Engineering, Harbin Institute of Technology, Harbin 150001, China; gaolina@hit.edu.cn
* Correspondence: zhouweidong@hrbeu.edu.cn

Abstract: A proper filtering method for jump Markov system (JMS) is an effective approach for tracking a maneuvering target. Since the coexisting of heavy-tailed measurement noises (HTMNs) and one-step random measurement delay (OSRMD) in the complex scenarios of the surface maneuvering target tracking, the effectiveness of typical interacting multiple model (IMM) techniques may decline severely. To solve the state estimation problem in JMSs with HTMN and OSRMD simultaneously, this article designs a novel robust IMM filter utilizing the variational Bayesian (VB) inference framework. This algorithm models the HTMNs as student's t-distribuitons, and presents a random Bernoulli variable to describe the OSRMD in JMSs. By transforming measurement likelihood function form from weighted summation to exponential product, this paper constructs hierarchical Gaussian state space models. Then, the state vectors, random Bernoulli vairable, and model probability are inferred jointly according to VB inference. The surface maneuvering target tracking simulation example result indicates that the presented IMM filter achieves superior target state estimation accuracy among existing IMM filters.

Keywords: variational Bayesian; surface maneuvering target tracking; random measurement delay; heavy-tailed measurement noise; interacting multiple model

Citation: Chen, C.; Zhou, W.; Gao, L. A Novel Robust IMM Filtering Method for Surface-Maneuvering Target Tracking with Random Measurement Delay. *J. Mar. Sci. Eng.* **2023**, *11*, 1047. https://doi.org/ 10.3390/jmse11051047

Academic Editor: Mohamed Benbouzid

Received: 17 April 2023
Revised: 9 May 2023
Accepted: 13 May 2023
Published: 14 May 2023

1. Introduction

State estimation is an important topic in maneuvering target tracking of traditional ships and surface autonomous ships. As a minimum mean square error estimator, the Kalman filter (KF) provides the optimal state estimation method for a linear Gaussian system (The abbreviations are summarized in Table 1). However, the system nonlinearity and system model uncertainty in jump Markov systems (JMSs) may lead to the KF's performances degrading dramatically. Additionally, the KF assumes that the measurement noises obey Gaussian distributions and that all measurements need to arrive in time. Both aspects may affect the filtering accuracy significantly when these assumptions are not satisfied. Thus, the extension for the KF under various assumptions has attracted considerable attention on account of its widely applied in engineering, for example, targets tracking, signal processing, integrated navigation, fault diagnosis, ect. [1–6].

The surface maneuvering target tracking is a problem of state estimation in JMSs. Since the state estimation in JMSs is well-known computational intractability and nondeterministic polynomial difficult, it is hard to find an optimal solution [7]. In the past few decades, some sub-optimal solutions were presented, for example, the interacting multiple model (IMM) approach [8], pseudo-Bayesian technique [9], particle filter [10], etc. Among these methods, the IMM approach is one of the most efficient since it reasonably balances the estimation performances and the computation complexity. In the process of IMM technique development, many researchers contributed remarkable achievements. The stability and performances of the IMM filter are analyzed by [11,12]. A series of executable pseudocode is provided in [13]. IMM algorithms are widely used in various practical engineering [14–16].

The typical IMM filters parallelly perform a bank of KFs to estimate mode-conditional system state and probability at each cycle. The weighted sum of sub-filter outputs obtains the final state estimates [17]. An effective method to develop the performances of sub-filters (KFs) is the variational Bayesian (VB) inference, which implements approximations for the conjugate exponential model with acceptable computational costs [18]. Based on the combination of VB theory and the KF, the development of the KF has made significant progress and remarkable achievements [19–22]. Extended to IMM filters for the state estimation in JMSs, some VB-based IMM methods were designed in past few years. Ref. [23] presented an IMM filter to adaptively estimate the unknown process and measurement noise covariance matrices in JMS. The authors selected the conjugate prior distribution of noise covariance matrix as the inverse-Wishart distribution, the conditional system state vectors and the noise parameters can be estimated simultaneously by VB inference. This method achieved remarkable estimation accuracy when the noise covariances were unknown. However, this technique requires linear Gaussian systems, and the measurements need to be arrived in time, which is usually not satisfied in actual tracking scenarios affected by measurement outliers and communication channel latency.

Table 1. Definitions of notations in this paper.

Notation	Definition	Notation	Definition
VB	Variational Bayesian	HGSSM	Hierarchical Gaussian state space model
KF	Kalman filter	OSRMD	One-step random measurement delay
PDF	Probability density function	$N(\cdot; \bar{\mu}, P)$	Gaussian distribution, $\bar{\mu}$-mean vector, P-scale matrix
JMS	Jump Markov system		
STD	student's t-distribution	$St(\cdot; \bar{\mu}, P, v)$	Student's t-distribution, $\bar{\mu}$-mean vector, P-scale matrix, v-degree of freedom parameter
RBV	Random Bernoulli variable		
KLD	Kullback-Leibler divergence		
$E[\cdot]$	Expectation computation	$G(\cdot; a, b)$	Gamma distribution, a-shape parameter, b-rate parameter
$tr(\cdot)$	Trace operation		
HTMN	Heavy-tailed measurement noise		

Aiming to solve state estimation problems in JMSs with measurement outliers, Ref. [24] modeled the measurement noises as students' t-distributions (STDs), and estimated the system states and mode probabilities jointly by VB inference. Compared with the typical IMM algorithm, the estimation performance was significantly developed. Ref. [25] developed the IMM filter in [24] by reasonably selecting the conjugate prior distributions of covariances and degree of freedom parameters as inverse Wishart and Gamma distributions, respectively. The state vectors and the degree of freedom parameters were inferred simultaneously by VB method. This algorithm improved the estimation accuracy and can be utilized for nonlinear JMSs. The methods mentioned in [24,25] show superior estimation accuracy in the scenarios of state estimation in JMSs containing heavy-tailed measurement noises (HTMNs). However, the performances of these approaches may lose efficacy when the random measurement delay happens since they assume the measurements can be obtained in real-time. To deal with the one-step random measurement delay (OSRMD), Ref. [26] designed a particle filter by defining two variables to extend the original system state vector. This filter can effectively handle the systems with delayed measurements. Although this algorithm provided a solution to measurement delay and achieved satisfactory estimation accuracy, the particle filters have problems of high computational burden and curse of dimensionality. Ref. [27] proposed a Gaussian approximate filter with OSRMD, and measurement noise vectors are inferred by the Bayesian rules. Ref. [28] developed the KF to deal with the OSRMD in nonlinear systems. In this method, the authors utilized a random Bernoulli variable (RBV) to describe the random measurement delay, and the system states were inferred by the VB technique. However, these methods failed to deal with the HTMN and system model jumping problems in JMSs. In complex JMSs engineering applications, the HTMNs and random measurement delay often exist simultaneously. It is necessary to propose a more general IMM filter for HTMNs coexist with random measurement delay.

This article designs a novel robust IMM filter to deal with the filtering problems in JMSs with HTMN and OSRMD. The HTMN and the OSRMD in JMSs are reasonably modeled in the designed filter. By transforming the measurement likelihood function to a new form, this algorithm constructed a new hierarchical Gaussian state space model (HGSSM). According to the VB inference, the state vectors, model probability, RBVs, and distribution parameters are inferred simultaneously. Target tracking experiment validates that the presented approach has better performance than existing IMM approaches.

The main contribution of our work is summarized as follows:

- The one-step predictive probability density function (PDF) and HTMN are assumed to obey Gaussian and STDs, respectively. This paper presents an RBV to characterize the OSRMD in JMSs. Aiming to introduce the VB method directly, this article converts measurement likelihood function form from weighted summation to exponential product, and constructs the HGSSM.
- To address the coupled state vectors and the noise covariance matrices, a novel IMM filter is designed by combining the VB theory with IMM method. In measurement update part, the mode conditional posterior PDFs are approximated recursively. The state vectors, RBVs, model probabilities, and unknown parameters are estimated through VB technique. Then, the final estimates are obtained by the weighted sum of sub-filters.
- Four parts of the surface target tracking simulation indicate that the presented method outperforms existing IMM filters on estimation accuracy. The presented algorithm provides a robust solution to the filtering problem in the scenarios of HTMNs coexisting with OSRMD.

The rest parts of this paper can be arranged as the following sections. Section 2 provides the problem statement. We construct a new HGSSM in Section 3. Furtherly, we summarize the derivations of our designed IMM method in Section 4. The performances of exsiting IMM algorithms and the proposed are verified in Section 5. Finally, the conclusions of this paper is in Section 6.

2. Problem Statement

Considering a state-space model of nonlinear jump Markov system:

$$x_s = f_{s-1}(x_{s-1}, M_{s-1}) + g_{s-1}, s \geq 1 \tag{1}$$

$$z_s = h_s(x_s, M_s) + \varepsilon_s, s \geq 1 \tag{2}$$

$$y_s = \begin{cases} (1 - \sigma_s)z_s + \sigma_s z_{s-1}, & s > 1 \\ z_s, & s = 1 \end{cases} \tag{3}$$

where s is time index, $x_s \in \mathbb{R}^n$ is the system state vector, $z_s \in \mathbb{R}^m$ and $y_s \in \mathbb{R}^m$ denote the ideal and real measurement vectors, m and n are their dimension numbers. $f_{s-1}(\cdot)$ and $h_s(\cdot)$ refer to the process function and measurement function. We denote $f_{s-1}(x_{s-1}, M_{s-1})$ as $f^i_{s-1}(x_{s-1})$ and $h_s(x_s, M_s)$ as $h^i_s(x_s)$ conditioned on $M_s = i$, respectively. The system mode M_s indicates the state of the Markov chains. It selects a value from a limited set

$\{1, 2, 3, \ldots, k\}$ according to the transition probability matrix $T = \left[\lambda_{ji}\right]_{k \times k}$, which satisfies the equations as follows:

$$\lambda_{ji} \overset{\Delta}{=} P\{M_s = i | M_s = j\} \tag{4}$$

$$\sum_{i=1}^{k} \lambda_{ji} = 1 \tag{5}$$

g_{s-1} represents the Gaussian process noise vector with nominal covariance matrix Q_s, while ε_s refers to the HTMN vector caused by measurement outliers with nominal covariance matrix R_s. $y_s \in \mathbb{R}^m$ denotes the real random delayed measurement vector. The RBV $\sigma_s \in \{0, 1\}$ with corresponding probability can be defined as follows:

$$p(\sigma_s = 1) = \varphi_s \tag{6}$$
$$p(\sigma_s = 0) = 1 - \varphi_s \tag{7}$$

where $\varphi_s \in (0, 1)$ denotes the probability of OSRMD. Note that random variables x_s, σ_s, g_{s-1}, and r_s are independent of each others.

Utilizing Equation (3) and Equations (6) and (7), the measurement likelihood PDF is as follows:

$$p(y_s | x_s, x_{s-1}, M_s = i) = \sum_{\sigma_s = 0}^{1} p(y_s, \sigma_s | x_s, x_{s-1}, M_s = i)$$

$$= p(\sigma_s = 1) p(y_s | x_s, x_{s-1}, \sigma_s = 1, M_s = i) + p(\sigma_s = 0) p(y_s | x_s, x_{s-1}, \sigma_s = 0, M_s = i)$$

$$= \varphi_s p(y_s | x_s, x_{s-1}, \sigma_s = 1, M_s = i) + (1 - \varphi_s) p(y_s | x_s, x_{s-1}, \sigma_s = 0, M_s = i) \tag{8}$$

Utilizing Equations (2) and (3) yields

$$p(y_s | x_s, x_{s-1}, \sigma_s = 1, M_s = i) = p_{\varepsilon_{s-1}}[y_s - h_{s-1}(x_{s-1})] \tag{9}$$
$$p(y_s | x_s, x_{s-1}, \sigma_s = 0, M_s = i) = p_{\varepsilon_s}[y_s - h_s(x_s)] \tag{10}$$

In Equation (8), replaced by Equations (9) and (10), we can obtain:

$$p(y_s | x_s, x_{s-1}, M_s = i) = \varphi_s p_{\varepsilon_{s-1}}[y_s - h_{s-1}(x_{s-1})] + (1 - \varphi_s) p_{\varepsilon_s}[y_s - h_s(x_s)] \tag{11}$$

However, the measurement likelihood PDF in Equation (11), which is formulated by a weighted sum form, has neither conjugate property nor closed property. Therefore, it is not easy to introduce the VB technique directly. Aiming at this difficulty, in the next section, the weighted sum form of measurement likelihood PDF is converted to exponential multiplication, which can be further utilized to design a novel adaptive estimation method for JMSs.

3. Construction of the HGSSM

3.1. Measurement Likelihood PDF Convertion

The relationship between RBV σ_s with fixed delay probability can be seen in Equations (6) and (7). The probability mass function $p(\sigma_s | \varphi_s)$ is defined as the following formula:

$$p(\sigma_s) = (1 - \varphi_s)^{(1 - \sigma_s)} \varphi_s^{\sigma_s} \tag{12}$$

according to Equations (11) and (12), reformulating Equation (8) by:

$$p(y_s|x_s, x_{s-1}, M_s = i) = \sum_{\sigma_s=0}^{1} p(y_s, \sigma_s | x_s, x_{s-1}, M_s = i)$$

$$= \sum_{\sigma_s=0}^{1} (\varphi_s)^{\sigma_s} p_{\varepsilon_{s-1}}(y_s - h_{s-1}(x_{s-1}))^{\sigma_s} (1 - \varphi_s)^{(1-\sigma_s)} p_{\varepsilon_s}(y_s - h_s(x_s))^{(1-\sigma_s)}$$

$$= \sum_{\sigma_s=0}^{1} p(\sigma_s) p_{\varepsilon_{s-1}}(y_s - h_s(x_s))^{\sigma_s} p_{\varepsilon_s}(y_s)^{(1-\sigma_s)} \tag{13}$$

then, transforming the conditional measurement likelihood PDF to exponential product form:

$$p(y_s|x_s, x_{s-1}, \sigma_s) = \left[p_{\varepsilon_{s-1}}(y_s - h_{s-1}(x_{s-1})) \right]^{\sigma_s} \left[p_{\varepsilon_s}(y_s - h_s(x_s)) \right]^{(1-\sigma_s)} \tag{14}$$

Equation (14) can be also denoted by the formula as follows:

$$p(y_s|\eta_s, \sigma_s) = \left[p_{\varepsilon_{s-1}}(y_s - h_{s-1}(x_{s-1})) \right]^{\sigma_s} \left[p_{\varepsilon_s}(y_s - h_s(x_s)) \right]^{(1-\sigma_s)} \tag{15}$$

where the extended state vector $\eta_s = \left[\begin{array}{cc} x_s^T & x_{s-1}^T \end{array} \right]^T$.

3.2. Prior PDFs Selection

First of all, the one-step predictive PDF of the extended system state η_s is formulated by

$$p(\eta_s|y_{1:s-1}) = N\left(\eta_s; \hat{\eta}_{s|s-1}, P^{\eta\eta}_{s|s-1} \right) \tag{16}$$

where $\hat{\eta}_{s|s-1}$ represents the predicted mean vector, $P^{\eta\eta}_{s|s-1}$ refers to the predicted covariance matrix, and they are expressed by:

$$\hat{\eta}_{s|s-1} = \left[\begin{array}{cc} \hat{x}_{s|s-1}^T & \hat{x}_{s-1|s-1}^T \end{array} \right]^T \tag{17}$$

$$P^{\eta\eta}_{s|s-1} = \left[\begin{array}{cc} P_{s|s-1} & P_{s-1,s|s-1} \\ \left(P_{s-1,s|s-1} \right)^T & P_{s-1|s-1} \end{array} \right] \tag{18}$$

where $\hat{\eta}_{s|s-1}$, $P_{s|s-1}$, and $P_{s-1,s|s-1}$ can be calculated by the standard Gaussian approximate filter [29], i.e.,

$$\hat{x}_{s|s-1} = \int f_{s-1}^i(x_{s-1}) N\left(x_{s-1}; \hat{x}_{s-1|s-1}, P_{s-1|s-1} \right) dx_{s-1} \tag{19}$$

$$P_{s|s-1} = \int f_{s-1}^i(x_{s-1}) f_{s-1}^{i(T)}(x_{s-1}) N\left(x_{s-1}; \hat{x}_{s-1|s-1}, P_{s-1|s-1} \right) dx_{s-1} \tag{20}$$

$$P_{s-1,s|s-1} = \int x_{s-1} f_{s-1}^{i(T)}(x_{s-1}) N\left(x_{s-1}; \hat{x}_{s-1|s-1}, P_{s-1|s-1} \right) dx_{s-1} - \hat{x}_{s-1|s-1} \hat{x}_{s|s-1}^T \tag{21}$$

Secondly, in the aspects of processing the measurement noises, since the STD has heavier tails compared with Gaussian distribution and is more robust to outlier [30–32], STD is selected to model HTMN, i.e.,

$$p(\varepsilon_{s-1}) = St(\varepsilon_{s-1}; 0, R_{s-1}, \tau_{s-1}) \tag{22}$$

$$p(\varepsilon_s) = St(\varepsilon_s; 0, R_s, \tau_s) \tag{23}$$

where the $p(\varepsilon_{s-1})$ and $p(\varepsilon_s)$ represent the PDFs of previous and current measurement noise vector. Based on the properties of the STDs, PDFs $p(\varepsilon_{s-1})$ and $p(\varepsilon_s)$ are formulated in hierarchical form as follows:

$$p(\varepsilon_{s-1}) \int N(\varepsilon_{s-1}; 0, R_{s-1}/v_{s-1}) p(v_{s-1}) dv_{s-1} \tag{24}$$

$$p(v_{s-1}) = G\left(v_{s-1}; \frac{a_{s-1}}{2}, \frac{a_{s-1}}{2}\right) \tag{25}$$

$$p(\varepsilon_s) \int N(\varepsilon_s; 0, R_s/v_s) p(v_s) dv_s \tag{26}$$

$$p(v_s) = G\left(v_s; \frac{a_s}{2}, \frac{a_s}{2}\right) \tag{27}$$

Utilizing Equations (15), (24)–(27), the conditional likelihood PDF $p(y_s|x_s, x_{s-1}, v_s, v_{s-1}, \sigma_s)$ can be obtained by:

$$p(y_s|x_s, x_{s-1}, v_s, v_{s-1}, \sigma_s) = [N(y_s; h_s(x_s), R_s/v_s)]^{(1-\sigma_s)} [N(y_{s-1}; h_{s-1}(x_{s-1}), R_{s-1}/v_{s-1})]^{\sigma_s} \tag{28}$$

Finally, according to Equation (28), the measurement vector y_s depends on the extended state vector η_s, auxiliary variables v_s and v_{s-1}, and the RBV σ_s, the joint prior PDFs $p(\eta_s, v_s, v_{s-1}, \sigma_s|y_{1:s-1})$ can be obtained as follows:

$$\begin{cases} p(\Xi|y_{1:s-1}) = p(\eta_s|y_{1:s-1}) p(v_s) p(v_{s-1}) p(\sigma_s) \\ = N\left(\eta_s; \hat{\eta}_{s|s-1}, P^{\eta\eta}_{s|s-1}\right) G\left(v_s; \frac{a_s}{2}, \frac{a_s}{2}\right) G\left(v_{s-1}; \frac{a_{s-1}}{2}, \frac{a_{s-1}}{2}\right) (1-\varphi_s)^{(1-\sigma_s)} \varphi_s^{\sigma_s} \\ \Xi \overset{\Delta}{=} \{\eta_s, v_s, v_{s-1}, \sigma_s\} \end{cases} \tag{29}$$

Then, the HGSSM consists Equations (15), (17)–(21) and (24)–(29) is constructed, based on which a novel adaptive estimation algorithm will be designed to infer the state vector η_s and BRV σ_s in JMSs with OSRMD and HTMN.

4. Design of the Proposed Filter

Based on the IMM filtering framework, the designed filter as an iterative algorithm mainly consists of four recursive steps, i.e., interacting/mixing, mode-conditioned filtering, mode probability updating, and combination.

Step 1: Interacting/Mixing

To interact the state vector and noise parameters, the mixing probability needs to be computed first utilizing the mode transition probability λ_{ji} as follows:

$$\mu^{ji}_{s-1} = \frac{1}{\bar{e}_i} p_{ji} \mu^{j}_{s-1} \tag{30}$$

where the mode transition probability has been defined in Equations (4) and (5), and the normalized constant $\bar{e}_i$ is defined by Equation (31):

$$\bar{e}_i = \sum_{j=1}^{k} \lambda_{ji} \mu^{j}_{s-1} \tag{31}$$

where μ^{j}_{s-1} represents the mode probability, and the mode index $j, i \in \{1, 2, 3, \ldots, k\}$.
Suppose that the posterior PDF at j-th mode can be obtained by

$$p(\eta_{s-1}|y_{1:s-1}, M_{s-1} = j) = N\left(\eta_{s-1}; \hat{\eta}^{j}_{s-1|s-1}, P^{j}_{s-1|s-1}\right) \tag{32}$$

Utilizing the total probability thorem and the Bayes rules to calculate the mixing PDFs $p(\eta_s|y_{1:s-1}, M_s = i)$, i.e.,

$$p(\eta_s|y_{1:s-1}, M_s = i) = \sum_{j=1}^{k} p(\eta_s|y_{1:s-1}, M_{s-1} = j)p(M_{s-1} = j|M_s = i, y_{1:s-1})$$

$$= \sum_{j=1}^{k} \mu_{s-1}^{ji} N\left(\eta_{s-1}; \hat{\eta}_{s-1}^{j}, P_{s-1}^{j}\right) \tag{33}$$

Using the Kullback-Leibler average fusion method [33], the summation term in Equation (33) is approximated as follows:

$$p(\eta_{s-1}|y_{1:s-1}, M_s = i) \approx N\left(\eta_{s-1}; \hat{\eta}_{s-1|s-1}^{0i}, P_{s-1|s-1}^{0i}\right) \tag{34}$$

where the mixing mean vector and covariance matrix of the distribution of η_{s-1} can be obtained as follows:

$$\begin{cases} \eta_{s-1|s-1}^{0i} = \sum_{j=1}^{k} \mu_{s-1|s-1}^{ji} \eta_{s-1|s-1}^{j} \\ P_{s-1|s-1}^{\eta\eta,0i} = \sum_{j=1}^{k} \mu_{s-1|s-1}^{ji} \left(P_{s-1|s-1}^{\eta\eta,j} + \left(\hat{x}_{s-1|s-1}^{j} - \hat{x}_{s-1|s-1}^{0i}\right)\left(\hat{x}_{s-1|s-1}^{j} - \hat{x}_{s-1|s-1}^{0i}\right)^{T}\right) \end{cases} \tag{35}$$

Step 2: Mode-Conditioned Filtering
The posterior PDFs at the i-th mode is formulated by:

$$p(\eta_s, v_s, v_{s-1}, \sigma_s|y_{1:s}, M_s = i) = \frac{p(\eta_s, v_s, v_{s-1}, \sigma_s|y_s, M_s = i)}{p(y_s|y_{1:s-1}, M_s = i)} \tag{36}$$

Since the presence of couplings between the state vectors and the noise covariances, the analytic solution can not be obtained for the mode conditional posterior PDFs $p(\eta_s, v_s, v_{s-1}, \sigma_s|y_s, M_s = i)$. However, by using the VB inference, the approximate solution can be available. Based on the VB technique, the abovementioned posterior PDFs is approximated as a multiplied factor form:

$$p(\eta_s, v_s, v_{s-1}, \sigma_s|y_{1:s}, M_s = i) \approx q\left(\eta_s^{i}\right)q\left(v_s^{i}\right)q\left(v_{s-1}^{i}\right)q\left(\sigma_s^{i}\right) \tag{37}$$

where $q(\cdot)$ is approximated posterior PDF of $p(\cdot)$. Using the VB technique to minimize the kullback-Leibler divergence (KLD) between the actual and approximated posterior PDFs, the optimal approximation PDFs can be obtained, i.e.,

$$q\left(\eta_s^{i}\right)q\left(v_s^{i}\right)q\left(v_{s-1}^{i}\right)q\left(\sigma_s^{i}\right)$$
$$= \arg\min KLD\left[q\left(\eta_s^{i}\right)q\left(v_s^{i}\right)q\left(v_{s-1}^{i}\right)q\left(\sigma_s^{i}\right)\middle\|p\left(\eta_s^{i}, v_s^{i}, v_{s-1}^{i}, \sigma_s^{i}|y_{1:s}, M_s = i\right)\right] \tag{38}$$

where $KLD(q(\cdot)\|p(\cdot)) \triangleq \int q(\cdot) \ln \frac{q(y)}{p(y)} dy$, Equation (38) is the optimal method for approximation:

$$\ln q(\phi_s) = E_{\Xi_s^{-\phi_s}}[\ln p(\Xi|y_{1:s})|M_s = i] + cst_{\phi_s} \tag{39}$$

where ϕ_s denotes one element from Ξ, $\Xi_s^{-\phi_s}$ represents the rest elements after removing element ϕ_s in Ξ_s. cst_{ϕ_s} refers to the constant about ϕ_s.
The joint PDF $p(\Xi, y_{1:s}|M_s = i)$ in Equation (39) can be formulated by

$$p(\Xi, y_{1:s} | M_s = i) = p(y_{1:s-1} | M_s = i) p(y_s | \eta_s, v_s, v_{s-1}, \sigma_s, M_s = i) p(\eta_s | y_{1:s-1}, M_s = i)$$
$$p(v_s | M_s = i) p(v_{s-1} | M_s = i) p(\sigma_s | M_s = i) \tag{40}$$

Substituting Equations (12) and (28), (29) into (40):

$$p(\Xi, y_{1:s} | M_s = i) = \left[N\left(y_s^i; h_s^i\left(x_s^i\right), R_s^i / v_s^i \right) \right]^{\left(1-\sigma_s^i\right)} \left[N\left(y_{s-1}^i; h_{s-1}^i\left(x_{s-1}^i\right), R_{s-1}^i / v_{s-1}^i \right) \right]^{\left(1-\sigma_s^i\right)}$$
$$\cdot N\left(\eta_s^i; \hat{\eta}_{s|s-1}^i, P_{s|s-1}^{\eta\eta,i} \right) G\left(v_s^i; \frac{a_s^i}{2}, \frac{a_s^i}{2} \right) G\left(v_{s-1}^i; \frac{a_{s-1}^i}{2}, \frac{a_{s-1}^i}{2} \right) \left(1 - \varphi_s^i\right)^{\left(1-\sigma_s^i\right)}$$
$$\cdot \left(\varphi_s^i \right)^{\sigma_s^i} p(y_{1:s-1}) \tag{41}$$

Utilizing Equation (41), $\ln p(\Xi, y_{1:s} | M_s = i)$ is further derived:

$$\ln p(\Xi, y_{1:s} | M_s = i) = 0.5\left(m\left(1 - \sigma_s^i\right) + a_s^i - 2 \right) \ln v_s^i + 0.5\left(m\sigma_s^i + a_{s-1}^i - 2 \right) \ln v_{s-1}^i$$
$$- 0.5\left(1 - \sigma_s^i\right) v_s^i \left(y_s^i - h_s^i\left(x_s^i\right) \right)^T R_s^{i(-1)} \left(y_s^i - h_s^i\left(x_s^i\right) \right) - 0.5\sigma_s^i v_{s-1}^i \left(y_s^i - h_{s-1}^i\left(x_{s-1}^i\right) \right)^T$$
$$\cdot R_{s-1}^{i(-1)} \left(y_s^i - h_{s-1}^i\left(x_{s-1}^i\right) \right) - 0.5\left(\eta_s^i - \hat{\eta}_{s|s-1}^i \right)^T P_{s|s-1}^{\eta\eta(-1)} \left(\eta_s^i - \hat{\eta}_{s|s-1}^i \right) + \sigma_s^i \ln \varphi_s^i$$
$$+ \left(1 - \sigma_s^i\right) \ln\left(1 - \varphi_s^i\right) - 0.5 v_{s-1}^i - 0.5 v_s^i + cst_{\eta,v,\sigma} \tag{42}$$

The approximated PDFs together with relevant expectations are obtained after N iterations:

$$q^{(N)}\left(\eta_s^i \right) \approx N\left(\eta_s^i; \hat{\eta}_{s|s}^{(N)i}, P_{s|s}^{(N)\eta\eta,i} \right) \tag{43}$$

$$q^{(N)}\left(v_s^i \right) \approx G\left(v_s^i; \alpha_{s|s}^{(N)i}, \beta_{s|s}^{(N)i} \right) \tag{44}$$

$$q^{(N)}\left(v_{s-1}^i \right) \approx G\left(v_{s-1}^i; \alpha_{s-1|s-1}^{(N)i}, \beta_{s-1|s-1}^{(N)i} \right) \tag{45}$$

$$E^{(N)}\left[v_s^i \right] = \frac{\alpha_s^{(N)i}}{\beta_s^{(N)i}} \tag{46}$$

$$E^{(N)}\left[v_{s-1}^i \right] = \frac{\alpha_{s-1}^{(N)i}}{\beta_{s-1}^{(N)i}} \tag{47}$$

$$E^{(N)}\left[\ln v_s^i \right] = \psi\left(\alpha_s^{(N)i} \right) - \ln\left(\beta_s^{(N)i} \right) \tag{48}$$

$$E^{(N)}\left[\ln v_{s-1}^i \right] = \psi\left(\alpha_{s-1}^{(N)i} \right) - \ln\left(\beta_{s-1}^{(N)i} \right) \tag{49}$$

$$E^{(N)}\left[\sigma_s^i \right] = \frac{\Pr^{(N)}\left(\sigma_s^i = 1 \right)}{\Pr^{(N)}\left(\sigma_s^i = 1 \right) + \Pr^{(N)}\left(\sigma_s^i = 0 \right)} \tag{50}$$

$$E^{(N)}\left[1 - \sigma_s^i \right] = 1 - E^{(N)}\left[\sigma_s^i \right] \tag{51}$$

where $\psi(\cdot)$ refers to digamma function, and the derivation details of Equations (43)–(51) can be found in Appendix A.

Step 3: Mode Probability Updating

In this step, the mode probability $\Pr\{M_s = i|y_{1:s}\}$ can be updated by combining the total probability equation with the Bayes rule, i.e.,

$$\mu_s^i = \Pr\{M_s = i|y_{1:s}\} = \frac{p(y_s|M_s = i, y_{1:s-1})\Pr(M_s = i|y_{1:s-1})}{\sum\limits_{j=1}^{k} p(y_s|M_s = j, y_{1:s-1})\Pr(M_s = j|y_{1:s-1})} = \frac{\bar{\mu}_{s-1}^i \Gamma_s^i}{\sum\limits_{j=1}^{k} \bar{\mu}_{s-1}^i \Gamma_s^i} \tag{52}$$

where

$$\bar{\mu}_{s-1}^i = \sum_{j=1}^{k} \lambda_{ji} \mu_{s-1}^j \tag{53}$$

Γ_s^i refers to the likelihood function, and the $\ln p(y_s|y_{1:s-1}, M_s = i)$ is rewritten by

$$\ln p(y_s|y_{1:s-1}, M_s = i) = L(\Lambda) + KLD(\Lambda\|p(\Xi|y_{1:s}, M_s = i)) \tag{54}$$

In Equation (54), $L(\Lambda)$ represents the evidence low bound of the function $\Lambda(\cdot)$. Utilizing the VB technique, the term of $KLD(\cdot)$ can be minimized and approximated to zero, therefore, we can obtain:

$$\Gamma_s^i \simeq \exp\{L(\Lambda)\} \tag{55}$$

where

$$L(\Lambda) = \ln \frac{p(y_{1:s}, \eta_s, v_s, v_{s-1}, \sigma_s|M_s = i)}{p(\eta_s, v_s, v_{s-1}, \sigma_s|y_{1:s}, M_s = i)} - \ln p(y_{1:s-1}|M_s = i) \tag{56}$$

and

$$p(\eta_s, v_s, v_{s-1}, \sigma_s|y_{1:s}, M_s = i) \simeq q\left(\eta_s^i\right) q\left(v_s^i\right) q\left(_{s-1}^i\right) q\left(\sigma_s^i\right) \tag{57}$$

After approximate operation, the $L(\Lambda)$ can be newly derived:

$$\begin{aligned} L(\Lambda) =& E_\Lambda[\ln p(y_{1:s}, \eta_s, v_s, v_{s-1}, \sigma_s|M_s = i)] - E_\Lambda\left[\ln q\left(\eta_s^i\right) q\left(v_s^i\right) q\left(v_{s-1}^i\right) q\left(\sigma_s^i\right)\right] \\ & - \ln p(y_{1:s-1}|M_s = i) \end{aligned} \tag{58}$$

where

$$\begin{aligned} p(y_{1:s}, \eta_s, v_s, v_{s-1}, \sigma_s|M_s = i) =& p(y_s|\eta_s, v_s, v_{s-1}, \sigma_s, M_s = i)p(\sigma_s|M_s = i)p(y_{1:s-1}|M_s = i) \\ & \cdot p(\eta_s|y_{1:s-1}, M_s = i)p(v_s|M_s = i)p(v_{s-1}|M_s = i) \end{aligned} \tag{59}$$

and Equation (60) is obtained:

$$\begin{aligned} L(\Lambda) =& E_\Omega[\ln p(y_s|\eta_s, v_s, v_{s-1}, \sigma_s, M_s = j)] + E_\Omega[\ln p(\eta_s|y_{1:s-1}, M_s = j)] \\ & + E_\Omega[\ln p(v_s|M_s = j)] + E_\Omega[\ln p(v_{s-1}|, M_s = j)] + E_\Omega[\ln p(\sigma_s|M_s = j)] \\ & - E_\Omega\left[\ln q\left(\eta_s^j\right)\right] - E_\Omega\left[\ln q\left(v_s^j\right)\right] - E_\Omega\left[\ln q\left(v_{s-1}^j\right)\right] - E_\Omega\left[\ln q\left(\sigma_s^j\right)\right] \end{aligned} \tag{60}$$

based on Equation (60), the likelihood function in Equation (55) can be calculated.

Step 4: Combination

The estimated system states and covariances of sub-filters are combined in this step and can be defined as follows:

$$\hat{x}_{s|s} = \sum_{i=1}^{k} \mu_s^i \hat{x}_{s|s}^i \tag{61}$$

$$P_{s|s} = \sum_{i=1}^{k} \mu_s^i \left(P_{s|s}^i + \left(\hat{x}_{s|s}^i - \hat{x}_{s|s} \right) \left(\hat{x}_{s|s}^i - \hat{x}_{s|s} \right)^T \right) \tag{62}$$

Based on the abovementioned derivations, the design of the proposed method consists of the time update Equations (16)–(21) and the recursive measurement update Equations (43)–(51). Executable pseudocode of the presented algorithm can be found in Algorithm 1.

Algorithm 1: One filtering iteration of the designed algorithm.

Input: $f_s^i(\cdot)$, $h_s^i(\cdot)$, y_s, $\hat{x}_{s-1|s-1}^i$, $\hat{P}_{s-1|s-1}^i$, μ_0, λ_{ji}, n, m, Q, R, ϕ_s, N, δ.

Step 1: Interaction/Mixing Process

Mixing weight μ_{s-1}^{ji} via Equation (30).

Mixing system state vector $\hat{\eta}_{s-1|s-1}^{0i}$ and covariance matrix $\hat{P}_{s-1|s-1}^{\eta\eta,0i}$ via Equation (35).

Step 2: Mode-conditioned Filtering Process

Time update:

Predict system state vector $\hat{\eta}_{s|s-1}^i$ and covariance matrix $\hat{P}_{s|s-1}^{\eta\eta,i}$ via Equations (17) and (18).

Measurement update:

Initial expectations by Equations (46)–(51).

for $L = 0, 1, 2, \cdots, N-1$ **do**

    Updating $q^{(L+1)}\left(\eta_s^i\right)$ by Gaussian distribution in Equation (43).

    Updating $q^{(L+1)}\left(v_s^i\right)$ by Gamma distribution in Equation (44).

    Updating $A_s^{(L+1)i}$ and $B_s^{(L+1)i}$ according to Equation (A16).

    Calculating $E^{(L+1)}\left[v_s^i\right]$ and $E^{(L+1)}\left[\log v_s^i\right]$ in Equations (46) and (48).

    Updating $q^{(L+1)}\left(v_{s-1}^i\right)$ by Gamma distribution in Equation (45).

    Calculating $E^{(L+1)}\left[v_{s-1}^i\right]$ and $E^{(L+1)}\left[\log v_{s-1}^i\right]$ in Equations (47) and (49).

    Updating $q^{(L+1)}\left(\sigma_s^i\right)$ by Bernoulli distribution in Equations (A25) and (A26).

    Calculating $E^{(L+1)}\left[\sigma_s^i\right]$ and $E^{(L+1)}\left[1-\sigma_s^i\right]$ in Equations (50) and (51).

    if $\dfrac{\left\|\hat{\eta}_{s|s}^{(L+1)}-\hat{\eta}_{s|s}^{(L)}\right\|}{\left\|\hat{\eta}_{s|s}^{(L)}\right\|} \leq \delta$, iteration stops.

Subfilter Outputs $\hat{x}_{s|s}^i = \hat{x}_{s|s}^{(N)i}$, $\hat{P}_{s|s}^i = \hat{P}_{s|s}^{(N)i}$.

Step 3: Mode Probability Update Process

Calculate μ_s^i via Equations (52)–(60).

Step 4: Combination Process

Calculate $\hat{x}_{s|s}$ and $\hat{P}_{s|s}$ via Equations (61) and (62).

Output: $\hat{x}_{s|s}$, $\hat{P}_{s|s}$.

---

5. Simulation Results of Maneuvering Target Tracking

To verify the estimation performances of the designed method, this paper utilizes moving target tracking simulation experiments. The target moves by following the coordinate turning (CT) and the constant velocity (CV) model alternately, where the dynamics of the CV model can be described as:

$$\mathbf{x}_s = \begin{bmatrix} 1 & D & 0 & 0 \\ 0 & 1 & 0 & 0 \\ 0 & 0 & 1 & D \\ 0 & 0 & 0 & 1 \end{bmatrix} \mathbf{x}_{s-1} + g_{s-1} \tag{63}$$

where D represents sampling interval, the system state in CV model $\mathbf{x}_s = (p_x, v_x, p_y, v_y)$ comprises the positions (p_x, p_y) and the velocities (v_x, v_y). The nominal process noise covariance matrix is provided as follows:

$$Q = \begin{bmatrix} b_1 U_{2\times2} & 0_{2\times2} \\ 0_{2\times2} & b_1 U_{2\times2} \end{bmatrix} \tag{64}$$

where the matrix

$$U_{2\times2} = \begin{bmatrix} \frac{D^3}{3} & \frac{D^2}{2} \\ \frac{D^2}{2} & D \end{bmatrix} \tag{65}$$

and the CT model can be described by:

$$\mathbf{x}_s = \begin{bmatrix} 1 & \frac{\sin(\omega D)}{\omega} & 0 & \frac{\cos(\omega D) - 1}{\omega} & 0 \\ 0 & \cos(\omega D) & 0 & -\sin(\omega D) & 0 \\ 0 & \frac{1 - \cos(\omega D)}{\omega} & 1 & \frac{\sin(\omega D)}{\omega} & 0 \\ 0 & \sin(\omega D) & 0 & \cos(\omega D) & 0 \\ 0 & 0 & 0 & 0 & 1 \end{bmatrix} \mathbf{x}_{s-1} + g_{s-1}, \ \omega \neq 0 \tag{66}$$

where the system state in CT model $\mathbf{x}_s = (p_x, v_x, p_y, v_y, \omega)$ comprises positions (p_x, p_y), the velocities (v_x, v_y), and the turning rate ω. The nominal process noise covariance matrix is in Equation (67)

$$Q = \begin{bmatrix} b_1 U_{2\times2} & 0_{2\times2} & 0 \\ 0_{2\times2} & b_1 U_{2\times2} & 0 \\ 0_{1\times2} & 0_{1\times2} & b_2 D \end{bmatrix} \tag{67}$$

where b_1 and b_2 represent the power spectral densities.

One sensor locates in (p_{x0}, p_{y0}) receives the noise ranges and bearing measurements by the following formula [25]:

$$h(\mathbf{x}_s) = \begin{bmatrix} \sqrt{(y_s - p_{y0})^2 + (x_s - p_{x0})^2} \\ \arctan((y_s - p_{y0})/(x_s - p_{x0})) \end{bmatrix} \tag{68}$$

According to [20,34], the HTMNs are generated by

$$\varepsilon_s \sim \begin{cases} N(0, R) & w.p.\ 0.9 \\ N(0, 100R) & w.p.\ 0.1 \end{cases} \tag{69}$$

where $w.p.$ denotes "with probabilities of", the nominal covariance matrices of the measurement noises are assumed to be $R = diag\left((10\ \text{m})^2, (0.1°)^2\right)$. The meaning of Equation (69) is that the measurement noise mostly obeys Gaussian distributions in a probability of 0.9, and follows Gaussian distributions with severely large covariances in a probability of 0.1.

The simulation time is from $0s$ to 1000 s. In detail, the moving target switches dynamic models alternately between the CT model and the CV model, i.e., the target executes the CV model with coordinate turn $-5°/s$ in 0–200 s, 401–600 s and 801–1000 s, while moves by the CT model in 201–400 s and 601–800 s. The sensor is positioned at $(0\text{ m}, 0\text{ m})$. Initialize the state vector $x_0 = (10{,}000\text{ m}, 20\text{ m/s}, 10{,}000\text{ m}, 20\text{ m/s}, -5°)/\text{s}$ and the covariance matrix $P_0 = \text{diag}(100\text{ m}^2, 50\text{ m}^2/\text{s}^2, 100\text{ m}^2, 50\text{ m}^2/\text{s}^2, 100\text{ m rad}^2/\text{s}^2)$.

Four IMM approaches, containing the IMM cubature KF (IMM-CKF) [8], the robust STD-based IMM unscented KF (IMM-RSTUKF) [25], the Gaussian approximate filter with OSRMD [27] in the IMM framework (IMM-DGAF), and the designed IMM filter are compared simultaneously. All these algorithms are implemented with MATLAB 2020b in Windows 10, and the simulation experiments are run on a computer with Intel Core i5-10400F CPU at 2.9 GHz and 16GB 2666MHz memory. The simulation parameters are summarized in Table 2. 1000 Monte Carlos are carried out for each filter. Two evaluation indicators are utilized to validate the performances of all algorithms, one is root mean square errors (RMSEs), the other one is averaged RMSEs (ARMSEs). According to [35,36], the RMSE and the ARMSE can be defined as follows:

$$RMSE_{pos}(s) = \sqrt{\frac{1}{K_{mc}} \sum_{j=1}^{K_{mc}} \left(\left(\hat{x}_s^{(j)} - x_s^{(j)} \right)^2 + \left(\hat{y}_s^{(j)} - y_s^{(j)} \right)^2 \right)} \tag{70}$$

$$ARMSE_{pos} = \frac{1}{t_s} \sum_{s=1}^{t_s} RMSE_{pos}(s) \tag{71}$$

where K_{mc} denotes the total Monte Carlo run times, $\left(x_s^j, y_s^j \right)$ and $\left(\hat{x}_s^j, \hat{y}_s^j \right)$ represent the real and estimated positions in j-th Monte Carlo cycle, and t_s refers to the total simulation time. Additionally, the RMSEs and the ARMSEs of the velocities and turning rates can be defined similarly to the abovementioned positions.

Table 2. Filtering parameters.

Index	Values
Probability of OSRMD φ_s	0.5
Sampling interval D	$1s$
Number of iterations N	10
Filtering parameter δ	10^{-16}
Simulation time t_s	1000 s
Power spectral densities $b1$	$0.1\text{ m}^2\text{s}^{-3}$
Power spectral densities $b2$	$1.75 \times 10^{-4}\text{ rad}^2\text{s}^{-3}$
Transition probability matrix T	$\begin{bmatrix} 0.99 & 0.01 \\ 0.01 & 0.99 \end{bmatrix}.$

The surface maneuvering target tracking simulation can be divided into four parts to evaluate the performances of different algorithms. The first part considers the designed algorithm and existing ones regarding their estimation accuracy. Figures 1–3 displays the RMSEs of position, velocity, and turning rate of all algorithms. The IMM-CKF performs poorly due to the coexistence of the HTMNs with random measurement delay, significantly affecting filtering accuracy. It can be seen in Figures 1–3 that the proposed method has smaller RMSEs than existing filters. When the measurement outliers occur, the RMSE fluctuation of the proposed method is relatively small. This is because the proposed method is robust to the measurement outliers and adaptive to random measurement delay. As illustrated in Figures 1–3, the designed algorithm outperforms existing algorithms in estimation accuracy. Figures 4 and 5 provide the estimated and actual model probabilities for the two models. The estimated model probabilities from the designed filter are closer to the actual than the other filters. Figures 4 and 5 illustrate that the proposed method can better match the system model. Table 3 shows that the designed filter has smaller ARMSEs in each period in the scenarios of HTMNs coexisting with OSRMD. Compared with IMM-RSTUKF, which utilizes VB approach, the accuracy of ARMSE in position, velocity and turning rate of the designed filter is improved by 41.46%, 47.96% and 6.92%, respectively. Moreover, the introduction of the VB inference increases computational costs.

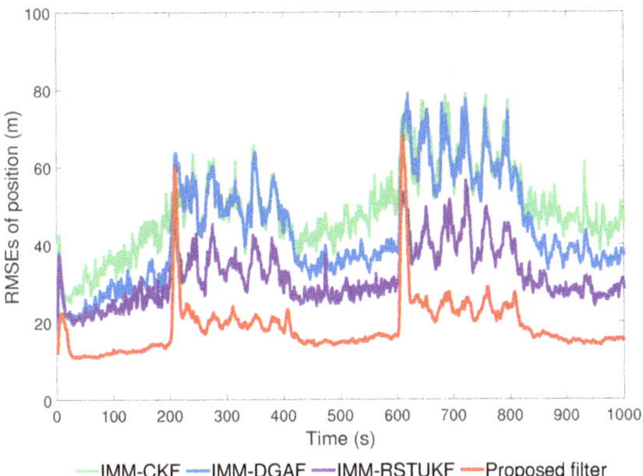

Figure 1. RMSEs of position for different filters.

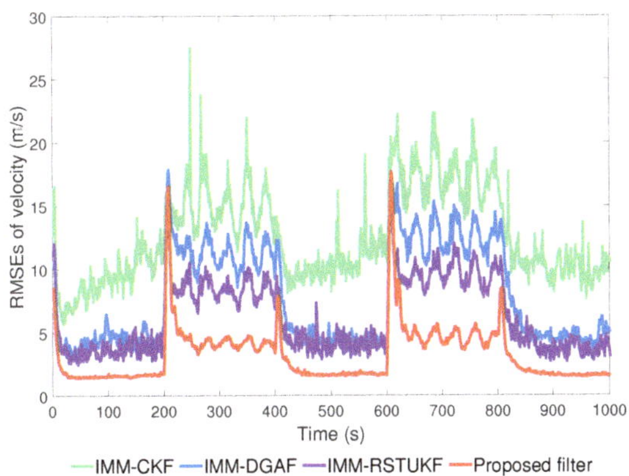

Figure 2. RMSEs of velocity for different filters.

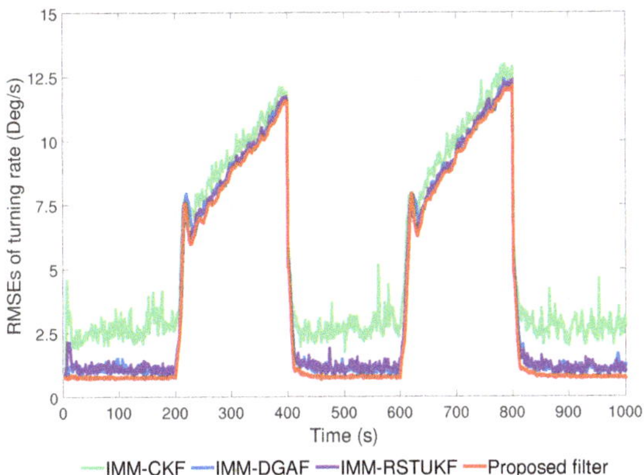

Figure 3. RMSEs of turning rate for different filters.

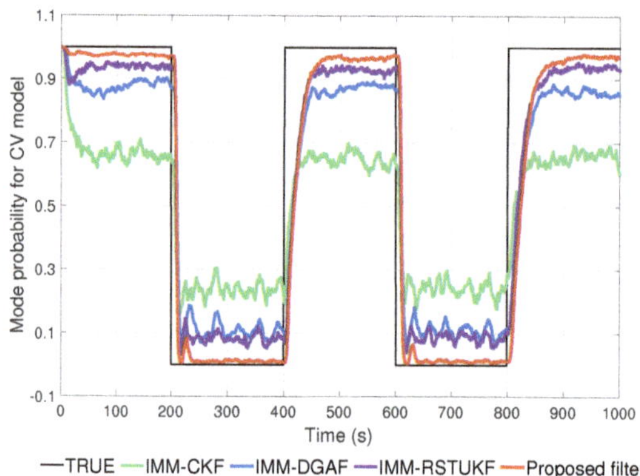

Figure 4. The CV model probability for different filters.

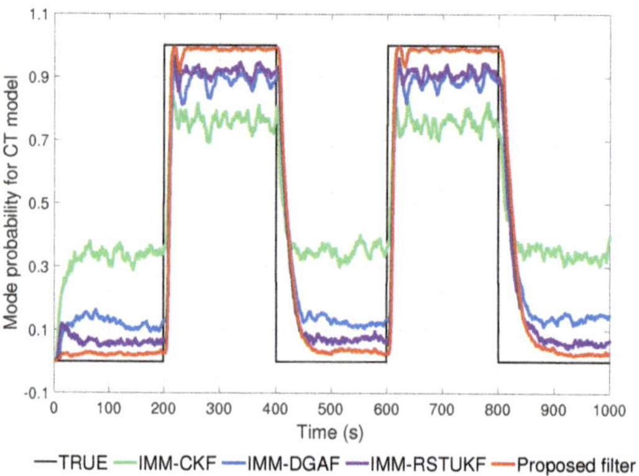

Figure 5. The CT model probability for different filters.

Table 3. ARMSEs and SSRTs from all algorithms in different time periods. Filter 1, 2, 3, and 4 respectively represent the IMM cubature KF [8], the robust STD-based IMM unscented KF [25], the GAF with OSRMD in IMM filtering framework [27], and our designed robust IMM filter with HTMNs and OSRMD.

	Time (s)	Filter 1	Filter 2	Filter 3	Filter 4
ARMSE_{pos} (m)	1~200	37.28	24.42	27.69	12.77
	201~400	53.33	35.49	52.29	21.60
	401~600	47.07	28.45	37.60	15.66
	601~800	65.01	41.03	61.88	26.29
	801~1000	48.90	29.13	39.99	16.47
ARMSE_{vel} (m/s)	1~200	10.45	3.87	4.55	1.75
	201~400	16.00	8.79	11.64	4.74
	401~600	11.65	4.37	5.28	2.06
	601~800	18.06	9.52	12.48	5.41
	801~1000	19.84	4.39	5.51	2.14
ARMSE_{omg} (deg/s)	1~200	3.33	1.12	1.10	0.77
	201~400	9.80	9.07	9.08	8.85
	401~600	3.19	1.30	1.27	0.91
	601~800	9.73	8.91	8.94	8.71
	801~1000	6.23	1.28	1.33	0.94
SSIT (ms)	1~1000	0.166	1.632	0.264	1.651

The second experiment part validates the impact of different OSRMD probabilities on estimation accuracy. The range of the measurement delay probability φ_s is set to be from 0.1 to 0.9. In Figures 6–8, the ARMSEs of the positions, velocities, and turning rate from the designed filtering method and existing algorithms are provided. The horizontal axises of Figures 6–8 denote different measurement delay probabilities, and the longitudinal axises of Figures 6–8 represent the value of ARMSEs. From Figures 6–8, it can be seen that the IMM-CKF shows the worst estimation performance among all filters. The reason is that IMM-CKF deals with neither measurement outliers nor random measurement delay, leading to unsatisfactory estimation performance. As the probability of delay increases, the ARMSEs of IMM-RSTUKF has a relatively large change, the reason is that the IMM-RSTUKF lacks of adaptive estimation ability to random measurement delay. For the IMM-DGAF, although it has good adaptability to random measurement delay, it still has large estimation error due to inaccurate modeling of HTMNs. Since the proposed method can address the HTMNs and OSRMD simultaneously, this algorithm shows better adaptability to different measurement delays and obtain higher estimation accuracy in the presence of measurement outliers than existing algorithms.

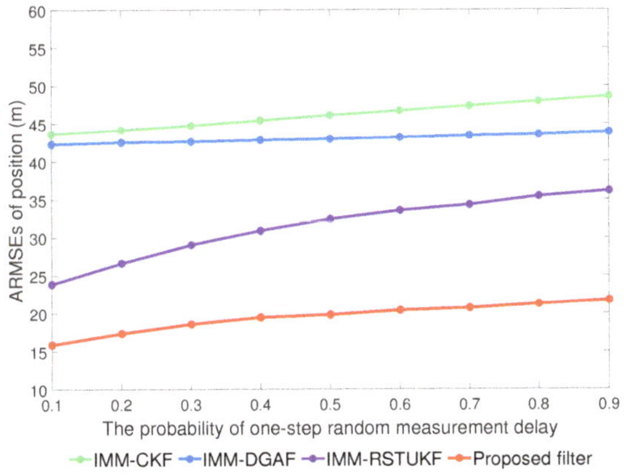

Figure 6. ARMSEs of position with different measurement delay probabilities.

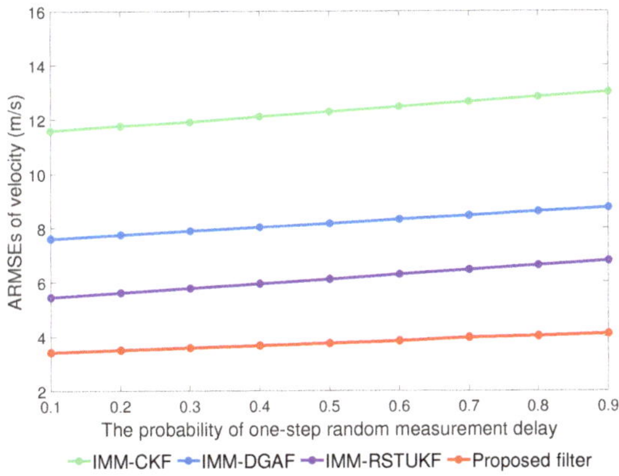

Figure 7. ARMSEs of velocity with different measurement delay probabilities.

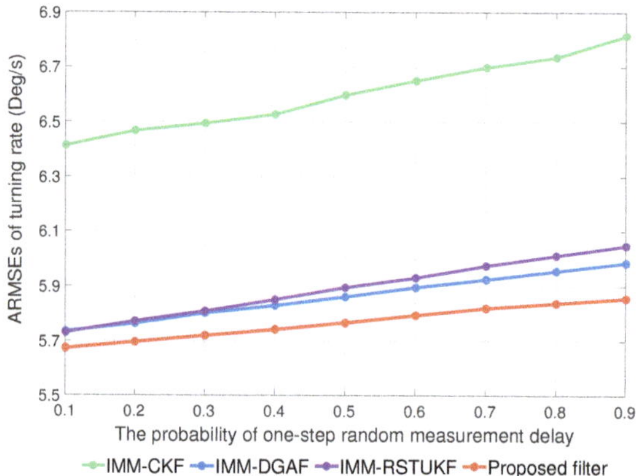

Figure 8. ARMSEs of turning rate with different measurement delay probabilities.

The third part of the experiment aimed to verify the performances of all filters under different measurement outlier probabilities. The fixed measurement outlier probability value in Equation (69) is transformed into a range from 0.05 to 0.15. Figures 9–11 exhibits the ARMSEs from all algorithms under different measurement outlier probabilities. It can be seen from Figures 9–11 that the IMM-CKF still performs poorly, the conclusions about the IMM-CKF in the second experiment are further verified. In addition, with the increasing of the measurement outlier probability, the ARMSE value of the IMM-DGAF shows great fluctuation, this is because the IMM-DGAF assumes the measurement noises to obey Gaussian distributions, however, the Gaussian distribution is sensitive to outliers. As a result, the IMM-DGAF shows serious performance degradation to the increase of the measurement outlier probability. The reason for the IMM-RSTUKF performs worse than the proposed method is that the IMM-RSTUKF assumes all the measurement vectors can arrive in time, which is not satisfied when the OSRMD exists. The results in Figures 9–11 indicate that the presented algorithm can achieve better adaptive estimation performances than existing algorithms under different measurement outlier probabilities.

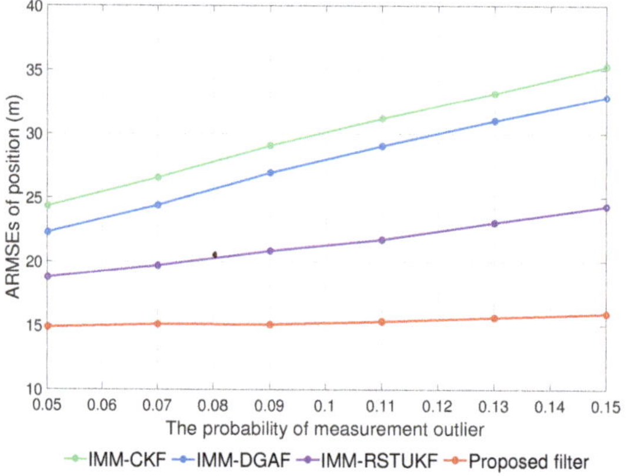

Figure 9. ARMSEs of position with different measurement outlier probabilities.

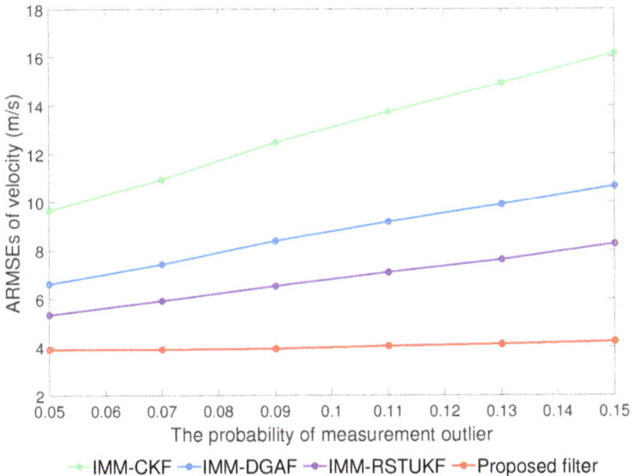

Figure 10. ARMSEs of velocity with different measurement outlier probabilities.

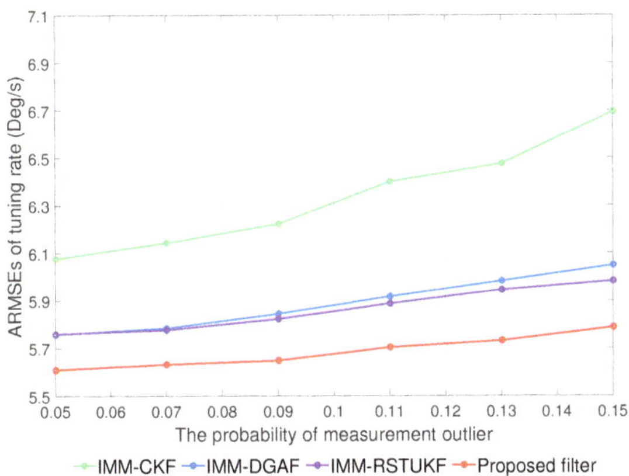

Figure 11. ARMSEs of turning rate with different measurement outlier probabilities.

In the fourth part of the experiment, the effects of the variational iterations on filtering accuracy are validated individually. In Figures 12–14, the ARMSEs from all filters under different variational iterations $N = 1, 2, \ldots, 15$ are shown. The simulation results in Figures 12–14 indicate that the presented algorithm converges when iteration number $N = 2$, and performs better than existing filters in estimation accuracy when $N \geq 2$. The proposed algorithm converges faster than the IMM-RSTUKF which is also based on the VB framework. We can see from Figures 12–14 that the estimation accuracy of the IMM-CKF and IMM-DGAF is relatively stable since these two algorithms do not depend on the variational iterations. As the variational iteration N increases, the estimation performance of the designed filter improves. However, computational efficiency must be taken into account. Considering the balance between computing efficiency and estimation performances, the recommended variational iteration number is 4 to 8.

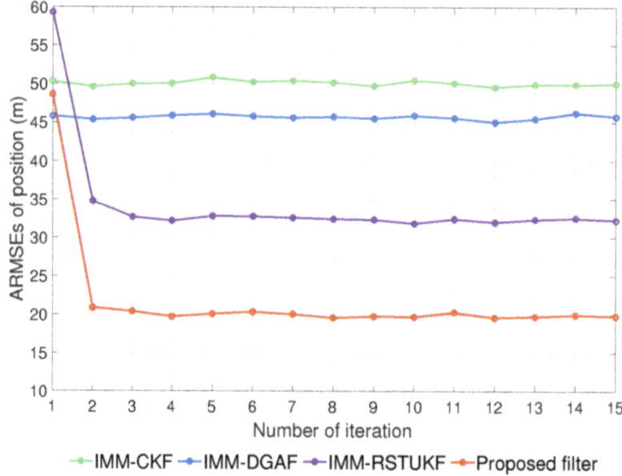

Figure 12. ARMSEs of position under different iteration numbers.

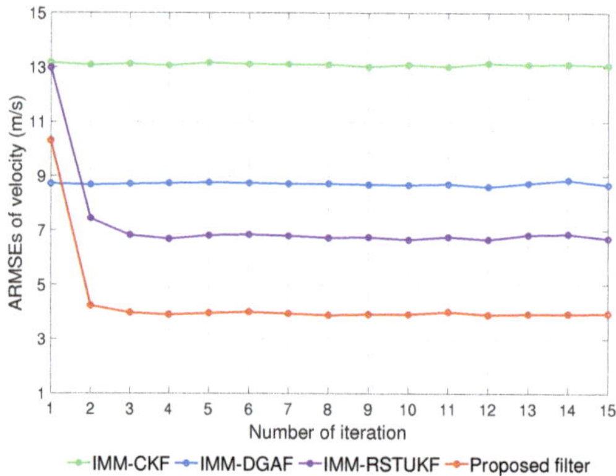

Figure 13. ARMSEs of velocity under different iteration numbers.

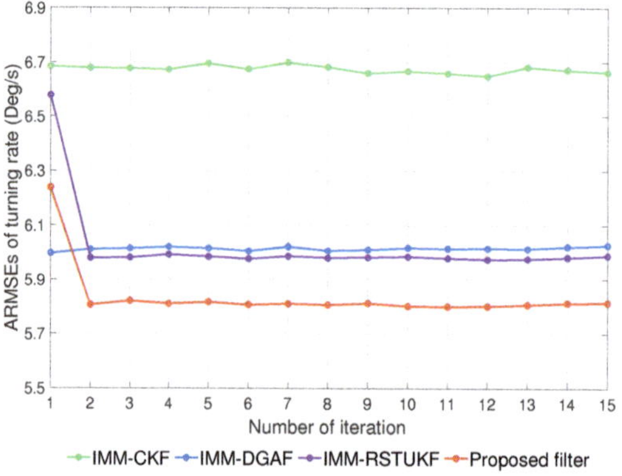

Figure 14. ARMSEs of turning rate under different iteration numbers.

6. Conclusions

In the applications of surface maneuvering target tracking, due to the adverse effects of unreliable sensors and communication channel latency, the measurement outliers and random measurement delay often occur simultaneously. The state estimation issue of surface maneuver target faces severe challenges. Aiming at solving the problem of state estimation in JMSs with HTMNs and OSRMD, this paper designed a novel robust IMM filtering method and applied to surface maneuvering target tracking. Firstly, the HTMNs are assumed to follow STDs, and we introduced an RBV to characterize the random measurement delay. Secondly, this paper established a new HGSSM to utilize VB method. Finally, by using VB inference, the system state, RBV, and model probability are estimated simultaneously. The surface maneuvering target tracking simulation results indicated that the designed filtering method has better estimation accuracy than existing algorithms. In addition, compared with existing filters, the proposed filtering method showed better robustness to different outlier probabilities, and achieved better adaptive estimation performances under various of random measurement delay probabilities. Although the computational costs of the designed method increased slightly, the accuracy of ARMSE in position, velocity and turning rate is improved by 41.46%, 47.96% and 6.92% than the state-of-the-art, respectively. In our future work, based on the theoretical content and results of the proposed filtering method, more complex measurement environments will be considered, including the cases of multi-step random measurement delay and random measurement loss with unknown probability. The effect of heavy-tailed process noises on filtering accuracy and computational efficiency are also focuses of our future research.

Author Contributions: Conceptualization, C.C.; methodology, C.C.; software, C.C. and L.G.; data curation, W.Z.; writing—original draft preparation, C.C.; writing—review and editing, C.C. and L.G.; visualization, C.C.; supervision, W.Z.; project administration, W.Z. All authors have read and agreed to the published version of the manuscript.

Funding: This work was supported in part by the National Natural Science Foundation of China (NSFC) under Grants 61573113.

Institutional Review Board Statement: Not applicable.

Informed Consent Statement: Not applicable.

Data Availability Statement: No new data was created or analyzed in this research work. Data sharing is not applicable to this paper.

Conflicts of Interest: The authors declare no conflict of interest.

Appendix A

Proposition A1. *Let* $\phi = \eta_s^i$ *in Equation (39), and* $\ln q^{(L+1)}\left(\eta_s^i\right)$ *is derived by:*

$$
\begin{aligned}
\ln q^{(L+1)}\left(\eta_s^i\right) = &-0.5\left(\eta_s^i - \hat{\eta}_{s|s-1}^i\right)^T \left(P_{s|s-1}^{\eta\eta,i}\right)^{-1}\left(\eta_s^i - \hat{\eta}_{s|s-1}^i\right) \\
&-0.5E^{(L)}\left[\sigma_s^i\right]E^{(L)}\left[v_s^i\ 1\right]\left(y_s^i - h_{s-1}^i\left(x_{s-1}^i\right)\right)^T R_{s-1}^{-1}\left(y_s^i - h_{s-1}^i\left(x_{s-1}^i\right)\right) \\
&-0.5E^{(L)}\left[1 - \sigma_s^i\right]E^{(L)}\left[v_s^i\right]\left(y_s^i - h_s^i\left(x_s^i\right)\right)^T R_s^{-1}\left(y_s^i - h_s^i\left(x_s^i\right)\right) + c_\eta
\end{aligned}
\tag{A1}
$$

where $q^{(L+1)}(\cdot)$ *is the approximate posterior PDF of* $p(\cdot)$ *when variational iteration is* $(L+1)$ *. The likelihood PDF is modified as:*

$$
p^{(L+1)}\left(\bar{y}_s^i \middle| \eta_s^i\right) = N\left(\bar{y}_s^i; \bar{h}_s^i\left(\eta_s^i\right), \bar{R}_s^{(L+1)i}\right)
\tag{A2}
$$

where $\bar{y}_s^i$, $\bar{h}_s^i\left(\eta_s^i\right)$ and $\bar{R}_s^{(L+1)i}$ represent the extended measurement vector, mean vector, and updated measurement noise covariances, they are defined as follows:

$$\bar{y}_s^i = \left[\begin{array}{c} y_s^i \\ y_s^i \end{array} \right] \tag{A3}$$

$$\bar{h}_s^i\left(\eta_s^i\right) = \left[\begin{array}{c} \bar{h}_s^i\left(x_s^i\right) \\ \bar{h}_{s-1}^i\left(x_{s-1}^i\right) \end{array} \right] \tag{A4}$$

$$\bar{R}_s^{(L+1)i} = \left[\begin{array}{cc} \dfrac{R_s^i}{E^{(L)}\left[1-\sigma_s^i\right]E^{(L)}\left[v_s^i\right]} & \mathbf{0}_{m\times m} \\ \mathbf{0}_{m\times m} & \dfrac{R_{s-1}^i}{E^{(L)}\left[\sigma_s^i\right]E^{(L)}\left[v_{s-1}^i\right]} \end{array} \right] \tag{A5}$$

Then, the $q^{(L+1)}\left(\eta_s^i\right)$ can be updated by

$$q^{(L+1)}\left(\eta_s^i\right) = N\left(\eta_s^i; \hat{\eta}_{s|s}^{(L+1)i}, P_{s|s}^{(L+1)\eta\eta,i}\right) \tag{A6}$$

where $\hat{\eta}_{s|s}^{(L+1)i}$ and $P_{s|s}^{(L+1)\eta\eta,i}$ are calculated through the following equations:

$$K_s^{(L+1)i} = P_{\eta\bar{y},s|s-1}^i\left(P_{\bar{y}\bar{y},s|s-1}^{(L+1)i}\right)^{-1} \tag{A7}$$

$$\hat{\eta}_{s|s}^{(L+1)i} = \hat{\eta}_{s|s-1}^i + K_s^{(L+1)i}\left(\bar{y}_s^i - \dot{y}_{s|s-1}^i\right) \tag{A8}$$

$$P_{s|s}^{(L+1)\eta\eta,i} = P_{s|s-1}^{\eta\eta,i} - K_s^{(L+1)i}P_{\bar{y}\bar{y},s|s-1}^{(L+1)i}\left(K_{s-1}^{(L+1)i}\right)^T \tag{A9}$$

where

$$\dot{y}_{s|s-1}^i = \int \bar{h}_s^i\left(\eta_s^i\right)N\left(\eta_s^i; \hat{\eta}_{s|s-1}^i, P_{s|s-1}^{\eta\eta,i}\right)d\eta_s^i \tag{A10}$$

$$P_{\bar{y}\bar{y},s|s-1}^i = \int \bar{h}_s^i\left(\eta_s^i\right)\left(\bar{h}_s^i\left(\eta_s^i\right)\right)^T N\left(\eta_s^i; \hat{\eta}_{s|s-1}^i, P_{s|s-1}^{\eta\eta,i}\right)d\eta_s^i - \dot{y}_{s|s-1}^i\left(\dot{y}_{s|s-1}^i\right)^T + \bar{R}_s^{(L+1)i} \tag{A11}$$

$$P_{\eta\bar{y},s|s-1}^i = \int \eta_s^i\left(\bar{h}_s^i\left(\eta_s^i\right)\right)^T N\left(\eta_s^i; \hat{\eta}_{s|s-1}^i, P_{s|s-1}^{\eta\eta,i}\right)d\eta_s^i - \eta_{s|s-1}^i\left(\dot{y}_{s|s-1}^i\right)^T \tag{A12}$$

Proposition A2. *Let* $\phi = v_s^i$ *in Equation (39), and* $\ln q^{(L+1)}\left(v_s^i\right)$ *is updated by Equation (A13):*

$$\ln q^{(L+1)}\left(v_s^i\right) = 0.5\left(mE^{(L)}\left[1-\sigma_s^i\right]+a_s^i - 2\right)\ln v_s^i$$

$$- 0.5\left\{E^{(L)}\left[1-\sigma_s^i\right]\left(y_s^i - h_s^i\left(x_s^i\right)\right)^T R_s^{-1}\left(y_s^i - h_s^i\left(x_s^i\right)\right)+a_s^i\right\}v_s^i + c_v \tag{A13}$$

and $q^{(L+1)}\left(v_s^i\right)$ *can be updated by*

$$q^{(L+1)}\left(v_s^i\right) = G\left(v_s^i; \alpha_s^{(L+1)i}, \beta_s^{(L+1)i}\right) \tag{A14}$$

where $\alpha_s^{(L+1)i}$ *and* $\beta_s^{(L+1)i}$ *are calculated as:*

$$\left\{ \begin{array}{l} \alpha_s^{(L+1)i} = 0.5\left(mE^{(L)}\left[1-\sigma_s^i\right]+a_s^i\right) \\ \beta_s^{(L+1)i} = 0.5\left\{E^{(L)}\left[1-\sigma_s^i\right]tr\left(A_s^{(L+1)}Y_1\left(B_s^{(L+1)}\right)^{-1}\right)+a_s^i\right\} \end{array} \right. \tag{A15}$$

where necessary matrices are defined as follows:

$$
\begin{cases}
A_s^{(L+1)i} = \int \left(y_s^i - h_s^i\left(x_s^i\right)\right)\left(y_s^i - h_s^i\left(x_s^i\right)\right)^T N\left(\eta_s^i; \hat{\eta}_{s|s}^{(L+1)i}, P_{s|s}^{(L+1)\eta\eta,i}\right) d\eta_s^i \\[6pt]
B_s^{(L+1)i} = \begin{bmatrix} \dfrac{R_s^i}{E^{(L+1)}\left[1-\sigma_s^i\right]} & 0_{m\times m} \\[10pt] 0_{m\times m} & \dfrac{R_{s-1}^i}{E^{(L+1)}\left[\sigma_s^i\right]} \end{bmatrix} \\[14pt]
Y_1 = \begin{bmatrix} I_{m\times m} & 0_{m\times m} \\ 0_{m\times m} & 0_{m\times m} \end{bmatrix}, Y_2 = \begin{bmatrix} 0_{m\times m} & 0_{m\times m} \\ 0_{m\times m} & I_{m\times m} \end{bmatrix}
\end{cases}
\tag{A16}
$$

$I_{m\times m}$ refers the identity matrix, and the necessary expectations are provided by:

$$
E^{(L+1)}\left[v_s^i\right] = \frac{\alpha_s^{(L+1)i}}{\beta_s^{(L+1)i}}
\tag{A17}
$$

$$
E^{(L+1)}\left[\ln v_s^i\right] = \psi\left(\alpha_s^{(L+1)i}\right) - \ln\left(\beta_s^{(L+1)i}\right)
\tag{A18}
$$

Proposition A3. *Let $\phi = v_{s-1}^i$ in Equation (39), and $\ln q^{(L+1)}\left(v_{s-1}^i\right)$ can be derived by the following equations:*

$$
\ln q^{(L+1)}\left(v_{s-1}^i\right) = 0.5\left(mE^{(L)}\left[\sigma_s^i\right] + a_{s-1}^i - 2\right)\ln v_{s-1}^i + c_v
$$
$$
- 0.5\left\{E^{(L)}\left[\sigma_s^i\right]\left(y_s^i - h_{s-1}^i\left(x_{s-1}^i\right)\right)^T R_{s-1}^{-1}\left(y_s^i - h_{s-1}^i\left(x_{s-1}^i\right)\right) + a_{s-1}^i\right\}v_{s-1}^i
\tag{A19}
$$

and $q^{(L+1)}\left(v_{s-1}^i\right)$ can be updated by

$$
q^{(L+1)}\left(v_{s-1}^i\right) = G\left(v_{s-1}^i; \alpha_{s-1}^{(L+1)i}, \beta_{s-1}^{(L+1)i}\right)
\tag{A20}
$$

where $\alpha_{s-1}^{(L+1)i}$ and $\beta_{s-1}^{(L+1)i}$ are calculated as:

$$
\begin{cases}
\alpha_{s-1}^{(L+1)i} = 0.5\left(mE^{(L)}\left[\sigma_s^i\right] + a_{s-1}^i\right) \\[6pt]
\beta_{s-1}^{(L+1)i} = 0.5\left\{E^{(L)}\left[\sigma_s^i\right]tr\left(A_s^{(L+1)}Y_2\left(B_s^{(L+1)}\right)^{-1}\right) + a_{s-1}^i\right\}
\end{cases}
\tag{A21}
$$

and the matrices uesd in Equation (A21) can be calculated by

$$
E^{(L+1)}\left[v_{s-1}^i\right] = \frac{\alpha_{s-1}^{(L+1)i}}{\beta_{s-1}^{(L+1)i}}
\tag{A22}
$$

$$
E^{(L+1)}\left[\ln v_{s-1}^i\right] = \psi\left(\alpha_{s-1}^{(L+1)i}\right) - \ln\left(\beta_{s-1}^{(L+1)i}\right)
\tag{A23}
$$

Proposition A4. *Let $\phi = \sigma_s^i$ in Equation (39), and $\ln q^{(L+1)}\left(\sigma_s^i\right)$ is updated as follows:*

$$
\ln q^{(L+1)}\left(\sigma_s^i\right) = -0.5\sigma_s^i E^{(L+1)}\left[v_{s-1}^i\right]\left(y_s^i - h_{s-1}^j\left(x_{s-1}^i\right)\right)^T\left(R_{s-1}^i\right)^{-1}\left(y_s^i - h_{s-1}^j\left(x_{s-1}^i\right)\right)
$$
$$
+ 0.5m\sigma_s^i E^{(L+1)}\left[\ln v_{s-1}^i\right] - 0.5\left(1-\sigma_s^i\right)E^{(L+1)}\left[v_s^i\right]\left(y_s^i - h_s^j\left(x_s^i\right)\right)^T\left(R_s^i\right)^{-1}\left(y_s^i - h_s^j\left(x_s^i\right)\right)
$$
$$
+ 0.5m\left(1-\sigma_s^i\right)E^{(L+1)}\left[\ln v_s^i\right] + \left(1-\sigma_s^i\right)\ln\left(1-\varphi_s^i\right) + \sigma_s^i \ln\varphi_s^i + cst_\sigma
\tag{A24}
$$

the RBV σ_s selecting values from $\{0,1\}$ is updated as follows:

$$Pr^{(L+1)}\left(\sigma_s^i = 1\right) = \Omega^{(L+1)i}\exp\left\{S_1^{(L+1)i}\right\} \tag{A25}$$

$$Pr^{(L+1)}\left(\sigma_s^i = 0\right) = \Omega^{(L+1)i}\exp\left\{S_1^{(L+1)i}\right\} \tag{A26}$$

where the $\Omega^{(L+1)i}$ denotes the normalized constant, and $S_1^{(L+1)i}$ and $S_2^{(L+1)i}$ are calculated as follows:

$$S_1^{(L+1)i} = E^{(L+1)}\left[\ln v_s^i\right] - 0.5tr\left(A_s^{(L+1)i}Y_1\left(C_s^{(L+1)i}\right)^{-1}\right) + \ln\left(1 - \varphi_s^i\right) \tag{A27}$$

$$S_1^{(L+1)i} = E^{(L+1)}\left[\ln v_{s-1}^i\right] - 0.5tr\left(A_s^{(L+1)i}Y_2\left(C_s^{(L+1)i}\right)^{-1}\right) + \ln\left(\varphi_s^i\right) \tag{A28}$$

and the matrix $C_s^{(L+1)i}$ is formulated as:

$$C_s^{(L+1)i} = \begin{bmatrix} \frac{R_s}{E^{(L+1)}\left[v_s^i\right]} & \mathbf{0}_{m\times m} \\ \mathbf{0}_{m\times m} & \frac{R_{s-1}}{E^{(L+1)}\left[v_{s-1}^i\right]} \end{bmatrix} \tag{A29}$$

the necessary expectations $E^{(L+1)}\left[\sigma_s^i\right]$ and $E^{(L+1)}\left[1 - \sigma_s^i\right]$ can be obtained by:

$$E^{(L+1)}\left[\sigma_s^i\right] = \frac{Pr^{(L+1)}\left(\sigma_s^i = 1\right)}{Pr^{(L+1)}\left(\sigma_s^i = 1\right) + Pr^{(L+1)}\left(\sigma_s^i = 0\right)} \tag{A30}$$

$$E^{(L+1)}\left[1 - \sigma_s^i\right] = 1 - E^{(L+1)}\left[\sigma_s^i\right] \tag{A31}$$

References

1. Zhang, W.; Zhao, X.; Liu, Z.; Liu, K.; Chen, B. Converted state equation Kalman filter for nonlinear maneuvering target tracking. *Signal Process.* **2023**, *202*, 108741. [CrossRef]
2. Huang, Y.; Zhang, Y.; Zhao, Y.; Shi, P.; Chambers, J.A. A novel outlier-robust Kalman filtering framework based on statistical similarity measure. *IEEE Trans. Autom. Control* **2020**, *66*, 2677–2692. [CrossRef]
3. Xu, B.; Hu, J.; Guo, Y. An Acoustic Ranging Measurement Aided SINS/DVL Integrated Navigation Algorithm Based on Multivehicle Cooperative Correction. *IEEE Trans. Instrum. Meas.* **2022**, *71*, 1–15. [CrossRef]
4. Demirci, M.; Gözde, H.; Taplamacioglu, M.C. Improvement of power transformer fault diagnosis by using sequential Kalman filter sensor fusion. *Int. J. Electr. Power Energy Syst.* **2023**, *149*, 109038. [CrossRef]
5. Lindner, L.; Sergiyenko, O.; Rivas-López, M.; Ivanov, M.; Rodríguez-Quiñonez, J.C.; Hernández-Balbuena, D.; Flores-Fuentes, W.; Tyrsa, V.; Muerrieta-Rico, F.N.; Mercorelli, P. Machine vision system errors for unmanned aerial vehicle navigation. In Proceedings of the 2017 IEEE 26th International Symposium on Industrial Electronics (ISIE), Edinburgh, UK, 19–21 June 2017; pp. 1615–1620.
6. Mercorelli, P.; Lehmann, K.; Liu, S. Robust flatness based control of an electromagnetic linear actuator using adaptive PID controller. In Proceedings of the 42nd IEEE International Conference on Decision and Control (IEEE Cat. No. 03CH37475), Maui, HI, USA, 9–12 December 2003; Volume 4, pp. 3790–3795.
7. Li, K.; Zhao, S.; Liu, F. Joint state estimation for nonlinear state-space model with unknown time-variant noise statistics. *Int. J. Adapt. Control Signal Process.* **2021**, *35*, 498–512. [CrossRef]
8. Li, X.R.; Jilkov, V.P. Survey of maneuvering target tracking. Part V. Multiple-model methods. *IEEE Trans. Aerosp. Electron. Syst.* **2005**, *41*, 1255–1321.
9. Freni, G.; Mannina, G.; Viviani, G. Urban runoff modelling uncertainty: Comparison among Bayesian and pseudo-Bayesian methods. *Environ. Model. Softw.* **2009**, *24*, 1100–1111. [CrossRef]
10. Djuric, P.M.; Kotecha, J.H.; Zhang, J.; Huang, Y.; Ghirmai, T.; Bugallo, M.F.; Miguez, J. Particle filtering. *IEEE Signal Process. Mag.* **2003**, *20*, 19–38. [CrossRef]
11. Seah, C.E.; Hwang, I. Algorithm for performance analysis of the IMM algorithm. *IEEE Trans. Aerosp. Electron. Syst.* **2011**, *47*, 1114–1124. [CrossRef]
12. Hwang, I.; Seah, C.E.; Lee, S. A study on stability of the interacting multiple model algorithm. *IEEE Trans. Autom. Control* **2016**, *62*, 901–906. [CrossRef]
13. Zhao, S.; Ahn, C.; Shi, P.; Shmaliy, Y.; Liu, F. Bayesian State Estimations for Markovian Jump Systems: Employing Recursive Steps and Pseudocodes. *IEEE Syst. Man, Cybern. Mag.* **2019**, *5*, 27–36. [CrossRef]

14. Li, J.; Yuan, G.; Duan, H. Consensus CIF-Based IMM Filtering for Multiple-UAV Target Tracking. In Proceedings of the 2022 International Conference on Guidance, Navigation and Control: Advances in Guidance, Navigation and Control, Tianjin, China, 23–25 October 2020; pp. 7060–7069.

15. Youn, W.; Ko, N.Y.; Gadsden, S.A.; Myung, H. A novel multiple-model adaptive Kalman filter for an unknown measurement loss probability. *IEEE Trans. Instrum. Meas.* **2020**, *70*, 1–11. [CrossRef]

16. Zubača, J.; Stolz, M.; Seeber, R.; Schratter, M.; Watzenig, D. Innovative Interaction Approach in IMM Filtering for Vehicle Motion Models With Unequal States Dimension. *IEEE Trans. Veh. Technol.* **2022**, *71*, 3579–3594. [CrossRef]

17. Blair, W.D.; Watson, G. Interacting multiple bias model algorithm with application to tracking maneuvering targets. In Proceedings of the 1992 31st IEEE Conference on Decision and Control, Tucson, AZ, USA, 16–18 December 1992; pp. 3790–3795.

18. Sarkka, S.; Nummenmaa, A. Recursive noise adaptive Kalman filtering by variational Bayesian approximations. *IEEE Trans. Autom. Control* **2009**, *54*, 596–600. [CrossRef]

19. Huang, Y.; Zhang, Y.; Wu, Z.; Li, N.; Chambers, J. A novel adaptive Kalman filter with inaccurate process and measurement noise covariance matrices. *IEEE Trans. Autom. Control* **2017**, *63*, 594–601. [CrossRef]

20. Huang, Y.; Zhang, Y.; Li, N.; Wu, Z.; Chambers, J.A. A novel robust Student's t-based Kalman filter. *IEEE Trans. Aerosp. Electron. Syst.* **2017**, *53*, 1545–1554. [CrossRef]

21. Zhu, H.; Zhang, G.; Li, Y.; Leung, H. An adaptive Kalman filter with inaccurate noise covariances in the presence of outliers. *IEEE Trans. Autom. Control* **2021**, *67*, 374–381. [CrossRef]

22. Jia, G.; Huang, Y.; Zhang, Y.; Chambers, J. A novel adaptive Kalman filter with unknown probability of measurement loss. *IEEE Signal Process. Lett.* **2019**, *26*, 1862–1866. [CrossRef]

23. Wang, G.; Wang, X.; Zhang, Y. Variational Bayesian IMM-filter for JMSs with unknown noise covariances. *IEEE Trans. Aerosp. Electron. Syst.* **2019**, *56*, 1652–1661. [CrossRef]

24. Shen, C.; Xu, D.; Huang, W.; Shen, F. An interacting multiple model approach for state estimation with non-Gaussian noise using a variational Bayesian method. *Asian J. Control* **2015**, *17*, 1424–1434. [CrossRef]

25. Li, D.; Sun, J. Robust Interacting Multiple Model Filter Based on Student'st-Distribution for Heavy-Tailed Measurement Noises. *Sensors* **2019**, *19*, 4830. [CrossRef] [PubMed]

26. Yan, B.; Zuo, J.; Chen, X.; Zou, H. Improved multiple model particle filter for maneuvering target tracking in the presence of delayed measurements. In Proceedings of the 2017 International Conference on Computer Systems, Electronics and Control (ICCSEC), Dalian, China, 25–27 December 2017; pp. 810–814.

27. Wang, X.; Liang, Y.; Pan, Q.; Zhao, C. Gaussian filter for nonlinear systems with one-step randomly delayed measurements. *Automatica* **2013**, *49*, 976–986. [CrossRef]

28. Tong, Y.; Zheng, Z.; Fan, W.; Li, Q.; Liu, Z. An improved unscented Kalman filter for nonlinear systems with one-step randomly delayed measurement and unknown latency probability. *Digit. Signal Process.* **2022**, *121*, 103324. [CrossRef]

29. Ito, K.; Xiong, K. Gaussian filters for nonlinear filtering problems. *IEEE Trans. Autom. Control* **2000**, *45*, 910–927. [CrossRef]

30. Bai, M.; Huang, Y.; Chen, B.; Yang, L.; Zhang, Y. A novel mixture distributions-based robust Kalman filter for cooperative localization. *IEEE Sens. J.* **2020**, *20*, 14994–15006. [CrossRef]

31. Huang, Y.; Zhang, Y.; Zhao, Y.; Chambers, J.A. A novel robust Gaussian–Student's t mixture distribution based Kalman filter. *IEEE Trans. Signal Process.* **2019**, *67*, 3606–3620. [CrossRef]

32. Huang, Y.; Zhang, Y.; Chambers, J.A. A novel Kullback–Leibler divergence minimization-based adaptive student's t-filter. *IEEE Trans. Signal Process.* **2019**, *67*, 5417–5432. [CrossRef]

33. Lu, C.; Feng, W.; Li, W.; Zhang, Y.; Guo, Y. An adaptive IMM filter for jump Markov systems with inaccurate noise covariances in the presence of missing measurements. *Digit. Signal Process.* **2022**, *127*, 103529. [CrossRef]

34. Roth, M.; Özkan, E.; Gustafsson, F. A Student's t filter for heavy tailed process and measurement noise. In Proceedings of the 2013 IEEE International Conference on Acoustics, Speech and Signal Processing, Vancouver, BC, Canada, 26–31 May 2013; pp. 5770–5774.

35. Jia, G.; Huang, Y.; Bai, M.B.; Zhang, Y. A novel robust Kalman filter with non-stationary heavy-tailed measurement noise. *IFAC-PapersOnLine* **2020**, *53*, 368–373. [CrossRef]

36. Fu, H.; Cheng, Y. A Novel Robust Kalman Filter Based on Switching Gaussian-Heavy-Tailed Distribution. *IEEE Trans. Circuits Syst. II Express Briefs* **2022**, *69*, 3012–3016. [CrossRef]

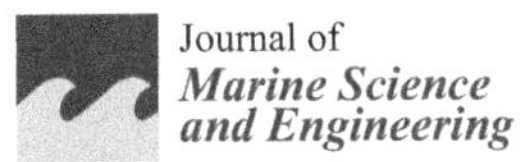
Journal of
*Marine Science
and Engineering*

Article

Three-Dimensional Trajectory Tracking of AUV Based on Nonsingular Terminal Sliding Mode and Active Disturbance Rejection Decoupling Control

Wei Zhang, Wenhua Wu *, Zixuan Li, Xue Du and Zheping Yan

College of Intelligent Systems Science and Engineering, Harbin Engineering University, Harbin 150001, China
* Correspondence: wenhuawuf@163.com

Abstract: This paper presents a nonsingular terminal sliding mode and active disturbance rejection decoupling control (NTSM-ADRDC) scheme for the three-dimensional (3D) trajectory tracking of autonomous underwater vehicles (AUV). Firstly, the AUV model is decoupled into five independent single input–single output (SISO) channels using ADRDC technology. Secondly, the NTSM-ADRDC controller is designed. The linear extended state observer (LESO) is used to observe the AUV state variables, and estimate the total disturbance of the system. In addition, to improve the system error convergence rate, the combination of exponential reaching rate and NTSM constitutes a nonlinear states error feedback control law for the controller. Finally, the stability of the proposed control law is proved using the Lyapunov theory. The simulation results demonstrate the effectiveness and robustness of the designed NTSM-ADRDC trajectory tracking approach.

Keywords: AUV; 3D trajectory tracking; LESO; nonsingular terminal sliding mode control

1. Introduction

Autonomous underwater vehicles (AUV) are increasingly being utilized in a variety of civil and military applications, such as intelligence collection, ocean mapping, pipeline inspection, and maritime rescue [1,2]. In order to better perform these tasks, it is essential that the AUV has the ability to accurately track three-dimensional (3D) trajectories in underwater space [3]. However, due to the highly nonlinear, strong coupling, complex hydrodynamic coefficient of the AUV model, the precise control of AUV has become a significant challenge [4]. Moreover, the sensitivity of AUVs to external disturbances further increases the difficulty in controller design [5]. Therefore, how to design an AUV trajectory tracking controller with good robustness has become a research hotspot.

The trajectory tracking of AUVs has been the subject of extensive research in recent years, and various control methods have been used in the design of AUV controllers. These control methods mainly include proportional–integral–derivative (PID) control [6,7], fuzzy logic control [8,9], adaptive control [10,11], neural network control (NNC) [12,13], and model predictive control [14–16]. In the literature [6], an intelligent PID controller is applied to AUV's horizontal plane path tracking control and vertical plane depth control. However, PID cannot provide accurate control in the presence of ocean current disturbances. In [8], a fuzzy dynamic surface control method was designed for solving the 3D trajectory tracking problem of the under-actuated AUV in the presence of model uncertainty and time-varying disturbance. To solve the dynamic trajectory tracking control of AUV in a three-dimensional underwater environment, a variable fuzzy predictor-based predictive control approach was proposed [9]. However, the membership functions and fuzzy rules of fuzzy logic control need to be determined based on rich experience. An adaptive controller based on Lyapunov's direct method and the back-stepping technique was proposed [10], which can address the issue of trajectory tracking control of underactuated AUV of six degrees of freedom. In the literature [11], an adaptive disturbance observer has been designed

Citation: Zhang, W.; Wu, W.; Li, Z.; Du, X.; Yan, Z. Three-Dimensional Trajectory Tracking of AUV Based on Nonsingular Terminal Sliding Mode and Active Disturbance Rejection Decoupling Control. *J. Mar. Sci. Eng.* **2023**, *11*, 959. https://doi.org/10.3390/jmse11050959

Academic Editor: Rafael Morales

Received: 24 March 2023
Revised: 24 April 2023
Accepted: 26 April 2023
Published: 30 April 2023

for AUV trajectory tracking control in the presence of unknown external disturbances, and the gain of the observer can be adjusted automatically by introducing an adaptive law. However, adaptive control can only achieve good control effects if the model of the controlled object is known, the parameters change slowly, and the uncertainty of the system is finite. A neural network-based tracking control method for underactuated AUV with model uncertainties was presented [12]. Simulation results demonstrated the effectiveness of the proposed control strategy. However, due to a large amount of calculation, NNC cannot meet the real-time requirement of AUV. The literature [14] presented a novel 3D underwater trajectory tracking approach for underwater robots based on model predictive control, considering actual constraints on system inputs and states. The main disadvantage of MPC is that it not only relies on an accurate mathematical model but also requires a high level of computational power of the AUV. Although the above control algorithm has achieved a better trajectory tracking effect to a certain extent, there are still some respective weaknesses.

The active disturbance rejection control (ADRC) algorithm can effectively solve the problem of system uncertainty (internal model and external disturbance uncertainty) [17,18]. The unique anti-interference capability of ADRC technology allows for a wide range of applications in engineering control fields [19,20]. ADRC was first proposed by Han in the 1990s [21,22]. As the core of ADRC, the extended state observer (ESO) can estimate the total disturbance including internal dynamics and external disturbance. Furthermore, the nonlinear states error feedback control law can compensate for the total disturbance of the system [23]. The control objective of the ADRC is to converge the system state error to zero so that the desired control effect can be achieved. However, the traditional ADRC has too many parameters, so it is difficult to set the parameters in engineering applications. To simplify the structure of ADRC, Professor Gao designed a linear active disturbance rejection controller (LADRC) [24]. The LADRC simplified the control parameters compared to ADRC, which is very convenient for engineering applications. In addition, the theory for the stability proof of LDARC was provided by Gao [25].

Sliding mode variable structure control has received extensive attention from scholars due to its ideal robustness [26,27]. For example, a second-order sliding mode controller is designed to addresses the problems of depth regulation control of AUV in wave circumstance [28].The simulation results show that this method can be effectively applied to robust tracking of AUV. However, chattering is the main drawback of sliding mode control. The literature [29] proposed a non-singular terminal sliding mode control (NTSMC) method, which can effectively suppress chattering and avoid singular. In the literature [30], an NTSMC method based on an exponential convergence law was proposed to improve the convergence speed for reaching non-singular terminal sliding surfaces. The simulation results showed that the designed control law can make the system converge to the equilibrium point in a short time.

In order to solve the problem of external disturbances and ocean current in AUV 3D trajectory tracking control, and also include the problem of tracking error convergence. A novel control scheme based on NTSM-ADRDC is proposed in this paper. The main idea is to combine the strong robustness of the NTSMC method with the LADRC controller's ability to suppress model uncertainty and external disturbance. Firstly, the AUV model is decoupled by taking advantage of active disturbance rejection decoupling control technology, in which a new virtual control vector is introduced. Then, the LESO is utilized to estimate the internal unmodeled dynamics and external disturbance of the AUV system as the total disturbance. After that, the NTSM nonlinear states error feedback control law is designed to compensate for the total disturbance of the system. Finally, the stability of the AUV system is proved by the Lyapunov theory.

The rest of this paper is organized as follows: the AUV model and its decoupling control process are given in Section 2. Section 3 illustrates the total structure of the AUV trajectory tracking control based on NTSM-ADRDC and the detailed proof of the controller's stability. Section 4 verifies the effectiveness of the designed controller through simulation. Finally, the conclusions are given in Section 5.

2. AUV Model and Decoupling

In this section, the kinematics and dynamics model of the AUV is first described. Then, the decoupling control of the AUV model based on the ADRC technique is introduced.

2.1. AUV Kinematics and Dynamics

Establishing the kinematics and dynamics model of the AUV is the prerequisite for studying its motion control. The inertial reference frame (I-frame) and body-fixed frame (B-frame) of the AUV are depicted in Figure 1. We assumed the roll of the AUV is passively stable, and its coupling nonlinear effect can be ignored. Based on the assumption, the kinematic and dynamic model of a five-degrees-of-freedom (5-DOF) full actuated AUV are stated as follows [31]:

$$\begin{cases} \dot{x} = u\cos\psi\cos\theta - v\sin\psi + w\sin\theta\cos\psi \\ \dot{y} = u\sin\psi\cos\theta + v\cos\psi + w\sin\theta\sin\psi \\ \dot{z} = -u\sin\theta + w\cos\theta \\ \dot{\theta} = q \\ \dot{\psi} = r/\cos\theta \end{cases} \tag{1}$$

$$\begin{cases} \dot{u} = \frac{m_{22}}{m_{11}}vr - \frac{m_{33}}{m_{11}}wq - \frac{d_{11}}{m_{11}}u + \frac{1}{m_{11}}\tau_u + \frac{1}{m_{11}}d_u \\ \dot{v} = -\frac{m_{11}}{m_{22}}ur - \frac{d_{22}}{m_{22}}v + \frac{1}{m_{22}}\tau_v \\ \dot{w} = \frac{m_{11}}{m_{33}}uq - \frac{d_{33}}{m_{33}}w + \frac{1}{m_{33}}\tau_w + \frac{1}{m_{33}}d_w \\ \dot{q} = \frac{m_{33}-m_{11}}{m_{55}}uw - \frac{d_{55}}{m_{55}}q + \frac{1}{m_{55}}\tau_q + \frac{1}{m_{55}}d_q \\ \dot{r} = \frac{m_{11}-m_{22}}{m_{66}}uv - \frac{d_{66}}{m_{66}}r + \frac{1}{m_{66}}\tau_r + \frac{1}{m_{66}}d_r \end{cases} \tag{2}$$

where $\eta = \begin{bmatrix} \eta_1 & \eta_2 \end{bmatrix}^T$ represent the position and Euler angles of the AUV in I-frame. $\eta_1 = \begin{bmatrix} x & y & z \end{bmatrix}^T$ indicate the north, east, and depth coordinates of the AUV. $\eta_2 = \begin{bmatrix} \theta & \psi \end{bmatrix}^T$ indicate the pitch and yaw angle of the AUV. $v = \begin{bmatrix} v_1 & v_2 \end{bmatrix}^T$ represent the linear velocity and angular velocity of the AUV in B-frame. $v_1 = \begin{bmatrix} u & v & w \end{bmatrix}^T$ denote the longitudinal, lateral, and vertical velocity of the AUV. $v_2 = \begin{bmatrix} q & r \end{bmatrix}^T$ denote the pitch and yaw angular velocity of the AUV. $\tau = \begin{bmatrix} \tau_1 & \tau_2 \end{bmatrix}^T$ are the input forces and moments of the AUV system. $\tau_1 = \begin{bmatrix} \tau_u & \tau_v & \tau_w \end{bmatrix}^T$ are the longitudinal, lateral, and vertical input forces. $\tau_2 = \begin{bmatrix} \tau_q & \tau_r \end{bmatrix}^T$ are the pitch and yaw input moments. m_{ii}, d_{jj} represent AUV hydrodynamic parameters and damping coefficients, $m_{11} = m - X_{\dot{u}}$, $m_{22} = m - Y_{\dot{v}}$, $m_{33} = m - Z_{\dot{w}}$, $m_{55} = I_y - M_{\dot{q}}$, $m_{66} = I_z - N_{\dot{r}}$, $d_{11} = -X_u - X_{u|u|}|u|$, $d_{22} = -Y_v - Y_{v|v|}|v|$, $d_{33} = -Z_w - Z_{w|w|}|w|$, $d_{55} = -M_q - M_{q|q|}|q|$. $d = \begin{bmatrix} d_u & d_v & d_w & d_q & d_r \end{bmatrix}^T$ represent bounded external disturbances. For detailed definitions of hydrodynamic parameters of AUV, readers can refer to the literature [32,33].

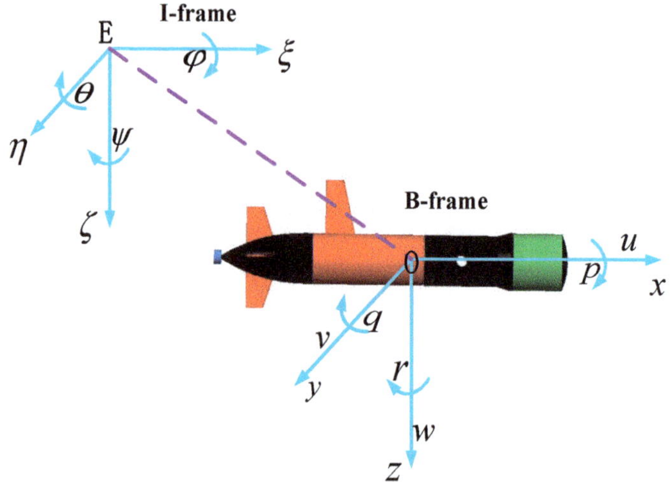

Figure 1. The AUV coordinate system.

2.2. Decoupling Control of AUV Model

From the above-established AUV multi-input and multi-output (MIMO) system, we can observe that the AUV is a nonlinear, multivariable, strongly coupled system. It brings a dramatic challenge to the subsequent design of the AUV tracking controller due to its characteristics. The ADRC technique is a good scheme for solving control problems of the coupled MIMO system. When using the ADRC technique for decoupling control of MIMO system, the coupling term between the different input and output channels can be seen as an external disturbance. The ADRC controller for each channel can estimate and compensate for external disturbances independently, thus achieving the decoupling of control for each channel by introducing virtual control variables. Next, the active disturbance rejection decoupling control technology will be introduced in detail [34].

Suppose there is an m-dimensional MIMO system as follows:

$$
\begin{cases}
\ddot{x}_1 = f_1(x_1, \dot{x}_1, \cdots, x_m, \dot{x}_m) + b_{11}u_1 + \cdots + b_{1m}u_m \\
\ddot{x}_2 = f_2(x_1, \dot{x}_1, \cdots, x_m, \dot{x}_m) + b_{21}u_1 + \cdots + b_{2m}u_m \\
\quad \vdots \\
\ddot{x}_m = f_m(x_1, \dot{x}_1, \cdots, x_m, \dot{x}_m) + b_{m1}u_1 + \cdots + b_{mm}u_m \\
y_1 = x_1, y_2 = x_2, \cdots, y_m = x_m
\end{cases}
\tag{3}
$$

where $x_i, y_i(i = 1, 2 \cdots m)$ are, respectively, expressed as system state variables and output; u_i represents control input; the amplification factor of the control variable $b_{ij} = b_{ij}(x, \dot{x}, t)$ is a function of state variables and time, which can be written in a matrix form as:

$$
B(x, \dot{x}, t) = \begin{bmatrix}
b_{11}(x, \dot{x}, t) & \cdots & b_{1m}(x, \dot{x}, t) \\
\vdots & \vdots & \vdots \\
b_{m1}(x, \dot{x}, t) & \cdots & b_{mm}(x, \dot{x}, t)
\end{bmatrix}
\tag{4}
$$

If the control variable coefficient matrix $B(x, \dot{x}, t)$ is invertible, the system Equation (3) can be simplified as:

$$
\begin{cases}
\ddot{x} = f(x, \dot{x}, t) + U \\
y = x
\end{cases}
\tag{5}
$$

where $x = \begin{bmatrix} x_1 & x_2 & \cdots & x_m \end{bmatrix}^T$, $y = \begin{bmatrix} y_1 & y_2 & \cdots & y_m \end{bmatrix}^T$, $f = \begin{bmatrix} f_1 & f_2 & \cdots & f_m \end{bmatrix}^T$, $u = \begin{bmatrix} u_1 & u_2 & \cdots & u_m \end{bmatrix}^T$, $U = B(x, \dot{x}, t)u$ is the newly introduced virtual control vector.

Therefore, the i−*th* channel in the system (5) can be expressed as:

$$\begin{cases} \ddot{x}_i = f_i(x_1, \dot{x}_1, \cdots, x_m, \dot{x}_m, t) + U_i \\ y_i = x_i \end{cases} \tag{6}$$

Essentially, the ADRC technology regards the $f_i(x_1, \dot{x}_1, \cdots, x_m, \dot{x}_m, t)$ of i-th channel as the external disturbance of the channel, while the ADRC controller of each channel can estimate and compensate for the external disturbance independently. Therefore, the virtual control variable U_i and the output variable y_i of each channel are in SISO relationship, that is, the system realizes the complete decoupling control by introducing virtual control vector. The decoupling control process of the MIMO system based on active disturbance rejection technology is shown in Figure 2. The desired input value $v_i(i = 1, 2 \cdots m)$ and the actual output value $y_i(i = 1, 2 \cdots m)$ of the system constitute a closed-loop channel. Each ADRC controller can achieve independent control of the corresponding channel so that the actual output of the system converges to the desired value.

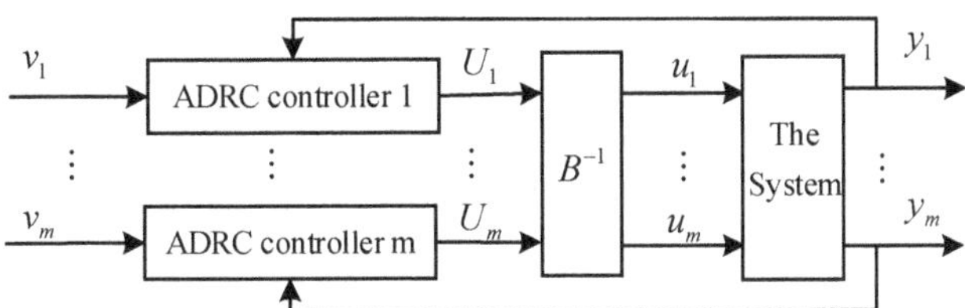

Figure 2. The active disturbance rejection decoupling control of MIMO system.

For the 5-DOF model of AUV in Equations (1) and (2), which is a strongly coupled MIMO system. In order to facilitate the design of the controller in the latter, it is necessary to decouple the control of the AUV model using the ADRC technique.

First, the derivation of Equation (1) is obtained:

$$\begin{cases} \ddot{x} = \dot{u}\cos\theta\cos\psi - u\cos\theta\sin\psi\dot{\psi} - u\sin\theta\cos\psi\dot{\theta} - \dot{v}\sin\psi \\ \quad -v\cos\psi\dot{\psi} + \dot{w}\sin\theta\cos\psi + w\cos\theta\cos\psi\dot{\theta} - w\sin\theta\sin\psi\dot{\psi} \\ \ddot{y} = \dot{u}\cos\theta\sin\psi - u\sin\theta\sin\psi\dot{\theta} + u\cos\theta\cos\psi\dot{\psi} + \dot{v}\cos\psi \\ \quad -v\sin\psi\dot{\psi} + \dot{w}\sin\theta\cos\psi + w\cos\theta\sin\psi\dot{\theta} + w\sin\theta\cos\psi\dot{\psi} \\ \ddot{z} = -\dot{u}\sin\theta - u\cos\theta\dot{\theta} + \dot{w}\cos\theta - w\sin\theta\dot{\theta} \\ \ddot{\theta} = \dot{q} \\ \ddot{\psi} = \dot{r}\sec\theta + r\tan\theta\sec\theta \end{cases} \tag{7}$$

Next, substitute Equation (2) into Equation (7):

$$\begin{cases} \ddot{x} = f_1(u,v,w,q,r,\theta,\psi) + \frac{1}{m_{11}}\cos\psi\cos\theta(\tau_u + d_u) - \frac{1}{m_{22}}\sin\psi(\tau_v + d_v) + \frac{1}{m_{33}}\cos\psi\sin\theta(\tau_w + d_w) \\ \ddot{y} = f_2(u,v,w,q,r,\theta,\psi) + \frac{1}{m_{11}}\sin\psi\cos\theta(\tau_u + d_u) + \frac{1}{m_{22}}\sin\psi(\tau_v + d_v) + \frac{1}{m_{33}}\sin\psi\sin\theta(\tau_w + d_w) \\ \ddot{z} = f_3(u,v,w,q,r,\theta) - \frac{1}{m_{11}}\sin\theta(\tau_u + d_u) + \frac{1}{m_{33}}\cos\theta(\tau_w + d_w) \\ \ddot{\theta} = f_4(u,w,q) + \frac{1}{m_{55}}(\tau_q + d_q) \\ \ddot{\psi} = f_5(u,v,r,\theta) + \frac{1}{m_{66}}\sec\theta(\tau_r + d_r) \end{cases} \tag{8}$$

where $f_i(i = 1,2,3,4,5)$ is the sum of other items except τ and d.

Here, we will introduce a new virtual control vector as follows:

$$\boldsymbol{U} = \boldsymbol{B} * \boldsymbol{\tau} \tag{9}$$

With

$$\boldsymbol{U} = \begin{bmatrix} u_x & u_y & u_z & u_\theta & u_\psi \end{bmatrix}^T \tag{10}$$

$$B = \begin{bmatrix} \dfrac{\cos\psi\cos\theta}{m_{11}} & -\dfrac{\sin\psi}{m_{22}} & \dfrac{\cos\psi\sin\theta}{m_{33}} & 0 & 0 \\[2mm] \dfrac{\sin\psi\cos\theta}{m_{11}} & \dfrac{\cos\psi}{m_{22}} & \dfrac{\sin\psi\sin\theta}{m_{33}} & 0 & 0 \\[2mm] -\dfrac{\sin\theta}{m_{11}} & 0 & \dfrac{\cos\theta}{m_{33}} & 0 & 0 \\[2mm] 0 & 0 & 0 & \dfrac{1}{m_{55}} & 0 \\[2mm] 0 & 0 & 0 & 0 & \dfrac{\sec\theta}{m_{66}} \end{bmatrix} \tag{11}$$

After calculation, $|B| = \sec\theta / m_{11}m_{22}m_{33}m_{55}m_{66}$ is reversible, the Equation (8) can be simplified as:

$$\begin{cases} \ddot{x} = f_1(u,v,w,q,r,\theta,\psi) + w_1 + u_x \\ \ddot{y} = f_2(u,v,w,q,r,\theta,\psi) + w_2 + u_y \\ \ddot{z} = f_3(u,v,w,q,r,\theta) + w_3 + u_z \\ \ddot{\theta} = f_4(u,w,q) + w_4 + u_q \\ \ddot{\psi} = f_5(u,v,r,\theta) + w_5 + u_r \end{cases} \tag{12}$$

where $w = \begin{bmatrix} w_1 & w_2 & w_3 & w_4 & w_5 \end{bmatrix}^T$, $w = B * d$.

The decoupling control process of the AUV system based on ADRC technology is shown in Figure 3. The ADRC technology regards $f_i + w_i (i = 1,2,3,4,5)$ as the total external disturbance of the i-th channel of the AUV system, while the ADRC controller of each channel can estimate and compensate for the total disturbance independently. Therefore, the virtual control variable U and the output variable η of each channel are in SISO relationship, that is, the AUV system realizes the complete decoupling control through virtual control vector U. The five channels of the AUV system can be individually designed with controllers. $\eta_d = \begin{bmatrix} x_d & y_d & z_d & \theta_d & \psi_d \end{bmatrix}^T$ are the reference signals of position and angle of AUV. Through the action of the control coefficient matrix B^{-1}, the virtual control vector U can generate the real control input τ acting on the AUV.

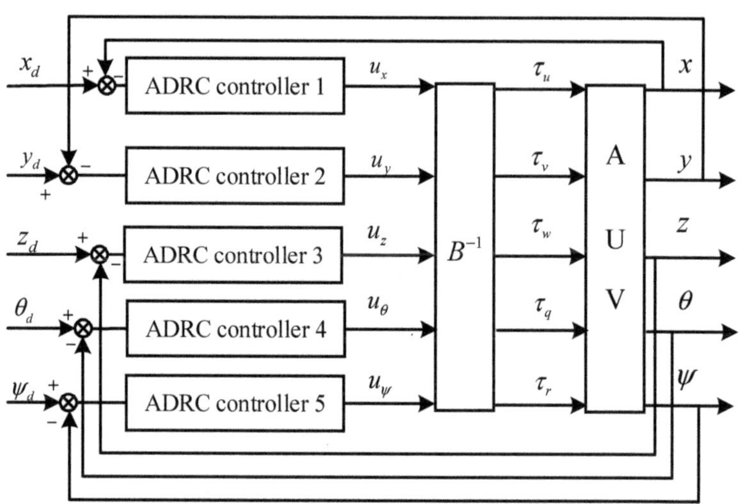

Figure 3. The ADRDC of AUV.

3. Three-Dimensional Trajectory Tracking Controller Design

After the above discussion, the basic framework of the decoupling control of the AUV system based on ADRC technology has been initially developed. In this section, we will introduce the design process of the AUV 3D trajectory tracking controller based on the NTSM-ADRDC. The total control framework of the proposed algorithm is depicted in Figure 4. The control frame is divided into five channels: $x_d - x$, $y_d - y$, $z_d - z$, $\theta_d - \theta$

and $\psi_d - \psi$, among which each channel is an independent SISO system. The LESO has strong ability to observe the total disturbance of AUV system, and the introduction of the exponential reaching law in NTSM allows for faster convergence of tracking errors. The close combination of the methods constitutes the NTSM-ADRDC controller.

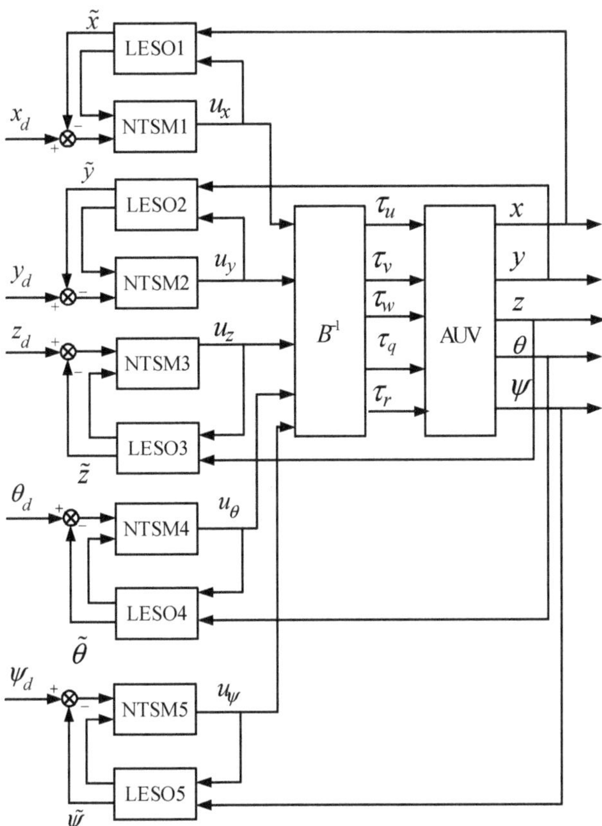

Figure 4. The control framework of the AUV system.

3.1. Linear Extended State Observer Design

As the core component of LADRC algorithm, the LESO can estimate the internal uncertain dynamics and external disturbances according to the input–output states of the system. Taking the position state x of AUV as an example, according to Equation (12), the $f_1(u,v,r,\psi) + w_1$ is regarded by LESO1 as the total disturbances of the $x_d - x$ channel in AUV system, which can be expanded into the new state variable x_3, namely $x_3 = f_1(u,v,r,\psi) + w_1$, and let $x_1 = x$, $\dot{x}_3 = h_1$. Where the h_1 denotes the derivative of the total disturbance observed by LESO1. Therefore, the equation of $\ddot{x}$ in the system (12) can be expanded into the following control system:

$$\begin{cases} \dot{x}_1 = x_2 \\ \dot{x}_2 = x_3 + u_x \\ \dot{x}_3 = h_1 \\ y = x_1 \end{cases} \tag{13}$$

Next, we can construct LESO1 according to Equation (13)

$$\begin{cases} \dot{\tilde{x}}_1 = x_2 - \beta_{01}(\tilde{x}_1 - x_1) \\ \dot{\tilde{x}}_2 = x_3 - \beta_{02}(\tilde{x}_1 - x_1) + u_x \\ \dot{\tilde{x}}_3 = -\beta_{03}(\tilde{x}_1 - x_1) \end{cases} \tag{14}$$

Here $\tilde{x}_1$, $\tilde{x}_2$, $\tilde{x}_3$ are the estimated values of x_1, x_2, x_3, respectively, and β_{01}, β_{02}, β_{03} are the gains of the LESO1. According to the characteristics of linear ESO, β_{01}, β_{02}, β_{03} have the following relationship:

$$\begin{bmatrix} \beta_{01} & \beta_{02} & \beta_{03} \end{bmatrix} = \begin{bmatrix} 3\omega_0 & 3\omega_0^2 & \omega_0^3 \end{bmatrix} \tag{15}$$

where ω_0 is often called the observer bandwidth. Note that there is a LESO1 estimation error between the estimated values $\tilde{x}_i$ and actual values $x_i (i = 1, 2, 3)$, but the estimation error is able to converge to 0 according to the literature [35,36]. The remaining three channels of LESO can be designed as follows:

$$\begin{cases} \dot{\tilde{y}}_1 = \tilde{y}_2 - \beta_{01}(\tilde{y}_1 - y_1) \\ \dot{\tilde{y}}_2 = \tilde{y}_3 - \beta_{02}(\tilde{y}_1 - y_1) + u_y \\ \dot{\tilde{y}}_3 = -\beta_{03}(\tilde{y}_1 - y_1) \end{cases} \tag{16}$$

$$\begin{cases} \dot{\tilde{z}}_1 = \tilde{z}_2 - \beta_{01}(\tilde{z}_1 - z_1) \\ \dot{\tilde{z}}_2 = \tilde{z}_3 - \beta_{02}(\tilde{z}_1 - z_1) + u_z \\ \dot{\tilde{z}}_3 = -\beta_{03}(\tilde{z}_1 - z_1) \end{cases} \tag{17}$$

$$\begin{cases} \dot{\tilde{\theta}}_1 = \tilde{\theta}_2 - \beta_{01}(\tilde{\theta}_1 - \theta_1) \\ \dot{\tilde{\theta}}_2 = \tilde{\theta}_3 - \beta_{02}(\tilde{\theta}_1 - \theta_1) + u_\theta \\ \dot{\tilde{\theta}}_3 = -\beta_{03}(\tilde{\theta}_1 - \theta_1) \end{cases} \tag{18}$$

$$\begin{cases} \dot{\tilde{\psi}}_1 = \tilde{\psi}_2 - \beta_{01}(\tilde{\psi}_1 - \psi_1) \\ \dot{\tilde{\psi}}_2 = \tilde{\psi}_3 - \beta_{02}(\tilde{\psi}_1 - \psi_1) + u_\psi \\ \dot{\tilde{\psi}}_3 = -\beta_{03}(\tilde{\psi}_1 - \psi_1) \end{cases} \tag{19}$$

where $\tilde{y}_i, \tilde{z}_i, \tilde{\theta}_i, \tilde{\psi}_i (i = 1, 2, 3)$ are the state estimates of $y_i, z_i, \theta_i, \psi_i (i = 1, 2, 3)$, respectively.

3.2. Design of NTSM Nonlinear States Error Feedback Control Law

Combined with LESO's observation ability to system state variables and total disturbance, the NTSM nonlinear states error feedback control law will be designed. Its key idea is to use LESO to improve the control law by real-time estimates of the AUV's internal uncertain dynamics and external disturbance. On the premise of ensuring the advantages of LADRC, it improves the robustness of the controller. In addition, the exponential reaching law is introduced to enhance the convergence speed of the system tracking error.

To ensure that the tracking error in the $x_d - x$ channel converges to zero in finite time, and to solve the singularity problem in terminal sliding mode control, we choose the following non-singular terminal sliding surface [29]:

$$S_x = e_x + \frac{1}{\beta_1} \text{sgn}(\dot{e}_x) |\dot{e}_x|^{\lambda_1} \tag{20}$$

where $e_x = x_1 - x_d$, $\dot{e}_x = x_2 - \dot{x}_d$. Since LESO1 can observe the estimated value of state variable x, we can rewrite it as $e_x \approx \tilde{x}_1 - x_d$, $\dot{e}_x \approx \tilde{x}_2 - \dot{x}_d$. β_1, and λ_1 are adjustable parameters, satisfying $\beta_1 > 0, 1 < \lambda_1 < 2$.

From the derivation of Equation (20),

$$\begin{aligned}
\dot{S}_x &= \ddot{e}_x + \frac{\lambda_1}{\beta_1}|\dot{e}_x|^{\lambda_1-1}\ddot{e}_x \\
&= \ddot{e}_x + \frac{\lambda_1}{\beta_1}|\dot{e}_x|^{\lambda_1-1}\left(f_1(u,v,r,\psi) + w_1 + u_x - \ddot{x}_d\right)
\end{aligned} \tag{21}$$

Here, we introduce the exponential reaching law,

$$\dot{S}_x = -k_1 S_x - \varepsilon_1 \tanh(S_x) \tag{22}$$

Among them, the first term is to use the exponential to shorten the convergence time, and the second term uses the function $\tanh(.)$ to weaken the system chattering [37], satisfying $k_1 > 0, \varepsilon_1 > 0$.

Combining Equations (21) and (22), we can conclude that

$$\ddot{e}_x + \frac{\lambda_1}{\beta_1}|\dot{e}_x|^{\lambda_1-1}\left(f_1(u,v,r,\psi) + w_1 + u_x - \ddot{x}_d\right) = -k_1 S_x - \varepsilon_1 \tanh(S_x) \tag{23}$$

Let $\rho(\dot{e}_x) = \frac{\lambda_1}{\beta_1}|\dot{e}_x|^{\lambda_1-1}$, according to the Equation (23), the NTSM control based on the exponential reaching law can be deduced as:

$$u_x = -\rho(\dot{e}_x)^{-1}(k_1 S_x + \varepsilon_1 \tanh(S_x) + \ddot{e}_x) - \left(f_1(u,v,r,\psi) + w_1 - \ddot{x}_d\right) \tag{24}$$

Considering that the control law (24) contains $\rho(\dot{e}_\xi) = \frac{\lambda_1}{\beta_1}|\dot{e}_\xi|^{\lambda_1-1}$ term that will cause a relatively large amount of calculation, we can simplify the control law on the basis of ensuring the reaching law and the simplified control law can be described as:

$$u_x = -\left(\frac{\beta_1}{\lambda_1}\mathrm{sgn}(\dot{e}_x)|\dot{e}_x|^{2-\lambda_1} + J_1 S_x + \sigma_{\mathrm{LESO1}}\tanh(S_x) + f_1(u,v,r,\psi) + w_1 - \ddot{x}_d\right) \tag{25}$$

where $J_1 > 0$ and $\sigma_{\mathrm{LESO1}} > 0$, σ_{LESO1} is the upper bound of the observation error of LESO1 to the uncertainty.

Since $\tilde{x}_3$ is LESO1's real-time estimated value of the sum of internal dynamics and external disturbances of the AUV channel $x_d - x$, it can be obtained that

$$\tilde{x}_3 = f_1(u,v,r,\psi) + w_1 \tag{26}$$

Substituting Equation (26) into Equation (25), the nonlinear states error feedback control law of channel $x_d - x$ can finally be expressed as:

$$u_x = -\left(\frac{\beta_1}{\lambda_1}\mathrm{sgn}(\dot{e}_x)|\dot{e}_x|^{2-\lambda_1} + J_1 S_x + \sigma_{\mathrm{LESO1}}\tanh(S_x) + \tilde{x}_3 - \ddot{x}_d\right) \tag{27}$$

Similarly, the sliding mode surfaces of the remaining three channels can be chosen as:

$$\begin{cases}
S_y = e_y + \frac{1}{\beta_2}\mathrm{sgn}(\dot{e}_y)|\dot{e}_y|^{\lambda_2} \\
S_z = e_z + \frac{1}{\beta_3}\mathrm{sgn}(\dot{e}_z)|\dot{e}_z|^{\lambda_3} \\
S_\theta = e_\theta + \frac{1}{\beta_4}\mathrm{sgn}(\dot{e}_\theta)|\dot{e}_\theta|^{\lambda_4} \\
S_\psi = e_\psi + \frac{1}{\beta_5}\mathrm{sgn}(\dot{e}_\psi)|\dot{e}_\psi|^{\lambda_5}
\end{cases} \tag{28}$$

where

$$\begin{cases}
e_y = \tilde{y}_1 - y_d \\
e_z = \tilde{z}_1 - z_d \\
e_\theta = \tilde{\theta}_1 - \theta_d \\
e_\psi = \tilde{\psi}_1 - \psi_d
\end{cases} \tag{29}$$

$$\begin{cases} \dot{e}_y = \tilde{y}_2 - \dot{y}_d \\ \dot{e}_z = \tilde{z}_2 - \dot{z}_d \\ \dot{e}_\theta = \tilde{\theta}_2 - \dot{\theta}_d \\ \dot{e}_\psi = \tilde{\psi}_2 - \dot{\psi}_d \end{cases} \tag{30}$$

The nonlinear error control law of the remaining four channels can be designed as follows:

$$\begin{cases} u_y = -\left(\frac{\beta_2}{\lambda_2} |\dot{e}_y|^{2-\lambda_2} + J_2 S_y + \sigma_{\text{LESO2}} \tanh(S_y) + \tilde{y}_3 - \ddot{y}_d \right) \\ u_z = -\left(\frac{\beta_3}{\lambda_3} |\dot{e}_z|^{2-\lambda_3} + J_3 S_z + \sigma_{\text{LESO3}} \tanh(S_z) + \tilde{z}_3 - \ddot{z}_d \right) \\ u_\theta = -\left(\frac{\beta_4}{\lambda_4} |\dot{e}_\theta|^{2-\lambda_4} + J_4 S_\theta + \sigma_{\text{LESO4}} \tanh(S_\theta) + \tilde{\theta}_3 - \ddot{\theta}_d \right) \\ u_\psi = -\left(\frac{\beta_5}{\lambda_5} |\dot{e}_\psi|^{2-\lambda_5} + J_5 S_\psi + \sigma_{\text{LESO5}} \tanh(S_\psi) + \tilde{\psi}_3 - \ddot{\psi}_d \right) \end{cases} \tag{31}$$

3.3. System Stability Analysis

Assumption 1. The reference trajectory of AUV system $\eta_d = \begin{bmatrix} x_d & y_d & z_d & \theta_d & \psi_d \end{bmatrix}^T$ are bounded.

Theorem 2. For the 5-DOF AUV system, considering the model (12), LESO (14)–(19), and the control laws Equations (27) and (31), then there exist control parameters satisfying $\beta_i > 0, 1 < \lambda_i < 2, J_i > 0$, and $\sigma_{\text{LESO}i} > 0$ $(i = 1, 2 \cdots, 5)$, such that the closed-loop system is stable.

Proof. Construct the Lyapunov function of the AUV system as:

$$V = V_x + V_y + V_z + V_\theta + V_\psi = \frac{1}{2}S_x^2 + \frac{1}{2}S_y^2 + \frac{1}{2}S_z^2 + \frac{1}{2}S_\theta^2 + \frac{1}{2}S_\psi^2 \tag{32}$$

According to Lyapunov stability theory, AUV system is stable if the condition of $\dot{V} = s\dot{s} \leq 0$ is satisfied. Therefore, the derivative of the Equation (21) is as follows:

$$\dot{V} = \dot{V}_x + \dot{V}_y + \dot{V}_z + \dot{V}_\theta + \dot{V}_\psi = S_x\dot{S}_x + S_y\dot{S}_y + S_z\dot{S}_z + S_\theta\dot{S}_\theta + S_\psi\dot{S}_\psi \tag{33}$$

where

$$\begin{aligned} \dot{V}_x = S_x\dot{S}_x &= S_x[\dot{e}_x + \tfrac{\lambda_1}{\beta_1}|\dot{e}_x|^{\lambda_1-1}(f_1(u,v,r,\psi) + w_1 + u_x - \ddot{x}_d)] \\ &= S_x\left\{ \dot{e}_x + \tfrac{\lambda_1}{\beta_1}|\dot{e}_x|^{\lambda_1-1}[f_1(u,v,r,\psi) + w_1 - \ddot{x}_d - (\tfrac{\beta_1}{\lambda_1}\text{sgn}(\dot{e}_x)|\dot{e}_x|^{2-\lambda_1} + J_1S_x + \sigma_{\text{LESO1}}\tanh(S_x) + \tilde{x}_3 - \ddot{x}_d)] \right\} \\ &= S_x[\dot{e}_x - \tfrac{\lambda_1}{\beta_1}|\dot{e}_x|^{\lambda_1-1}(\tfrac{\beta_1}{\lambda_1}\text{sgn}(\dot{e}_x)|\dot{e}_x|^{2-\lambda_1} + J_1S_x + \sigma_{\text{LESO1}}\tanh(S_x)] = S_x[-\tfrac{\lambda_1}{\beta_1}|\dot{e}_x|^{\lambda_1-1}(J_1S_x + \sigma_{\text{LESO1}}\tanh(S_x))] \\ &= -\tfrac{\lambda_1}{\beta_1}|\dot{e}_x|^{\lambda_1-1}(J_1S_x^2 + \sigma_{\text{LESO1}}S_x\tanh(S_x)) \end{aligned} \tag{34}$$

Since $\beta_1 > 0$ and $1 < \lambda_1 < 2$, so $\frac{\lambda_1}{\beta_1}|\dot{e}_x|^{\lambda_1-1} \geq 0$. In addition, because of $S_x\tanh(S_x) \geq 0$, $S_x^2 \geq 0$, we can conclude that $\dot{V}_x \leq 0$.

In a similar way, we can derive the following results:

$$\dot{V} = \dot{V}_x + \dot{V}_y + \dot{V}_z + \dot{V}_\theta + \dot{V}_\psi \leq 0 \tag{35}$$

$\square$

4. Simulation Results and Analysis

To verify the effectiveness of the NTSM-ADRDC algorithm proposed in this paper, the AUV 3D trajectory tracking simulation is studied in this section. By comparing with the LADRC method, the proposed control strategy has better performance in control accuracy, anti-disturbance, and robustness. The control laws for the NTSM-ADRDC algorithm are Equations (27) and (31). A detailed introduction to the LDARC method can be referred to in the literature [24]. The detailed hydrodynamic parameter values of AUV as shown in Table 1.

Table 1. Hydrodynamic parameters and damping coefficient of AUV.

Hydrodynamic Parameters	Damping Coefficients
$m_{11} = 215$ kg	$d_{11} = (70 + 100|u|)$ kg/s
$m_{22} = 265$ kg	$d_{22} = (100 + 200|v|)$ kg/s
$m_{33} = 265$ kg	$d_{33} = (100 + 100|w|)$ kg/s
$m_{55} = 80$ kg.m^2	$d_{55} = (50 + 100|q|)$ kg.m^2/s
$m_{66} = 80$ kg.m^2	$d_{66} = (50 + 100|r|)$ kg.m^2/s

The control parameters of the NTSM-ADRDC controller in the simulation are selected as: $h = 0.05$, $\omega_0 = 15$; $\beta_1 = 0.60$, $\lambda_1 = 15/13$, $J_1 = 0.60$, $\sigma_{\text{LESO1}} = 1.00$; $\beta_2 = 0.75$, $\lambda_2 = 13/11$, $J_2 = 0.65$, $\sigma_{\text{LESO2}} = 0.85$; $\beta_3 = 0.55$, $\lambda_3 = 15/13$, $J_3 = 0.75$, $\sigma_{\text{LESO3}} = 0.80$; $\beta_4 = 0.25$, $\lambda_4 = 11/9$, $J_4 = 0.50$, $\sigma_{\text{LESO4}} = 0.52$; $\beta_5 = 0.20$, $\lambda_5 = 17/15$, $J_5 = 0.50$, $\sigma_{\text{LESO5}} = 0.55$; AUV initial state $\eta_0 = \begin{bmatrix} 0 & 0 & 0 & 0 & 0 \end{bmatrix}^T$. In the simulation, the method of determining the bandwidth ω_0 of the LESO can be found in the literature [38]. The control parameter λ_i can be determined first since it has little effect on the control effect as long as it satisfies $1 < \lambda_i < 2$. Then, the value range of β_i is constrained to be between 0 and 1. Finally, the larger the parameters of $\sigma_{\text{LESO}i}$ and J_i, the faster the convergence rate. However, the "chattering" phenomenon may be generated as the values of $\sigma_{\text{LESO}i}$ and J_i increase. Therefore, it is necessary to trade off between suppressing "chattering" and speeding up the convergence rate. Usually, we choose a slightly larger value of $\sigma_{\text{LESO}i}$ than J_i to weaken the "chattering".

AUV three-dimensional tracking reference trajectory is defined as:

$$\begin{cases} x_d = 5 + 10\sin(0.03t) \\ y_d = -5 + 10\cos(0.03t) \\ z_d = -0.05t \\ \theta_d = 0.03 \\ \psi_d = 0.03t \end{cases} \tag{36}$$

Figure 5 shows the three-dimensional trajectory curves of two different control algorithms without disturbance. The cylindrical spiral is used as the reference trajectory to simulate the tracking of the spiral diving target by the AUV. The black dotted line represents the reference trajectory. The blue solid line represents the AUV tracking result under the LADRC algorithm. The red solid line represents the simulation result of the designed NTSM-ADRDC. It can be seen from the figure that the three-dimensional trajectory curves of the two control algorithms without disturbance are consistent with the reference trajectory. It shows that although the initial position of the AUV is far from the reference trajectory, both methods can track the reference trajectory more accurately and quickly. However, there is an overshoot in NTSM-ADRDC compared to LADRC. The position tracking performance of the AUV is shown in Figure 6. We can observe that both control algorithms show good trajectory tracking performance.

The position tracking performance of the AUV is shown in Figure 6. We can observe that the two control algorithms can follow the corresponding reference trajectory curve well. Through the comparison of the first 20 s, it can be clearly found that the NTSM-ADRDC algorithm can approach the desired position faster. By comparing the position error of the two control methods in Figure 7, can we discover NTSM-ADRDC has a faster error convergence rate than LADRC.

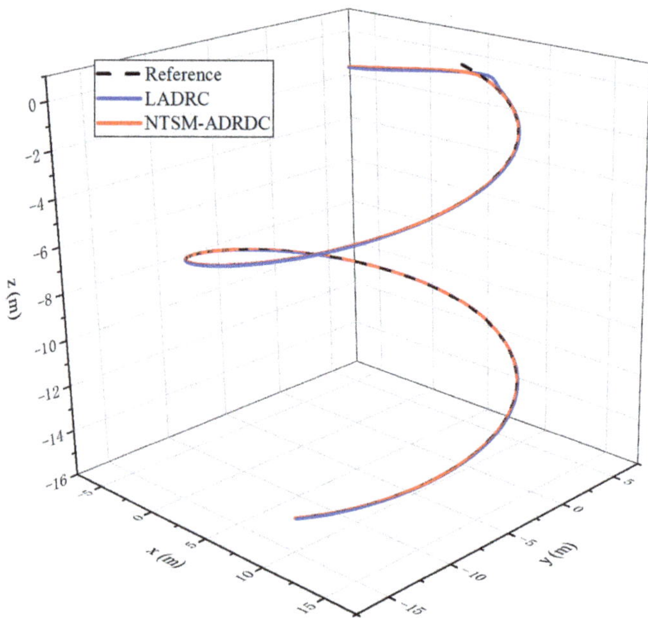

Figure 5. AUV 3D trajectory tracking curve without disturbance.

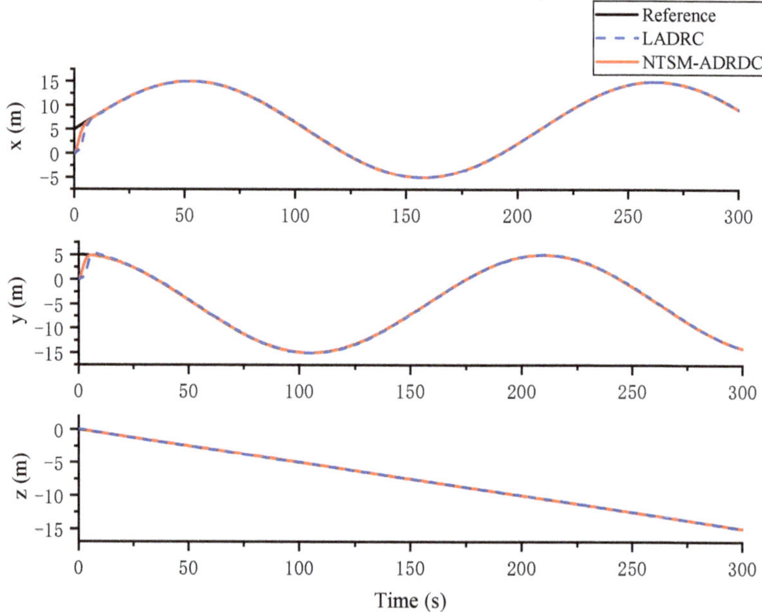

Figure 6. The position-tracking performance of the AUV without disturbance.

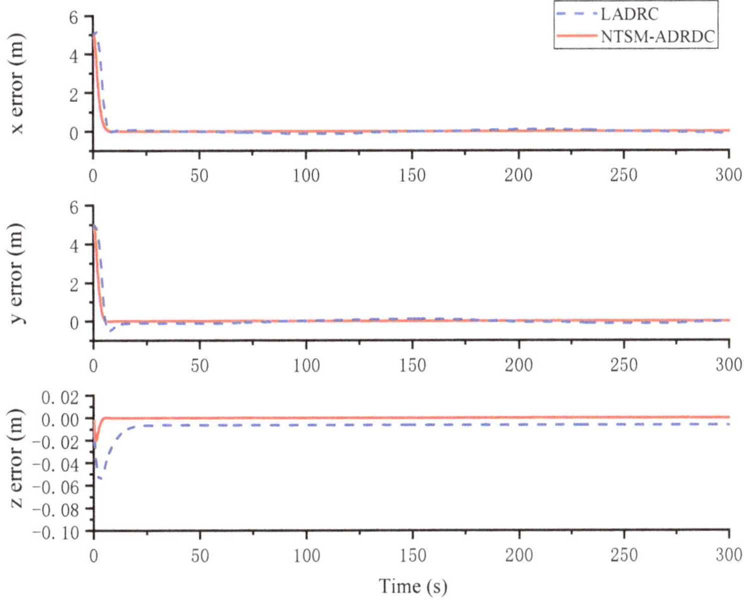

Figure 7. The position errors of the AUV without disturbance.

In order to reflect the trajectory tracking accuracy of different controllers, we define the AUV trajectory tracking error in the 3D space as follows:

$$E_\eta(t) = \sqrt{(x(t) - x_d(t))^2 + (y(t) - y_d(t))^2 + (z(t) - z_d(t))^2} \qquad (37)$$

where $x(t)$, $y(t)$, $z(t)$ are the actual position values of the AUV at the time t, and $x_d(t)$, $y_d(t)$, $z_d(t)$ represent the desired position values at the time t.

Figure 8 shows the comparison curve of AUV trajectory tracking error. Since the initial position of the AUV is far from the starting point of the reference position, it takes a certain time for the AUV to make the trajectory tracking error tend to 0. The AUV tracking error tends to zero after approximately 8 s under the ATSM-ADRDC algorithm, while LADRC is about 20 s. This shows that the error convergence time of the former is shorter than that of the latter. From the partial enlargement, we can clearly see that the trajectory tracking error of ATSM-ADRDC converges more smoothly compared to LADRC.

To further evaluate the tracking error accuracy of the AUV under the two controllers, we introduce three indicators related to the tracking error: the maximum tracking error (Max), the minimum tracking error (Min), and the average tracking error (Avg). The tracking error measurement values after 20 s are shown in Table 2. Taken together, it can be concluded that the control accuracy of NTSM-ADRDC is much higher than that of LADRC. As can be seen from Figure 9, the proposed NTSM-ADRC scheme has almost no "chattering" in the control inputs. It also indicates that the use of the tanh function to replace the sign function can reduce "chattering".

Table 2. AUV trajectory tracking error measurement values without disturbance.

Methods	Max (m)	Min (m)	Avg (m)
LADRC	0.12163	0.07649	0.10533
NTSM-ADRDC	0.00158	0.00131	0.00146

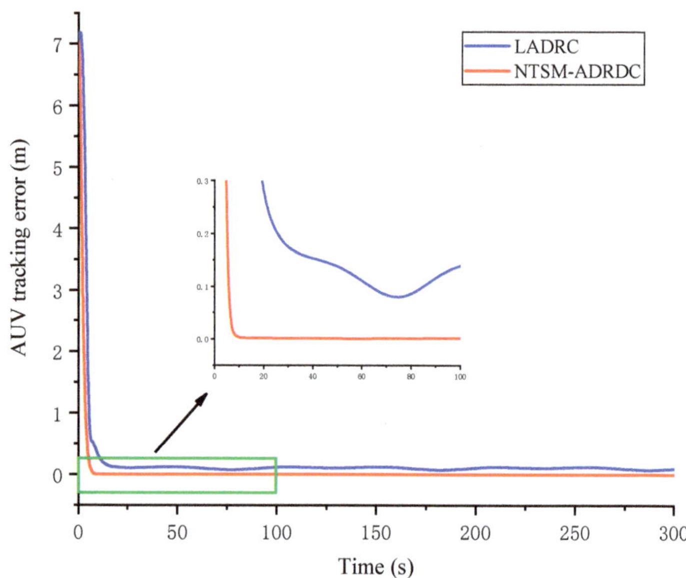

Figure 8. The AUV trajectory tracking errors without disturbance.

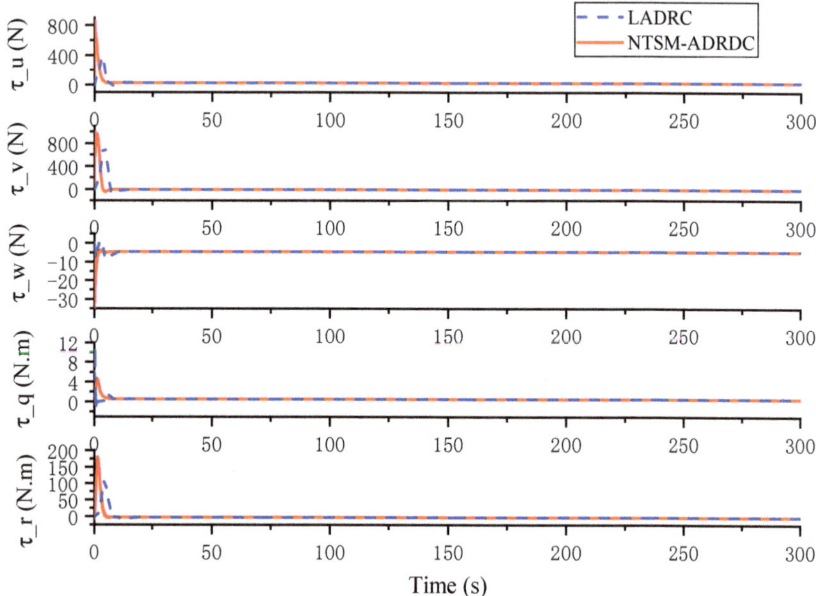

Figure 9. The AUV of control input without disturbance.

Based on the above analysis, both controllers can track the reference trajectory well in the absence of disturbance. In comparison, the trajectory following performance of NTSM-ADRDC is significantly better than LADRC. The main reason is that the former combines the exponential approach law (22), which ensures the rapid error convergence of the designed algorithm in a finite time.

In order to analyze the anti-disturbance performance of the designed algorithm, ocean current and bounded disturbances are added to the simulation. Ocean current disturbance in I-frame: $u_c^I = 0.25$ m/s, $v_c^I = 0.25$ m/s. External bounded disturbances:

$$
\begin{aligned}
d_u &= 0.02m_{11}[1 + \sin(0.05t)], & d_v &= 0.02m_{22}[1 + \sin(0.05t)], \\
d_w &= 0.02m_{33}[1 + \sin(0.05t)], & d_q &= 0.01m_{55}[1 + \cos(0.01t)], \\
d_r &= 0.01m_{66}[1 + \cos(0.01t)].
\end{aligned}
$$

Figure 10 shows the AUV 3D trajectory curves of the two control algorithms with the disturbance. We can find that both control approaches can still track the reference trajectory relatively well in the presence of disturbances. Nevertheless, the LADRC produces significant overshoot. From Figure 11 we can observe that the NTSM-ADRDC algorithm not only allows a faster approach to the desired position, but also without overshoot.

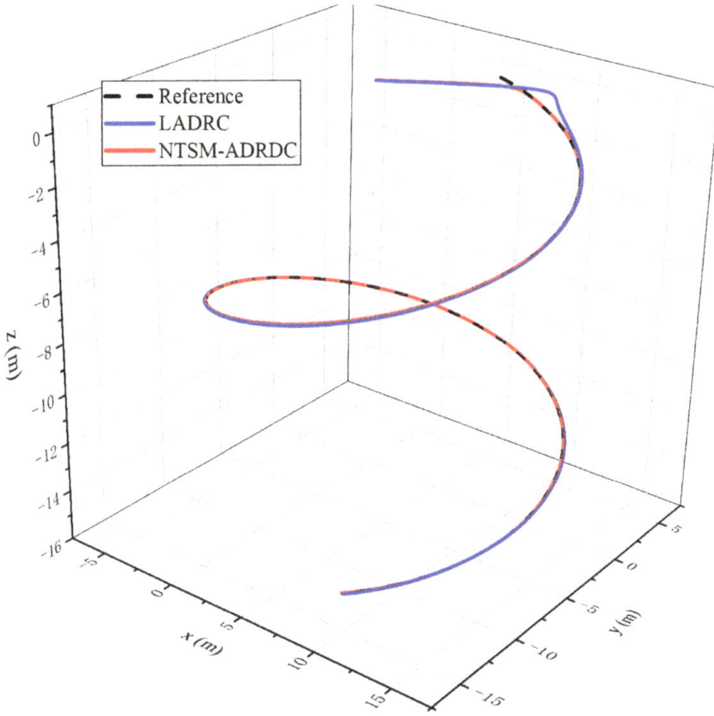

Figure 10. AUV 3D trajectory tracking curve with disturbance.

By comparing the position error under the two control algorithms, we can find from Figure 12 that the algorithm designed in this paper still has a faster error convergence rate with the disturbance, which shows that the exponential reaching law introduced into the NTSM-ADRDC algorithm is effective in the presence of disturbance. From Figure 13, we can clearly find that the AUV tracking error tracking convergence time is approximately 10 s under the ATSM-ADRDC algorithm, while LADRC is about 26 s. This shows that the proposed NTSM-ADRDC method has a faster error convergence compared to the LADRC. Additionally, the partial enlargement shows that there are larger fluctuations in the LADRC method compared to NTSM-ADRDC. It shows that the stability of NTSM-ADRDC is much better than that of the latter.

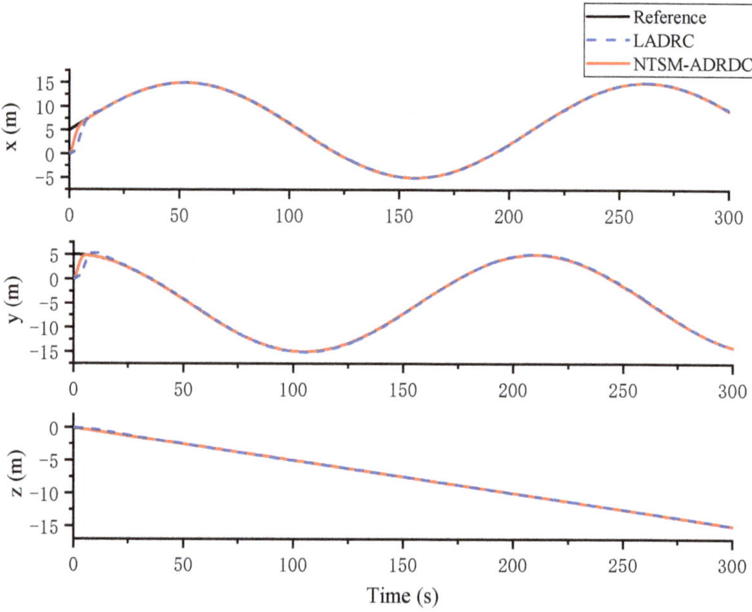

Figure 11. The position-tracking performance of the AUV with disturbance.

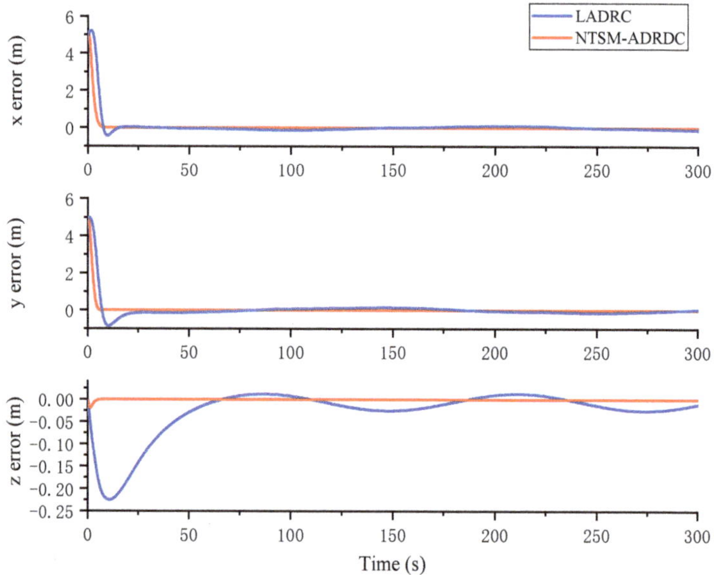

Figure 12. The position errors of the AUV with disturbance.

The tracking error measurement values after 26 s are shown in Table 3. In the presence of disturbance, the average of the tracking error of NTSM-ADRDC is 0.00162 m, while LADRC is about 0.12461 m. This indicates that the designed controller has high control accuracy in the presence of ocean currents and external disturbances. Figure 14 shows the control input of AUV with disturbance. We can see that there is some fluctuation in the control input due to the presence of ocean currents and external disturbances.

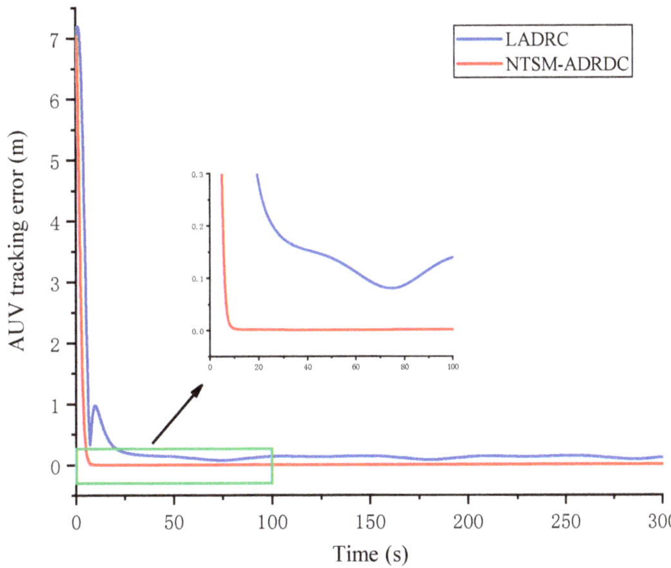

Figure 13. The AUV trajectory tracking errors with disturbance.

Table 3. AUV trajectory tracking error measurement values with disturbance.

Methods	Max (m)	Min (m)	Avg (m)
LADRC	0.20466	0.08036	0.12461
NTSM-ADRDC	0.00227	0.00143	0.00162

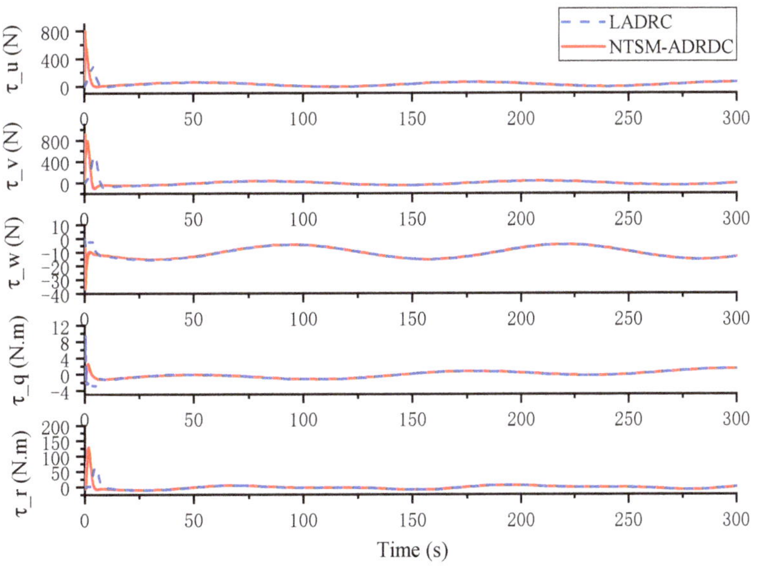

Figure 14. The AUV of control input with disturbance.

Based on the above analysis, the designed algorithm has good robustness and anti-disturbance. There are two main reasons: one is that ADRC technology can effectively suppress model uncertainty and external disturbance, and the other is that the combination of NTSM and the exponential reaching law not only retains strong robustness but also can ensure the rapid convergence of tracking errors in a finite time.

5. Conclusions

Aiming at the problem of AUV trajectory tracking control, this paper designs a novel tracking control method based on NTSM-ADRDC. Firstly, the AUV 5-DOF model is decoupled by introducing the ADRC technology. Secondly, the 3D trajectory tracking controller based on NTSM-ADRDC is designed. The controller uses LESO to observe the state variable values of the AUV and estimate the sum of the unmodeled dynamics and external disturbances of the system. By introducing the exponential reaching law into NTSM, a nonlinear error feedback law is designed to compensate for the total disturbance of the system. Combining NTSMC and ADRC technology can retain the advantages of the two control algorithms to the maximum. The NTSMC strategy can make the AUV quickly approach the reference trajectory, and the ADRC can suppress model uncertainty and external disturbance. Finally, the simulation verifies the effectiveness of the designed controller by comparing it with LADRC.

In future work, we will investigate the theory of optimization of the control parameters of the NTSM-ADRDC algorithm to improve the engineering applicability of the designed algorithm. Meanwhile, considering the input and state constraints existing in the AUV system, we will combine other methods, such as model predictive control, in the design process of the ADRC controller. In addition, in order to match the configuration of AUV actuators in actual applications, thrust allocation schemes will be investigated for the design of the controller.

Author Contributions: Methodology, W.W.; Validation, Z.Y.; Investigation, Z.L.; Data curation, X.D.; Writing—original draft, W.W.; Writing—review & editing, W.W.; Funding acquisition, W.Z. All authors have read and agreed to the published version of the manuscript.

Funding: This work was supported by the National Natural Science Foundation of China under grant E1102/52071108, and in part by National Natural Science Foundation of China under grant 5217110332, and the Natural Science Foundation of Heilongjiang Province under grant JJ2021JQ0075.

Institutional Review Board Statement: Not applicable.

Informed Consent Statement: Not applicable.

Data Availability Statement: Not applicable.

Conflicts of Interest: The authors declare no conflict of interest.

References

1. Wynn, R.B.; Huvenne, V.A.; Le Bas, T.P.; Murton, B.J.; Connelly, D.P.; Bett, B.J. Autonomous Underwater Vehicles (AUVs): Their past, present and future contributions to the advancement of marine geoscience. *Mar. Geol.* **2014**, *352*, 451–468. [CrossRef]
2. Palomeras, N.; Vallicrosa, G.; Mallios, A.; Bosch, J.; Vidal, E.; Hurtos, N.; Carreras, M.; Ridao, P. AUV homing and docking for remote operations. *Ocean Eng.* **2018**, *154*, 106–120. [CrossRef]
3. An, L.; Li, Y.; Cao, J.; Jiang, Y.; He, J.; Wu, H. Proximate time optimal for the heading control of underactuated autonomous underwater vehicle with input nonlinearities. *Appl. Ocean Res.* **2020**, *95*, 102002. [CrossRef]
4. Yan, Z.; Yu, H.; Zhang, W. Globally finite-time stable tracking control of underactuated UUVs. *Ocean Eng.* **2015**, *107*, 106–120. [CrossRef]
5. Yan, Z.; Gong, P.; Zhang, W.; Wu, W.H. Model predictive control of autonomous underwater vehicles for trajectory tracking with external disturbances. *Ocean Eng.* **2020**, *217*, 107884. [CrossRef]
6. Li, Y.; Jiang, Y.Q.; Wang, L.F.; Cao, J.; Zhang, G.C. Intelligent PID guidance control for AUV path tracking. *J. Cent. South Univ.* **2015**, *22*, 3440–3449. [CrossRef]

7. Robert, S.M.; Brett, W.H.; Lance, M. Docking control system for a 54-cm-diameter (21-in) AUV. *IEEE J. Oceanic. Eng.* **2009**, *33*, 550–562.
8. Liang, X.; Qu, X.; Wang, N. Three-Dimensional Trajectory Tracking of an Underactuated AUV based on Fuzzy Dynamic Surface Control. *IET Intell. Transp. Syst.* **2019**, *14*, 364–370. [CrossRef]
9. Yin, J.C.; Wang, N. Predictive Trajectory Tracking Control of Autonomous Underwater Vehicles Based on Variable Fuzzy Predictor. *Int. J. Fuzzy Syst.* **2020**, *23*, 1809–1822. [CrossRef]
10. Rezazadegan, F.; Shojaei, K.; Sheikholeslam, F.; Chatraei, A. A novel approach to 6-DOF adaptive trajectory tracking control of an AUV in the presence of parameter uncertainties. *Ocean Eng.* **2015**, *107*, 246–258. [CrossRef]
11. Guerrero, J.; Torres, J.; Creuze, V.; Chemori, A. Adaptive disturbance observer for trajectory tracking control of underwater vehicles. *Ocean Eng.* **2020**, *200*, 107080. [CrossRef]
12. Park, B.S. Neural Network-Based Tracking Control of Underactuated Autonomous Underwater Vehicles with Model Uncertainties. *J. Dyn. Syst.-T Asme.* **2015**, *137*, 021004. [CrossRef]
13. Eski, K.; Yldrm, S. Design of Neural Network Control System for Controlling Trajectory of Autonomous Underwater Vehicles. *Int. J. Adv. Robot. Syst.* **2014**, *11*, 1–17. [CrossRef]
14. Zhang, Y.D.; Liu, X.F.; Liu, M.Z.; Yang, C.G. MPC-based 3-D trajectory tracking for an autonomous underwater vehicle with constraints in complex ocean environments. *Ocean Eng.* **2019**, *189*, 106309. [CrossRef]
15. Chao, S.; Yang, S.; Buckham, B. Path-Following Control of an AUV: A Multiobjective Model Predictive Control Approach. *IEEE Trans. Control Syst. Technol.* **2019**, *27*, 1334–1342.
16. Chao, S.; Yang, S.; Buckham, B. Integrated Path Planning and Tracking Control of an AUV: A Unified Receding Horizon Optimization Approach. *IEEE ASME Trans. Mechatron.* **2017**, *22*, 1163–1173.
17. Przybyła, M.; Kordasz, M.; Madoński, R.; Herman, P.; Sauer, P. Active Disturbance Rejection Control of a 2DOF manipulator with significant modeling uncertainty. *Bull. Pol. Acad. Sci.* **2012**, *60*, 509–520. [CrossRef]
18. Tian, S.; Zhang, Y. Active disturbance rejection control based robust output feedback autopilot design for airbreathing hypersonic vehicles. *ISA Trans.* **2018**, *74*, 7–15. [CrossRef]
19. Yang, H.J.; Chen, L. Active Disturbance Rejection Attitude Control for a Dual Closed-Loop Quadrotor Under Gust Wind. *IEEE Trans. Contr. Syst. Technol.* **2018**, *26*, 1400–1405. [CrossRef]
20. Dong, W.; Gu, G.Y.; Zhu, X.Y. A high-performance flight control approach for quadrotors using a modified active disturbance rejection technique. *Robot. Auton. Syst.* **2016**, *83*, 177–187. [CrossRef]
21. Han, J. From PID to active disturbance rejection control. *IEEE Trans. Ind. Electron.* **2009**, *56*, 900–906. [CrossRef]
22. Han, J. The "extended state observer" of a class of uncertain systems. *Control Decis.* **1995**, *10*, 85–88.
23. Han, J.Q. Auto-disturbance rejection controller and it's applications. *Control Decis.* **1998**, *13*, 19–23.
24. Gao, Z. Active disturbance rejection control: A paradigm shift in feedback control system design. In Proceedings of the 2006 American Control Conference, Minneapolis, MN, USA, 14–16 June 2006; pp. 2399–2405.
25. Gao, Z.Q.; Guo, B. Active disturbance rejection control approach to stabilization of lower triangular systems with uncertainty. *Int. J. Robust. Nonlin.* **2015**, *26*, 2314–2337.
26. Chen, M.; Chen, W.H. Sliding mode control for a class of uncertain nonlinear system based on disturbance observer. *Int. J. Adapt Control* **2010**, *24*, 51–64. [CrossRef]
27. Besnard, L.; Shtessel, Y.; Landrum, B. Quadrotor vehicle control via sliding mode controller driven by sliding mode disturbance observer. *J. Frankl. Inst.* **2012**, *349*, 658–684. [CrossRef]
28. Ismail, Z.H.; Putranti, W.E. Second Order Sliding Mode Control Scheme for an Autonomous Underwater Vehicle with Dynamic Region Concept. *Math. Probl. Eng.* **2015**, *2*, 11–13. [CrossRef]
29. Feng, Y.; Yu, X.; Man, Z. Non-singular terminal sliding mode control of rigid manipulators. *Automatica* **2002**, *38*, 2159–2167. [CrossRef]
30. Zhang, W.W.; Wang, J. Nonsingular Terminal sliding model control based on exponential reaching law. *Control Decis.* **2012**, *27*, 909–913.
31. Sahu, B.K.; Subudhi, B.J.I.; Jo, A. Adaptive Tracking Control of an Autonomous Underwater Vehicle. *Int. J. Autom. Comput.* **2014**, *11*, 299–307. [CrossRef]
32. Fossen, T.I. *Handbook of Marine Craft Hydrodynamics and Motion Control*; John Wiley & Sons: Hoboken, NJ, USA, 2011.
33. Fossen, T.I. *Guidance and Control of Ocean Vehicles*; John Wiley & Sons Inc.: Chichester, UK, 1994.
34. Han, J. *Active Disturbance Rejection Control Technology—The Control Technology for Estimating and Compensating the Uncertainties*; China Defensive Industry Press: Beijing, China, 2008.
35. Chen, Z.Q.; Sun, M.W.; Yang, R.G. On the stability of linear disturbance rejection control. *Acta Autom. Sin.* **2013**, *39*, 574–580. [CrossRef]
36. Zhang, Y.; Chen, Z.Q.; Sun, M.W. Trajectory tracking control of a quadrotor UAV based on sliding mode active disturbance rejection control. *Nonlinear Anal.-Model.* **2019**, *24*, 545–560. [CrossRef]

37. Wu, Y.; Wang, L.F.; Zhang, J.Z. Path Following Control of Autonomous Ground Vehicle Based on Nonsingular Terminal Sliding Mode and Active Disturbance Rejection Control. *IEEE Trans. Veh. Technol.* **2019**, *68*, 6379–6390. [CrossRef]
38. Gao, Z. Scaling and parameterization based controller tuning. In Proceedings of the 2003 American Control Conference, Denver, CO, USA, 4–6 June 2003; pp. 4989–4996.

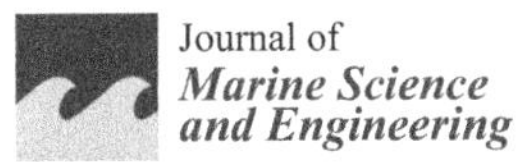

Journal of
Marine Science and Engineering

Article

Multi-AUV Formation Predictive Control Based on CNN-LSTM under Communication Constraints

Juan Li [1,2], Zhenyang Tian [2], Gengshi Zhang [2,*] and Wenbo Li [2]

[1] Key Laboratory of Underwater Robot Technology, Harbin Engineering University, Harbin 150001, China; lijuan041@hrbeu.edu.cn
[2] College of Intelligent Systems Science and Engineering, Harbin Engineering University, Harbin 150001, China; heu9436@hrbeu.edu.cn (Z.T.); liwenbo049@hrbeu.edu.cn (W.L.)
* Correspondence: zgengshi@163.com

Abstract: For the problem of hydroacoustic communication constraints in multi-AUV leader follower formation, this paper designs a formation control method combining CNN-LSTM prediction and backstepping sliding mode control. First, a feedback linearization method is used to transform the AUV nonlinear model into a second-order integral model; then, the influence of hydroacoustic communication constraints on the multi-AUV formation control problem is analyzed, and a sliding window-based formation prediction control strategy is designed; for the characteristics of AUV motion trajectory with certain temporal order, the CNN-LSTM prediction model is selected to predict the trajectory state of the leader follower and compensate the effect of communication delay on formation control, and combine the backstepping method and sliding mode control to design the formation controller. Finally, the simulation experimental results show that the proposed CNN-LSTM prediction and backstepping sliding mode control can improve the effect of hydroacoustic communication constraints on formation control.

Keywords: formation control; communication constraints; feedback linearization; CNN-LSTM prediction; backstepping slide control

Citation: Li, J.; Tian, Z.; Zhang, G.; Li, W. Multi-AUV Formation Predictive Control Based on CNN-LSTM under Communication Constraints. *J. Mar. Sci. Eng.* **2023**, *11*, 873. https://doi.org/10.3390/jmse11040873

Academic Editor: Rafael Morales

Received: 27 March 2023
Revised: 17 April 2023
Accepted: 18 April 2023
Published: 20 April 2023

1. Introduction

With the further exploration of the ocean, Autonomous Underwater Vehicles (AUVs) have started to play an important role in various marine activities, and AUVs are commonly used in tasks such as marine ecosystem detection, underwater inspection and surveillance, and subsea pipeline laying [1–3]. As the complexity of AUV missions increases, the operating environment of AUVs will become more and more complex. Due to constraints, such as the limited energy carried by them, AUVs start to look overwhelmed when facing some more demanding tasks. Therefore, multi-AUV collaboration, information sharing, and joint mission accomplishment have become the new direction of AUV development today. Multi-AUV collaboration can accomplish difficult tasks faster and better for single AUVs, especially in data acquisition [4], target search [5–7] and path planning [8,9], etc. Therefore, multi-AUV collaborative operation is the future development trend of AUVs to deal with complex problems in complex environments.

In the actual application, the multi-AUV formation will inevitably be affected by the actual environment, there will be a communication delay when multi-AUVs communicate with each other, and it takes some time to fuse and calculate the information of each sensor, so the real-time information sharing between multi-AUVs cannot be achieved in the actual application and the multi-AUV formation control will produce large control errors [10,11]. Therefore, the study of multi-AUV formation control under communication delay is helpful to apply the theory to practice and promote the development of multi-AUV formation technology.

For the multi-AUV formation control problem, different authors have proposed different solutions. Kang [12] used fuzzy control theory to coordinate the behavior of multiple AUV members, and the fuzzy control scheme inputs for the leader AUV in a multi-AUV formation were the yaw angle during obstacle avoidance and the yaw angle during target finding maneuvers, and the fuzzy control scheme for the follower consisted of the yaw angle deviations during obstacle avoidance and formation keeping. Borhaug [13] proposed a time-varying smooth feedback control law for multiple non-complete AUVs to maintain formation. An integral backstepping method was used to cooperatively park the follower AUV in its desired docking position and orientation relative to the leader, and the above control law was applied to a real AUV formation system to investigate the implementation problem and singularity avoidance problem of the physical AUV system. Ding [14] proposed a multi-AUV 3D formation control and obstacle avoidance method based on backstepping control and a bio-inspired neural network model. The followers track the virtual AUVs, during which the backstepping control method is guided to achieve 3D underwater formation control. The formation of AUVs was transformed using a bio-neural network model in order to avoid obstacles and pass through the area of obstacles. For the problem of leader failure in multi-AUV leader-following formations, Juan [15] proposed a solution to the problem of leader failure in multi-AUV leader-following formations by using the Hungarian algorithm to reconstruct the failed formation with the lowest cost. The Hungarian algorithm was improved to solve the nonstandard assignment problem. To address the issue of increased leader communication pressure after formation reconstruction, an event trigger mechanism was applied to reduce unnecessary communication. The efficiency of the event trigger mechanism was improved by increasing the event trigger condition of the sampling error threshold. Zheping [10] considered the presence of bounded communication delay and non-convex control input constraints in multi-AUV formation under weak communication conditions. They proposed a formation consistency constrained controller algorithm for discrete-time leaderless multi-AUV systems with dual independent communication topologies by introducing a constraint operator. For the problem of hydroacoustic communication constraints between multiple AUVs, Yuepeng [16] proposed a consensus control algorithm for multi-AUVs combined with the leader-following method under communication time delay, using graph theory to describe the communication topology of multi-AUVs and introducing a hybrid communication topology to accommodate large formation control. The consensus theory was combined with the leader-following method to construct distributed control laws. Suryendu [17] designed a time-lag estimator based on the gradient descent method to estimate the communication delay, and the actual delay was significantly reduced because the time tagging of the leader AUV state packets was avoided in the formulation of the estimator. Shibin [18] investigated the leader-following consistency problem for a multi-intelligent body system with input delays. A distributed state observer was designed to estimate the states of neighbors using the output information between neighboring intelligences, and a consistency algorithm was proposed using the estimated state information. Sufficient conditions for stability were constructed using Lyapunov theory and solved by a set of linear matrix inequalities with iterative parameters.

Based on the above research results, this paper proposes a formation control method combining CNN-LSTM prediction and backstepping sliding mode control. The specific contributions of this paper are summarized as follows:

1. A multi-AUV formation control method combining CNN-LSTM prediction and backstepping sliding mode control is proposed, the stability of the control method is demonstrated, and the effectiveness of the control method is verified by simulation.

2. Combining the advantages of CNN feature extraction, filtering noise and LSTM temporal memory, a CNN-LSTM prediction model is built for predicting the state information of navigators.

3. Applying the feedback linearization method, the AUV nonlinear model is transformed into a second-order integral model, and the controller is designed by combining

the backstepping method and sliding mode control, which improves the robustness of the controller.

2. AUV Nonlinear Model Building and Feedback Linearization

2.1. AUV Nonlinear Model

To study the motion of the AUV, the fixed coordinate system {E} and the motion coordinate system {O} established in this section are shown in Figure 1.

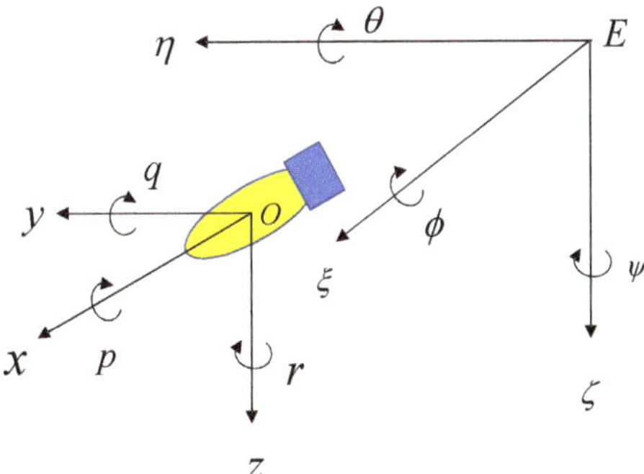

Figure 1. Coordinate system diagram, where $E - \xi\eta\zeta$ is the fixed coordinate system, ξ points due north, η points due east, $O - xyz$ is the motion coordinate system, and O coincides with the center of gravity of the AUV, where the x-axis points to the bow of the vehicle.

A fixed point at sea level is usually chosen as the origin of the fixed coordinate system, where the ξ axis points to due north and the η axis points to due east. In order to simplify the nonlinear model of the AUV, the center of gravity of the AUV is chosen as the origin of the motion coordinate system {O}, where the x axis is located in the longitudinal mid-profile and points to the bow of the AUV, and the y axis is perpendicular to the longitudinal mid-profile and points to the starboard side of the AUV.

In model building, it may be assumed that the AUV studied in this paper is a rigid body with a certain mass distribution, and the effect of its transverse rocking motion is not considered when the AUV is operating underwater, i.e., the transverse rocking attitude angle and angular velocity are kept as desired values. In the following, the nonlinear model of the AUV and the feedback linearization process are based on this assumption.

For the purpose of the following study, the following motion variables are defined:
The position vector in a fixed coordinate system is $\eta = [x\, y\, z\, \theta\, \psi]^T \in R^3 \times S^2$.
The position is $\eta_1 = [x\, y\, z]^T \in R^3$, The attitude angle is $\eta_2 = [\theta\, \psi]^T \in S^2$.
The velocity vector in the motion coordinate system is $v = [u\, v\, w\, q\, r]^T \in R^5$.
The linear velocity in the motion coordinate system is $v_1 = [u\, v\, w]^T \in R^3$.
The angular velocity in the motion coordinate system is $v_2 = [q\, r]^T \in R^2$.
The forces and moments in the motion coordinate system are $T = [X\, Y\, Z\, M\, N]^T \in R^6$.
The force in the motion coordinate system is $T_1 = [X\, Y\, Z]^T \in R^3$.
The moment in the motion coordinate system is $T_2 = [M\, N]^T \in R^2$.
Where R^3 denotes the three-dimensional Euclidean space and S^3 denotes the three-dimensional torus, i.e., there exist three angles in the range $[0, 2\pi]$.

Combining the AUV kinematic model and dynamics model, the AUV nonlinear mathematical model vector expression can be obtained as:

$$\dot{\eta} = J(\eta)v$$
$$M_R\dot{v} + M_A\dot{v} + C_R(v)v + Y(v) + g(\eta) = T + \lambda \tag{1}$$

The kinematic and kinetic mathematical model derivation process and model parameters of the AUV are shown in the literature [19] shown.

2.2. AUV Feedback Linearization Model

As can be seen from Equation (1), the nonlinear model of the AUV is still very complicated even if it is written in vector form. In this subsection, we simplify the AUV nonlinear model by using the transformation method to make the complex problem simple. By coordinate transformation, we can transform the nonlinear model of the AUV in the motion coordinate system to a specific coordinate system, in which the nonlinear model will realize the decoupling of each control channel and transform into a second-order integral model.

According to the literature [20], the AUV model is transformed appropriately:

$$\begin{cases} \dot{\eta} = J(\eta)v \\ \dot{v} = M^{-1}N(\eta, v) + M^{-1}T \end{cases} \tag{2}$$

where $M = M_R + M_A$ is the sum of the inertia matrix and the additional inertia matrix. T denotes the control input forces and moments. Synthesizing the three terms of the model $C_R(v)v$, $Y(v)$, $g(\eta)$ into a column vector $N(\eta, v)$, then Equation (2) can be transformed into:

$$\begin{bmatrix} \dot{\eta} \\ \dot{v} \end{bmatrix} = \begin{bmatrix} I & 0 \\ 0 & M^{-1} \end{bmatrix} \begin{bmatrix} J(\eta)v \\ N(\eta, v) \end{bmatrix} + \begin{bmatrix} 0 \\ M^{-1} \end{bmatrix} T \tag{3}$$

In Equation (3), a mathematical model with three axial thrusters and two rudders is considered, replacing the controller input T in Equation (3) with the thrust of the axial thrusters X_{prop}, Y_{prop}, Z_{prop} and the rudder angles δ_r, δ_s. The vector $\xi = [\eta^T, v^T]^T$ will be formed by η and v. The two matrices in Equation (3) are taken to be $M_1 = \begin{bmatrix} I & 0 \\ 0 & -M^{-1} \end{bmatrix} \in R^{10 \times 10}$ and $M_2 = \begin{bmatrix} 0 \\ M^{-1} \end{bmatrix} \in R^{10 \times 5}$, respectively. The above Equation (3) is transformed into the following vector form for model linearization:

$$\dot{\xi} = f(\xi) + M_2 g'(\xi)\hat{u} \tag{4}$$

Among them $f(\xi) = M_1 \begin{bmatrix} J(\eta)v \\ N(\eta, v) \end{bmatrix} \in R^{10 \times 1}$, $g'(\xi) = \left[g'_{ij}(\xi) \right] \in R^{5 \times 5}$, $\hat{u} = \left[X_{prop}, Y_{prop}, Z_{prop}, \delta_s, \delta_r \right]^T$.

Vector field: the nonlinear first-order model is taken as the following equation:

$$\begin{aligned} \dot{x} &= f(x) + g(x)u \\ y &= h(x) \end{aligned} \tag{5}$$

where $f(x)$, $g(x)$, $h(x)$ is smooth enough over the definition domain $D \in R^n$, the mapping $f : D \to R^n$ and $g : D \to R^n$ are vector fields over the domain of definition D.

Lie derivative: derivative of y in Equation (5).

$$\dot{y} = \frac{\partial h}{\partial x}[f(x) + g(x)u] = L_f h(x) + L_g h(x)u \tag{6}$$

where $L_f h(x) = \frac{\partial h}{\partial x} f(x)$, $L_g h(x) = \frac{\partial h}{\partial x} g(x)$, is said to be the Lie derivative of h along the smooth vector field f.

Define the output function $\zeta = h(\xi)$, then the dynamics of the AUV are modeled as:

$$\begin{aligned} \dot{\xi} &= f(\xi) + M_2 g'(\xi)\hat{u} \\ \zeta &= h(\xi) \end{aligned} \tag{7}$$

The basic idea of feedback linearization is to find an appropriate coordinate transformation and a control rate after the coordinate transformation.

Select the coordinate transformation $z = \varphi(x)$.

$$
\begin{aligned}
z_1 &= [h_1(x), h_2(x), h_3(x), h_4(x), h_5(x)]^T \\
z_2 &= \left[L_f h_1(x), L_f h_2(x), L_f h_3(x), L_f h_4(x), L_f h_5(x)\right]^T
\end{aligned}
\tag{8}
$$

From transforming the coordinates, we have:

$$
\begin{aligned}
z_1 &= h(x) \\
z_2 &= L_f h(x)
\end{aligned}
\tag{9}
$$

The transformation gives:

$$
\begin{aligned}
\dot{z}_1 &= z_2 \\
\dot{z}_2 &= L_f^2 h(x) + L_g L_f h(x)\hat{u}
\end{aligned}
\tag{10}
$$

In a given coordinate system, to obtain a simpler form, we might as well allow u to equal $L_f^2 h(x) + L_g L_f h(x)\hat{u}$. So, we can obtain the following equation:

$$
u = B(x) + \Gamma(x)\hat{u} = L_f^2 h(x) + L_g L_f h(x)\hat{u}
\tag{11}
$$

Then, the second-order integral model of the AUV in the new coordinate system after transformation can be obtained under the action of Equations (6) and (10).

$$
\hat{u} = \Gamma^{-1}(x)(u - B(x))
\tag{12}
$$

The AUV linearized mathematical model can be obtained as:

$$
\begin{aligned}
\dot{z}_1 &= z_2 \\
\dot{z}_2 &= u
\end{aligned}
\tag{13}
$$

where, z_1 is the position information of the AUV after the coordinate transformation, z_2 is the speed information of the AUV after the coordinate transformation, and u is the control input of the AUV after the coordinate transformation.

3. CNN-LSTM Prediction Model

3.1. Pre-Requisite Knowledge

3.1.1. Convolutional Neural Network

In underwater formations of multiple AUVs, the transmitted track data from the leader to the follower may be subject to both delay and noise interference caused by various factors such as oceanic noise. To enable accurate trajectory prediction, the data must be filtered prior to analysis. In this study, a convolutional neural network is employed to filter the data and extract the relevant trajectory data features. The basic structure diagram of the network is illustrated in Figure 2.

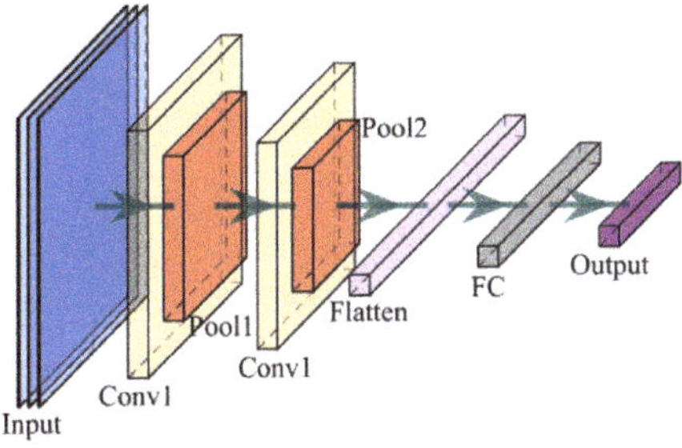

Figure 2. CNN structure schematic.

Figure 2 shows the structure of a convolutional neural network (CNN), which consists of an input layer, a convolutional layer, a ReLU layer, a pooling layer, and a fully connected layer. CNNs differ from traditional neural networks in two main ways:

1. CNNs use a common filter for different regions, which reduces parameters, improves training speed, and prevents overfitting;
2. The output of a CNN is related to only a portion of the input data due to the convolutional layers, which allows for the extraction of exclusive features for each input, whereas a traditional neural network is fully connected and outputs are related to all input units.

3.1.2. Long Short-Term Memory

For problems related to time series, such as AUV formation tracking, traditional neural network algorithms such as CNNs are not fully applicable. Long short-term memory (LSTM) networks are better suited for these problems due to their memory effect. LSTM networks use memory modules instead of traditional storage units, which are interconnected recursive subnetworks. The memory module contains gates that control the flow of information, allowing for memory information to affect neuronal nodes at longer time intervals. The three gates of an LSTM cell are the input gate, output gate, and forgetting gate, which control the storage and inflow of information as well as the core cell unit. The cell structure of LSTM is shown in Figure 3. The activation function plays an important role in the neural network by introducing nonlinear factors into the model, enabling it to perform well on problems where the linear model is not suitable.

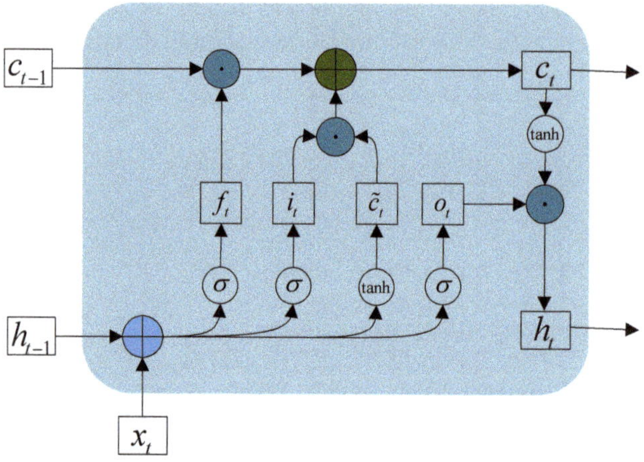

Figure 3. LSTM cell structure diagram.

In Figure 3, the symbol "f_t" represents the forgetting gate, "i_t" represents the input gate, and "o_t" represents the output gate. "x_t" denotes the input to the input layer at time "t", "h_t" denotes the output at time "t", "C_t" denotes the state value of the memory cell at time "t", and "σ" represents the sigmoid function. The mathematical expressions for "σ" and "tanh" in the figure are as follows:

$$\sigma(z) = \frac{1}{1 - e^{-z}} \tag{14}$$

$$\tanh(x) = \frac{e^x - e^{-x}}{e^x + e^{-x}} \tag{15}$$

The LSTM processes the data internally as follows:

$$f_t = \sigma\left(W_{xf}x_t + W_{hj}h_{t-1} + b_f\right) \tag{16}$$

$$i_t = \sigma(W_{ri}x_t + W_{hi}h_{t-1} + b_i) \tag{17}$$

$$o_f = \sigma(W_{xv}x_t + W_{hu}h_{t-1} + b_o) \tag{18}$$

$$c_t = f_t \cdot c_{t-1} + i_t \cdot \tanh(W_{xc}x_t + W_{hc}h_{t-1} + b_c) \tag{19}$$

$$h_i = o_t \cdot \tanh(c_t) \tag{20}$$

where, W is the weight matrix, $\cdot$ is the product of point pairs, and b is the deviation.

From Equations (14)–(18), it can be seen that the LSTM is computed by first calculating the values of the forgetting gate, input gate, output gate, and candidate state h_{t-1} and the input at the current moment based on the external state. Next, the internal state c_{t-1} is used to compute the values of the forgetting gate, the input gate and the candidate state in order to update the internal state c_t. Finally, the information is passed to the external state h_t via the current internal state and output gates.

3.2. CNN-LSTM Prediction Model Building

This paper proposes a neural network prediction model that combines the advantages of CNN feature extraction and noise filtering with LSTM temporal memory. The model is designed by connecting the CNN and LSTM layers in series, and its structure is depicted in Figure 4.

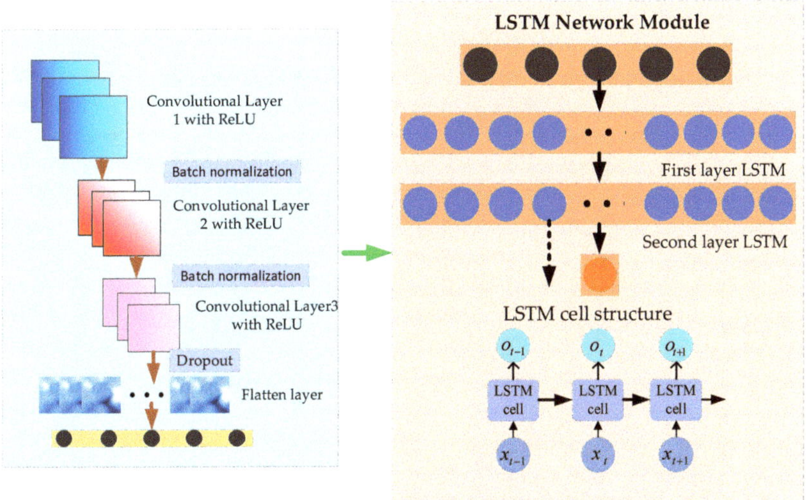

Figure 4. CNN-LSTM prediction model diagram.

The proposed structure is composed of two main modules: the data processing module and the model prediction module. Upon receiving the navigator state information, the data are first preprocessed and then fed into the prediction model. As illustrated in Figure 4, the CNN module is composed of three convolutional layers: a BatchNorm layer, a dropout layer, an expansion layer, and a fully connected layer, which is responsible for receiving the preprocessed data and extracting data features. The LSTM module, on the other hand, consists of two LSTM layers, which analyze the features extracted by the CNN, explore the time series relationships in the data, and predict multiple future points.

The overall prediction process is as follows: the navigator state information is pre-processed by the data processing module, and the processed data are passed to the CNN module for filtering and spatial feature learning. The CNN generates a sequence of high-level features representing the capture and passes it to the tensor processing module. The tensor processing layer then reshapes the output of the CNN so that it can be accepted by the LSTM sub-module. Finally, the LSTM module learns the time-series dependencies of the delayed data and outputs the predicted values for the current moment.

4. Predictive Control of Multi-AUV Formations Based on CNN-LSTM Models

4.1. Multi-AUV Formation Controller Design under Ideal Communication Conditions

It may be assumed that there are five AUVs in the formation: one leader and four followers. The formation that the formation wants to form and maintain is an isosceles triangle (the specific formation is shown in Figure 5 below), and the AUVs are required to maintain the formation even when making a spiral dive.

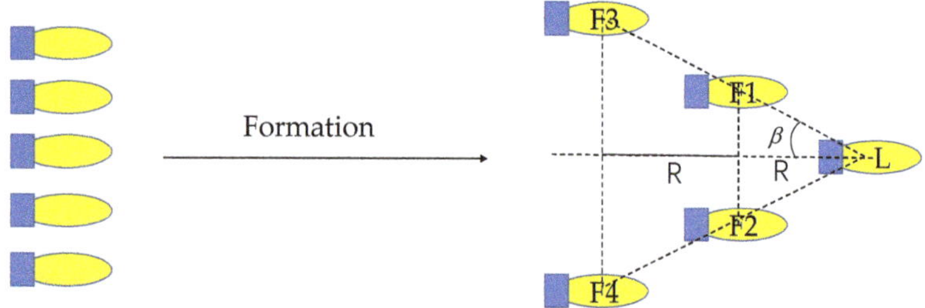

Figure 5. Formation diagram.

As shown in Figure 5, L denotes the leader, and F1, F2, F3 and F4 are all follower AUVs. According to the formation that we want to achieve, we introduce the variables R and β to constrain the formation, where the distance from the leader to the followers F1 and F2 line is R, the distance from the follower F1 and F2 line to the follower F3 and F4 line is also R, and the attitude angle of the formation hold is β. The formation constraints proposed in this paper are:

$$
\begin{cases}
\eta_{F1} + d_1 = \eta_L \\
\eta_{F2} + d_2 = \eta_L \\
\eta_{F3} + d_3 = \eta_L \\
\eta_{F4} + d_4 = \eta_L
\end{cases}
\qquad
\begin{cases}
\dot{\eta}_{F1} + dv_1 = \dot{\eta}_L \\
\dot{\eta}_{F2} + dv_2 = \dot{\eta}_L \\
\dot{\eta}_{F3} + dv_3 = \dot{\eta}_L \\
\dot{\eta}_{F4} + dv_4 = \dot{\eta}_L
\end{cases}
\tag{21}
$$

where $d_1, d_2, d_3, d_4, dv_1, dv_2, dv_3$ and dv_4 are denoted as:

$$
\begin{cases}
d_1 = \left(-(\cos\beta)^{-1}R\cos(\psi_L - \beta - \frac{\pi}{2}), (\cos\beta)^{-1}R\cos(\psi_L + \beta - \frac{\pi}{2}), 0, 0, 0\right)^T \\
d_2 = \left(-(\cos\beta)^{-1}R\cos(\psi_L + \beta - \frac{\pi}{2}), -(\cos\beta)^{-1}R\cos(\psi_L - \beta - \frac{\pi}{2}), 0, 0, 0\right)^T \\
d_3 = \left(-(\cos\beta)^{-1}2R\cos(\psi_L - \beta - \frac{\pi}{2}), (\cos\beta)^{-1}2R\cos(\psi_L + \beta - \frac{\pi}{2}), 0, 0, 0\right)^T \\
d_4 = \left(-(\cos\beta)^{-1}2R\cos(\psi_L + \beta - \frac{\pi}{2}), -(\cos\beta)^{-1}2R\cos(\psi_L - \beta - \frac{\pi}{2}), 0, 0, 0\right)^T
\end{cases}
\tag{22}
$$

$$
\begin{cases}
dv_1 = J_\eta(r_L \tan(-\beta) \times R, 0, 0, 0, 0)^T \\
dv_2 = J_\eta(r_L \tan(\beta) \times R, 0, 0, 0, 0)^T \\
dv_3 = J_\eta(r_L \tan(-\beta) \times 2R, 0, 0, 0, 0)^T \\
dv_4 = J_\eta(r_L \tan(\beta) \times 2R, 0, 0, 0, 0)^T
\end{cases}
$$

In a leader-follower formation control with five AUVs, the motion state vector of the i th follower AUV at the moment of t is $\varepsilon_i(t) = \eta_i(t)$ and the motion state vector of the leader is $\varepsilon_L(t) = \eta_L(t)$. If the formation satisfies Equation (23), it is said that the formation can achieve formation maintenance and stability convergence.

$$\begin{aligned} \lim_{t\to\infty}\left|\varepsilon_i(t) - \varepsilon_L(t) + d_i\right| &= 0 \\ \lim_{t\to\infty}\left|\dot{\varepsilon}_i(t) - \dot{\varepsilon}_L(t) + dv_i\right| &= 0 \end{aligned} \qquad i = 1,2,3,4 \tag{23}$$

Let the attitude vector of the i th follower AUV at the time of t and the attitude vector of the leader AUV at the time of z_{1d} in the lead follower formation control of the AUV be z_{1i}.

Define the trajectory tracking error of the i th follower AUV as $z_{i1e} = z_{i1} - z_{1d} + d_i$, then $\dot{z}_{i1e} = z_{i2} - \dot{z}_{1d}$.

Define the following Lyapunov function:

$$V_{i1} = \frac{1}{2}z_{i1e}{}^2 \tag{24}$$

Define $z_{i2} = z_{i2e} + \dot{z}_{1d} - c_{i1}z_{i1e}$, where c_{i1} is the positive constant and z_{i2e} is the intermediate virtual control item. We can get $z_{i2e} = z_{i2} - \dot{z}_{1d} + c_{i1}z_{i1e}$, and the derivation gives $\dot{z}_{i1e} = z_{i2} - \dot{z}_{i1d} = z_{i2e} - c_{i1}z_{i1e}$.

The derivative of V_{i1} gives:

$$\dot{V}_{i1} = z_{i1e}\dot{z}_{i1e} = z_{i1e}z_{i2e} - c_{i1}z_{i1}{}^2 \tag{25}$$

Define the switching function as:

$$\sigma_i = k_{i1}z_{i1e} + z_{i2e} \tag{26}$$

Among them, $k_{i1} > 0$.

Because of $\dot{z}_{i1e} = z_{i2e} - c_{i1}z_{i1e}$, we can derive:

$$\sigma_i = k_{i1}z_{i1e} + z_{i2e} = k_{i1}z_{i1e} + \dot{z}_{i1e} + c_{i1}z_{i1e} = (k_{i1} + c_{i1})z_{i1e} + \dot{z}_{i1e} \tag{27}$$

Because of $k_{i1} + c_{i1} > 0$, there is $\sigma_i = 0$ only when $z_{i1e} = 0$, $z_{i2e} = 0$ and $\dot{V}_{i1} \leq 0$. For this, the next design step is needed.

Define the following Lyapunov function.

$$V_{i2} = V_{i1} + \frac{1}{2}\sigma_i{}^2 \tag{28}$$

The derivative of V_{i2} gives:

$$\begin{aligned} \dot{V}_2 &= \dot{V}_{i1} + \sigma_i\dot{\sigma}_i \\ &= z_{i1e}z_{i2e} - c_{i1}z_{i1e}{}^2 + \sigma_i\dot{\sigma}_i \\ &= z_{i1e}z_{i2e} - c_{i1}z_{i1e}{}^2 + \sigma_i(k_{i1}\dot{z}_{i1e} + \dot{z}_{i2e}) \\ &= z_{i1e}z_{i2e} - c_{i1}z_{i1e}{}^2 + \sigma_i(k_{i1}(z_{i2e} - c_{i1}z_{i1e}) + \dot{z}_{i2} - \ddot{z}_{1d} + c_{i1}\dot{z}_{i1e}) \\ &= z_{i1e}z_{i2e} - c_{i1}z_{i1e}{}^2 + \sigma_i(k_{i1}(z_{i2e} - c_{i1}z_{i1e}) + U_i + F - \ddot{z}_{1d} + c_{i1}\dot{z}_{i1e}) \end{aligned} \tag{29}$$

where U_i is the expression of the controller to be designed. F is the total uncertainty of the system.

The design of the i follower controller is shown below.

$$U_i = -k_{i1}(z_{i2e} - c_{i1}z_{i1e}) - \bar{F}\tanh(\sigma_i) + \ddot{z}_{1d} - c_{i1}\dot{z}_{i1e} - h_i(\sigma_i + \beta_i\tanh(\sigma_i)) \tag{30}$$

where h_i and β_i are positive constants.

Substituting Equation (30) into $\dot{V}_{i2}$ yields:

$$\begin{aligned}
\dot{V}_{i2} &= z_{i1e}z_{i2e} - c_{i1}z_{i1e}^2 - h_i\sigma_i^2 - h_i\beta_i|\sigma_i| + F\sigma_i - \overline{F}\sigma_i \\
&\leq -c_{i1}z_{i1e}^2 + z_{i1e}z_{i2e} - h_i\sigma_i^2 - h_i\beta_i|\sigma_i|
\end{aligned} \tag{31}$$

Let Q_i be equal to the following matrix.

$$Q_i = \begin{bmatrix} c_{i1} + h_i k_{i1}^2 & h_i k_{i1} - \frac{1}{2} \\ h_i k_{i1} - \frac{1}{2} & h_i \end{bmatrix} \tag{32}$$

Due to

$$\begin{aligned}
z_{ie}{}^{\mathrm{T}} Q_i z_{ie} &= \begin{bmatrix} z_{i1e} & z_{i2e} \end{bmatrix} \begin{bmatrix} c_{i1} + h_i k_{i1}^2 & h_i k_{i1} - \frac{1}{2} \\ h_i k_{i1} - \frac{1}{2} & h_i \end{bmatrix} \begin{bmatrix} z_{i1e} & z_{i2e} \end{bmatrix}^{\mathrm{T}} \\
&= c_{i1}z_{i1e}^2 - z_{i1e}z_{i2e} + h_i k_{i1}^2 z_{i1e}^2 + 2h_i k_{i1} z_{i1e}z_{i2e} + h_i z_{i2e}^2 \\
&= c_{i1}z_{i1e}^2 - z_{i1e}z_{i2e} + h_i\sigma_i^2
\end{aligned} \tag{33}$$

Among them, $z_{ie}{}^{\mathrm{T}} = \begin{bmatrix} z_{i1e} & z_{i2e} \end{bmatrix}$.

If you want to guarantee that Q_i is a positive definite matrix, then

$$\dot{V}_{i2} = -z_{ie}{}^{\mathrm{T}} Q_i z_{ie} - h_i\beta_i\Big|\sigma_i\Big| \leqslant 0 \tag{34}$$

Due to

$$\left| Q_i \right| = h_i\left(c_{i1} + h_i k_{i1}^2\right) - \left(h_i k_{i1} - \frac{1}{2}\right)^2 = h_i(c_{i1} + k_{i1}) - \frac{1}{4} \tag{35}$$

Therefore, it is possible to guarantee $\dot{V}_{i2} \leq 0$ by taking the values of h_i, c_{i1} and k_{i1} such that $|Q_i| > 0$, i.e., Q_i is a positive definite matrix.

By taking the values of h, c_1 and k_1, you can make $|Q| > 0$. Thus, it can be deduced that Q is a positive definite matrix and that $\dot{V}_2 \leq 0$ is guaranteed.

According to LaSalle's invariance principle, when $\dot{V}_{i2} \equiv 0$ is taken, it can be deduced that $z_{ie} \equiv 0, \sigma_i \equiv 0$. When $t \to \infty$, since $z_{i1e} \to 0$, $z_{i2e} \to 0$, it can be deduced that $z_{i2e} \to 0$, $\dot{z}_{i1} \to \dot{z}_{1d}$.

In summary, it can be seen that the Lyapunov functions V_{i1} and V_{i2} are positive definite, and the values of V_{i1}, c_{i1} and k_{i1} can be reasonably chosen to ensure that $\dot{V}_{i1}$ and $\dot{V}_{i2}$ are negative definite, so the designed AUV formation controller (30) is stable and convergent.

4.2. Sliding Window-Based Predictive Control of Multi-AUV Formations under Communication Constraints

In the previous section, the backstepping sliding mode control method was used and the formation controller was designed according to the formation constraint relationship. The controller for the follower AUV in the formation with time-lag state is presented below due to the communication delay between the leader and the follower and the limitations of the hydroacoustic sonar in transmitting high-frequency signals, resulting in a longer communication interval between them. As a consequence, the follower may not receive the real-time status information of the leader.

$$U_i = -k_{i1}(z_{i2e} - c_1 z_{i1e}) - \overline{F}\tanh(\sigma_i) + \ddot{z}_{1d} - c_1\dot{z}_{i1e} - h(\sigma_i + \beta\tanh(\sigma_i)) \tag{36}$$

where, $z_{i1e} = z_{i1} - z_{1d}(t - \tau) + d_i$, $z_{i2e} = z_{i2} - z_{2d}(t - \tau) + c_1 z_{i1e} + dv_i$, $\sigma_i = k_1 z_{i1e} + z_{i2e}$, and τ are the communication delay times between the navigator and the follower.

To illustrate the effect of communication delay on formation control while preparing for a new predictive control strategy, the following assumptions are made about the communication delay between the leader and the follower and the hydroacoustic sonar occurrence interval:

Assumption 1. *The distance between the navigator and the follower is close, and the speed of acoustic wave transmission in the water is 1500 m/s, so the communication time delay caused by the communication transmission is small, where it is assumed that the delay time between the broadcast of the navigator sending the status information and the follower receiving the information and measuring the settlement is 1 s.*

Assumption 2. *Due to the limitation of communication bandwidth, the navigator cannot send too many beats of historical status data to the follower at one time; so, suppose the navigator can send five beats of status data to the follower at one time.*

Assumption 3. *The hydroacoustic sonar is unable to sound at high frequencies and the sounding time is affected by the size of the data sent, assuming that the communication interval of the hydroacoustic sonar is 6–9 s.*

To solve the communication delay and communication interval problem between the leader and the follower, this section proposes a formation control strategy based on a sliding window to achieve multi-step prediction, which iterates the historical state information of the leader to predict the current state information of the leader step by step, which saves computational efficiency and has better adaptability compared with the observer-based iterative prediction method. The specific principle of the strategy is described below.

At the M time, the navigator sends its own status data $\{Z_1, Z_2, \cdots\cdots Z_{M-1}, Z_M\}$ from the previous M time to the follower in the formation. Due to the communication transmission delay τ_{tran} and the hydroacoustic sonar sounding time consuming τ_{inter}, a fixed time delay $\tau_{once\_tal} = \tau_{tran} + \tau_{inter}$ is defined, and the follower receives the status information of the navigator at the $M + \tau_{once\_tal}$ time, and the status information of the navigator received by the follower at this time is the status information of the navigator at the M time. So, the follower needs to predict the state information of the leader at the $M + \tau_{once\_tal}$ time as the tracking target based on the state information of the leader at the M time.

The second sounding of the sonar starts immediately after the first sounding. Since the transmission delay after the first sounding is included in the second sonar sounding elapsed time, the follower needs time τ_{inter} to receive the information of the navigator for the second time. Therefore, after the follower receives the status information of the navigator at the $M + \tau_{once\_tal}$ time, the follower firstly has to predict the status information of the navigator at the M time as the tracking target; secondly, since the follower cannot receive the status information of the navigator at the τ_{inter} time in the future, the follower needs to then predict the status information of the navigator at the τ_{inter} time in the future. A schematic diagram of the information transfer process is shown in Figure 6.

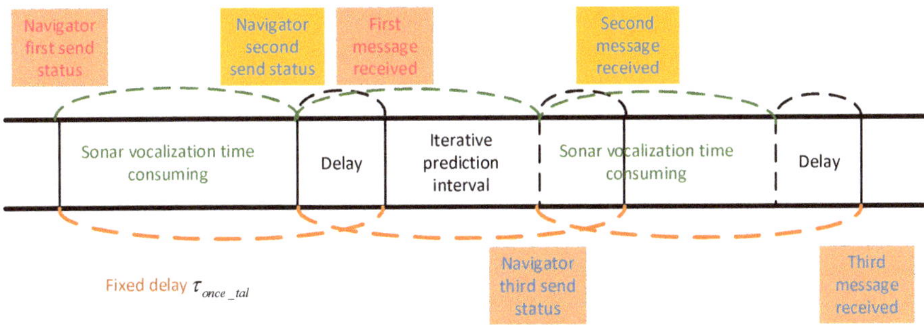

Figure 6. Information transmission diagram.

This paper focuses on predicting the real-time state of the leader in AUV formation by using delayed state data received by the followers as input to the prediction model. The delayed data are in the form of a time series, and to achieve continuous prediction, a sliding window approach is designed where the delay information is fed into the window as input and the real information as output, as illustrated in Figure 7. To evaluate the model's performance, a delay time of 10 s is set, and the size of the sliding window, which corresponds to the time step of the input data, is set to 5. The prediction equation is given as follows:

$$z(t) = f(\{z(t-14), \cdots , z(t-11), z(t-10)\}) \tag{37}$$

where $z = [z_1, z_2]$ denotes the position vector of the navigator in time. $z_1 = [x, y, depth, \theta, \psi]^T$, where x, y and $depth$ represent the displacement in three coordinate directions; θ and ψ represent the pitch and heading angles. $z_2 = [u, v, w, q, r]^T$, where u, v and w are the longitudinal, lateral and vertical velocities respectively; q and r are the longitudinal and bow angular velocities.

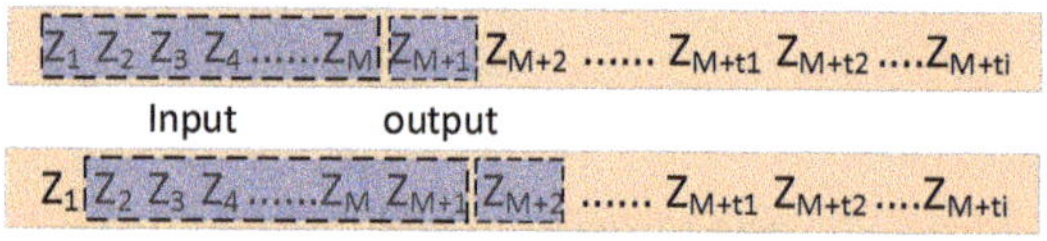

Z_1 Z_2 Z_3 Z_4 ......Z_M Z_{M+1} Z_{M+2} ...... Z_{M+t1} Z_{M+t2} ....Z_{M+ti}

Input   output

Z_{11} Z_2 Z_3 Z_4 ......Z_M Z_{M+1} Z_{M+2} ...... Z_{M+t1} Z_{M+t2} ....Z_{M+ti}

Figure 7. Schematic diagram of the sliding window.

The prediction strategy designed in this paper has two main phases: fixed delay prediction and communication interval prediction. In the fixed delay prediction stage, the follower puts the state information sent by the navigator into the designed sliding window and uses the prediction model to predict the state quantity $\hat{Z}_{M+1}$ at the $M+1$ th time based on the data in the first M times of the sliding window. Put $\hat{Z}_{M+1}$ into the sliding window, and then the sliding window moves forward to obtain $\hat{Z}_{M+2}$ using $\hat{Z}_{M+1}$ and the historical state quantity prediction, and finally obtain the state prediction $\hat{Z}_{M+\tau_{once\_tal}}$ at $M + \tau_{once\_tal}$ moments through continuous iterative prediction.

At the same time, due to the effect of hydroacoustic sonar sounding time consumption, the follower will only receive the next status data from the navigator at the moment of $M + \tau_{once\_tal} + \tau_{inter}$, so the follower will continue to make iterative predictions based on the status quantity $\hat{Z}_{M+\tau_{once\_tal}}$ obtained from the prediction compensation during this period, obtain $\hat{Z}_{M+\tau_{once\_tal}+1+1}$, $\hat{Z}_{M+\tau_{once\_tal}+2} \cdots \cdots \hat{Z}_{M+\tau_{once\_tal}+\tau_{inter}}$, and output in turn until it receives the time delay status data from the navigator again. Based on the above strategy, the follower will get the predicted value of the current moment of the leader; the controller of the follower in the AUV formation at this time is shown below.

$$U_i = -k_{i1}(\hat{z}_{i2e} - c_1\hat{z}_{i1e}) - \bar{F}\tanh(\hat{\sigma}_i) + \ddot{z}_{1d} - c_1\dot{z}_{i1e} - h(\hat{\sigma}_i + \beta\tanh(\hat{\sigma}_i)) \tag{38}$$

where, $\hat{z}_{i1e} = z_{i1} - \hat{z}_{1d}(t) + d_i$, $\hat{z}_{i2e} = z_{i2} - \hat{z}_{2d}(t) + c_1\hat{z}_{i1e} + dv_2$, $\hat{\sigma}_i = k_1\hat{z}_{i1e} + \hat{z}_{i2e}$, $\hat{z}_{1d}(t)$ and $\hat{z}_{1d}(t)$ are the predicted values of CNN-LSTM model.

The block diagram of CNN-LSTN-based multi-AUV formation prediction control under communication constraints is shown in Figure 8. Based on the pilot-follower formation control strategy, there is a communication delay when the follower AUV receives the position and speed information from the pilot due to the influence of hydroacoustic communication. In this paper, a CNN-LSTM prediction model is established to make predictions based on the historical information of the pilot, which can well offset the effects of noise and communication delay on formation control. The prediction information and feedback information are used as the input of the AUV formation controller to finally realize the AUV formation prediction control.

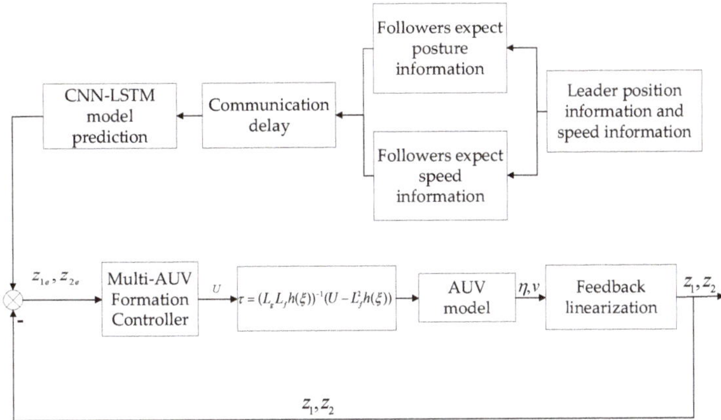

Figure 8. AUV formation prediction control block diagram.

5. Simulation Verification and Analysis

5.1. Simulation Results and Analysis of CNN-LSTM Model

The trajectory data of a small AUV, consisting of longitude and latitude measurements from multiple positioning systems, as well as values from GPS, bathymetry, and Doppler measurements with a maximum depth of 20 m, were selected as the training set for this study. The relevant information of the training set is shown in Table 1.

Table 1. AUV status information.

Sample Size	Maximum Depth (m)	Lon	Lat	U (m/s)
39,875	20	119.18° E	29.56° N	1–3

These training data were obtained from the trajectory data of an AUV on-lake experiment, and some of its trajectories are shown in Figure 9. The raw data were preprocessed and used for the training of the CNN-LSTM model.

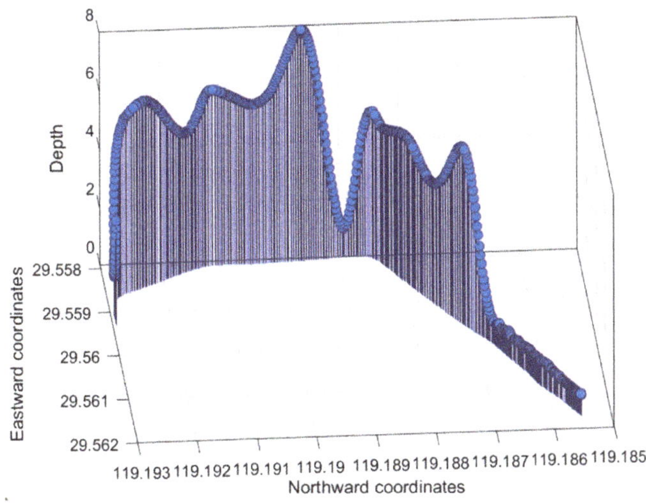

Figure 9. AUV partial trajectory data.

The CNN model designed in this paper contains three convolutional layers with filter sizes of (2, 1), (3, 1) and (3, 1) for each layer, and a dropout layer is added to prevent overfitting. The processed features were passed to the two-layer LSTM model, and the predicted data were output by the last LSTM layer. Through continuous debugging, it was found that the network with 125 and 128 neurons in each layer was trained well. Additionally, to prevent the overfitting of the network, a discard layer with probability 0.3 is built after the hidden layer. The Adam algorithm is used for optimization, and the design learning rate decline period is 100, the learning rate is 0.012, the learning rate decline coefficient is 0.8, and finally, the gradient threshold is set to 1 in order to prevent gradient explosion.

After processing the delayed data according to the aforementioned data processing steps, they are fed into the CNN-LSTM model using the sliding window format. The performance of the model is then evaluated by computing the mean square error (*MSE*) and maximum absolute error (MAXERR) between the predicted values and the actual trajectory data. The evaluation metrics can be formulated as follows:

$$MSE = \frac{1}{N} \sum_{t=1}^{N} \left(\text{observed}_t - \text{predicted}_i \right)^2 \tag{39}$$

where N indicates the number of samples.

$$\text{Maxerr} = \max \left| \frac{\text{observed}_t - \text{predicted}_t}{\text{observed}} \right| \tag{40}$$

According to Assumption 2, the leader broadcasts the data of the past five beats to each follower at a time, so the size of the sliding window is set to 5, and the prediction effect of the prediction model is verified under the fixed delay of 2 s and the communication interval of 7 s. The selected navigator trajectory is a spiral dive trajectory, and Gaussian white noise with an amplitude of 0.003 is superimposed on the trajectory data, and the LSTM prediction model is selected for simulation comparison. The parameters of the two model designs are shown in Table 2.

Table 2. AUV status information.

Models	Structural Layer	Parameter Setting	Learning Rate
LSTM model	Hidden layer neurons	[10, 10, 10, 10]	0.02
	activation function	ReLU	
	Optimizers	Adam	
	Epochs	30	
	Batch size	128	
CNN-LSTM model	Filter 1	×16 size (2, 1)	0.012
	Filter 2	×16 size (3, 1)	
	Filter 3	×16 size (3, 1)	
	Dropout ratio	0.3	
	Optimizers	Adam	
	LSTM cells 1	10	
	LSTM cells 2	10	
	Activation function	ReLU	

Since the velocity quantities in the selected trajectories are kept constant, in order to objectively compare the advantages and disadvantages of the two prediction models, only the navigator state quantities $z = [x, y, depth, \theta, \psi]$ are compared for prediction, and the simulation results are shown in Figure 10.

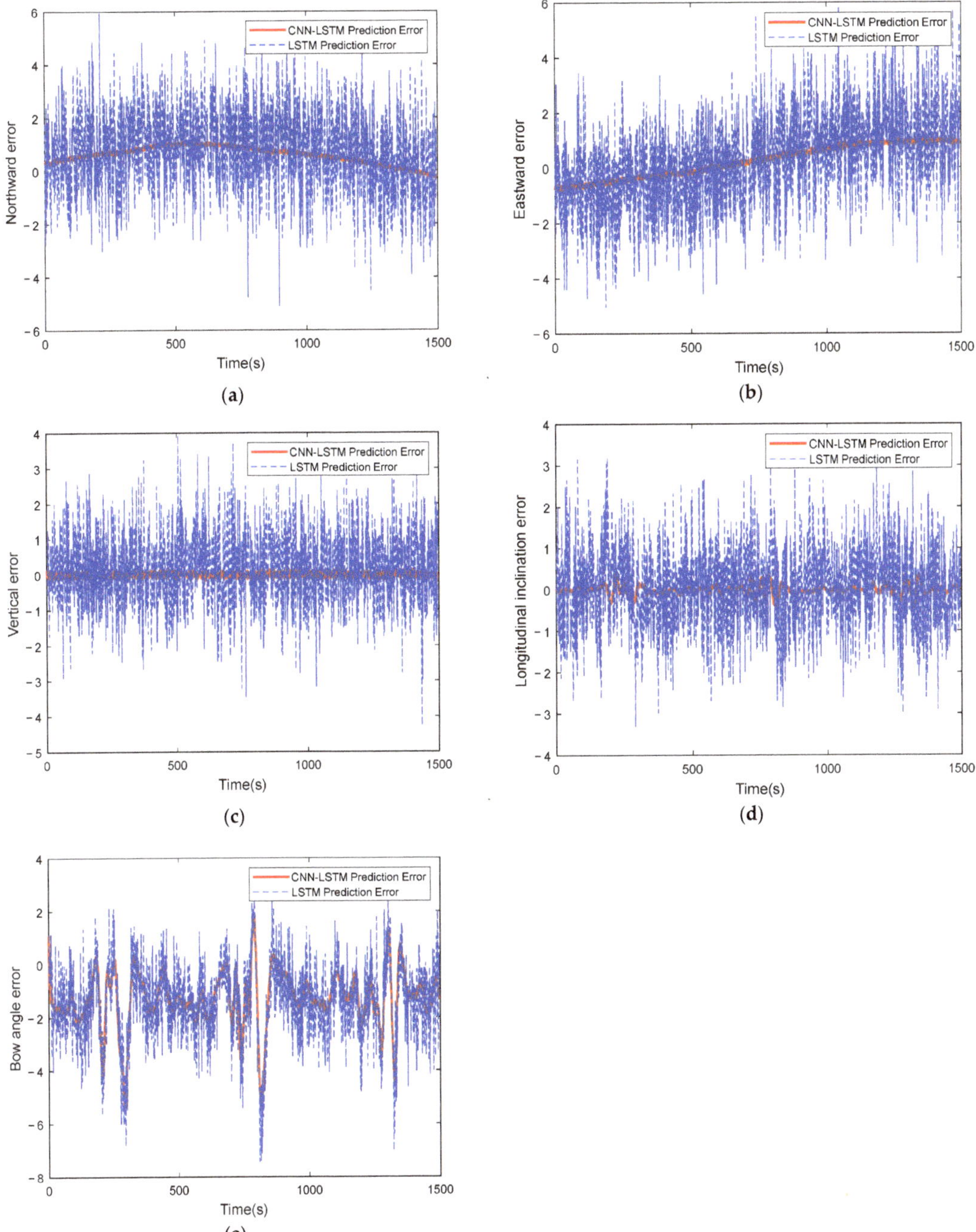

Figure 10. AUV trajectory prediction error: (**a**) northward trajectory prediction error, (**b**) eastward trajectory prediction error, (**c**) vertical trajectory prediction error, (**d**) longitudinal inclination angle prediction error and (**e**) bow angle prediction error.

Based on Figures 10 and 11, it can be observed that the CNN-LSTM model predicts a trajectory that is closer to the actual value, with a smoother prediction curve and lower error fluctuations. These results demonstrate that the CNN-LSTM model provides higher accuracy and stability. The *MSE* values for the predicted states by the LSTM model are 1.7911, 1.7947, 1.1921, 1.6871, and 0.2564, while the CNN-LSTM model predicts the state with lower *MSE* values of 0.6868, 0.6315, 0.0664, 1.3078, and 0.1139. These *MSE* values are smaller compared to those of the pure LSTM model, indicating that the CNN-LSTM model provides better prediction results.

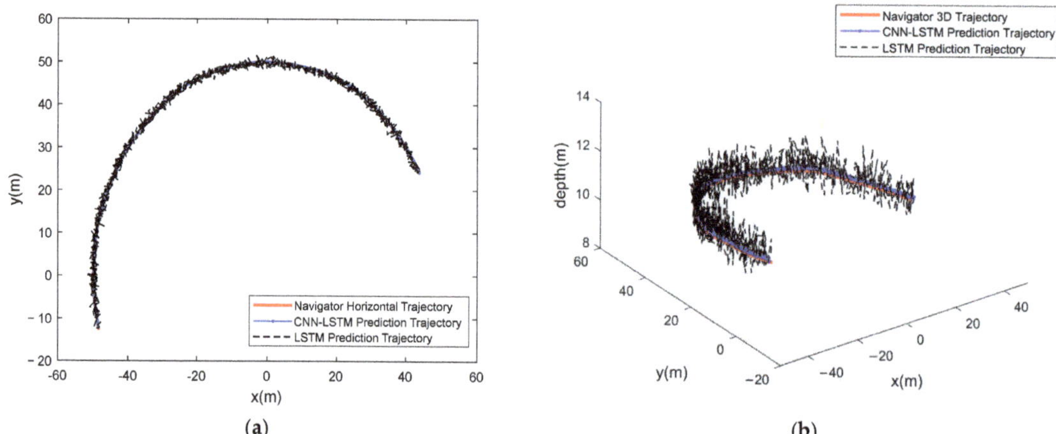

Figure 11. AUV trajectory prediction: (**a**) horizontal plane trajectory prediction and (**b**) 3D trajectory prediction.

5.2. Formatting of Mathematical Components

To verify the prediction effectiveness of the CNN-LSTM prediction model in the formation and formation holding phases of the multi-AUV formation, the communication transmission delay τ_{tran} is set to 1 s and the hydroacoustic sonar sounding delay τ_{inter} is set to 4 s, i.e., the fixed time delay τ_{once_tal} is defined to be 5 s and the maximum total time delay is 9 s. The formation design is consistent with Figure 5.

The navigational track of the navigator is

$$\begin{cases} x_p = 60\cos(2\pi t/1000) \\ y_p = 60\sin(2\pi t/1000) \qquad 0 \leq t \leq 2000 \\ z_p = -0.3t \end{cases} \tag{41}$$

The initial state of the AUV is as follows: initial position $x(0)$ is randomly taken in the range of $[55, 65]$ m, $y(0)$ is randomly taken in the range of $[-10, 10]$ m, $x(0)$ is 65 m, depth is 0 m, initial attitude $\theta(0)$ is 0 rad, bow angle $\psi(0)$ is $4\pi/3$ rad, longitudinal velocity $u(0)$ is 0.5 m/s, all other velocities are initialized to 0 m/s, and controller parameters are $h = 1$, $k_1 = 0.3$, $c_1 = 0.3$, $F = 0.02$, $\beta = 0.5$.

The simulation results are shown in Figures 12 and 13.

In Figure 12, (a) to (e) are the simulation plots of AUV formation position information, from which it can be seen that the leader and the follower always keep the same position, pitch angle and bow angle during the spiral dive under the action of the formation controller; (f) to (j) are the simulation plots of AUV formation speed information, from which it can be seen that the bow speed, lateral speed and vertical speed of the follower in the formation have some fluctuations, but the overall velocity remains stable. Figure 13 shows the 3D trajectory of the AUV formation and its projection on the horizontal plane, from which it can be seen that the followers can follow the leader more accurately and can realize the multi-AUV formation control in a 3D environment. The simulation results illustrate that the formation control method combining CNN-LSTM prediction and backstepping sliding mode control designed in this paper can better realize the three-dimensional predictive control of multi-AUV formation under the communication constraints.

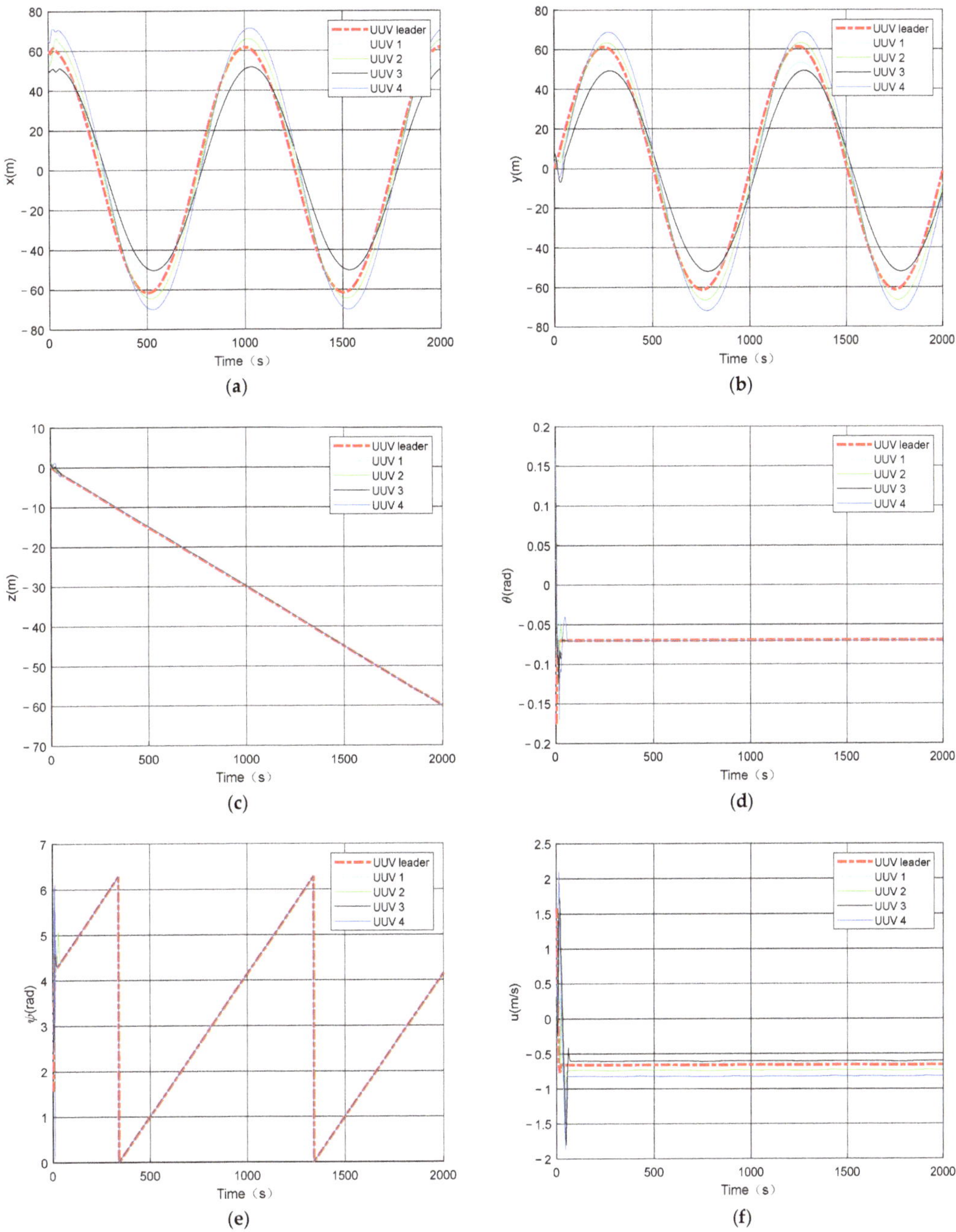

Figure 12. *Cont.*

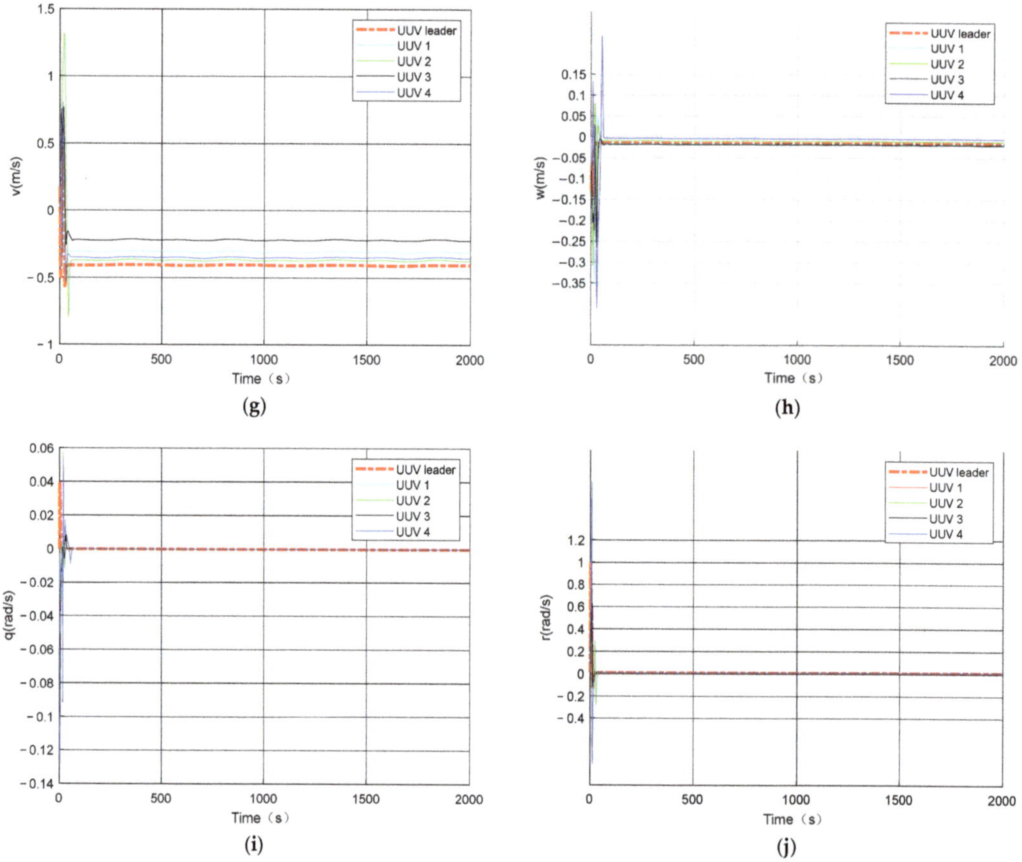

Figure 12. Simulation diagrams of formation position and velocity information: (**a**) AUV northward trajectory, (**b**) AUV eastward trajectory, (**c**) AUV vertical trajectory, (**d**) AUV longitudinal inclination angle state, (**e**) AUV bow angle state, (**f**) AUV longitudinal velocity, (**g**) AUV lateral velocity, (**h**) AUV vertical velocity, (**i**) AUV longitudinal inclination angle velocity and (**j**) AUV bow angle velocity.

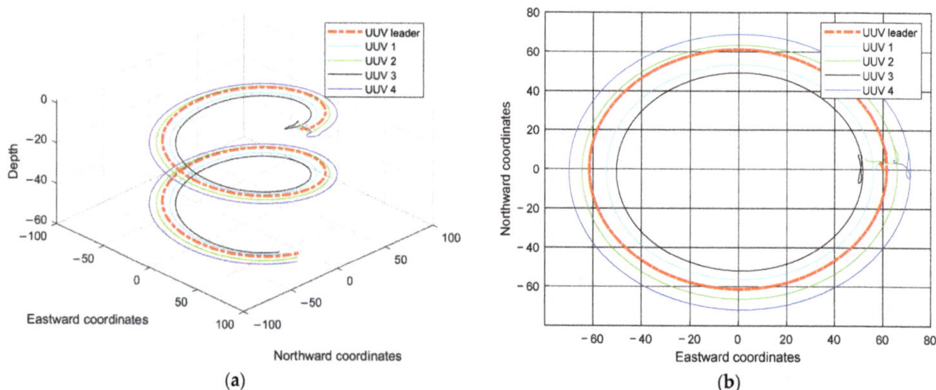

Figure 13. AUV formation 3D trajectory diagram and its horizontal projection: (**a**) 3D trajectory diagram and (**b**) horizontal projection diagram.

Figure 14 shows the position and attitude errors of the AUV formation under CNN-LSTM prediction. Figure 15 shows the position and attitude error of AUV formation under communication delay. From Figures 14a,b and 15a,b, it can be seen that the northward and eastward errors of the AUV formation under predictive control are much smaller than the control errors under delay, indicating that the CNN-LSTM prediction-based AUV formation control method can better overcome the effect of communication delay on formation control.

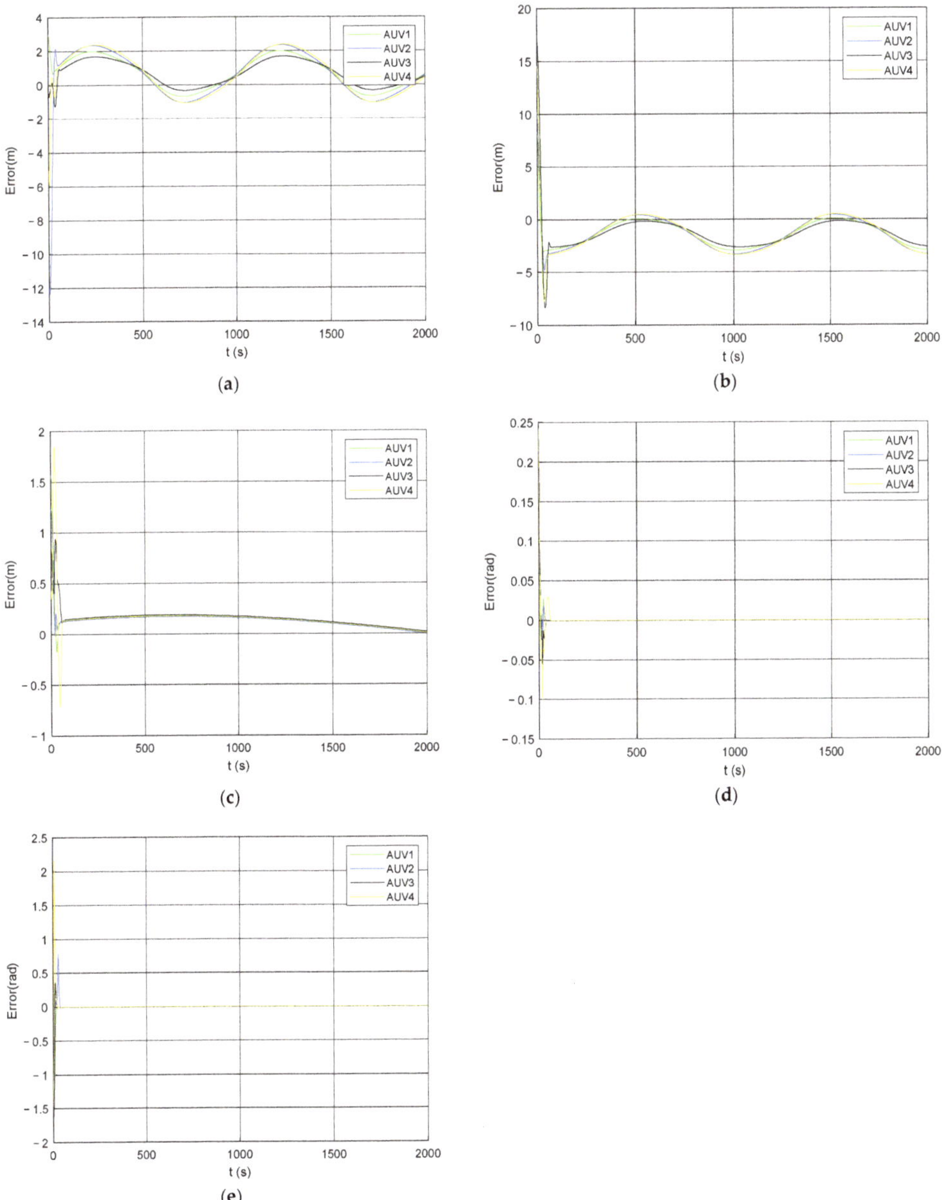

Figure 14. Errors of follower AUV under CNN-LSTM model prediction: (**a**) northward error, (**b**) eastward error, (**c**) vertical error, (**d**) longitudinal inclination angle error and (**e**) bow angle error.

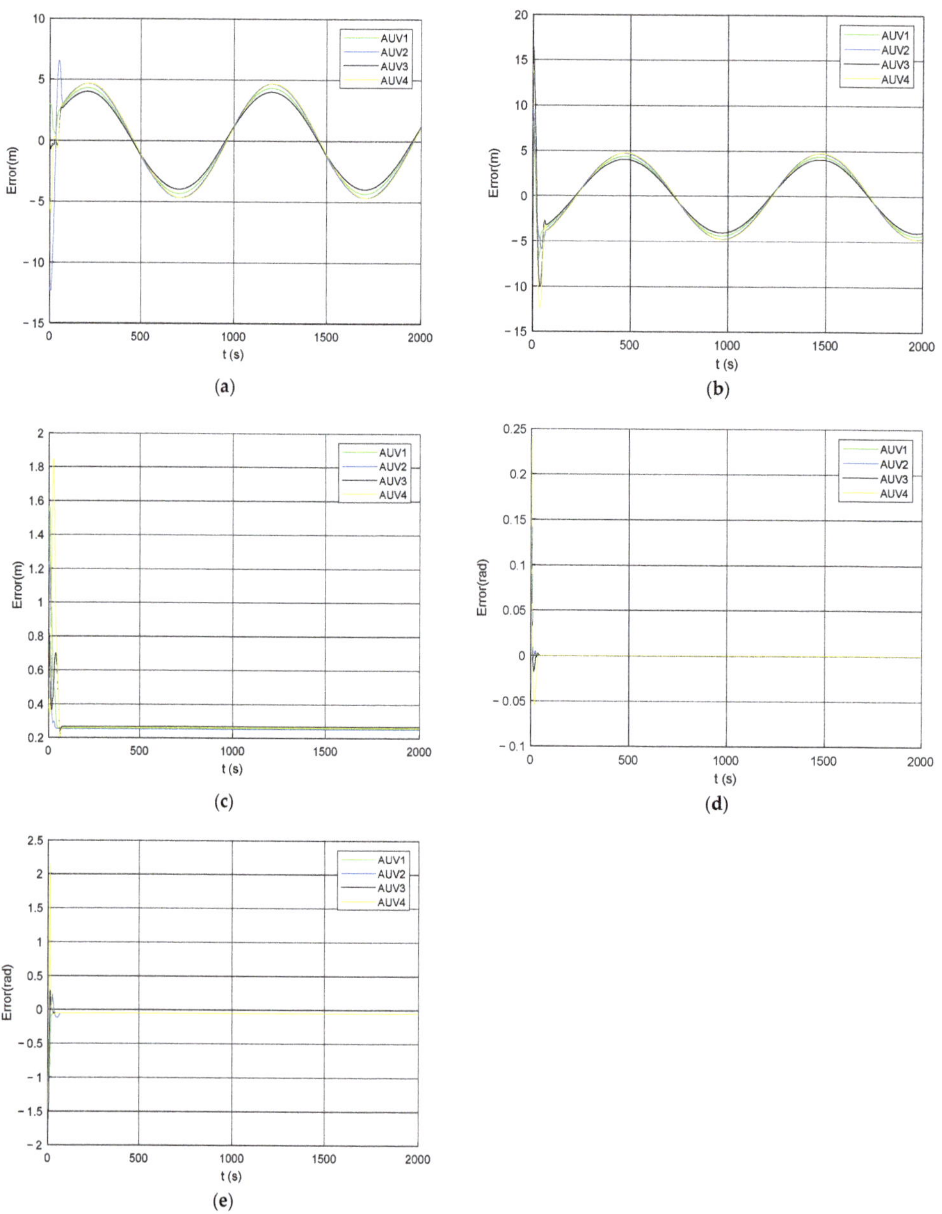

Figure 15. Errors of follower AUV under time delay: (**a**) northward error, (**b**) eastward error, (**c**) vertical error, (**d**) longitudinal inclination angle error and (**e**) bow angle error.

6. Conclusions

This paper focuses on the multi-AUV formation control problem under communication constraints. Firstly, a five-degree-of-freedom nonlinear model of the AUV is established and processed by using feedback linearization to obtain a second-order integral model of the AUV. A sliding-window-based formation prediction control strategy is designed to iteratively predict the current state information of the leader by the historical state information of the leader step by step. The method saves computational efficiency and has better adaptability. The CNN-LSTM prediction model is chosen to predict the trajectory state of the navigator for the characteristics of AUV motion trajectory with certain temporality, which compensates for the influence of communication delay on the formation control; and the backstepping method and sliding mode control are combined to design the formation controller, which improves the robustness of the controller. The stability of the control is proved based on Lyapunov stability theory. The effectiveness of the CNN-LSTM prediction model and the designed controller are verified by simulation.

Author Contributions: Conceptualization, J.L. and Z.T.; methodology, J.L.; software, J.L.; validation, J.L. and Z.T.; formal analysis, J.L.; investigation, J.L. and Z.T.; resources, J.L.; data curation, G.Z.; writing—original draft preparation, J.L.; writing—review and editing, W.L. and Z.T.; visualization, G.Z. and Z.T.; supervision, J.L.; project administration, J.L.; funding acquisition, J.L. All authors have read and agreed to the published version of the manuscript.

Funding: This research was funded by the National Natural Science Foundation of China (Grant No. 5217110503), the Research Fund from Science and Technology on Underwater Vehicle Technology (Grant No. JCKYS2021SXJQR-09) and the Natural Science Foundation of Shandong Province (Grant No. ZR202103070036).

Institutional Review Board Statement: Not applicable.

Informed Consent Statement: Not applicable.

Data Availability Statement: Not applicable.

Conflicts of Interest: The authors declare no conflict of interest.

References

1. Glaviano, F.; Esposito, R.; Di Cosmo, A.; Esposito, F.; Gerevini, L.; Ria, A.; Molinara, M.; Bruschi, P.; Costantini, M.; Zupo, V. Management and Sustainable Exploitation of Marine Environments through Smart Monitoring and Automation. *J. Mar. Sci. Eng.* **2022**, *10*, 297. [CrossRef]
2. Wolek, A.; McMahon, J.; Dzikowicz, B.R.; Houston, B.H. The Orbiting Dubins Traveling Salesman Problem: Planning inspection tours for a minehunting AUV. *Auton. Robot.* **2021**, *45*, 31–49. [CrossRef]
3. Ru, J.; Yu, H.; Liu, H.; Liu, J.; Zhang, X.; Xu, H. A Bounded Near-Bottom Cruise Trajectory Planning Algorithm for Underwater Vehicles. *J. Mar. Sci. Eng.* **2023**, *11*, 7. [CrossRef]
4. Khan, J.U.; Cho, H.-S. Data-Gathering Scheme Using AUVs in Large-Scale Underwater Sensor Networks: A Multihop Approach. *Sensors* **2016**, *16*, 1626. [CrossRef] [PubMed]
5. Liu, H.; Xu, B.; Liu, B. An Automatic Search and Energy-Saving Continuous Tracking Algorithm for Underwater Targets Based on Prediction and Neural Network. *J. Mar. Sci. Eng.* **2022**, *10*, 283. [CrossRef]
6. Chen, T.; Qu, X.; Zhang, Z.; Liang, X. Region-Searching of Multiple Autonomous Underwater Vehicles: A Distributed Cooperative Path-Maneuvering Control Approach. *J. Mar. Sci. Eng.* **2021**, *9*, 355. [CrossRef]
7. Li, J.; Zhai, X.; Xu, J.; Li, C. Target Search Algorithm for AUV Based on Real-Time Perception Maps in Unknown Environment. *Machines* **2021**, *9*, 147. [CrossRef]
8. Mao, S.; Yang, P.; Gao, D.; Bao, C.; Wang, Z. A Motion Planning Method for Unmanned Surface Vehicle Based on Improved RRT Algorithm. *J. Mar. Sci. Eng.* **2023**, *11*, 687. [CrossRef]
9. Chen, Y.; Wu, W.; Jiang, P.; Wan, C. An Improved Bald Eagle Search Algorithm for Global Path Planning of Unmanned Vessel in Complicated Waterways. *J. Mar. Sci. Eng.* **2023**, *11*, 118. [CrossRef]
10. Yan, Z.; Yue, L.; Zhou, J.; Pan, X.; Zhang, C. Formation Coordination Control of Leaderless Multi-AUV System with Double Independent Communication Topology and Nonconvex Control Input Constraints. *J. Mar. Sci. Eng.* **2023**, *11*, 107. [CrossRef]
11. Yu, H.; Zeng, Z.; Guo, C. Coordinated Formation Control of Discrete-Time Autonomous Underwater Vehicles under Alterable Communication Topology with Time-Varying Delay. *J. Mar. Sci. Eng.* **2022**, *10*, 712. [CrossRef]
12. Kang, X.; Xu, H.; Feng, X. Fuzzy Logic Based Behavior Fusion for Multi-AUV Formation Keeping in Uncertain Ocean Environment. In Proceedings of the OCEANS 2009, Biloxi, MS, USA, 26–29 October 2009; pp. 1–7.
13. Borhaug, E.; Pavlov, A.; Pettersen, K.Y. Straight line path following for formations of underactuated underwater vehicles. In Proceedings of the 2007 IEEE 46th IEEE Conference on Decision and Control, New Orleans, LA, USA, 12–14 December 2007.
14. Ding, G.; Zhu, D.; Sun, B. Formation control and obstacle avoidance of multi-AUV for 3-D underwater environment. In Proceedings of the 2014 33rd Chinese Control Conference, Nanjing, China, 28–30 July 2014.
15. Li, J.; Zhang, Y.; Li, W. Formation Control of a Multi-Autonomous Underwater Vehicle Event-Triggered Mechanism Based on the Hungarian Algorithm. *Machines* **2021**, *9*, 346. [CrossRef]

16. Chen, Y.; Guo, X.; Luo, G.; Liu, G. A Formation Control Method for AUV Group Under Communication Delay. *Front. Bioeng. Biotechnol.* **2022**, *10*, 848641. [CrossRef] [PubMed]
17. Suryendu, C.; Subudhi, B. Formation Control of Multiple Autonomous Underwater Vehicles Under Communication Delays. In Proceedings of the 2020 IEEE Transactions on Circuits and Systems II: Express Briefs, Goa, India, 28 February 2020.
18. Yang, S.; Chen, J.; Liu, F. Observer-Based Consensus Control of Multi-Agent Systems with Input Delay. In Proceedings of the 2018 IEEE 4th International Conference on Control Science and Systems Engineering (ICCSSE), Wuhan, China, 21–23 August 2018.
19. Liu, Y.B. Research on Coordination Control of Multiple Underwater Vehicles for Ocean Exploratio. Ph.D. Thesis, Harbin Engineering University, Harbin, China, 2017. (In Chinese)
20. He, Y.Y.; Yan, M.D. *Nonlinear Control Theory and Application*; Electronic Science and Technology University Press: Xi'an, China, 2007; pp. 115–119. (In Chinese)

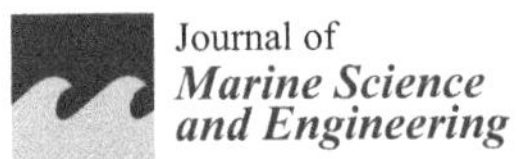
Journal of
Marine Science and Engineering

Article

Output Feedback Tracking Control with Collision Avoidance for Dynamic Positioning Vessel under Input Constraint

Benwei Zhang [1,2,*] and Guoqing Xia [1]

[1] College of Intelligent Systems Science and Engineering, Harbin Engineering University, Harbin 150001, China; xiaguoqing@hrbeu.edu.cn
[2] China Classification Society, Beijing 100007, China
* Correspondence: bwzhang@ccs.org.cn

Abstract: This dissertation presents a fresh control strategy for dynamic positioning vessels exposed to model uncertainty, various external disturbances, and input constraint. The vessel is supposed to work in a particular situation surrounding a lighthouse or a submerged reef, where collision avoidance must be prevented. The control strategy involves making the vessel navigate under the action of modified artificial potential functions (MAPFs) along a smooth trajectory. To achieve this goal, we put forward a collision-avoidance control strategy, which consists of the backstepping technique, an extended state observer (ESO), and an active dynamic positioning control technique. The MAPFs, together with a strategy, are applied to realize collision avoidance. To address the input constraint problem, an auxiliary dynamic system (ADS) is constructed. Entire related signals of the control system could converge to a small neighboring zone of the equilibrium state via Lyapunov deduction. Simulation outcomes verify the effectiveness of the presented control strategy.

Keywords: dynamic positioning vessels; collision avoidance; input constraint

1. Introduction

As an increasingly widely used tool for exploring and exploiting marine resources, the motion control strategy of dynamic positioning vessels has attracted increasing attention [1–4]. The study and investigation on the control method of dynamic positioning vessels have been sufficient. However, due to model uncertainty and the effects of environmental disturbances, undertaking research remains challenging [5–7]. To deal with this challenge, several control methods have been proposed for the tracking control of dynamic positioning vessels. In general, control schemes used for compensating the model uncertainty and disturbances can be classified as passive and active approaches. The active approaches exploit the estimation of the model uncertainty and disturbances to compensate for these effects [8], while the passive approaches rely on the robustness property of the designed controller [9–11]. To reduce the ship roll motion in waves, a self-tuning proportional integral derivative (PID) controller is used to adjust optimal stabilizer fin angles in [12]. However, the controller has poor performance to deal with the high magnitude of the disturbance values. In order to increase the performance, fuzzy logic systems and intelligent control techniques have been proposed in [13,14]. The fuzzy logic systems depend on a good human understanding of the dynamical behavior of the system. Model predictive control is employed to ensure optimal system performance [15], and the method algorithm requires a large number of account operations and resources. To solve this problem and enhance robustness property, sliding mode control is proposed for the tracking control of dynamic positioning vessels [16]. To reduce the effects of uncertainty and disturbances, an active approach is designed based on observers. In order to estimate uncertainty and disturbances, the disturbance observer is proposed for the tracking control of vessels [17]; the key features of disturbance observer are based on velocity measurements. The position and heading of vessels can be acquired by using GPS and electronic compass, although

Citation: Zhang, B.; Xia, G. Output Feedback Tracking Control with Collision Avoidance for Dynamic Positioning Vessel under Input Constraint. *J. Mar. Sci. Eng.* **2023**, *11*, 811. https://doi.org/10.3390/jmse11040811

Academic Editor: Gerasimos Theotokatos

Received: 10 March 2023
Revised: 30 March 2023
Accepted: 5 April 2023
Published: 11 April 2023

it is difficult to obtain their velocities. To avoid using velocity measurements, a nonlinear extended state observer (ESO) is employed for dynamic positioning vessel in [18].

Besides model uncertainty and disturbances, another challenge to tracking control for dynamic positioning vessel is input constraint. Due to physical limitations, the actuator outputs are constantly bounded or constrained [19]. Input constraint may degrade the system performance and lead to an imbalance [20]. To deal with this challenge, a Gaussian error function is used to approximate input saturation in [21]. The hyperbolic tangent function is used to handle the input saturation in [22], and a trajectory tracking control law is designed based on the hyperbolic tangent function. However, the hyperbolic tangent function and Gaussian error function limit system performance, even if the control inputs are not saturated. To avoid this, an anti-windup compensator is incorporated into controller design to solve trajectory tracking of dynamic positioning vessel with input saturation in [23]. To solve the input saturation in trajectory tracking control of vessels, an auxiliary dynamic system (ADS) is applied to controller design in [24]. Ref. [25] introduces an auxiliary control to solve the formation control of autonomous underwater vehicles with input saturation.

However, few papers take the problem of collision-avoidance control into account. In the view of engineering practice, it is inevitable for unmanned vessels to encounter fixed or moving obstacles such as lighthouses and submerged reefs. Collision avoidance is crucial to navigation safety. The existing achievements often use a path-planning algorithm to solve the collision-avoidance problem [26–29]. The method depends on an optimization algorithm, which is complex and takes a long time to obtain a feasible solution, so it is not conducive to lowering the construction cost of dynamic positioning vessels. In order to address this problem, some scholars design the control system of unmanned vessels with collision avoidance [30,31]. However, input saturation is neglected in the controller-design process, and the performance of the actuator has a great impact on collision-avoidance ability. Therefore, input constraint also needs to be considered when proposing the collision-avoidance control algorithm.

Motivated by the above research background, this paper investigates the trajectory tracking control of surface vessels subject to input saturation, uncertainty, and environmental disturbances. Even more significantly, both static obstacles and unknown non-cooperative ships are taken into consideration. Inspired by the works in [30], an improved output feedback controller is proposed. Firstly, an ESO is used to estimate the velocity, model uncertainty, and disturbances. To avoid collisions with obstacle and unknown non-cooperative ships, modified artificial potential functions (MAPFs) are incorporated into the controller-design process. In particular, MAPFs can render the vessel bypass obstacle and unknown non-cooperative ships smoothly, which greatly reduces the actuator performance requirements. An ADS is introduced to solve the input saturation. Then, an output feedback controller is designed based on the ESO, the MAPFs, and the ADS. Finally, the stability of closed-loop system is proved. Simulations demonstrate the proposed control strategy.

Three salient features of this paper are summarized as follows. First, MAPFs are designed to solve the problem of automatic collision avoidance system for dynamic positioning vessels. Second, both obstacle and unknown non-cooperative ships are considered in the controller-design process. The embedded processor of the vessels carries out reading computations using only the position and yaw, and the transmission loads are saved. Third, the designed strategy can achieve collision avoidance, even if the control inputs are saturated.

Compared with the existing related results, the novelty of this paper is summarized as follows. First, compared with the control strategy [20–23], this paper designs a strategy to achieve collision avoidance between a vessel and an obstacle. Second, unlike the existing control approaches [30,31], input saturation is considered in the controller-design process; the proposed controller has become more attractive in practice engineering. Third, compared with the collision-avoidance algorithm proposed in [18], a non-cooperative ship is also considered.

The other sections are organized as follows. The problem is formulated in Section 2. Section 3 provides a collision-avoidance strategy, Section 4 presents an observer design, and Section 5 presents a controller design and stability analysis. Simulation outcomes are described in Section 6. Conclusions are given in Section 7.

2. Problem Formulation

The kinematic (position and orientation) and dynamic (linear velocity and angular velocity) models of dynamic positioning vessel are [32]

$$\dot{\eta} = R(\psi)v \tag{1}$$

$$M\dot{v} + Cv + Dv = \tau + d, \tag{2}$$

where $\eta = [x, y, \psi]^T$, x, and y are vertical and horizontal coordinates in the earth-fixed reference frame, respectively. ψ is the yaw angle in the earth-fixed reference frame. $v = [u, v, r]^T$ denotes the velocity vector containing linear velocity u, v and the angular velocity r. $M \in \mathbb{R}^{3 \times 3}$ denotes the inertia matrix including added mass, which is positive definite, invertible, and constant. $C \in \mathbb{R}^{3 \times 3}$ represents a skew-symmetric matrix of Coriolis and centripetal term. $D \in \mathbb{R}^{3 \times 3}$ denotes the damping matrix, which is positive definite. $\tau = [\tau_u, \tau_v, \tau_r]^T$ is the input signal. $d = [d_1, d_2, d_3]^T$ are the environmental disturbances. $R(\psi)$ is the rotation matrix and given by

$$R(\psi) = \begin{bmatrix} \cos(\psi) & -\sin(\psi) & 0 \\ \sin(\psi) & \cos(\psi) & 0 \\ 0 & 0 & 1 \end{bmatrix}.$$

In practical application, due to the limited engine power, the input signals are limited by maximum force and moment. The input limitation is given:

$$\tau = \begin{cases} \tau_{max} & \tau_c > \tau_{max} \\ \tau_c & \tau_{min} \leq \tau_c \leq \tau_{max} \\ \tau_{min} & \tau_c < \tau_{min} \end{cases}$$

where τ_{max} and $\tau_{min} \in \mathbb{R}^3$ are the maximum and minimum control force and moment, respectively. The mismatch function is defined as $\varpi = \tau_c - \tau$, $\tau_c = [\tau_{uc}, \tau_{vc}, \tau_{rc}]^T$, τ_{uc}, τ_{vc} and τ_{rc} are control force and moment calculated by the proposed controller, respectively. The saturated input signal can be attained by $\tau = \tau_c - \varpi$.

3. Collision Avoidance Strategy

The collision-avoidance ability of the control system plays an important role in ship-collision avoidance and ensuring the safety of ship navigation, and artificial potential functions are used for achieving collision avoidance [33,34]. Here, the MAPFs are designed; they consist of a repulsion function, and they do not work when the vessel is sufficiently far away from obstacle and other ship. The MAPFs repel the vessel when the vessel approaches the obstacle and the other ship, which keeps a safe distance from the obstacle and the other ship. The obstacle is modeled as a circle-shaped object [35], and the MAPFs are given as follows:

$$p_o = \begin{cases} \frac{1}{2}\alpha_o(\frac{1}{p_v} - \frac{1}{\bar{R}_o})^2, & \text{if } p_v \leq \bar{R}_o \\ 0, & \text{otherwise} \end{cases} \tag{3}$$

where $p_v = \|p - p_o\|$ is the distance between the vessel and the obstacle , $p = [x, y]^T$ is the position of the vessel, $p_o = [x_o, y_o]^T$ is the center of the obstacle, and α_o is a positive constant. $\bar{R}_o = \max\{R_o, R_u\}$, R_o is the collision avoidance range with respect to the obstacle, and R_u is the collision avoidance range with respect to the vessel.

To avoid collision with an obstacle, the MAPFs are introduced to the control objective. Additionally, the control objective is given as follows:

$$\lim_{t \to +\infty} \|\eta - \eta_d - \eta_o\| < c_1 \tag{4}$$

where η_d is the desired trajectory, $\eta_o = [p_o, p_o, 0]^{\mathrm{T}}$.

Hypothesis 1. *η_d and its derivative $\dot{\eta}_d$ are bounded.*

Remark 1. *The MAPFs do not work when the vessel is sufficiently far away from obstacle . This means that $\eta_o = [0, 0, 0]^{\mathrm{T}}$ if $p_v > \bar{R}_o$, and the control objective is a trajectory tracking task. The MAPFs work and repel the vessel when it approaches the obstacle. The design of η_o ensures that the MAPFs can render the vessel bypass obstacle smoothly, which leads to conservative input signals.*

4. Observer Design

In this section, we develop a nonlinear observer to estimate the unknown term containing environmental disturbances d, the damping matrix D, and the Coriolis and centripetal term C; the model of USV can be rewritten as

$$\dot{\eta} = R(\psi)v \tag{5}$$

$$\dot{v} = M^{-1}\tau + \zeta, \tag{6}$$

where $\zeta = M^{-1}(-Cv - Dv + d)$.

Hypothesis 2. *There is a positive constant ζ^* such that $\|\dot{\zeta}\| \le \zeta^*$.*

Inspired by [36], an ESO is given to provide the estimations of model uncertainties and disturbances, which is shown as:

$$\dot{\hat{\eta}} = - K_{o1}(\hat{\eta} - \eta) + R(\psi)\hat{v}, \tag{7}$$

$$\dot{\hat{v}} = - K_{o2}R^{\mathrm{T}}(\psi)(\hat{\eta} - \eta) + \hat{\zeta} + M^{-1}\tau, \tag{8}$$

$$\dot{\hat{\zeta}} = - K_{o3}R^{\mathrm{T}}(\psi)(\hat{\eta} - \eta) \tag{9}$$

where $\hat{\eta} = [\hat{x}, \hat{y}, \hat{\psi}]^{\mathrm{T}} \in \mathbb{R}^3$, $\hat{v} = [\hat{u}, \hat{v}, \hat{r}]^{\mathrm{T}} \in \mathbb{R}^3$, $\hat{\zeta} = [\hat{\zeta}_1, \hat{\zeta}_2, \hat{\zeta}_3]^{\mathrm{T}} \in \mathbb{R}^3$, $\hat{x}, \hat{y}, \hat{\psi}, \hat{u}, \hat{v}, \hat{r}, \hat{\zeta}_1, \hat{\zeta}_2, \hat{\zeta}_3$ are the estimates of $x, y, \psi, u, v, r, \zeta_1, \zeta_2$ and ζ_3. $K_{o1} \in \mathbb{R}^{3\times3}$, $K_{o2} \in \mathbb{R}^{3\times3}$, and $K_{o3} \in \mathbb{R}^{3\times3}$ are gain matrices.

From (5)–(9), the error dynamics of the ESO are given as

$$\dot{\tilde{\eta}} = -K_{o1}\tilde{\eta} + R(\psi)\tilde{v},$$
$$\dot{\tilde{v}} = -K_{o2}R^{T}(\psi)\tilde{\eta} + \tilde{\zeta}, \tag{10}$$
$$\dot{\tilde{\zeta}} = -K_{o3}R^{T}(\psi)\tilde{\eta} - \dot{\zeta},$$

where $\tilde{\eta} = \hat{\eta} - \eta$, $\tilde{v} = \hat{v} - v$ and $\tilde{\zeta} = \hat{\zeta} - \zeta$ are estimation errors.

The observer error dynamics (10) can be rewritten as

$$\dot{X} = AX + B\phi \tag{11}$$

$$\tilde{\eta} = C_o X, \tag{12}$$

where $X = [\tilde{\eta}^T, \tilde{v}^T, \tilde{\zeta}^T]^T \in \mathbb{R}^{9\times1}$,

$$A = \begin{bmatrix} -K_{o1} & R(\psi) & 0_3 \\ -K_{o2}R^T(\psi) & 0_3 & I_3 \\ -K_{o3}R^T(\psi) & 0_3 & 0_3 \end{bmatrix}, \tag{13}$$

$$B = \begin{bmatrix} 0_3 \\ 0_3 \\ I_3 \end{bmatrix}, \tag{14}$$

$$C_{oi} = \begin{bmatrix} I_3 & 0_3 & 0_3 \end{bmatrix}, \tag{15}$$

$\phi = -\dot{\zeta}$. 0_3 is a 3×3 dimensional zero matrix. I_3 is a 3×3 dimensional identity matrix.

According to Hypothesis 2, we can devise a hypothesis that $\|\phi\| \leq \bar{\phi}$ and $\bar{\phi}$ is a positive constant.

A nonlinear term $J(\psi)$ makes it difficult to conduct the stability analysis of the ESO. To solve the difficulty, a transformation $\chi = TX$ with $T = \mathrm{diag}\{J^T(\psi), I_3, I_3\}$ is introduced, thus

$$\dot{\chi} = (A_0 + rS_T)\chi + B\phi, \tag{16}$$

where $S_T = \mathrm{diag}\{S_1, 0_3, 0_3\}$,

$$S_1 = \begin{bmatrix} 0 & 1 & 0 \\ -1 & 0 & 0 \\ 0 & 0 & 0 \end{bmatrix}, \tag{17}$$

$$A_0 = \begin{bmatrix} -K_{o1} & I_3 & 0_3 \\ -K_{o2} & 0_3 & I_3 \\ -K_{o3} & 0_3 & 0_3 \end{bmatrix}. \tag{18}$$

Lemma 1. *The error of the ESO X is bounded, if there are symmetric definite positive matrices* $Q, P \in \mathbb{R}^{9 \times 9}$ *such that*

$$\begin{aligned} A_0^T P + PA_0 + PBB^T P + Q \\ + \bar{r}(S_T^T P + PS_T) \leq 0, \end{aligned} \tag{19}$$

$$\begin{aligned} A_0^T P + PA_0 + PBB^T P + Q \\ - \bar{r}(S_T^T P + PS_T) \leq 0, \end{aligned} \tag{20}$$

$\bar{r} \in \mathbb{R}$ is the upper bound of r.

Chose a Lyapunov function as follows:

$$V_o = \frac{1}{2}\chi^T P\chi. \tag{21}$$

We can obtain that $\dot{V}_o = \frac{1}{2}(\chi^T P\dot{\chi} + \dot{\chi}^T P\chi)$. Form (16), we can obtain that

$$\dot{V}_o = \frac{1}{2}(\chi^T P((A_0 + rS_T)\chi + B\phi)) + \frac{1}{2}((A_0 + rS_T)\chi + B\phi)^T P\chi \tag{22}$$

By using $((A_0 + rS_T)\chi)^T P\chi = \chi^T(A_0^T p + rS_T^T P)\chi$ and $(B\phi)^T P\chi = \chi^T PB\phi$, differentiating V_o with respect to time is given:

$$\begin{aligned} V_o &= \frac{1}{2}\chi^T(PA_0 + A_0^T P + rPS_T + rS_T^T P)\chi \\ &\quad + \chi^T PB\phi \\ &\leq \frac{1}{2}\chi^T(PA_0 + A_0^T P + rPS_T + rS_T^T P \\ &\quad + PBB^T P)\chi + \frac{1}{2}\phi^T\phi \\ &\leq -c_{o_1}V_o + c_{o_2}, \end{aligned} \tag{23}$$

where $c_{o_1} = \frac{\lambda_{\min}(Q)}{\lambda_{\max}(P)}$, $c_{o_2} = \frac{1}{2}\bar{\phi}^2$. Since c_{o_2} is symmetric definite positive, the state χ_i is bounded. Using $\|T^{-1}\| = 1$ and $X = T^{-1}\chi$, the estimation error signal X is bounded .

5. Controller Design

In order to clearly describe the process of controller design, the diagram of the proposed output feedback controller is given in Figure 1. The designed ESO provides estimations of velocities, model uncertainty, and disturbances by using position of the vessel. A filter is used to give the velocity estimations of obstacle or non-cooperative ship. An ADS is designed based on the mismatch function $\varpi = \tau_c - \tau$. The proposed kinetic control law is designed by using the ADS and the ESO.

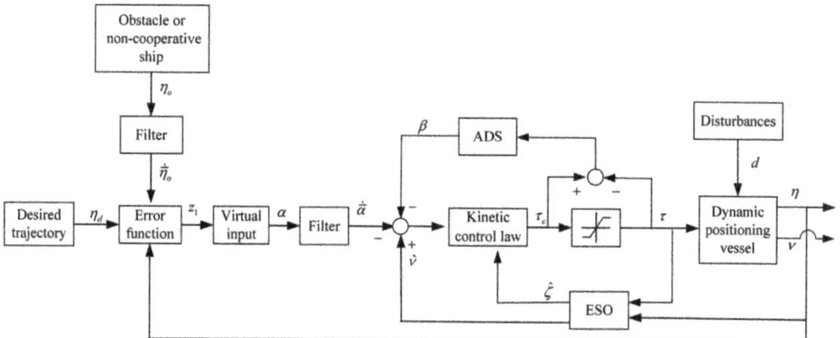

Figure 1. The diagram of the proposed output feedback controller.

An output feedback controller design for vessel with collision avoidance is presented step by step.

Step 1: A modified error function based on the information of desired trajectory and obstacle is defined as

$$z_1 = \eta - \eta_d - \eta_o. \tag{24}$$

Along with the kinematics (5), the time derivative of (24) is given as

$$\dot{z}_1 = R(\psi)v - \dot{\eta}_d - \dot{\eta}_o. \tag{25}$$

Choosing α as a virtual input, the kinematic control law α is proposed as

$$\alpha = R^{\mathrm{T}}(\psi)(\dot{\eta}_d + \dot{\bar{\eta}}_o - K_1 z_1), \tag{26}$$

where $K_1 \in \mathbb{R}^{3\times3}$ is a diagonal matrix, $\dot{\eta}_d$ is known time derivative of η_d, $\dot{\bar{\eta}}_o$ is the signal originated from the filter, and $l_o\dot{\bar{\eta}}_o + \bar{\eta}_o = \eta_o$, l_o is a positive constant.

Remark 2. *The virtual input (26) is a virtual velocity signal. If z_1 and $\dot{\bar{\eta}}_o \to [0,0,0]^T$, $\alpha \to R^{\mathrm{T}}(\psi)\dot{\eta}_d$, which is the desired velocity in the hull coordinate system. If $z_1 \to [0,0,0]^T$ and $\dot{\bar{\eta}}_o \neq [0,0,0]^T$, $\alpha \to R^{\mathrm{T}}(\psi)(\dot{\eta}_d + \dot{\bar{\eta}}_o)$, the velocity of the vessel converges to the vector piled up by the desired velocity and the velocity of obstacle. The term $-K_1 z_1$ ensures the convergence of the signal α, the therms $\dot{\eta}_d + \dot{\bar{\eta}}_o$ give equilibrium points of the control system, and the $R^{\mathrm{T}}(\psi)$ is the transform matrix between the earth-fixed reference frame and the hull coordinate system.*

Remark 3. *Narrow channels and coastal waters form regions with heavy maritime traffic. Small sea vehicles do not have navigational aids such as VHF radio or AIS [37]. Users are unable to gain the velocity information. We refer to these actors as non-cooperative ships. Non-cooperative ships need to be taken into account in controller design. To estimate the velocity information of non-cooperative ships, a filter is introduced to the process of controller design. Errors are caused* *when the velocity estimation is used. Errors are bounded due to position-bounded η_o. This shows that enough safety distance between dynamic positioning vessels and non-cooperative ships or obstacles is necessary.*

Step 2: In this step, a dynamic controller at the kinetic level is developed. Define the second error vector as

$$z_2 = \hat{v} - \bar{\alpha} - \beta, \tag{27}$$

where β is signal of ADS; it is designed to solve input saturation. $\bar{\alpha}$ is introduced to avoid the calculation of the time derivative of α; a first-order filter is used instead of the time derivative of α, $l\dot{\bar{\alpha}} + \bar{\alpha} = \alpha$ with time constant $l > 0$. The update law of β is given as

$$\dot{\beta} = -K_2\beta - M^{-1}\omega \tag{28}$$

where $K_2 \in \mathbb{R}^{3\times3}$ is a diagonal matrix.

The time derivative of z_2 is derived as

$$\dot{z}_2 = -K_{o2}R^{\mathrm{T}}(\psi)\tilde{\eta} + \hat{\zeta} + M^{-1}(\tau_c - \omega) - \dot{\bar{\alpha}} - \dot{\beta}. \tag{29}$$

To stabilize z_2, a kinetic control law is designed as

$$\tau_c = M(-K_3 z_2 - \hat{\zeta} + \dot{\bar{\alpha}} + K_2\beta - R^T(\psi)z_1), \tag{30}$$

where $K_3 \in \mathbb{R}^{3\times3}$ is a diagonal matrix.

Lemma 2. *Consider the system consisting of the dynamic positioning vessel dynamics (1) and (2); the kinetic control law (30); the observer (7), (8), and (9); and the ADS (28) with input saturation, environmental disturbances, and model uncertainty. If Hypotheses 1–2 are satisfied, the proposed output feedback-control scheme guarantees that all of the signals in the closed-loop system are bounded.*

Proof. Choose a Lyapunov function candidate as follows:

$$V_1 = \frac{1}{2}z_1^T z_1. \tag{31}$$

Using (25), the time derivative of V_1 is given

$$\dot{V}_1 = z_1^{\mathrm{T}}[R(\psi)v - \dot{\eta}_d - \dot{\eta}_o]. \tag{32}$$

By applying (27) to (32), the time derivative of V_1 is rewritten as

$$\dot{V}_1 = z_1^{\mathrm{T}}[R(\psi)(z_2 + \alpha + \beta + \bar{\alpha} - \tilde{v}) - \dot{\eta}_d - \dot{\eta}_o]. \tag{33}$$

Substituting (26) into (33), the time derivative of V_1 can be rewritten as

$$\dot{V}_1 = -z_1^{\mathrm{T}}K_1 z_1 + z_1^{\mathrm{T}}[R(\psi)(z_2 + \beta + \bar{\alpha} - \tilde{v})] + \Delta_v, \tag{34}$$

where Δ_v is a positive constant containing estimate error of the velocity information of non-cooperative ships.

Define a another Lyapunov function

$$V_2 = V_1 + \frac{1}{2}z_2^T z_2. \tag{35}$$

Form (8), (27), and (34), the time derivative of V_2 can be attained

$$\dot{V}_2 = -z_1^T K_1 z_1 + z_1^T [R(\psi)(z_2 + \beta + \tilde{\alpha} - \tilde{v})] \\ + z_2^T [-K_{o2} R^T(\psi)\tilde{\eta} + \hat{\zeta} + M^{-1}(\tau_c - \omega) \\ -\dot{\tilde{\alpha}} - \dot{\beta}] + \Delta_v \tag{36}$$

Using equations (30) and (28),

$$\dot{V}_2 = -z_1^T K_1 z_1 + z_1^T [R(\psi)(\beta + \tilde{\alpha} - \tilde{v})] \\ - z_2^T K_{o2} R^T(\psi)\tilde{\eta}. \tag{37}$$

Define the last Lyapunov function

$$V_3 = V_2 + \frac{1}{2}\beta^T \beta + V_o. \tag{38}$$

Using (28) and (37) yields, the time derivative of V_3 is given

$$\dot{V}_3 = -z_1^T K_1 z_1 - z_2^T K_3 z_2 + z_1^T [R(\psi)(\beta + \tilde{\alpha} - \tilde{v})] \\ - z_2^T K_{o2} R^T(\psi)\tilde{\eta} - \beta^T K_2 \beta - \beta M^{-1}\omega + \Delta_v + \dot{V}_o. \tag{39}$$

Using Young's inequality, the inequalities are given as

$$z_1^T [R(\psi)(\beta + \tilde{\alpha} - \tilde{v})] \leq \frac{1}{2}(3z_1^T z_1 + \beta^T \beta + \tilde{\alpha}^T \tilde{\alpha} + \tilde{v}^T \tilde{v}), \tag{40}$$

$$-z_2^T K_{o2} R^T(\psi)\tilde{\eta} \leq \lambda_{\min}(K_{o2})/2(z_2^T z_2 + \tilde{\eta}^T \tilde{\eta}), \tag{41}$$

$$-\beta M^{-1}\omega \leq 1/2(\beta^T \beta + \Delta), \tag{42}$$

where $\Delta = M^{-1}\omega^T \omega M^{-1}$.

Substituting (40)–(42) into (39), the time derivative of V_3 is rewritten as

$$\dot{V}_3 \leq -[\lambda_{\min}(K_1) - 3/2]z_1^T z_1 - [\lambda_{\min}(K_3) - \lambda_{\min}(K_{o2})/2]z_2^T z_2 \\ - [\lambda_{\min}(K_2) - 1]\beta^T \beta - c_{o3} V_o + c_{o4}, \tag{43}$$

where $c_{o3} = [\lambda_{\min}(Q) - \max\{\lambda_{\min}(K_{o2}), 1\}]/\lambda_{\max}(P)$, $c_{o4} = c_{o2} + \tilde{\alpha}^T \tilde{\alpha} + \Delta/2 + \Delta_v$.
The inequality (43) becomes

$$\dot{V}_3 \leq -c_1 V_3 + c_2, \tag{44}$$

where $c_1 = \min\{2[\lambda_{\min}(K_1) - 3/2], 2[\lambda_{\min}(K_3) - \lambda_{\min}(K_{o2})/2], 2[\lambda_{\min}(K_2) - 1], c_{o3}\}$, $c_2 = c_{o4}$.

From the definition of V_3, it can be concluded that all of the signals in the closed-loop are bounded. $\square$

6. Simulation Results

In this section, numerical simulations are designed to show the effectiveness of the proposed controller (PC). The reference signal η_d is used to generate desired trajectory. In simulations, the model of dynamic positioning ship Cybership II is used [38]. The control forces and moment are constrained as $\tau_{\max} = [2\,\text{N}, 2\,\text{N}, 1.5\,\text{Nm}]$, $\tau_{\min} = [-2\,\text{N}, -2\,\text{N}, -1.5\,\text{Nm}]$. Environmental disturbances are modeled as the sum of some sinusoidal signals. The parameters of observer are chosen as $K_{o1} = \text{diag}[15, 15, 15]$, $K_{o2} = \text{diag}[15, 15, 15]$, $K_{o3} = \text{diag}[15, 15, 15]$. The design parameters of the controller are $K_1 = \text{diag}[5, 5, 5]$, $K_2 = \text{diag}[10, 10, 10]$, $K_3 = \text{diag}[5, 5, 5]$. The avoidance ranges $\bar{R}_o = 8\,\text{m}$, $\alpha_o = 20$.

6.1. Trajectory Tracking Control with Obstacle Avoidance

The formation tracking results for case 1, case 2, case 3, and case 4 are shown in Figures 2 and 3. For all cases, the initial postures of desired trajectory are given in $[0\ \text{m}, 0\ \text{m}, 0\ \text{rad}]^T$. The trajectory is generated by time-varying velocity vector v_d, and $v_d(t)$ is given in Table 1. The center of the obstacle is presented in Table 1, and the safe collision-avoidance radii is 1.5 m. The initial postures of dynamic positioning vessel $\eta(0)$ are chosen in Table 1.

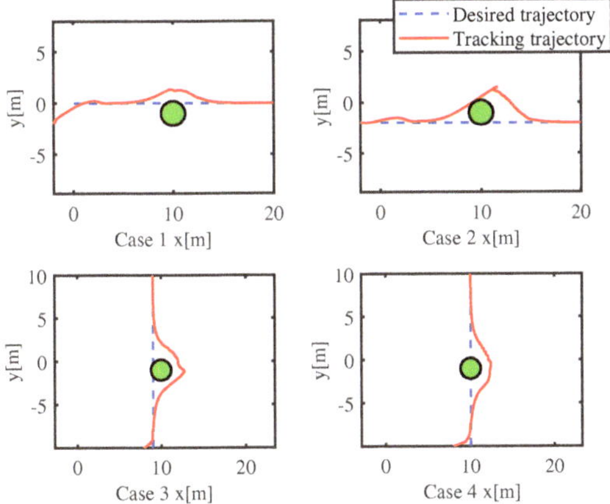

Figure 2. Trajectory tracking control under four cases (the blue circle denotes obstacle).

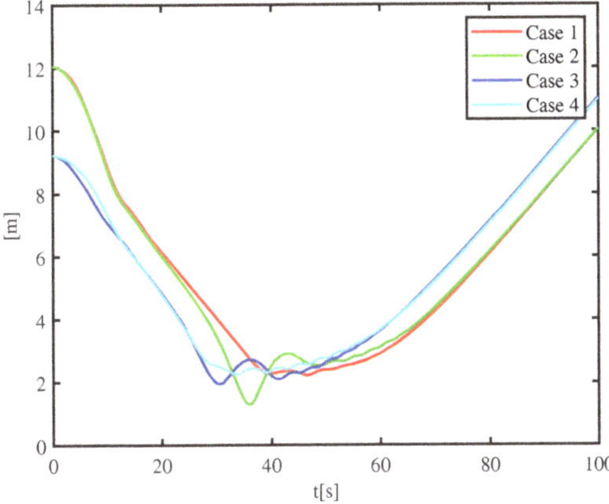

Figure 3. The distances between dynamic positioning vessel and obstacle under four cases.

Table 1. Design simulations.

Simulation Scenario	$\eta(0)$	$v_d(t)$	P_o
Case 1	$[-2\ \text{m},\ -2\ \text{m},\ 0\ \text{rad}]^T$	$[0.2\ \text{m/s},\ 0\ \text{m/s},\ 0\ \text{rad/s}]^T$	$[10\ \text{m},\ -1\ \text{m}]^T$
Case 2	$[-2\ \text{m},\ -2\ \text{m},\ 0\ \text{rad}]^T$	$[0.2\ \text{m/s},\ 0\ \text{m/s},\ 0\ \text{rad/s}]^T$	$[10\ \text{m},-1\ \text{m}]^T$
Case 3	$[-8\ \text{m},\ -10\ \text{m},\ 0\ \text{rad}]^T$	$[0\ \text{m/s},\ 0.2\ \text{m/s},\ 0\ \text{rad/s}]^T$	$[10\ \text{m},\ -1\ \text{m}]^T$
Case 4	$[-8\ \text{m},\ -10\ \text{m},\ 0\ \text{rad}]^T$	$[0\ \text{m/s},\ 0.2\ \text{m/s},\ 0\ \text{rad/s}]^T$	$[10\ \text{m},\ -1\ \text{m}]^T$

Figure 2 shows trajectory tracking control under four cases. In all cases, the dynamic positioning vessel overtakes the obstacle on its portside, and it makes a detour around the obstacle. Figure 3 shows the distances between the dynamic positioning vessel and the obstacle. The results show all of the distances are greater than the safe collision-avoidance radii. This explains why there are no safety risks during voyaging: collisions between the dynamic positioning vessel and the obstacle are avoided . To describe the collision-avoidance process, Figure 4 shows the time-varying error function $z_1^T z_1$ containing the tracking error z_1. In 0 s–25 s, the tracking error z_1 gradually converges to the neighboring zone of zero, and the MAPFs are ineffective. During 25 s–45 s, the dynamic positioning vessel is moving toward the obstacle, and the MAPFs start to work, which leads to the actual operation deviating from the predetermined trajectory. At 45 s–60 s, the dynamic positioning vessel is moving away from the obstacle, the error function $z_1^T z_1$ decreases continuously, and the dynamic positioning vessel gets back on the predetermined trajectory. At 60 s–100 s, the dynamic positioning vessel keeps a safe distance from the obstacle, and the MAPFs do not work.

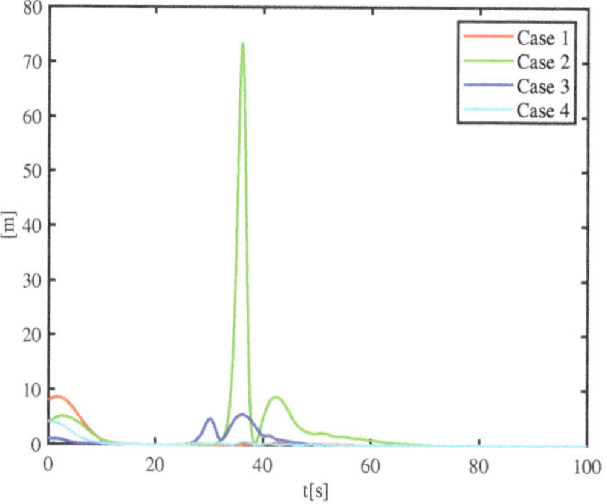

Figure 4. $z_1^T z_1$ under four cases.

Figure 5 describes the velocity recovery performance of the observer. At 25 s–60 s, owing to the effect of the MAPFs, the lateral speed and longitudinal speed are changing zigzag. Figure 6 shows that the uncertainty and disturbances can be estimated by the proposed observer, and the estimated initial value is set as zero. Figure 7 shows that the input signals of system are constrained by the input saturation. At 25 s–60 s, , to ensure the safety navigation of the dynamic positioning vessel, the MAPFs work and affect the input signals, which generates terrible twitter and saturated input signals.

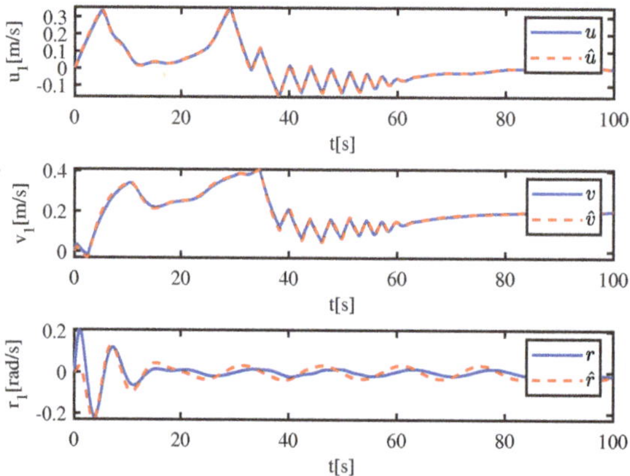

Figure 5. Velocity estimation.

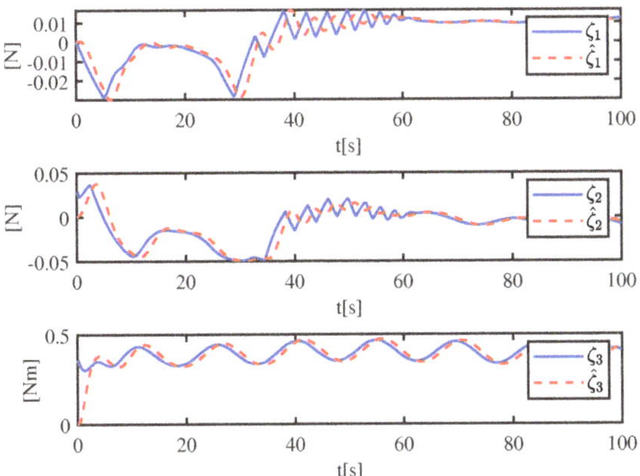

Figure 6. Unknown function and estimation.

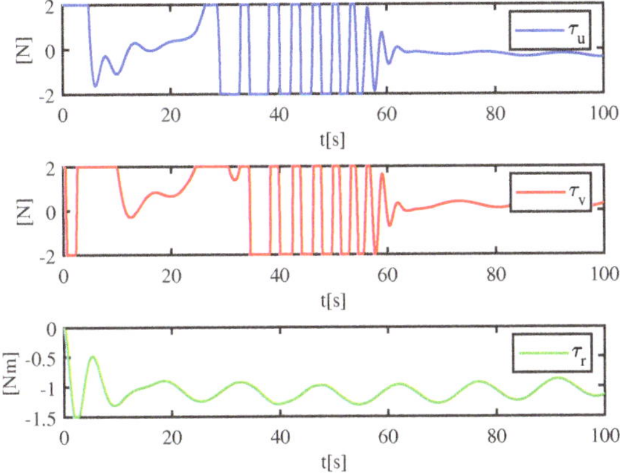

Figure 7. Control inputs.

6.2. Trajectory Tracking Control with Non-Cooperative Ship

To show the effectiveness of the collision-avoidance ability of the PC, a non-cooperative ship is considered in trajectory tracking control. The initial position of non-cooperative ship is set as $[4\text{ m}, -1\text{ m}, 0\text{ rad}]^T$, and the constant velocity non-cooperative ship is set as $[0.1\text{ m/s}, 0\text{ m/s}, 0\text{ rad/s}]^T$. Figure 8 shows the trajectory-tracking results. At $t = 10$ s, the dynamic positioning vessel tracks the desired signal successfully. The non-cooperative ship comes from port side of the dynamic positioning vessel. At $t = 50$ s, the non-cooperative ship takes no actions to avoid collisions, and the dynamic positioning vessel turns to avoid a collision with the non-cooperative ship, which leads to large formation tracking errors. The dynamic positioning vessel overtakes the non-cooperative ship from its starboard. At $t = 60$ s, the dynamic positioning vessel keeps a safe distance between the non-cooperative ship, and tracking trajectory gradually converges to the desired trajectory. The non-cooperative ship is moving out from behind the dynamic positioning vessel. At $t = 100$ s, the dynamic positioning vessel successfully tracks the desired trajectory again. The dynamic positioning vessel moves away from the non-cooperative ship, and the non-cooperative ship sails on the starboard of the dynamic positioning vessel . Figure 9 demonstrates the distance between the dynamic positioning vessel and the non-cooperative ship. Figure 10 depicts the bounded control force and moment. Under the effect of the ADS, the rapid saturation of the input signals is fast convergence.

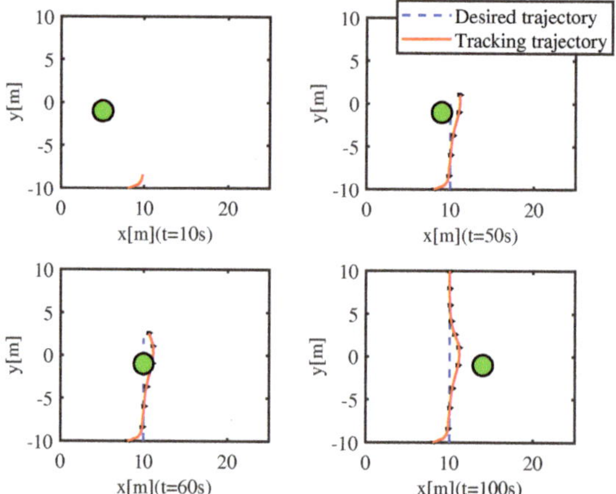

Figure 8. Four moments of trajectory tracking control.

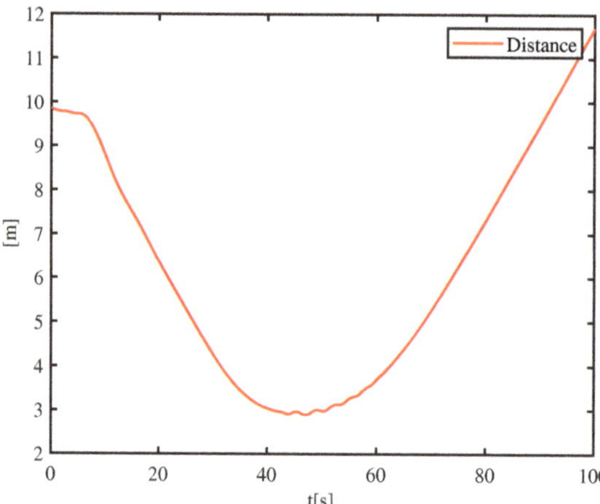

Figure 9. The distance between dynamic positioning vessel and non-cooperative ship.

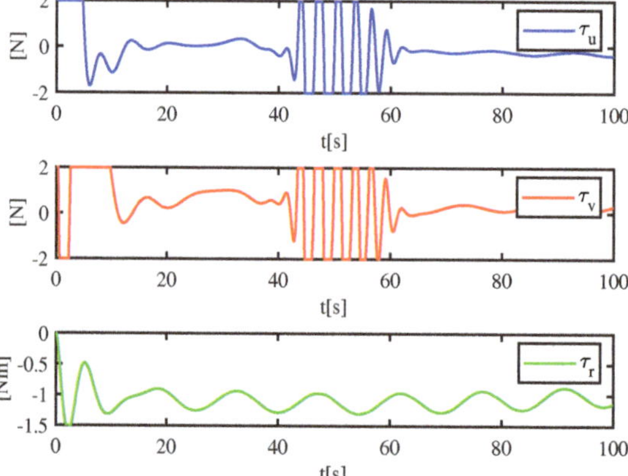

Figure 10. Control inputs under non-cooperative ship.

6.3. Comparison Study

To further appraise the performance of the PC, an output feedback tracking controller (OFC) without an auxiliary dynamic system is given by

$$z_1 = \eta - \eta_d - \eta_o \tag{45}$$

$$z_2 = \hat{v} - \bar{\alpha}, \tag{46}$$

$$\tau_c = M(-K_3 z_2 - \hat{\zeta} + \dot{\bar{\alpha}} - R^T(\psi)z_1), \tag{47}$$

where the parameters K_1, K_2, and K_3 are same as the proposed controller, and the parameters and structure of the observer are also the same. In the same environmental interference, a non-cooperative ship is considered in trajectory tracking control. The initial position of the non-cooperative ship is set as $[4\,\mathrm{m}, -1\,\mathrm{m}, 0\,\mathrm{rad}]^T$, and the constant velocity non-cooperative ship is set as $[0.1\,\mathrm{m/s}, 0\,\mathrm{m/s}, 0\,\mathrm{rad/s}]^T$. The simulation results are shown as Figures 11–13. Figure 11 shows the four moments of the trajectory-tracking results, and the dynamic positioning vessel takes evasive action under OFC. At $t = 10$ s, the dynamic positioning vessel tracks the desired signal successfully. The non-cooperative ship comes from port side of the dynamic positioning vessel . At $t = 50$ s, the non-cooperative ship takes no actions to avoid collisions, and the dynamic positioning vessel makes turns to avoid collisions. However, at $t = 60$ s, the dynamic positioning vessel overtakes the non-cooperative ship from its portside. However, unlike the PC, the trajectory of the dynamic positioning vessel has a coincidence with the non-cooperative ship navigation curve. At $t = 100$ s, the dynamic positioning vessel moves away from the non-cooperative ship, and the non-cooperative ship sails on the starboard of the dynamic positioning vessel.

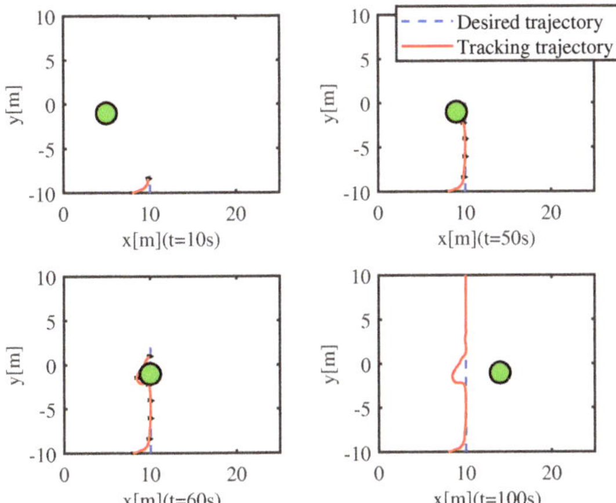

Figure 11. Four moments of trajectory tracking control under OFC.

Figure 12 shows the time-varying distances between the dynamic positioning vessel and the non-cooperative ship under different controllers. The safe collision-avoidance radii is set to 1.5 m; collisions may happen between the dynamic positioning vessel and non-cooperative ship under OFC. Figure 13 shows that the control inputs suffer from sudden jumps, and the proposed scheme can decrease jitters.

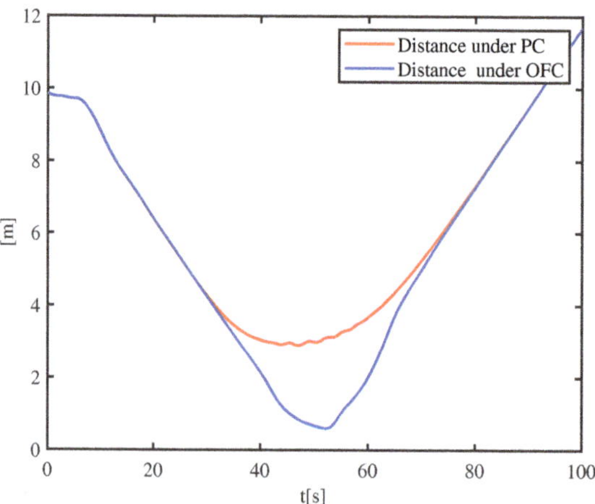

Figure 12. The distances between dynamic positioning vessel and non-cooperative ship under different controllers.

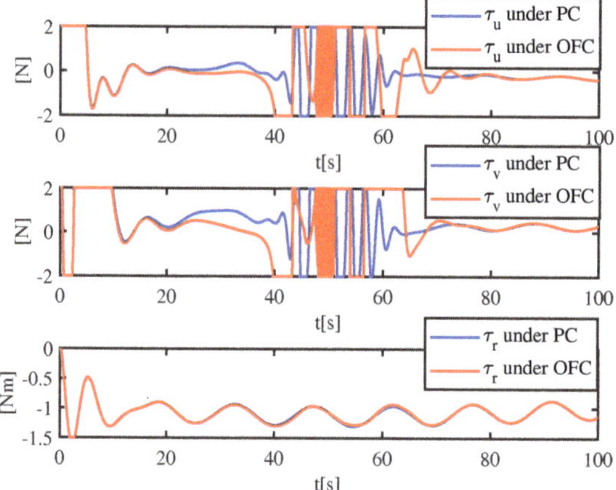

Figure 13. Input signals under different controllers.

7. Conclusions

This article suggests an output feedback controller for a dynamic positioning vessel with model uncertainty, unknown environmental disturbances, and input constraints. Static obstacles and unknown non-cooperative ships are also considered. An ESO is given, and unknown model dynamics and velocity are simultaneously estimated. The controller is designed based on the ADS and MAPFs. Finally, a mathematical analysis is undertaken to prove that all of the error signals of the system are bounded. Simulation experiments affirm the tracking performance of the proposed controller. This paper proposes an anti-collision controller for the dynamic positioning vessel, and simulations prove that the vessel can avoid the obstacle and navigate itself to the desired trajectory. However, there is no mathematical analysis that shows that the controller can guarantee safety. Finding the best way to solve this problem can be considered as a part of future works.

Author Contributions: Conceptualization, G.X.; methodology, B.Z.; software, B.Z.; validation, B.Z.; and writing—review and editing, B.Z. All of the authors have read and agreed to the published version of the manuscript.

Funding: This research received no external funding.

6.3. Comparison Study

To further appraise the performance of the PC, an output feedback tracking controller (OFC) without an auxiliary dynamic system is given by

$$z_1 = \eta - \eta_d - \eta_o \tag{45}$$

$$z_2 = \hat{v} - \bar{\alpha}, \tag{46}$$

$$\tau_c = M(-K_3 z_2 - \hat{\zeta} + \dot{\bar{\alpha}} - R^T(\psi)z_1), \tag{47}$$

where the parameters K_1, K_2, and K_3 are same as the proposed controller, and the parameters and structure of the observer are also the same. In the same environmental interference, a non-cooperative ship is considered in trajectory tracking control. The initial position of the non-cooperative ship is set as $[4\text{ m}, -1\text{ m}, 0\text{ rad}]^T$, and the constant velocity non-cooperative ship is set as $[0.1\text{ m/s}, 0\text{ m/s}, 0\text{ rad/s}]^T$. The simulation results are shown as Figures 11–13. Figure 11 shows the four moments of the trajectory-tracking results, and the dynamic positioning vessel takes evasive action under OFC. At $t = 10$ s, the dynamic positioning vessel tracks the desired signal successfully. The non-cooperative ship comes from port side of the dynamic positioning vessel . At $t = 50$ s, the non-cooperative ship takes no actions to avoid collisions, and the dynamic positioning vessel makes turns to avoid collisions. However, at $t = 60$ s, the dynamic positioning vessel overtakes the non-cooperative ship from its portside. However, unlike the PC, the trajectory of the dynamic positioning vessel has a coincidence with the non-cooperative ship navigation curve. At $t = 100$ s, the dynamic positioning vessel moves away from the non-cooperative ship, and the non-cooperative ship sails on the starboard of the dynamic positioning vessel.

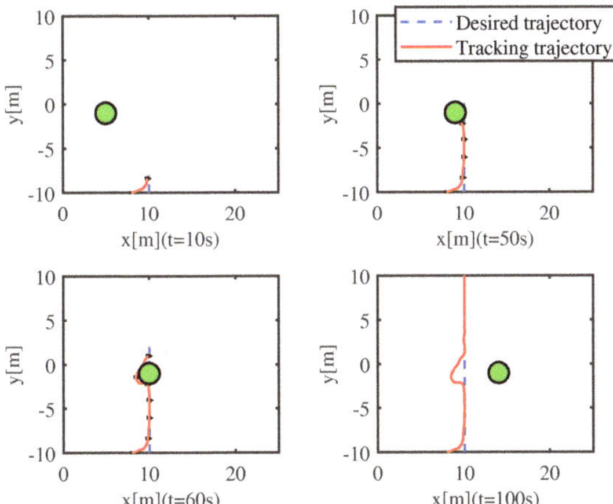

Figure 11. Four moments of trajectory tracking control under OFC.

Figure 12 shows the time-varying distances between the dynamic positioning vessel and the non-cooperative ship under different controllers. The safe collision-avoidance radii is set to 1.5 m; collisions may happen between the dynamic positioning vessel and non-cooperative ship under OFC. Figure 13 shows that the control inputs suffer from sudden jumps, and the proposed scheme can decrease jitters.

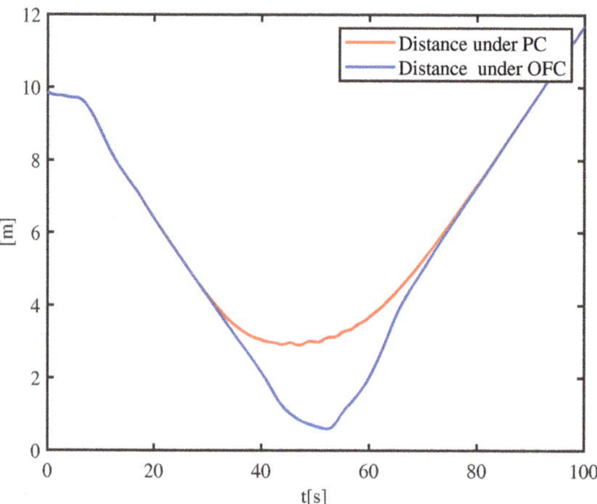

Figure 12. The distances between dynamic positioning vessel and non-cooperative ship under different controllers.

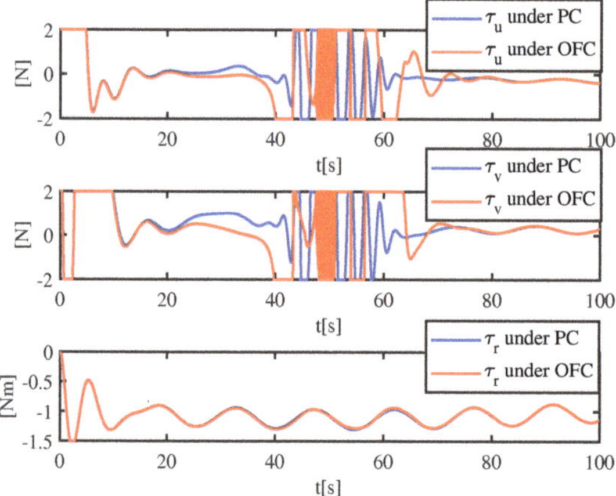

Figure 13. Input signals under different controllers.

7. Conclusions

This article suggests an output feedback controller for a dynamic positioning vessel with model uncertainty, unknown environmental disturbances, and input constraints. Static obstacles and unknown non-cooperative ships are also considered. An ESO is given, and unknown model dynamics and velocity are simultaneously estimated. The controller is designed based on the ADS and MAPFs. Finally, a mathematical analysis is undertaken to prove that all of the error signals of the system are bounded. Simulation experiments affirm the tracking performance of the proposed controller. This paper proposes an anti-collision controller for the dynamic positioning vessel, and simulations prove that the vessel can avoid the obstacle and navigate itself to the desired trajectory. However, there is no mathematical analysis that shows that the controller can guarantee safety. Finding the best way to solve this problem can be considered as a part of future works.

Author Contributions: Conceptualization, G.X.; methodology, B.Z.; software, B.Z.; validation, B.Z.; and writing—review and editing, B.Z. All of the authors have read and agreed to the published version of the manuscript.

Funding: This research received no external funding.

Institutional Review Board Statement: This article does not contain any studies with human or animal subjects.

Informed Consent Statement: The study did not involve humans.

Data Availability Statement: The data of this paper are unavailable.

Conflicts of Interest: The authors declare no conflict of interest.

References

1. Dai, S.L.; Wang, M.; Wang, C. Neural learning control of marine dynamic positioning vessels with guaranteed transient tracking performance. *IEEE Trans. Ind. Electron.* **2016**, *63*, 1717–1727. [CrossRef]
2. Gao, S.; Liu, C.; Tuo, Y.; Chen, K.; Zhang, T. Augmented model-based dynamic positioning predictive control for underactuated unmanned surface vessels with dual propellers. *Ocean Eng.* **2022**, *266*, 112885. [CrossRef]
3. Sørensen, A.J. A survey of dynamic positioning control systems. *Annu. Rev. Control.* **2011**, *35*, 123–136. [CrossRef]
4. Xia, G.Q.; Zhang, B.W. Finite-Time Control of Dynamic Positioning Vessel Based on Disturbance Observer. *Math. Probl. Eng.* **2022**, *2022*, 9262457 . [CrossRef]
5. Zhang, Z.Y.; Wu, H.T.; Liu, W.X. Effects of mooring line hydrodynamic coefficients and wave parameters on the floating production storage and offloading motions. *Desalin. Water Treat.* **2021** , *239*, 278–288. [CrossRef]
6. Liu, X.; Miao, Q.; Wang, X.; Xu, S.; Fan, H. A novel numerical method for the hydrodynamic analysis of floating bodies over a sloping bottom. *J. Mar. Sci. Technol.* **2021** , *26*, 1198–1216. [CrossRef]
7. Fan, H.Q.; Miao, Q.M.; Allan, R.M. Wave Loads on the Large Vertical Cylinder with the Conformal Mapping and Series Expansion Method. In Proceedings of the Fourteenth (2020) ISOPE Pacific-Asia Offshore Mechanics Symposium, Dalian, China, 22–25 November 2020.
8. Van M.; Do, V.T.; Khyam, M.O.; Xuan P.D. Tracking control of uncertain dynamic positioning vessels with global finite-time convergence. *J. Adv. Res.* **2021**, *241*, 109974.
9. Zhu, Y.; Zhang, H.; Li, H.; Zhang, J.; Zhang, D. Optimal Jamming Strategy Against Two-state Switched System. *IEEE Commun. Lett.* **2022**, *13*, 1767–1775. [CrossRef]
10. Gao, S.; Xue, J.J. Nonlinear vector model control of underactuated air cushion vehicle based on parameter reduction algorithm, *Trans. Inst. Meas. Control.* **2021**, *43*, 1202–1211. [CrossRef]
11. Li, H.; Xu, W.; Zhang, H.; Zhang, J.; Liu, Y. Polynomial regressors based data-driven control for autonomous underwater vehicles. *Peer-to-Peer Netw. Appl.* **2020**, *13*, 1767–1775. [CrossRef]
12. Fang, M.C.; Zhuo, Y.Z.; Lee, Z.Y. The application of the self-tuning neural network PID controller on the ship roll reduction in random waves. *Ocean. Eng.* **2010**, *37*, 529–538. [CrossRef]
13. Larrazabal, J.M.; Penas, M.S. Intelligent rudder control of an unmanned surface vessel. *Expert Syst. Appl.* **2016**, *55*, 106–117. [CrossRef]
14. Ishaque, K.; Abdullah, S.; Ayob, S.; Salam, Z. A simplified approach to design fuzzy logic controller for an underwater vehicle. *Ocean Eng.* **2011**, *38*, 271–284. [CrossRef]
15. Abdelaal, M.; Fränzle, M.; Hahn, A. Nonlinear model predictive control for trajectory tracking and collision avoidance of underactuated vessels with disturbances. *Ocean Eng.* **2018**, *160*, 168–180. [CrossRef]
16. Ashrafiuon, H.; Muske, K.; McNinch, L. Sliding-mode tracking control of dynamic positioning vesselS *IEEE Trans. Ind. Electron.* **2008**, *55*, 4004–4012. [CrossRef]
17. Xia, G.; Xia, X.; Zhao, B.; Sun, C.; Sun, X. A solution to leader following of underactuated dynamic positioning vessel s with actuator magnitude and rate limits. *Int. J. Adapt. Control. Signal Process.* **2021**, *35*, 1860–1878. [CrossRef]
18. Xia, G.Q.; Xia, X.M.; Sun, X.X. Formation tracking control for underactuated surface vehicles with actuator magnitude and rate saturations. *Ocean Eng.* **2022**, *260*, 111935. [CrossRef]
19. Zhu, H.; Yu, H.M.; Guo, C. Finite time PAILOS based path following control of underactuated marine surface vessel with input saturation. *ISA Trans.* **2022**, *135*, 66–77. [CrossRef]
20. Xia, G.; Xia, X.; Bo, Z.; Sun, X.; Sun, C. Event-Triggered Controller Design for Autopilot with Input Saturation. *Math. Probl. Eng.* **2020**, *2020*, 5362895. [CrossRef]
21. Zhu, G.; Du, J. Global Robust Adaptive Trajectory Tracking Control for dynamic positioning ships Under Input Saturation. *IEEE J. Ocean. Eng.* **2020**, *45*, 442–450. [CrossRef]
22. Qin, H.; Li, C.; Sun, Y.; Li, X.; Du, Y.; Deng, Z. Finite-time trajectory tracking control of unmanned surface vessel with error constraints and input saturations. *J. Frankl. Inst.* **2020**, *357*, 11472–11495. [CrossRef]
23. Qin, H.; Li, C.; Sun, Y.; Wang, N. Adaptive trajectory tracking algorithm of unmanned dynamic positioning vessel based on anti-windup compensator with full-state constraints. *Ocean Eng.* **2020**, *200*, 106906. [CrossRef]
24. Xia, G.; Xia, X.; Zhao, B.; Sun, C.; Sun, X. Distributed Tracking Control for Connectivity-Preserving and Collision-Avoiding Formation Tracking of Underactuated Surface Vessels with Input Saturation. *Appl. Sci.* **2020**, *10*, 3372. [CrossRef]
25. Zeng, Z.; Yu, H.; Guo, C.; Yan, Z. Finite-time coordinated formation control of discrete-time multi-AUV with input saturation under alterable weighted topology and time-varying delay. *Ocean Eng.* **2022**, *266*, 112881. [CrossRef]

26. Tam, C.K.; Bucknall, R. Cooperative path-planning algorithm for marine dynamic positioning vessels. *Ocean Eng.* **2013**, *57*, 25–33. [CrossRef]

27. Li, Y.; Zheng, J. Real-time collision avoidance planning for unmanned surface vessels based on field theory. *ISA Trans.* **2020**, *106*, 233–242. [CrossRef]

28. Wang, W.M.; Du, J.L.; Tao, Y.H. A dynamic collision avoidance solution scheme of unmanned surface vessels based on proactive velocity obstacle and set-based guidance. *Ocean Eng.* **2022**, *248*, 110794.

29. Park, J.W. Improved Collision Avoidance Method for Autonomous Surface Vessels Based on Model Predictive Control Using Particle Swarm Optimization. *Int. J. Fuzzy Log. Intell. Syst.* **2021**, *21*, 378–390. [CrossRef]

30. Peng, Z.; Wang, D.; Li, T.; Han, M. Output-Feedback Cooperative Formation Maneuvering of Autonomous Surface Vehicles With Connectivity Preservation and Collision Avoidance. *IEEE Trans. Cybern.* **2020**, *50*, 2527–2535. [CrossRef]

31. Park, B.S.; Yoo, S.J. An error transformation approach for connectivity-preserving and collision-avoiding formation tracking of networked uncertain underactuated dynamic positioning vessels. *IEEE Trans. Cybern.* **2019**, *49*, 353–359. [CrossRef]

32. Fossen, T.I.; Strand, J.P. Passive nonlinear observer design for ships using Lyapunov methods: Full-Scale experiments with a supply vessel. *Automatica* **1999**, *35*, 3–16. [CrossRef]

33. Liang, X.; Qu, X.; Wang, N.; Li, Y.; Zhang, R. Swarm control with collision avoidance for multiple underactuated surface vehicles. *Ocean Eng.* **2019**, *191*, 106516. [CrossRef]

34. Xia, G.Q.; Xia, X.M.; Sun, X.X. Formation control with collision avoidance for underactuated surface vehicles. *Asian J. Control.* **2022**, *24*: 2244–2257. [CrossRef]

35. Kowalczyk, W.; Michaek, M.; Kozowski, K. Trajectory tracking control and obstacle avoidance for a differentially driven mobile robot. *IFAC-World Congr.* **2011**, *41*, 1058–1063. [CrossRef]

36. Xia, G.; Sun, C.; Zhao, B.; Xia, X.; Sun, X. Neuroadaptive Distributed Output Feedback Tracking Control for Multiple Marine Surface Vessels With Input and Output Constraints. *IEEE Access* **2019**, *7*, 123076–123085. [CrossRef]

37. Kowalczyk, W.; Michaek, M.; Kozowski, K. Collaborative collision avoidance for Maritime Autonomous dynamic positioning ships: A review. *IFAC-World Congr.* **2011**, *41*, 1058–1063.

38. Skjetne, R.; Fossen, T.I.; Kokotovi, P. V. Adaptive maneuvering, with experiments, for a model ship in a marine control laboratory. *Automatica* **2005**, *41*, 289–298. [CrossRef]

Journal of
*Marine Science
and Engineering*

Article

Horizontal Trajectory Tracking Control for Underactuated Autonomous Underwater Vehicles Based on Contraction Theory

Caipeng Ma [1], Jinjun Jia [1], Tiedong Zhang [1,2], Shaoqun Wu [3] and Dapeng Jiang [1,2,*]

[1] School of Ocean Engineering and Technology, Sun Yat-sen University, Zhuhai 519082, China; macp3@mail2.sysu.edu.cn (C.M.); jiajj5@mail2.sysu.edu.cn (J.J.); zhangtd5@mail.sysu.edu.cn (T.Z.)

[2] Southern Marine Science and Engineering Guangdong Laboratory (Zhuhai), Zhuhai 519000, China

[3] YaGuang Technology Group Company Limited, Yuanjiang 413100, China; wsq0927yt@163.com

* Correspondence: jiangdp5@mail.sysu.edu.cn; Tel.: +86-131-3675-7927

Abstract: In this paper, contraction theory is applied to design a control law to address the horizontal trajectory tracking problem of an underactuated autonomous underwater vehicle. Suppose that the vehicle faces challenges such as model uncertainties, external environmental disturbances, and actuator saturation. Firstly, a coordinate transformation is introduced to solve the problem of underactuation. Then, a disturbance observer is designed to estimate the total disturbances, which are composed of model uncertainties and external environmental disturbances. Next, a saturated controller is designed based on singular perturbation theory and contraction theory. Meanwhile, contraction theory is used to analyse the convergence properties of the observer and the full singular perturbation system, and make quantitative analysis of the estimation error and the tracking error. Finally, the results of numerical simulations prove that the method in this paper enables the vehicle to track the desired trajectory with relatively high accuracy, while the control inputs do not exceed the limitations of the actuators.

Keywords: underactuated autonomous underwater vehicle; trajectory tracking; actuator saturation; singular perturbation system; contraction theory

Citation: Ma, C.; Jia, J.; Zhang, T.; Wu, S.; Jiang, D. Horizontal Trajectory Tracking Control for Underactuated Autonomous Underwater Vehicles Based on Contraction Theory. *J. Mar. Sci. Eng.* **2023**, *11*, 805. https://doi.org/10.3390/jmse11040805

Academic Editor: Sergei Chernyi

Received: 15 March 2023
Revised: 4 April 2023
Accepted: 5 April 2023
Published: 10 April 2023

1. Introduction

Underactuated AUV is a kind of AUV which has fewer independent control inputs than the DOF to be controlled. Compared with the fully actuated AUV, it has more advantages in saving costs, reducing consumption, and improving system reliability, so it has a wide range of applications [1]. However, due to the challenges such as nonlinearity, model uncertainties, time-varying external disturbance, actuator saturation, etc., the control of the underactuated AUV becomes difficult. Currently, the precise motion control of the underactuated AUV is one of the research hotspots.

As a powerful system design and analysis tool, CT [2] has been applied in many fields; however, to our knowledge, there is currently no literature on the application of CT to underactuated AUVs. Based on this, this paper focuses on the horizontal trajectory tracking control of an underactuated AUV in the presence of model uncertainties, time-varying environmental disturbances, and actuator saturation. CT and its application in SPS [3] are used to design the controller and give quantitative analyses of various errors. In general, the contributions of this paper are as follows:

(1) A coordinate transformation is introduced to cope with the problem of underactuation and a disturbance observer is designed to estimate the total disturbances.

(2) CT and SPT are used to construct a saturated controller such that the reference states defined by the coordinate transformation converge to the desired states asymptotically, while the error between the actual states and the desired states converges to the region near zero, and the control inputs do not exceed the limitations of the actuator.

Compared with existing methods, the controller designed in this paper has a simple form and is easy to apply. In solving input saturation problems especially, there is no need to design auxiliary systems or run complex algorithms such as NN or FL.

(3) CT is applied to analyse the convergence properties of the DO and the SPS, and the upper bounds of estimation error, tracking error, and the error between the ideal controller and actual controller are given.

The structure of this paper will be arranged as follows. Section 1 introduces the research background, purpose, contributions, and structure of this paper. Section 2 reviews the literature related to the work of this paper. Section 3 establishes the mathematical model for an underactuated 3-DOF AUV in the presence of model uncertainties, external disturbances, and actuator saturation. A coordinate transformation is introduced to deal with the problem of underactuation. The DO and the saturated controller are designed in Section 4. Section 5 gives quantitative analysis of the control variable error and the tracking error. Section 6 conducts numerical simulations and analyses the performance of the controller and DO. Finally, Section 7 gives conclusions. In addition, Appendix A summarizes the abbreviations used in this paper and Appendix B briefly introduces the theoretical basis of this paper.

2. Literature Review

Underactuation is the first problem to be solved in the motion control of a USV or underactuated AUV. One of the common methods to tackle this problem is to derive the tracking error model in SF frame [4–7], and then design the control law on each controllable DOF. In [4], a 3D-path-following error model for an underactuated 5-DOF AUV was established based on virtual guidance method in SF frame, and then backstepping and sliding mode control were applied to design controllers in surge, pitch, and yaw directions. In [6], the path-following error dynamics were derived and several reduced-order ESOs were designed to estimate various disturbances. Coordinate transformation is also a common method [8–12]. In [8,9], the model of a USV was converted into a cascade system and the control problem of the USV was transformed into the stabilization analysis of two small subsystems. In [10,11], the output variables of the USV were redefined via a coordinate transformation and then the other control methods were applied to design the controller such that the new output variables could track their desired values. In addition, the system order of the underactuated AUV can be reduced by constructing an SPS [13–15]. In [13], SPT was used to decompose the full system of a 4-DOF underactuated AUV into two time-scale subsystems and then the independent controller design was carried out on each subsystem.

Disturbances are a ubiquitous challenge for the motion control of an AUV. It may come from model uncertainties, unknown system parameters, or external environment disturbances, or more likely the superposition of these factors. The estimation and compensation of disturbances are very important for AUVs' motion control. For mechanical systems including AUVs, a common practice is to combine various disturbances into total disturbances, and then design a DO to estimate it. The observer based on auxiliary variables and the ESO [16] are two common types of DO. Readers can find a detailed overview of the first type of DO in [17,18]. In [19], an auxiliary variable was introduced to design a nonlinear DO and then a backstepping finite-time sliding mode controller was constructed for the trajectory tracking control of a 5-DOF AUV. In [20], a reduced-order observer was proposed to estimate the total disturbances. ESO estimates the total disturbances as another state of the system. In [14], a high gain ESO based on SPT was designed to estimate the total disturbances. In [21], a proportional-integral velocity variable based third-order fast finite-time ESO was designed to estimate the lumped uncertainties and their first derivatives. In addition, with the continuous maturity of artificial intelligence, more and more researchers use NN [22] and FL [23] to approximate disturbances and uncertainties.

In practical applications, the problem of actuator saturation is almost unavoidable since the force/torque provided by the actuators is limited. Ignoring this problem may reduce the performance of the controller or even cause instability. At present, the auxiliary variable system [24–27], adaptive method [28], FL [29], NN [30], MPC [31,32], and DAI [33,34] are common means used to address actuator saturation. In [27], the saturation effects of rudder angle in diving control of AUVs were compensated by a modified auxiliary system with time-varying nonlinear gains. In [28], a smooth dead zone based model was designed to linearize the actuator model, so that the adaptive law could be applied to eliminate input saturation and actuator failure. In [31,32], a 3D trajectory tracking controller based on MPC was proposed such that the state and control input of the AUV did not exceed their respective physical constraints. In [33], DAI was recently applied to guarantee saturation avoidance and the techniques were applied to DC motor control for UUVs [34]. However, most of the methods used to solve input saturation are relatively complex, and some require additional design of new systems. In addition, methods such as NN and FL require strong computing power, which is a significant challenge for AUVs.

In the above literature, the Lyapunov method occupies an absolute position in the design and stability analysis of the system. However, for a complex nonlinear system, it is not easy to find a suitable Lyapunov function and prove that its first derivative is UND. In recent years, the continuous development of CT [2] provides researchers with new ideas. It studies the convergence properties of system trajectory, which is very applicable to tracking control problems. At present, CT has been widely used in many fields, such as controller and observer design [35–39], cooperative control [40,41], SPS [3,42], iterative learning control [43,44], convex optimization [45,46], and so on.

Compared with the extensive application in other fields, the application of CT in AUVs is still rare, and the model of AUVs in some rare applications is relatively simple. In [35], CT was used to design the position and velocity observer for an AUV with 1-DOF and analyse the convergence properties. In [47], after omitting the Coriolis and centripetal terms in the dynamic equation, CT was applied to design an UGES observer for a 6-DOF AUV. Unfortunately, the simulation results of the observer were not shown. In [48,49], combining the CT and the backstepping method, the author designed a speed stabilization controller and a trajectory tracking controller for a simplified AUV, and discussed the incremental stability of the AUV system. In [37], CT was applied to construct a trajectory tracking controller for an openframe AUV under the pH framework.

3. AUV Modelling

3.1. AUV Modelling in the Horizontal Plane

This section establishes a horizontal 3-DOF kinematics and dynamics model for an underactuated AUV. To describe the motion of the AUV, we first define two reference frames: earth-fixed frame and body-fixed frame, as shown in Figure 1. Here, $(x, y, z)^T$ and $(\varphi, \theta, \psi)^T$ denote the position vector and the attitude vector relative to the earth-fixed frame, respectively. The terms $(u, v, w)^T$ and $(p, q, r)^T$ represent the linear velocity vector and angular velocity vector with respect to the body-fixed frame, respectively.

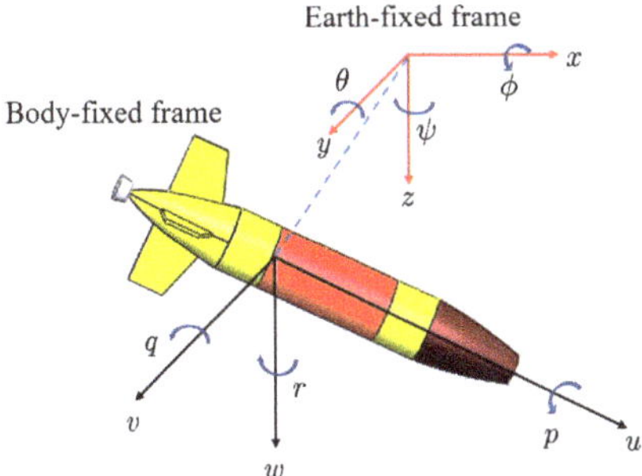

Figure 1. Schematic diagram of coordinate system.

Since the motion considered here is the motion of an underactuated 3-DOF AUV in the horizontal plane, the dynamics in heave, roll, and pitch directions are all neglected and the AUV is supposed to be neutrally buoyant. The kinematics model is such that [50]

$$\begin{cases} \dot{x} = u\cos\psi - v\sin\psi \\ \dot{y} = u\sin\psi + v\cos\psi \\ \dot{\psi} = r \end{cases} \tag{1}$$

and the dynamics model

$$\begin{cases} m_{11}\dot{u} - m_{22}vr - X_u u - X_{|u|u}|u|u & = \tau_u + \tau_{d1} \\ m_{22}\dot{v} + m_{11}ur - Y_v v - Y_{|vs.|v}|vs.|v & = \tau_{d2} \\ m_{33}\dot{r} + (m_{22} - m_{11})uv - N_r r - N_{|r|r}|r|r & = \tau_r + \tau_{d3} \end{cases} \tag{2}$$

where $m_{11} = m - X_{\dot{u}}$, $m_{22} = m - Y_{\dot{v}}$ and $m_{33} = I_z - N_{\dot{r}}$. m is the AUV's mass; I_z is the moment of inertia about the yaw rotation; $X_{\dot{u}}$, $Y_{\dot{v}}$, and $N_{\dot{r}}$ are the hydrodynamic added mass terms in the surge, the sway, and the yaw directions, respectively; X_u, Y_v, and N_r are the linear damping terms, and $X_{|u|u}$, $Y_{|vs.|v}$, and $N_{|r|r}$ are the second-order damping terms. τ_u and τ_r are control inputs for the surge force and the yaw torque. τ_{d1}, τ_{d2}, and τ_{d3} represent the total disturbances in three directions, they are composed of model uncertainties and time-varying environmental disturbances.

To facilitate subsequent design, (2) is simplified as follows:

$$\begin{cases} \dot{u} = f_u + m_{11}^{-1}\tau_u + m_{11}^{-1}\tau_{d1} \\ \dot{v} = f_v + m_{22}^{-1}\tau_{d2} \\ \dot{r} = f_r + m_{33}^{-1}\tau_r + m_{33}^{-1}\tau_{d3} \end{cases} \tag{3}$$

with $f_u = m_{11}^{-1}(m_{22}vr + X_u u + X_{|u|u}|u|u)$, $f_v = m_{22}^{-1}(-m_{11}ur + Y_v v + Y_{|vs.|v}|vs.|v)$, and $f_r = m_{33}^{-1}((m_{11} - m_{22})uv + N_r r + N_{|r|r}|r|r)$.

This paper also considers the problem of actuator saturation. In order to deal with this problem, we design a control law $\tau_i = \sigma(\tau_{ci}, \tau_{max}), i = u, r$, where τ_{ci} is the force/torque calculated by the controller, τ_{max} represents the maximum force/torque that the actuator can provide, τ_i is the actual output force/torque of the actuator, and $\sigma(\cdot)$ is a bounded smooth function, which satisfies

$$\begin{cases} \sigma(0, \tau_{max}) = 0 \\ s\sigma(s, \tau_{max}) > 0, \forall\, s \neq 0 \\ \lim\limits_{s \to +\infty} \sigma(s, \tau_{max}) = \tau_{max}, \lim\limits_{s \to -\infty} \sigma(s, \tau_{max}) = -\tau_{max} \\ \frac{\partial\sigma(s, \tau_{max})}{\partial s} > 0, \forall s \in \mathbb{D}_s \in \mathbb{R} \end{cases} \tag{4}$$

As we know, many functions including the Gaussian error function and hyperbolic tangent function satisfy the properties in (4). Here we choose the hyperbolic tangent function (Figure 2), so the form of the controller is $\boldsymbol{\tau} = \sigma(\boldsymbol{\tau}_c, \tau_{max}) = \tau_{max}\tanh(\boldsymbol{\tau}_c/\tau_{max})$, where $\boldsymbol{\tau} = [\tau_u\ \tau_r]^T$, $\boldsymbol{\tau}_c = [\tau_{cu}\ \tau_{cr}]^T$.

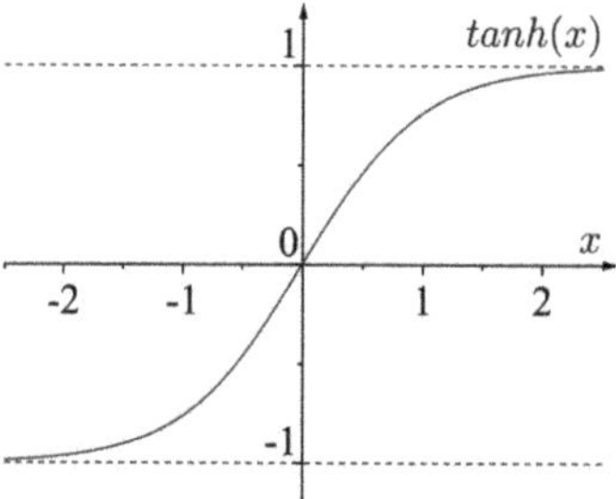

Figure 2. Schematic diagram of *tanh(x)*.

The dynamics model considering actuator saturation is obtained:

$$\begin{cases} \dot{u} = f_u + m_{11}^{-1}\sigma(\tau_{cu}, \tau_{max}) + m_{11}^{-1}\tau_{d1} \\ \dot{v} = f_v + m_{22}^{-1}\tau_{d2} \\ \dot{r} = f_r + m_{33}^{-1}\sigma(\tau_{cr}, \tau_{max}) + m_{33}^{-1}\tau_{d3} \end{cases} \tag{5}$$

The following Assumptions are made to facilitate the subsequent design and analyses:

Assumption 1. *τ_{d1}, τ_{d2}, and τ_{d3} are all unknown and time-varying, and their first and second time derivatives are bounded.*

Remark 1. *The disturbances are bounded since the energy of the environmental disturbance and the state of AUVs are finite. Therefore, Assumption 1 is reasonable.*

3.2. Coordinate Transformation

To cope with the problem of underactuation, the following coordinate transformation is introduced [10,12]:

$$\begin{cases} x_r = x + l\cos\psi \\ y_r = y + l\sin\psi \end{cases} \tag{6}$$

where $l > 0$ is a constant, $x_r \to x, y \to y$ as $l \to 0$. The first order derivatives of x_r and y_r with respect to time are given:

$$\begin{cases} \dot{x}_r = u\cos\psi - v\sin\psi - lr\sin\psi \\ \dot{y}_r = u\sin\psi + v\cos\psi + lr\cos\psi \end{cases} \tag{7}$$

Subsequently, the second derivatives of x_r and y_r are obtained:

$$\begin{cases} \ddot{x}_r = \dot{u}\cos\psi - (\dot{v} + l\dot{r})\sin\psi - f_x(\mathbf{z}) \\ \ddot{y}_r = \dot{u}\sin\psi + (\dot{v} + l\dot{r})\cos\psi + f_y(\mathbf{z}) \end{cases} \tag{8}$$

where $\mathbf{z} = [u\ v\ r\ \psi]^T$,

$$\begin{cases} f_x(\mathbf{z}) = ur\sin\psi + (vr + lr^2)\cos\psi \\ f_y(\mathbf{z}) = ur\cos\psi - (vr + lr^2)\sin\psi \end{cases} \tag{9}$$

In the light of (5) and (8), we have

$$\begin{cases} \ddot{x}_r = F_x(\mathbf{z}) - f_x(\mathbf{z}) + \Delta x + m_{11}^{-1}\cos\psi\sigma(\tau_{cu}, \tau_{max}) - lm_{33}^{-1}\sin\psi\sigma(\tau_{cr}, \tau_{max}) \\ \ddot{y}_r = F_y(\mathbf{z}) + f_y(\mathbf{z}) + \Delta y + m_{11}^{-1}\sin\psi\sigma(\tau_{cu}, \tau_{max}) + lm_{33}^{-1}\cos\psi\sigma(\tau_{cr}, \tau_{max}) \end{cases} \tag{10}$$

where

$$\begin{cases} F_x(\mathbf{z}) = f_u\cos\psi - (f_v + lf_r)\sin\psi \\ \Delta x = m_{11}^{-1}\tau_{d1}\cos\psi - (m_{22}^{-1}\tau_{d2} + lm_{33}^{-1}\tau_{d3})\sin\psi \end{cases} \tag{11}$$

$$\begin{cases} F_y(\mathbf{z}) = f_u\sin\psi + (f_v + lf_r)\cos\psi \\ \Delta y = m_{11}^{-1}\tau_{d1}\sin\psi + (m_{22}^{-1}\tau_{d2} + lm_{33}^{-1}\tau_{d3})\cos\psi \end{cases} \tag{12}$$

Let $\boldsymbol{\eta}_1 = [x_r\ y_r]^T$ be the reference trajectory and $\boldsymbol{\eta}_2 = [\dot{x}_r\ \dot{y}_r]^T$, (10) can be rewritten as follows:

$$\begin{cases} \dot{\boldsymbol{\eta}}_1 = \boldsymbol{\eta}_2 \\ \dot{\boldsymbol{\eta}}_2 = \boldsymbol{\Phi}(\mathbf{z}) + \boldsymbol{\Delta} + \boldsymbol{g}(\psi)\sigma(\boldsymbol{\tau}_c, \tau_{max}) \end{cases} \tag{13}$$

where $\boldsymbol{\Phi}(\mathbf{z}) = [F_x(\mathbf{z}) - f_x(\mathbf{z})\ F_y(\mathbf{z}) + f_y(\mathbf{z})]^T$, $\boldsymbol{\Delta} = [\Delta x\ \Delta y]^T$ is the combined disturbance vector and

$$\boldsymbol{g}(\psi) = \begin{bmatrix} m_{11}^{-1}\cos\psi & -lm_{33}^{-1}\sin\psi \\ m_{11}^{-1}\sin\psi & lm_{33}^{-1}\cos\psi \end{bmatrix} \tag{14}$$

It is obvious that the matrix $\boldsymbol{g}(\psi)$ is nonsingular for any ψ. The control objective is to design a controller $\boldsymbol{\tau}_c$ such that the reference trajectory $\boldsymbol{\eta}_1 = [x_r\ y_r]^T$ converges to the desired trajectory $\boldsymbol{\eta}_d = [x_d \; y_d]^T$ asymptotically, and the error between the actual trajectory $\boldsymbol{\eta} = [x \; y]$ and the desired trajectory is small enough. In addition, considering that the disturbance $\boldsymbol{\Delta}$ is unknown, a DO is needed.

Assumption 2. *According to Assumption 1, (11) and (12), the disturbance vector $\boldsymbol{\Delta}$ and its first derivative are also bounded. For the convenience of subsequent analysis, we assume that $\|\dot{\boldsymbol{\Delta}}\| \leq \overline{\Delta}$.*

Assumption 3. *The desired trajectory $\boldsymbol{\eta}_d = [x_d \; y_d]^T$ and its first two time derivatives $\dot{\boldsymbol{\eta}}_d$, $\ddot{\boldsymbol{\eta}}_d$ are bounded. Furthermore, the desired yaw angle $\psi_d = \arctan\frac{\dot{y}_d(t)}{\dot{x}_d(t)}$ is also bounded.*

4. DO and Saturated Controller Design

4.1. DO Design

Here, we introduce an auxiliary variable to design a DO to estimate $\boldsymbol{\Delta}$, and its expression is as follows:

$$\begin{cases} \hat{\boldsymbol{\Delta}} & = \boldsymbol{\xi} + K_1 \boldsymbol{\eta}_2 \\ \dot{\boldsymbol{\xi}} & = K_1(-\boldsymbol{\Phi}(\mathbf{z}) - g\psi\sigma(\tau_c, \tau_{max})) - K_1(\boldsymbol{\xi} + K_1\boldsymbol{\eta}_2) \end{cases} \tag{15}$$

where $\hat{\boldsymbol{\Delta}}$ is the estimated value of $\boldsymbol{\Delta}$, $\boldsymbol{\xi}$ is the auxiliary variable and $K_1 > 0$ is the observer gain. Define the estimation error $e_1 = \hat{\boldsymbol{\Delta}} - \boldsymbol{\Delta}$. According to (13) and (15), the dynamic of e_1 is obtained as follows:

$$\begin{aligned} \dot{e}_1 &= \dot{\hat{\boldsymbol{\Delta}}} - \dot{\boldsymbol{\Delta}} \\ &= K_1(-\boldsymbol{\Phi}(\mathbf{z}) - f(\mathbf{z}) - g\psi\sigma(\tau_c, \tau_{max})) - K_1\hat{\boldsymbol{\Delta}} + K_1\dot{\boldsymbol{\eta}}_2 - \dot{\boldsymbol{\Delta}} \\ &= -K_1 e_1 - \dot{\boldsymbol{\Delta}} \end{aligned} \tag{16}$$

according to Assumption 2, the dynamic of e_1 can be regarded as the combination of the nominal dynamic $\dot{e}_0 = -K_1 e_0$ and the bounded disturbance $\dot{\boldsymbol{\Delta}}$. Obviously, the nominal dynamic is contracting with respect to e_0 with a contracting rate $\lambda_1 = K_1$ and a transformation matrix $\boldsymbol{\Theta}_1$.

According to triangular inequality, we obtain $\|e_1\| \leq \|e_1 - e_0\| + \|e_0\|$. Since $\dot{e}_0 = -K_1 e_0$ is contracting, there is $\|e_0(t)\| \leq \|e_0(0)\|$ and $e_0(0)$ is the initial value of e_0. Then, applying the robust properties of contracting system, we can obtain

$$\begin{aligned} \|e_1\| &\leq \|e_1 - e_0\| + \|e_0\| \\ &\leq \chi_1 \|e_1(0) - e_0(0)\| exp^{-\lambda_1 t} + \frac{\chi_1 \overline{\Delta}}{\lambda_1} + \|e_0(0)\| \\ &= v_1 \end{aligned} \tag{17}$$

where χ_1 is the upper bound of the condition number of $\boldsymbol{\Theta}_1$.

We can see that v_1 consists of three parts. The first part $\chi_1 \|e_1(0) - e_0(0)\| exp^{-\lambda_1 t}$ will converge to zero exponentially, so the size of v_1 will ultimately depend on the last two items. Obviously, $\|e_1\|$ can be infinitely close to zero by selecting the appropriate observer gain and the initial estimation values of the disturbances.

4.2. Saturated Controller Design

In this subsection, we apply CT and SPT to design a saturated controller. Firstly, we define a tracking error $e_2 = \boldsymbol{\eta}_1(t) - \boldsymbol{\eta}_d(t)$ and then construct a new variable based on e_2:

$$S = \dot{e}_2 + K_2 e_2 \tag{18}$$

here $K_2 > 0$ is the gain. From the definition of S, we know that
$$S \to 0 \Rightarrow e_2 \to 0 \tag{19}$$

In the light of (13) and (18), we have

$$\dot{S} = F(\mathbf{z}, \boldsymbol{\tau}_c, \Delta, e_2) = \Phi(\mathbf{z}) + \Delta + g(\psi)\sigma(\boldsymbol{\tau}_c, \tau_{max}) - \ddot{\eta}_d + K_2 \dot{e}_2 \tag{20}$$

The present objective is to design a controller $\boldsymbol{\tau}_c$ to make S converge to 0. Applying SPT, we construct a dynamic process for $\boldsymbol{\tau}_c$:

$$\mu \dot{\boldsymbol{\tau}}_c = H(\mathbf{z}, \boldsymbol{\tau}_c, \Delta, e_2, \mu) = -g^T(\psi)(K_3 S + \Phi(\mathbf{z}) + \Delta + g(\psi)\sigma(\boldsymbol{\tau}_c, \tau_{max}) - \ddot{\eta}_d + K_2 \dot{e}_2) \tag{21}$$

where $0 < \mu \ll 1$ is the singular perturbation parameter and $K_3 > 0$ is the controller gain.

Equations (20) and (21) constitute a standard SPS [51], where (20) is the slow subsystem, and (21) is the fast subsystem. Since $\mu \ll 1$, the dynamic process of (21) is much faster than (20).

Now, we apply CT to analyse the properties of the controller. By solving the partial derivative of $\boldsymbol{\tau}_c$, the Jacobian of (21) is

$$J_b = -\frac{1}{\mu} g^T(\psi) g(\psi) \begin{bmatrix} \frac{\partial \sigma(\cdot)}{\partial \tau_u} & \\ & \frac{\partial \sigma(\cdot)}{\partial \tau_r} \end{bmatrix} \tag{22}$$

According to the properties of $\sigma(\cdot)$ and $g(\psi)$, it is obvious that J_b is UND. Therefore, for any $\mathbf{z}, \Delta, e_2$, and μ, $H(\mathbf{z}, \boldsymbol{\tau}_c, \Delta, e_2, \mu)$ is partially contracting with respect to $\boldsymbol{\tau}_c$, we assume that the contracting rate is $\frac{\lambda_2}{\mu}$ and the transformation matrix is Θ_2. Then, according to the results in [3], the algebraic equation $H(\mathbf{z}, \boldsymbol{\tau}_c, \Delta, e_2, 0) = 0$ can be equivalently written as $\boldsymbol{\tau}_d = \vartheta(\mathbf{z}, \Delta, e_2)$, i.e., there is a unique, global mapping between $\boldsymbol{\tau}_c$ and $\mathbf{z}, \Delta, e$, where $\boldsymbol{\tau}_d$ is the root of the above algebraic equation, which is also called the quasi-steady state of the fast subsystem (21). According to the properties of contracting system, $\boldsymbol{\tau}_c$ converges to its quasi-steady state $\boldsymbol{\tau}_d$ exponentially. Now, solving the algebraic equation $H(\mathbf{z}, \boldsymbol{\tau}_d, \Delta, e_2, 0) = 0$, we obtain

$$\sigma(\boldsymbol{\tau}_d, \tau_{max}) = g^{-1}(\psi)(-K_3 S - \Phi(\mathbf{z}) - \Delta + \ddot{\eta}_d - K_2 \dot{e}_2) \tag{23}$$

according to SPT, the slow subsystem can be simplified via the quasi-steady state of the fast subsystem, therefore, bring (23) into (20), and the simplified slow subsystem is obtained:

$$\dot{S} = -K_3 S \tag{24}$$

since $K_3 > 0$, the dynamic of S is contracting with a transformation metric I and a contracting rate $\lambda_s = K_3$. Therefore, S converges to 0 exponentially. Based on the above analysis, we know that

$$\mu \to 0 \Rightarrow \boldsymbol{\tau}_c \to \boldsymbol{\tau}_d \Rightarrow S \to 0 \tag{25}$$

Therefore, when μ is small enough and K_3 is large enough, $\boldsymbol{\tau}_c \to \boldsymbol{\tau}_d$ and $S \to 0$ quickly. At the same time, due to the existence of $\sigma(\cdot)$, the control input will not exceed the limit of the actuator. Compared with the literature in the introduction, the method proposed in this paper is simple in form and convenient in application when dealing with actuator saturation.

Since it has been proven in Section 3.1 that $\hat{\Delta}$ approaches Δ infinitely, we replace Δ in (21) with its estimated value, and the practical controller can be obtained:

$$\begin{aligned} \mu \dot{\boldsymbol{\tau}}_c &= -g^T(\psi)(K_3 S + \Phi(\mathbf{z}) + \hat{\Delta} + g(\psi)\sigma(\boldsymbol{\tau}_c, \tau_{max}) - \ddot{\eta}_d + K_2 \dot{e}_2 \\ \tau &= \tau_{max} tanh(\boldsymbol{\tau}_c / \tau_{max}) \end{aligned} \tag{26}$$

In order to ensure the fast convergence of $\boldsymbol{\tau}_c$ to $\boldsymbol{\tau}_d$ and e_2 to 0, K_2, K_3 and μ should be reasonably selected. At the same time, K_1 should be large enough to ensure the rapid convergence of the estimation error. Of course, due to the complexity of the system, each parameter has an impact on the performance of the controller. After many simulations, we find that three parameters l, μ, and K_2 have a significant impact on the performance of the controller. For l, when l is too small, the controller may diverge, while too large l will reduce the tracking accuracy. Reducing μ can improve the convergence speed, but it also makes the changes of the control input less smooth. For μ , the smaller the μ , the faster the convergence speed will be, but it will lead to drastic changes in the control input, while an excessively large μ will reduce the convergence speed. At the same time, the larger the K_2 , the faster the convergence speed, but this will cause oscillation. In addition, K_1 mainly affects the performance of DO, and K_3 has a limited impact on the results. Based on the above analysis, it is necessary to make a balance between tracking accuracy, convergence speed, and smooth changes in control input when selecting parameters.

5. Error Analysis

5.1. Error Analysis of Control Variables

As analysed above, when τ_c converges exponentially to its quasi-steady state τ_d, the contracting reduced-order slow subsystem will be obtained. However, there is always an error between τ_c and τ_d, because μ can only be be as small as possible and cannot be equal to zero. Here, we make a quantitative analysis of the error between τ_c and τ_d.

Defining the control variable error $\tau_e = \tau_c - \tau_d$. According to SPT, the fast boundary layer dynamics in a new time scale $t_1 = \frac{t}{\mu}$ can be derived as

$$\frac{d\tau_e}{dt_1} = H(z, \tau_e + \tau_d, \hat{\Delta} - e_1, e_2, \mu) - \mu \dot{\tau}_d \tag{27}$$

The unperturbed error dynamic of (27) is

$$\frac{d\tau_e}{dt_1} = H(z, \tau_e + \tau_d, \hat{\Delta} - e_1, e_2, \mu) \tag{28}$$

Because the Jacobian J_b in (22) is UND, it implies that (28) is also partially contracting with respect to τ_e, and the contracting rate is $\frac{\lambda_2}{\mu}$ and the transformation matrix is denoted as Θ_2 with a supermum condition number χ_2. Assume that $\dot{\tau}_d$ is Lipschitz continuous in τ_e and e_1 [42], i.e.,

$$\|\dot{\tau}_d(\cdot)\| \le c_1 \|\tau_e\| + c_2 \|e_1\| + c_3 \tag{29}$$

where c_1, c_2, c_3 are all positive constants.

According to the robust properties of contracting system and ignoring the initial value of τ_e,

$$\|\tau_e(t)\| = \|\tau_c - \tau_d\| \le \chi_2 \|\tau_c(0) - \tau_d(0)\| exp^{-\frac{\lambda_2}{\mu^2}t} + \frac{\mu\chi_2(c_1\|\tau_e\| + c_2\|e_1\| + c_3)}{\lambda_2} \tag{30}$$

then, we can obtain

$$\|\tau_e(t)\| \le \frac{\lambda_2\chi_2}{\lambda_2 - \mu\chi_2 c_1} \|\tau_c(0) - \tau_d(0)\| exp^{-\frac{\lambda_1}{\mu^2}t} + \frac{\mu\chi_2(c_2v_1 + c_3)}{\lambda_2 - \mu\chi_2 c_1} \tag{31}$$

it can be observed that τ_e depends not only on μ, but also on λ_2 and χ_2, which are related to the values of observer gain and controller gain. Therefore, τ_e can be very small by reasonably selecting parameters.

5.2. Analysis of Tracking Error

The existence of e_1 and τ_e causes S to fail to converge to zero. In this subsection, a quantitative analysis of S is conducted. By performing some simple transformations on (20), we obtain a new dynamic form of S

$$\begin{aligned} \dot{S} =& \Phi(z) + g(\psi)\sigma(\tau_d, \tau_{max}) + \Delta - \ddot{\eta}_d + K_2\dot{e}_2 + \\ & \Phi(z) + g(\psi)\sigma(\tau_c, \tau_{max}) + \Delta - \ddot{\eta}_d + K_2\dot{e}_2 - \\ & (\Phi(z) + g(\psi)\sigma(\tau_d, \tau_{max}) + \Delta - \ddot{\eta}_d + K_2\dot{e}_2) \end{aligned} \tag{32}$$

By substituting (23) in (32), we obtain

$$\dot{S} = -K_3 S + g(\psi)\sigma(\tau_d, \tau_{max})(\sigma(\tau_c, \tau_{max}) - \sigma(\tau_d, \tau_{max})) \tag{33}$$

Similarly, the dynamic of S can be regarded as the coupling of the contracting nominal dynamic $\dot{S}_0 = -K_3 S_0$ and the bounded disturbance $g(\psi)(\sigma(\tau_c, \tau_{max}) - \sigma(\tau_d, \tau_{max}))$. Suppose the contracting rate is λ_3 and the transformation matrix is Θ_3. For the convenience of analysis, it is further assumed that $g(\psi)(\sigma(\tau_c, \tau_{max}) - \sigma(\tau_d, \tau_{max}))$ is Lipschitz continuous in τ_e with constant L, i.e.,

$$\|g(\psi)(\sigma(\tau_c, \tau_{max}) - \sigma(\tau_d, \tau_{max}))\| \leq L\|\tau_c - \tau_d\| = L\|\tau_e\| \tag{34}$$

Continuing to apply the triangular inequality and the robustness property of the contracting system, and we can obtain:

$$\|S\| \leq \|S - S_0\| + \|S_0(0)\|$$
$$\leq \chi_3 \|S(0) - S_0(0)\| exp^{-\lambda_3 t} + \frac{\chi_3 L \|\tau_e\|}{\lambda_3} + \|S_0(0)\| \tag{35}$$

where χ_3 is the upper bound of the condition number of Θ_3 and $S_0(0)$ is the initial value of S_0. If we replace τ_e in (35) with (31), a more detailed expression of S can be obtained.

From the above analysis, we can see that the estimation error and tracking error can finally converge to a small range by reasonably selecting the controller gain, the observer gain, and the singular perturbation parameters.

6. Numerical Simulations

In this section, based on Matlab Simulink, numerical simulations on the Remus AUV [52] are conducted to illustrate the effectiveness of the proposed method; the parameters in (2) are given as follows: $m = 30.58$, $I_z = 3.45$, $X_{\dot{u}} = -0.93$, $Y_{\dot{v}} = -35.5$, $N_{\dot{r}} = -4.88$, $X_u = -13.5$, $Y_v = -66.6$, $N_r = -6.87$, $X_{|u|u} = -1.62$, $Y_{|v|v} = -131$ and $N_{|r|r} = -188$.

In order to highlight the advantages of this method, the simulation results obtained based on the methods in [53] are compared with the results in this paper. In [53], the sliding mode control method was applied to deal with the trajectory tracking problem of an underactuated AUV by introducing a first-order sliding surface in terms of surge tracking errors and a second-order surface in terms of lateral motion tracking errors. The sliding surfaces are defined as

$$\begin{cases} S_1 = e_u + \lambda_1 \int_0^t e_u(\tau)d\tau \\ S_2 = \dot{e}_v + \lambda_3 e_v + \lambda_2 \int_0^t e_v(\tau)d\tau \end{cases} \tag{36}$$

where $\lambda_1, \lambda_2, \lambda_3 > 0$. $e_u = u - u_d$ and $e_v = v - v_d$ are tracking errors for surge and sway velocity, respectively. u_d and v_d are desired surge velocity and sway velocity, and they are defined as follows:

$$\begin{bmatrix} u_d \\ v_d \end{bmatrix} = \begin{bmatrix} cos\psi & sin\psi \\ -sin\psi & cos\psi \end{bmatrix} \begin{bmatrix} \dot{x}_d + l_x tanh(-\frac{k_x}{l_x}x_e) \\ \dot{y}_d + l_y tanh(-\frac{k_y}{l_y}y_e) \end{bmatrix} \tag{37}$$

where $k_x, k_y > 0$ are controller gains and $l_x, l_y > 0$ are saturation constants. $x_e = x - x_d$ and $y_e = y - y_d$ are position tracking errors.

Then, the control law is given by

$$\begin{cases} \tau_u = \tau_{u,eq} + \tau_{u,sw} \\ \tau_u = \tau_{r,eq} + \tau_{r,sw} \end{cases} \tag{38}$$

where

$$\tau_{u,eq} = -m_{22}vr - X_u u - X_{|u|u}|u|u + m_{11}\dot{u}_d - m_{11}\lambda_1 e_u$$

$$\tau_{u,sw} = m_{11}(-k_1 S_1 - W_1 sign(S_1))$$

$$\tau_{r,eq} = -(m_{11} - m_{22})uv - N_r r - N_{|r|r}|r|r + (m_{11}m_{22}^{-1}\dot{u}r - m_{22}^{-1}(Y_v + 2sign(v)Y_{|v|v}v)\dot{v})/b$$
$$+ (\Gamma - \lambda_3 \dot{e}_v - \lambda_2 e_v)/b$$

$$\tau_{r,sw} = (-k_2 S_2 - W_2 sign(S_2))/b$$

$$b = m_{33}^{-1}(u_d - m_{11}m_{22}^{-1})$$

$$\Gamma = -\ddot{x}_d sin\psi + \ddot{y}_d cos\psi - \ddot{x}_d r cos\psi - \ddot{y}_d r sin\psi - \dot{u}_d r + \gamma_1 r cos\psi + \gamma_2 r sin\psi$$
$$+ \dot{\gamma}_1 sin\psi - \dot{\gamma}_2 cos\psi$$

$$\gamma_1 = k_x \dot{x}_e sech^2(-\frac{k_x}{l_x}x_e)$$

$$\gamma_2 = k_y \dot{y}_e sech^2(-\frac{k_y}{l_y}y_e)$$

The first derivatives of the desired velocity is given by

$$\begin{bmatrix} \dot{u}_d \\ \dot{v}_d \end{bmatrix} = r \begin{bmatrix} -sin\psi & cos\psi \\ -cos\psi & -sin\psi \end{bmatrix} \begin{bmatrix} \dot{x}_d + l_x tanh(-\frac{k_x}{l_x}x_e) \\ \dot{y}_d + l_y tanh(-\frac{k_y}{l_y}y_e) \end{bmatrix} + \begin{bmatrix} cos\psi & sin\psi \\ -sin\psi & cos\psi \end{bmatrix} \begin{bmatrix} \ddot{x}_d - \gamma_1 \\ \ddot{y}_d - \gamma_2 \end{bmatrix} \quad (39)$$

There are two types of trajectories that need to be tracked by the AUV, one is a sinusoidal curve, and the other is a combination of straight lines and circles.

Trajectory 1: Sinudoidal curve

$$\begin{cases} x_d(t) = 0.5t \\ y_d(t) = 20cos(0.02t) \end{cases}$$

The total simulation time is 200π s. To facilitate the distinction, we mark the simulation based on the method in this paper as Case 1, while the simulation based on the literature [53] is marked as Case 2. The various settings for Case 1 and Case 2 simulations have been given in Table 1.

Table 1. Settings in Case 1 and Case 2.

Terms	Case 1	Case 2
parameters	$l = 0.35, K_1 = 10, K_2 = 0.024$ $K_3 = 500, \mu = 0.01$	$l_x = l_y = 1, k_x = k_y = 0.5$ $l_1 = l_2 = l_3 = 1$ $k_1 = k_2 = 1, W_1 = W_2 = 0.1$
initial conditions	$[x(0)\ y(0)\ \psi(0)]^T = [0\ -2\ 0]^T$ $[u(0)\ v(0)\ r(0)]^T = [0.2\ 0\ 0]^T$ $\xi(0) = 0, \tau_c(0) = 0$	$[x(0)\ y(0)\ \psi(0)]^T = [0\ -2\ 0]^T$ $[u(0)\ v(0)\ r(0)]^T = [0.2\ 0\ 0]^T$
actuator limitation	$\tau_{u,max} = \tau_{r,max} = 30$	

Trajectory 2: A combination of straight lines and circles

$$\begin{cases} x_d(t) = 0.4t, y_d(t) = 20, t < 100 \\ x_d(t) = 40 + 20cos(0.02t + 1.5\pi - 2), y_d(t) = 40 + 20sin(0.02t + 1.5\pi/ -2), t \geq 100 \end{cases}$$

The total simulation time is $100\pi + 100$ s. Similarly, two simulations are called Case 3 and Case 4, respectively. The various settings required for them are shown in Table 2.

Table 2. Settings in Case 3 and Case 4.

Terms	Case 3	Case 4
parameters	$l = 0.25, K_1 = 50, K_2 = 0.08$ $K_3 = 500, \mu = 0.02$	$l_x = l_y = 1, k_x = k_y = 0.5$ $l_1 = l_2 = l_3 = 1$ $k_1 = k_2 = 1, W_1 = W_2 = 0.1$
initial conditions	$[x(0)\ y(0)\ \psi(0)]^T = [-1\ -18\ 0]^T$ $[u(0)\ v(0)\ r(0)]^T = [0.2\ 0\ 0]^T$ $\xi(0) = 0, \tau_c(0) = 0$	$[x(0)\ y(0)\ \psi(0)]^T = [-1\ -18\ 0]^T$ $[u(0)\ v(0)\ r(0)]^T = [0.2\ 0\ 0]^T$
actuator limitation	$\tau_{u,max} = 30, \tau_{r,max} = 25$	

The following disturbances are considered in these simulations:

$$\begin{cases} \tau_{d1} = 0.1 f_u - 3 + 2sin(0.2t) + 2\varepsilon(5) \\ \tau_{d2} = 0.2 f_v + 5 + 0.2sin(0.2t) + 5\varepsilon(5) \\ \tau_{d3} = 0.15 f_r + 2 + sin(0.1t) + 3\varepsilon(5) \end{cases} \tag{40}$$

here, $\varepsilon(5)$ is the zero-mean white noise with power intensity of 5%. Specifically, the first term denotes degrees of model uncertainties. The second, third, and forth terms, respectively, account for the constant, periodic unknown disturbances, and Gaussian white noise.

Simulation results for Case 1 and Case 2 are shown in Figures 3–6. Figure 3 illustrates the trajectory of the AUV. It can be seen that both the method in this paper and the sliding mode control method in the literature have high tracking accuracy. The reference trajectory in Case 1 and the actual trajectory in Case 2 all converge to the desired trajectory. Of course, the error between the actual trajectory and the desired trajectory in Case 1 is relatively large, which can be more clearly seen from Figure 4.

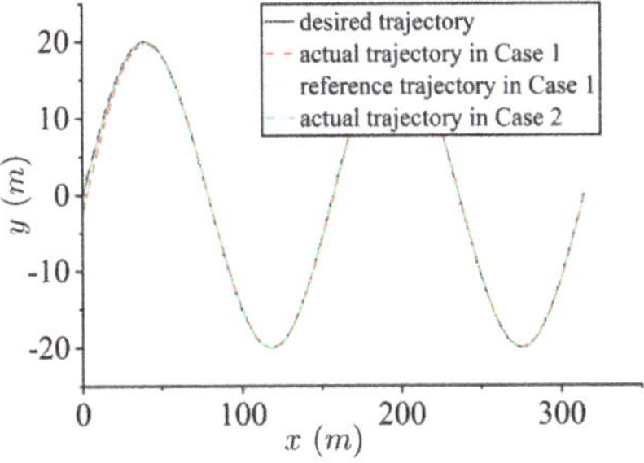

Figure 3. The comparison of trajectories in Case 1 and Case 2.

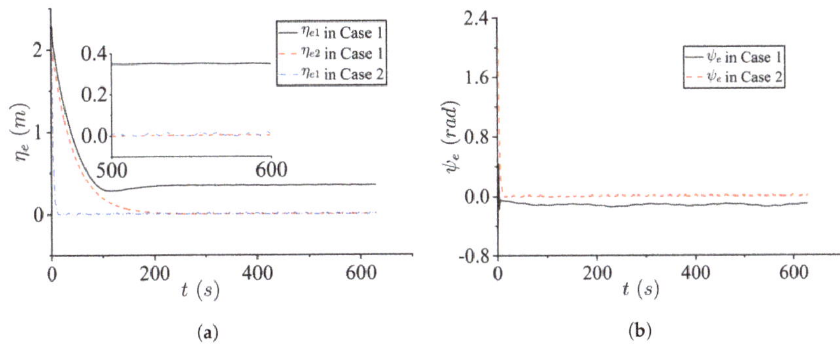

Figure 4. The tracking errors in Case 1 and Case 2: (**a**) position error, (**b**) heading angle error.

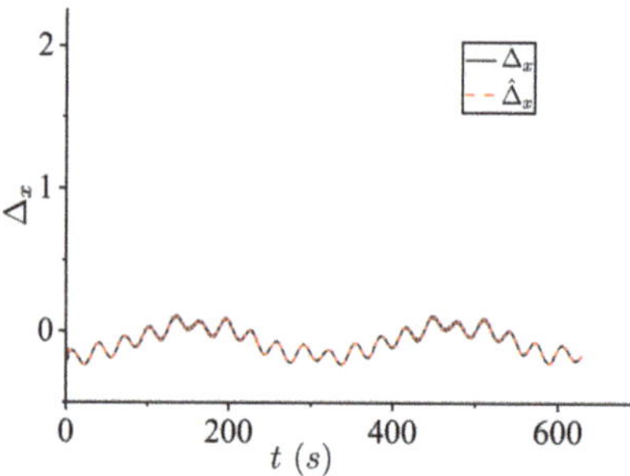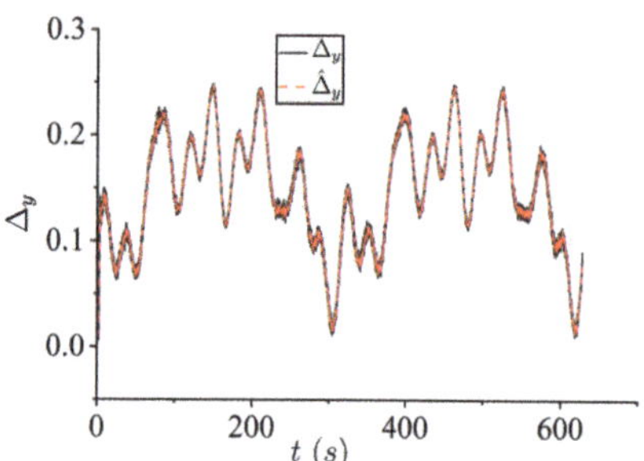

Figure 5. The estimation results of DO in Case 1.

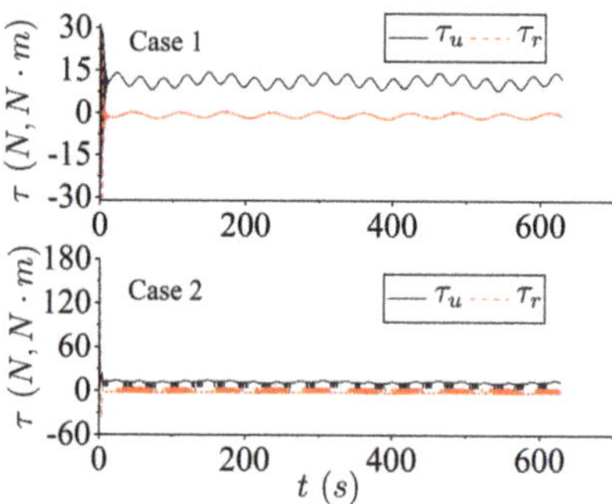

Figure 6. The control inputs in Case 1 and Case 2.

Figure 4 shows the trajectory tracking error and heading angle error, where $\eta_{e1} = \sqrt{(x - x_d)^2 + (y - y_d)^2}$ and $\eta_{e2} = \sqrt{(x_r - x_d)^2 + (y_r - y_d)^2}$ denote the trajectory tracking error, $\psi_e = \psi - \psi_d$ represents the heading angle error. It can be seen from Figure 4a that the trajectory error η_{e2} in Case 1 converges after about 200s, and it is almost 0, indicating that the reference state defined via coordinate transformation can accurately be traced to the desired state. However, η_{e1} in Case 1 does not converge to 0 but remains around 0.35, which is just the value of l. Reviewing Equation (6), we know that there is an inherent error between the reference state and the actual state. The results here just confirm this phenomenon. A very natural idea is that the value of l should be as small as possible to reduce η_{e1}. However, we find in the simulation that too small l will lead to slower convergence and even divergence. Therefore, the value of l should be balanced between tracking accuracy and convergence speed. In contrast, the sliding mode control methods in the literature have high tracking accuracy and convergence speed regardless of trajectory tracking error shown in Figure 4a or heading angle error shown in Figure 4b, which is a disadvantage of the method in this paper.

Figure 5 provides the results of the DO in Case 1. It reveals that the unknown disturbance including model uncertainties and time-varying environmental disturbance can be accurately estimated by the DO designed in this paper.

Figure 6 shows the control inputs of surge force and yaw torque in Case 1 and Case 2. It can be seen that the method in this paper considers the problem of input saturation, so neither τ_u nor τ_r exceed the limitation of the actuator, and the control input changes smoothly, which is conducive to the stable operation of the actuator. However, the methods in the literature do not solve the problem of input saturation. We can see that at the initial stage of simulation, both τ_u and τ_r are large, which can easily exceed the limitation of the actuator. Moreover, due to the use of symbolic functions in the controller, there is significant chattering in the control input curve, which is detrimental to the stable operation of the actuator.

In general, the sliding mode control methods in the literature excel in tracking accuracy and convergence speed. The advantage of the method in this paper lies in the simple form of the controller, and it easily solves the problem of input saturation, ensuring stable changes in control inputs and stable operation of the actuator. Of course, improving tracking accuracy and convergence speed is the work we must do in the next stage.

Simulation results for Case 3 and Case 4 are shown in Figures 7–10. It can be seen that the conclusions obtained from the analysis of Case 1 and Case 2 are still applicable here. Although the desired trajectory becomes more complex, the AUV still tracks the desired trajectory accurately. At the same time, even if the input saturation is more complex, the saturated controller proposed in this paper enables the control input in each direction to not exceed the limit of the actuator. Similarly, it can be seen that the performance of the sliding mode control method in the literature in Case 4 is the same as in Case 2.

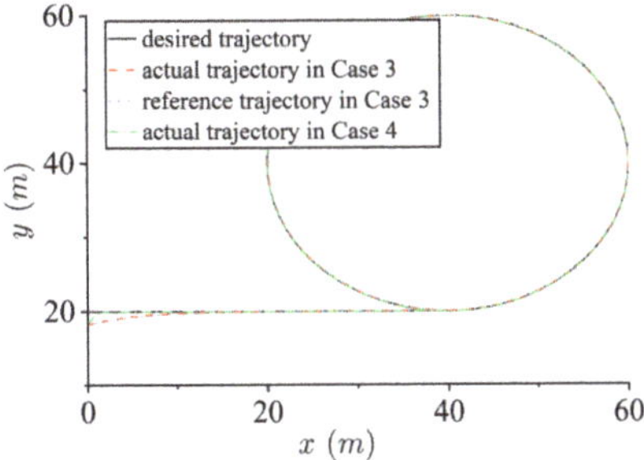

Figure 7. The comparison of trajectories in Case 3 and Case 4.

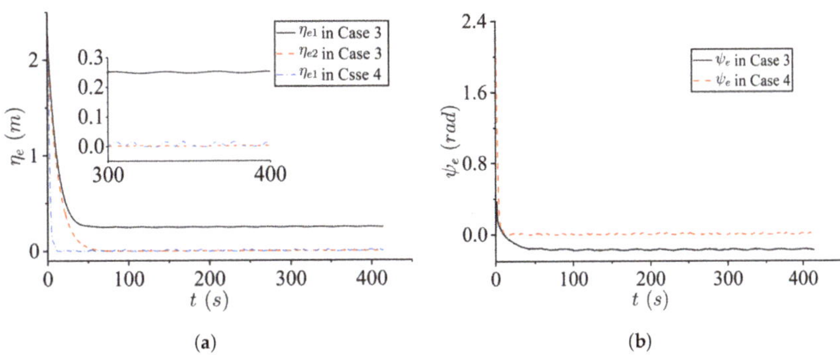

(a) (b)

Figure 8. The tracking errors in Case 3 and Case 4: (**a**) position error, (**b**) heading angle error.

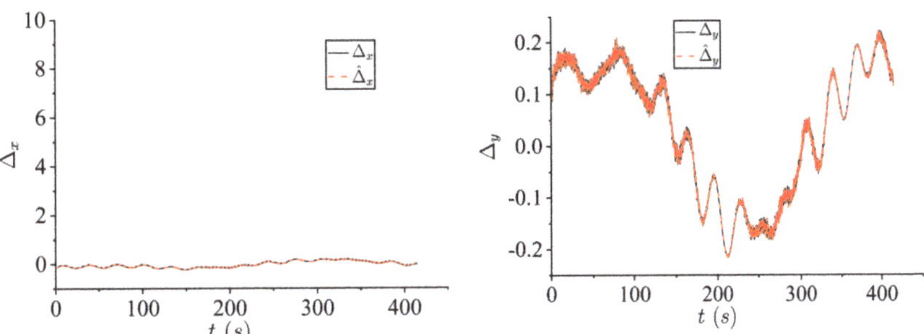

Figure 9. The estimation results of DO in Case 3.

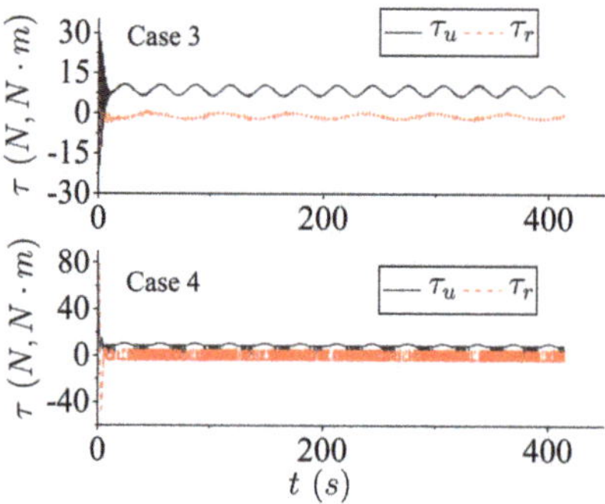

Figure 10. The control inputs in Case 3 and Case 4.

In all of above simulation results, the proposed controller works well for the horizontal trajectory tracking of underactuated AUVs in the presence of unknown internal and external disturbances. Therefore, we reach a conclusion that the validity and efficacy of the proposed method and proposed control scheme are sufficiently demonstrated.

7. Conclusions

This paper focusses on the horizontal trajectory tracking control of a 3-DOF underactuated AUV in the face of model uncertainties, time-varying external disturbance, and input saturation. A coordinate transformation is introduced to tackle the problem of underactuation and a DO is designed to estimate the total unknown disturbance. Applying CT and SPT, we design a saturation controller and perform quantitative analysis for the estimation error and the tracking error. Simulation results show that the controller proposed in this paper makes the AUV track the desired trajectory well and avoid the problem of input saturation. Of course, compared to the methods in the literature, the method in this paper still needs to be improved in terms of tracking accuracy and convergence speed. Therefore, the next research direction is to improve the tracking accuracy and convergence speed on the basis of this paper, while expanding the research results to the 3D trajectory tracking of underactuated AUVs.

Author Contributions: C.M.: Conceptualization, Methodology, Software, Validation, Investigation, Writing—original draft; J.J.: Methodology, Validation, Writing—Review and Editing; T.Z.: Writing—Review and Editing, Supervision; S.W.: Writing—Review and Editing; D.J.: Conceptualization, Resources, Writing—Review and Editing, Supervision, Project administration. All authors have read and agreed to the published version of the manuscript.

Funding: This research received no external funding.

Institutional Review Board Statement: No applicable.

Informed Consent Statement: No applicable.

Data Availability Statement: Not applicable.

Conflicts of Interest: The authors declare no conflict of interest. The funders had no role in the design of the study; in the collection, analysis, or interpretation of data; in the writing of the manuscript, or in the decision to publish the results.

Appendix A

Due to the use of many abbreviations in this paper, for the sake of standardization, they are summarized in the Table A1 according to the order in which they appear in the paper.

Table A1. The abbreviations used in this paper.

Abbreviation	Full Name
AUV	autonomous underwater vehicle
CT	contraction theory
DO	disturbance observer
SPT	singular perturbation theory
SPS	singular perturbation system
NN	neural networks
FL	fuzzy logic
DOF	degrees-of-freedom
USV	unmanned surface vehicle
SF	Serrent–Frenet
ESO	extended state observer
3D	three-dimensional
MPC	model predictive control
DAI	deterministic artificial intelligence
UUV	unmanned underwater vehicles
UND	uniformly negative definite
UGES	uniformly global exponential stable
pH	port-Hamiltonian

Appendix B

Appendix B.1. Contraction Theory

For a nonlinear system

$$\dot{x} = f(x,t),\ x(t_0) = x_0,\ \forall t \geq t_0 \geq 0 \tag{A1}$$

where $x(t) \in \mathbb{R}^n$ is the state vector, t is the time and $f : \mathbb{R} \times \mathbb{R}_{t \geq t_0} \to \mathbb{R}^n$ is a nonlinear smooth function, meaning that all required derivatives and partial derivatives exist and are continuous. If there is a positive scalar λ and a uniformly positive matrix Θ, such that

$$(F + F^T)/2 \leq -\lambda I_n \tag{A2}$$

then (A1) is said to be contracting, where I_n denotes the identity matrix with dimension n, $F = (\dot{\Theta} + \Theta^T \frac{\partial f(x,t)}{\partial x})\Theta^{-1}$ is the generalized Jacobian. For a contracting system, the trajectories starting from any initial condition will converge together exponentially. If $\lambda = 0$, (A1) is called semi-contracting and all its trajectories converge together asymptotically [2].

Partial contraction is a very important concept in CT [40]. Let the auxiliary system, called virtual system

$$\dot{\varsigma} = f(\varsigma, x, t) \tag{A3}$$

associated with (A1) through $f(x, x, t) = f(x, t)$. Assume that (A3) is contracting with respect to ς, i.e., the Jacobian $\frac{\partial f(\varsigma, x, t)}{\partial \varsigma}$ is UND for any ς and x.

If a particular solution of the virtual system verifies a smooth specific property, then all trajectories of the original x-system verify the same property exponentially. The original system is called partial contracting.

When a contracting system is subject to bounded disturbance, the error between the trajectory of the system after the disturbance and the original system trajectory is very small, that is, the contracting system is robust.

Consider the perturbed system:

$$\dot{x}_p = f(x_p, t) + d(x_p, t) \tag{A4}$$

where $d(x_p, t)$ is bounded, i.e., $\exists\, d_0 \geq 0, \forall x_p, \forall t \geq 0, \|d(x_p, t)\| \leq d_0$. Then the error between $x(t)$ and $x_p(t)$ satisfies

$$\|x_p(t) - x(t)\| \leq \chi_p \|x_p(0) - x(0)\| exp^{-\lambda t} + \frac{\chi_p d_0}{\lambda} \tag{A5}$$

where χ_p is the upper bound of the condition number of Θ.

Appendix B.2. Contraction Analysis of Singular Perturbation System

For a standard singular perturbation system (SPS) [51]:

$$\begin{cases} \dot{x} & = f(x, z) \\ \epsilon \dot{z} & = g(x, z, \epsilon) \end{cases} \tag{A6}$$

$0 < \epsilon \ll 1$ is the singular perturbation parameter. For any z, if the virtual system $\dot{y}_x = f(y_x, z)$ is contracting with respect to y_x, then the system (A6) is called to be partially contracting with respect to x. Similarly, it is partially contracting in z when the virtual system $\epsilon \dot{y}_z = g(x, y_z, \epsilon)$ is contracting for any x [3,40].

If the system (A6) is partially contracting in z, there exist a unique, global mapping between x, z and ϵ [3], i.e., the algebraic equation $g(x, z, \epsilon) = 0$ can be equivalently written as $z = h(x, \epsilon)$, here $h(x, \epsilon)$ is called slow manifold or the quasi-steady state of the z-subsystem. According to SPT, the x-subsystem can be simplified by introducing the slow manifold into the x-subsystem:

$$\dot{x}_{re} = f(x_{re}, h(x_{re}, \epsilon)) \tag{A7}$$

To analyse the convergence behavior between z and the slow manifold $h(x, \epsilon)$, we define a error variable $y = z - h(x, \epsilon)$, and the dynamic for y can be expressed as:

$$\frac{dy}{d\tau} = g(x, y + h(x, \epsilon), \epsilon) - \epsilon \frac{h(\cdot)}{dt} \tag{A8}$$

where $\tau = \frac{t}{\epsilon}$ is a new time scale. Then, the dynamic behavior of the whole SPT (A6) can be determined by analysing the behavior of the reduced order state variables x_{re} and z.

References

1. Wang, F.; Wan, L.; Li, Y.; Su, Y.; Xu, Y. A survey on development of motion control for underactuated AUV. *Shipbuild. China* **2010**, *51*, 227–241.
2. Lohmiller, W.; Slotine, J.J.E. On contraction analysis for non-linear systems. *Automatica* **1998**, *34*, 683–696. [CrossRef]
3. Del Vecchio, D.; Slotine, J.-J.E. A contraction theory Approach to Singularly Perturbed Systems. *IEEE Trans. Autom. Control* **2013**, *58*, 752–757. [CrossRef]
4. Liang, X.; Qu, X.; Wan, L.; Ma, Q. Three-Dimensional Path Following of an Underactuated AUV Based on Fuzzy Backstepping Sliding Mode Control. *Int. J. Fuzzy Syst.* **2017**, *20*, 640–649. [CrossRef]
5. Miao, J.; Wang, S.; Tomovic, M.M.; Zhao, Z. Compound line-of-sight nonlinear path following control of underactuated marine vehicles exposed to wind, waves, and ocean currents. *Nonlinear Dyn.* **2017**, *89*, 2441–2459. [CrossRef]
6. Miao, J.; Deng, K.; Zhang, W.; Gong, X.; Lyu, J.; Ren, L. Robust Path-Following Control of Underactuated AUVs with Multiple Uncertainties in the Vertical Plane. *J. Mar. Sci. Eng.* **2022**, *10*, 238. [CrossRef]
7. Zhang, J.; Xiang, X.; Lapierre, L.; Zhang, Q.; Li, W. Approach-angle-based three-dimensional indirect adaptive fuzzy path following of under-actuated AUV with input saturation. *Appl. Ocean. Res.* **2021**, *107*, 102486. [CrossRef]
8. Ghommam, J.; Mnif, F.; Benali, A.; Derbel, N. Asymptotic Backstepping Stabilization of an Underactuated Surface Vessel. *IEEE Trans. Control Syst. Technol.* **2006**, *14*, 1150–1157. [CrossRef]
9. Ding, F.; Wu, J.; Wang, Y. Stabilization of an Underactuated Surface Vessel Based on Adaptive Sliding Mode and Backstepping Control. *Math. Probl. Eng.* **2013**, *2013*, 324954. [CrossRef]
10. Qin, J.; Du, J. Adaptive fast nonsingular terminal sliding mode control for underactuated surface vessel trajectory tracking. In Proceedings of the 19th International Conference on Control, Automation and Systems (ICCAS), Jeju, Republic of Korea, 15–18 October 2019; pp. 1365–1370. [CrossRef]
11. Qin, J.; Du, J. Robust adaptive asymptotic trajectory tracking control for underactuated surface vessels subject to unknown dynamics and input saturation. *J. Mar. Sci. Technol.* **2021**, *27*, 307–319. [CrossRef]

12. Shojaei, K.; Arefi, M.M. On the neuro-adaptive feedback linearising control of underactuated autonomous underwater vehicles in three-dimensional space. *IET Control. Theory Appl.* **2015**, *9*, 1264–1273. [CrossRef]
13. Yi, B.; Qiao, L.; Zhang, W. Two-time scale path following of underactuated marine surface vessels: Design and stability analysis using singular perturbation methods. *Ocean. Eng.* **2016**, *124*, 287–297. [CrossRef]
14. Lei, M.; Li, Y.; Pang, S. Extended state observer-based composite-system control for trajectory tracking of underactuated AUVs. *Appl. Ocean. Res.* **2021**, *112*, 102694. [CrossRef]
15. Lei, M.; Li, Y.; Pang, S. Robust singular perturbation control for 3D path following of underactuated AUVs. *Int. J. Nav. Archit. Ocean. Eng.* **2021**, *13*, 758–771. [CrossRef]
16. Han, J. From PID to active disturbance rejection control. *IEEE Trans. Ind. Electron.* **2009**, *56*, 900–906. [CrossRef]
17. Radke, A.; Gao, Z. A survey of state and disturbance observers for practitioners. In Proceedings of the 2006 American Control Conference, Minneapolis, MN, USA, 14–16 June 2006. [CrossRef]
18. Chen, W.-H.; Yang, J.; Guo, L.; Li, S. Disturbance-Observer-Based Control and Related Methods—An Overview. *IEEE Trans. Ind. Electron.* **2016**, *63*, 1083–1095. [CrossRef]
19. Liu, S.; Liu, Y.; Wang, N. Nonlinear disturbance observer-based backstepping finite-time sliding mode tracking control of underwater vehicles with system uncertainties and external disturbances. *Nonlinear Dyn.* **2016**, *88*, 465–476. [CrossRef]
20. Peng, Z.; Wang, J.; Wang, J. Constrained Control of Autonomous Underwater Vehicles Based on Command Optimization and Disturbance Estimation. *IEEE Trans. Ind. Electron.* **2019**, *66*, 3627–3635. [CrossRef]
21. Ali, N.; Tawiah, I.; Zhang, W. Finite-time extended state observer based nonsingular fast terminal sliding mode control of autonomous underwater vehicles. *Ocean. Eng.* **2020**, *218*, 108179. [CrossRef]
22. Liu, J.; Du, J. Composite learning tracking control for underactuated autonomous underwater vehicle with unknown dynamics and disturbances in three-dimension space. *Appl. Ocean. Res.* **2021**, *112*, 102686. [CrossRef]
23. Duan, K.; Fong, S.; Chen, C. Fuzzy observer-based tracking control of an underactuated underwater vehicle with linear velocity estimation. *IET Control. Theory Appl.* **2020**, *14*, 584–593. [CrossRef]
24. Galeani, S.; Tarbouriech, S.; Turner, M.; Zaccarian, L. A Tutorial on Modern Anti-windup Design. *Eur. J. Control* **2009**, *15*, 418–440. [CrossRef]
25. Chen, M.; Ge, S.S.; Ren, B. Adaptive tracking control of uncertain MIMO nonlinear systems with input constraints. *Automatica* **2011**, *47*, 452–465. [CrossRef]
26. Cui, R.; Zhang, X.; Cui, D. Adaptive sliding-mode attitude control for autonomous underwater vehicles with input nonlinearities. *Ocean. Eng.* **2016**, *123*, 45–54. [CrossRef]
27. Chu, Z.; Xiang, X.; Zhu, D.; Luo, C.; Xie, D. Adaptive Fuzzy Sliding Mode Diving Control for Autonomous Underwater Vehicle with Input Constraint. *Int. J. Fuzzy Syst.* **2017**, *20*, 1460–1469. [CrossRef]
28. Zhu, C.; Huang, B.; Su, Y.; Zheng, Y.; Zheng, S. Finite-time rotation-matrix-based tracking control for autonomous underwater vehicle with input saturation and actuator faults. *Int. J. Robust Nonlinear Control* **2021**, *32*, 2925–2949. [CrossRef]
29. Yu, C.; Xiang, X.; Zhang, Q.; Xu, G. Adaptive Fuzzy Trajectory Tracking Control of an Under-Actuated Autonomous Underwater Vehicle Subject to Actuator Saturation. *Int. J. Fuzzy Syst.* **2017**, *20*, 269–279. [CrossRef]
30. Xia, Y.; Xu, K.; Huang, Z.; Wang, W.; Xu, G.; Li, Y. Adaptive energy-efficient tracking control of a X rudder AUV with actuator dynamics and rolling restriction. *Appl. Ocean. Res.* **2022**, *118*, 102994. [CrossRef]
31. Yan, Z.; Gong, P.; Zhang, W.; Wu, W. Model predictive control of autonomous underwater vehicles for trajectory tracking with external disturbances. *Ocean. Eng.* **2020**, *217*, 107884. [CrossRef]
32. Gong, P.; Yan, Z.; Zhang, W.; Tang, J. Lyapunov-based model predictive control trajectory tracking for an autonomous underwater vehicle with external disturbances. *Ocean. Eng.* **2021**, *232*, 109010. [CrossRef]
33. Sands, T. Development of Deterministic Artificial Intelligence for Unmanned Underwater Vehicles (UUV). *J. Mar. Sci. Eng.* **2020**, *8*, 578. [CrossRef]
34. Shah, R.; Sands, T. Comparing Methods of DC Motor Control for UUVs. *Appl. Sci.* **2021**, *11*, 4972. [CrossRef]
35. Jouffroy, J. A relaxed criterion for contraction theory: Application to an underwater vehicle observer. In Proceedings of the 2003 European Control Conference (ECC), Cambridge, UK, 1–4 September 2003; pp. 2999–3004. [CrossRef]
36. Jouffroy, J.; Fossen, T.I. Tutorial on Incremental Stability Analysis using contraction theory. *Model. Identif. Control. Nor. Res. Bull.* **2010**, *31*, 93–106. [CrossRef]
37. Reyes-Báez, R. Virtual Contraction and Passivity Based Control of Nonlinear Mechanical Systems: Trajectory Tracking and Group Coordination. Ph.D. Thesis, University of Groningen, Groningen, The Netherlands, 2019.
38. Espindola-Lopez, E.; Tang, Y. Global exponential attitude tracking for spacecraft with gyro bias estimation. *ISA Trans.* **2021**, *116*, 46–57. [CrossRef]
39. Su, Y.; Ren, L.; Sun, C. Event-triggered robust distributed nonlinear model predictive control using contraction theory. *J. Frankl. Inst.* **2022**, *359*, 4874–4892. [CrossRef]
40. Wang, W.; Slotine, J.J. On partial contraction analysis for coupled nonlinear oscillators. *Biol. Cybern.* **2005**, *92*, 38–53. [CrossRef]
41. Chung, S.-J.; Slotine, J.J.E. Cooperative Robot Control and Concurrent Synchronization of Lagrangian Systems. *IEEE Trans. Robot.* **2009**, *25*, 686–700. [CrossRef]
42. Rayguru, M.M.; Mohan, R.E.; Parween, R.; Yi, L.; Le, A.V.; Roy, S. An Output Feedback Based Robust Saturated Controller Design for Pavement Sweeping Self-Reconfigurable Robot. *IEEE/ASME Trans. Mechatron.* **2021**, *26*, 1236–1247. [CrossRef]

43. Kong, F.H.; Manchester, I.R. Contraction analysis of nonlinear noncausal iterative learning control. *Syst. Control. Lett.* **2020**, *136*, 104599. [CrossRef]
44. Tsukamoto, H.; Chung, S.-J.; Slotine, J.-J.E. Neural Stochastic Contraction Metrics for Learning-Based Control and Estimation. *IEEE Control. Syst. Lett.* **2021**, *5*, 1825–1830. [CrossRef]
45. Tsukamoto, H.; Chung, S.-J. Neural Contraction Metrics for Robust Estimation and Control: A Convex Optimization Approach. *IEEE Control. Syst. Lett.* **2021**, *5*, 211–216. [CrossRef]
46. Tsukamoto, H.; Chung, S.-J. Robust Controller Design for Stochastic Nonlinear Systems via Convex Optimization. *IEEE Trans. Autom. Control* **2020**, *66*, 4731–4746. [CrossRef]
47. Jouffroy, J.; Fossen, T.I. On the combination of nonlinear contracting observers and UGES controllers for output feedback. In Proceedings of the 43rd IEEE Conference on Decision and Control (CDC), (IEEE Cat. No. 04CH37601), Paradise Island, Bahamas, 14–17 December 2004; Volume 5, pp. 4933–4939. [CrossRef]
48. Mohamed, M.; Swaroop, S. Incremental Input to State Stability of Underwater Vehicle. *IFAC-PapersOnLine* **2016**, *49*, 41–46. [CrossRef]
49. Mohamed, M.; Su, R. Contraction based tracking control of autonomous underwater vehicle. *IFAC-PapersOnLine* **2017**, *50*, 2665–2670. [CrossRef]
50. Fossen, T.I. *Handbook of Marine Craft Hydrodynamics and Motion Control*; John Wiley & Sons: Hoboken, NJ, USA, 2011.
51. Khalil, H.K. *Nonlinear Systems*, 3rd ed.; Prentice Hall: Upper Saddle River, NJ, USA, 2002.
52. Prestero, T. Verification of a Six-Degree of Freedom Simulation Model for the REMUS Autonomous Underwater Vehicle. Ph.D. Thesis, Massachusetts Institute of Technology, Cambridge, MA, USA, 2001.
53. Elmokadem, T.; Zribi, M.; Youcef-Toumi, K. Trajectory tracking sliding mode control of underactuated AUVs. *Nonlinear Dyn.* **2016**, *84*, 1079–1091. [CrossRef]

Journal of
*Marine Science
and Engineering*

Article

An Algorithm of Complete Coverage Path Planning for Unmanned Surface Vehicle Based on Reinforcement Learning

Bowen Xing [1,†], Xiao Wang [1,2], Liu Yang [1,†], Zhenchong Liu [3] and Qingyun Wu [1,*]

1   College of Engineering Science and Technology, Shanghai Ocean University, Shanghai 201306, China; bwxing@shou.edu.cn (B.X.)
2   Shanghai Investigation Design & Research Institute, Shanghai 200335, China
3   Shanghai Zhongchuan NERC-SDT Co., Ltd., Shanghai 201114, China
*   Correspondence: qywu@shou.edu.cn
†   These authors contributed equally to this work.

Abstract: A deep reinforcement learning method to achieve complete coverage path planning for an unmanned surface vehicle (USV) is proposed. This paper firstly models the USV and the workspace required for complete coverage. Then, for the full-coverage path planning task, this paper proposes a preprocessing method for raster maps, which can effectively delete the blank areas that are impossible to cover in the raster map. In this paper, the state matrix corresponding to the preprocessed raster map is used as the input of the deep neural network. The deep Q network (DQN) is used to train the complete coverage path planning strategy of the agent. The improvement of the selection of random actions during training is first proposed. Considering the task of complete coverage path planning, this paper replaces random actions with a set of actions toward the nearest uncovered grid. To solve the problem of the slow convergence speed of the deep reinforcement learning network in full-coverage path planning, this paper proposes an improved method of deep reinforcement learning, which superimposes the final output layer with a dangerous actions matrix to reduce the risk of selection of dangerous actions of USVs during the learning process. Finally, the designed method validates via simulation examples.

Keywords: environment modeling; raster map; screening matrix; DQN; reward function

Citation: Xing, B.; Wang, X.; Yang, L.; Liu, Z.; Wu, Q. An Algorithm of Complete Coverage Path Planning for Unmanned Surface Vehicle Based on Reinforcement Learning. *J. Mar. Sci. Eng.* **2023**, *11*, 645. https://doi.org/10.3390/jmse11030645

Academic Editors: Rafael Morales and Tieshan Li

Received: 27 January 2023
Revised: 20 February 2023
Accepted: 16 March 2023
Published: 19 March 2023

1. Introduction

In recent years, the field of USV based on sensor technology and intelligent algorithms has made great progress [1]. During the working process, USV first uses sensors to model the working environment, estimates its own state, and then makes decisions based on the surrounding environment and its own state to complete the corresponding tasks. Compared with humans, USVs are more suitable for performing some highly dangerous or repetitive tasks, such as target reconnaissance [2], hydrographic mapping [3], and water patrol [4]. To accomplish these tasks in unpredictable water environments, path planning techniques are crucial for USVs.

As an important aspect of robotics, coverage path planning (CPP), is special path planning. full-coverage path planning requires that the generated USV motion trajectory can cover all areas in the workspace, except obstacles, to the greatest extent. The full-coverage path planning algorithm has a wide range of applications, such as unmanned surface vehicles [5–7], intelligent sweeping robots [8], gardening electric tractors [9], cleaning robots [10,11], tile robots [12], rescue robots [13], window cleaning robot [14], automatic lawn mower [15], etc. In practice, there are many optimization algorithms that USV can use, such as the Taguchi method, ant colony algorithm, fuzzy logic, and TLBO, which provide many options for the design of path planning. The Taguchi method is an analytical approach for a design algorithm, first proposed to enhance the quality of products in manufacturing. Wang et al. [16] designed optimal bridge-type compliant mechanism flexure hinges with low stress by using a flexure joint. Sun et al. [17] proposed a new binary fully convolutional neural network (B-FCN) based on Taguchi method sub-optimization for the segmentation of robotic floor regions, which can precisely distinguish floor regions in complex indoor environments. The ant colony optimization algorithm is a bionic intelligent algorithm inspired by the foraging behavior of ant colonies. A global path planning algorithm based on an improved quantum ant colony algorithm (IQACA) is proposed by Xia et al. [18]. Fuzzy logic has gained much attention due to its ease of implementation and simplicity. The concept behind it is to smooth the classical Boolean logic of true (1 or yes) and false (0 or no) to a partial value located between these values, coming in a term of the degree of membership. Abdullah Alomari et al. [19] proposed a novel, distributed, range-free movement mechanism for mobility-assisted localization based on fuzzy-logic approach. Teaching a learning-based optimization (TLBO) algorithm is a distinguished, nature-inspired, population-based meta-heuristic, which is basically designed for unconstrained optimization. Wang et al. [20] proposed a novel approach to improve centrifugal pump performance based on TLBO.

In recent years, due to the new demand for USVs with full-coverage path planning in various fields, such as civil, military, patrol [21], search and rescue [22–25], cybersecurity industry [26], and inspection [27,28], scholars and experts from various countries have proposed various methods to complete the task of full-coverage path planning.

The full-coverage path planning algorithm based on the intelligent algorithm means that USV first models the environment through sensors and relies on the modeled map to perform full-coverage path planning [29]. The grid map is a common expression of the environment, and it divides the map into several square grids, which can be used for a full-coverage route plan. The full-coverage path planning algorithm, based on the traditional algorithm, includes the full-coverage path planning algorithm of boustrophedon or the inner spiral full-coverage path planning [30,31]. The boustrophedon method drives the USV to go back-and-forth along a path, similar to that of an ox plowing a field. In the work area that needs to be covered, the USV performs full-coverage path planning work with certain direction rules. Once an obstacle is encountered, the USV changes direction to cover in another direction. Lee et al. [32] proposed a full-coverage path planning algorithm based on the inner spiral. This algorithm is an online full-coverage path planning algorithm, which ensures complete coverage in unstructured planar environments. However, this kind of algorithm does not introduce global control [33], and the USV can only perceive the regional state within a certain range, instead of global information, so it will lead to a large cumulative error.

The most classic full-coverage path planning algorithm based on the exact cell decomposition method was proposed by Choset [34] in 2000. Choset developed an accurate cell decomposition method for full-coverage path planning, which is essentially a generalization of trapezoidal decomposition. It can be used in work environments with non-polygonal obstacles and is more efficient than trapezoidal decomposition. Zhu et al. [35] proposed the Glasius bionic neural network (GBNN) algorithm and improved this algorithm in view of the shortcomings of the bionic neural network algorithm, such as high computational complexity and long path planning time. This algorithm constructs a grid graph by discretizing the two-dimensional underwater environment and then establishes a corresponding dynamic neural network on the grid graph. Luo et al. [36] first proposed a complete coverage neural network (CCNN) algorithm for ultrasonic motor path planning. By simplifying the calculation process of neural activities, this algorithm can significantly reduce the calculation time. In order to improve coverage efficiency and generate more standardized paths, the author establishes the optimal next-step location decision formula, combined with the coverage direction term.

However, the above studies mainly focus on the coverage rate, and a systematic study on the coverage path repetition rate is rare. In this study, a new full-coverage path planning algorithm based on reinforcement learning is proposed. In the decision phase, USV can use the model learning capability of reinforcement learning to select actions based on the historical information of the area coverage. In the training phase, the information of area coverage is used to train the neural network, which makes the neural network model fit the best motion action. This algorithm improves the defects of the traditional full-coverage path planning algorithm with higher efficiency. Both ensure the coverage rate and significantly reduce the coverage path repetition rate. The feasibility and efficiency of the algorithm are verified in simulation experiments. This contribution is expected to provide a scientific reference for the future research about the coverage path repetition rate.

2. USV Models and Environment Modeling

2.1. USV Model

When the USV moves in the workspace, in order to avoid the USV colliding with the obstacles or the map boundary, it is necessary to model according to the size and shape of the USV. In path planning, there are two ways to model the USV. The first one is to model the USV in real-time and the second one is to inflate obstacles to a width the size of the vehicle's radius. Since the first method needs to model the USV in real-time, it can move the USV irregularly with a better effect and stronger robustness, but it consumes too many computing resources and is difficult to implement. The obstacle avoidance effect is very good, but the efficiency of the path planning is too low. In the second method, an irregular USV can be regarded as a circular USV with a diameter equal to the maximum width of the USV, as shown in Figure 1. The model diagram of the USV is a subset of the approximate model of the USV, that is, the area of the approximate model of the USV is greater than or equal to the model diagram of the USV, so as long as the approximate model of the USV does not collide with obstacles, the USV can be guaranteed. During the movement, it is possible not to collide with obstacles. This will not lose too much USV modeling accuracy, but also has the advantages of good real-time performance, convenience, reliability, and high efficiency. Since the grid size selected in this paper is equal to the diameter of the approximate image of the USV, the target point moved by the USV each time can be set as the midpoint of the next grid, and when the USV needs to move obliquely, it is very likely that, as shown in Figure 2, in order to simplify the kinematic model of the USV, this paper sets the movement direction of the USV to only the horizontal and vertical directions and does not support the oblique direction. The red arrows in Figure 2 indicate the direction of usv's oblique motion.

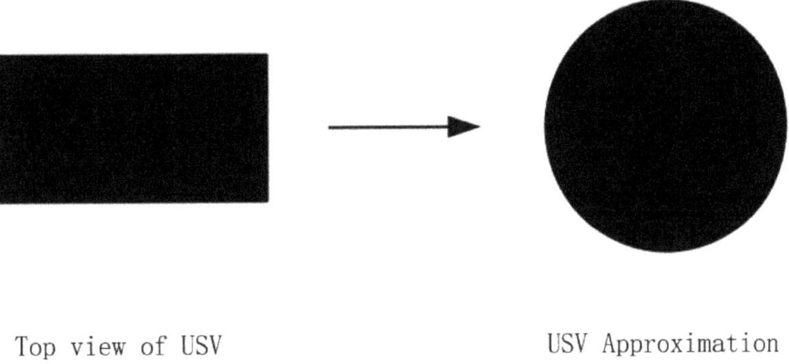

Figure 1. USV approximation.

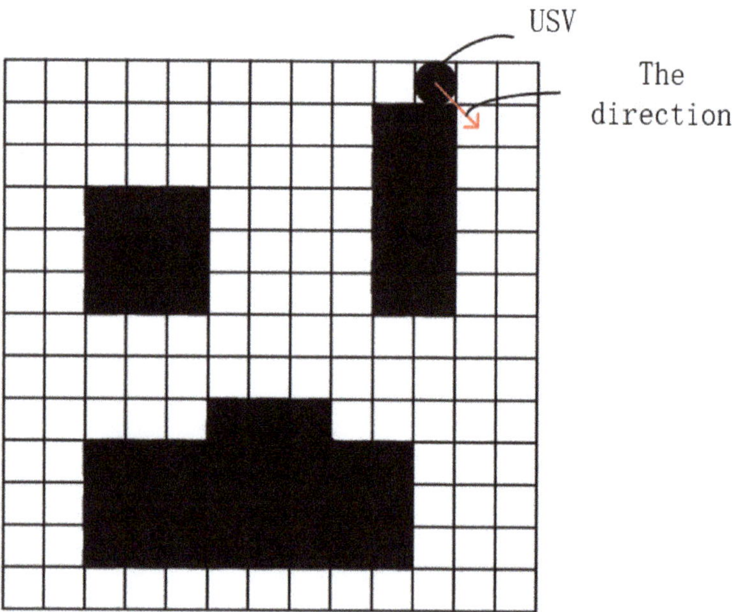

Figure 2. Oblique motion diagram of USV.

2.2. Workspace Modeling and Preprocessing

When the USV uses the grid map modeling method for full-coverage path planning, the resolution of the grid map has a great influence on the efficiency of its algorithm. If the resolution is too high, the map information will be redundant, will take up a lot of memory, and will even lead to repeated traversal, which will reduce the efficiency of the full-coverage path planning algorithm. When the resolution is too low, although the memory footprint of the grid map will become very small, it cannot represent some details of the environmental map, and there is a high probability that the blank area around the obstacle will be represented as the obstacle area, so using low-resolution raster maps results in lower coverage for USVs. Therefore, when modeling the working environment of the USV, in order to reduce the memory occupied by the map and represent some details of the environment, this paper adopts the coverage size of the USV as the resolution size, that is, the size of each grid is the size that the USV can use once. The size of the coverage area. When the USV reaches the middle position of the grid, it is considered that the grid has been covered by the USV.

The construction method of the grid map in this paper is to first divide the working environment of the USV into several grids of the same size, set the grid that the USV can reach and cover as white with a value of 0, and set the grid where the obstacles are located. The grid is set to black, with a value of -1. However, there is only a part of obstacles in some grids, that is, there are both white areas and black areas in this grid. Here, in order to prevent the USV from being damaged by collisions when performing full-coverage tasks, this paper uses such grids. Unpassable areas are uniformly set to black. The construction method of the grid map is shown in Figure 3.

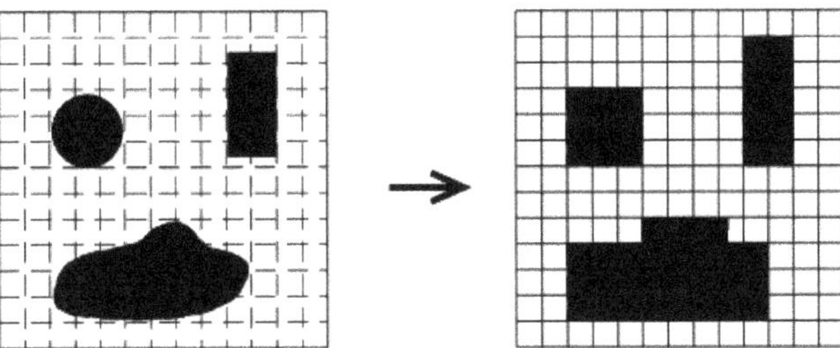

Figure 3. The way to build a grid map.

In order to realize the full-coverage path planning of the USV, it is first necessary to determine the area that the USV needs to cover. In the grid map, the general area that the USV needs to cover is all the blank grids in the grid map, but there are some special For example, if an obstacle surrounds a blank grid, although there is no obstacle in the surrounded grid, the USV cannot pass through the obstacle to cover the grid, as shown in Figure 4. Therefore, it is necessary to deal with this situation, that is, the non-coverable grid becomes the grid with obstacles.

Figure 4. Unreachable grid.

The difficulty of this task lies in how to use the algorithm to automatically find the non-coverable grids. The non-coverable grids can be easily found by human judgment, but for the computer, the map is just a series of two-dimensional matrices inputs. For this reason, this paper proposes a raster map preprocessing method to remove non-overridable rasters.

Since the memory occupied by the matrix is much smaller than the memory occupied by the image, and each element in the matrix of the same dimension as the grid map can corresponds to the grid in the grid map, the grid map preprocessing method is proposed in this paper. First, the grid map is converted into a two-dimensional state matrix form. The element value of the blank grid in the corresponding position in the state matrix is 0, and the element value of the black grid representing the obstacle in the corresponding position in the state matrix is −1. The element value of the corresponding position of the covered grid in the state matrix is 1, and the element value of the corresponding position of the grid where the USV is located in the state matrix is 7. In the preprocessing stage of the grid map, the location of the USV does not need to be considered, and the USV cannot appear in the grid where obstacles exist, so the element value corresponding to the grid where the USV is located is recorded as 0 in the preprocessing stage. The grid map shown in Figure 4 is converted into a two-dimensional matrix, as shown in matrix 1.

$$\begin{bmatrix} -1 & -1 & -1 \\ 0 & -1 & 0 \\ -1 & 0 & 0 \end{bmatrix} \tag{1}$$

Because the surrounding of the map is the map boundary, the map boundary is regarded as an obstacle, and the USV should not go out of the map boundary, so firstly, add -1 around the two-dimensional matrix representing the grid map, as shown in matrix 2.

$$\begin{bmatrix} -1 & -1 & -1 & -1 & -1 \\ -1 & -1 & -1 & -1 & -1 \\ -1 & 0 & -1 & 0 & -1 \\ -1 & -1 & 0 & 0 & -1 \\ -1 & -1 & -1 & -1 & -1 \end{bmatrix} \tag{2}$$

Then, set a 3×3 matrix and denote it as a screening matrix M_f, which is matrix 3.

$$\begin{bmatrix} 0 & 1 & 0 \\ 1 & 4 & 1 \\ 0 & 1 & 0 \end{bmatrix} \tag{3}$$

Decompose matrix 2 into $(n-2) \times (n-2)$ 3×3 matrices, where n is the dimension of the matrix after adding -1, and then do a dot product operation with matrix 3. That is, multiply and sum the elements of each position of the two 3×3 matrices, and finally, obtain a 3×3 matrix, and write the obtained result into a matrix of the same size as matrix 1. The calculation process is shown in Figure 5. The final result is matrix 4.

$$\begin{bmatrix} -7 & -7 & -3 \\ -4 & -5 & -2 \\ -6 & -3 & -2 \end{bmatrix} \tag{4}$$

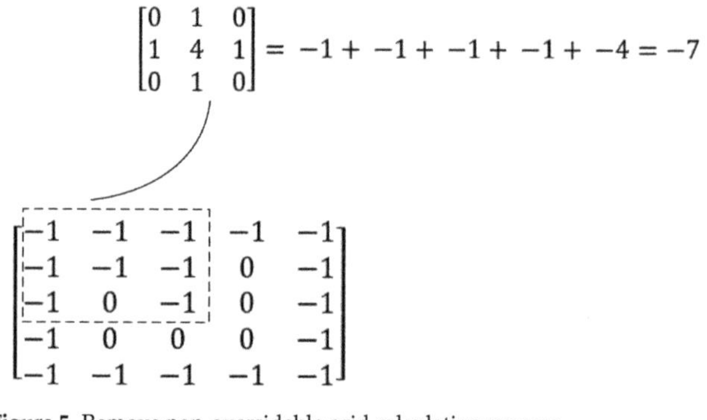

$$\begin{bmatrix} 0 & 1 & 0 \\ 1 & 4 & 1 \\ 0 & 1 & 0 \end{bmatrix} = -1 + -1 + -1 + -1 + -4 = -7$$

$$\begin{bmatrix} -1 & -1 & -1 & -1 & -1 \\ -1 & -1 & -1 & 0 & -1 \\ -1 & 0 & -1 & 0 & -1 \\ -1 & 0 & 0 & 0 & -1 \\ -1 & -1 & -1 & -1 & -1 \end{bmatrix}$$

Figure 5. Remove non-overridable grid calculation process.

Then, filter the generated matrix. If the value of a certain position is less than or equal to -4, the grid representing the position is an uncoverable point or an obstacle, and the value corresponding to the element of its position should be -1. The other grids are grids that need to be fully covered, and the element value corresponding to their position is 0. The two-dimensional matrix corresponding to the filtered raster map is shown in matrix 5. Use this method to correctly and efficiently remove non-overridable blank rasters, turning them into black rasters.

$$\begin{bmatrix} -1 & -1 & -1 \\ -1 & -1 & 0 \\ -1 & 0 & 0 \end{bmatrix} \tag{5}$$

In the process of performing the full-coverage task, use this matrix to record the grid where the USV is located and the grid that the USV has covered. The element value of the corresponding matrix position is changed from 0 to 1, and the element value of the matrix position corresponding to the rest of the grids remains unchanged. The matrix corresponding to Figure 6 is shown in matrix 6.

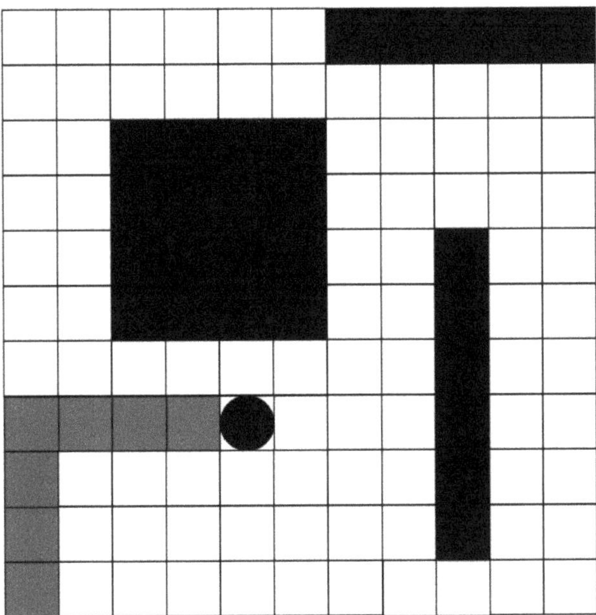

Figure 6. A state in the process.

$$
\begin{bmatrix}
0 & 0 & 0 & 0 & 0 & 0 & -1 & -1 & -1 & -1 & -1 \\
0 & 0 & 0 & 0 & 0 & 0 & 0 & 0 & 0 & 0 & 0 \\
0 & 0 & -1 & -1 & -1 & -1 & 0 & 0 & 0 & 0 & 0 \\
0 & 0 & -1 & -1 & -1 & -1 & 0 & 0 & 0 & 0 & 0 \\
0 & 0 & -1 & -1 & -1 & -1 & 0 & 0 & -1 & 0 & 0 \\
0 & 0 & -1 & -1 & -1 & -1 & 0 & 0 & -1 & 0 & 0 \\
0 & 0 & 0 & 0 & 0 & 0 & 0 & 0 & -1 & 0 & 0 \\
1 & 1 & 1 & 1 & 7 & 0 & 0 & 0 & -1 & 0 & 0 \\
1 & 0 & 0 & 0 & 0 & 0 & 0 & 0 & -1 & 0 & 0 \\
1 & 0 & 0 & 0 & 0 & 0 & 0 & 0 & -1 & 0 & 0 \\
1 & 0 & 0 & 0 & 0 & 0 & 0 & 0 & 0 & 0 & 0
\end{bmatrix}
\tag{6}
$$

In order to increase the difference between elements, make the features more obvious, improve the efficiency of the neural network, and multiply the input matrix M of the preprocessed raster map by a coefficient β as the input of DQN, that is, the input is $\beta \times M$.

The size of the grid map affects the speed and success rate of the algorithm learning process. Bian, T et al. [37] have investigated the correlation between the size of the grid map and the learning process by using several different algorithms. When the size of grid map increases, the time required for the algorithm to learn increases and the success rate decreases to different degrees.

3. Deep Reinforcement Learning Models and the Improvements

3.1. Q Learning

Different from the supervised learning commonly used in machine learning, reinforcement learning is a process in which an agent continuously explores the environment and tends to an optimal solution. The goal of reinforcement learning is to find the optimal strategy for a continuous time series. Inspired by psychology, it focuses on how the agent can take different actions in the environment to obtain the highest reward.

Reinforcement learning is mainly composed of agents, environmental states, actions, rewards, and punishments. After the agent performs an action, the action will affect the environment, so that the state will change to a new state. For the new state, a reward and punishment signal will be given according to the completion of the task. Subsequently, the agent performs new actions according to a certain strategy and according to the new environment state and the feedback reward and punishment.

Q-learning is a reinforcement learning method using the temporal difference method, which needs to be updated step-by-step in the process of interaction between the agent and environment, so that the global state transition probability cannot be known. The temporal difference method combines the Monte Carlo sampling method and the dynamic programming method, making it suitable for model-free algorithms, and it can be updated in a single step.

The core idea of Q-learning is to use the greedy algorithm to select the behavior according to a certain probability through the agent, according to the current state s, and then obtain the corresponding reward r. At the same time, the environment is updated according to the current state s and the action a selected by the agent. The environment state is $s(t+1)$, and the next action is selected based on $s(t+1)$, until the end of the task (including the successful completion of the task and the failure of the task). The parameter update formula of Q-learning is Equation (7).

$$[Q(s,a) < -Q(s,a) + \alpha * (r + \gamma maxQ(s',a') - Q(s,a))] \tag{7}$$

Among them, α is the learning rate, α is a number less than 1, γ is the discount factor, and $Q(s,a)$ is essentially a table, which stores the Q value corresponding to each action in various environments. This table is called the Q-value table. Q-learning uses the Q value of each action in the Q-value table to decide what action to choose in the corresponding environment. However, in order to avoid Q-learning, it can only find the local optimal solution. Using the greedy selection method to select behavior a, the agent has a certain probability to randomly select the action, instead of according to the Q-value table, and then updates the Q-value table according to the reward obtained by the agent after taking the action. An example of the Q-value table is shown in Table 1.

Table 1. Q-value table.

Q-Value Table	a_1	a_2	a_3
s_1	$Q(s_1, a_1)$	$Q(s_1, a_2)$	$Q(s_1, a_3)$
s_2	$Q(s_2, a_1)$	$Q(s_2, a_2)$	$Q(s_2, a_3)$
s_3	$Q(s_3, a_1)$	$Q(s_3, a_2)$	$Q(s_3, a_3)$
...	...	...	...

3.2. Deep Reinforcement Learning Models

Q-learning is a very classic algorithm in reinforcement learning, but it has a big flaw. Its actions are selected based on the Q-value table. On the one hand, the capacity of the Q-value table is very small, and when the environment becomes more complicated, the number of states will increase exponentially. If the Q-value table needs to completely record the Q-values corresponding to all states and actions, it will consume a lot of storage space, and when the state and action spaces are high-dimensional and continuous, it becomes more difficult to use the Q-value table to store the action space and state. Another aspect is that the Q-value table is just a table. If there is a state that has never appeared in the Q-value table, then Q-learning will be helpless, that is, Q-learning has no generalization ability.

Therefore, the deep mind team proposed the deep Q network (DQN) based on an artificial neural network to make up for the defects of Q-learning. Both DQN and Q-learning are value-based algorithms in nature. DQN uses a neural network to calculate the Q-value of the corresponding action in a state and generates the Q-value corresponding to the action in a certain environment by fitting a function, instead of the Q-value table, thus avoiding the Q-value table occupying a lot of memory and querying the Q-value. The slowness of the table and other shortcomings, this method of using a deep neural network to fit a Q-value table is called deep reinforcement learning. The basic framework of DQN is shown in Figure 7. In the figure, S is the current state, A is the current action, R is the reward, S' is the next state, A' is the next action, θ is the parameter of the Q network, θ' is the network parameters used to compute the target.

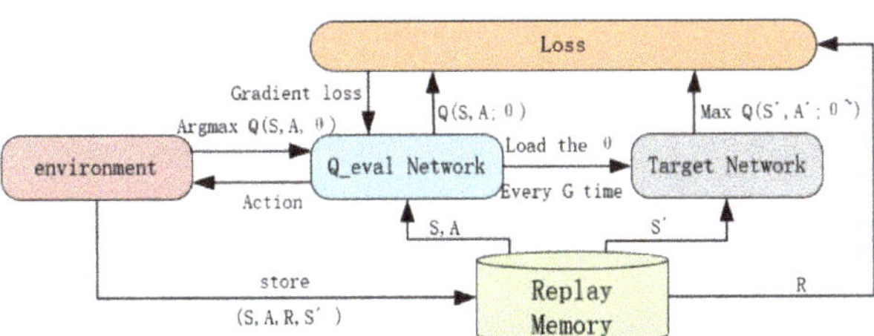

Figure 7. The structure of DQN.

DQN is an improvement to Q-learning. It uses the characteristics of the artificial neural network to fit the Q-value table. On this basis, DQN also adds an experience replay pool to solve the problem of too high a correlation between samples. The experience replay pool stores A large amount of historical data, DQN does not update parameters immediately after performing an action like Q-learning, but stores a sample corresponding to the action $< s_t, a_t, r_t, a\ (t+1) >$ in Experience playback pool. When the acquired samples accumulate to a certain number, the training of the neural network starts. During training, a fixed number of samples are randomly selected from the experience replay pool for training. Since it is randomly selected, the probability of data drawn in the same round is very small, and the strong correlation between samples is broken, thus avoiding the need for neural networks. The local optimum problem arises from the instability. At the same time, labels are constructed through reward and punishment functions for neural network training. In addition, DQN also uses two neural networks with the same structure, one of which is a neural network with hysteresis parameters, namely the current value network and the target value network. The parameters of the Q-target (target value network) are the historical version of the Q-eval (current value network) neural network. Only the Q-eval neural network is trained. After a certain step, Q-target updates parameters that are the same as the parameters of Q-eval neural network, so as to reduce the connection between the current Q-value and the target Q-value, as well as to improve the convergence speed and stability of the neural network. The specific neural network parameter update formula is shown in Equation (8).

$$\left[loss = (r_t + \gamma maxQ(s_{t+1}, a'; \theta^-) - Q(s_t, a_t; \theta))^2\right] \tag{8}$$

In Equation (8), $Q(*; \theta^-)$ is the Q-target neural network. It does not need to train parameters through the Loss function, but only needs to update the parameters after a certain step size—$Q(*; \theta)$ is the Q-eval neural network, and it directly gives the action that the agent needs to make through the environment. DQN uses the gradient descent method to train neural network parameters.

There are two artificial neural networks with the same structure, but different parameters in DQN. $Q(s_t, a_t; \theta)$ represents the value output by the current network, that is, Q-eval, which is used to output the optimal action in the current state. Additionally, $Q(s_{t+1}, a'; \theta^-)$ represents the output result of the target value network Q-target, so when the agent takes an action, it can update the parameters of Q-eval according to Equation (8), and after a certain step, iterate several times and copy the parameters of Q-eval to Q-target, thus completing a learning process.

Three existing improved DQN are combined to achieve the improvements proposed by this manuscript, including double DQN, dueling DQN, and prioritized experience replay (DQN).

3.3. Sampling Method

In DQN, the training samples in the playback memory pool are generated by the interaction between the agent and the environment and do not need to be provided by humans. During the interaction between the agent and the environment, new samples will be continuously generated and stored in the playback memory pool. When the number of memories stored in the playback memory pool reaches the size of the storage memory pool, the oldest sample will be deleted, and the new round is stored in the playback memory pool. Figure 8 shows the update method of the playback memory pool, which can ensure that the maximum capacity of the playback memory pool remains unchanged.

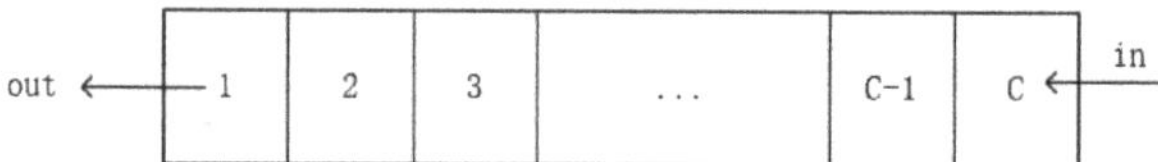

Figure 8. Replay memory pool update method.

In DQN, the random sampling method is used to extract data from the playback memory pool for training, while in the full-coverage path planning, the repeated data that has less effect on the neural network update will have a high probability of being extracted as the neural network training data. It will slow down the convergence speed of the neural network. In this paper, the method of sumtree is used to extract data for learning.

Sumtree is a binary tree structure, and its tree structure is only used to store the priority, and the extra data block is the data needed to be stored in the playback memory pool. The structure of sumtree is shown in Figure 9.

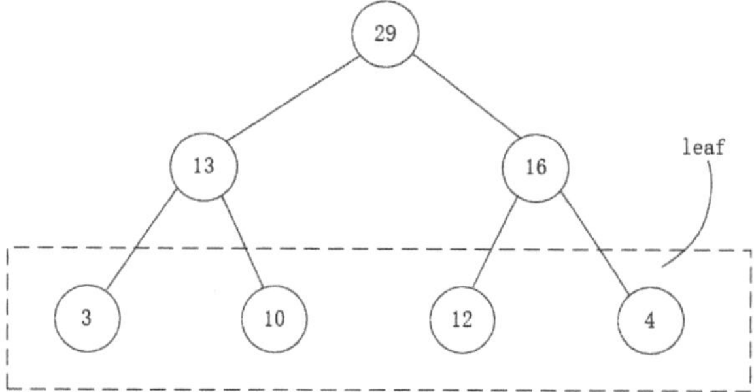

Figure 9. The framework of sumtree.

The leaf node stores the training priority value of each state, and this priority value is determined according to the value of its td-error. The larger the value of td-error, the higher the priority of selecting the memory. Its parent node is the sum of its two child nodes, and the value of the top-level node is the sum of the values of all the child nodes of the next layer. Each memory sample corresponds to a leaf node, and a certain amount of memory is extracted from the leaf nodes for the training each time. Saving data in a tree structure can greatly facilitate the sampling process, and it also ensures that each leaf node has a probability of being drawn, which greatly increases the convergence speed of the neural network.

In addition to replaying the memory pool, it is also necessary to optimize the loss function of the current value network to eliminate bias. The improved sample priority function is:

$$\left[loss = 1/m \sum_{j=1} w_j (y_j - Q(s_j, a_j; \theta))^2 \right] \tag{9}$$

The definition of w_j is shown in Equation (10).

$$[w_j = (1/N * 1/P(j))^\beta] \tag{10}$$

Due to the use of priority to select samples, this will cause the distribution of samples to change, which will cause the sample model to converge to different values, that is, the neural network will be unstable. Therefore, importance sampling is used in this paper, which not only ensures that the probability of each sample being selected for training is different, in order to improve the training speed, but also ensures that its influence on the gradient descent of the neural network is the same, thereby ensuring that the neural network receives the same the result of. In Equation (10), N is the number of samples in the playback memory pool, and β is a hyperparameter that needs to be set manually to offset the impact of changes in sample distribution.

3.4. Improvements Based on Action Selection

The USV explores the map under the strategy of the initial neural network. Since the parameters of the initial neural network are not trained or trained only a few times, the initial neural network is likely to give a strategy to collide with obstacles or go out of the map boundary. When the USV collides with an obstacle or walks out of the map boundary, the full-coverage task fails, and the USV returns to the initial position to try the next full-coverage path planning task, so the USV needs to undergo much training to avoid collision obstacles. The worst behavior is to avoid objects or go outside the boundaries of the map. Moreover, during the training process, even if the USV is in the same position, the state matrix of the map is different, due to the different previous behaviors. When the USV does the worst behavior, the reinforcement learning model will only make the state corresponding to this time. The Q-value of the action is reduced without extending it to all similar states, as shown in Figure 10.

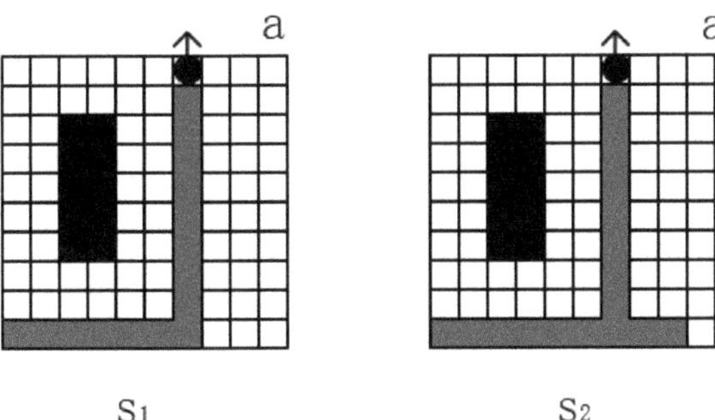

Figure 10. Dangerous actions in different states.

As shown in Figure 10, state s_1 and state s_2 are two different states, black squares are obstacles, white squares are movable paths, gray squares are the paths traveled by USV, black circle is USV, and a represents the action taken by USV. When the USV makes a dangerous action only in state s_1, deep reinforcement learning will only train the neural network, so that $Q\text{--}eval(s_1, a)$ becomes smaller without making the value of $Q\text{--}eval(s_2, a)$ smaller, so that when the USV encounters the s_2 state, there is still a probability to choose the dangerous action a, which leads to the failure of the task.

When humans perform similar tasks, they will actively avoid the above dangerous actions, so this paper proposes a method that can greatly reduce the probability of such dangerous actions. The policies for USVs are derived from current networks in deep reinforcement learning. The output of the current network is the Q-value of each action in the current environment. If you want to reduce the selection of dangerous actions, you only need to subtract a large number from the Q-value of the dangerous action. The specific implementation method is as follows.

Step 1: Assuming the size of the state map is n × n, first add −1 around the state map to represent the map boundary, as mentioned in the previous chapter. Then, convert the state map to the initial state, that is, change the values of all covered grids to 0.

Step 2: The position value in the state matrix of the USV is 7—find the position of the USV, and then multiply it by the surrounding elements based on the action value. For example, if the action value is up, it is multiplied by the element directly above the USV to obtain the danger of the action in this state. The four actions derive their corresponding dangers.

Step 3: Multiply the risk by the risk coefficient *t*, and then add it to the final output layer of DQN to form the final output. Its structure diagram is shown in Figure 11.

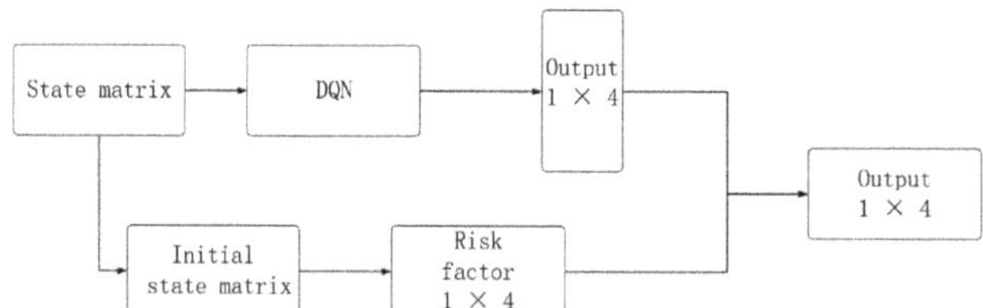

Figure 11. Framework for reducing the probability of choosing a dangerous action.

3.5. Design of the Reward Function

In order to improve the efficiency of full-coverage path planning, the USV should try to avoid repeatedly covering the covered area during the full-coverage path planning task and try to select the area that can cover the uncovered area every time an action is selected. So, the reward function initially defined by the text is shown in Equation (11).

$$
r = \begin{cases}
1, m(i,j) = 0 \\
0, m(i,j) = 1 \ and \ m_{-1}(i \pm 1, j \pm 1) = 1 \\
-1, m(i,j) = 1 \ and \ \exists m_{-1}(i\pm, j \pm 1) = 0 \\
-20, m(i,j) = -1 \ or \ i,j < 0 \ or \ i,j > N \\
20, done
\end{cases}
\tag{11}
$$

where $m(i,j)$ represents the position in the grid map after the USV takes an action, and $m_{-1}(i \pm 1, j \pm 1)$ is the grid situation around the grid before the USV takes an action. When the USV enters an uncovered grid, it will obtain reward 1. When there is no uncovered grid around the USV, that is, all the grids around the USV have been covered or are obstacles, the USV takes no collision obstacles or actions that go out of the map's borders and are rewarded with 0. When there is an uncovered grid around the USV, the vehicle still takes action to make itself enter the covered grid, and the reward is −1. When the vehicle takes an action and collides with an obstacle or goes out of the map boundary, it will obtain a reward of −20.

4. Simulation Results and Discussion

4.1. Simulation Platform

The simulation experiment in this paper runs in the Ubuntu 20.04 environment, using Python as the development language, and the parameters of the simulation platform are shown in Table 2.

Table 2. Simulation platform

Os	Language	CPU	GPU	RAM
Ubuntu 20.04	Python 3.8	intel10900	GTX2070	6 Gb

Based on the simulation platform we use, the duration of the proposed algorithm learning process usually takes one to two days. The duration of the learning process is limited by simulation platform, and a better simulation platform can significantly reduce learning time of the proposed algorithm. In order to verify the algorithm proposed in this paper, the size of the working environment of the USV in this paper is 20×20 m^2, which is divided into 11×11 grid maps.

4.2. Algorithm Simulation

In the simulation process, the above environmental parameters and DQN network model parameters are initialized first, and then a preprocessed grid map is constructed to record the grid map information and location information covered by the USV. Before each training starts, the position of the USV appears randomly on the grid map, and it is ensured that the position of the USV is not the position of the obstacle. During training, when the USV task is moved, the training will terminate, and the next training will begin. The conditions for task termination are that the USV collides with the map boundary or the obstacle grid, completes the full-coverage task, and the USV's step size exceeds the set maximum step size. During the training process, each action will generate a sample, and the sample will be stored in the playback memory pool. When the number of samples reaches a certain level, the artificial neural network will be trained. When the training round reaches the set maximum training round, the training will end.

The simulation diagrams are shown in Figures 12–15. Both the improved DQN proposed in this paper and the traditional boustrophedon method can complete the full-coverage path planning task well, and the coverage rate is 100%. The DQN and inner spiral coverage cannot complete the full-coverage path planning task. Their coverage rates are only 13% and 86%. As shown in Figures 13 and 15, the coverage repetition rate of the improved DQN was 0.04%, while that of the boustrophedon method was 0.13%. Its final composite scores were 0.96 and 0.87 points, respectively. The effect of using the improved DQN-based full-coverage path planning algorithm is due to the traditional boustrophedon method.

Figure 16 shows the coverage of the full-coverage path planning algorithm using boustrophedon and the full-coverage path planning algorithm using the improved DQN. The red curve is the real-time coverage using boustrophedon, and the blue curve is the real-time coverage using the improved DQN full-coverage path planning. The horizontal axis is the step size, and the vertical axis is the coverage. From the real-time coverage map, it can be concluded that the improved DQN-based full-coverage path planning algorithm is more efficient than the boustrophedon method.

The loss function diagram of the improved DQN is shown in Figure 17, and its ordinate is the distance between the path and the optimal path, which can be obtained from Figures 4–7. When the number of training times reaches 1000, the neural network has converged. The 1000 times at this point is not that the agent has performed 1000 tasks, but that 1000 times of memory are extracted from the playback memory pool for playback training, and the overall efficiency is high.

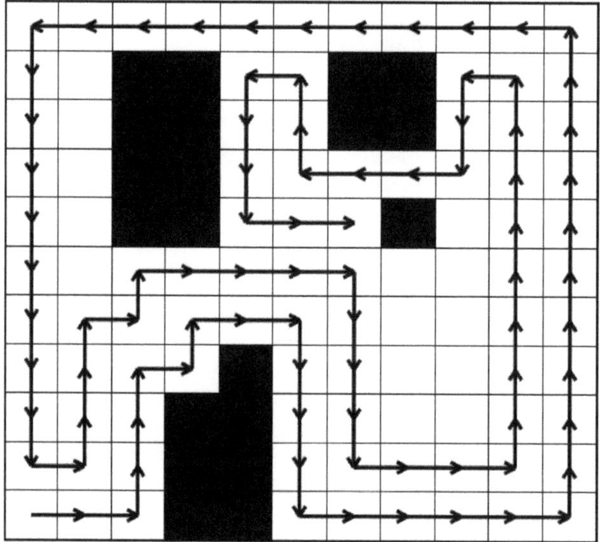

Figure 12. Path plan of inner spiral coverage.

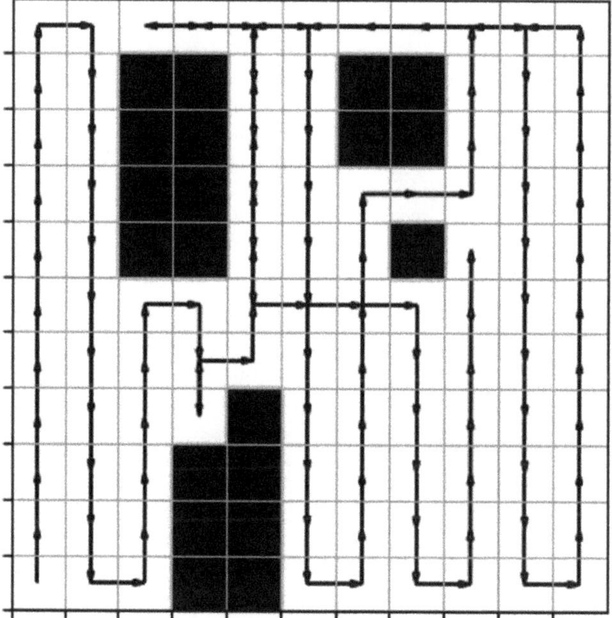

Figure 13. Path plan of boustrophedon.

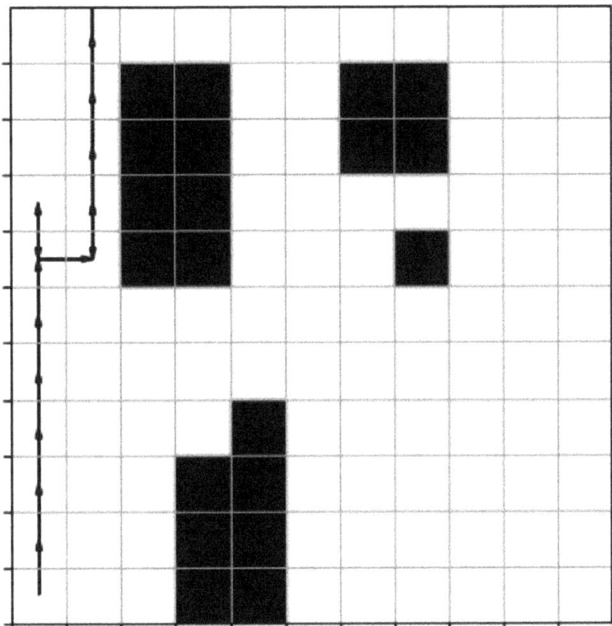

Figure 14. Path plan of DQN.

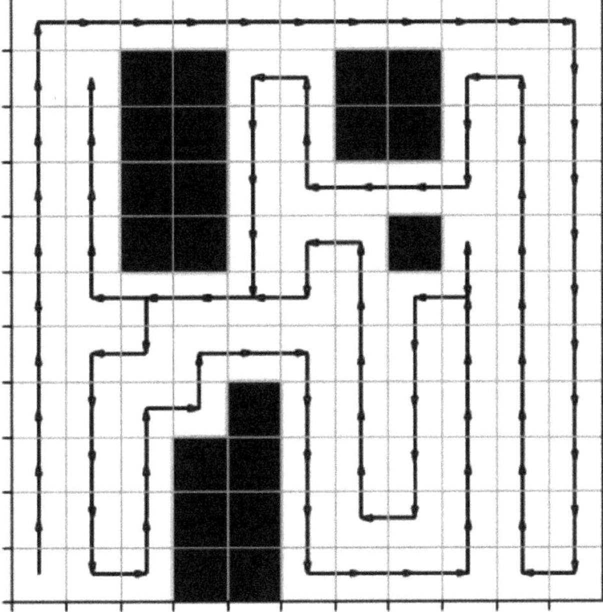

Figure 15. Path plan of improved DQN.

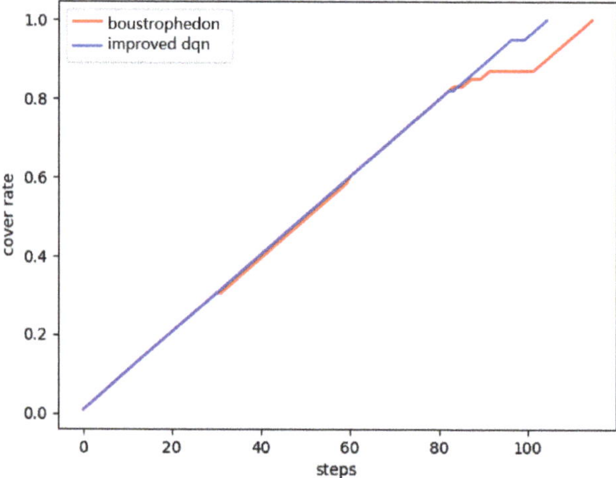

Figure 16. Coverage map.

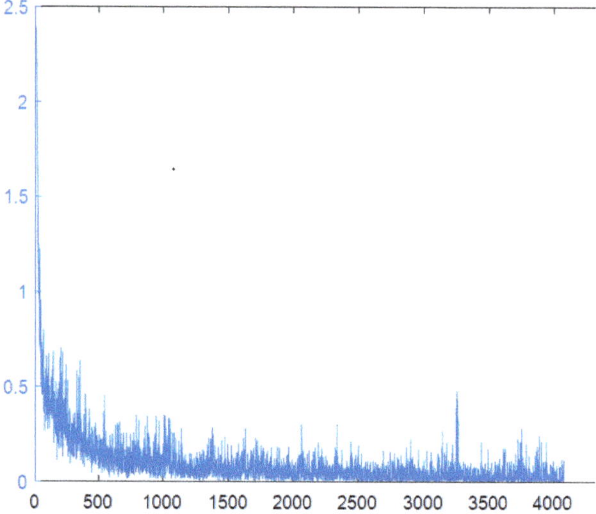

Figure 17. Training graph.

5. Conclusions

5.1. Main Conclusions and Findings

This paper proposes a raster map preprocessing method. Combined with the description and analysis of typical full-coverage path planning task scenarios, and further according to the characteristics of full-coverage path planning, we found the grids that cannot be covered by USVs and set them as obstacle grids.

To solve the problem of the low learning efficiency of traditional DQN, an improved action selection mechanism is proposed. This makes the failure rate of full-coverage path planning in the initial training phase lower. At the same time, the method of extracting memory training is improved, so that valuable data has a greater probability of being trained, and the efficiency of full-coverage of the area is improved.

Simulation experiments show that the algorithm can effectively complete the full-coverage path planning task. The results show that, compared with the traditional full-coverage path planning algorithm, the proposed algorithm has good coverage rate. Compared with the boustrophedon and DQN algorithms, the performance of this algorithm on the coverage repetition rate is better, indicating that the performance of this algorithm is better.

5.2. Main Limitation of the Research

In modeling phase, the dynamical constraints and performance of the USV were not fully considered. The actual model of USV is more complex than the one in this study. This full-coverage path planning algorithm for USV might not have good performance in some situations.

5.3. Future Research Prospects

Considering economic factors, we usually want USVs to achieve full-coverage tasks efficiently at a low cost. To achieve this goal, the efficiency of path planning algorithms and their adaptability to different environments need to be improved.

In future work, the focus will be on the USV model and reward function, since the full-coverage path planning algorithm is affected by USV model and reward function when using simulation platform. At the same time, how to extend this algorithm to a complex actual environment with disturbances to obtain application value is also worth studying.

Author Contributions: Methodology, B.X.; formal analysis, Q.W.; data curation, Z.L.; data analysis, L.Y. and X.W.; software, X.W.; writing—original draft preparation, L.Y.; writing—review and editing, X.W.; supervision, Q.W. and B.X.; project administration, B.X. All authors have read and agreed to the published version of the manuscript.

Funding: This research was funded by Shanghai Science and Technology Committee (STCSM) Local Universities Capacity-building Project (No. 22010502200), Scientific Research Project of China Three Gorges Corporation (No. 202003111).

Institutional Review Board Statement: Not applicable.

Informed Consent Statement: Not applicable.

Data Availability Statement: The data are available on request.

Acknowledgments: The authors would like to express their gratitude for the support of Fishery Engineering and Equipment Innovation Team of Shanghai High-level Local University.

Conflicts of Interest: The authors declare no conflict of interest.

References

1. Zhang, H.; Hu, B.; Xu, Z. Visual Navigation and Landing Control of an Unmanned Aerial Vehicle on a Moving Autonomous Surface Vehicle via Adaptive Learning. *IEEE Trans. Neural Netw. Learn. Syst.* **2021**, *32*, 5345–5355. [CrossRef] [PubMed]
2. Jin, J.; Zhang, J.; Liu, D.; Shi, J.; Wang, D.; Li, F. Vision-Based Target Tracking for Unmanned Surface Vehicle Considering Its Motion Features. *IEEE Access* **2020**, *8*, 132655–132664. [CrossRef]
3. Pyrchla, K.; Pyrchla, J.; Kantak, T. Hydrographic Multisensory Unmanned Watercraft. In Proceedings of the 2018 Baltic Geodetic Congress (BGC Geomatics), Olsztyn, Poland, 21–23 June 2018; pp. 231–235.
4. Xie, X.; Wang, Y.; Wu, Y.; You, M.; Zhang, S. Random Patrol Path Planning for Unmanned Surface Vehicles in Shallow Waters. In Proceedings of the 2022 IEEE International Conference on Mechatronics and Automation (ICMA), Guilin, China, 7–10 August 2022; pp. 1860–1865.
5. Xu, F.; Xie, Y.; Liu, X.; Chen, X.; Han, W. Research Status and Key Technologies of Intelligent Technology for Unmanned Surface Vehicle System. In Proceedings of the 2020 International Conference on Sensing, Diagnostics, Prognostics, and Control (SDPC), Beijing, China, 5–7 August 2020; pp. 229–233.
6. Zhou, X.; Wu, P.; Zhang, H.; Guo, W.; Liu, Y. Learn to Navigate: Cooperative Path Planning for Unmanned Surface Vehicles Using Deep Reinforcement Learning. *IEEE Access* **2019**, *7*, 165262–165278. [CrossRef]
7. Fan, J.; Li, Y.; Liao, Y.; Jiang, W.; Wang, L.; Jia, Q.; Wu, H. Second Path Planning for Unmanned Surface Vehicle Considering the Constraint of Motion Performance. *J. Mar. Sci. Eng.* **2019**, *7*, 104. [CrossRef]
8. Zhang, H.; Hong, W.; Chen, M. A Path Planning Strategy for Intelligent Sweeping Robots. In Proceedings of the 2019 IEEE International Conference on Mechatronics and Automation (ICMA), Tianjin, China, 4–7 August 2019; pp. 11–15.

9. Zhang, M.; Guo, W.; Wang, L.; Li, D.; Hu, B.; Wu, Q. Modeling and optimization of watering robot optimal path for ornamental plant care. *Comput. Ind. Eng.* **2021**, *157*, 107263. [CrossRef]

10. Miao, X.; Lee, J.; Kang, B. Scalable Coverage Path Planning for Cleaning Robots Using Rectangular Map Decomposition on Large Environments. *IEEE Access* **2018**, *6*, 38200–38215. [CrossRef]

11. Miao, X.; Lee, H.; Kang, B. Multi-Cleaning Robots Using Cleaning Distribution Method Based on Map Decomposition in Large Environments. *IEEE Access* **2020**, *8*, 97873–97889. [CrossRef]

12. Le, A.; Parween, R.; Kyaw, P.; Mohan, R.; Minh, T.; Borusu, C. Reinforcement Learning-Based Energy-Aware Area Coverage for Reconfigurable hRombo Tiling Robot. *IEEE Access* **2020**, *8*, 209750–209761. [CrossRef]

13. Takeda, T.; Ito, K.; Matsuno, F. Path generation algorithm for search and rescue robots based on insect behavior—Parameter optimization for a real robot. In Proceedings of the 2016 IEEE International Symposium on Safety, Security, and Rescue Robotics (SSRR), Lausanne, Switzerland, 23–27 October 2016; pp. 270–271.

14. Farsi, M.; Ratcliff, K.; Johnson, J.P.; Allen, C.R.; Karam, K.Z.; Pawson, R. Robot control system for window cleaning. In Proceedings of the 1994 American Control Conference—ACC '94, Baltimore, MD, USA, 29 June–1 July 1994; pp. 994–995.

15. Aponte-Roa, D.A.; Collazo, X.; Goenaga, M.; Espinoza, A.A.; Vazquez, K. Development and Evaluation of a Remote Controlled Electric Lawn Mower. In Proceedings of the 2019 IEEE 9th Annual Computing and Communication Workshop and Conference (CCWC), Las Vegas, NV, USA, 7–9 January 2019; pp. 1–5.

16. Wang, C.-N.; Yang, F.-C.; Nguyen, V.T.T.; Nguyen, Q.M.; Huynh, N.T.; Huynh, T.T. Optimal Design for Compliant Mechanism Flexure Hinges: Bridge-Type. *Micromachines* **2021**, *12*, 1304. [CrossRef]

17. Sun, C.-C.; Ahamad, A.; Liu, P.-H. SoC FPGA Accelerated Sub-Optimized Binary Fully Convolutional Neural Network for Robotic Floor Region Segmentation. *Sensors* **2020**, *20*, 6133. [CrossRef]

18. Xia, G.; Han, Z.; Zhao, B.; Liu, C.; Wang, X. Global Path Planning for Unmanned Surface Vehicle Based on Improved Quantum Ant Colony Algorithm. *Math. Probl. Eng.* **2019**, *2019*, 2902170. [CrossRef]

19. Alomari, A.; Phillips, W.; Aslam, N.; Comeau, F. Dynamic Fuzzy-Logic Based Path Planning for Mobility-Assisted Localization in Wireless Sensor Networks. *Sensors* **2017**, *17*, 1904. [CrossRef] [PubMed]

20. Wang, C.-N.; Yang, F.-C.; Nguyen, V.T.T.; Vo, N.T.M. CFD Analysis and Optimum Design for a Centrifugal Pump Using an Effectively Artificial Intelligent Algorithm. *Micromachines* **2022**, *13*, 1208. [CrossRef] [PubMed]

21. Xie, J.; Zhou, R.; Luo, J.; Peng, Y.; Liu, Y.; Xie, S.; Pu, H. Hybrid Partition-Based Patrolling Scheme for Maritime Area Patrol with Multiple Cooperative Unmanned Surface Vehicles. *J. Mar. Sci. Eng.* **2020**, *8*, 936. [CrossRef]

22. Jorge, V.A.M.; Granada, R.; Maidana, R.G.; Jurak, D.A.; Heck, G.; Negreiros, A.P.F.; dos Santos, D.H.; Gonçalves, L.M.G.; Amory, A.M. A Survey on Unmanned Surface Vehicles for Disaster Robotics: Main Challenges and Directions. *Sensors* **2019**, *19*, 702. [CrossRef]

23. Mendonça, R.; Marques, M.; Marques, F. A cooperative multi-robot team for the surveillance of shipwreck survivors at sea. In Proceedings of the OCEANS 2016 MTS/IEEE Monterey, Monterey, CA, USA, 19–23 September 2016; pp. 1–6.

24. Kang, C.; Yeh, L.; Jie, S.; Pei, T.; Nugroho, H. Design of USV for Search and Rescue in Shallow Water. In *Intelligent Robotics and Applications*; Springer: Cham, Switzerland, 2020; pp. 351–363.

25. Ozkan, M.; Carrillo, L.; King, S. Rescue Boat Path Planning in Flooded Urban Environments. In Proceedings of the 2019 IEEE International Symposium on Measurement and Control in Robotics (ISMCR), Houston, TX, USA, 19–21 September 2019; pp. B2-2-1–B2-2-9.

26. Wang, C.-N.; Yang, F.-C.; Vo, N.T.M.; Nguyen, V.T.T. Wireless Communications for Data Security: Efficiency Assessment of Cybersecurity Industry—A Promising Application for UAVs. *Drones* **2022**, *6*, 363. [CrossRef]

27. Paravisi, M.; Santos, D.H.; Jorge, V.; Heck, G.; Gonçalves, L.M.; Amory, A. Unmanned Surface Vehicle Simulator with Realistic Environmental Disturbances. *Sensors* **2019**, *19*, 1068. [CrossRef]

28. Osen, O.; Havnegjerde, A.; Kamsvåg, V.; Liavaag, S.; Bye, R. A low cost USV for aqua farm inspection. In Proceedings of the 2016 Techno-Ocean (Techno-Ocean), Kobe, Japan, 6–8 October 2016; pp. 291–298.

29. Han, S.; Wang, L.; Wang, Y.; He, H. An efficient motion planning based on grid map: Predicted Trajectory Approach with global path guiding. *Ocean Eng.* **2021**, *238*, 109696. [CrossRef]

30. Long, Y.; He, H. Robot path planning based on deep reinforcement learning. In Proceedings of the 2020 IEEE Conference on Telecommunications, Optics and Computer Science (TOCS), Shenyang, China, 11–13 December 2020; pp. 151–154.

31. Choi, S.; Lee, S.; Viet, H. B-Theta*: An Efficient Online Coverage Algorithm for Autonomous Cleaning Robots. *Intell Robot Syst* **2017**, *87*, 265–290. [CrossRef]

32. Lee, T.-K.; Baek, S.-H.; Choi, Y.-H.; Oh, S.-Y. Smooth coverage path planning and control of mobile robots based on high-resolution grid map representation. *Robot. Auton. Syst.* **2011**, *59*, 801–812. [CrossRef]

33. Yildirim, O.; Diepold, K.; Vural, R.A. Decision Process of Autonomous Drones for Environmental Monitoring. In Proceedings of the 2019 IEEE International Symposium on Innovations in Intelligent Systems and Applications (INISTA), Sofia, Bulgaria, 3–5 July 2019; pp. 1–6.

34. Choset, H. Coverage of Known Spaces: The Boustrophedon Cellular Decomposition. *Auton. Robot.* **2000**, *9*, 247–253. [CrossRef]

35. Sun, B.; Zhu, D.; Tian, C.; Luo, C. Complete Coverage Autonomous Underwater Vehicles Path Planning Based on Glasius Bio-Inspired Neural Network Algorithm for Discrete and Centralized Programming. *IEEE Trans. Cogn. Dev. Syst.* **2019**, *11*, 73–84. [CrossRef]

36. Luo, J. Complete Coverage Path Planning of an Unmanned Surface Vehicle Based on a Complete Coverage Neural Network Algorithm. *J. Mar. Sci. Eng.* **2019**, *9*, 1163.
37. Bian, T.; Xing, Y.; Zolotas, A. End-to-End One-Shot Path-Planning Algorithm for an Autonomous Vehicle Based on a Convolutional Neural Network Considering Traversability Cost. *Sensors* **2022**, *22*, 9682. [CrossRef] [PubMed]

Journal of
*Marine Science
and Engineering*

Article

Autonomous Underwater Vehicle Path Tracking Based on the Optimal Fuzzy Controller with Multiple Performance Indexes

Qunhong Tian [1,2,3,*], Tao Wang [4], Yuming Song [4], Yunxia Wang [4] and Bing Liu [4]

[1] Qingdao Ship Science and Technology Co., Ltd., Harbin Engineering University, Qingdao 266000, China
[2] College of Intelligent Systems Science and Engineering, Harbin Engineering University, Harbin 150001, China
[3] College of Ocean Science and Engineering, Shandong University of Science and Technology, Qingdao 266590, China
[4] College of Mechanical and Electronic Engineering, Shandong University of Science and Technology, Qingdao 266590, China
* Correspondence: tianqunhong@sdust.edu.cn

Abstract: Autonomous underwater vehicles (AUVs) are increasingly being used in missions involving submarine cable detection, underwater archaeology, pipeline inspection, military reconnaissance, and so on. It is very important to realize AUV path tracking to accomplish these missions. In this paper, a fuzzy controller based on the established kinematic and dynamic models of AUV systems is presented to solve the AUV path-tracking problem. In order to design the fuzzy controller to exhibit good performance, we select the path length, smoothness, and cross-track position error as the multiple optimization performance indexes for the fuzzy controller. We propose the particle swarm optimization (PSO) algorithm to determine the parameters of the membership functions. Different scenarios are presented to test the performance of the proposed algorithm, including the straight line, sine curve, half-moon shape, Archimedean spiral, and practical paths. The results are given to illustrate the effectiveness and feasibility of the fuzzy controller with the optimization of multiple performance indexes.

Keywords: AUV; path tracking; fuzzy controller; optimization; multiple performance indexes

Citation: Tian, Q.; Wang, T.; Song, Y.; Wang, Y.; Liu, B. Autonomous Underwater Vehicle Path Tracking Based on the Optimal Fuzzy Controller with Multiple Performance Indexes. *J. Mar. Sci. Eng.* **2023**, *11*, 463. https://doi.org/10.3390/jmse11030463

Academic Editor: Sergei Chernyi

Received: 8 January 2023
Revised: 10 February 2023
Accepted: 13 February 2023
Published: 21 February 2023

1. Introduction

The ocean is an important strategic space for economic and social development. Due to the increasing exploitation and utilization of marine resources, autonomous underwater vehicles (AUVs) have been rapidly developed for use in the fields of pipeline inspection, seawater quality detection, marine geological exploration, submarine cable detection, and so on [1–3]. It is important to control the AUV to complete missions quickly and accurately [4–6]; however, complex underwater working environments increase the difficulty of controlling AUVs [7]. Therefore, as the premise to realize AUV missions, path tracking becomes an important problem to be solved.

There has been much research on AUV path tracking based on traditional design methods, including proportional integral differential (PID) control [8,9], sliding mode control [10,11], predictive control [12], and so on. PID is one of the most widely used algorithms for controller design; to guide an AUV to track an underwater cable, it forms a straight-line path-tracking control problem with magnetic sensing. Previous research presented a PID controller to track the desired guidance profiles in light of the linearizing feedback method [8]. The PID algorithm can give a line-of-sight guidance law with drift angle compensation to solve the problem of AUV tracking in a complex marine environment, and the PID algorithm can be used to guide AUVs to track the desired path [9]. However, PID is a linear control method, which is difficult to use in complex nonlinear systems related to AUVs in practice. To solve the under-actuated AUV horizontal path-tracking control problem, the authors of another study utilized a sliding mode control method with line-of-sight guidance and cross-track errors [10]; however, chattering was inevitable in the tracking control process. This necessitated a nonlinear second-order sliding mode controller for the purpose of improving the stability of the control system by eliminating the chattering motion [11]. In light of the backstepping algorithm, a model predictive control method was given for AUV path tracking [12]. In summary, the above traditional methods [8–12] are highly dependent on the models of AUV control systems, which are difficult to apply in engineering practice because of the difficulty of obtaining accurate models in complex underwater environments.

Compared with the traditional control methods of path tracking, intelligent algorithms have attracted more and more attention because their application of intelligent control is independent of an accurate kinetic model. Reinforcement learning refers to agent learning according to "trial and error", which interacts with the environment to obtain rewards to guide behavior for path tracking [13,14]. A line-of-sight method of guidance was proposed by reinforcement learning algorithms, which can also utilize long short-term memory neural networks to pre-train the reinforcement learning framework to speed up the learning process [13]. Considering the uncertainty of ocean currents and according to the data on ocean currents obtained by the observer, one study presented a novel reward function to improve the learning ability of a reinforcement learning architecture, and thus, deep reinforcement learning was developed for path tracking [14]. Fuzzy control is a method considering the theory of fuzzy mathematics, which has been used for AUV path tracking [15–17]. In fuzzy control, a nonlinear single-input fuzzy controller was used for path tracking, resulting in reductions in the computation complexity [15]. For an AUV with an unknown actuator saturation and environmental disturbance, the authors of one study presented a kinematic controller by the backstepping technique, whose law was designed in light of an indirect adaptive fuzzy logic control system [16]. To improve the performance of fuzzy controllers, the authors of another study proposed a fuzzy controller optimized by the genetic algorithm [17]. In summary, although there are many different intelligent control methods for AUV path tracking, they are focused on improving the tracking error; in practice, the path length and smoothness are also important for path-tracking performance. Therefore, a new method is needed to solve the AUV path-tracking problem with multiple good performance indexes.

It should be noted that the existing fuzzy theory and the related research have made some contributions regarding the collision avoidance problem in marine engineering. In one study, fuzzy theory was used to reason the degree of the collision risk, A*search was used to make an avoidance action plan, and the action space searched by the ship was formed in the expert system using marine traffic rules [18]. In order to trigger the prompt alert of a potential collision, the authors of another study utilized evidential research theory to calculate the risk of collision according to the relevant navigation information [19]. Based on the need to prevent collisions according to sea rules, the authors of another study proposed a collision risk inference system using fuzzy rules to enable autonomous surface ships to avoid collisions [20]. Furthermore, on the basis of [20], the authors of another study proposed a local route planning method to avoid collisions with maritime autonomous surface ships [21]. Although the above works can provide some reference to fuzzy theory [18–21], they are difficult to directly use in AUV path tracking.

Path tracking is very important to perform missions in complex underwater environments. To solve the path-tracking problem for AUVs according to the established kinematic and dynamic models, a fuzzy controller is presented to realize path tracking. We select the path length, smoothness and cross-track position error as the performance indexes for the fuzzy controller, which determines the AUV tracking path energy, traveling time and tracking accuracy. The fuzzy controller is optimized with the optimization of membership function by the particle swarm optimization (PSO) algorithm. The designed fuzzy controller makes a path-tracking system with strong robustness and anti-interference ability.

The rest of this paper is arranged as follows. Section 2 gives the problem description and definition for AUV path tracking; Section 3 presents mathematical models for AUV path tracking; Section 4 proposes a fuzzy control method based on the PSO algorithm for AUV path tracking; Section 5 gives the results and discussion for the proposed algorithm; Section 6 provides some conclusive remarks.

2. Problem Description and Definition for AUV Path Tracking

The goal of AUV path tracking is to follow the desired path with the performance indexes. In the tracking process, once the coordinates of the current and target locations are given, the desired attitude angles can be obtained with the desired AUV velocity, the essence is to eliminate the error between the practical and desired angles. In order to solve the AUV path-tracking problem, we select the path length, smoothness and cross-track position error as the path-tracking performance indexes. The reason is that the indexes determine the path tracking energy consumption, travelling time, etc. [22]. The fuzzy control can solve the control problem of nonlinear system with strong robustness effectively; therefore, the fuzzy controller is presented for path tracking in this paper. Figure 1 gives the technology roadmap for AUV path tracking. The optimization performances for the fuzzy controller are formed according to the established kinematic and dynamic models, the deviations and their derivatives are taken as input parameters. We present the PSO algorithm to optimize the membership function to realize optimal path tracking. The AUV tracks with the desired path based on the optimized fuzzy controller.

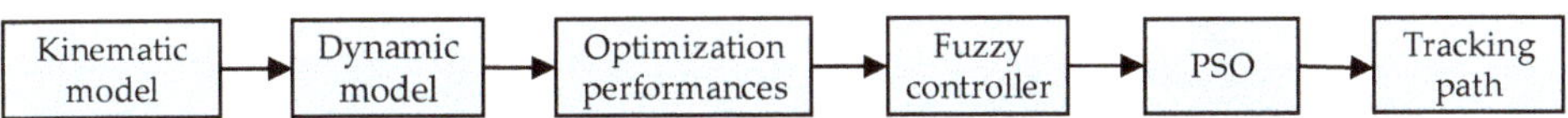

Figure 1. The technology roadmap for AUV path tracking.

We select the cross-track position and heading angle errors as the important factors to illustrate the path-tracking effect [10]. Figure 2 shows the definitions of the AUV's cross-track position and heading angle errors, where $P_{pr} = P_{dx}(i-1), P_{dy}(i-1)$, $P_{cu} = P_{dx}(i), P_{dy}(i)$ and $P_{ne} = P_{dx}(i+1), P_{dy}(i+1)$ are the adjacent discrete points of the desired path. $P_{ce} = (P_x(t), P_y(t))$ is the AUV's center.

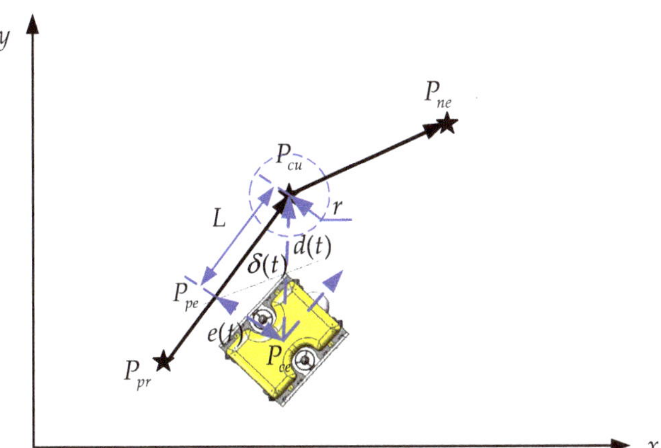

Figure 2. Cross-track position and heading angle errors for the AUV.

The cross-track position error $e(t)$ is the vertical distance from the AUV's center to the adjacent segment of the desired path. In light of the points P_{pr}, P_{cu} and P_{ce}, the cross-track position error can be obtained as follows:

$$e(t) = d(t) \times \sin(\delta(t)) \tag{1}$$

where $d(t)$ is the distance from the AUV's center to the target point.

$$d(t) = \sqrt{(P_y(t) - P_{dy}(i))^2 + (P_x(t) - P_{dx}(i))^2} \tag{2}$$

The cross-track heading angle error $\delta(t)$ is an angle between the $\overrightarrow{P_{ce}P_{cu}}$ and $\overrightarrow{P_{pr}P_{cu}}$ vectors, which can be given as follows:

$$\delta(t) = \arctan\left(P_{dy}(i) - P_y(t)\right)/(P_{dx}(i) - P_x(t)) - \arctan\left(P_{dy}(i) - P_{dy}(i-1)\right)/(P_{dx}(i) - P_{dx}(i-1)) \tag{3}$$

In this paper, the smaller the cross-track position and heading angle errors, the better the tracking effect.

In Figure 2, the straight line $P_{ce}P_{pe}$ is perpendicular to $P_{pr}P_{cu}$, P_{pe} is the corresponding crossing point. Whether the AUV drives into the next segment path can be evaluated by the following decision factor $f_d(t)$:

$$L = d(t)\cos(\delta(t)) \tag{4}$$

$$f_d(t) = \begin{cases} 1 & (L \leq 0 \ \text{or} \ d(t) \leq r) \\ 0 & (L > 0 \ \text{and} \ d(t) > r) \end{cases} \tag{5}$$

where L denotes the length of straight line $P_{pe}P_{cu}$. r and P_{cu} are the circle radius and center, respectively. If $L \leq 0$ or $d(t) \leq r$, then $f_s(t) = 1$, that is, the AUV executes path tracking in the next segment. If $L > 0$ and $d(t) > r$, then $f_s(t) = 0$, and the AUV executes path tracking in the current segment.

3. Mathematical Models for AUVs

3.1. AUV Kinematic Model

Figure 3 shows the platform of our self-made "Ocean Star" AUV, six thrusters provide forces to control the AUV. The detailed AUV's mechanical structure and size can be found in the Section 5. We give the earth ($\{O - X, Y, Z\}$) and body-fixed frames ($\{O_1 - X_1, Y_1, Z_1\}$) to express the AUV kinematic model, the AUV kinematic model can be written as follows [23–27]:

$$\dot{\eta} = \begin{bmatrix} J_1 & 0_{3\times3} \\ 0_{3\times3} & J_2 \end{bmatrix} \mathbf{v} \tag{6}$$

$$\begin{bmatrix} \dot{x} & \dot{y} & \dot{z} & \dot{\phi} & \dot{\theta} & \dot{\varphi} \end{bmatrix}^T = \begin{bmatrix} J_1 & 0_{3\times3} \\ 0_{3\times3} & J_2 \end{bmatrix} \begin{bmatrix} u & v & w & p & q & r \end{bmatrix}^T \tag{7}$$

where $\eta = \begin{bmatrix} x & y & z & \phi & \theta & \psi \end{bmatrix}^T$, x, y, z represent the AUV's location in the Earth-frame, ϕ, θ, ψ are AUV's roll, pitch and yaw angles in the Earth-frame frame; $v = \begin{bmatrix} u & v & w & p & q & r \end{bmatrix}$, u, v, w are the surge, sway and heave in the body-fixed frame, p, q, r are the roll, pitch and yaw rates in the body-fixed frame; J_1 and J_2 denote the coordinate transformation matrixes, whose mathematical expressions can be presented as follows:

$$J_1 = \begin{bmatrix} \cos\theta\cos\psi & \sin\theta\sin\phi\cos\psi - \cos\phi\sin\psi & \sin\theta\sin\psi + \sin\theta\cos\phi\cos\psi \\ \cos\theta\sin\psi & \sin\theta\sin\phi\sin\psi + \cos\phi\cos\psi & \sin\theta\cos\phi\sin\psi - \sin\phi\cos\psi \\ -\sin\theta & \sin\phi\cos\theta & \cos\phi\cos\theta \end{bmatrix} \tag{8}$$

$$J_2 = \begin{bmatrix} 1 & \sin\phi\tan\theta & \cos\phi\tan\theta \\ 0 & \cos\phi & -\sin\phi \\ 0 & \sin\phi/\cos\theta & \cos\phi/\cos\theta \end{bmatrix} \tag{9}$$

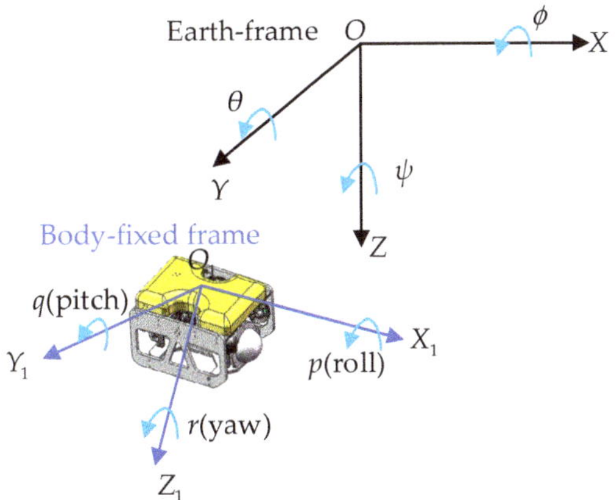

Figure 3. Two coordinate systems for the AUV.

3.2. AUV Dynamic Model

The dynamic equation of the AUV's motion can be given as follows [11,25,28]:

$$M\dot{v} + C(v)V + D(v)v + g(\eta) = \tau \tag{10}$$

where $M \in R^{6\times6}$ is the inertial matrix, which includes the added mass; $D \in R^{6\times6}$ is the damping matrix; $g \in R^{6\times6}$ is the gravitational terms matrix; $C \in R^{6\times6}$ is the matrix of Coriolis and centripetal terms; v is the position and orientation vector; τ is the control forces vector. The AUV's gravitational and buoyancy centers are located at one point. The translational and rotational motions can be described with six equations, which denote surge, sway, heave, roll, pitch, and yaw, respectively as follows:

$$m\left(\dot{u} - vr + wq - x_g\left(q^2 + r^2\right) + z_g\left(pr + q\right)\right)$$
$$= X_{HS} + X_u|u| + X_{\dot{u}}u + X_{wq}wq + X_{qq}qq + X_{vr}vr + X_{rr}rr + X_{prop} \tag{11}$$

$$m\left(\dot{v} - wp + ur - z_g(qr - \dot{p}) + x_g(pq + \dot{r})\right)$$
$$= Y_{HS} + Y_{v|v|}v|v| + Y_{r|r|}r|r| + Y_{\dot{v}}\dot{v} + Y_{\dot{r}}\dot{r} + Y_{ur}ur + Y_{wp}wp + Y_{pq}pq + Y_{uv}uv + Y_{uu\delta_r}u^2\delta_r \tag{12}$$

$$m\left(\dot{w} - uq + vp - z_g\left(q^2 + p^2\right) + x_g(rp + \dot{q})\right)$$
$$= Z_{HS} + Z_{w|w|}w|w| + Z_{q|q|}q|q| + Z_{\dot{w}}\dot{w} + Z_{\dot{q}}\dot{q} + Z_{uq}uq + Z_{vp}vp + Z_{pr}pr + Z_{uw}uw + Z_{uu\delta_s}u^2\delta_s \tag{13}$$

$$I_{xx}\dot{p} + (I_{zz} - I_{yy})qr + m\left|-z_g\left(\dot{v} - wp + ur\right)\right|$$
$$= K_{HS} + K_{p|p|}p|p| + k_{p|p|}p|p| + k_{\dot{p}}\dot{p} + k_{prop} \tag{14}$$

$$I_{yy}\dot{q} + (I_{xx} - I_{zz})pr + m\left|z_g\left(\dot{u} - vr + wq\right) - x_g\left(\dot{w} - uq + vp\right)\right| =$$
$$M_{HS} + M_{w|w|}q|q| + M_{q|q|}q|q| + M_{\dot{w}}\dot{w} + M_{\dot{q}}\dot{q} + M_{uq}uq + M_{vp}vp + M_{rp}rp + M_{uw}uw + M_{uu\delta_s}u^2\delta_s \tag{15}$$

$$I_{zz}\dot{r} + (I_{yy} - I_{xx})qp + m\left|x_g\left(\dot{v} - wp + ur\right)\right| =$$
$$N_{HS} + N_{v|v|}v|v| + N_{r|r|}r|r| + N_{\dot{v}}\dot{v} + N_{\dot{r}}\dot{r} + N_{ur}ur + N_{wp}wp + N_{pq}pq + N_{uv}uv + N_{uu\delta_r}u^2\delta_r \tag{16}$$

where m is the AUV's mass; x_g and z_g are the location parameters of the AUV's gravitational centers; I_{xx}, I_{yy}, I_{zz} are the AUV's inertia mass moments; X_u, Y_v, Z_q, N_r, N_v and M_q are the speed coefficients; $X_{u|u|}$, $Y_{v|v|}$, $Z_{q|q|}$, $M_{w|w|}$, $M_{|w|q}$, $N_{r|r|}$, $M_{q|q|}$ are the second-order speed coefficients; X_{wq} and X_{qq} are the second-order speed coefficients; δ_s and δ_r are the vertical angle and horizontal fins, respectively; X_{prop} is the propeller thrust; and k_{prop} is the propeller torque.

It should be noted that one can select several equations from Equations (11) to (16) for path-tracking design according to the actual engineering.

4. AUV Path Tracking with the Fuzzy Controller

4.1. The Design of the Fuzzy Controller

Fuzzy control is a human-like intelligent control method embodying the human control experience and strategy [29]. Fuzzy control does not depend on the precise models of the nonlinear control system, which can be used in the control process with strong robustness and good anti-interference performance [30–32]. A fuzzy controller contains four basic elements: fuzzification, knowledge base, fuzzy inference and defuzzification.

In this paper, we select triangle membership function for fuzzification, the reasons are given as follows: (1) Consisting of simple straight-line segments, triangular membership function is very easy to apply in fuzzy control; (2) Triangular membership function can obtain good performance for AUV control [30]. The methodologies to use for membership function optimization are given as follows: the membership function is assumed as an isosceles triangle, the function includes three parameters; the first and third parameters are optimized by the PSO algorithm, the middle is obtained as the average of the first and third values. Input parameters $e(t)$ and $\dot{e}(t)$ are transformed into fuzzy information based on the given triangle membership function. Based on the results of different fuzzy control rules with a trial-and-error method, Table 1 can be obtained with the fuzzy control rules for AUV path tracking. Combined with the input parameters, the output parameters attitude angles are obtained according to the fuzzy control rules, which are changed with $e(t)$ and $\dot{e}(t)$. The center of gravity method has smooth output inference control, whose output changes in response to small changes in the input value. Therefore, we select the center of gravity method for defuzzification in this paper.

Table 1. Fuzzy control rules for the AUV.

$\dot{e}(t)$ \ $e(t)$	NB	NM	NS	ZO	PS	PM	PB
NB	NB	NB	NM	NM	NS	ZO	ZO
NM	NB	NB	NM	NS	NS	ZO	PS
NS	NM	NM	NM	NS	ZO	PS	PS
ZO	NM	NM	NS	ZO	PS	PM	PM
PS	NS	NS	ZO	PS	PS	PM	PM
PM	NS	ZO	PS	PM	PM	PM	PB
PB	ZO	ZO	PM	PM	PM	PB	PB

Remark 1. *NB denotes negative big, NM denotes negative middle, NS denotes negative small, ZO denotes zero, PS denotes positive small, PM denotes positive middle, PB denotes positive big.*

In the control process for AUV path tracking, once the coordinates of the current and target locations are given, the desired attitude angles can be obtained at the desired AUV velocity [17]. Therefore, we select not only the cross-track position and attitude angle errors but also their derivatives as the input parameters. Figures 4 and 5 show the membership functions of cross-track position errors and corresponding derivatives, respectively, which are like the ones for cross-track attitude angle errors and their derivatives. Input and output variables need to be divided into different fuzzy sets. The more numbers of the fuzzy sets, the more detailed and flexible for the constituted fuzzy control rules, this may increase the programming complexity and computing time. Conversely, the less numbers of the fuzzy sets, the less simple the fuzzy control rules, this may lead to difficulties for the controller to achieve the expected effect. We divide input and output variables into seven fuzzy sets based on the experiment, which are good enough to balance the calculation time and control accuracy. The fuzzy states of cross-track position errors and their derivatives are given as follows: (1) Cross-track position error ($e(t)$) is divided into seven fuzzy states as follows, $NB \in [\alpha_1^1, \alpha_3^1]$, $NM \in [\alpha_2^1, \alpha_5^1]$, $NS \in [\alpha_4^1, \alpha_7^1]$, $ZO \in [\alpha_6^1, \alpha_9^1]$, $PS \in [\alpha_8^1, \alpha_{11}^1]$, $PM \in [\alpha_{10}^1, \alpha_{13}^1]$, $PB \in [\alpha_{12}^1, \alpha_{14}^1]$; (2) Cross-track position error derivative is divided into seven fuzzy states as follows: $NB \in [\alpha_1^2, \alpha_3^2]$, $NM \in [\alpha_2^2, \alpha_5^2]$, $NS \in [\alpha_4^2, \alpha_7^2]$, $ZO \in [\alpha_6^2, \alpha_9^2]$, $PS \in [\alpha_8^2, \alpha_{11}^2]$, $PM \in [\alpha_{10}^2, \alpha_{13}^2]$, $PB \in [\alpha_{12}^2, \alpha_{14}^2]$. Similar to Figures 4 and 5, the fuzzy states of the attitude angles and their derivatives can be given as follows: (1) Cross-track attitude angle error is divided into seven fuzzy states as follows: $NB \in [\alpha_1^3, \alpha_3^3]$, $NM \in [\alpha_2^3, \alpha_5^3]$, $NS \in [\alpha_4^3, \alpha_7^3]$, $ZO \in [\alpha_6^3, \alpha_9^3]$, $PS \in [\alpha_8^3, \alpha_{11}^3]$, $PM \in [\alpha_{10}^3, \alpha_{13}^3]$, $PB \in [\alpha_{12}^3, \alpha_{14}^3]$; (2) Cross-track attitude angle error derivative is divided into seven fuzzy states as follows: $NB \in [\alpha_1^4, \alpha_3^4]$, $NM \in [\alpha_2^4, \alpha_5^4]$, $NS \in [\alpha_4^4, \alpha_7^4]$, $ZO \in [\alpha_6^4, \alpha_9^4]$, $PS \in [\alpha_8^4, \alpha_{11}^4]$, $PM \in [\alpha_{10}^4, \alpha_{13}^4]$, $PB \in [\alpha_{12}^4, \alpha_{14}^4]$.

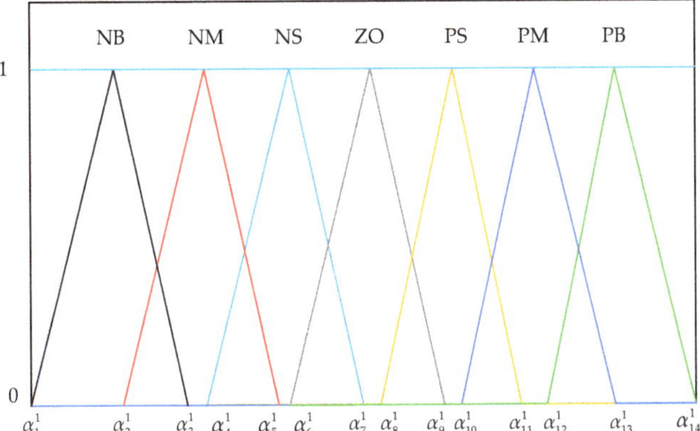

Figure 4. Membership function of the cross-track position error.

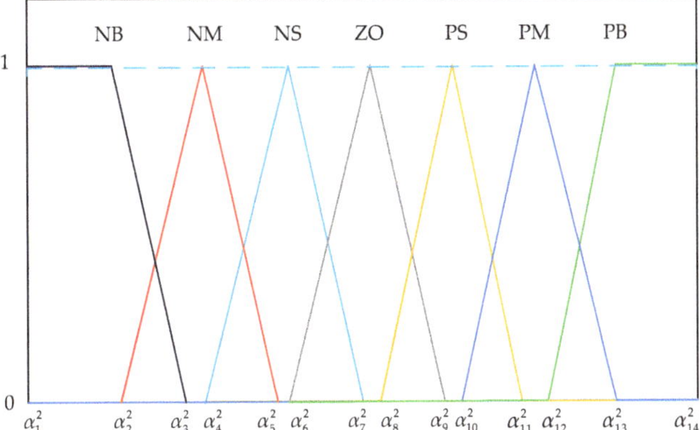

Figure 5. Membership function of the corresponding derivative.

Remark 2. *Cross-track attitude angle errors include the cross-track roll, pitch, and yaw (heading) angle errors in the three-dimensional environment, cross-track attitude angle error is the heading angle error if just considering path tracking in the XOY plane, the definition is given in Section 2.*

4.2. The Optimization for the Fuzzy Controller

In this paper, fuzzy control converts a linguistic control strategy into an automatic controller capable of managing the AUV path-tracking system. The membership function greatly affects the performance of the fuzzy controller, whose parameters are usually obtained by the cut-and-trial method with many experiments. This takes a lot of time and increases the difficulty to obtain the optimal value. Therefore, an algorithm needs to optimize the controller to ensure the control performance.

PSO is a popular and efficient method to solve many different kinds of engineering optimization problems [33,34]. The PSO algorithm has been proven to be simple for formulation, easy for programming and has good convergence [35–38], this conclusion is verified for fuzzy controller optimization of AUV path tracking. The PSO algorithm simulates the foraging behavior of a flock of flying birds, the result is achieved through cooperation and competition between individuals in the flock. In the PSO algorithm, flying "birds" are simulated by particles without weight or volume, each potential solution is represented by one particle in the search space, and its "flight state" is denoted with both the velocity and position of the particle. The global optimal search is realized by the particles, which continuously learns from the group and neighborhood optimal solutions. The new particle position and velocity can be obtained as follows [39,40]:

$$v_p(k+1) = w_p v_p(k) + c_1 r_1 (B_p - x_p(k)) + c_2 r_2 (B_g - x_p(k)) \tag{17}$$

$$x_p(k+1) = x_p(k) + v_p(k+1)) \tag{18}$$

$$w_p(k) = w_{max} - \frac{w_{max} - w_{min}}{N_{max}} N_{iter}(k) \tag{19}$$

where $v_p(k)$ is the current velocity of one particle; $v_p(k+1)$ is the new velocity of one particle; k is the k-th iteration; w_p is the inertia weight; c_1 is the social acceleration for the local best position; c_2 is the cognitive acceleration for the global best position; r_1 and r_2 are random numbers between 0 and 1; B_p and B_g are the best solutions found by a particle and in a population, respectively; $x_p(k)$ is the current position; $x_p(k+1)$ is the new position; N_{max} is the maximum number of iterations; $N_{iter}(k)$ is the current number of iteration; and w_{max} and w_{min} are the initial and final inertia weight, respectively.

In this paper, the following function is given to optimize the fuzzy controller to improve the path-tracking performance.

$$F_{TM} = f_L + f_S + f_E \tag{20}$$

where F_{TM} is the optimization performance index including f_L, f_S and f_E; f_L is the length of the tracking path; f_S is the smoothness of the tracking path; and f_E is the average cross-track position error. To solve the multiple objective functions for path tracking, a weight coefficient method is presented to transform Equation (20) as follows:

$$F_{TM} = \alpha \lfloor f_L \rfloor + \beta \lfloor f_S \rfloor + \gamma \lfloor f_E \rfloor \tag{21}$$

where α, β, γ are the weight coefficients; and $\lfloor f_L \rfloor$, $\lfloor f_S \rfloor$, $\lfloor f_E \rfloor$ are the normalized values of f_L, f_S, f_E respectively.

(1)   Tracking path length

The start and target points are (x_0, y_0, z_0) and (x_T, y_T, z_T) for the path, respectively, and T is the total number of the segments of the tracking path. The path can be expressed as follows: $A = [(x_0, y_0, z_0), \cdots, (x_t, y_t, z_t), \cdots, (x_T, y_T, z_T)]$.

The total length of the tracking path can be obtained as follows [22,41]:

$$f_L = \sum_{t=0}^{T} d(q_t, q_{t+1}) \tag{22}$$

where $q_t = (x_t, y_t, z_t)$ is one point of the tracking path at t, and t is the corresponding serial segment number. $d(q_t, q_{t+1})$ is the length from the adjacent point q_t to q_{t+1}, which can be calculated as follows:

$$d(q_t, q_{t+1}) = \sqrt{(x_{t+1} - x_t)^2 + (y_{t+1} - y_t)^2 + (z_{t+1} - z_t)^2} \tag{23}$$

(2)   Tracking path smoothness

A virtual triangle is used to describe the path smoothness, which is composed of three adjacent points q_{t-1}, q_t, and q_{t+1}. The cosine function is given to obtain the intersection angle for the AUV tracking path as follows.

$$\cos \alpha_s = \frac{\widetilde{b}^2 + \widetilde{c}^2 - \widetilde{a}^2}{2\widetilde{b}\widetilde{c}} \tag{24}$$

$$f_S(A) = \sum_{r=1}^{R} \frac{1}{\alpha_s}, \qquad \alpha_{min} < \alpha_s < \alpha_{max} \tag{25}$$

where α_s is the intersection angle with the corner vertex of q_t (rad). $\widetilde{a}$, $\widetilde{b}$, and $\widetilde{c}$ are the lengths of triangle sides with the angles of α_s, γ_s, and β_s, respectively. β_s is the intersection angle with the corner vertex of q_{t-1} (rad); and γ_s is the intersection angle with the corner vertex of q_{t+1} (rad). $f_S(A)$ is the AUV tracking path smoothness, the smaller $f_S(A)$, the better the path smoothness. $r = 1, 2, 3 \cdots R$, R is the serial number of the corner for each turning point.

(3)   Cross-track position error

The average of the total absolute of the cross-track position errors is given as a performance index, which is referred to simply as the average cross-track position error in this paper, calculated as follows:

$$f_E = \frac{1}{T} \sum_{t=0}^{T} |e_t| \tag{26}$$

where f_E is the average of the total $|e_t|$; e_t is one cross-track position error; and $|e_t|$ denotes the absolute value of e_t.

Figure 6 shows the flowchart of the path tracking in light of the fuzzy controller with the PSO algorithm, algorithm 1 presents the path-tracking steps in detail.

Algorithm 1: The path-tracking algorithm

1: Establishing the AUV kinematic model (Equation (7)) and the dynamic model (Equation (10)), obtaining the optimization function (Equation (21)) for the fuzzy controller for path tracking;
2: Defining the rules for the fuzzy controller, setting the basic initial parameters of the triangle membership function;
3: Initializing the parameters including c_1, c_2, and so on;
4: Setting the AUV start point as q_0, $t = 0$;
5. Obtaining the target point of the desired path $(\bar{q}_t)$;
6: Calculating the errors and these derivatives of the input parameters of the fuzzy controller, executing fuzzification optimization;
7: Realizing the fuzzy inference based on the membership functions, executing the defuzzification operation;
8: Calculating the fitness functions of each particle based on Equation (21), obtaining the optimization particle position;
9: Updating the current velocity of each particle based on Equations (17)–(19);
10: If N_{iter} is smaller than N_{max}, go to the next step, else if, obtain the parameters of membership functions, setting $N_{iter} = N_{iter} + 1$, go to step 7;
11: Obtaining the tracking point q_t based on the AUV kinematic and dynamic models, if t is smaller than T, setting $t = t + 1$, go to step 4, else if, go to the next step;
12: Obtain the optimal tracking path.

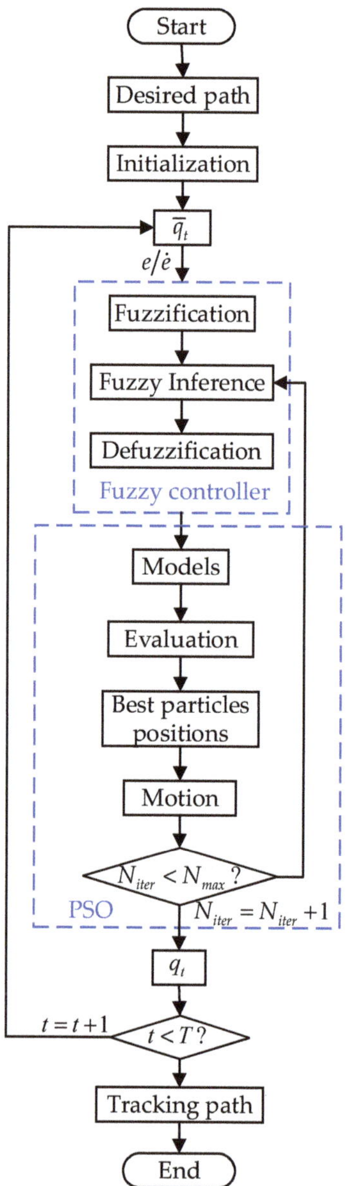

Figure 6. Flowchart for path tracking with the fuzzy controller.

5. Results and Discussion

In this section, a self-made "Ocean Star" AUV is used to illustrate the effectiveness of the proposed algorithm. Figure 7 gives the "Ocean Star" AUV and its configuration, the length is 400 mm, the width is 295 mm, and the high is 255 mm. Six thrusters are given to provide forces to control the AUV, Figure 7b shows the corresponding configuration.

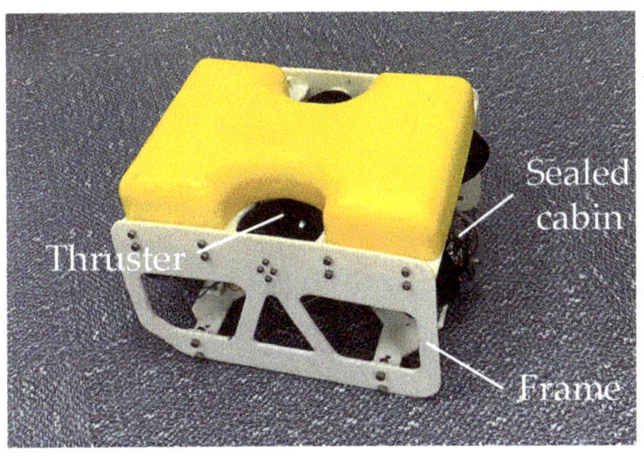

(**a**) "Ocean Star" AUV

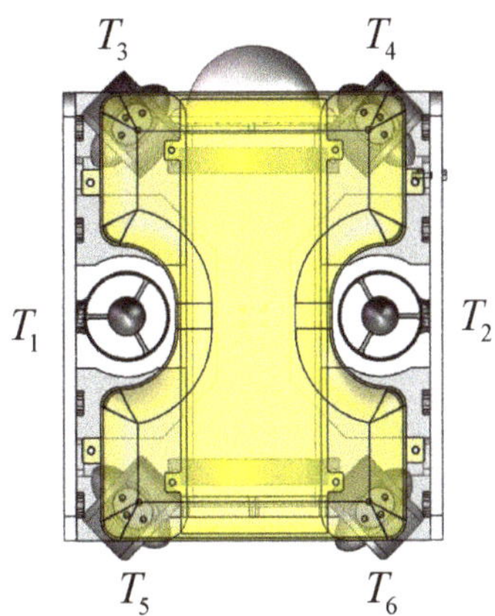

(**b**) Thruster configuration

Figure 7. The mechanical structure and physical drawing.

Different scenarios are given including straight line, sine, half-moon shape, Archimedean spiral, and practical paths tracking. For the PSO algorithm, the initial value of the inertia weight $w_p = 0.8$, the social and cognitive accelerations are $c_1 = 0.5$ and $c_2 = 0.2$, respectively. In general, the fewer particles, the more iterations are required to obtain good optimization results. For example, the algorithm needs more than 100 iterations with 15 particles and 160 iterations with 10 particles to guarantee the optimization results. In this paper, the number of particles and the maximum number of iterations are good enough for the proposed algorithm with the settings as follows: the number of the particles is 20 for the group, the maximum number of iterations is 50. We set the weight coefficients as $\alpha = 0.3$, $\beta = 0.1$, and $\gamma = 0.6$ for the optimization objective functions.

A desired straight line is given as follows: $x(t_1) = t_1 + 10$, $y(t_1) = 2t_1 + 10$, where $t_1 \in [0, 60]$ (as shown in Figure 8). The straight line is discretized by the sequential discrete points $[(10, 10, 0), (11.7, 13.3, 0), \cdots, (70, 130, 0)]$, and the corresponding start point is $(10, 10, 0)$ [m]. The initial position, Euler angle, and velocity are set to $(x, y, z) = (0, 0, 0)$ [m], $(\varphi, \theta, \phi) = (0°, 0°, 0°)$, $(u, v, w) = (0, 0, 0)$ [kn], respectively. The desired velocity is set to $(u, v, w) = (0.18, 0, 0)$ [kn]. Figure 8 shows the desired and tracking path curves, we can see that the tracking path coincides well with the desired path. The tracking path length is 141.46 m, smoothness is 982.4, and average cross-track position error is 0.078 m. Figure 9 shows the cross-track position error, Figure 10 shows the heading angle error. The cross-track position and heading angle errors become smaller and smaller over time. The cross-track position error is limited in the range of $[-0.084 \text{ m}, 0.037\text{m}]$, and the standard deviation is 0.007 m for path cross-track position errors calculated from the start to the target tracking points. The cross-track heading angle error is limited in the range of $[-0.008 \text{ rad}, 0.19 \text{ rad}]$, and the standard deviation is 0.0045 rad for the cross-track heading angle errors calculated from the start to the target tacking points. The proposed fuzzy controller with the PSO algorithm can deal with straight line path tracking practically and effectively.

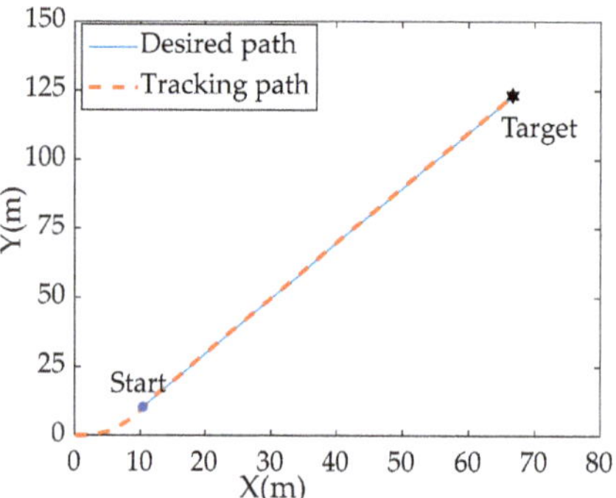

Figure 8. Path tracking results for the straight line.

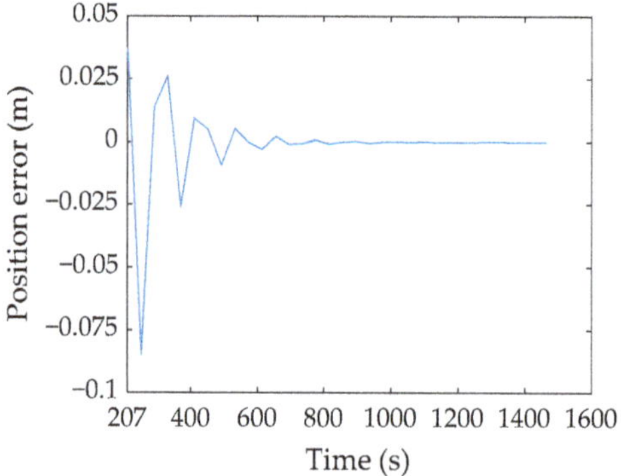

Figure 9. Cross-track position error for the straight line.

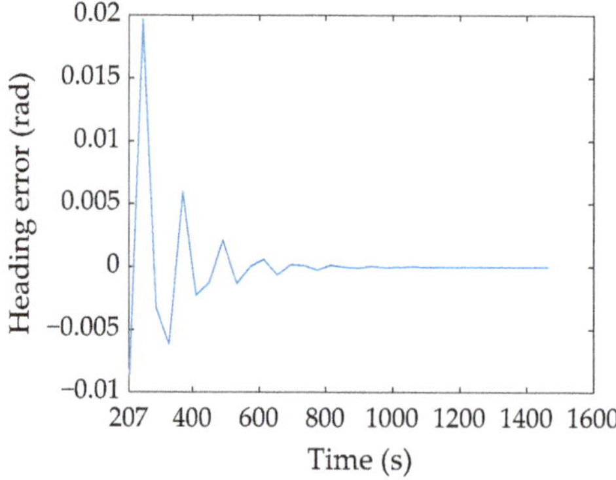

Figure 10. Cross-track heading angle error for the straight line.

The desired sine curve path is given as follows: $x(t_2) = t_2$, $y(t_2) = 5\sin(t_2\pi/40)$, where $t_2 \in [0, 240]$, Figure 11 shows the corresponding curve. The sine curve is discretized by the sequential discrete points $[(0,0,0), (2,0.78,0), \cdots, (240,0,0)]$, and the corresponding start point is $(0,0,0)$ [m]. The initial position, Euler angle, and velocity are set to $(x,y,z) = (-10,-1,0)$ [m], $(\varphi,\theta,\phi) = (0^\circ,0^\circ,0^\circ)$, and $(u,v,w) = (0,0,0)$ [kn], respectively. The desired velocity is set to $(u,v,w) = (0.18,0,0)$ [kn]. Figure 11 shows the tracking path by the proposed algorithm, the tracking path length is 259.52 m, smoothness is 1802.2, and average cross-track position error is 0.18 m. Figure 12 shows the cross-track position error, Figure 13 shows the cross-track heading angle error. The cross-track position error is limited in the range of $[-0.79 \text{ m}, 0.66 \text{ m}]$, and the standard deviation is 0.25 m for the path-tracking position. The cross-track heading angle error is limited in the range of $[-0.31 \text{ rad}, 0.22 \text{ rad}]$, and the standard deviation is 0.08 rad for the cross-track heading angle error. Although larger values are at the peaks and valleys of the cross-track position and heading angle errors, the tracking path still coincides well with the desired curve. The proposed fuzzy controller with the PSO algorithm can deal with sine curve path tracking practically and effectively.

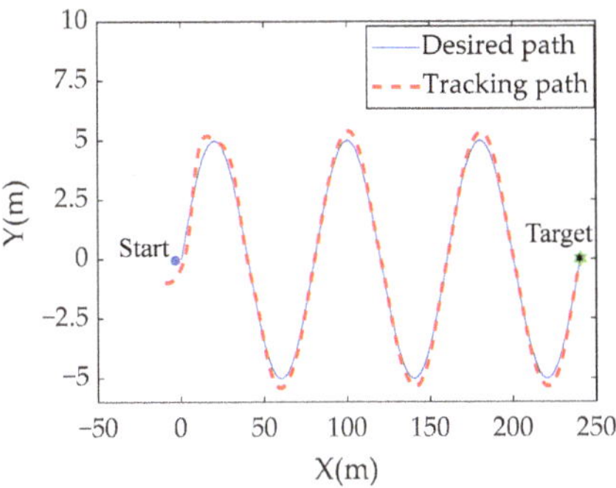

Figure 11. Path-tracking results for the sine curve.

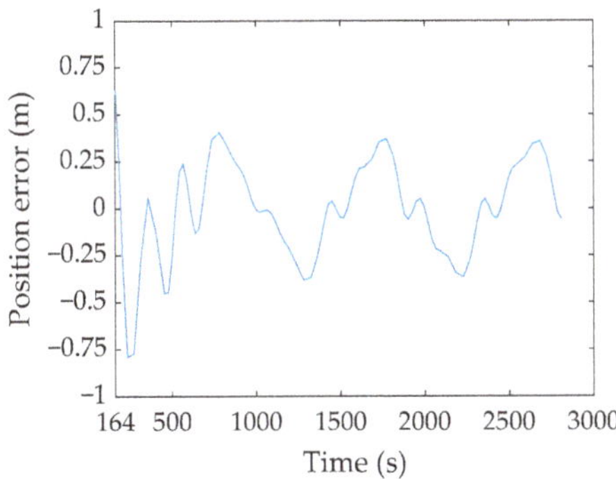

Figure 12. Cross-track position error for the sine curve.

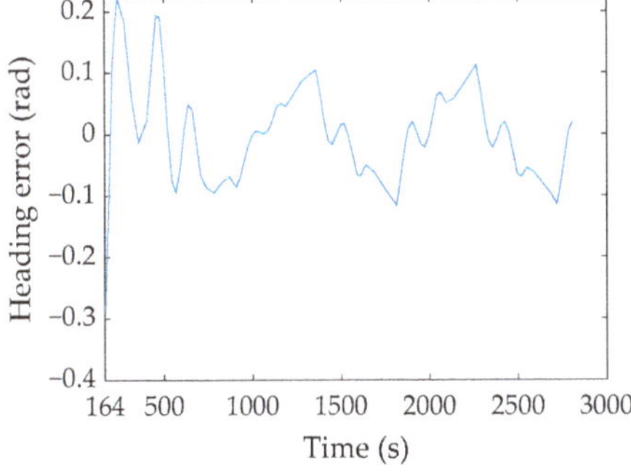

Figure 13. Cross-track heading angle error for the sine curve.

The desired half-moon shape path is given as follows: $x(t_3) = 50\cos(t_3)$, $y(t_3) = 50\sin(t_3)$, where $t_3 \in [-\pi/2, \pi/2]$. $x(t_4) = 100\cos(t_4) + 100\cos(5\pi/6)$, $y(t_4) = 100\sin(t_4)$, where $t_4 \in [\pi/6, -\pi/6]$. Figure 14 shows the half-moon shape path, which is discretized by the sequential discrete points $[(0, -50, 0), (0.7854, -49.9938, 0), \cdots, (0, -50, 0)]$. The start and target points are all at $(0, -50, 0)$ [m] for the desired path, the AUV moves along the half-moon shape in an anti-clockwise direction. The initial position, Euler angle, and velocity are set to $(x, y, z) = (-10, -60, 0)$ [m], $(\varphi, \theta, \phi) = (0°, 0°, 0°)$, and $(u, v, w) = (0, 0, 0)$ [kn], respectively. The desired velocity is set to $(u, v, w) = (0.18, 0, 0)$ [kn]. Figure 14 shows the tracking path by the proposed algorithm, the red curve coincides well with the desired path. The tracking path length is 278.38 m, smoothness is 1931.5, and average cross-track position error is 0.17 m. Figures 15 and 16 show the cross-track position and heading angle errors, respectively. The standard deviation is 0.38 m for the path cross-track position calculated from the start to the target tracking points. The standard deviation is 0.11 rad for the cross-track heading angle error calculated from the start to the target tracking point. The proposed fuzzy controller algorithm can deal with the half-moon shape path tracking practically and effectively.

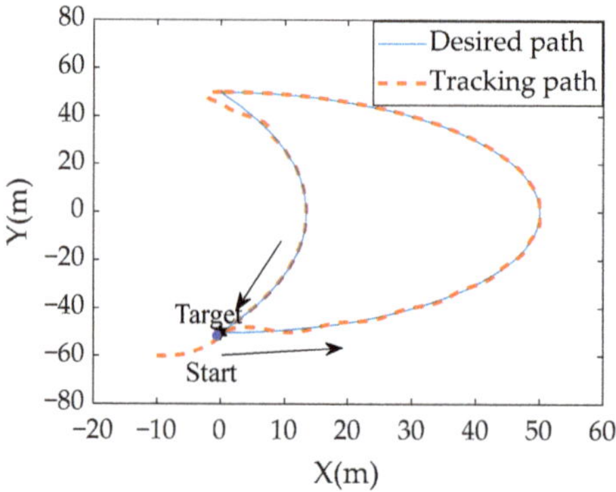

Figure 14. Tracking results for the half-moon shape.

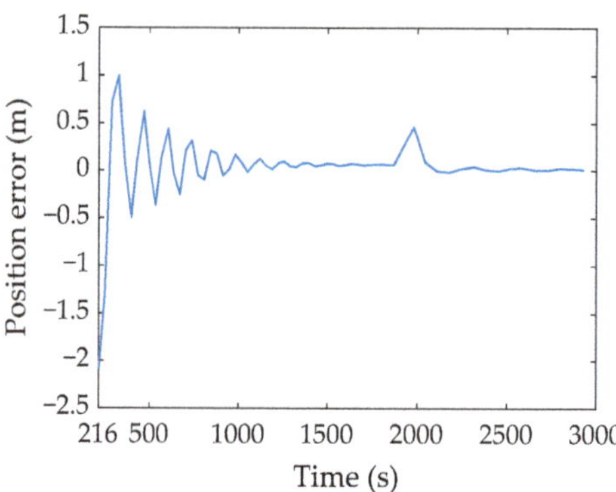

Figure 15. Cross-track position error for the half-moon shape.

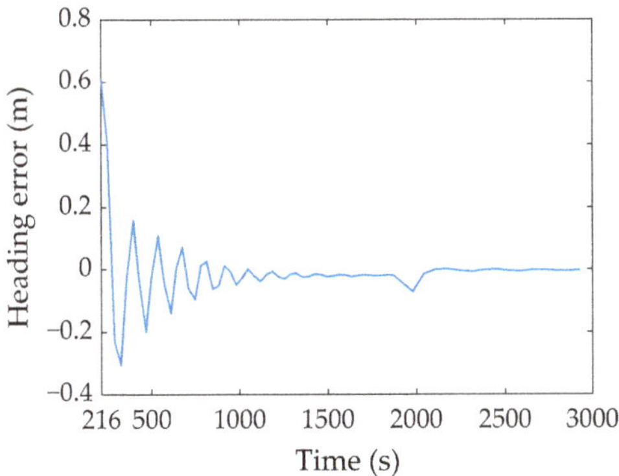

Figure 16. Cross-track heading angle error for the half-moon shape path.

Some comparison results are given to make an in-depth analysis of the path-tracking simulation results. A desired Archimedean spiral path is given as follows: $x(t_5) = (5t_5 \cos(t_5))$, $y(t_5) = 40 + (5t_5 \sin(t_5))$, where $t_5 \in [1.5\pi : 4.1\pi]$. Figure 17a shows the Archimedean path, which is discretized by the sequential discrete points $[(0, 16.44, 0)\ [m],\ (2.26, 16.0, 0)\ [m], \cdots, (61.25, 59.9, 0)\ [m]]$, the corresponding start point is $(0, 16.44, 0)\ [m]$. It should be noted that Figure 17b is a partial enlargement of Figure 17a. The initial position, Euler angle, and velocity are set to $(x, y, z) = (-3, 16, 0)\ [m]$, $(\varphi, \theta, \phi) = (0°, 0°, 0°)$, and $(u, v, w) = (0, 0, 0)\ [kn]$, respectively. The desired velocity is set to $(u, v, w) = (0.34, 0, 0)\ [kn]$. Figures 18 and 19 show the comparison tracking results between the proposed algorithm, fuzzy control method (using only fuzzy logic), and the traditional method (PID algorithm). Table 2 gives the corresponding comparison with some performance indexes. The tracking path length is 360.9 m, smoothness is 1353, and average cross-track position error is 0.05 m by the proposed algorithm. The tracking path length is 361.2 m, smoothness is 1355, and average cross-track position error is 0.06 m by the fuzzy control method. The tracking path length is 361.8 m, smoothness is 1357, and average cross-track position error is 0.09 m by the traditional method. Figure 18 shows the cross-track position error for the Archimedean spiral. The fluctuating range of the error curve by the proposed algorithm is smaller than the fuzzy control and traditional methods. The standard deviations of the cross-track position errors are 0.065 m, 0.078 m, and 0.113 m by the proposed, fuzzy control, and traditional methods, respectively. The proposed algorithm obtains better tracking results compared with the fuzzy control and traditional methods.

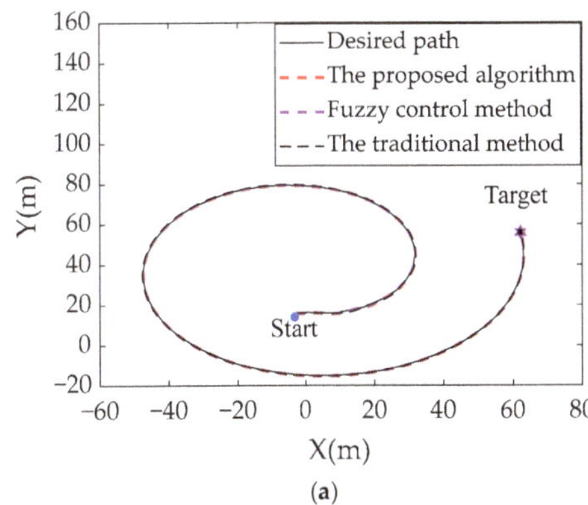

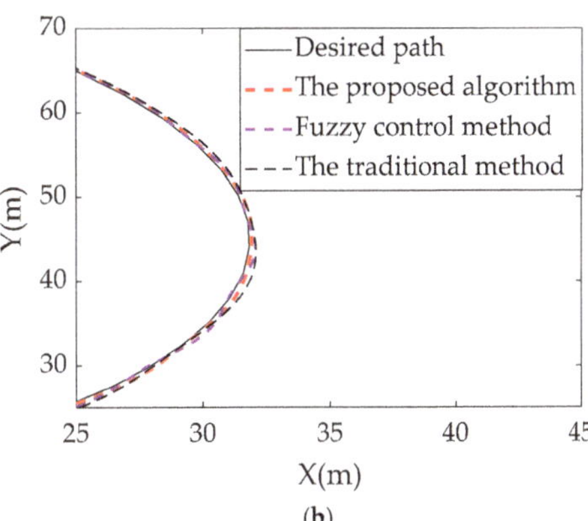

Figure 17. (**a**). Path-tracking results for Archimedean spiral. (**b**) Partial enlargement for the tracking path.

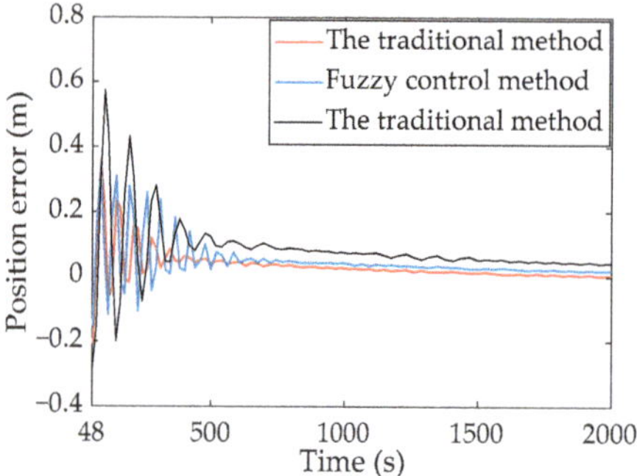

Figure 18. Cross-track position error for the Archimedean spiral.

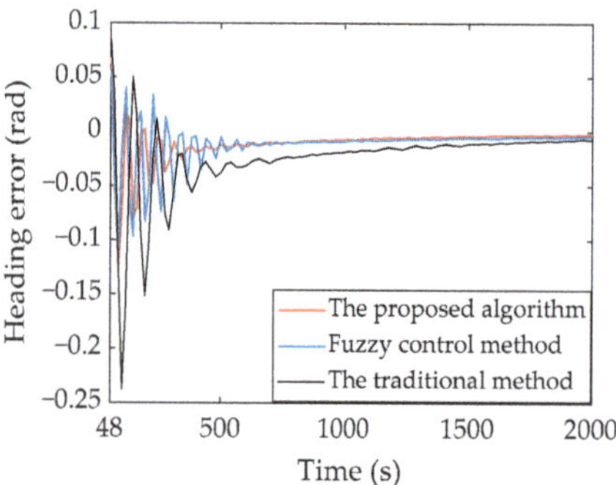

Figure 19. Cross-track heading angle error for the Archimedean spiral.

Table 2. Comparison of the proposed and existing algorithms.

Method	Tracking Path Length (m)	Path Smoothness (–)	Average Cross-Track Position Error (m)	Standard Deviation (m)
The proposed algorithm	360.9	1353	0.05	0.065
The fuzzy control method	361.2	1355	0.06	0.078
The traditional method	361.8	1357	0.09	0.113

Figures 20 and 21 show a desired path in the real environment of our university. The length and width are obtained for a man-made river with an irregular polygon shape based on a diastimeter. The desired path is the line with a distance of 1 m from the riverbank to the AUV's center. As shown in Figure 21, we establish the coordinate system (XOY), and select a point of the bottom left corner of the river as the origin. The AUV start and target positions are at the same point $(x, y, z) = (44.7, 1, 0)$ [m], and the AUV moves along the path in a clockwise direction. The initial Euler angle is $(\varphi, \theta, \phi) = (180°, 0°, 0°)$, the initial velocity is $(u, v, w) = (0, 0, 0)$ [kn], and the desired velocity is $(u, v, w) = (0.18, 0, 0)$ [kn]. Figure 21 shows the tracking results, the tracking path length is 99.14 m, smoothness is 633.59, and average cross-track position error is 0.05 m. Figure 22 shows the cross-track position error, and the standard deviation is 0.1 m of the cross-track position errors. Figure 23 shows the cross-track heading angle error, and the standard deviation is 0.04 of the cross-track heading angle errors. Results show that the AUV can follow the path effectively by the proposed algorithm.

Figure 20. The desired path in a real man-made river.

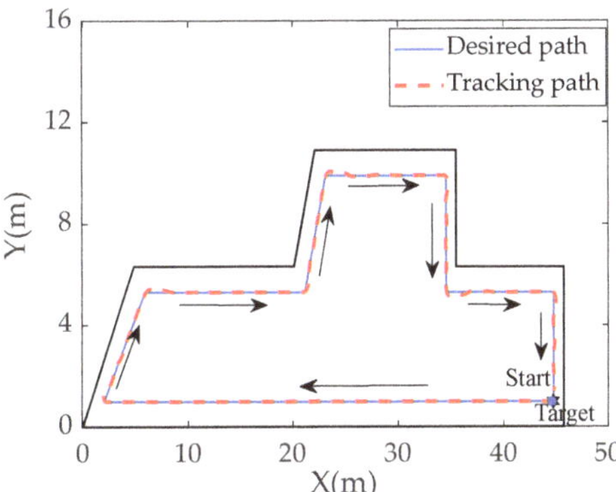

Figure 21. The path tracking in the real man-made river.

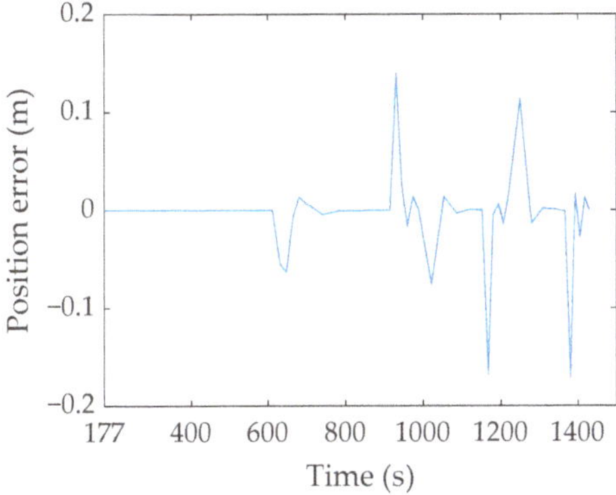

Figure 22. Cross-track position error for the AUV.

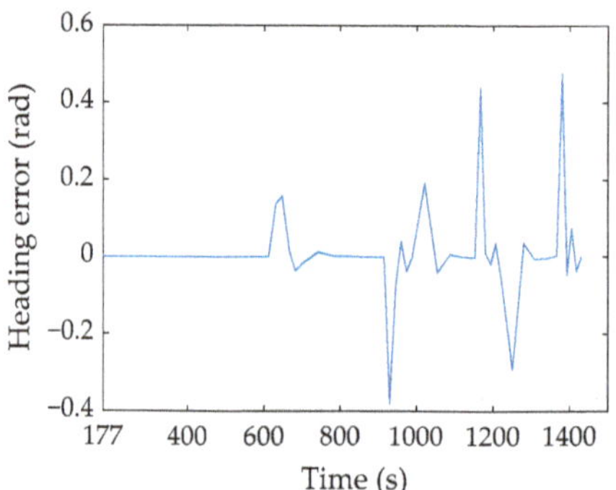

Figure 23. Cross-track heading angle error for the AUV.

6. Conclusions

AUV path tracking is important to realize for AUV application in numerous missions for the exploration of marine resources. In this paper, a fuzzy controller was designed with multiple performance optimizations to complete path tracking. Based on the established AUV kinematic and dynamic models, we selected the path length, smoothness and cross-track position error as the objective functions of the optimization problem for the fuzzy controller. The PSO algorithm was used to optimize the membership function. We gave the straight line, sine curve, half-moon shape, Archimedean spiral, and practical paths to test the fuzzy controller's performance for path tracking. The test results showed that the AUV effectively tracked the desired path using the proposed algorithm. Compared to the traditional algorithm, the designed controller made the path-tracking system perform well with better tracking accuracy, better smoothness and a shorter length using the proposed algorithm.

In the near future, we are interested in improving the path-tracking performance by incorporating salient features such as sliding mode control, deep reinforcement learning, and so on. Collaborative work on multiple AUVs is an important way to efficiently accomplish these missions, of which the key factor is how to accurately track multiple paths. Therefore, providing a method for multiple AUV path tracking is also our next work.

Author Contributions: Conceptualization, Q.T. and T.W.; methodology, Q.T.; software, T.W.; validation, Q.T., Y.W., Y.S. and B.L.; formal analysis, Q.T.; investigation, Q.T. and T.W.; resources, Y.W.; data curation, Y.S.; writing—original draft preparation, Q.T.; writing—review and editing, Y.W.; visualization, Y.S.; supervision, B.L.; project administration, B.L.; funding acquisition, Y.W. All authors have read and agreed to the published version of the manuscript.

Funding: This research was funded by the China Post-doctoral Science Foundation (2022M710934), Post-doctoral Applied Research Project of Qingdao City, and the Project of Shandong Province Higher Educational Young Innovative Talent Introduction and Cultivation Team (Intelligent Transportation Team of Offshore Products).

Institutional Review Board Statement: Not applicable.

Informed Consent Statement: Not applicable.

Data Availability Statement: Not applicable.

Conflicts of Interest: The authors declare no conflict of interest.

References

1. Hien, N.V.; Truong, V.-T.; Bui, N.-T.; de Oña, R. A Model-Driven Realization of AUV Controllers Based on the MDA/MBSE Approach. *J. Adv. Transp.* **2020**, *2020*, 8848776. [CrossRef]
2. Marvian Mashhad, A.; Mousavi Mashhadi, S.K.-e.-d. PD like fuzzy logic control of an autonomous underwater vehicle with the purpose of energy saving using H∞ robust filter and its optimized covariance matrices. *J. Mar. Sci. Technol.* **2018**, *23*, 937–949. [CrossRef]
3. Tian, Q.; Wang, T.; Liu, B.; Ran, G. Thruster Fault Diagnostics and Fault Tolerant Control for Autonomous Underwater Vehicle with Ocean Currents. *Machines* **2022**, *10*, 582. [CrossRef]
4. Tran, H.N.; Nhut Pham, T.N.; Choi, S.H. Robust depth control of a hybrid autonomous underwater vehicle with propeller torque's effect and model uncertainty. *Ocean Eng.* **2021**, *220*, 108257. [CrossRef]
5. Bejarbaneh, E.Y.; Masoumnezhad, M.; Armaghani, D.J.; Pham, B.T. Design of robust control based on linear matrix inequality and a novel hybrid PSO search technique for autonomous underwater vehicle. *Appl. Ocean Res.* **2020**, *101*, 102231. [CrossRef]
6. Yao, X.; Wang, X. Integral vector field control for three-dimensional path following of autonomous underwater vehicle. *J. Mar. Sci. Technol.* **2020**, *26*, 159–173. [CrossRef]
7. Yang, X.; Yan, J.; Hua, C.; Guan, X. Trajectory Tracking Control of Autonomous Underwater Vehicle With Unknown Parameters and External Disturbances. *IEEE Trans. Syst. Man Cybern. Syst.* **2021**, *51*, 1054–1063. [CrossRef]
8. Yu, C.Y.; Xiang, X.B.; Zuo, M.J.; Liu, H. Underwater cable tracking control of under-actuated AUV. In Proceedings of the 2016 IEEE/OES Autonomous Underwater Vehicles, Tokyo, Japan, 6–9 November 2016; IEEE: Piscataway, NJ, USA, 2016; pp. 324–329.
9. Liu, F.; Shen, Y.; He, B.; Wang, D.; Wan, J.; Sha, Q.; Qin, P. Drift angle compensation-based adaptive line-of-sight path following for autonomous underwater vehicle. *Appl. Ocean Res.* **2019**, *93*, 101943. [CrossRef]
10. Nie, W.; Feng, S. Planar path-following tracking control for an autonomous underwater vehicle in the horizontal plane. *Optik* **2016**, *127*, 11607–11616. [CrossRef]
11. Zhang, G.; Huang, H.; Qin, H.; Wan, L.; Li, Y.; Cao, J.; Su, Y. A novel adaptive second order sliding mode path following control for a portable AUV. *Ocean Eng.* **2018**, *151*, 82–92. [CrossRef]
12. Jia, C.; Chen, Z.; Zhang, G.; Shi, L.; Sun, Y.; Pang, S. AUV Tunnel Tracking Method Based on Adaptive Model Predictive Control. *IOP Conf. Ser. Mater. Sci. Eng.* **2018**, *428*, 012070. [CrossRef]
13. Wang, D.; He, B.; Shen, Y.; Li, G.; Chen, G. A Modified ALOS Method of Path Tracking for AUVs with Reinforcement Learning Accelerated by Dynamic Data-Driven AUV Model. *J. Intell. Robot. Syst.* **2022**, *49*, 104. [CrossRef]
14. Sun, Y.; Zhang, C.; Zhang, G.; Xu, H.; Ran, X. Three-Dimensional Path Tracking Control of Autonomous Underwater Vehicle Based on Deep Reinforcement Learning. *J. Mar. Sci. Eng.* **2019**, *7*, 443. [CrossRef]
15. Yu, C.; Xiang, X.; Lapierre, L.; Zhang, Q. Nonlinear guidance and fuzzy control for three-dimensional path following of an underactuated autonomous underwater vehicle. *Ocean Eng.* **2017**, *146*, 457–467. [CrossRef]
16. Zhang, J.; Xiang, X.; Lapierre, L.; Zhang, Q.; Li, W. Approach-angle-based three-dimensional indirect adaptive fuzzy path following of under-actuated AUV with input saturation. *Appl. Ocean Res.* **2021**, *107*, 102486. [CrossRef]
17. Chen, J.; Zhu, H.; Zhang, L.; Sun, Y. Research on fuzzy control of path tracking for underwater vehicle based on genetic algorithm optimization. *Ocean Eng.* **2018**, *156*, 217–223. [CrossRef]
18. Lee, H.J.; Rhee, K.P. Development of collision avoidance system by using expert system and search algorithm. *Int. Shipbuild. Prog.* **2001**, *48*, 197–212.
19. Zhao, Y.; Li, W.; Shi, P. A real-time collision avoidance learning system for Unmanned Surface Vessels. *Neurocomputing* **2016**, *182*, 255–266. [CrossRef]
20. Namgung, H.; Kim, J.S. Collision risk inference system for maritime autonomous surface ships using COLREGs rules compliant collision avoidance. *IEEE Access* **2021**, *9*, 7823–7835. [CrossRef]
21. Namgung, H. Local route planning for collision avoidance of maritime autonomous surface ships in compliance with COLREGs rules. *Sustainability* **2021**, *14*, 198. [CrossRef]
22. Ataei, M.; Yousefi-Koma, A. Three-dimensional optimal path planning for waypoint guidance of an autonomous underwater vehicle. *Robot. Auton. Syst.* **2015**, *67*, 23–32. [CrossRef]
23. Wang, L.; Liu, L.; Qi, J.; Peng, W. Improved Quantum Particle Swarm Optimization Algorithm for Offline Path Planning in AUVs. *IEEE Access* **2020**, *8*, 143397–143411. [CrossRef]
24. Cao, X.; Sun, H.; Jan, G.E. Multi-AUV cooperative target search and tracking in unknown underwater environment. *Ocean Eng.* **2018**, *150*, 1–11. [CrossRef]
25. Taheri, E.; Ferdowsi, M.H.; Danesh, M. Closed-loop randomized kinodynamic path planning for an autonomous underwater vehicle. *Appl. Ocean Res.* **2019**, *83*, 48–64. [CrossRef]
26. MahmoudZadeh, S.; Yazdani, A.M.; Sammut, K.; Powers, D.M.W. Online path planning for AUV rendezvous in dynamic cluttered undersea environment using evolutionary algorithms. *Appl. Soft Comput.* **2018**, *70*, 929–945. [CrossRef]
27. Tian, Q.; Wang, T.; Wang, Y.; Li, C.; Liu, B. Robust Optimization Design for Path Planning of Bionic Robotic Fish in the Presence of Ocean Currents. *J. Mar. Sci. Eng.* **2022**, *10*, 1109. [CrossRef]
28. Karkoub, M.; Wu, H.; Hwang, C. Nonlinear trajectory-tracking control of an autonomous underwater vehicle. *Ocean Eng.* **2017**, *145*, 188–198. [CrossRef]

29. Ran, G.; Chen, H.; Li, C.; Ma, G.; Jiang, B. A Hybrid Design of Fault Detection for Nonlinear Systems Based on Dynamic Optimization. *IEEE Trans. Neural Netw. Learn. Syst.* **2022**, *18*, 1–11. [CrossRef]

30. Xiang, X.; Yu, C.; Lapierre, L.; Zhang, J.; Zhang, Q. Survey on Fuzzy-Logic-Based Guidance and Control of Marine Surface Vehicles and Underwater Vehicles. *Int. J. Fuzzy Syst.* **2017**, *20*, 572–586. [CrossRef]

31. Ran, G.; Liu, J.; Li, C.; Lam, H.; Li, D.; Chen, H. Fuzzy-Model-Based Asynchronous Fault Detection for Markov Jump Systems with Partially Unknown Transition Probabilities: An Adaptive Event-Triggered Approach. *IEEE Trans. Fuzzy Syst.* **2022**, *11*, 4679–4689. [CrossRef]

32. Ran, G.; Shu, Z.; Lam, H.; Liu, J.; Li, C. Dissipative Tracking Control of Nonlinear Markov Jump Systems With Incomplete Transition Probabilities: A Multiple-Event-Triggered Approach. *IEEE Trans. Fuzzy Syst.* **2022**, *20*, 1–11. [CrossRef]

33. Bangyal, W.; Hameed, A.; Alosaimi, W.; Alyami, H. A New Initialization Approach in Particle Swarm Optimization for Global Optimization Problems. *Comput. Intell. Neurosci.* **2021**, *2021*, 1–17. [CrossRef]

34. Bai, B.; Guo, Z.; Zhou, C.; Zhang, W.; Zhang, J. Application of adaptive reliability importance sampling-based extended domain PSO on single mode failure in reliability engineering. *Inf. Sci.* **2021**, *546*, 42–59. [CrossRef]

35. Bonyadi, M.R.; Michalewicz, Z. Particle swarm optimization for single objective continuous space problems: A review. *Evol. Comput.* **2017**, *25*, 1–54. [CrossRef] [PubMed]

36. Yarat, S.; Senan, S.; Orman, Z. A Comparative Study on PSO with Other Metaheuristic Methods. *Appl. Part. Swarm Optim.: New Solut. Cases Optim. Portf.* **2021**, *83*, 49–72.

37. Askarzadeh, A. Comparison of particle swarm optimization and other metaheuristics on electricity demand estimation: A case study of Iran. *Energy* **2014**, *72*, 484–491. [CrossRef]

38. Ren, M.; Huang, X.; Zhu, X.; Shao, L. Optimized PSO algorithm based on the simplicial algorithm of fixed point theory. *Appl. Intell.* **2020**, *50*, 2009–2024. [CrossRef]

39. Fattahi, P.; Hajipour, V.; Nobari, A. A bi-objective continuous review inventory control model: Pareto-based meta-heuristic algorithms. *Appl. Soft Comput.* **2015**, *32*, 211–223. [CrossRef]

40. Pluhacek, M.; Senkerik, R.; Davendra, D.; Kominkova Oplatkova, Z.; Zelinka, I. On the behavior and performance of chaos driven PSO algorithm with inertia weight. *Comput. Math. Appl.* **2013**, *66*, 122–1234. [CrossRef]

41. Tian, Q.; Wang, T.; Wang, Y.; Wang, Z.; Liu, C. A two-level optimization algorithm for path planning of bionic robotic fish in the three-dimensional environment with ocean currents and moving obstacles. *Ocean. Eng.* **2022**, *266*, 112829. [CrossRef]

Journal of
*Marine Science
and Engineering*

Article

Finite Time Trajectory Tracking with Full-State Feedback of Underactuated Unmanned Surface Vessel Based on Nonsingular Fast Terminal Sliding Mode

Donghao Xu [1,2], Zipeng Liu [1], Jiuzhen Song [3] and Xueqian Zhou [2,4,*]

[1] College of Automation, Harbin University of Science and Technology, Harbin 150080, China
[2] College of Shipbuilding Engineering, Harbin Engineering University, Harbin 150001, China
[3] The 714 Research Institute of CSSC, Beijing 100101, China
[4] International Joint Laboratory of Naval Architecture and Offshore Technology between Harbin Engineering University and University of Lisbon, Harbin 150001, China
* Correspondence: xueqian.zhou@hrbeu.edu.cn; Tel.: +86-451-82519902

Abstract: Marine transportation and operations have attracted the attention of more and more countries and scholars in recent years. A full-state finite time feedback control scheme is designed for the model parameters uncertainty, unknown ocean environment disturbances, and unmeasured system states in the underactuated Unmanned Surface Vessel (USV) trajectory tracking control. The external wind, wave and current environmental disturbances and model parameters perturbation are extended by Nonlinear Extended State Observer (NESO) to the state of the system, namely complex disturbances. The complex disturbances, positions and velocities of USV can be observed by NESO and feedback to USV control system. Next, the underactuated USV error model is obtained by operating the obtained feedback information and the virtual ship model. According to the error model, a Nonsingular Fast Terminal Sliding Model surface (NFTSM) is constructed to realize finite-time control. The control law is deduced through the Lyapunov stability theory to ensure the stability of the system. The results of MATLAB numerical simulations under different disturbances show that the trajectory tracking algorithm has fast responses, and a good convergence of the errors is observed, which verifies the effectiveness of the designed scheme.

Keywords: nonsingular fast terminal sliding model; nonlinear extended state observer; finite time control; full-state feedback; trajectory tracking

Citation: Xu, D.; Liu, Z.; Song, J.; Zhou, X. Finite Time Trajectory Tracking with Full-State Feedback of Underactuated Unmanned Surface Vessel Based on Nonsingular Fast Terminal Sliding Mode. *J. Mar. Sci. Eng.* **2022**, *10*, 1845. https://doi.org/10.3390/jmse10121845

Academic Editor: Alessandro Ridolfi

Received: 22 October 2022
Accepted: 21 November 2022
Published: 1 December 2022

Publisher's Note: MDPI stays neutral with regard to jurisdictional claims in published maps and institutional affiliations.

1. Introduction

In recent years, with the continuous development of unmanned driving, artificial intelligence, image processing, Internet space technology, etc., the whole industry has made great progress in digitalization and intelligence, and the USV is also rapidly developing towards intelligence [1]. The research on USV has also become a research hotspot. USV has a wide application prospect [2] because of its high flexibility and cost-effective characteristics, carrying specific equipment to complete a variety of dangerous specific tasks. In the military, the USV can be used for counter-reconnaissance and other intelligence collection, while in the civilian field, it can be used for water quality detection/monitoring, maritime patrol and ocean mapping [3–6]. To accomplish such missions, USVs require good motion control ability, and trajectory tracking control is the solution. Trajectory tracking is different from path tracking as for the former, the trajectory is parameterized and depends on time [7], which requires the ship to reach specified coordinates along a specific trajectory at a specified time. The under-actuation of a USV increases the control difficulty because the number of independent control inputs is less than its degrees of freedom [8] and is thus a nonlinear system [9].

In practice, there are many factors in the system to deal with, such as nonlinearity, parameter uncertainty, unmeasurable noises, unmeasurable speeds and external environment disturbances, which bring great challenges to the design of the control system. A neural network can approximate any nonlinear function and can be used to solve the control problems in uncertain models. In [10], the existing model uncertainties and external disturbances were treated by radial basis function neural network and disturbance observer, respectively. Similarly, neural network techniques and adaptive techniques were used to compensate for the uncertainty of the model [11]. In [12], a wavelet neural network was used to approximate completely unknown dynamic and external disturbances. In [13], an adaptive neural network was proposed to approximate uncertain nonlinear dynamics and external environmental disturbances. Combined with the backstepping method, adaptive technology was used to approximate the disturbed boundary, and a neural network was used to approximate the uncertain function to achieve trajectory tracking [14]. The difficulty of neural network radial basis function approximating arbitrary function lies in how to properly select the center points, the number of nodes and the appropriate width of radial basis function to generate hidden layers. Based on the stability theory of immersion and invariance, the asymptotic stability was guaranteed, and the adaptive law was derived to deal with the parameter uncertainty [15]. An online constructive fuzzy approximator was designed to dynamically adjust fuzzy rules to deal with time-varying uncertainties [16]. Combined with dynamic surface technology, the desired signal is smooth and bounded, and the control output realizes trajectory tracking.

In [17], the external disturbances and model uncertainty were compensated by a disturbance observer, and trajectory tracking was realized by sliding mode control. A nonlinear disturbance observer was constructed to deal with the disturbances problem [18]. For the unmeasured velocities problem, [19] constructed a hyper-distortion observer to observe the velocity. The unknown disturbance is estimated by RBF neural network combined with adaptive method. In order to ensure sufficient safety, the state of the system is subject to certain boundary constraints to perform a specific task; Li et al. [20] used the barrier Lyapunov function to constrain all states of the system to achieve trajectory tracking control. A pre-defined performance tracking control law based on quantization state was proposed to achieve tracking control without requiring system structural parameters and function approximators for a long time [21]. In the trajectory tracking control of mobile robot [22], state feedback control and disturbance feedforward compensation control are designed to solve the problems of unmeasurable speed and disturbance. Given that the disturbances are bounded, the disturbances suppression and tracking error reduction were guaranteed by finite time convergence.

Finite time control has better convergence performance and is more desirable in practical applications. It is able to complete the trajectory tracking task within finite time and has fast response characteristics [23–28]. Sliding mode control has been widely used in nonlinear uncertain systems because of its strong robustness, simple design and easy implementation. Traditional sliding mode control does not have the characteristics of finite time control. The terminal sliding mode can control the convergence rate near or away from the equilibrium point to realize finite time control because it introduces nonlinear function into sliding hyperplane design. It is widely used to solve finite time control problems [29–31]. In [32], a finite time adaptive sliding mode controller is designed, and the adaptive law is obtained by Lyapunov function. In [33], a finite time control strategy is designed based on proportional integral-differential sliding mode control to make the error converge in finite time.

The above studies deal with the problems of external environmental disturbances by designing disturbance observers to compensate for the disturbance or neural network approximation and using state observers for velocity estimation. In this study, perturbation and velocity unmeasurable problems are considered and dealt with together. NESO and NFTSM are combined to realize finite time underactuated USV trajectory tracking control. The overall structure block diagram of this study is shown in Figure 1. The underactuated

USV mathematical model is taken as the research plant, and then NESO and NFTSM are designed and verified by numerical simulation. The main contributions are as follows:

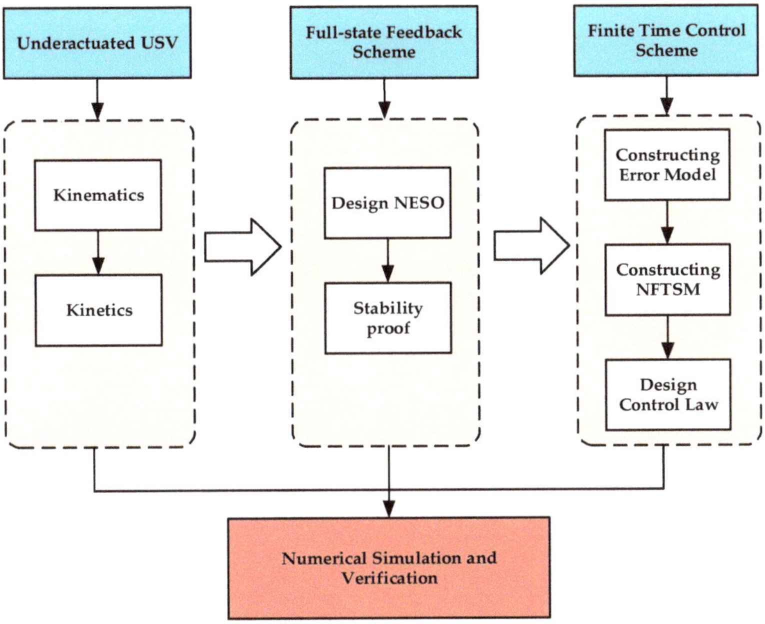

Underactuated USV → Full-state Feedback Scheme → Finite Time Control Scheme

- Underactuated USV: Kinematics → Kinetics
- Full-state Feedback Scheme: Design NESO → Stability proof
- Finite Time Control Scheme: Constructing Error Model → Constructing NFTSM → Design Control Law
- Numerical Simulation and Verification

Figure 1. Structure diagram of research.

A NESO is designed to achieve full-state feedback, which eliminates the need for a separate disturbance observer to compensate the disturbance, thus reducing the design difficulty. Uncertainty of model parameters and external disturbances are regarded as complex disturbances and expanded as part of the state variables of the system, and the observation values of complex disturbances feedback to the controller to simplify the design process. The observed values of velocity and position are obtained and feedback to the system, which is particularly useful in the case of sensor malfunction.

NFTSM can realize the convergence of tracking error in a shorter time, which is used to solve the problem of slow convergence of trajectory tracking error. According to Lyapunov stability theory, the corresponding surge force and yaw moment are derived to ensure the stability of the system. The numerical simulation by MATLAB verifies that the controller has fast response characteristics, and the chattering phenomenon of the controller is also reduced.

2. Definition

For first-order nonlinear systems, NTSM and NFTSM were defined [34] as follows:

$$
\begin{aligned}
&\sigma(t) = x + k\,\mathrm{sign}^a \dot{x} = 0, k > 0,\ 1 < a < 2, \\
&\sigma(t) = x + k_1\mathrm{sign}^{a_1} x + k_2\mathrm{sign}^a \dot{x} = 0,\ k_1 > 0, k_2 > 0, 1 < a < 2, a_1 > a
\end{aligned}
\tag{1}
$$

where $x \in R$, $\mathrm{sign}^a x := \mathrm{sign}\,x \cdot |x|^a$, k, k_1 and k_2 are coefficients greater than 0, $a = p/q$, p and q are positive integers $p > 0, q > 0$.

3. Underactuated USV Modeling and Problem Formulation

Assumption 1: *Only three degrees of freedom of the underactuated USV are considered, namely, surge, sway and yaw.*

Assumption 2: The underactuated USV has a uniform mass distribution and a left-right symmetry.

Assumption 3: The (Center of Gravity) CoG and Center of Buoyancy (COB) of the underactuated USV are located on the z-axis of the body-fixed coordinate system.

On the premise of the above assumptions, the kinematic and dynamic equations of USV can be obtained as follows:

$$\dot{\boldsymbol{\eta}} = \mathbf{R}(\psi)\mathbf{v}$$
$$\mathbf{M}\dot{\mathbf{v}} = -\mathbf{C}(\mathbf{v})\mathbf{v} - \mathbf{D}(\mathbf{v})\mathbf{v} + \boldsymbol{\tau} + \boldsymbol{\tau}_E \tag{2}$$

where $\boldsymbol{\eta} = [x, y, \psi]^T$ denotes the actual positions and heading angle of underactuated USV in the Earth-fixed coordinate also called North-East-Down coordinate; $\mathbf{v} = [u, v, r]^T$ denotes the actual velocities and angle velocity; $\mathbf{R}(\psi)$ is the transformation matrix between the Earth-fixed coordinate system and the Body-fixed coordinate system, as shown in Figure 2. $\mathbf{M}$ is inertial matrix; $\mathbf{C}(\mathbf{v})$ is the Coriolis and centripetal matrix; $\mathbf{D}(\mathbf{v})$ is the hydrodynamic damping parameters matrix; $\boldsymbol{\tau} = [\boldsymbol{\tau}_u, 0, \boldsymbol{\tau}_r]^T$ denotes the surge force, sway force and yaw moment, where sway force is 0; $\boldsymbol{\tau}_E = [\boldsymbol{\tau}_{Eu}, \boldsymbol{\tau}_{Ev}, \boldsymbol{\tau}_{Er}]$ is used to represent the disturbances caused by the wind, wave and current environment of the USV. The above parameter matrices satisfy the following properties:

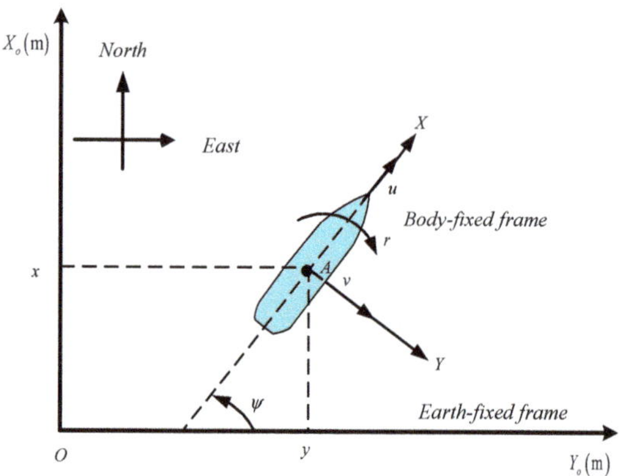

Figure 2. Earth-fixed $O - X_o Y_o$ and Body-fixed $A - XY$ coordinate frames of a USV.

Property 1: The rotation matrix is orthogonal to satisfy $\|\mathbf{R}(\psi)\| = 1$ and

$$\mathbf{R}^T(\psi) = \mathbf{R}^{-1}(\psi)$$
$$\dot{\mathbf{R}}(\psi) = \mathbf{R}(\psi)\mathbf{S}(r) \tag{3}$$
$$\mathbf{R}(\psi)\mathbf{S}(r)\mathbf{R}^T(\psi) = \mathbf{R}(\psi)^T\mathbf{S}(r)\mathbf{R}(\psi)$$

where

$$\mathbf{R}(\psi) = \begin{bmatrix} \cos(\psi) & -\sin(\psi) & 0 \\ \sin(\psi) & \cos(\psi) & 0 \\ 0 & 0 & 1 \end{bmatrix}, \mathbf{S}(r) = \begin{bmatrix} 0 & -r & 0 \\ r & 0 & 0 \\ 0 & 0 & 0 \end{bmatrix} \tag{4}$$

Property 2: The symmetric positive definite of the inertia matrix satisfies:

$$\mathbf{M} = \mathbf{M}^T > 0 \tag{5}$$

Property 3: The Coriolis and centripetal force matrices are obliquely symmetric and satisfy:

$$\mathbf{C}(\mathbf{v}) = -\mathbf{C}^T(\mathbf{v}), \forall \mathbf{v} \in \mathbb{R}^3. \tag{6}$$

Property 4: Hydrodynamic damping matrix is an asymmetric and strictly positive definite real matrix, which satisfies:

$$\mathbf{D}(\mathbf{v}) > 0, \forall \mathbf{v} \in \mathbb{R}^3. \tag{7}$$

The expressions of matrix $\mathbf{M}$, $\mathbf{C}(\mathbf{v})$, and $\mathbf{D}(\mathbf{v})$ are, respectively:

$$\mathbf{M} = \begin{bmatrix} m_{11} & 0 & 0 \\ 0 & m_{22} & 0 \\ 0 & 0 & m_{33} \end{bmatrix} = \begin{bmatrix} m - X_{\dot{u}} & 0 & 0 \\ 0 & m - Y_{\dot{v}} & 0 \\ 0 & 0 & I_z - N_{\dot{r}} \end{bmatrix} \tag{8}$$

$$\mathbf{C}(\mathbf{v}) = \begin{bmatrix} 0 & 0 & -m_{22}v \\ 0 & 0 & m_{11}u \\ m_{22}v & -m_{11}u & 0 \end{bmatrix} \tag{9}$$

$$\mathbf{D}(\mathbf{v}) = \begin{bmatrix} d_{11} & 0 & 0 \\ 0 & d_{22} & 0 \\ 0 & 0 & d_{33} \end{bmatrix} = \begin{bmatrix} -X_u & 0 & 0 \\ 0 & -Y_v & 0 \\ 0 & 0 & -N_r \end{bmatrix} \tag{10}$$

Define auxiliary quantity $\boldsymbol{\omega} = \mathbf{R}(\psi)\mathbf{v}$, there are:

$$\begin{aligned} \dot{\boldsymbol{\omega}} &= \ddot{\boldsymbol{\eta}} \\ &= \dot{\mathbf{R}}(\psi)\mathbf{v} + \mathbf{R}(\psi)\dot{\mathbf{v}} \\ &= f(\boldsymbol{\eta}, \mathbf{v}, \boldsymbol{\tau}_E) + \mathbf{B}(\psi)\boldsymbol{\tau} \end{aligned} \tag{11}$$

where $f(\boldsymbol{\eta}, \mathbf{v}, \boldsymbol{\tau}_E) = \mathbf{R}(\psi)\mathbf{M}^{-1}[\boldsymbol{\tau}_E - \mathbf{C}(\mathbf{v})\mathbf{v} - \mathbf{D}(\mathbf{v})\mathbf{v}] + \dot{\mathbf{R}}(\psi)\mathbf{v}$, $\mathbf{B}(\psi) = \mathbf{R}(\psi)\mathbf{M}^{-1}$.

Therefore, the underactuated USV mathematical model can be written in the following form:

$$\begin{aligned} \dot{\boldsymbol{\eta}} &= \boldsymbol{\omega} \\ \dot{\boldsymbol{\omega}} &= f(\boldsymbol{\eta}, \mathbf{v}, \boldsymbol{\tau}_E) + \mathbf{B}(\psi)\boldsymbol{\tau} \end{aligned} \tag{12}$$

In this study, a reference trajectory is generated by providing a predefined input to the virtual underactuated USV, as shown in Figure 3.

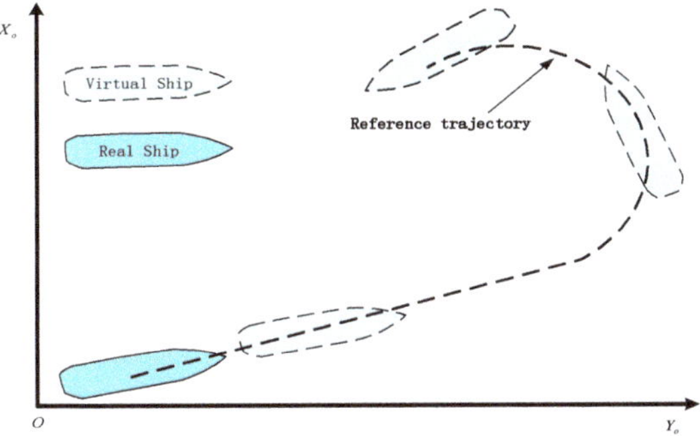

Figure 3. Virtual USV generation reference estimation schematic.

Similarly for virtual underactuated USV:

$$\begin{aligned} \dot{\boldsymbol{\eta}}_d &= \mathbf{R}_d(\psi_d)\mathbf{v}_d \\ \mathbf{M}_d\dot{\mathbf{v}}_d &= -\mathbf{C}_d(\mathbf{v}_d)\mathbf{v}_d - \mathbf{D}_d(\mathbf{v})\mathbf{v}_d + \boldsymbol{\tau}_d \end{aligned} \tag{13}$$

Defining an auxiliary quantity $\omega_d = R_d(\psi_d)v_d$, there are:

$$
\begin{aligned}
\dot{\omega}_d &= \ddot{\eta}_d \\
&= \dot{R}_d(\psi_d)v_d + R_d(\psi_d)\dot{v}_d \\
&= f_d(\eta_d, v_d) + B_d(\psi_d)\tau_d
\end{aligned}
\tag{14}
$$

where $f_d(\eta_d, v_d) = R_d(\psi_d)M_d^{-1}[-C_d(v_d)v_d - D_d(v)v_d] + \dot{R}_d(\psi_d)v_d$, $B_d(\psi_d) = R_d(\psi_d)M_d^{-1}$.

According to Equations (11) and (14), an error model can be obtained.

$$
\begin{aligned}
\dot{\eta}_e &= \dot{\eta} - \dot{\eta}_d \\
\ddot{\eta}_e &= \ddot{\eta} - \ddot{\eta}_d \\
&= f(\eta, v, \tau_E) + B(\psi)\tau - f_d(\eta_d, v_d) - B_d(\psi_d)\tau_d
\end{aligned}
\tag{15}
$$

4. Design of NESO

In the actual navigation process, it is inevitable that the USV is subjected to the disturbances of the external environment: wind, wave and current. At the same time, considering the parameters perturbation, that is, the system inertia matrix, Coriolis centripetal force matrix and hydrodynamic damping matrix, this part of the influence will be regarded as whole complex disturbances, and it will be expanded into part of the system states to deal with together. Further considering the situation that the system states cannot be measured, NESO will be used to observe the system states, that is, the velocities, positions and composite disturbances are feedback to the underactuated USV system.

The following NESO can be designed for the underactuated USV model (12).

$$
\begin{cases}
e_1 = z_1 - x_1, e_2 = z_2 - x_2, e_3 = z_3 - x_3 \\
\dot{z}_1 = z_2 + \lambda_1 e_1 \\
\dot{z}_2 = z_3 + B\tau + \lambda_2 \text{fal}(e_1, \beta_1, \delta_1) \\
\dot{z}_3 = \lambda_3 \text{fal}(e_1, \beta_2, \delta_2)
\end{cases}
\tag{16}
$$

where $x_1 = \eta$, $x_2 = \dot{\eta} = \omega$, the derivative of complex disturbances $x_3 = f(\eta, v, \tau_E)$ is unknown but bounded, that is, $f(\eta, v, \tau_E) <= \bar{f} > 0$, z_1 is position observed value of η, z_2 is the observed value of $\dot{\eta}$, z_3 is complex disturbances observed value of $f(\eta, v, \tau_E)$; λ_1, λ_2, λ_3 are gain matrix of NESO, $e_i = [e_{i,j}]^T (i = 1, 2, 3, j = 1, 2, 3)$ are the corresponding approximation errors. The function of $\text{fal}(\bullet)$ expression is as follow:

$$
\text{fal}(x, \beta, \delta) = \begin{cases}
|x|^\beta \text{sign}(x) & , |x| > \delta \\
\frac{s}{\delta^{1-\beta}} & , |x| \leq \delta
\end{cases}
\tag{17}
$$

where x is independent variable of the function. $\beta \in (0, 1)$, δ an arbitrary small positive number.

Combining Equations (12) and (16) to calculate the derivatives of e_1, e_2, e_3, the error system of NESO is as follows:

$$
\begin{cases}
\dot{e}_1 = e_2 - \lambda_1 e_1 \\
\dot{e}_2 = e_3 - \lambda_2 e_1 \\
\dot{e}_3 = f(\eta, v, \tau_E) - \lambda_3 e_1
\end{cases}
\tag{18}
$$

The error state equation of NESO can be expressed as:

$$
\dot{e} = \overline{A}e + \overline{B}\dot{f}(\eta, v, \tau_E)
\tag{19}
$$

where

$$\overline{\mathbf{A}} = \begin{bmatrix} -\lambda_1 & \mathbf{I}_{3\times3} & 0_{3\times3} \\ -\lambda_2 & 0_{3\times3} & \mathbf{I}_{3\times3} \\ -\lambda_3 & 0_{3\times3} & 0_{3\times3} \end{bmatrix}, \ \overline{\mathbf{B}} = \begin{bmatrix} 0_{3\times3} \\ 0_{3\times3} \\ \mathbf{I}_{3\times3} \end{bmatrix} \tag{20}$$

The eigenequation of matrix $\overline{\mathbf{A}}$ is

$$\left| \mathbf{sI} - \overline{\mathbf{A}} \right| = \begin{bmatrix} \mathbf{s} + \lambda_1 & -\mathbf{I}_{3\times3} & 0_{3\times3} \\ \lambda_2 & \mathbf{s} & -\mathbf{I}_{3\times3} \\ \lambda_3 & 0_{3\times3} & \mathbf{s} \end{bmatrix} \tag{21}$$

The characteristic polynomial of Equation (19) is

$$\mathbf{s}_i^3 + \alpha_1 \mathbf{s}_i^2 + \alpha_2 \mathbf{s}_i^1 + \alpha_3 = 0 \tag{22}$$

where $\mathbf{s}_i = [\mathbf{s}_{1i}, \mathbf{s}_{2i}, \mathbf{s}_{3i}]^T, i = 1, 2, 3$. Appropriate choice of parameter matrixes, $\lambda_i, i = 1, 2, 3$, makes $\overline{\mathbf{A}}$ satisfy the Hurwitz stability condition. By Lyapunov's second method, for arbitrary given positive definite matrix $\mathbf{Q}$ and $\mathbf{P}$, the following Lyapunov equation is satisfied.

$$\overline{A}^T P + P\overline{A} + Q = 0 \tag{23}$$

Define the Lyapunov function regarding NESO as

$$V = e^T P e \tag{24}$$

The differential Equation (24) can be obtained.

$$\begin{aligned} \dot{V} &= \dot{e}^T P e + e^T P \dot{e} \\ &= e^T \overline{A}^T P e + \left[\overline{B} \dot{f}(\mathbf{\eta}, \mathbf{v}, \mathbf{\tau}_{\mathrm{E}}) \right]^T P e + e^T P \overline{A} e + e^T P \overline{B} \dot{f}(\mathbf{\eta}, \mathbf{v}, \mathbf{\tau}_{\mathrm{E}}) \\ &= e^T \left(\overline{A}^T P + P\overline{A} \right) e + 2 e^T P \overline{B} \dot{f}(\mathbf{\eta}, \mathbf{v}, \mathbf{\tau}_{\mathrm{E}}) \\ &\le e^T (Q) e + 2\|e\| \cdot \|P\overline{B}\| \cdot \|\dot{f}(\mathbf{\eta}, \mathbf{v}, \mathbf{\tau}_{\mathrm{E}})\| \\ &\le \lambda_{\min}(Q) \|e\|^2 + 2\overline{f}\|e\| \cdot \|P\overline{B}\| \end{aligned} \tag{25}$$

When $\dot{V} \le 0$, it can be proved that NESO is convergent, and the convergence conditions are:

$$\|e\| \le \frac{2\overline{f}\|P\overline{B}\|}{\lambda_{\min}(Q)} \tag{26}$$

From Equation (16), the observed value $\hat{\mathbf{\eta}} = \mathbf{z}_1$ of position, the observed value $\hat{\mathbf{v}} = \mathbf{R}^{-1}(\psi)\mathbf{z}_2$ of velocity, and the observed value $\hat{f}(\mathbf{\eta}, \mathbf{v}, \mathbf{\tau}_{\mathrm{E}}) = \mathbf{z}_3$ of complex disturbance can be obtained. Using the observations obtained by NESO to replace the position states and velocity states corresponding to the actual USV, and the complex disturbances composed of unknown internal disturbances, namely model parameter disturbances and external unknown wind, wave and current environment disturbances, can simplify the design of the controller and reduce the complex operation process.

5. Controller Design

The feedback value of the underactuated USV system states obtained by NESO has been mentioned before. After redefining the error model, the sliding mode surface is designed for controller design. The control scheme structure diagram is shown in Figure 4.

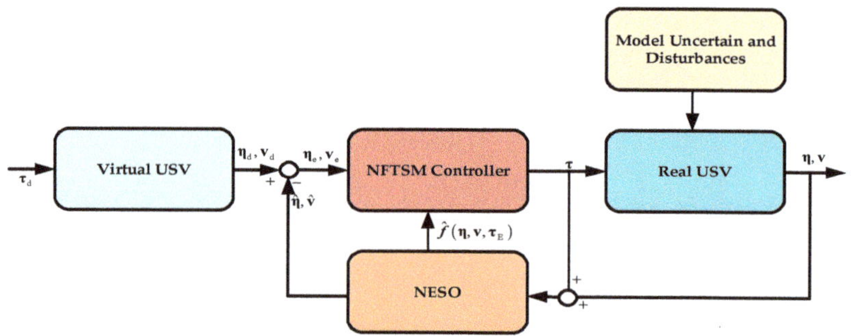

Figure 4. Control scheme structure diagram.

According to the design idea, the error model is redefined by the error model (15) and NESO (16).

$$
\begin{aligned}
\dot{\eta}_e &= \dot{\hat{\eta}} - \dot{\eta}_d \\
\ddot{\eta}_e &= \ddot{\hat{\eta}} - \ddot{\eta}_d \\
&= \hat{f}\left(\hat{\eta}, \hat{v}, \hat{\tau}_E\right) + \mathbf{B}(\hat{\psi})\tau - f_d(\eta_d, v_d) - \mathbf{B}_d(\psi_d)\tau_d
\end{aligned}
\tag{27}
$$

From the definition of NFTSM, the following sliding surface can be obtained:

$$
s = \eta_e + k_1 \operatorname{sign}^{a_1}\eta_e + k_2 \operatorname{sign}^{a}\dot{\eta}_e
\tag{28}
$$

From Equations (27) and (28), the differential equation of the sliding mode surface can be obtained:

$$
\dot{s} = \dot{\eta}_e + a_1 k_1 \operatorname{diag}\left(\operatorname{sign}^{a_1-1}\tau_e\right)\dot{\eta}_e + a k_2 \operatorname{diag}\left(\operatorname{sign}^{a-1}\dot{\eta}_e\right)\ddot{\eta}_e
\tag{29}
$$

Since $\ddot{\eta}_e$ contains the required control τ, the Lyapunov function about the sliding surface is constructed, and the desired controller output τ can be obtained by selecting the appropriate τ to satisfy the Lyapunov stability condition.

Define the following Lyapunov function,

$$
V_1 = \frac{1}{2}s^T s
\tag{30}
$$

and obtain the differential equation:

$$
\begin{aligned}
\dot{V}_1 &= s^T \dot{s} \\
&= s^T\left[\dot{\eta}_e + a_1 k_1 \operatorname{diag}\left(\operatorname{sign}^{a_1-1}\eta_e\right)\dot{\eta}_e + a k_2 \operatorname{diag}\left(\operatorname{sign}^{a-1}\dot{\eta}_e\right)\ddot{\eta}_e\right]
\end{aligned}
\tag{31}
$$

If the underactuated USV input is selected as:

$$
\begin{aligned}
\tau = \mathbf{M}\mathbf{R}^{-1}(\hat{\psi})\Big\{ &-\hat{f}\left(\hat{\eta}, \hat{v}, \hat{\tau}_E\right) + \mathbf{R}_d(\psi_d)\mathbf{M}d^{-1}\tau_d + f_d(\eta_d, v_d) \\
&-\tfrac{1}{ak_2}\left[ak_1\operatorname{diag}\left(\operatorname{sign}^{a_1-1}\eta_e\right) + \mathbf{I}\right]\operatorname{sign}^{2-a}\dot{\eta}_e - k_3 s - k_4 \operatorname{sign} s\Big\}
\end{aligned}
\tag{32}
$$

Substituting Equation (32) into Equation (31) can be obtained:

$$\begin{aligned}
\dot{V}_1 &= s^T \dot{s} \\
&= s^T \left[\dot{\boldsymbol{\eta}}_e + a_1 k_1 \mathrm{diag}\left(\mathrm{sign}^{a_1-1}\boldsymbol{\eta}_e\right)\dot{\boldsymbol{\eta}}_e + ak_2\mathrm{diag}\left(\mathrm{sign}^{a}\dot{\boldsymbol{\eta}}_e\right)\ddot{\boldsymbol{\eta}}_e \right] \\
&= ak_2 s^T \mathrm{diag}\left(\mathrm{sign}^{a-1}\dot{\boldsymbol{\eta}}_e\right)\left(-k_3 s - k_4 \mathrm{sign} s\right) \\
&\leq -ak_2 \sum_{i=1}^{3} k_{3i}\left|\mathrm{sign}^{a-1}\dot{\eta}_{ei}\right|s_i^2 - ak_2 \sum_{i=1}^{3} k_{4i}\left|\mathrm{sign}^{a-1}\dot{\eta}_{ei}\right|\left|s_i\right|
\end{aligned} \qquad (33)$$

The Lyapunov function $V_1 \geq 0$ is selected, and the design $\boldsymbol{\tau}$ makes $\dot{V}_1 \leq 0$, so it can be concluded that the underactuated USV system is stable.

6. Numerical Simulation and Analysis

In order to verify the effectiveness of using NESO for full-state feedback with the combination of NFTSM for finite-time trajectory tracking control, the BAYCLASS long-range patrol ship in reference [35] is used for simulation, and the parameters of underactuated USV are shown in Table 1. The virtual USV, actual USV, NESO and controller in the design process are modeled using the Simulink simulation tool in MATLAB, and the solver is chosen ode45 with variable step size. The work of the simulation is divided into two stages. First of all, the simulation verifies whether the mentioned NESO has a good approximation effect on the complex disturbances, positions and velocities state of the underactuated USV system. Finally, the simulation incorporates parameters perturbation using NESO to obtain the three-part state feedback values of positions, velocities, and expansion of the underactuated USV system combined with NFTSM to verify the simulation and compare it with the NTSM in [36] in order to verify the design of the control.

Table 1. USV model parameters.

Parameter	Definition	Value	Units
m	Mass of USV	1.18×10^3	kg
L	Length of USV	38	m
m_{11}	Parameter of inertia matrix	1.2×10^5	kg/s
m_{22}	Parameter of inertia matrix	1.779×10^5	kg/s
m_{33}	Parameter of inertia matrix	6.36×10^7	kg/s
d_{11}	Parameter of damping matrix	2.15×10^4	kg/s
d_{22}	Parameter of damping matrix	1.47×10^5	kg/s
d_{33}	Parameter of damping matrix	8.02×10^6	kg/s

The virtual underactuated USV used to generate the reference trajectory uses the same model parameters as the actual underactuated USV model. The initial values of position and velocity of underactuated USV are $\boldsymbol{\eta}_0 = [0,0,0]^T$ and $\mathbf{v}_0 = [0,0,\pi/6]^T$. The initial values of positions and velocities of virtual underactuated USV are $\boldsymbol{\eta}_{d0} = [5,3.5,\pi/6]^T$ and $\mathbf{v}_{d0} = [0,0,0]^T$. To reflect the parameter perturbation, $\mathbf{M} = \pm1.1\mathbf{M}$, $\mathbf{C}(\mathbf{v}) = \pm1.1\mathbf{C}(\mathbf{v})$, $\mathbf{D}(\mathbf{v}) = \pm1.1\mathbf{D}(\mathbf{v})$ are used in the simulation. Four periods with higher probability of disturbance occurrence were chosen according to the scatter diagram of the sea states. The disturbances $\boldsymbol{\tau}_E$ caused by wind, wave and current are assumed to be:

$$\boldsymbol{\tau}_E = \begin{bmatrix} 8 \times 10^n \sin(\omega \cdot t) \\ 6 \times 10^n \sin(\omega \cdot t) \\ 7 \times 10^n \sin(\omega \cdot t) \end{bmatrix} \qquad (34)$$

where $n = 3,4$, $T = 5,6,7,8s$, four different periods can be obtained when the angular frequency as $\omega = 1.2566, 1.0472, 0.8976, 0.7854$.

The reference trajectory is designed as the complex trajectory obtained by the combination of straight line and circle, and the control input τ_d of the virtual underactuated USV is selected as follows:

$$\tau_d = \begin{cases} \begin{bmatrix} 1 & 0 & 0 \end{bmatrix}^T \times 10^5, 0 \leq t \leq 100 \\ \begin{bmatrix} 1 & 0 & 2 \end{bmatrix}^T \times 10^5, 100 < t \leq 300 \end{cases} \tag{35}$$

The parameters selected for Equation (16) NESO are: $\lambda_{1i} = 6, \lambda_{2i} = 9, \lambda_{3i} = 15 (i = 1, 2, 3)$, $\beta_i = 0.5, \delta_i = 0.001 (i = 1, 2)$. Regarding the design form Equation (32) of the controller, the chosen parameters are $k_1 = 3, k_2 = 1, k_3 = k_4 = \text{diag}(10 \quad 10 \quad 10)$, $a = 9/5, a_1 = 5/3$. It is worth noting that the symbolic function $k_4 \text{sign} s$ in Equation (32) will cause obvious chattering in the controller in the actual simulation, so the saturation function is used to replace the symbolic function in the simulation, and the form of the saturation function is shown below.

$$\text{Sat}(s_i) = \begin{cases} \text{sign}(s_i) & , |s_i| > \omega \\ |s_i|^{\xi} \text{sign}(s_i) / \omega^{\xi} & , |s_i| \leq \omega \end{cases} , (i = 1, 2, 3) \tag{36}$$

The parameters used are $\omega = 7, \xi = 0.2$.

For the disturbances in the form of Equation (34), there are two different amplitudes. For the first amplitude, the disturbances as in Equation (37) are chosen for simulation.

$$\tau_E = \begin{bmatrix} 8 \times 10^3 \sin(\omega \cdot t) \\ 6 \times 10^3 \sin(\omega \cdot t) \\ 7 \times 10^3 \sin(\omega \cdot t) \end{bmatrix} \tag{37}$$

Figure 5 shows the tracking curves of the complex disturbances $f(\eta, v, \tau_E)$ for four periods under the disturbances of Equation (37). The complex disturbances are related to the USV's own velocity, position information and external environmental disturbances, which are non-periodic disturbances. The red curve, which is the observed value, is very close to the actual value (the blue curve), as can be seen in the figures below.

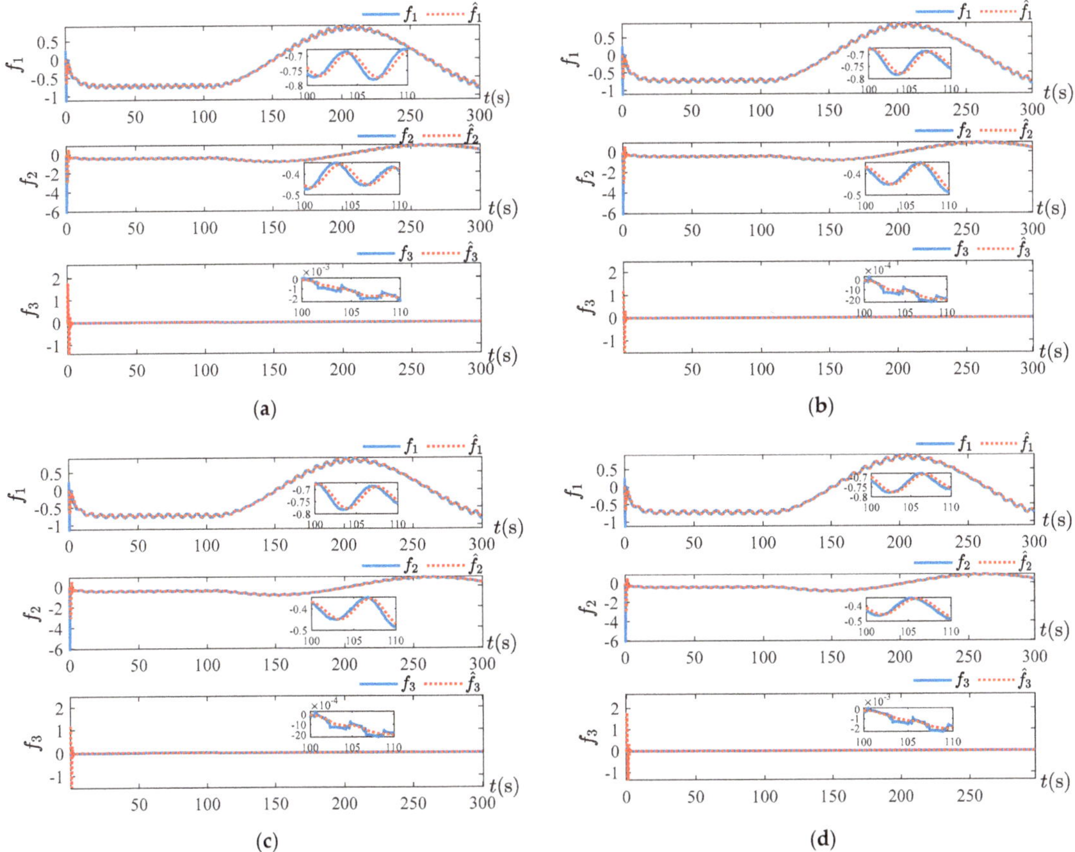

Figure 5. Simulation results of complex disturbances under four periods. (**a**) Complex disturbances tracking curve graph at period $T = 5s$; (**b**) Complex disturbances tracking curve graph at period $T = 6s$; (**c**) Complex disturbances tracking curve graph at period $T = 7s$; (**d**) Complex disturbances tracking curve graph at period $T = 8s$.

Figure 6 shows the observation error curves of complex disturbances $f(\mathbf{\eta}, \mathbf{v}, \mathbf{\tau}_E)$, positions $\mathbf{\eta}_e$ and velocities $\mathbf{v}_e$. It can be seen in the three figures that for the three different state tracking error curves converge in a minimal neighborhood of zero. The composite disturbance error is less than 10^{-1}, and the position and velocity observation error is less than 10^{-3}. Despite the change in the period of the disturbance, the NESO observations are still valid within the error tolerance. There is an initial error in the estimation error of the vessel positions and velocities state at the initial stage. As time goes on, the error converges very quickly. From the convergence range of the error, it can be concluded that the estimated vessel speed and position states can replace the actual measured states information and feedback to the system. Therefore, when the states of the underactuated USV system cannot be measured or is disturbed by the external environment, the structural information of the system itself can be used to estimate the states information of the underactuated USV.

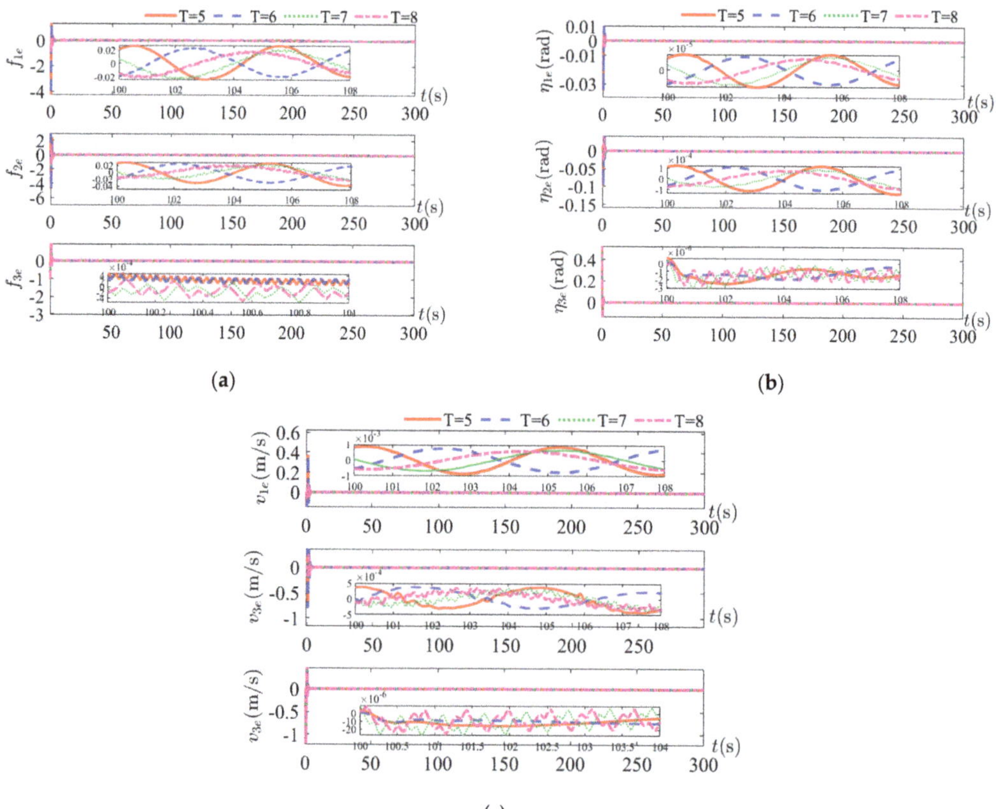

Figure 6. Observation error results at four periods. (**a**) Graph of complex disturbances tracking error under four periods; (**b**) Graph of positions tracking error under four periods; (**c**) Graph of velocities tracking error under four periods.

Figures 7–10 show the simulation results under four periods of disturbance $T = 5s, 6s, 7s, 8s$. The trajectory tracking curve, position tracking error curve, velocity tracking error curve and surge force and yaw moment curve are given for each period. For the trajectory tracking curves under four periods, it can be seen that, overall, both control methods can track the reference trajectory, but from the local zoom in, it can be seen that the NFTSM control method used in this study approaches the reference trajectory earlier than the NTSM method and the approximation error is smaller. The small error indicates that the safety is higher and the risk is lower when performing the task using this method. As seen from the velocity tracking error curve, there is not much difference between the two methods because, overall, both methods can make the USV track on the reference trajectory, while achieving the position tracking needs to satisfy the velocity tracking first. For the simulation curves of surge force and yaw moment, it can be seen that the output curve of the controller is smoother than the control method used in this study, and the output curve of the NTSM control method is more strongly chattered. When the periods are $T = 7s$ and $T = 8s$, it is obvious from the position tracking Figures 9b and 10b under the two sets of periods that the control method NFTSM of this paper is better than the NTSM control scheme.

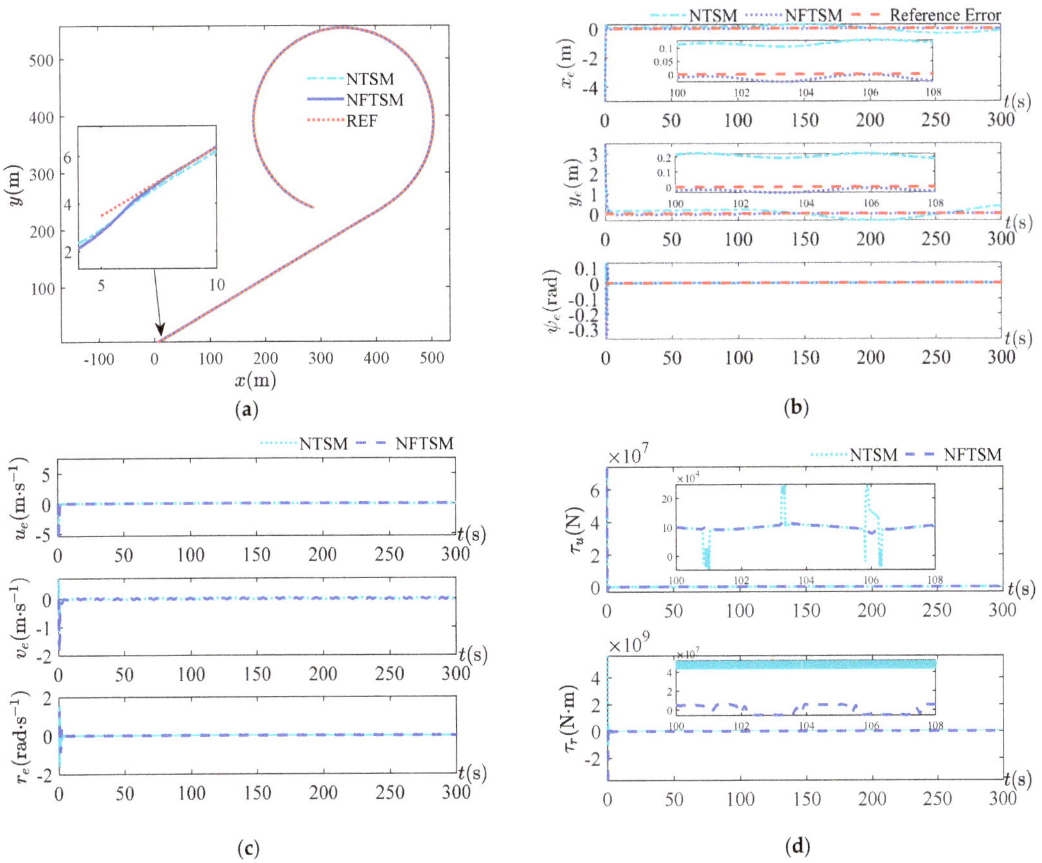

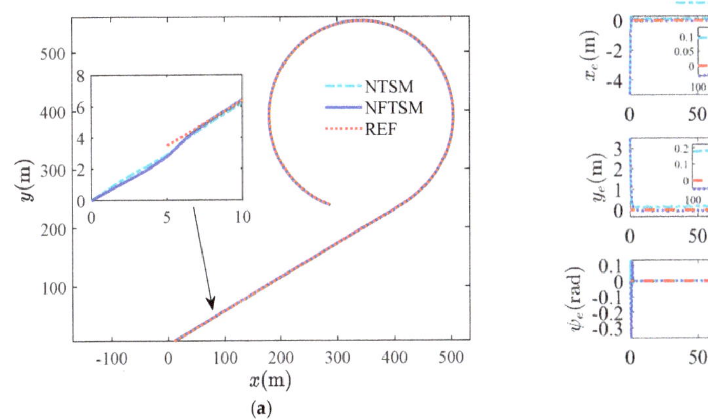

Figure 7. Simulation results at disturbance period of $T = 5s$. (**a**) Trajectory tracking curves; (**b**) Position tracking error curves; (**c**) Velocity tracking error curves; (**d**) Surge force and yaw moment curves.

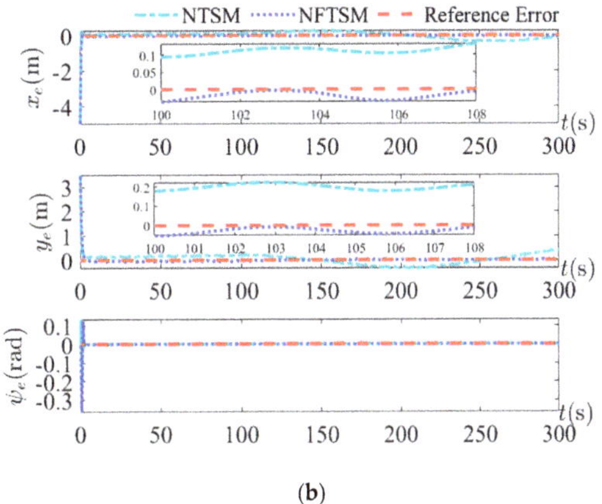

Figure 8. *Cont.*

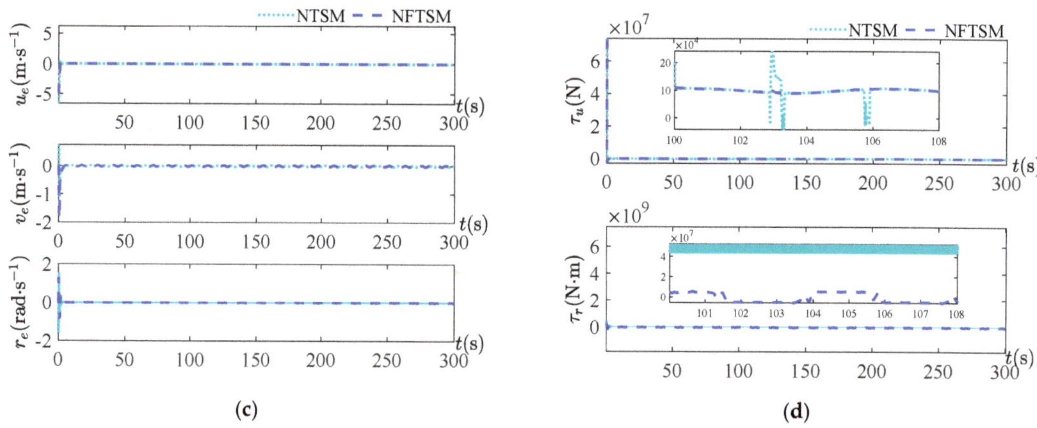

Figure 8. Simulation results at disturbance period of $T = 6s$. (**a**) Trajectory tracking curves; (**b**) Position tracking error curves; (**c**) Velocity tracking error curves; (**d**) Surge force and yaw moment curves.

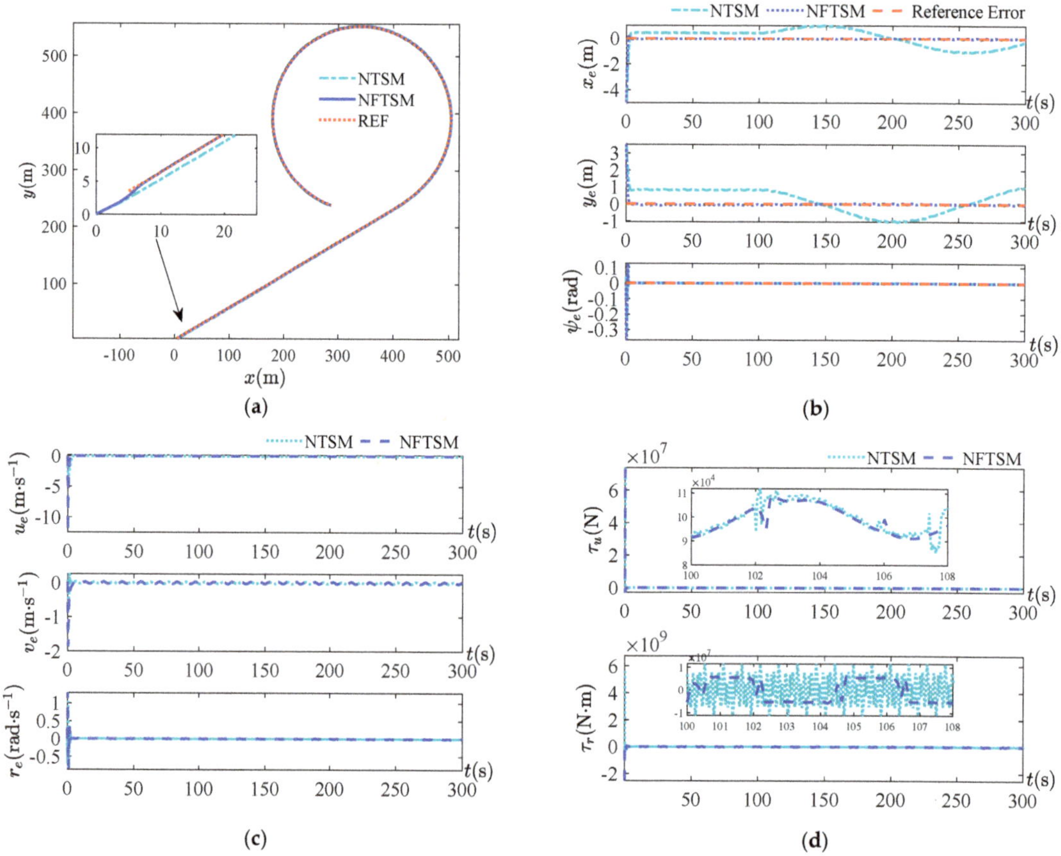

Figure 9. Simulation results at disturbance period of $T = 7s$. (**a**) Trajectory tracking curves; (**b**) Position tracking error curves; (**c**) Velocity tracking error curves; (**d**) Surge force and yaw moment curves.

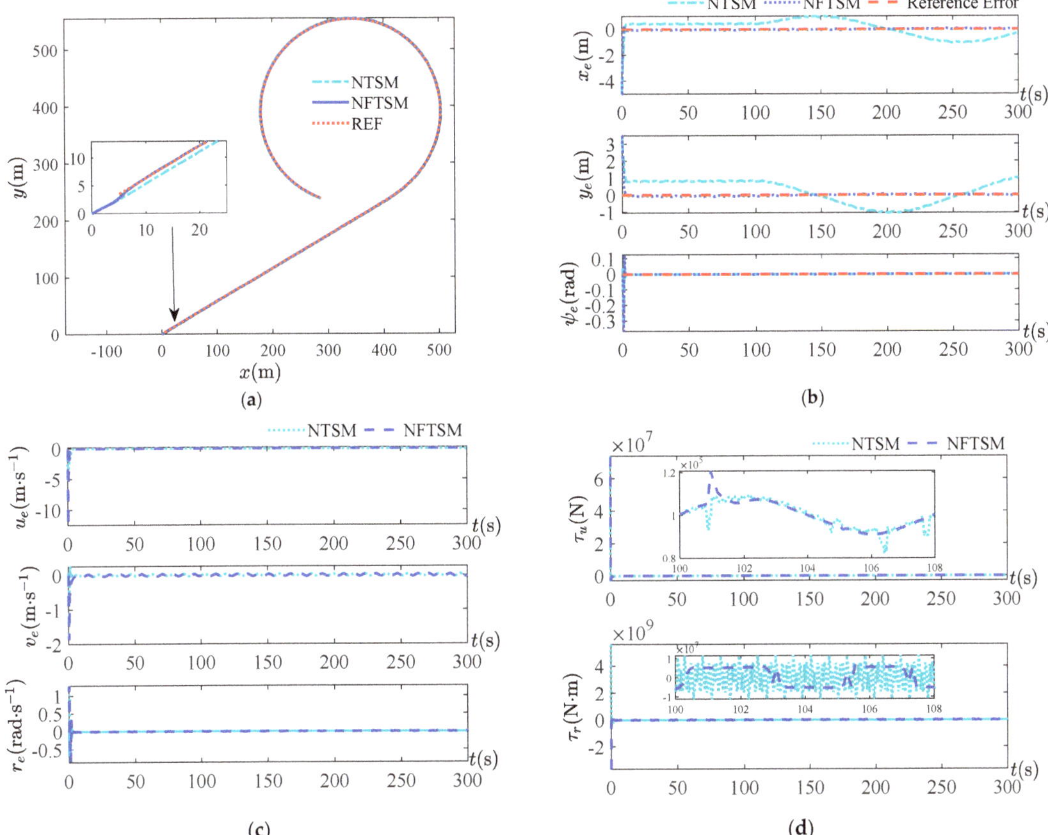

Figure 10. Simulation results at disturbance period of $T = 8s$. (**a**) Trajectory tracking curves; (**b**) Position tracking error curves; (**c**) Velocity tracking error curves; (**d**) Surge force and yaw moment curves.

The second disturbances amplitude is 10 times the first disturbances amplitude, and the disturbances as in Equation (38) are chosen for simulation.

$$\boldsymbol{\tau}_{\mathrm{E}} = \begin{bmatrix} 8 \times 10^4 \sin(\omega \cdot t) \\ 6 \times 10^4 \sin(\omega \cdot t) \\ 7 \times 10^4 \sin(\omega \cdot t) \end{bmatrix} \tag{38}$$

Figure 11 presents the observation error curves of complex disturbances, position and velocity under disturbances as shown in Equation (38). For disturbances with amplitude 11 times larger, the observations of the complex disturbances as shown in Figure 11a are not as good as in the case of small disturbances, but the observation errors of the position and velocity states are within 10^{-2}. In the case of period variation, there is no significant difference between the simulation graphs under the four periods, and there is good adaptability for the period variation.

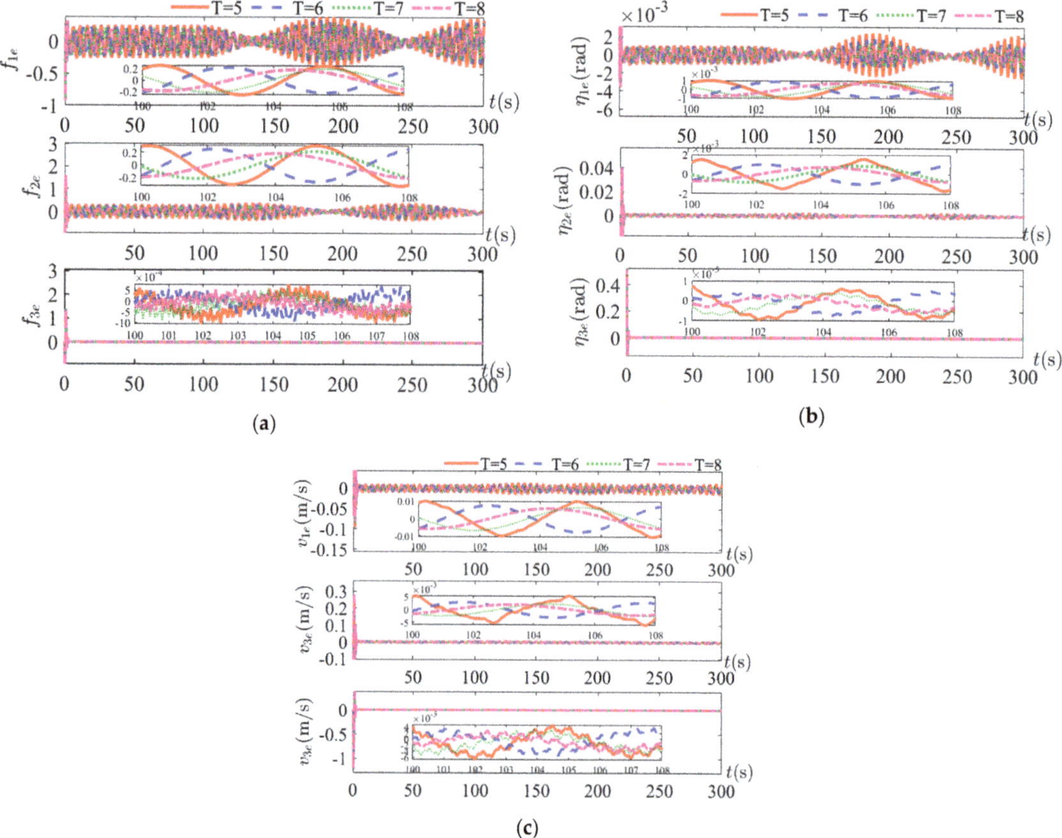

Figure 11. Observation error results at four periods. (**a**) Graph of complex disturbances tracking error under four periods; (**b**) Graph of positions tracking error under four periods; (**c**) Graph of velocities tracking error under four periods.

The simulation results shown in Figures 12–15 are obtained by varying the period $T = 5, 6, 7, 8s$ of the disturbances. Since the disturbances amplitude is 10 times the disturbances in Equation (37), the errors of the observed value obtained by NESO are relatively large compared to the small amplitude disturbances, which have an impact on the position tracking results and lead to a larger error. In the presence of observation errors, both control methods can achieve trajectory tracking control as seen in the trajectory tracking result graph, but the control effects are different. From the simulation results of Figure 12 at the disturbance period of $T = 5s$, it is obvious from Figure 12b that the control effect of the blue curve NFTSM of the control method used in this paper has a smaller error, and Figure 12d shows that the chattering of the yaw moment is smaller. When the period becomes larger, as seen from the position error curves of the four periods, the tracking error of the control method in this paper is smaller, and its error curves are closer to the reference error curve of zero value. The simulation graph of the yaw moment at four periods shows that the control scheme proposed in this study has a smaller frequency of control output variation and less chattering.

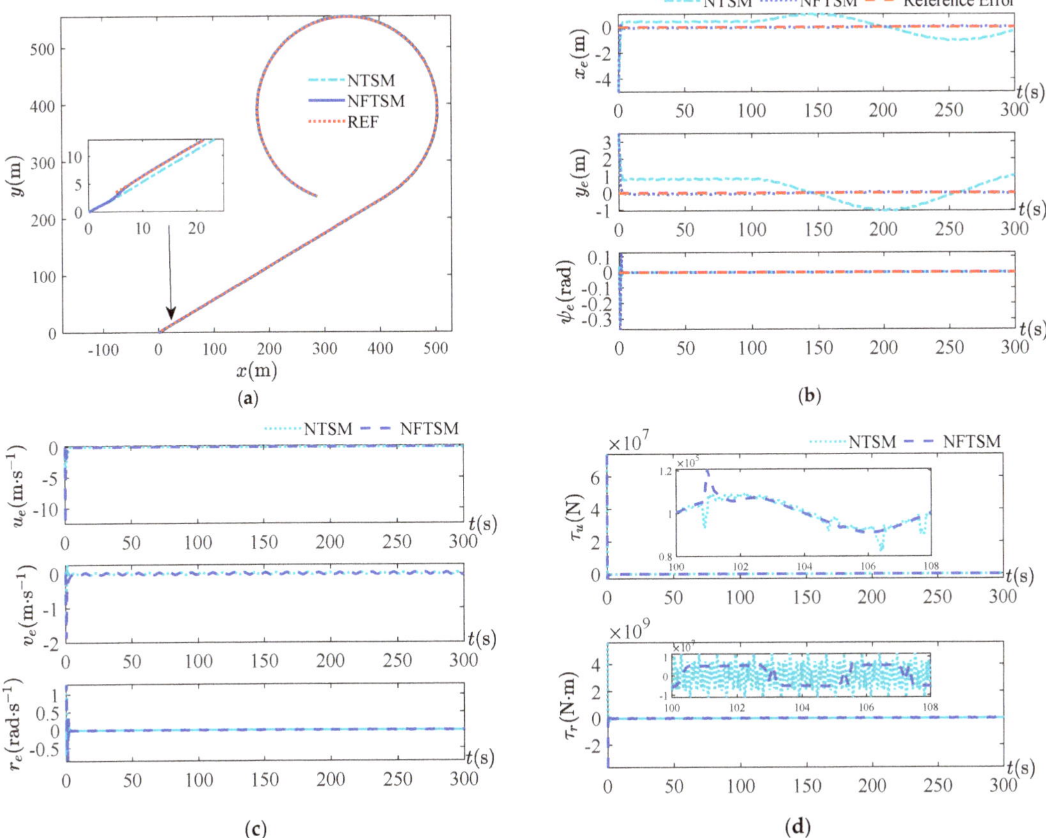

Figure 10. Simulation results at disturbance period of $T = 8s$. (**a**) Trajectory tracking curves; (**b**) Position tracking error curves; (**c**) Velocity tracking error curves; (**d**) Surge force and yaw moment curves.

The second disturbances amplitude is 10 times the first disturbances amplitude, and the disturbances as in Equation (38) are chosen for simulation.

$$\boldsymbol{\tau}_E = \begin{bmatrix} 8 \times 10^4 \sin(\omega \cdot t) \\ 6 \times 10^4 \sin(\omega \cdot t) \\ 7 \times 10^4 \sin(\omega \cdot t) \end{bmatrix} \tag{38}$$

Figure 11 presents the observation error curves of complex disturbances, position and velocity under disturbances as shown in Equation (38). For disturbances with amplitude 11 times larger, the observations of the complex disturbances as shown in Figure 11a are not as good as in the case of small disturbances, but the observation errors of the position and velocity states are within 10^{-2}. In the case of period variation, there is no significant difference between the simulation graphs under the four periods, and there is good adaptability for the period variation.

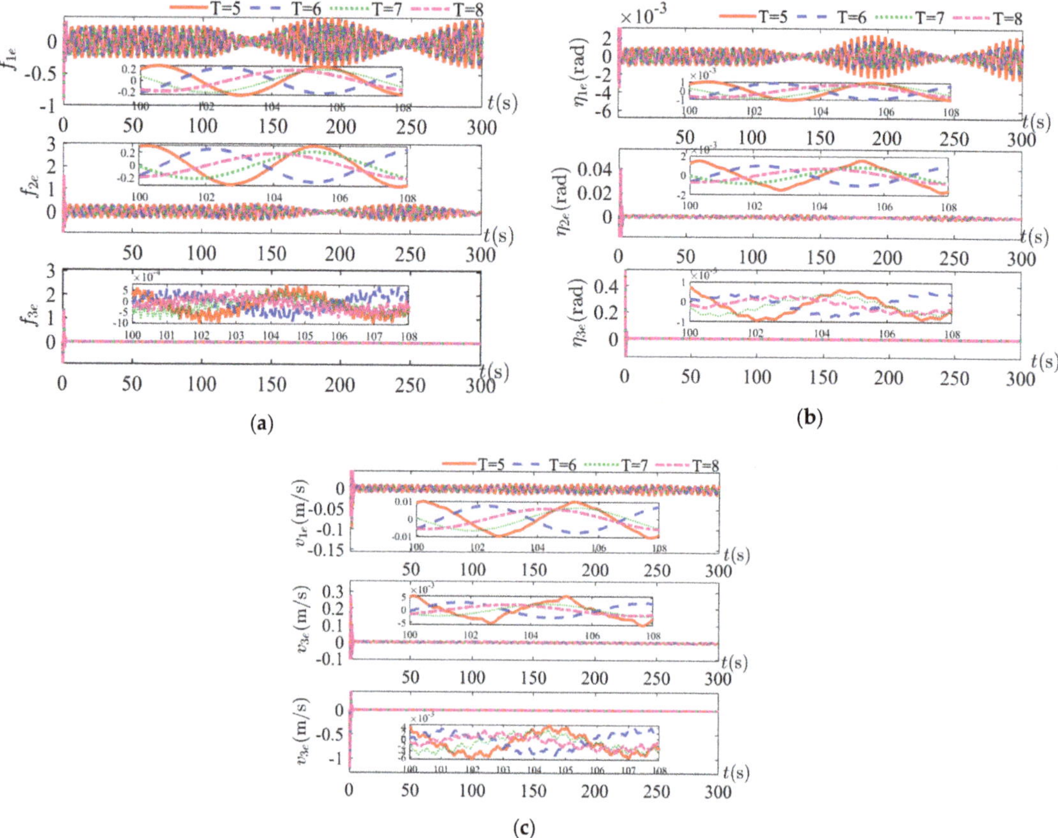

Figure 11. Observation error results at four periods. (**a**) Graph of complex disturbances tracking error under four periods; (**b**) Graph of positions tracking error under four periods; (**c**) Graph of velocities tracking error under four periods.

The simulation results shown in Figures 12–15 are obtained by varying the period $T = 5, 6, 7, 8s$ of the disturbances. Since the disturbances amplitude is 10 times the disturbances in Equation (37), the errors of the observed value obtained by NESO are relatively large compared to the small amplitude disturbances, which have an impact on the position tracking results and lead to a larger error. In the presence of observation errors, both control methods can achieve trajectory tracking control as seen in the trajectory tracking result graph, but the control effects are different. From the simulation results of Figure 12 at the disturbance period of $T = 5s$, it is obvious from Figure 12b that the control effect of the blue curve NFTSM of the control method used in this paper has a smaller error, and Figure 12d shows that the chattering of the yaw moment is smaller. When the period becomes larger, as seen from the position error curves of the four periods, the tracking error of the control method in this paper is smaller, and its error curves are closer to the reference error curve of zero value. The simulation graph of the yaw moment at four periods shows that the control scheme proposed in this study has a smaller frequency of control output variation and less chattering.

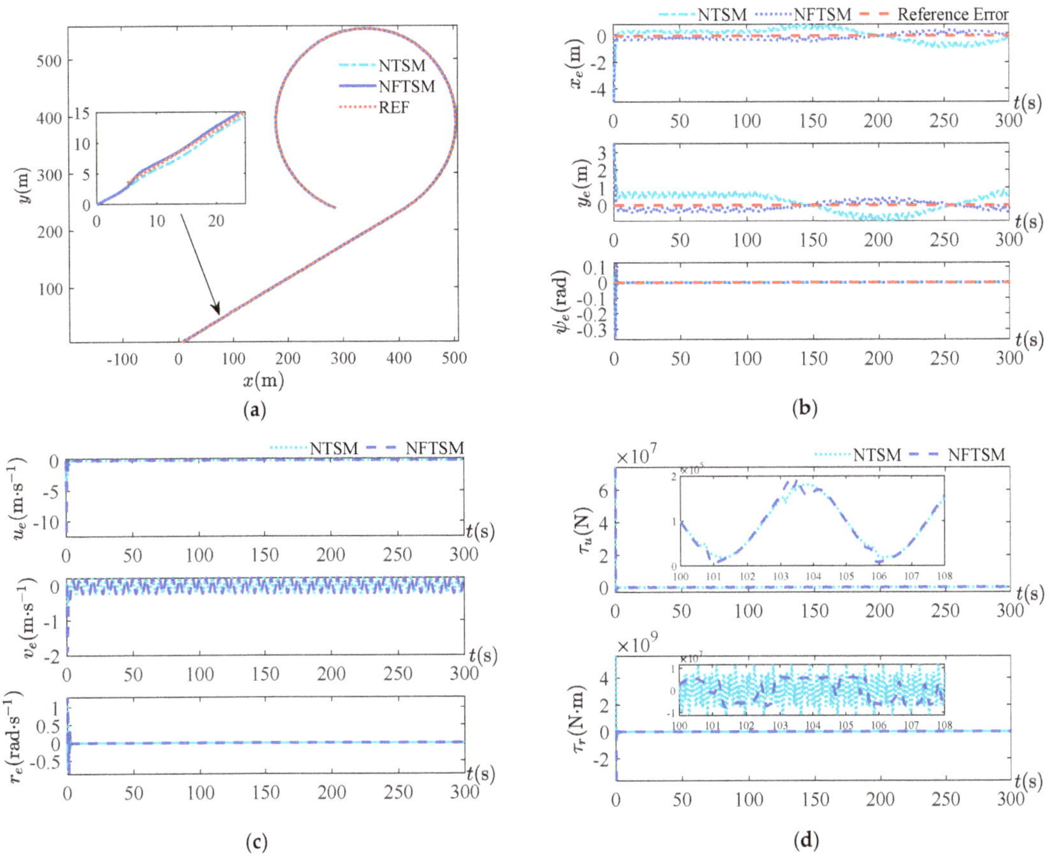

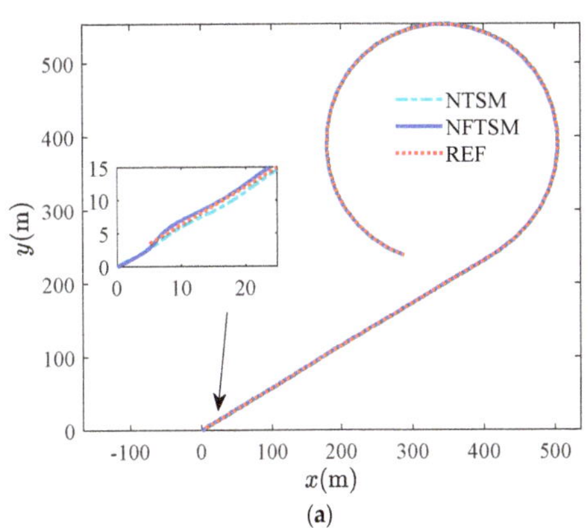

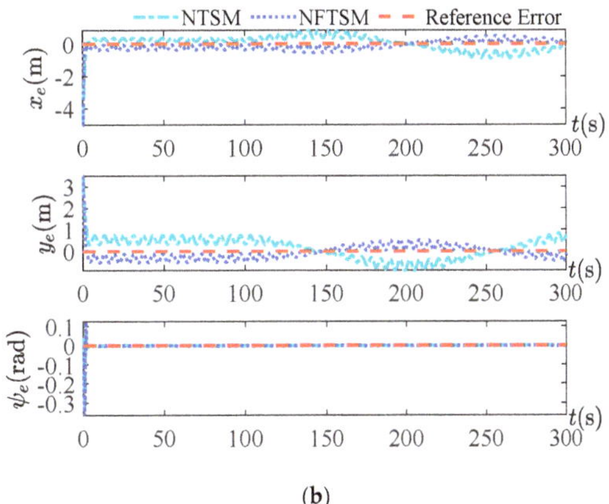

Figure 12. Simulation results at disturbance period of $T = 5s$. (**a**) Trajectory tracking curves; (**b**) Position tracking error curves; (**c**) Velocity tracking error curves; (**d**) Surge force and yaw moment curves.

Figure 13. *Cont.*

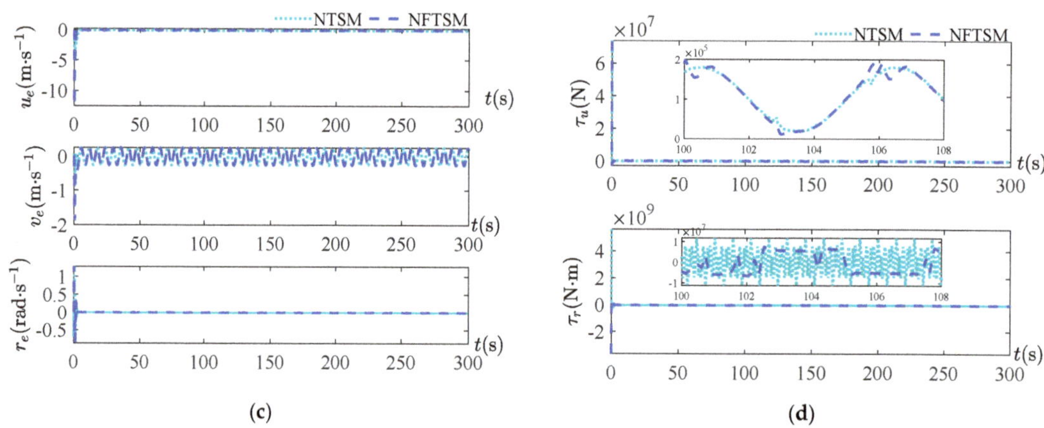

Figure 13. Simulation results at disturbance period of $T = 6s$. (**a**) Trajectory tracking curves; (**b**) Position tracking error curves; (**c**) Velocity tracking error curves; (**d**) Surge force and yaw moment curves.

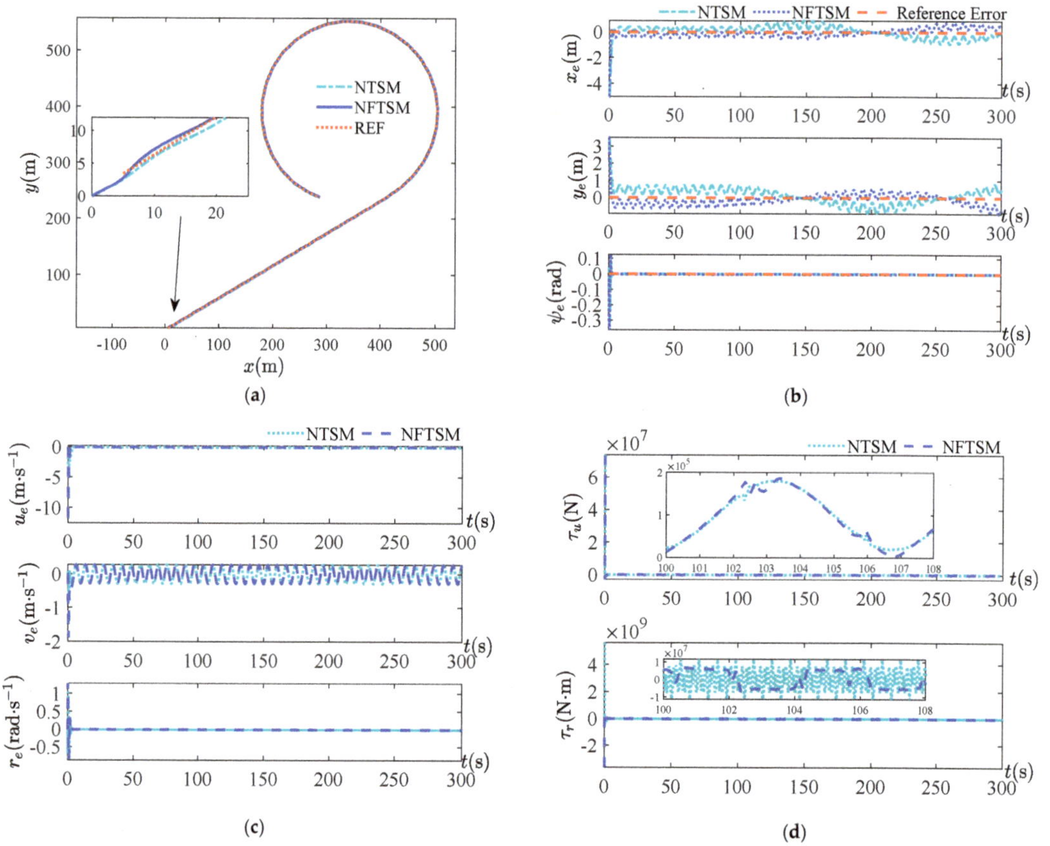

Figure 14. Simulation results at disturbance period of $T = 7s$. (**a**) Trajectory tracking curves; (**b**) Position tracking error curves; (**c**) Velocity tracking error curves; (**d**) Surge force and yaw moment curves.

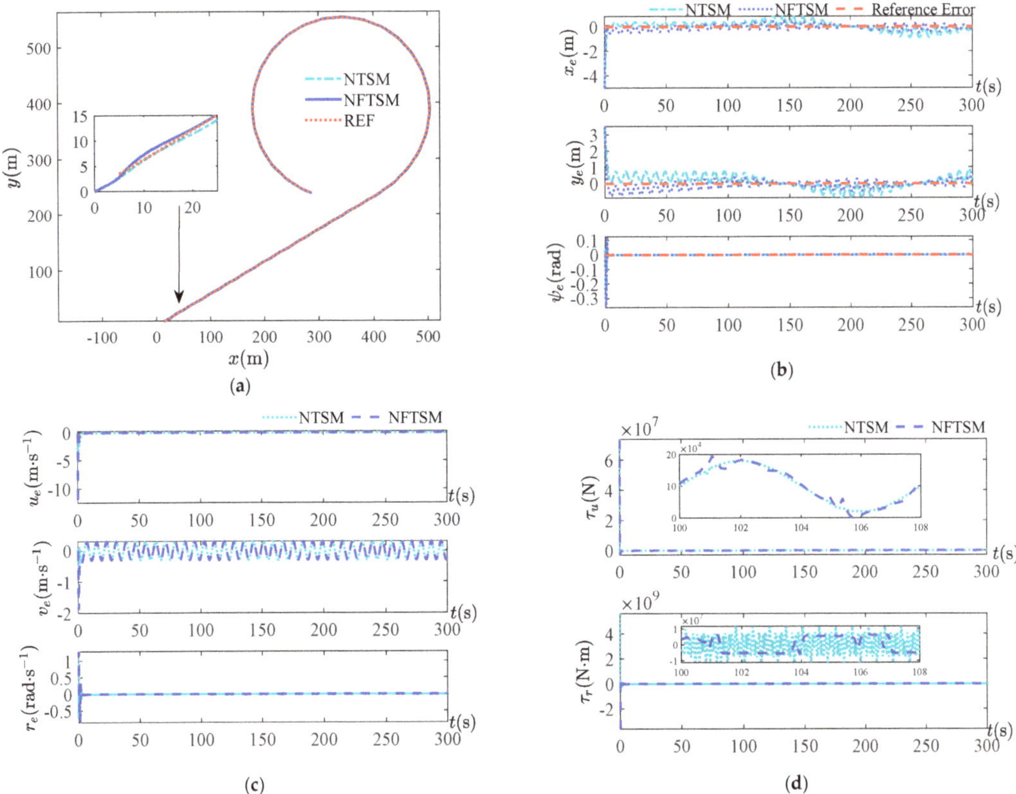

Figure 15. Simulation results at disturbance period of $T = 8s$. (**a**) Trajectory tracking curves; (**b**) Position tracking error curves; (**c**) Velocity tracking error curves; (**d**) Surge force and yaw moment curves.

The simulation results of the above two amplitudes and four periods show that the observation errors of position and velocity are less than 10^{-3} and the complex disturbance less than 2×10^{-2} in the case of small disturbance amplitude. In the case of a large disturbance of 10 times the small amplitude disturbance, the observation errors of position and velocity are less than 1×10^{-2}, and the observation errors of the complex disturbance are less than 0.3. Under the disturbance of small amplitude and large amplitude disturbances, the position, velocity and composite disturbance observations with a small enough error can be obtained, which shows that the designed full state observer is feasible. The comparison of different periods of the two control methods highlights that the control method NFTSM in this paper has a better trajectory tracking effect, a smaller error and less controller chattering.

7. Conclusions

In this study, a method combining NESO and NFTSM is presented for trajectory control of an underactuated USV under the disturbances of wind, wave and current in external environment, in the presence of perturbation of the parameters of an underactuated USV model and in the absence of accurate information about the state of the system. Based on theoretical analysis, it is proved that the design process of NESO and NFTSM satisfies the Lyapunov stability condition. Two types of disturbances with 10 times difference in amplitude and four different periods are designed. The four period disturbances with the higher probability of generating wind and wave currents are chosen according to the scatter diagram of the sea states, and the numerical simulation verifies the effectiveness of NESO for all of the states observed. Finally, the NTSM and NFTSM are compared with the position, velocity and complex perturbation observations obtained by NESO, and the following conclusions can be drawn: (1) The stability demonstration and simulation results of NESO show that the designed observer can quickly approximate the actual complex disturbances and the underactuated USV system state, and the vessel position and vessel velocity observation errors for different periods of the two disturbances are less than 10^{-3}, thus it is feasible and easier to implement the observer as a measurement module of the system to feedback the underactuated USV system state. (2) The simulation results show that the actual trajectory has a good approximation for the combined linear and circular trajectory, and the error curve shows the control scheme of NFTSM with a smaller error, and the error finally converges to the small neighborhood of zero, thus the probability of accidents is lower. (3) The simulation results show that under the conditions of environmental disturbances, model parameter uncertainty and state unpredictability, the control method of NFTSM can achieve rapid trajectory tracking control with less tracking error and less chattering in the process.

In future work, the focus will be on adaptive NESO, since the control accuracy is affected by the observer accuracy when using state feedback.

Author Contributions: Conceptualization, X.Z., Z.L. and D.X.; Data curation, J.S.; Formal analysis, X.Z., Z.L. and D.X.; Funding acquisition, X.Z. and D.X.; Methodology, D.X.; Resources, X.Z., Z.L. and D.X.; Writing—original draft, D.X.; Writing—review and editing, X.Z., D.X. and J.S. All authors have read and agreed to the published version of the manuscript.

Funding: This research was supported by the Science Fund Project of Heilongjiang Province (LH2021E087), the National Natural Science Foundation of China (51779055), and Fundamental Research Funds for the Central Universities (GK2010260347).

Institutional Review Board Statement: Not applicable.

Informed Consent Statement: Not applicable.

Data Availability Statement: Not applicable.

Acknowledgments: The authors would like to thank the fund for supporting, the Editor and Reviewers for their constructive comments.

Conflicts of Interest: The authors declare no conflict of interest.

References

1. Huang, Z.; Sui, B.; Wen, J.; Jiang, G. An Intelligent Ship Image/Video Detection and Classification Method with Improved Regressive Deep Convolutional Neural Network. *Complexity* **2020**, *2020*, 1520872. [CrossRef]
2. Yi, G.; Liu, Z.; Zhang, J.-Q.; Dong, J. Research on Underactuated USV Path Following Algorithm. In Proceedings of the 2020 IEEE 4th Information Technology, Networking, Electronic and Automation Control Conference (ITNEC), Chongqing, China, 12–14 June 2020; pp. 2141–2145.
3. Zeng, J.; Xie, Y.; Guo, X.; Liu, X.; Han, W. Transiting Target Value Based Path Following Control of Unmanned Surface Vessel. In Proceedings of the 2020 International Conference on Sensing, Diagnostics, Prognostics, and Control (SDPC), Beijing, China, 5–7 August 2020; pp. 252–256.
4. von Ellenrieder, K.D. Dynamic surface control of trajectory tracking marine vehicles with actuator magnitude and rate limits. *Automatica* **2019**, *105*, 433–442. [CrossRef]
5. Yu, H.; Guo, C.; Shen, Z.; Yan, Z. Output Feedback Spatial Trajectory Tracking Control of Underactuated Unmanned Undersea Vehicles. *IEEE Access* **2020**, *8*, 42924–42936. [CrossRef]
6. Kim, J. Target Following and Close Monitoring Using an Unmanned Surface Vehicle. *IEEE Trans. Syst. Man Cybern. Syst.* **2020**, *50*, 4233–4242. [CrossRef]
7. Paliotta, C.; Lefeber, E.; Pettersen, K.Y.; Pinto, J.; Costa, M. Trajectory Tracking and Path Following for Underactuated Marine Vehicles. *IEEE Trans. Control. Syst. Technol.* **2019**, *27*, 1423–1437. [CrossRef]
8. Do, K.D. Synchronization motion tracking control of multiple underactuated ships with collision avoidance. *IEEE Trans. Ind. Electron.* **2016**, *63*, 2976–2989. [CrossRef]
9. Ashrafiuon, H.; Nersesov, S.; Clayton, G. Trajectory Tracking Control of Planar Underactuated Vehicles. *IEEE Trans. Autom. Control* **2017**, *62*, 1959–1965. [CrossRef]

10. Chen, Z.; Zhang, Y.; Nie, Y.; Tang, J.; Zhu, S. Adaptive Sliding Mode Control Design for Nonlinear Unmanned Surface Vessel Using RBFNN and Disturbance-Observer. *IEEE Access* **2020**, *8*, 45457–45467. [CrossRef]
11. Ghommam, J.; Saad, M.; Mnif, F.; Zhu, Q.M. Guaranteed Performance Design for Formation Tracking and Collision Avoidance of Multiple USVs With Disturbances and Unmodeled Dynamics. *IEEE Syst. J.* **2020**, *15*, 4346–4357. [CrossRef]
12. Liang, X.; Qu, X.; Hou, Y.; Li, Y.; Zhang, R. Distributed coordinated tracking control of multiple unmanned surface vehicles under complex marine environments. *Ocean Eng.* **2020**, *205*, 107328. [CrossRef]
13. Liu, H.; Li, Y.; Tian, X. Practical Finite-Time Event-Triggered Trajectory Tracking Control for Underactuated Surface Vessels with Error Constraints. *IEEE Access* **2022**, *10*, 43787–43798. [CrossRef]
14. Qin, H.; Chen, X.; Sun, Y. Adaptive state-constrained trajectory tracking control of unmanned surface vessel with actuator saturation based on RBFNN and tan-type barrier Lyapunov function. *Ocean Eng.* **2022**, *253*, 110966. [CrossRef]
15. Taghavifar, H.; Qin, Y.; Hu, C. Adaptive immersion and invariance induced optimal robust control of unmanned surface vessels with structured/unstructured uncertainties. *Ocean Eng.* **2021**, *239*, 109792. [CrossRef]
16. Wang, S.; Fu, M.; Wang, Y.; Tuo, Y.; Ren, H. Adaptive Online Constructive Fuzzy Tracking Control for Unmanned Surface Vessel with Unknown Time-Varying Uncertainties. *IEEE Access* **2018**, *6*, 70444–70455. [CrossRef]
17. Chen, Z.; Zhang, Y.; Zhang, Y.; Nie, Y.; Tang, J.; Zhu, S. Disturbance-Observer-Based Sliding Mode Control Design for Nonlinear Unmanned Surface Vessel with Uncertainties. *IEEE Access* **2019**, *7*, 148522–148530. [CrossRef]
18. Xu, D.; Liu, Z.; Zhou, X.; Yang, L.; Huang, L. Trajectory Tracking of Underactuated Unmanned Surface Vessels: Non-Singular Terminal Sliding Control with Nonlinear Disturbance Observer. *Appl. Sci.* **2022**, *12*, 3004. [CrossRef]
19. Zou, L.; Liu, H.; Tian, X. Robust Neural Network Trajectory-Tracking Control of Underactuated Surface Vehicles Considering Uncertainties and Unmeasurable Velocities. *IEEE Access* **2021**, *9*, 117629–117638. [CrossRef]
20. Li, L.; Dong, K.; Guo, G. Trajectory tracking control of underactuated surface vessel with full state constraints. *Asian J. Control.* **2020**, *23*, 1762–1771. [CrossRef]
21. Park, B.S.; Yoo, S.J. Robust trajectory tracking with adjustable performance of underactuated surface vessels via quantized state feedback. *Ocean Eng.* **2022**, *246*, 110475. [CrossRef]
22. Shi, S.; Yu, X.; Khoo, S. Robust finite-time tracking control of nonholonomic mobile robots without velocity measurements. *Int. J. Control* **2016**, *89*, 411–423. [CrossRef]
23. Wang, N.; Su, S.F. Finite-Time Unknown Observer-Based Interactive Trajectory Tracking Control of Asymmetric Underactuated Surface Vehicles. *IEEE Trans. Control Syst. Technol.* **2021**, *29*, 794–803. [CrossRef]
24. Liang, X.; Qu, X.; Hou, Y.; Li, Y.; Zhang, R. Finite-time unknown observer based coordinated path-following control of unmanned underwater vehicles. *J. Frankl. Inst.* **2021**, *358*, 2703–2721.
25. Shen, D.; Tang, L.; Hu, Q.; Guo, C.; Li, X.; Zhang, J. Space manipulator trajectory tracking based on recursive decentralized finite-time control. *Aerosp. Sci. Technol.* **2020**, *102*, 105870. [CrossRef]
26. Fu, M.; Gao, S.; Wang, C.; Li, M. Human-Centered Automatic Tracking System for Underactuated Hovercraft Based on Adaptive Chattering-Free Full-Order Terminal Sliding Mode Control. *IEEE Access* **2018**, *6*, 37883–37892. [CrossRef]
27. Wang, N.; Lv, S.; Liu, Z. Global finite-time heading control of surface vehicles. *Neurocomputing* **2016**, *175*, 662–666. [CrossRef]
28. Zou, A.-M. Finite-Time Output Feedback Attitude Tracking Control for Rigid Spacecraft. *IEEE Trans. Control Syst. Technol.* **2014**, *22*, 338–345. [CrossRef]
29. Zhang, J.; Yu, S.; Wu, D.; Yan, Y. Nonsingular fixed-time terminal sliding mode trajectory tracking control for marine surface vessels with anti-disturbances. *Ocean Eng.* **2020**, *217*, 108158. [CrossRef]
30. Ali, N.; Tawiah, I.; Zhang, W. Finite-time extended state observer based nonsingular fast terminal sliding mode control of autonomous underwater vehicles. *Ocean Eng.* **2020**, *218*, 108179. [CrossRef]
31. Elmokadem, T.; Zribi, M.; Youcef-Toumi, K. Terminal sliding mode control for the trajectory tracking of underactuated Autonomous Underwater Vehicles. *Ocean Eng.* **2017**, *129*, 613–625.
32. Rodriguez, J.; Castañeda, H.; Gonzalez-Garcia, A.; Gordillo, J.L. Finite-time control for an Unmanned Surface Vehicle based on adaptive sliding mode strategy. *Ocean Eng.* **2022**, *254*, 111255. [CrossRef]
33. Yu, H.; Guo, C.; Yan, Z. Globally finite-time stable three-dimensional trajectory-tracking control of underactuated UUVs. *Ocean Eng.* **2019**, *189*, 106329.
34. Yang, L.; Yang, J. Nonsingular fast terminal sliding-mode control for nonlinear dynamical systems. *Int. J. Robust Nonlin.* **2011**, *21*, 1865–1879. [CrossRef]
35. Do, K.D.; Pan, J.; Jiang, Z.P. Robust and adaptive path following for underactuated autonomous underwater vehicles. *Ocean Eng.* **2004**, *31*, 1967–1997. [CrossRef]
36. Wang, N.; Lv, S.; Zhang, W.; Liu, Z.; Er, M.J. Finite-time observer based accurate tracking control of a marine vehicle with complex unknowns. *Ocean Eng.* **2017**, *145*, 406–415. [CrossRef]

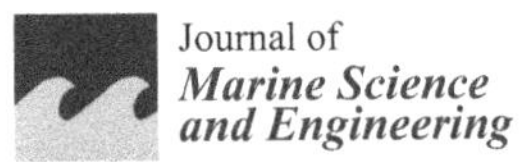

Journal of
Marine Science and Engineering

Article

Distributed Robust Fast Finite-Time Formation Control of Underactuated ASVs in Presence of Information Interruption

Guoqing Zhang *, Jun Han, Jiqiang Li and Xianku Zhang

Navigation College, Dalian Maritime University, Dalian 116026, China
* Correspondence: zhanggq87@dlmu.edu.cn

Abstract: To adapt to complex navigation conditions, this paper addresses the coordination formation of autonomous surface vehicles (ASVs) with the constraint of information interruption. For this purpose, a distributed robust fast finite-time formation control algorithm is proposed by fusion of the directed graph and neural network method. In the strategy, the graph theory is utilized for the channel of information transmission to maintain the stability of the formation system. In addition, the radial basic function (RBF) neural network is employed to approximate the structure uncertainty. Due to the merits of the robust neural damping technique, only two adaptive parameters are designed to compensate the perturbation from the model uncertainty and external environmental. Furthermore, an improved dynamic surface control (DSC) technology is developed for constituting the exponential term of the Lyapunov function. It is proven that the proposed scheme is able to achieve consensus tracking in finite time quickly, and the errors rapidly approach a small region around the origin. Finally, the feasibility and effectiveness of the algorithm are verified by two numerical simulations.

Keywords: fast finite-time; formation control; autonomous surface vehicles; adaptive control

Citation: Zhang, G.; Han, J.; Li, J.; Zhang, X. Distributed Robust Fast Finite-Time Formation Control of Underactuated ASVs in Presence of Information Interruption. *J. Mar. Sci. Eng.* **2022**, *10*, 1775. https://doi.org/10.3390/jmse10111775

Academic Editor: Sergei Chernyi

Received: 25 October 2022
Accepted: 15 November 2022
Published: 18 November 2022

Publisher's Note: MDPI stays neutral with regard to jurisdictional claims in published maps and institutional affiliations.

1. Introduction

In the past decades, there has been increased attention aimed at the development of autonomous surface vehicles (ASVs) in the field of marine cybernetics [1]. This attention primarily originates from the practical applicability of ASVs in various civilian and military missions, e.g., the emergency search and rescue, the offshore supply and the cooperative transportation [2]. The formation control is a class of effective solutions to implement the coordinated and compensated operation of ASVs. In the existing work, the control protocols almost entirely use uniformly ultimately-bounded convergence performance. In practice, the possible delay or tardiness may cause the inaccuracy of the attitude of ASV and even lead to invalidation of the whole engineering operation. Therefore, the investigation of the formation control of ASVs, with consideration to the fast response and information loss, requires more efforts and holds significance in the marine industry.

The research on vehicle formation has been developed to some extent [3–6]. Its control objective is to drive a group of ASVs converged to an anticipated geometric formation with a desired heading angle. To obtain the predefined formation, the mainstream formation control methods in the existing literature are the leader-follower approach [7,8], the virtual structure approach [9], the behavioral-based approach [10] and the graph theory-based approach [11]. In addition, the above formation methods present a mixed development state in recent years, which effectively promote the further development of formation control. In [12], novel sliding mode control laws are proposed to control multiple surface vehicles with arbitrary formations. Xiang et al. [13] developed a nonlinear controller for formation coordination to avoid collision with a certain formation. Moreover, a leader-follower formation tracking approach with limited torque was developed for tackling actuator saturation. Considering accurate formation in [14], Xiao et al. [15] designed a novel disturbance estimation scheme for the formation controller based on a terminal sliding mode observer. In [4], a practical coordinated path-following algorithm was presented for a group of underactuated surface vehicles to achieve and maintain the desired formation pattern. For the multi-input and multi-output system, a traditional adaptive control approach was designed in the presence of the amplitude and rate saturation in [9].

Aside from the proposed formation control methods, the consensus control for multi-agent system (MAS) is also a method to solve the vehicle formation coordination problem. It has the superiorities of autonomy, distribution, coordination and certain autonomous learning ability. Moreover, it holds strong robustness to the external influence and high tolerance to the fault of internal single agent. The consensus problem for MAS is a basic research direction in multi-agent cooperation; it makes each individual reach a common state. The existing papers on consensus control are mainly divided into two general categories, e.g., leader-following consensus problem and leaderless consensus problem. Especially, leader-following consensus problem (also called distributed tracking control) has already become an active research area owing to its dependability. The leader is an individual different from a follower. On the one hand, the state information of the leader can be directly or indirectly transmitted to all followers, and affects the behaviors of followers. On the other hand, the leader can be a virtual, ideal individual which is not affected by external disturbance. Based on these proposed advantages, the consensus control of MAS has been widely used with the development of vehicle formation control, such as in papers [11,16]. Based on a modular design approach, a new cooperative control scheme is proposed for the dynamic positioning of multiple offshore vessels. In [16], the paper designs a leader-follower cooperative formation control algorithm for ASV with uncertain dynamics and external disturbances.

The control algorithms mentioned above deal with the asymptotic consensus for the system. Each system state can reach the equilibrium only when the time tends to infinity theoretically. Moreover, the convergence speed is an important performance index, which affects the real-time performance of formation systems. In the existing literature, there is little fast finite-time stability theory applied to the formation control of ASVs. Actually, the fast finite time stability analysis is used in [17–20] to achieve rapid stability. In [17], a tracking control scheme for the first-order MAS is proposed in the sense of finite time convergence. The works in [18] address the tracking control of second-order MASs, and the tracking control strategies are developed, which guarantee the achievement of fast tracking in finite time. Subsequently, a fast finite-time consensus proposal with a virtual leader is presented for heterogeneous MASs in [19]. In [20], the leader-follower consensus tracking problems are discussed for high-order MASs.

Motivated by the aforementioned discussions, an improved distributed fast finite-time formation of ASVs control scheme is developed for ASVs with information interruption. The objective is to propose a consensus formation control proposal for underactuated ASVs based on fast finite-time stability criterion; the main contributions are summarized as follows:

(1) Considering the possible information interruptions during the navigation, the directed theory has been adopted to improve the fault-tolerant performance of the vehicle formation. Unlike the traditional leader-follower method, the proposed distributed control algorithm is designed to communicate between the networked underactuated vehicles which only required the information of partial neighbors, which can effectively enhance the robustness and reduce the over-dependence on the leader vehicle.

(2) In this paper, the proposed formation control scheme is firstly developed for the underactuated ASVs based on the fast finite-time stable criterion and an improved DSC technique. Different from the existing finite time protocols, the proposed strategy has the superiority of a faster convergence rate and higher control accuracy. Moreover, the convergence time of the errors has been independent with the initial condition, and is only related to the design parameters, which are more in accordance with the practical requirements.

2. Problem for Formulation and Preliminaries

Throughout the paper, h is a constant with $h = \frac{4z-1}{4z+1}$ $(z \in Z_+)$. $\widehat{(\cdot)}$ is the estimate of $(\cdot)$, and the estimation error $\widetilde{(\cdot)} = \widehat{(\cdot)} - (\cdot)$. The variables u and r are represented by ς.

2.1. The Mathematical Model for Underactuated ASVs

In this paper, the control plant is the underactuated vehicle with h propellers and rudders for surge and yaw motions only. Based on the Newtonian and Lagrangian mechanics, the mathematical dynamic model of ASVs is given as Equations (1) and (2), following [21].

$$\begin{cases} \dot{x}_i = u_i \cos(\psi_i) - v_i \sin(\psi_i) \\ \dot{y}_i = u_i \sin(\psi_i) + v_i \cos(\psi_i) \\ \dot{\psi}_i = r_i \\ \dot{u}_i = \frac{m_v}{m_u} v_i r_i - f_u(u_i) + \frac{1}{m_u}\tau_{ui} + d_{wui} \\ \dot{v}_i = -\frac{m_u}{m_v} u_i r_i - f_v(v_i) + d_{wvi}, \\ \dot{r}_i = \frac{m_u - m_v}{m_r} u_i v_i - f_u(u_i) + \frac{1}{m_r}\tau_{ri} + d_{wri} \end{cases} \tag{1}$$

with

$$\begin{cases} f_u(u_i) = \frac{d_u}{m_u} u_i + \frac{d_{u2}}{m_u}|u_i|u_i + \frac{d_{u3}}{m_u} u_i^3 \\ f_v(v_i) = \frac{d_v}{m_v} v_i + \frac{d_{v2}}{m_v}|v_i|v_i + \frac{d_{v3}}{m_v} v_i^3 \\ f_r(r_i) = \frac{d_r}{m_r} r_i + \frac{d_{r2}}{m_r}|r_i|r_i + \frac{d_{r3}}{m_r} r_i^3 \end{cases} \tag{2}$$

where x_i, y_i ψ_i denote the position and heading angle of the ith vehicle based on the earth centered fixed coordinate frame. u_i, v_i, r_i are the vehicle velocities in surge, sway and yaw in the body-fixed frame. d_{wui}, d_{wvi}, d_{wri} are the interference effect of marine environment on vehicles, such as wind, wave and ocean current. τ_{ui}, τ_{ri} are control input of vehicle model on surge and yaw. m_u, m_v, m_r, d_u, d_v, d_r, d_{u2}, d_{v2}, d_{r2}, d_{u3}, d_{v3}, d_{r3} are all considered as unknown parameters, and described as the vehicle's inertia, hydrodynamic damping and nonlinear damping terms. $f_u(u_i)$, $f_v(v_i)$, $f_r(r_i)$ denote the high-order hydrodynamic effects.

2.2. Directed Graph Theory

In a formation system with M ASVs, each ASV can be regarded as the node of the directed graph G, and the channel among ASVs can be regarded as the edge of the directed graph G. $v = \{v_1, v_2, \cdots, v_M\}$ is called the vertex set of graph G, and its elements are nodes; $e = \{e_1, e_2, \cdots, e_M\}$ is called the edge set of graph G, whose elements are edges. The M-order matrix A is called adjacency matrix with the elements $a_{ij} = 1$, representing the adjacency relationship between the nodes of the whole system. When $a_{ij} = 1$, it means that the information flow can be directly transmitted from jth vehicle to ith vehicle; when $a_{ij} = 0$, it means that there is no direct connection between the two agents. Diagonal matrix $D = diag(d_1, d_2, \cdots, d_M)$ is the in-degree matrix of graph G, with $d_i = \sum_{j=1}^{M} a_{ij}$. When the autonomous vehicle sent a message to the jth vehicle (called its neighbor), thus, the neighbor set of jth vehicle is expressed as $N_j = \{v_k | (v_k, v_j)\}$. Taking the leader into consideration, the augmented graph $\overline{G}$ is used to denote the communication topology of the leader-following system for ASVs.

2.3. RBF NNs

Based on the existing results [22,23], the neural networks and fuzzy logic systems are effective approximation tools to model the model uncertainty and extraneous disturbances. This paper adopts the radial basis function neural networks (RBF NNs) to address the uncertainty of the above mathematical model.

Lemma 1 ([24]). *For any real continuous function $f(x)$ with $f(0) = 0$, there is always one result like Equation (3)*

$$f(x) = S(x)A(x) + \varepsilon \tag{3}$$

where $S(x) = [s_1, s_2, \cdots, s_n]$ with $s_i(x)$ is called the basis function and chosen as in Gaussian form:

$$s_i = \frac{1}{\sqrt{2\pi\eta_i}} \exp\left(-\frac{(x - \mu_i)^{\mathrm{T}}(x - \mu_i)}{2\eta_i}\right), i = 1, 2, \cdots, l \tag{4}$$

l is the node number of NNs, ε is the approximation error with unknown upper bound $\bar{\varepsilon}$, n is the dimension number of $\bar{x}$, and

$$A = \begin{bmatrix} w_{11} & w_{12} & \cdots & w_{1n} \\ w_{21} & w_{22} & \cdots & w_{1n} \\ \vdots & \vdots & \ddots & \vdots \\ w_{l1} & w_{l2} & \cdots & w_{ln} \end{bmatrix}$$

is an optimal weight matrix.

Assumption 1. *The formation system consists of a leader and its followers, and each follower just communicates with its neighbors. Thus, a directed graph $\overline{G}$ is formed mathematically from the cooperative formation, contained a directed spanning tree when the leader is considered as the root.*

Remark 1. *As for directed graph $\overline{G}$, if only the Laplacian matrix $L = D - A$ has only one zero eigenvalue and other eigenvalues have positive real parts, then the directed graph is strongly connected or has a directed spanning tree. That is, all the control information is available for every agent.*

Assumption 2. *Assume that the unstructured uncertainty terms satisfy $d_{wui} \leq d_{umax}$, $d_{wvi} \leq d_{vmax}, d_{wri} \leq d_{rmax}$, where $d_{umax}, d_{vmax}, d_{rmax}$ are unknown positive constants.*

Assumption 3. *The leader is an ideal vehicle with no inertia and damping, it is not affected by the marine environmental disturbance. x_0, y_0, ψ_0 represent its status information, the reference path of leader is generated by following Equation (5).*

$$\begin{cases} \dot{x}_0 = u_0 \cos \psi_0 \\ \dot{y}_0 = u_0 \sin \psi_0 \\ \dot{\psi}_0 = r_0 \end{cases} \tag{5}$$

where u_0 and r_0 are set values according to the needs.

Lemma 2 ([25]). *Supposing a continuous function $V(x, t)$ satisfies V is positive define.*

There exist scalars $\beta > 0, \gamma > 0, 0 < \alpha < 1, 0 < \rho < \infty$ and an open neighborhood Ω of origin such that $\dot{V}(x, t) \leq -\beta V(x, t)^{\alpha} - \gamma V(x, t) + \rho, x \in \Omega$. Then there exists a finite time $T, 0 < \eta < \gamma$ such that for any $t \geq T$, one has $V(x, t) \leq \frac{\rho}{\gamma - \eta}$, where

$$T = \frac{1}{\eta(1 - \alpha)} \ln\left(\frac{\frac{\beta}{\eta} + V(0)^{1-\alpha}}{\frac{\beta}{\eta} + \left(\frac{\rho}{\gamma - \eta}\right)^{1-\alpha}}\right).$$

Lemma 3. *For $s_l \in R, l = 1, 2, \cdots, k, 0 < \mu \leq 1$, the following inequality is correct.*

$$\left(\sum_{l=1}^{k}|s_l|\right)^{\mu} \leq \sum_{l=1}^{k}|s_l|^{\mu} \leq k^{1-\mu}\left(\sum_{l=1}^{k}|s_l|\right)^{\mu}$$

3. Roust Fast Finite-Time Controller Design

As shown in Figure 1, the purpose of this paper is to provide an algorithm such that the vehicle formation would converge to the desired planning trajectory in a predetermined geometric pattern. In this part, the algorithm for vehicle formation is fully explained and proved through mathematical derivation and theoretical analysis.

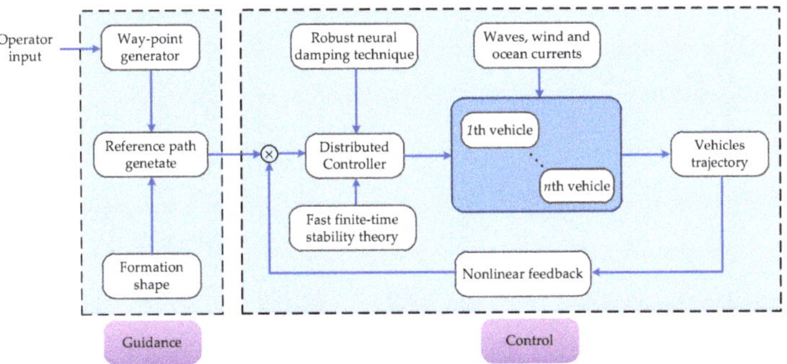

Figure 1. Conceptual signal flow box diagram for guidance, and control.

3.1. Distributed Robust Fast Finite-Time Controller Design

In this paper, the vehicle formation consists of a virtual leader and its followers. The function of the leader is to guide the vehicle formation to navigate along the planned path. The ith vehicle in the formation system just communicates with its neighbors or leader according to the designed directed graph. Based on the information from its own and other vehicles, the robust fast finite-time formation controller of ith ASV is designed by a backstepping approach and directed graph theory. In this part, the construction method of the controller is described in the following two steps.

Step 1: According to the relative position of the vehicles in Figure 2, the position errors of ith are defined based on directed graph as Equations (6) and (7).

$$x_{ei} = \sum_{j=1}^{M} a_{ij}(x_i - x_j - \Delta x_{ij}) + b_i(x_i - x_0) \tag{6}$$

$$y_{ei} = \sum_{j=1}^{M} a_{ij}(y_i - y_j - \Delta y_{ij}) + b_i(y_i - y_0) \tag{7}$$

where $\Delta x_{ij}, \Delta y_{ij}$ are constants maintained the geometry for vehicle formation.

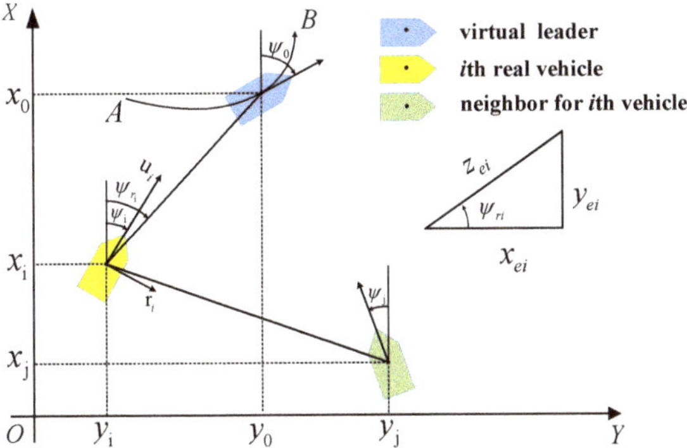

Figure 2. The schematic diagram of the proposed control principle.

Remark 2. $B = [b_1, b_2, \cdots, b_M]^{\mathrm{T}}$ *is a column of vectors, which is used to describe the existence of information connectivity between ith vehicle and leader. When the ith vehicle can receive the information from the leader, $b_i = 1$, otherwise $b_i = 0$. Here, the leader and vehicles in the formation are constructed into a directed graph $\overline{G}$.*

Based on the geometrical relation between the current position of *i*th vehicle and the reference signal in Figure 2, the error z_{ei} and ψ_{ei} are calculated as Equation (8).

$$
\begin{aligned}
z_{ei} &= \sqrt{x_{ei}^2 + y_{ei}^2} \\
\psi_{ei} &= \psi_i - \psi_{ri}
\end{aligned}
\tag{8}
$$

Note that $\psi_{ri} \in (-\pi, \pi)$ is the expected angle of the ith ship, which can be calculated as Equation (9).

$$
\psi_{ri} = 0.5[1 - \operatorname{sign}(x_{ei})]\operatorname{sign}(y_{ei})\pi + \arctan\left(\frac{y_{ei}}{x_{ei}}\right)
\tag{9}
$$

The mathematical relationship between position errors is calculated according to the trigonometric function as Equation (10).

$$
x_{ei} = z_{ei} \cos \psi_{ri}, \qquad y_{ei} = z_{ei} \sin \psi_{ri}
\tag{10}
$$

In view of the Equations (1) and (2), the time derivative of x_{ei} and y_{ei} are calculated as follows:

$$
\begin{aligned}
\dot{x}_{ei} &= (d_i + b_i)(u_i \cos \psi_i - v_i \sin \psi_i) - \sum_{i=1}^{M} a_{ij}(u_j \cos \psi_j - v_j \sin \psi_j + \Delta \dot{x}_{ij}) - b_i \dot{x}_0 \\
\dot{y}_{ei} &= (d_i + b_i)(u_i \sin \psi_i + v_i \cos \psi_i) - \sum_{i=1}^{M} a_{ij}(u_j \sin \psi_j + v_j \cos \psi_j + \Delta \dot{y}_{ij}) - b_i \dot{y}_0
\end{aligned}
\tag{11}
$$

Combining Equations (8)–(11), we can simplify the time derivation of z_{ei} as

$$
\begin{aligned}
\dot{z}_{ei} &= \dot{x}_{ei} \cos \psi_{ri} + \dot{y}_{ei} \sin \psi_{ri} \\
&= (d_i + b_i)[u_i \cos \psi_{ei} - v_i \sin \psi_{ei}] - \sum_{j=1}^{M} a_{ij}\zeta - b_i \cos \psi_{ri} \dot{x}_0 - b_i \sin \psi_{ri} \dot{y}_0
\end{aligned}
\tag{12}
$$

with $\zeta = u_j \cos(\psi_j - \psi_{rj}) + v_j \sin(\psi_j - \psi_{rj})$.

According to the Equations (2) and (8), the time derivative of the heading angle error ψ_{ei} is calculated as (13).

$$
\dot{\psi}_{ei} = \dot{\psi}_i - \dot{\psi}_{ri} = r_i - \dot{\psi}_{ri}
\tag{13}
$$

To stabilize the kinematic error dynamics (8), the virtual control laws α_{ui} and α_{ri} can be respectively designed for u_i and r_i as Equation (14).

$$
\begin{aligned}
\alpha_{ui} &= [(d_i + b_i) \cos \psi_{ei}]^{-1}[-k_{z1}z_{ei} - k_{z2}z_{ei}^{2h-1} + (d_i + b_i)v_i \sin \psi_{ei} \\
&\quad + \sum_{i=1}^{M} a_{ij}\zeta + b_i \cos \psi_{ri} \dot{x}_0 + b_i \sin \psi_{ri} \dot{y}_0] \\
\alpha_{ri} &= -k_{\psi1}\psi_{ei} - k_{\psi2}\psi_{ei}^{2h-1} + \dot{\psi}_{ri}
\end{aligned}
\tag{14}
$$

where $k_{z1}, k_{z2}, k_{\psi1}, k_{\psi2}$ are positive design parameters.

Remark 3. *For control law Equation (14), note that α_{ui} is not defined at $\psi_{ei} = \pm 0.5\pi$, thus, we assume in the control design that $|\psi_{ei} < 0.5\pi|$, and conceive that if $\psi_{ei} = \pm 0.5\pi$, then $\psi_{ei} = 0.5\pi \mp 0.01\pi$. Moreover, in the algorithm, the real vehicle follows a certain route generated by a virtual vehicle, and the heading angle error ψ_{ei} will not be equal to 0.5π owing to the small tracking error.*

In this part, we introduce a novel DSC technology by applying first-order nonlinear filter β_{ui} and β_{ri} instead of the traditional first-order linear filter to avoid repeatedly differentiating $\alpha_{\zeta i}$ as Equation (15). This effectively simplifies the process of controller design and solves the problem of the so-called "explosion of complexity".

$$\varepsilon_{\zeta i}\dot{\beta}_{\zeta i} = sig(\alpha_{\zeta i} - \beta_{\zeta i})^{2h-1} + sig(\alpha_{\zeta i} - \beta_{\zeta i}) \tag{15}$$

where $\varepsilon_{\zeta i}$ are time constants, $sig(\cdot)^{\kappa} = |\cdot|^{\kappa}sig(\cdot)$.

Defining the errors of filters as Equation (16).

$$q_{ui} = \beta_{ui} - \alpha_{ui}, \qquad q_{ri} = \beta_{ri} - \alpha_{ri} \tag{16}$$

From the above result, $q_{\zeta i}\dot{q}_{\zeta i}$ can be simplified as Equation (17)

$$\begin{aligned}
q_{\zeta i}\dot{q}_{\zeta i} &= q_{\zeta i}(\dot{\beta}_{\zeta i} - \dot{\alpha}_{\zeta i}) = -\tfrac{1}{\varepsilon_{\zeta i}}q_{\zeta i}^{2h} - \tfrac{1}{\varepsilon_{\zeta i}}q_{\zeta i}^{2} - q_{\zeta i}\dot{\alpha}_{\zeta i} \\
&\leq -\tfrac{1}{\varepsilon_{\zeta i}}q_{\zeta i}^{2h} - \tfrac{1}{\varepsilon_{\zeta i}}q_{\zeta i}^{2} + q_{\zeta i}^{2} + \tfrac{1}{4}\eta_{\zeta i}^{2}(\cdot)
\end{aligned} \tag{17}$$

where $\eta_{\zeta i}(\cdot) \geq \left|-\dot{\alpha}_{\zeta i}\right|$ is a nonnegative continuous function.

Step 2: At this step, the control inputs $\tau_{\varsigma i}$ and the derivative of adaptive laws $\dot{\hat{\lambda}}_{\varsigma i}$ for the ith vehicle are obtained according to the deduction of kinetic errors. In addition, the RBF NNs is used to approximate the model uncertainty part, which effectively enhances the stability of the system.

Firstly, one defines the kinetic errors as Equation (18).

$$\begin{aligned}
u_{ei} &= u_i - \beta_{ui} \\
r_{ei} &= r_i - \beta_{ri}
\end{aligned} \tag{18}$$

The time derivative of u_{ei} and r_{ei} can be calculated along Equations (2) and (18) as Equation (19).

$$\begin{aligned}
\dot{u}_{ei} &= \dot{u}_i - \dot{\beta}_{ui} = \tfrac{m_v}{m_u}v_i r_i - f_u(u_i) + \tfrac{1}{m_u}\tau_{ui} + d_{wui} - \dot{\beta}_{ui} \\
\dot{r}_{ei} &= \dot{r}_i - \dot{\beta}_{ri} = \tfrac{m_u - m_v}{m_r}u_i v_i - f_r(r_i) + \tfrac{1}{m_r}\tau_{ri} + d_{wri} - \dot{\beta}_{ri}
\end{aligned} \tag{19}$$

Here, $f_\varsigma(\varsigma_i)$ is an unknown smooth function in vehicle model which is used to describe hydrodynamic effects. According to Lemma 1, it can be approximated by NNs as

$$\begin{aligned}
f_\varsigma(\varsigma_i) &= S_{\varsigma i}(\varsigma)A_{\varsigma i}\varsigma_i + \varepsilon_{\varsigma i} \\
&= S_{\varsigma i}(\varsigma)A_{\varsigma i}\beta_{\varsigma i} + S_{\varsigma i}(\varsigma)A_{\varsigma i}\varsigma_{ei} + \varepsilon_{\varsigma i} \\
&= S_{\varsigma i}(\varsigma)A_{\varsigma i}\beta_{\varsigma i} + b_{\varsigma i}S_{\varsigma i}W_{\varsigma i} + \varepsilon_{\varsigma i}
\end{aligned} \tag{20}$$

where $b_\varsigma = \left\|A_\varsigma\right\|$, $A_\varsigma^m = (A_\varsigma/b_\varsigma)$, $W_\varsigma = A_\varsigma^m \varsigma e = (A_\varsigma/b_\varsigma)\varsigma_e, b_\varsigma W_\varsigma = A_\varsigma \varsigma_e$.

According to Equation (20), the simplified $\dot{u}_{ei}$ and $\dot{r}_{ei}$ can be obtained as Equation (21).

$$\begin{aligned}
\dot{u}_{ei} &= v_{ui} - b_{ui}S_{ui}W_{ui} + \tfrac{1}{m_u}\tau_{ui} - \dot{\beta}_{ui} \\
\dot{r}_{ei} &= v_{ri} - b_{ri}S_{ri}W_{ri} + \tfrac{1}{m_r}\tau_{ri} - \dot{\beta}_{ri}
\end{aligned} \tag{21}$$

To promote the understanding, one defines the variables $v_{\varsigma i}$, $\vartheta_{\varsigma i}$ and $\varphi_{\varsigma i}$ to describe simplistically the controller construction.

$$
\begin{aligned}
v_{ui} &= \frac{m_v}{m_u} v_i r_i - S_{ui} A_{ui} \beta_{ui} + d_{wui} - \varepsilon_{ui} \\
&\leq d_u \frac{v_i^2 + r_i^2}{2} + S_{ui} A_{ui} \beta_{ui} + d_{u\,max} + \bar{\varepsilon}_{ui} \\
&\leq \max\{\|A_{ui}\beta_{ui}\|, d_u, d_{u\,max} + \bar{\varepsilon}_{ui}\}\left(1 + \|S_{ui}(u)\| + \frac{v_i^2 + r_i^2}{2}\right) \\
&\leq \vartheta_{ui} \varphi_{ui}
\end{aligned}
\tag{22}
$$

where $\vartheta_{ui} = \max\{\|A_{ui}\beta_{ui}\|, d_u, d_{u\,max} + \bar{\varepsilon}_{ui}\}$ and $\varphi_{ui} = \left(1 + \|S_{ui}(u)\| + \frac{v_i^2 + r_i^2}{2}\right)$

Similar to the previous process, we can obtain

$$
v_{ri} \leq \vartheta_{ri} \varphi_{ri}
\tag{23}
$$

where $\vartheta_{ri} = \max\{\|A_{ri}\beta_{ri}\|, d_r, d_{r\,max} + \bar{\varepsilon}_{ri}\}$ and $\varphi_{ri} = \left(1 + \|S_{ri}(r)\| + \frac{u_i^2 + v_i^2}{2}\right)$, with d_u and d_r are an unknown positive constant.

Now, the controller inputs of τ_{ui}, τ_{ri} and parameter update laws $\dot{\hat{\lambda}}_{ui}$, $\dot{\hat{\lambda}}_{ri}$ are designed as follows Equations (24) and (25).

$$
\begin{aligned}
\tau_{ui} &= m_u \dot{\beta}_{ui} - \hat{\lambda}_{ui} \Phi_{ui}(\cdot) u_{ei} - k_{u1} u_{ei} - k_{u2} u_{ei}^{2h-1} \\
\tau_{ri} &= m_r \dot{\beta}_{ri} - \hat{\lambda}_{ri} \Phi_{ri}(\cdot) r_{ei} - k_{r1} r_{ei} - k_{r2} r_{ei}^{2h-1}
\end{aligned}
\tag{24}
$$

$$
\begin{aligned}
\dot{\hat{\lambda}}_{ui} &= \Gamma_{ui}\left[\Phi_{ui}(\cdot) u_{ei}^2 - \delta_{u1}(\hat{\lambda}_{ui} - \hat{\lambda}_{ui}^0) - \delta_{u2}(\hat{\lambda}_{ui} - \hat{\lambda}_{ui}^0)^{2h-1}\right] \\
\dot{\hat{\lambda}}_{ri} &= \Gamma_{ri}\left[\Phi_{ri}(\cdot) r_{ei}^2 - \delta_{r1}(\hat{\lambda}_{ri} - \hat{\lambda}_{ri}^0) - \delta_{r2}(\hat{\lambda}_{ri} - \hat{\lambda}_{ri}^0)^{2h-1}\right]
\end{aligned}
\tag{25}
$$

with

$$
\begin{aligned}
\Phi_{\varsigma i}(\cdot) &= (S_{\varsigma i}(\varsigma) S_{\varsigma i}(\varsigma)^{\mathrm{T}}/(4\gamma_{i1}^2)) + (\varphi_{\varsigma i}^2/(4\gamma_{\varsigma 2}^2)) \\
\lambda_\varsigma &= m_\varsigma \max\{b_\varsigma^2, \vartheta_\varsigma^2\} \quad \varsigma = u, r
\end{aligned}
$$

where $k_{\varsigma 1}, k_{\varsigma 2}, \Gamma_{\varsigma i}, \delta_{\varsigma 1}, \delta_{\varsigma 2}$ are positive parameters for controller. $\hat{\lambda}_{\varsigma i}$ is the estimate of λ_ς, and $\hat{\lambda}_\varsigma^0$ is the initial value of $\hat{\lambda}_\varsigma$.

3.2. Distributed Robust Fast Finite-Time Control Algorithm Stability Analysis

In this section, the stability analysis for the vehicle formation system, which is equipped with the proposed distributed robust fast finite-time controller, is carried out based on the Lemma 2 and Lyapunov approach.

The main result is summarized as Theorem 1.

For any $B_1 > 0$ and $B_2 > 0$, the sets

$$
\Pi_1 := \left\{(x_0, \dot{x}_0, \ddot{x}_0, y_0, \dot{y}_0, \ddot{y}_0, \dot{\psi}_0, \ddot{\psi}_0) : x_0^2 + \dot{x}_0^2 + \ddot{x}_0^2 + y_0^2 + \dot{y}_0^2 + \ddot{y}_0^2 + \dot{\psi}_0^2 + \ddot{\psi}_0^2 \leq B_1\right\}
$$

$$
\Pi_2 := \left\{z_{ei}, \psi_{ei}, u_{ei}, r_{ei}, q_{ui}, q_{ri}, \tilde{\lambda}_{ui}, \tilde{\lambda}_{ri} : z_{ei}^2 + \psi_{ei}^2 + u_{ei}^2 + r_{ei}^2 + q_{ui}^2 + q_{ri}^2 + \tilde{\lambda}_{ui}^2 + \tilde{\lambda}_{ri}^2 \leq B_2\right\}
$$

Therefore, the nonnegative continuous function $\eta_{\varsigma i}(\cdot)$ has a maximum value $M_{\varsigma i}$ on compact set $\Pi_1 \times \Pi$.

Theorem 1. *Assume the close-loop system consists of the vehicle dynamic (1) satisfied Assumption 2, the virtual controllers (14), the distributed robust controllers (24) and the adaptive laws (25). The proposed algorithm for the closed-loop will guarantee that: 1. all signals remain bounded uniformly; 2. the tracking errors converge to a small region in finite time T.*

Proof. To make the reasoning process clear, we divide it into the four steps according to the system state.

Step 1: Defining the Lyapunov function $V_1 = \frac{1}{2}z_{ei}^2$, we can get the derivative of V_1 as Equation (26) according to Equations (12), (16) and (18).

$$
\begin{aligned}
\dot{V}_1 = \; & z_{ei}\{(d_i + b_i)[(q_{ui} + \alpha_{ui} + u_{ei})\cos\psi_{ei} - v_i\sin\psi_{ei}] \\
& - \sum_{j=1}^{M} a_{ij}\zeta - b_i\cos\psi_{ri}\dot{x}_0 - b_i\sin\psi_{ri}\dot{y}_0\}
\end{aligned}
\tag{26}
$$

On account of Young's inequality, the time derivative of V_1 is derived along (14) and (26) as Equation (27)

$$
\begin{aligned}
\dot{V} \; & \leq -k_{z1}z_{ei}^2 - k_{z2}z_{ei}^{2h} + (d_i + b_i)\cos\psi_{ei}(q_{ui} + u_{ei})z_{ei} \\
& \leq -\left[k_{z1} - \frac{1}{2}(d_i + b_i)\right]z_{ei}^2 - k_{z2}z_{ei}^{2h} \\
& \quad + (d_i + b_i)q_{ui}^2 + (d_i + b_i)u_{ei}^2
\end{aligned}
\tag{27}
$$

Step 2: Constructing the Lyapunov function $V_2 = \frac{1}{2}\psi_{ei}^2$, one can obtain the $\dot{V}_2$ by substituting (13), (16) and (18).

$$
\dot{V}_2 = \psi_{ei}(q_{ri} + \alpha_{ri} + r_{ei} - \dot{\psi}_r)
\tag{28}
$$

Substituting the virtual law α_{ri} (14) into (28), $\dot{V}_2$ can be simplified as

$$
\begin{aligned}
\dot{V}_2 \; & \leq -k_{\psi 1}\psi_{ei}^2 - k_{\psi 2}\psi_{ei}^{2h} + q_{ri}\psi_{ei} + r_{ei}\psi_{ei} \\
& \leq -(k_{\psi 1} - \frac{1}{2})\psi_{ei}^2 - k_{\psi 2}\psi_{ei}^{2h} + q_{ri}^2 + r_{ei}^2
\end{aligned}
\tag{29}
$$

Step 3: In order to stabilize kinetic errors ς_{ei} and adaptive parameters λ_{ς_i}, here we allow

$$
V_3 = \frac{1}{2}\varsigma_{ei}^2 + \frac{1}{2m_\varsigma}\Gamma_{\varsigma_i}^{-1}\tilde{\lambda}_{\varsigma_i}^2 + \frac{1}{2}q_{\varsigma_i}^2
\tag{30}
$$

The time derivative of V_3 can be derived by invoking from Equations (21) and (17) as Equation (31)

$$
\begin{aligned}
\dot{V}_3 \; & = \varsigma_{ei}(v_{\varsigma_i} - b_{\varsigma_i}S_{\varsigma_i}W_{\varsigma_i} + \frac{1}{m_\varsigma}\tau_{\varsigma_i} - \dot{\beta}_{\varsigma_i}) + \frac{\Gamma_{\varsigma_i}^{-1}}{m_\varsigma}\tilde{\lambda}_{\varsigma_i}\dot{\hat{\lambda}}_{\varsigma_i} + q_{\varsigma_i}\dot{q}_{\varsigma_i} \\
& \leq v_{\varsigma_i}\varsigma_{ei} - b_{\varsigma_i}S_{\varsigma_i}W_{\varsigma_i}\varsigma_{ei} + \frac{1}{m_\varsigma}\tau_{\varsigma_i}\varsigma_{ei} - \dot{\beta}_{\varsigma_i}\varsigma_{ei} + \frac{\Gamma_{\varsigma_i}^{-1}}{m_\varsigma}\hat{\lambda}_{\varsigma_i}\dot{\hat{\lambda}}_{\varsigma_i} \\
& \quad - \frac{\Gamma_{\varsigma_i}^{-1}}{m_\varsigma}\lambda_{\varsigma_i}\dot{\hat{\lambda}}_{\varsigma_i} - \frac{1}{\varepsilon_{\varsigma_i}}q_{\varsigma_i}^{2h} - \frac{1}{\varepsilon_{\varsigma_i}}q_{\varsigma_i}^2 + q_{\varsigma_i}^2 + \frac{1}{4}M_{\varsigma_i}^2
\end{aligned}
\tag{31}
$$

Based on Young's inequality, one simplifies the partial variables of Equation (31).

$$
\begin{aligned}
& v_{\varsigma_i}\varsigma_{ei} - b_{\varsigma_i}S_{\varsigma_i}W_{\varsigma_i}\varsigma_{ei} \\
& \leq \frac{b_{\varsigma_i}^2 S_{\varsigma_i}^T S_{\varsigma_i}\varsigma_{ei}^2}{4\gamma_{\varsigma_1}^2} + \gamma_{\varsigma_1}^2 W_{\varsigma_i}^T W_{\varsigma_i} + \frac{\varphi_{\varsigma_i}^2\theta_{\varsigma_i}^2\varsigma_{ei}^2}{4\gamma_{\varsigma_2}^2} + \gamma_{\varsigma_2}^2 \\
& \leq (\frac{S_{\varsigma_i}^T S_{\varsigma_i}}{4\gamma_{\varsigma_1}^2} + \frac{\varphi_{\varsigma_i}^2}{4\gamma_{\varsigma_2}^2})m_\varsigma max\{b_{\varsigma_i}^2, \theta_{\varsigma_i}^2\}\frac{1}{m_\varsigma}\varsigma_{ei}^2 + \gamma_{\varsigma_1}^2 W_{\varsigma_i}^T W_{\varsigma_i} + \gamma_{\varsigma_2}^2 \\
& \leq \frac{1}{m_\varsigma}\lambda_{\varsigma_i}\Phi_{\varsigma_i}\varsigma_{ei}^2 + \gamma_{\varsigma_1}^2\varsigma_{ei}^2 + \gamma_{\varsigma_2}^2
\end{aligned}
\tag{32}
$$

with

$$
W_{\varsigma_i} = A_{\varsigma_i}^m \varsigma_{ei} = \frac{A_{\varsigma_i}}{\|A_{\varsigma_i}\|}\varsigma_{ei} = \frac{1}{\|A_{\varsigma_i}\|}
\begin{bmatrix}
w_{\varsigma_1}\varsigma_e \\
w_{\varsigma_2}\varsigma_e \\
\vdots \\
w_{\varsigma_L}\varsigma_e
\end{bmatrix}
$$

$$
W_{\varsigma_i}^T W_{\varsigma_i} = \frac{1}{\|A_{\varsigma_i}\|}(w_{\varsigma_1}^2 + w_{\varsigma_2}^2 + \cdots + w_{\varsigma_L}^2)\varsigma_{\varsigma_i}^2 = \varsigma_{\varsigma_i}^2
$$

Then, $\dot{V}_3$ is calculated by virtue of Equation (32).

$$
\begin{aligned}
\dot{V}_3 \;\leq\;& \tfrac{1}{m_\varsigma}\Phi_{\varsigma i}\lambda_{\varsigma i}\varsigma_{ei}^2 + \gamma_{\varsigma 1}^2\varsigma_{ei}^2 + \gamma_{\varsigma 2}^2 + \tfrac{1}{m_\varsigma}\tau_{\varsigma i}\varsigma_{ei} - \dot{\beta}_{\varsigma i}\varsigma_{ei} \\
&+ \tfrac{\Gamma_{\varsigma i}^{-1}}{m_\varsigma}\hat{\lambda}_{\varsigma i}\dot{\hat{\lambda}}_{\varsigma i} - \tfrac{\Gamma_{\varsigma i}^{-1}}{m_\varsigma}\lambda_{\varsigma i}\dot{\hat{\lambda}}_{\varsigma i} - \tfrac{1}{\varepsilon_{\varsigma i}}q_{\varsigma i}^{2h} - \tfrac{1}{\varepsilon_{\varsigma i}}q_{\varsigma i}^2 + q_{\varsigma i}^2 + \tfrac{1}{4}M_{\varsigma i}^2 \\
\leq\;& \tfrac{\varsigma_{ei}}{m_\varsigma}\left(\Phi_{\varsigma i}\hat{\lambda}_{\varsigma i}\varsigma_{ei} - m_\varsigma\dot{\beta}_{\varsigma i} + \tau_{\varsigma i}\right) + \tfrac{\tilde{\lambda}_{\varsigma i}}{m_\varsigma}\left(\Gamma_{\varsigma i}^{-1}\dot{\hat{\lambda}} - \Phi_{\varsigma i}\varsigma_{ei}^2\right) \\
&- \tfrac{1}{\varepsilon_{\varsigma i}}q_{\varsigma i}^{2h} - \tfrac{1}{\varepsilon_{\varsigma i}}q_{\varsigma i}^2 + q_{\varsigma i}^2 + \tfrac{1}{4}M_{\varsigma i}^2 + \gamma_{\varsigma 1}^2\varsigma_{ei}^2 + \gamma_{\varsigma 2}^2
\end{aligned}
\tag{33}
$$

Substituting control inputs Equation (24) and adaptive laws (25) into Equation (33), one can get

$$
\begin{aligned}
\dot{V}_3 \;\leq\;& -\tfrac{k_{\varsigma 1}}{m_u}\varsigma_{ei}^2 - \tfrac{k_{\varsigma 2}}{m_u}\varsigma_{ei}^{2h} - \tfrac{\delta_{\varsigma 1}}{m_u}\tilde{\lambda}_{\varsigma i}(\hat{\lambda}_{\varsigma i} - \hat{\lambda}_{\varsigma i}^0) \\
&- \tfrac{\delta_{\varsigma 2}}{m_u}\tilde{\lambda}_{\varsigma i}(\hat{\lambda}_{\varsigma i} - \hat{\lambda}_{\varsigma i}^0)^{2h-1} + \tfrac{1}{4}\eta_{\varsigma i}^2 - \tfrac{1}{\varepsilon_{\varsigma i}}q_{\varsigma i}^{2h} \\
&- (\tfrac{1}{\varepsilon_{\varsigma i}} - 1)q_{\varsigma i}^2 + \gamma_{\varsigma 1}^2\varsigma_{ei}^2 + \gamma_{\varsigma 2}^2
\end{aligned}
\tag{34}
$$

where

$$
\begin{aligned}
&-\tilde{\lambda}_{\varsigma i}(\hat{\lambda}_{\varsigma i} - \hat{\lambda}_{\varsigma i}^0)^{2h-1} \\
=\;& -\tilde{\lambda}_{\varsigma i}(\hat{\lambda}_{\varsigma i} - \lambda_{\varsigma i} + \lambda_{\varsigma i} - \hat{\lambda}_{\varsigma i}^0)^{2h-1} \\
=\;& -\left[\tilde{\lambda}_{\varsigma i}^{\frac{2h}{2h-1}} + \tilde{\lambda}_{\varsigma i}^{\frac{1}{2h-1}}(\lambda_{\varsigma i} - \hat{\lambda}_{\varsigma i}^0)\right]^{2h-1} \\
\leq\;& -\left[\tilde{\lambda}_{\varsigma i}^{\frac{2h}{2h-1}} - \tfrac{1}{2h}\tilde{\lambda}_{\varsigma i}^{\frac{2h}{2h-1}} - \tfrac{2h-1}{2h}(\lambda_{\varsigma i} - \hat{\lambda}_{\varsigma i}^0)^{\frac{2h}{2h-1}}\right]^{2h-1} \\
\leq\;& -\left(\tfrac{2h-1}{2h}\tilde{\lambda}_{\varsigma i}^{\frac{2h}{2h-1}}\right)^{2h-1} - \left[-\tfrac{2h-1}{2h}(\lambda_{\varsigma i} - \hat{\lambda}_{\varsigma i}^0)^{\frac{2h}{2h-1}}\right]^{2h-1} \\
\leq\;& -a\tilde{\lambda}_{\varsigma i}^{2h} + a(\lambda_{\varsigma i} - \hat{\lambda}_{\varsigma i}^0)^{2h}
\end{aligned}
\tag{35}
$$

with $a = \left(\tfrac{2h-1}{2h}\right)^{2h-1}$, $-\tilde{\lambda}_{\varsigma i}(\hat{\lambda}_{\varsigma i} - \hat{\lambda}_{\varsigma i}^0) \leq -\tfrac{1}{2}\tilde{\lambda}_{\varsigma i}^2 + \tfrac{1}{2}(\lambda_{\varsigma i} - \hat{\lambda}_{\varsigma i}^0)^2$, when $h = 1$.

Step 4: Based on the above argument, we construct the Lyapunov function V for5the close-loop control system.

$$
\begin{aligned}
V \;=\;& V_1 + V_2 + V_3 \\
=\;& \tfrac{1}{2}z_{ei}^2 + \tfrac{1}{2}\psi_{ei}^2 + \tfrac{1}{2}\varsigma_{ei}^2 + \tfrac{1}{2m_\varsigma}\Gamma_{\varsigma i}^{-1}\tilde{\lambda}_{\varsigma i}^2 + \tfrac{1}{2}q_{\varsigma i}^2
\end{aligned}
\tag{36}
$$

By combining (27), (29), (34) and (35), the time derivative of V can be derived as Equation (37)

$$
\begin{aligned}
\dot{V} \;=\;& z_{ei}\dot{z}_{ei} + \psi_{ei}\dot{\psi}_{ei} + \varsigma_{ei}\dot{\varsigma}_{ei} + \tfrac{\Gamma_{\varsigma i}^{-1}}{m_\varsigma}\tilde{\lambda}_{\varsigma i}\dot{\hat{\lambda}}_{\varsigma i} + q_{\varsigma i}\dot{q}_{\varsigma i} \\
\leq\;& -\left[k_{z1} - \tfrac{1}{2}(d_i + b_i)\right]z_{ei}^2 - \left(k_{\psi 1} - \tfrac{1}{2}\right)\psi_{ei}^2 \\
&- \left[\tfrac{k_{\varsigma 1}}{m_\varsigma} - \gamma_{\varsigma 1}^2 - (d_i + b_i)\right]\varsigma_{ei}^2 - \left[\tfrac{1}{\varepsilon_{ui}} - 1 - (d_i + b_i)\right]q_{ui}^2 \\
&- \left[\tfrac{1}{\varepsilon_{ri}} - 2\right]q_{ri}^2 - \tfrac{\delta_{\varsigma 1}}{2m_\varsigma}\tilde{\lambda}_{\varsigma i}^2 - k_{z2}z e_{ei}^{2h} - k_{\psi 2}\psi_{ei}^{2h} \\
&- \tfrac{k_{\varsigma 2}}{m_\varsigma}\varsigma_{ei}^{2h} - \tfrac{1}{\varepsilon_{\varsigma i}}q_{\varsigma i}^{2h} - \tfrac{a\delta_{\varsigma 2}}{m_\varsigma}\tilde{\lambda}_{\varsigma i}^{2h} + \rho
\end{aligned}
\tag{37}
$$

with

$$
\rho = \frac{\delta_{\varsigma 1}}{2m_\varsigma}(\lambda_{\varsigma i} - \hat{\lambda}_{\varsigma i}^0)^2 + \frac{a\delta_{\varsigma 2}}{m_\varsigma}(\lambda_{\varsigma i} - \hat{\lambda}_{\varsigma i}^0)^{2h} + \gamma_{\varsigma 2}^2 + \frac{1}{4}M_{\varsigma i}^2
$$

Now we can choose

$$\gamma = \min\left\{2k_{z1} - (d_i + b_i), 2k_{\psi_1} - 1, \frac{2k_\varsigma}{m_\varsigma} - \gamma_{\varsigma_1}^2 - 2(d_i + b_i),\right.$$
$$\left.\frac{2}{\varepsilon_{ui}} - 1 - (d_i + b_i), \frac{2}{\varepsilon_{ri}} - 4, \frac{\delta_{\varsigma_1}}{\Gamma_{\varsigma_i}}\right\} \tag{38}$$
$$\beta = \min\left\{2k_{z2}, 2k_{\psi 2}, \frac{k_{2\varsigma_2}}{m_\varsigma}, \frac{2}{\varepsilon_{\varsigma_i}}, \frac{a\delta_{\varsigma_2}}{\Gamma_{\varsigma_i}}\right\}$$

then $\dot{V} \leq -\gamma V - \beta V^h + \rho.$ $\square$

Remark 4. *By applying Lemma 2, this inequality implies that all the signals in the formation of ASVs are bounded uniformly. Moreover, $V(x) \leq \frac{\rho}{\gamma - \eta}$ always holds for $t > T$, with $\alpha = h$ according to the proof process of Lemma 2 in [25]. Compared with the system satisfied $\dot{V} \leq -\alpha V + \rho$ or $\dot{V} \leq -\alpha V^\alpha + \rho$, the proposed control strategy is with the merits of fast convergence and practicability.*

4. Numerical Simulation

In this section, two examples are given to verify the performance and feasibility of the proposed distributed robust fast finite-time control scheme. The simulation plant are the underactuated vehicles (length of 38 m, mass of 118×10^3 kg), which is the same as it in [26], and model parameters of simulation vehicles as follows: $m_u = 120 \times 10^3$ kg, $m_v = 177.9 \times 10^3$ kg, $m_r = 636 \times 10^5$ kg. Moreover, the personal computer (the PC (Ryzen 7 4700U CPU @2GHz RAM:16GB)) was chosen to simulate the proposed algorithm.

4.1. Search and Rescue Formation in Presence of Partial Information Interruption

In the field of maritime search and rescue, we expect that the task should be conducted with the advantages of a wide search range and high efficiency, hence multi-vehicle cooperative formations are more suitable for this task. In this part, we suppose the vehicle formation consists of a virtual leader and three analogous followers. They are arranged horizontally to carry out the search task under the disturbance of ocean environment. The processes of information communication between vehicles are designed as Figure 3.

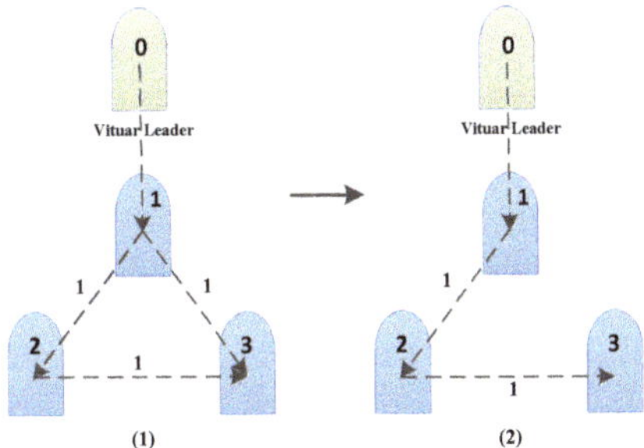

Figure 3. The augmented graph of information transfer in the ASVs.

Considering the possible delay or information interruption, the experiment simulates the interruption of a message during navigation by invalidating one of the information transmission channels at 600 s. Moreover, this above situation is described by graph theory as: the corresponding adjacency matrix and Laplacian matrix are expressed in the topological relationship of the vehicle formation, respectively,

$$A = \begin{bmatrix} 0 & 0 & 0 \\ 1 & 0 & 0 \\ 1 & 1 & 0 \end{bmatrix} \rightarrow \begin{bmatrix} 0 & 0 & 0 \\ 1 & 0 & 0 \\ 0 & 1 & 0 \end{bmatrix},$$

$$L = \begin{bmatrix} 0 & 0 & 0 \\ -1 & 1 & 0 \\ -1 & -1 & 2 \end{bmatrix} \rightarrow \begin{bmatrix} 0 & 0 & 0 \\ -1 & 1 & 0 \\ 0 & -1 & 1 \end{bmatrix}.$$

Remark 5. *Before or after the information channel interruption, the Laplacian matrix L from graph always satisfies the condition in Assumption 1, and the information needed to achieve control tasks is always globally accessible, hence the result of formation control will not be affected by the interruption of a specific signal.*

Additionally, $B = [1,0,0]$ is a used to describe the communication ability between the vehicle and leader; following Remark 2, the leader's path is composed of path planning based on waypoints, with details such as in article [26]. The geometric formation is constructed by the distances in X and Y directions between vehicles, and the relative distances between vehicles are given as $\Delta x_{21} = -1$, $\Delta y_{21} = 299$; $\Delta x_{32} = -1$, $\Delta y_{32} = 299$; $\Delta x_{31} = -2$, $\Delta y_{31} = 598$. Moreover, the initial positions and orientations of the three ASVs are designed as Equation (39).

$$\begin{bmatrix} x_i(0) \\ y_i(0) \\ \psi_i(0) \end{bmatrix} = \begin{bmatrix} -1m & -1m & -1m \\ 0m & -300m & -600m \\ 0^\circ & 0^\circ & 0^\circ \end{bmatrix} \tag{39}$$

The RBF NNs for $f_u(u_i)$ and $f_r(r_i)$ include 32 nodes, $l = 35$, the centers spaced in $[-10 \text{ m/s}, 10 \text{ m/s}]$ for $f_u(u_i)$ and $[-2.5 \text{ rad/s}, 2.5 \text{ rad/s}]$ for $f_r(r_i)$ and widths $\eta_j = 3(j = 1, 2, \ldots, l)$. For the external disturbances, the ASVs inevitably encounter the disturbance of wind and wind-generated waves in the course of navigation. This paper adopts a physical-based mathematical model to reflect the actual environment. The NORSOK wind and the JONSWAP wave spectrums are adapted to produce these two disturbances, which have been defined as an International Towing Tank Conference standard. Figure 4 describes the wind conditions (e.g., the speed is about 12.25 m/s and the direction is about 45 deg) and wind-generated waves in the simulation experiment. As for the current perturbation, the speed $v_{vurrent} = 0.8$ m/s and the direction $\psi_{current} = 60^\circ$. These disturbances exert influence on the underactuated ASVs through a certain mechanism, and the detailed descriptions of the impact method are illustrated in book [27]. For adjusting the parameters more conveniently in the application, the three ASVs in this simulation use the same set of controller parameters. The main controller parameters adjusted properly are selected as Table 1.

Table 1. Controller parameters for the controller.

Indexes	Items
Control gains	$k_{z1} = 5, k_{z2} = 0.9, k_{\psi1} = 1.2, k_{\psi2} = 10,$ $k_{u1} = 8.6 \times 10^5, k_{u2} = 20,$ $k_{r2} = 1.52 \times 10^9, k_{r2} = 20.$
Other parameters	$\delta_{u1} = 4.5, \delta_{u2} = 4.5, \delta_{r1} = 4.5, \delta_{r2} = 4.5,$ $\gamma_{u1} = 1.2, \gamma_{u2} = 4.3, \gamma_{r1} = 1.2, \gamma_{r2} = 4.3,$ $\Gamma_{u1} = 8.5, \Gamma_{u2} = 9.0, \Gamma_{r1} = 8.5, \Gamma_{r2} = 10.0,$ $\Gamma_u = 0.08, \Gamma_r = 0.08, \varepsilon_u = 0.01, \varepsilon_r = 0,01.$

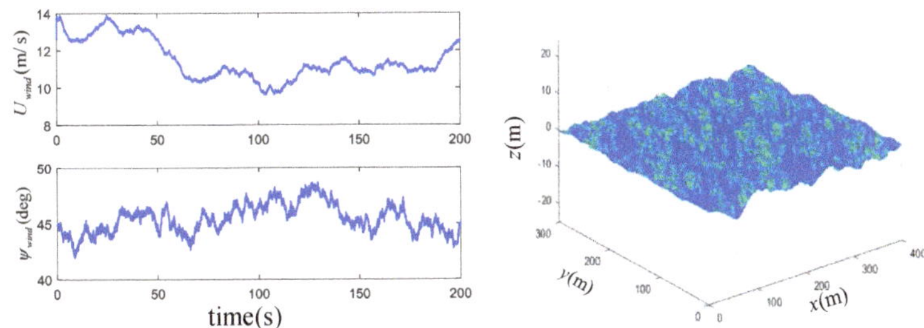

Figure 4. The simulated environment with physical-based mathematical model.

Based on the above configuration, the numerical simulation results are shown in Figures 5–10. Figure 5 is the navigation chart of ASVs formation under the proposed control scheme. The navigation path for ASVs formation is expanded by concentric square, the circles (with a radius of 150 m) in figure is the search radius. Figure 6 demonstrates the tracking errors x_{ei}, y_{ei} and ψ_{ei}, the fluctuation of the errors curve in the figure is caused by the turning of the ASVs formation. When the information is interrupted at 600 s, the positions errors x_{e3} and y_{e3} have a sudden change, which is consistent with the description of Equation (6) combined with the ASVs' position at 600 s in Figure 5. In Figure 6, other places are continuous except that the error at the signal interrupt is discontinuous. During the turning, the surge speed of the vehicle outside the circle is larger than that of the inner circle to achieve the formation turning behavior; the action is described in Figure 7. In addition, there exists a decline in the surge velocity of second and third vehicles at the time 180 s, which is different from at 600 s. For Figure 8, we can see that one of the changes in the adaptive parameters related to surge and yaw velocity, and the fluctuation time is the same as that of surge and yaw velocity, which is reasonable for this phenomenon to appear in the course of a formation system turning. Figure 9 shows the control efforts τ_u and τ_r of the three ASVs with the proposed algorithm.

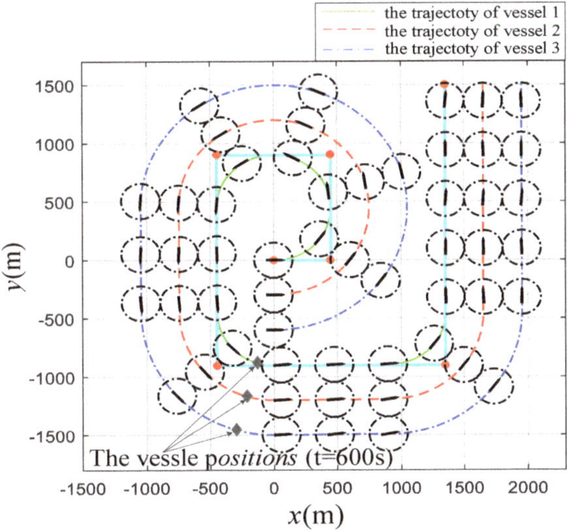

Figure 5. The position and orientation errors for the formation.

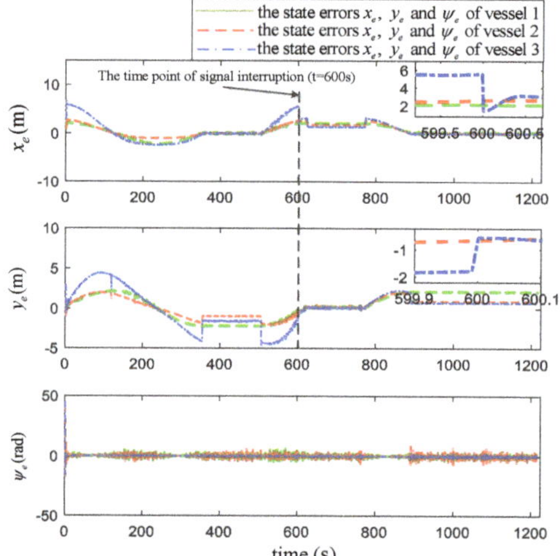

Figure 6. The position and orientation errors for the formation.

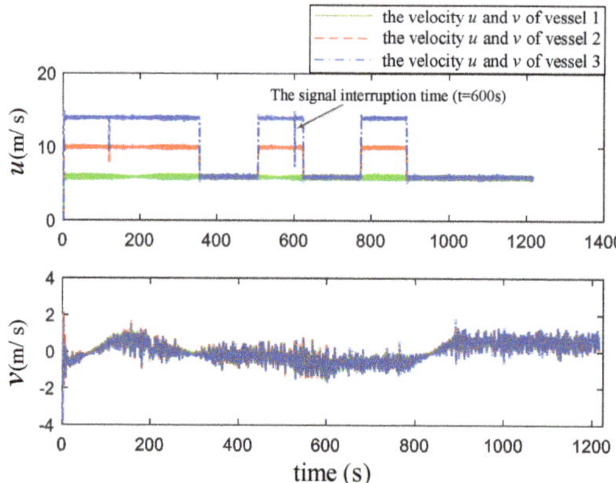

Figure 7. The change in surge velocity and sway velocity in vehicle navigation.

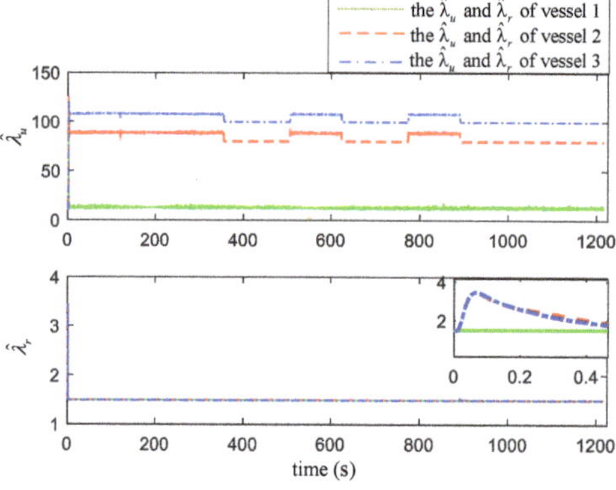

Figure 8. The changes in adaptive laws λ_u and λ_r during simulation.

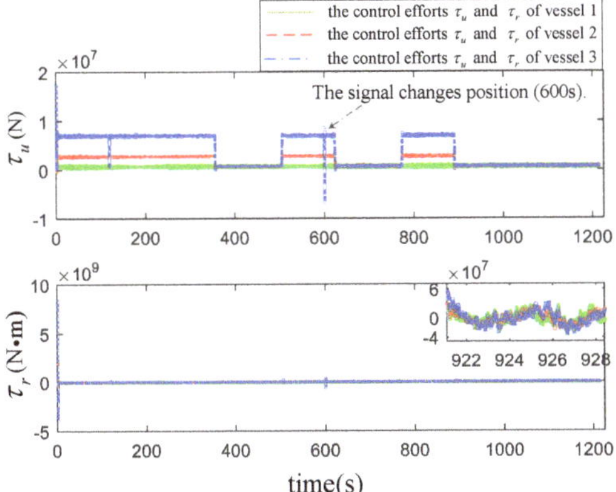

Figure 9. The control inputs τ_u and τ_r for ASVs.

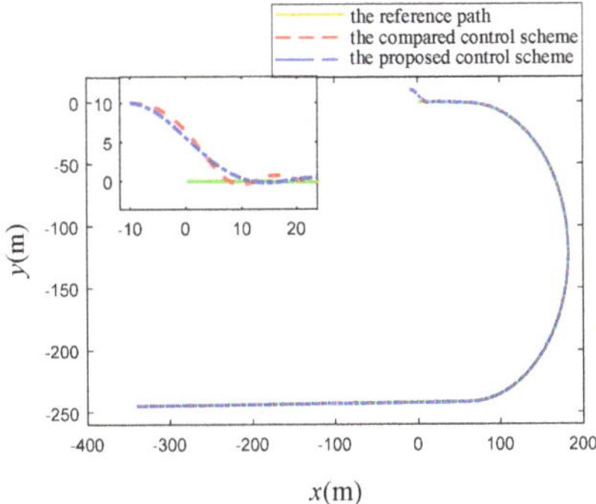

Figure 10. Comparison of the tracking path: the proposed scheme (dot dash line) and the result in [28] (dashed line).

4.2. Comparison Simulation for Individual Ship

In this section, a comparative experiment is given to demonstrate the superiority of the proposed robust fast finite-time control algorithm with a single ship. The proposed approach is compared with the finite-time path-following control algorithm in [28]. The vehicle model used in this experiment are same as above simulation, and the time-varying disturbances in the comparison simulation are as follows:

$$d_{wu} = (11/22)(1 + 0.35\sin(0.2t) + 0.15\cos(0.5t))$$
$$d_{wv} = (26/17.79)(1 + 0.3\cos(0.4t) + 0.2\sin(0.1t))$$
$$d_{wr} = -(950/636)(1 + 0.3\cos(0.3t) + 0.1\sin(0.5t))$$

The reference path is generated by a virtual vehicle, as shows in Equation (40).

$$r_d = \begin{cases} -\exp(\frac{0.005t}{300}), & 0s \leq t < 10s \\ -0.05, & 10s \leq t < 72s \\ 0, & t \geq 72. \end{cases} \tag{40}$$

$$\begin{aligned} MAE &= \frac{1}{t_\infty - t_0}\int_{t_0}^{t_\infty}|e(t)|dt \\ MAI &= \frac{1}{t_\infty - t_0}\int_{t_0}^{t_\infty}|\tau(t)|dt \end{aligned} \tag{41}$$

Figures 10–12 and Table 2 show the results of the comparison simulation. To better reflect the comparison results, three performance indexes in Table 2 are employed to display the advantages and disadvantages of the two algorithms, e.g., the mean absolute error (MAE), the mean absolute control input (MAI) and the adjustment time (AT). MAE is used for evaluating the control accuracy of control algorithm. MAI is for properties of energy consumption and smoothness; the two indicators are calculated as Equation (41). In addition, AT represents the time from the initial state to the steady state of the errors, which can directly reflect the convergence performance.

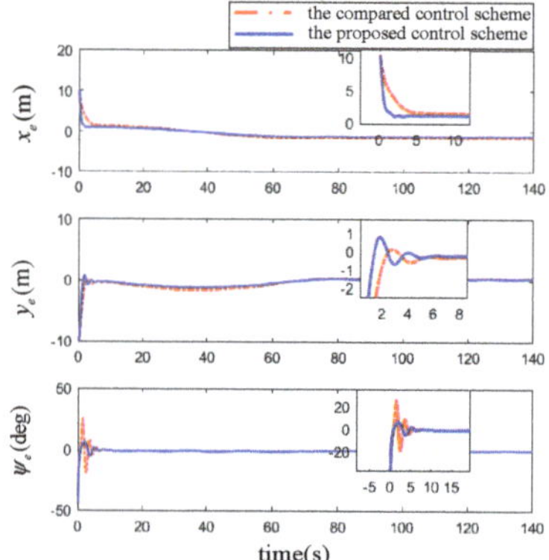

Figure 11. The error x_e, y_e, ψ_e of comparison experiment: the proposed scheme (solid line) and the result in [28] (dot dash line).

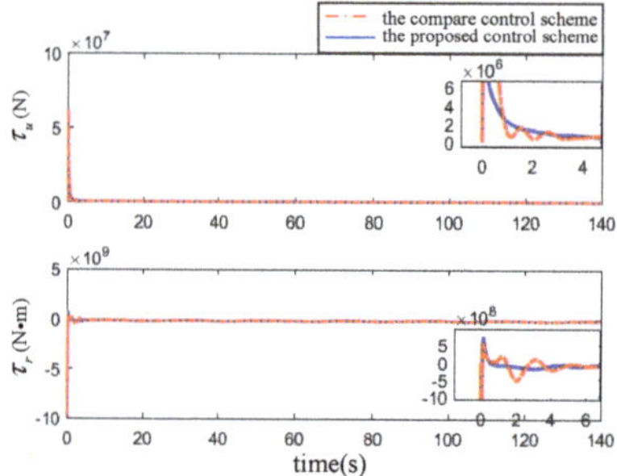

Figure 12. The control inputs τ_u and τ_r of comparison experiment: the proposed scheme (solid line) and the result in [28] (dot dash line).

Table 2. The main comparison results for two algorithms.

Indexes	Items	The Proposed Approach	The Scheme in [28]
	$x_e(m)$	0.6129	0.5000
MAE	$y_e(m)$	0.2712	0.3748
	$\psi_e(\deg)$	0.0084	0.0479
MAI	$\tau_u(N)$	6.4942×10^5	7.0622×10^5
	$\tau_r(N \cdot m)$	9.4191×10^7	9.4224×10^7
	$x_e(s)$	2.25	3.86
AT	$y_e(s)$	4.82	5.26
	$\psi_e(s)$	6.42	6.87

5. Conclusions

In this paper, a fast finite-time robust adaptive control method-based directed graph theory is proposed for underactuated vehicle formation, which has a superior faster convergence rate and better robust performance. In addition, the effectiveness of the algorithm mentioned above has been verified through the simulation results. It is meaningful for improving the coordination performance of vehicle formation. However, the proposed strategy of vehicle formation is suitable for the ideal sea area, which has limitations on areas with restricted formation deployment. In navigation, it is expected that the formation shape can be adjusted according to the changes in obstacles or coast. Hence, we will spend more energy on the research of variable vehicle formation of in the following research.

Author Contributions: Conceptualization, G.Z. and J.H.; methodology, G.Z. and J.H.; software, G.Z. and J.H.; validation, G.Z. and J.H.; formal analysis, J.H.; investigation, J.L.; writing—original draft preparation, J.H.; writing—review and editing, G.Z.; supervision, G.Z.; project administration, G.Z. and X.Z.; funding acquisition, G.Z. and X.Z. All authors have read and agreed to the published version of the manuscript.

Funding: The paper is partially supported by the National Natural Science Foundation of China (No. 52171291, 51909018), the Science and Technology Innovation Foundation of Dalian City (No. 2019J12GX026), the Liaoning BaiQianWan Talents Program (No. 2021BQWQ64), the Dalian Innovation Team Support Plan in the Key Research Field (No. 2020RT08), and the Fundamental Research Funds for the Central Universities (No. 3132022128). The authors would like to thank anonymous reviewers for their valuable comments.

Institutional Review Board Statement: Not applicable.

Informed Consent Statement: Not applicable.

Data Availability Statement: The data that support the findings of this study are available from the corresponding author upon reasonable request.

Conflicts of Interest: The authors declare no conflict of interest.

References

1. Liu, Z.; Zhang, Y.; Yu, X.; Yuan, C. Unmanned surface vehicles: An overview of developments and challenges. *Annu. Rev. Control.* **2016**, *41*, 71–93. [CrossRef]
2. Nussbaum, R.D. Some remarks on a conjecture in parameter adaptive control. *Syst. Control. Lett.* **1983**, *3*, 243–246. [CrossRef]
3. Lu, Y.; Zhang, G.; Sun, Z.; Zhang, W. Robust adaptive formation control of underactuated autonomous surface vessels based on MLP and DOB. *Nonlinear Dyn.* **2018**, *94*, 1–17. [CrossRef]
4. Zhang, G.; Huang, C.; Li, J.; Zhang, X. Constrained coordinated path-following control for underactuated surface vessels with the disturbance rejection mechanism. *Ocean Eng.* **2020**, *196*, 106725.1–106725.10. [CrossRef]
5. Huang, C.; Zhang, X.; Zhang, G. Improved decentralized finite-time formation control of underactuated USVs via a novel disturbance observer. *Ocean Eng.* **2019**, *174*, 117–124. [CrossRef]
6. He, Z.; Liu, C.; Chu, X.; Negenborn, R.R.; Wu, Q. Dynamic anti-collision A-star algorithm for multi-ship encounter situations. *Appl. Ocean Res.* **2022**, *118*, 102995. [CrossRef]
7. Xiao, H.; Li, Z.; Chen, C.L.P. Formation control of leader-follower mobile robots systems using model predictive control based on neural-dynamic optimization. *IEEE Trans. Ind. Electron.* **2016**, *63*, 5752–5762. [CrossRef]
8. Peng, Z.; Wang, D.; Li, T.; Wu, Z. Leaderless and leader-follower cooperative control of multiple marine surface vehicles with unknown dynamics. *Nonlinear Dyn.* **2013**, *74*, 95–106. [CrossRef]
9. Xu, B.; Shou, Y. Composite Learning Control of MIMO Systems With Applications. *IEEE Trans. Ind. Electron.* **2018**, *65*, 6414–6424. [CrossRef]
10. Balch, T.; Arkin, R.C. Behavior-based formation control for multi-robot teams. *IEEE Trans. Robot. Autom.* **1998**, *14*, 926–939. [CrossRef]
11. Peng, Z.; Dan, W.; Wang, J. Cooperative Dynamic Positioning of Multiple Marine Offshore Vessels: A Modular Design. *IEEE/ASME Trans. Mechatron.* **2016**, *21*, 1210–1221. [CrossRef]
12. Fahimi, F. Sliding-Mode Formation Control for Underactuated Surface Vessels. *IEEE Trans. Robot.* **2007**, *23*, 617–622. [CrossRef]
13. Xiang, X.; Lapierre, L.; Jouvencel, B. Guidance Based Collision Free and Obstacle Avoidance of Autonomous Vehicles under Formation Constraints. *IFAC Proc. Vol.* **2010**, *43*, 599–604. [CrossRef]
14. Shojaei, K. Leader-follower formation control of underactuated autonomous marine surface vehicles with limited torque. *Ocean Eng.* **2015**, *105*, 196–205. [CrossRef]

15. Xiao, B.; Yang, X.; Huo, X. A Novel Disturbance Estimation Scheme for Formation Control of Ocean Surface Vessels. *IEEE Trans. Ind. Electron.* **2017**, *64*, 4994–5003. [CrossRef]
16. Lu, Y.; Zhang, G.; Sun, Z.; Zhang, W. Adaptive cooperative formation control of autonomous surface vessels with uncertain dynamics and external disturbances. *Ocean Eng.* **2018**, *167*, 36–44. [CrossRef]
17. Chao, S.; Hu, G.; Xie, L. Fast finite-time consensus tracking for first-order multi-agent systems with unmodelled dynamics and disturbances. In Proceedings of the 2014 11th IEEE International Conference on Control & Automation (ICCA), Taichung, Taiwan, 18–20 June 2014; IEEE: Piscataway, NJ, USA, 2014.
18. Xiao, Q.; Wu, Z.; Li, P. Fast Finite-Time Consensus Tracking of First-Order Multi-Agent Systems with a Virtual Leader. *Appl. Mech. Mater.* **2014**, *596*, 552–559. [CrossRef]
19. Xiao, Q.; Wu, Z.; Li, P. Fast finite-time consensus tracking of heterogeneous multi-agent systems with a virtual leader. *Appl. Mech. Mater.* **2014**, *687*, 580–586. [CrossRef]
20. Khoo, S.; Trinh, H.M.; Man, Z.; Shen, W. Fast finite-time consensus of a class of high-order uncertain nonlinear systems. In Proceedings of the 2010 5th IEEE Conference on Industrial Electronics and Applications (ICIEA), Taichung, Taiwan, 15–17 June 2010; IEEE: Piscataway, NJ, USA; 2010.
21. Zhang, G.; Zhang, C.; Zhang, X.; Deng, Y. ESO-based path following control for underactuated vehicles with the safety prediction obstacle avoidance mechanism. *Ocean Eng.* **2019**, *188*, 106259. [CrossRef]
22. Chen, W.; Jiao, L. Adaptive Tracking for Periodically Time-Varying and Nonlinearly Parameterized Systems Using Multilayer Neural Networks. *IEEE Trans. Neural Netw.* **2010**, *21*, 345–351. [CrossRef]
23. Li, Y.; Shuai, S.; Tong, S. Adaptive Fuzzy Control Design for Stochastic Nonlinear Switched Systems With Arbitrary Switchings and Unmodeled Dynamics. *IEEE Trans. Cybern.* **2017**, *47*, 403–414. [CrossRef] [PubMed]
24. Li, j.; Zhang, G.; Shan, Q.; Zhang, W. A Novel Cooperative Design for USV-UAV Systems: 3D Mapping Guidance and Adaptive Fuzzy Control. *IEEE Trans. Control Netw. Syst.* **2022**. [CrossRef]
25. Shang, Y.; Chen, B.; Lin, C. Fast finite-time adaptive neural control of multi-agent systems. *J. Frankl. Inst.* **2020**, *357*, 10432–10452. [CrossRef]
26. Zhang, G.; Zhang, X. Concise Robust Adaptive Path-Following Control of Underactuated Ships Using DSC and MLP. *IEEE J. Ocean. Eng.* **2014**, *39*, 685–694. [CrossRef]
27. Sorensen, A.J. *Marine Control Systems Propulsion and Motion Control of Ships and Ocean Structures Lecture Notes*; Department of Marine Technology, NTNU: Trondheim, Norway, 2012.
28. Ghommam, J.; Iftekhar, L.; Saad, M. Adaptive Finite Time Path-Following Control of Underactuated Surface Vehicle With Collision Avoidance. *J. Dyn. Syst. Meas. Control.* **2019**, *141*, 121008. [CrossRef]

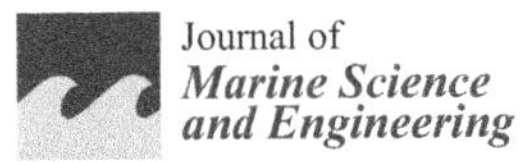

Journal of
Marine Science and Engineering

Review

A Review of Path Planning for Unmanned Surface Vehicles

Bowen Xing [1,†], Manjiang Yu [1,†], Zhenchong Liu [2], Yinchao Tan [3], Yue Sun [4] and Bing Li [5,*]

1   College of Engineering Science and Technology, Shanghai Ocean University, Shanghai 201306, China;
    bwxing@shou.edu.cn (B.X.); ymj19992022@163.com (M.Y.)
2   Shanghai Zhongchuan NERC-SDT Co., Ltd., Shanghai 201114, China; lzc_just@163.com
3   Department of Ship Engineering, Weihai Ocean Vocational College, Weihai 264315, China;
    tanyinchao@whovc.edu.cn
4   Shanghai Marine Diesel Engine Research Institute, Shanghai 201108, China; sunyuework711@163.com
5   College of Intelligent System Science and Engineering, Harbin Engineering University, Harbin 150001, China
*   Correspondence: libing265@hrbeu.edu.cn
†   These authors contributed equally to this work.

Abstract: With the continued development of artificial intelligence technology, unmanned surface vehicles (USVs) have attracted the attention of countless domestic and international specialists and academics. In particular, path planning is a core technique for the autonomy and intelligence process of USVs. The current literature reviews on USV path planning focus on the latest global and local path optimization algorithms. Almost all algorithms are optimized by concerning metrics such as path length, smoothness, and convergence speed. However, they also simulate environmental conditions at sea and do not consider the effects of sea factors, such as wind, waves, and currents. Therefore, this paper reviews the current algorithms and latest research results of USV path planning in terms of global path planning, local path planning, hazard avoidance with an approximate response, and path planning under clustering. Then, by classifying USV path planning, the advantages and disadvantages of different research methods and the entry points for improving various algorithms are summarized. Among them, the papers which use kinematic and dynamical equations to consider the ship's trajectory motion planning for actual sea environments are reviewed. Faced with multiple moving obstacles, the literature related to multi-objective task assignment methods for path planning of USV swarms is reviewed. Therefore, the main contribution of this work is that it broadens the horizon of USV path planning and proposes future directions and research priorities for USV path planning based on existing technologies and trends.

Keywords: unmanned surface vehicle (USV); path planning; hazard avoidance; hybrid algorithms; multi-objective task assignment

Citation: Xing, B.; Yu, M.; Liu, Z.; Tan, Y.; Sun, Y.; Li, B. A Review of Path Planning for Unmanned Surface Vehicles. *J. Mar. Sci. Eng.* **2023**, *11*, 1556. https://doi.org/10.3390/jmse11081556

Academic Editors: Rafael Morales and Tieshan Li

Received: 23 May 2023
Revised: 13 July 2023
Accepted: 3 August 2023
Published: 6 August 2023

1. Introduction

With the rapid development of science and technology, the USV is an indispensable means to accomplish tasks at sea. In addition, it is the outstanding advantage of USV that it is intelligent [1]. Individual USVs can perform intelligence acquisition, surface search and rescue, and marine resource exploration, while cluster of USVs can perform cooperative sensing and formation, intelligent escorting, and other operational tasks [2,3]. USV path planning is one of the most significant aspects of its safe navigation in the working environment, which directly affects the safety and economy of USVs during navigation [4]. Path planning is also one of the prominent technologies for the automation and intelligence of USVs and for performing complex tasks [5]. The evaluation criteria for USV path planning are to seek a safe and feasible optimal path from a defined starting point to an endpoint in an obstacle-ridden working environment. The aim of the path planning algorithm is the optimal selection of routes to maximize efficiency. An optimal path is predicted by analyzing the path length, smoothness, safety, and other indicators to save time and energy consumption.

The research method for this paper is the literature research method. New findings are obtained through extensive reading of the latest current literature to gain a comprehensive and correct understanding of the problems associated with USV path planning. In the past five years, several review papers [2–4,6–12] have been published summarizing the advancement of research on path planning. The following is a list of the most relevant review papers:

Ref. [2] describes the progress of USV path planning research based on multi-modality constraints in three stages: route planning, trajectory planning, and motion planning. Ref. [3] provides a comprehensive review of the development of USV from target tracking, trajectory tracking, path tracking, and cooperative formation control. This study focuses on intelligent motion control with less description of path planning. Ref. [4] addresses the USV local path planning problem and describes the characteristics of various algorithms at two levels of path search and trajectory optimization. Ref. [6] reviews recent advances and new breakthroughs in AUV path planning and obstacle avoidance methods, and compares constraints and marine environmental impacts of AUV from global and local path planning algorithms. Ref. [7] explores a path planning algorithm for autonomous maritime vehicles and its collision regulation correlation. This study focuses on USVs and COLREGs from the perspective of the safety of navigation.

Next, we will present the purpose and contribution of our review paper and emphasize the necessity of this work in comparison to current review papers. We present an up-to-date review of USV path planning. Not only are traditional graph-based search and sampling methods covered, but recent developments in reinforcement learning, neural networks, and swarm-intelligence-based optimization algorithms are also included. The innovation of this paper is to point out the limitations of current path planning methods, namely, most of them ignore the effects of winds, waves, and currents at sea on the ship. However, this paper systematically and comprehensively introduces new developments in current research into path planning algorithms in the face of unknown complex maritime situations. Secondly, in the face of multiple moving obstacles, this paper introduces the relevant algorithms to accomplish multi-objective task assignment planning and cooperation using USV clusters to achieve autonomous obstacle avoidance of vehicles. The rest of the paper is structured as follows: Section 2 presents and compares conventional as well as evolutionary algorithms under global path planning; Section 3 presents algorithms related to local path planning; Section 4 presents types of and methods for hazard avoidance in proximity response; Section 5 presents path planning algorithms under clustering; Section 6 gives conclusions and analyzes valuable future research directions in this area.

2. Global Path Planning

Global path planning is a large-scale offline path planning method based on provided information about the marine environment (electronic charts) to obtain information about static obstacles in the area that USV passes through. The global path planning algorithm acquires information about the entire environment, modeling the environment based on the obtained information pairs and performing the preliminary planning for a given path [13]. Global path planning is a viable path from the starting point to the ending point of the USV in a known operating environment. Once in sophisticated maritime environments, or when obstacles suddenly appear in the route, it can easily lead to a local optimization situation. Currently, the main global path planning methods are traditional algorithms such as the Dijkstra algorithm and A start (*) heuristic search algorithm, and evolutionary algorithms such as the genetic algorithm and neural network algorithm, as shown in Figure 1.

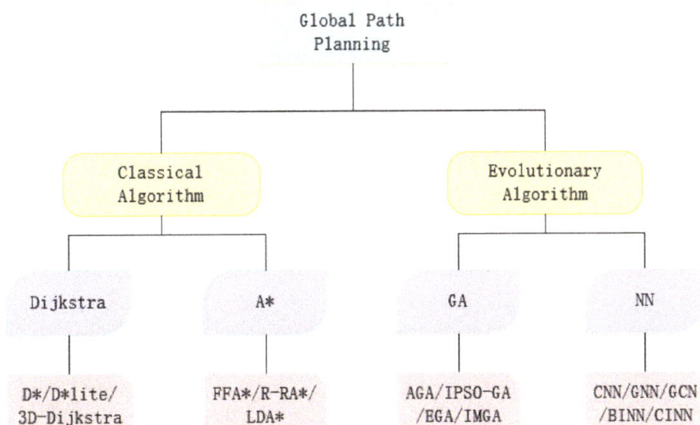

Figure 1. Global path planning algorithm.

2.1. Dijkstra Algorithm

The Dijkstra algorithm, a classical shortest-path search algorithm, was formulated by E.W. Dijkstra in 1959 [14]. By searching the graph and choosing any starting point among the schema, it is possible to calculate the closest path to all vertices. Due to the Dijkstra algorithm computing all vertices during the search process, it is less efficient to run. To address the problem of low operational efficiency, ref. [15] proposed an improved Dijkstra algorithm to add key nodes and divide regions, which can effectively reduce the computation time and improve the operational effectiveness of the algorithm. In [16], another improved Dijkstra algorithm was proposed, which needs to select the nearest node. As a result, the computation of non-critical nodes decreased, which saves time and speeds up the path-planning operation. Nevertheless, the method does not select an optimal route for the appearance of multiple paths with the same shortest distance. In [17], a running time calculation function was introduced to calculate the optimal route by running time when several paths of the same length occur. Once the data are heavy, the method consumes a lot of time. To address the inefficiency caused by a large amount of data, ref. [18] proposed a hierarchical Dijkstra improvement algorithm. It saves the location information that has been searched to be synchronized to avoid repeated searches for the same location. The algorithm can quickly find a more suitable path when there are large amounts of data. In a way, it saves time and improves efficiency, even if it is not the best path. For considering the effect of under static obstacles, ref. [19] raised a distance-seeking Dijkstra algorithm based on electronic nautical charts to solve the global path planning problem for USVs. By finding the node with the shortest path, the algorithm speeds up planning, optimizes the global route, and makes the planned path smoother. In [20], a three-dimensional Dijkstra optimization algorithm was proposed to generate globally optimal routes. Compared with the two-dimensional Dijkstra algorithm, this algorithm has a high global search capability, finds the globally optimal path more precisely, and saves time and fuel costs. Considering the effect of dynamic obstacles on global paths, ref. [21] presented a D*Lite algorithm. The prediction of dynamic path planning using dynamic information from the first computed path enables a bi-directional variable search in an unknown environment. If the map changes too much, it will calculate duplicate nodes and result in a slow planning efficiency. Moreover, the results are not convergent and the algorithm becomes stuck in a dead loop. In [22], a path planning algorithm was raised, based on the improved D*Lite algorithm was by enhancing the path cost function and reducing the expansion range of nodes. As a result, it avoids double-node computation and raises computational efficiency.

2.2. A* Algorithm

The A* algorithm, proposed by Cove in 1967 [23], was a heuristic search algorithm for discovering the shortest route between two nodes. The A* algorithm is simple in principle, and quicker than Dijkstra's algorithm. Using an optimal search approach ensures that the path has the lowest cost and enhances the efficiency of the operation. However, it relies more on heuristic functions. Once the heuristic functions are complex or invalid, it produces poor smoothness and continuity of the paths, which are not detrimental to the navigation of the vessel [24]. At present, the main improvements of the A* algorithm in academia include: firstly, expanding the number of neighboring points to be searched to improve the smoothness; secondly, optimizing the heuristic function to reduce the computation time; and thirdly, reducing the computation of the raster to hence efficiency. In [25], a finite angle FFA* algorithm was proposed by introducing a safety distance parameter. This algorithm expands the search range and increases the number of adjacent points, thus improving the smoothness of the generated route and increasing the safety of ship navigation. However, adding branches leads to more computing time and less efficiency. In [26], a limited destructive A* (LDA*) was raised, based on the problem of the slow running of the FFA* algorithm. By optimizing the heuristic function, the shortest path from the starting point to the endpoint is found in the grid environment to save time. The method is fast for static obstacles and generates feasible routes, but the performance is not optimal. In [27], another constrained A* algorithm was proposed. In a simulated closed ocean environment, the effects of static and dynamic obstacles are considered to generate a safer route by maintaining a safe distance. It reduces the computation time and improves operational efficiency by optimizing the heuristic function, which adapts to the globally optimal path planning. In [28], an R-RA* algorithm was proposed. As only a fraction of the grid map, it is possible to significantly reduce the length of the route, saving computational time and improving the operational efficiency of the algorithm. However, the generated paths are not globally optimal. In [29], a rectangular grid instead of a hexagonal grid was presented. With the reduction in intermediate nodes, it makes the path smoother. The optimization is also performed in the path length to ensure the robustness and safety of the ship.

2.3. Genetic Algorithm (GA)

The genetic algorithm (GA) was proposed by John Holland, and adheres to the principles of genetics and natural selection [30]. The main benefit of GA is that it is available for complex problems and is a general approach to solving search methods. However, classical GA suffered from low convergence and function-dependent optimization [31]. The slow processing speed makes it unsuitable for real-time route planning in dynamic and unknown situations. GA improvements have focused on improving the crossover and variation operators, choosing better coding methods, and adjusting genetic parameters. In [32], a new crossover operator to solve the complicated case of the USV path planning problem was utilized. With the aim of minimizing the overall flight cost, the algorithm randomly selects the child paths from the parent and generates paths to find the globally optimal solution and improve the convergence speed. In [33], an artificial adaptation function and two custom genetic operators were proposed, dealing with the path planning problem of USVs in the static case. It is shown that the method enables fast optimal solutions in a static environment but is hard to obtain in dynamic obstacle avoidance. In [34], a hybrid repetition-free string selection operator and heuristic multiple variables operators was proposed to find optimal paths and extend the population search. It obtains smooth flight paths and reduces the computational time of the algorithm. In [35], an adaptive genetic algorithm (AGA) was proposed that focuses on solving complex optimization problems. The hierarchical approach to selecting operators reduces time and speeds up convergence. The algorithm can shorten the length of the global path, improve quality, and find globally optimal routes in static environments. In [36], an improved particle swarm genetic algorithm (IPSO-GA) was proposed for UAV path planning, which optimizes continuously by swarm mating variation operations. The algorithm saves time and improves efficiency compared to greedy algorithms. It is more suitable for computationally intensive and complex situations to achieve collision avoidance of dynamic obstacles, but the quality of generating globally optimal paths in complex environments. In [37], a multi-objective enhanced genetic algorithm (EGA) was developed. Using crossing and variational operators, it takes full account of path smoothness and path length in ensuring that the generated paths are safe and feasible, and improves running time and efficiency. However, the method needs further improvement for dynamic obstacle avoidance. In [38], an improved multi-objective genetic algorithm (IMOGA) was proposed. Decreasing the path length is achieved by removing the redundant nodes of the running path. The optimized path quality is achieved by improving the mobility of the mobile robot, reducing the number of turning movements, and improving smoothness. However, the method does not consider the dynamic obstacle avoidance problem in the underwater environment.

2.4. Neural Network (NN)

The Neural Network (NN), introduced by McCulloch Pitts in 1944 [39], also becomes the Artificial Neural Network (ANN). It also describes the NN method to navigate in an unknown environment [40]. In [41], an NN based the COLREG rule and the risk of avoiding collisions with static obstacles was considered, to ensure the safety of MASS. However, the generated global path is not optimal by the limited environmental information. In [42], a CNN to extract information and collect visual data through two-dimensional images was proposed. Furthermore, considering COLREG, the algorithm changes the motion vector to avoid collisions. However, the data do not update in real-time and the path planning for dynamic obstacles is not accurate enough. In [43], a residual convolutional neural network (R-CNN) algorithm was investigated. It improves the capability of real-time path planning path quality by collecting previous global information, training by reinforcement, and avoiding dynamic obstacles in a real-time environment. At the same time, it can generate optimal paths on global paths. In [44], a GNN model to use features extracted from graphs by NN was proposed. With consideration of the structure and node characteristics of the graph, it improves the smoothness and speeds up the computation. However, using the GNN model directly on large-scale graphs can be computationally challenging [45]. In [46], a graph convolution network (GCN) was proposed. It is a semi-supervised learning method based on graphical data. Using a model of GCN, nodes with previously unknown data are created to capture global information to predict traffic states [47] and to achieve global path planning. The method has few weight parameters and fast convergence. However, it has a poor performance when it is applied to maritime USVs and works for planning globally optimal paths on the ground. In [48], a bio-inspired neural network (BINN) algorithm was proposed and applied to full-coverage path planning for robots on land. It is able to find the shortest path by a full-coverage approach while avoiding static obstacles. Applied to USV, it generates paths with too large turning angles and not smooth enough paths. To adapt full-coverage path planning for USV over the sea, ref. [49] proposed a full-coverage neural network (CCNN) algorithm for USV path planning. It can greatly reduce the computation time and improve coverage efficiency. The turning angle is further reduced to make the path smoother and to avoid all dynamic obstacles to obtain the globally optimal path.

2.5. Summary of Global Path Planning Algorithms

The above research uses tools such as shortening the path length and smoothing the generated paths when designing path planning algorithms to make the planned paths more compatible with navigation practice. The use of hierarchical processing of data in constraints to avoid repeated searches and the introduction of genetic crossover operators make the paths more convergent and efficient while reducing the solution time of the algorithm to meet practical needs. The safety distance parameter is entered, following COLREGs, to improve the safety and robustness of the ship. The global path algorithm is dependent on perceived environmental information and cannot optimally deal with unknown and dynamic obstacles. As shown in Table 1, it compares different algorithms for global path planning and considers various factors for optimal paths.

Table 1. Characteristics of different algorithms for global path planning.

References	Methods	Length	Smooth	Safety	COLREG	Time	Duplicate	DOA	Efficiency
[14]	Dijkstra	F	F	F	F	T	F	F	F
[15,16]	improved Dijkstra	T	F	F	F	T	F	F	T
[18]	hierachial Dijkstra	T	F	T	T	T	F	F	T
[20]	3D-Dijkstra	T	F	T	F	T	F	F	T
[21,22]	D*Lite	T	F	F	F	T	T	T	T
[23]	A*	T	F	F	F	T	T	F	T
[25]	FFA*	T	F	T	F	T	T	T	T
[26]	LDA*	T	F	T	F	T	F	F	T
[28,29]	R-RA*	T	T	F	F	T	T	F	T
[35]	AGA	T	T	F	F	T	F	F	T
[36]	improved GA	T	T	F	F	T	T	F	T
[37]	EGA	T	T	T	F	T	F	F	T
[38]	IMGA	T	T	T	F	T	T	F	T
[42]	CNN	T	F	T	T	T	F	F	T
[44]	GNN	T	T	T	F	T	F	F	T
[46]	GCN	T	T	T	F	T	F	F	T
[48]	BINN	T	T	T	F	T	F	F	T
[49]	CINN	T	T	T	F	T	F	T	T

Note: consider (T), no consider (F), dynamic obstacle avoidance (DOA), * (start).

3. Local Path Planning

In contrast to the global path planning algorithm, local path planning is the optimal path from the current node to a targeted node. It uses sensors to obtain data from the surrounding location environment, senses the obstacle distribution in the region, and performs local path planning in a real-time path search. Due to the lack of global environment information, it is easy for the pathway to fall into local optimum. Typical local path planning algorithms include artificial potential field (APF), fast extended random tree (RRT), velocity obstacle (VO), and dynamic window method (DWA). As shown in Figure 2, the typical local algorithms are analyzed for their advantages and disadvantages and the corresponding optimization algorithms.

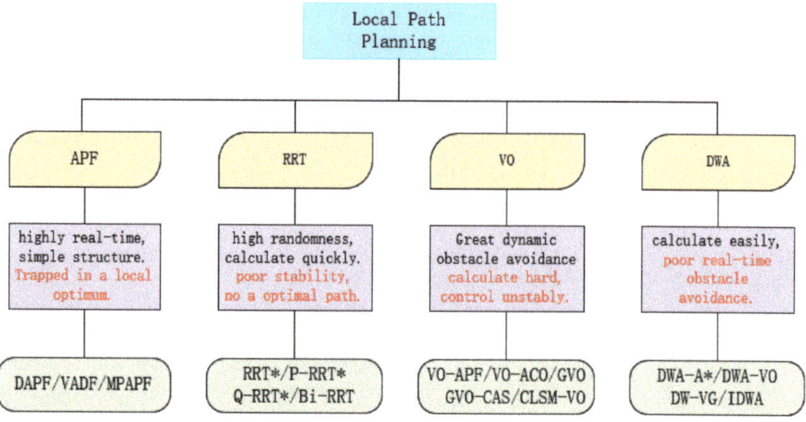

Figure 2. Local path planning algorithm.

3.1. Artificial Potential Field (APF)

The artificial potential field (APF) method is a well-known USV local path planning algorithm proposed by Khatib [50]. Although the APF method has the advantages of a single structure, fast speed, and good real-time performance, it suffers from problems such as local optimum and poor path smoothing. In [51], the APF method is applied to the local path planning of USVs to solve the problem of avoiding static obstacles during navigation and ensuring the safety of the ship's trajectory. However, when the target points near obstacles are difficult to detect, there is poor performance of the generated global paths. To solve the problem of dead loops, ref. [52] proposed an improved APF approach using map expansion to reprogram local routes. When an infinite loop occurs, the map is transformed and starts searching for travelable paths, but does not take into account the problem of unreachability due to obstacles being too close to the target point. In [35], a vector artificial potential field method (VAPF) was proposed that improves the use of the space vector method to calculate the total force of AUVs. It allows USVs to reach static targets and track dynamic targets. The improved algorithm improves computational efficiency and reduces the cumbersome time in AUV dynamic avoidance. It can avoid dynamic obstacles effectively in real-time and reduces the cost of AUVs. However, when the target points near obstacles are difficult to detect, there is poor performance of the generated global paths. In [53], the discrete artificial potential field (DAPF) method was proposed. The results show that the algorithm not only efficiently and quickly computes safe routes in static and dynamic obstacle environments but also solves local optimization methods. Using the path optimization algorithm can reduce the path length and improve the smoothness. In [54], the Ant Colony Potential Field Algorithm (APF-ACO) global obstacle avoidance scheme was used, where the real-time information from a LiDAR sensor can avoid local obstacles, and a method to create a potential field without local extrema was proposed. The length of the path is optimized to reduce steering during driving and to ensure the robustness and safety of the ship itself. In [55], a model prediction strategy and an artificial potential field (MPAPF) was proposed. It overcomes the drawbacks of the traditional APF and other algorithms and solves the difficulty of becoming stuck in the local optimum. At the same time, it considers the head-on and cross-encounter situations of ships, effectively avoiding dynamic obstacles in real-time and ensuring the safety of USVs under complex conditions.

3.2. Rapidly Expanding Random Tree (RRT)

The rapidly expanding Random Tree (RRT), formulated by Lavalle [56], is a sampling-based path planning algorithm. As the creation of new nodes is random, node creation will stop once the node determines the target location. However, the RRT algorithm provides a preferred path that generates many nodes, which consumes time and generates a not sufficiently smooth path. In [57], the algorithm reduces the randomness of the final path by deselecting the parent node. The shortest path is found by increasing the number of iterations, but it can fall into local optimality. In [58], RRT* trajectory optimization is used to create smoother paths and reduce path lengths. However, convergence to the optimal path solution is slow and consumes considerable memory and time. In [59], a Q-Learning-based RRT* (Q-RRT*) method was proposed that reduces the path length and accelerates convergence to the optimal solution by extending the search. However, it yields paths that are not globally optimal and obstacle avoidance for dynamic obstacles requires further consideration. In [60], the RRT-Connect algorithm was proposed, which is a simple and efficient stochastic algorithm. The greedy heuristic algorithm needs to speed up the search capability and the convergence process by searching from the starting and target points but is prone to local optima. To address the case of falling into a local optimum, ref. [61] proposed a path extension-based heuristic sampling RRT* algorithm (EP-RRT*). Based on RRT-Connect, the sampling-based asymptotically optimal path planning algorithm is proposed, which can quickly explore the global environment and find feasible paths. In [62], a novel probabilistic smoothed bi-directional RRT (PSBi-RRT) algorithm was presented. A greedy algorithm and biasing strategy require searching for nearby points. As a result, it avoids collisions in the environment, significantly reduces running time, and decreases the likelihood of local optimization. In [63], a method combining APF and RRT* (P-RRT*) was presented. By introducing randomized gradient descent (RGD), the algorithm sufficiently reduces the number of iterations and the time and converges to the optimal path solution more quickly. The algorithm solves the local minima problem, but the real-time performance needs to be enhanced. In [64], an improved heuristic Bi-RRT algorithm based on the Bi-RRT algorithm was presented. Not only can dynamic obstacles in the real-time environment be avoided, but further improvements are also achieved in the efficiency of algorithm operation and path length, and smoothness.

3.3. Velocity Obstacle (VO)

To address the dynamic obstacle avoidance of USVs, Fiorini and Shiller [65] proposed the velocity obstacle (VO) approach in 1998 and incorporated the velocity characteristics of obstacles into the scope. However, the algorithm failed to consider the dynamic changes in the velocity of the obstacle and the dynamic performance of the USV. In [66,67], a VO-based dynamic avoidance algorithm was designed for USV to avoid moving obstacles. Since COLREGs are not considered, the experiments verify that USV is ineffective for moving and unknown obstacle avoidance, and the safety of the generated paths needs further enhancement. In [68], an enhanced velocity obstacle-based method with a particular triangular obstacle geometry model for autonomous dynamic obstacle avoidance of USVs was proposed. COLREGs should be considered to optimize the heading angle and improve the portability of the ship. It also reduces the USV path length and fuel cost by ensuring accurate obstacle avoidance. In [69], a hybrid VO-APF algorithm for obstacle avoidance of dynamic obstacles was raised, which enhances the real-time performance of obstacle avoidance and integrates COLREGS rules and path optimization functions. However, the algorithm can easily fall into a local optimum and cannot generate a globally optimal path. To tackle the situation of falling into a local optimum, in [70], a path planning method, based on the COLREGs, was displayed, combining the VO algorithm with the ACO algorithm to realize dynamic obstacle transitions relative to the stationary state of the USV at a given instant. The method can effectively avoid dynamic obstacles and obtain a reasonable risk avoidance strategy that provides the globally optimal path. However, owing to the uncertainty of the obstacle motion state, the predicted obstacle information i nevitably appears inaccurate. Based on the Generalized Velocity Obstacle (GVO) algorithm [71,72] proposed a collision avoidance system (GVO-CAS). The method, based on COLREG, visualizes the collision course and velocity for collision avoidance of multiple dynamic obstacles at close range. The algorithm takes into account the ship dynamics to improve collision avoidance performance and provides the globally optimal path. In [73], a novel efficient path planning algorithm was presented, the Constraint Locked Sweep Method and Velocity Obstacles (CLSM-VO). The algorithm optimizes global search performance, increases computational efficiency, and enhances generated path smoothness. It is suitable for complex environments with multiple dynamic obstacles and enables fast and effective dynamic obstacle avoidance by providing USV initial yaw angle and constraint layer optimization.

3.4. Dynamic Window Approach (DWA)

The Dynamic Window Approach (DWA) is a commonly used path planning method, originally proposed by FOX et al. [74,75] in 1997. The algorithm samples multiple velocity values in velocity space and simulates the trajectory generated by the robot at each velocity. To avoid the influence of dynamic obstacles, in [76], a "dynamic collision model" was introduced for the conventional DWA to predict possible future collisions, taking into account obstacle movements. The DWA algorithm for real-time path planning reduces wasted space and time by simulating the obstacle shape based on the obstacle. In [77], a local virtual target dynamic window (DW-VG) obstacle avoidance method was proposed. Introducing virtual target points selects the optimal amount of motion control. As a result, the algorithm can avoid all dynamic obstacles and predict an optimal path. In [78], the DWA-VO algorithm was proposed. The algorithm can speed up the search and improve the efficiency of the operation. It only considers local moving and unknown obstacles, which solves the information error caused by dynamic obstacles encountered by USV. It fails to identify accurately the global obstacles and leads to a local optimum. To address the condition of being trapped in a local minimum, ref. [79] proposed a path planning system with DWA and A* algorithms, where USV can avoid unknown obstacles by selecting the optimal speed. USV can adapt to global and local maritime obstacles by choosing the best velocity to avoid unknown obstacles and avoid local optimum problems. In [80], an improved dynamic window approach (IDWA) was proposed to improve the efficiency of USV navigation. Combined with the idea of non-uniform theta*, the reverse search is performed from the target node to reduce the path length and improve the navigation efficiency and the real-time performance of the USV. The algorithm prevents the USV from falling into local minima and enhances the ability of the USV to avoid dynamic obstacles at close distances. In [81], the ant colony optimization (ACO) and dynamic window approach (DWA) were combined to solve local path planning. The global path length is optimized by enhancing the search capability and removing redundant points through a dual population heuristic function. Adaptive navigation strategies are used to achieve obstacle avoidance for unknown dynamic obstacles and to solve the local optimum problem.

3.5. Summary of Local Path Planning Algorithms

The above research focuses on solving local optimization problems. Using the sensor's ability to sense the maritime environment, it can efficiently and quickly avoid dynamic obstacles, making the generated paths smoother, and can calculate safe routes. By using space vectors, employing extended search, and introducing greedy algorithms, the length of the path is reduced while reducing the solution time of the algorithm to meet practical needs. Using a dynamic collision model and considering the speed variation of dynamic obstacles, it can avoid all obstacles in real time. However, local paths ignore the effects of wind and wave currents in real sea conditions. There is no good solution to the problem in the face of complex situations. Table 2 compares different algorithms for local path planning and considers the factors of the optimal path.

Table 2. Characteristics of different algorithms for local path planning.

References	Methods	Length	Smooth	Efficiency	DOA	GOA	COLREG	Real-Time	Optimal
[35]	VAPF	T	T	T	T	T	F	T	F
[51]	APF	T	F	F	F	F	F	T	F
[52]	improved APF	T	T	T	T	T	T	T	F
[53]	DAPF	T	T	T	T	F	F	T	T
[54]	APF-ACO	T	T	T	T	T	F	T	T
[55]	MPAPF	T	T	T	T	F	T	T	T
[56]	RRT	T	F	T	F	F	F	F	F
[58]	RRT*	T	T	F	T	F	F	T	F
[59]	Q-RRT*	T	T	F	T	T	F	T	F
[60]	RRT-Connect	T	T	T	T	F	F	T	T
[63]	P-RRT*	T	T	F	T	F	F	F	T
[64]	improved BI-RRT	T	T	T	T	T	F	T	T
[61]	PSBI-RRT	T	T	T	T	T	F	T	F
[66]	VO	T	F	F	T	T	T	F	F
[68]	improved VO	T	T	T	T	T	T	T	F
[69]	VO-APF	T	F	T	T	F	T	T	F
[70]	VO-ACO	T	T	T	T	F	T	T	T
[72]	GVO-CAS	T	T	T	T	F	T	T	F
[73]	CLSM-VO	T	T	T	T	T	T	T	F
[76]	DWA	F	F	T	T	F	F	T	F
[79]	DWA-A*	T	T	T	T	F	T	T	F

Table 2. *Cont.*

References	Methods	Length	Smooth	Efficiency	DOA	GOA	COLREG	Real-Time	Optimal
[78]	DWA-VO	T	T	T	T	T	T	T	F
[80]	IDWA	T	T	T	T	F	F	T	T
[81]	DWA-ACO	T	T	T	T	T	F	T	T

Note: consider (T), no consider (F), dynamic obstacle avoidance (DOA), global obstacle avoidance (GOA), * (start).

4. Proximity Responsive Hazard Avoidance

Proximity-responsive hazard avoidance is where a USV, following a set course, needs to adjust its response in time to effectively avoid a collision with an obstacle if the uncertainty of a dynamic obstacle is detected suddenly. In contrast, close-range collision avoidance requires consideration of the vessel's shape and turning angle. Hazard avoidance and multi-vessel collision avoidance in complex marine conditions are also hot topics of current research.

4.1. Hazard Avoidance in a Complex Environment

For USV path planning in complex environments, an increasing number of hybrid algorithms are used to deal with the above problem. In [82], a multi-layer path planner (MPP) was proposed, which combines the fast marching method (FMM) [83] and the SDCE model to simulate a static environment. Collision Avoidance (CA) decisions are used to introduce safety parameters to achieve obstacle avoidance for dynamic and unknown obstacles and reefs in the coastal environment. The method does not consider the effects of wind, waves, and currents. In [84], a combined GA and APF approach was presented. It can perform path planning for various complex dynamic obstacle situations and find the best path among the obstacles on three sides around the starting point. However, the method only plans for simple geometries and coastlines, which does not work well in actual maritime environments. To consider realistic factors at sea, ref. [85] improved the multi-objective route planning method of PSO-GA and proposed a risk assessment model for wind and waves. Meanwhile, the length of the path, smoothness, and speed of the ship are taken into account. The genetic algorithm is used for cross-calculation, which improves operational efficiency. However, it does not consider the effect of ocean currents. If you want to consider the impact of ocean currents on path planning, building a kinematic USV model is a key choice, as shown in Figure 3.

In [86], an improved imperial competition algorithm (ICA) was proposed. Based on a three-dimensional current model and the influence of static obstacles at sea, the locally optimal solution is solved effectively and the AUV can obtain the global optimal path. However, the method cannot be adapted to dynamic obstacles. In [87], the steering angle of the ship is incorporated into the cost by building a cost model of the effect of ocean currents on the energy consumption of the AUV. The algorithm uses an improved D* algorithm and considers the kinematic and dynamical equations to plan the path with the minimum energy consumption, but the path is not optimal. In [88], a VF-RRT* algorithm was proposed, which improves its effectiveness in the current environment by introducing a virtual field sampling function and an ocean current constraint function. As a result, it greatly reduces the navigation time and energy consumption of the USV. In [89], an improved sparrow search algorithm (SSA) was designed, which considers the effects of time-varying characteristics of ocean currents, wind and wave factors, and dynamic obstacles on the navigation of AUVs in complex ocean environments. Convergence speed and search capability are improved by building a 3D underwater model and optimizing the coefficient weights. The algorithm can avoid all the obstacles underwater and find the globally optimal path in the complex ocean environment.

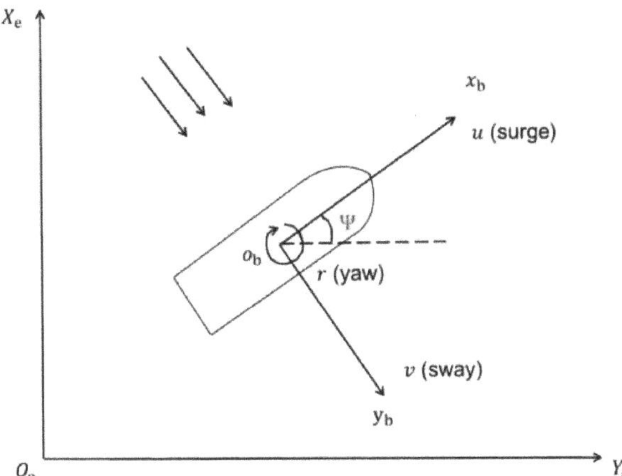

Figure 3. USV kinematic model.

Using machine learning algorithms is another way to deal with path planning in complex environments. In [90,91], a deep Q network (DQN) is developed for ship path tracking and heading control in calm seas without considering the actual sea environment. In [92], an improved DQN algorithm for vehicle path tracking using a nonlinear dynamic model was shown. It also ensures the safety and robustness of ship navigation in a windy and rough marine environment. Due to the complex environment of wind, waves, and currents, USV need to constantly change their speed and direction, considering the dynamics and kinematics of USV. In [93], a combination of Q-learning and neural networks (Q-NN) was proposed for path planning in unknown and harsh environments. It is able to avoid obstacles in real time and considers the kinematics of USVs, but it does not consider the effect of ocean currents. In [94], harsh environments such as wind, waves, and currents on USV path planning were considered. The optimization using the least squares strategy considers the dynamics and kinematics of the USV. However, the method is mainly applicable to static obstacle avoidance and textcolorred needs further breakthroughs for dynamic and unknown obstacles. In [95], an improved deep reinforcement learning algorithm (DRL) was presented. A two-dimensional ocean current model and a planar kinematic model are developed with energy cost as an important constraint. The global optimal path is solved by utilizing an improved reward function, action set, and state set and using B spline smoothing optimization.

4.2. Multi-Vessel Collision Hazard Avoidance

The multi-vessel collision problem is a common traffic accident at sea [96,97]. Currently, AIS ship trajectory data analysis is mainly used to require compliance with COLREG to prevent ship collisions as much as possible. The distance to the point of closest approach (DCPA) and the time to the point of closest approach (TCPA) are important indicators of a ship's collision risk [98]. Figure 4 shows the positional relationship between the ship and the target ship. The DCPA Collision Hazard Model is used to determine the distance from the vessel to the hazardous vessel for collision hazard assessment. In [99] DCPA and TCPA were combined to calculate hazard classes and analyses collision hazard classes in the context of ship safety domains, ship domains, and dynamic boundaries. A full-coverage algorithm is used to enable collision avoidance for multiple vessels. In [100], a full-coverage path planning algorithm based on deep reinforcement Q-learning neural network (DQN) was proposed, which uses convolutional layers to perceive the raster map and construct and train deep reinforcement learning. However, the traditional DQN suffers from low learning efficiency and low coverage. In [101,102], a method to improve its action selection mechanism was proposed. Training data is selected based on the priority of the task. A dynamic reward mechanism ensures that valuable data is trained and reduces the risk of dangerous USV action selection during the learning process. The method improves the convergence and efficiency of full-area coverage, and can effectively avoid other vessels. As shown in Figure 5, the improved DQN algorithm can achieve full coverage, which is able to avoid all obstacles and has fewer repeated paths. It is also feasible to apply this algorithm to collision avoidance of multiple vessels. In [103], a sliding window algorithm (SWA) was raised for accurately detecting ship collision avoidance behavior from AIS trajectory data, but the real-time performance is not strong enough. In [104], AIS and spatial-temporal was used to analyze the massive data of vessel traffic conditions in past waters, and used simulated vessel traffic demand and temporal characteristics to predict vessel trajectories. The method achieves global ship collision avoidance and cannot monitor the risk of ship collision in real-time. For the issue of weak real-time monitoring of the risk of ship collision, in [105], using AIS data and ACO-APF algorithm, real-time dynamic avoidance is performed to plan a safe route. It achieves real-time dynamic collision avoidance for multiple vessels and effectively reduces the risk of collision vessels. In [106], a practical rule-aware time-varying conflict risk (R-TCR) for multi-ship collision avoidance was raised, considering vessel maneuverability and COLREGs. In [107], a synergy ship domain (SSD) was proposed to construct a collision risk model in a two-ship situation, which can identify the risk of ship collision in real time. By simulating frontal collision, side collision, and cross-collision, it identifies the dynamic changes in collision risk. For the local optimal problem in multi-objective complex situations, ref. [108] investigated collision avoidance considering COLREGs for solving multi-ship encounter problems. To obtain an optimal decision to avoid ship collisions, multiple objective decision-making (MODM) was combined with CGA. The algorithm keeps the ship moving safely as it avoids dynamic obstacles. In the case of multiple objectives, it takes a global view rather than just considering optimizing one target. In [109], a RACO combination of the ant colony algorithm (ACO) with the rolling window method (RWM) [110] was proposed. To address the complexity of collisions, the algorithm considers three possible approaches: frontal collision, side collision, and random collision. By introducing secondary safety distances and complying with COLREGs, it achieves real-time collision avoidance safely and effectively and finds the globally optimal path.

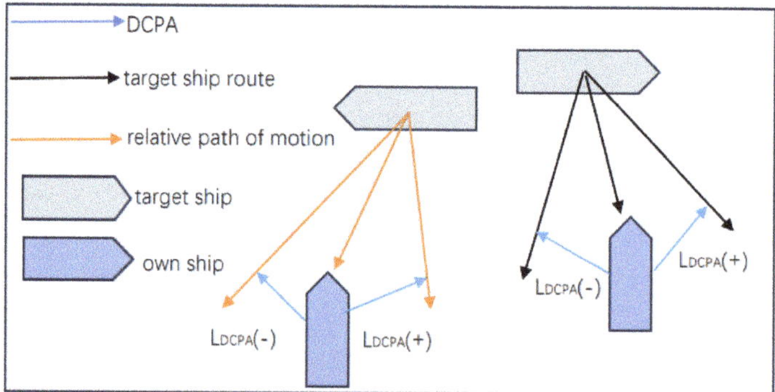

Figure 4. Vessel collision avoidance route map based on DCPA.

4.3. Summary of Hazard Avoidance Algorithm

The above study considers the problem of reactive avoidance of proximity in the design of path-planning algorithms. In the face of complex environments, a hybrid algorithm using hierarchical processing and a grid-based model prevents repeated searches and makes paths more efficient. In realistic marine environments, the effects of wind, waves, currents, and other factors are also considered to ensure safety and to produce globally optimal paths. Based on AIS data, COLREGs and machine learning algorithms combine historical data with real-time data for route planning to ensure safe navigation and avoid multi-vessel head-on collisions, cross-collisions, and random collisions. Table 3 compares different algorithms for hazard avoidance and considers the factors of the optimal path.

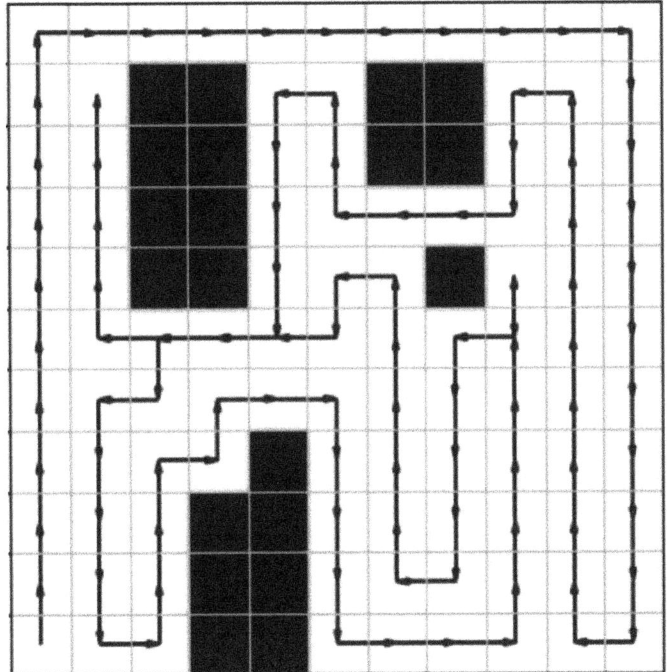

Figure 5. Path planning of improved DQN [102].

Table 3. Characteristics of different algorithms for hazard avoidance.

Reference	Method	Efficiency	DOA	COLREG	Real-Time	AIS	Wind	Current	Dynamics
[82]	MPP	F	T	F	T	F	F	F	F
[84]	GA-APF	T	F	T	T	F	F	F	F
[85]	PSO-GA	T	T	T	T	F	T	F	F
[86]	ICA	T	T	T	T	F	T	T	T
[89]	SSA	T	T	T	T	F	T	T	T
[87]	improved D*	T	T	T	F	F	T	T	T
[88]	VF-RRT*	T	T	T	T	F	T	T	T
[91]	DQN	T	F	F	T	F	F	F	T
[92]	IDQN	T	T	F	T	F	T	F	T
[90]	RL	F	F	T	T	F	F	F	F
[93]	Q-NN	T	F	F	T	F	T	F	T
[95]	DRL	T	F	F	T	F	T	T	T
[102]	CCPP	T	T	F	T	F	T	T	T
[103]	SWA	T	F	T	F	T	F	F	T
[105]	ACO-APF	T	T	T	T	T	F	F	T
[107]	SSD	F	T	T	T	T	F	F	T
[108]	MODM-CGA	T	T	F	T	T	F	F	F
[109]	ACO-RWM	T	T	T	T	F	F	F	T

Note: consider (T), no consider (F), DOA (dynamic obstacle avoidance), * (start).

5. Cluster Path Planning

USVs cluster path planning technology is rapidly developing as a core key technology. USVs cluster path planning is the key to cluster path generation, obstacle avoidance, collision avoidance, and other navigation and coordination tasks assignment. It is essentially a multi-constrained combinatorial optimization algorithm. The cluster intelligence optimization algorithm focusing on the bionic approach and the multi-task assignment strategy are essential assets to solve the cluster path planning problem [111].

5.1. Bionic Algorithm

For large-scale, high-dimensional, and non-linear USV cluster path planning problems, it is a good choice to use a bionic optimization algorithm. In [112], a firefly-based Approach (FA) for robot cluster path planning is proposed, where Firefly social behavior optimizes group behavior. Since the path planning problem is an NP complexity problem, multi-objective evolutionary algorithms are an effective way to solve this problem. In [113], a cluster path-planning algorithm based on ant colony optimization (ACO) [114] in a dynamic environment is proposed to perform path-planning optimization and establish the path-planning optimization objective function in a multi-tasking scenario. In [115], based artificial bee colony (ABC) algorithm [116], an efficient artificial bee colony (EABC) algorithm is proposed. It solves online path planning collision avoidance for multiple mobile robots by selecting appropriate objective functions for the target, obstacles, and robots. It utilizes elite individuals to maintain good evolution, improve performance and shorten path length. The method improves the quality of path planning for clusters, but it tends to fall into local optimality. For USV clusters caught in a local optimum, ref. [117] proposed an improved particle swarm optimization(PSO) algorithm based on an adaptive sensitivity decision operator that is more adapted to 3D path. It solves the drawback that traditional PSO is prone to fall into local optimality [118], improves the convergence accuracy and cooperative operation of clusters, and predicts the globally optimal planning. In [119], it proposed a decision support algorithm to use an artificial fish swarming algorithm (AFSA) [120] based on AIS data for path-planning decisions in USV clusters. It can calculate the optimal avoidance turn to time and avoidance angle, and the algorithm possesses excellent robustness and converges quickly. In [121], a hybrid improved artificial fish swarm algorithm (HIAFSA) was proposed to address the problem of falling into local optima. It combines the A* algorithm with AFSA for further optimization in global suboptimal paths. It introduces decay functions to enhance the visual range and motion steps to improve the convergence speed. The algorithm offers access to local optimum avoidance, convergence speed, and accuracy. In [122], the salp swarm algorithm (SSA) [123] optimization algorithm was used to divide the task area using Voronoi diagrams to achieve collaborative path planning for USV clusters. This optimization algorithm avoids repeated searches, improves search efficiency and accuracy, and effectively avoids collisions between obstacles and USVs. However, the method does not consider the influence of COLREGs and the actual environment at sea. To consider USV clusters in realistic environmental scenarios at sea, ref. [124] proposed a simulated annealing-bacterial foraging optimization algorithm (SA-BFO). The bacterial foraging optimization algorithm (BFO) is a population intelligence optimization algorithm with good search efficiency and robustness [125]. The hybrid approach, based on COLREGs, can perform real-time avoidance of dynamic obstacles using USV cooperative control. It is able to plan collision-free paths efficiently, moving away from local optima and converging to global optima.

5.2. Multi-Objective Task Assignment Algorithm

When faced with multiple target missions, it is difficult for a single USV to perform complex tasks in unknown underwater environments. Then, USV clustering can improve the performance of the system, shorten the mission time and increase the probability of the search success. Multi-objective search task assignment is the key to USV clustering path collaboration [126,127]. As shown in Figure 6, multiple search tasks are assigned to multiple specific USVs. Currently, reinforcement learning algorithms (RL) are more suitable for task assignment, and path planning [128,129]. In [130], combining HER with DQN and introducing a reward mechanism can increase the validity of the sample and the speed of convergence. In [131], a pheromone mechanism using ACO on the classical Q-learning approach is investigated. It solves the information-sharing problem in reinforcement learning systems and improves the efficiency of robot cluster path planning. In [132], a hexagonal area search (HAS-DQN) was proposed. It solves the collaborative path planning problem for UAVs, maximizes the data collected by UAVs, minimizes the total energy consumption, and extends the lifetime. However, due to the effects of offshore winds, waves, currents, etc., the above multi-task and multi-objective optimization decisions based on reinforcement learning do not work well when applied to USVs. In [133], an improved self-organizing map (SOM) [134] with collision avoidance capability was proposed, based on a fast marching square (FMS) [135] path planning algorithm. It is adapted to multi-task assigned USVs cluster for complex tasks, such as maritime patrol search and rescue and environmental detection. The algorithm first assigns tasks to each USV and enables the function of fast task assignment and optimized execution sequences. Not only does it achieve avoidance of all static and dynamic obstacles, but it also takes into account COLREGs. However, the algorithm does not consider the complex terrain and time-varying currents under the sea. In [136], an improved the K-means algorithm to accommodate unsupervised learning of competing strategies was proposed. The algorithm first assigns different tasks to multiple USVs and performs the task assignment by the SOM algorithm. It can autonomously undertake complex maritime missions in a limited environment and verifies its effectiveness by simulating it in a real marine environment. In [137], an improved SOM combined with spectral clustering (SC), is used to solve collaborative path planning for USVs cluster and multi-task allocation. The method allows for the selection of the globally optimal path with minimal energy consumption in a collision-free manner. To make the generated routes smoother, a dual smoothing strategy with B-samples and indirect adaptive disturbance observer-based line-of-sight (IADO-LOS) [138] are used to achieve the precise path following of the USV. The method also accounts for the effect of ocean currents on ship driving and is more in line with the realistic environment at sea.

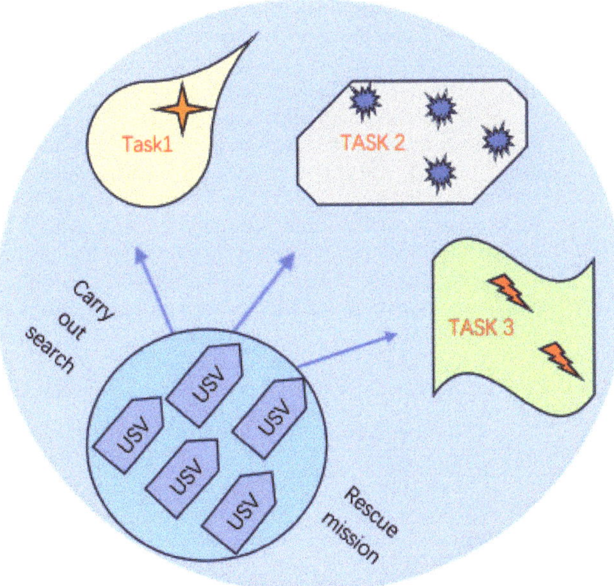

Figure 6. USV cluster multi-objective task assignment.

5.3. Summary of Cluster Path Planning Algorithm

The above study uses the bionic optimization concept for path planning algorithms in USV clusters to optimize multi-objective paths for multi-task, multi-constraint models. Using AIS big data, COLREG and machine learning algorithms are used to combine historical and real-time data for path planning to ensure safe navigation while completing multi-tasking assignment strategies at sea. Table 4 compares different algorithms for the USV cluster and considers the factors of the optimal path.

Table 4. Characteristics of different algorithms for clusters.

References	Methods	Mass	Efficiency	Convergence	COLREG	Real-Time	Muti	Environment
[112]	FA	T	T	T	F	F	T	F
[113]	ACO	F	T	F	F	F	F	T
[115]	EABC	T	T	T	F	T	F	F
[117]	improved PSO	T	T	T	F	T	T	F
[119]	AFSA	F	T	T	F	T	F	F
[121]	HIAFSA	T	T	T	F	T	T	F
[123]	SSA	T	T	T	F	T	T	F
[124]	SA-BFO	T	T	T	T	T	T	F
[128]	RL	T	T	T	F	F	T	T
[130]	HER-DQN	T	T	T	F	T	F	F
[132]	HAS-DQN	T	T	T	F	T	F	F
[133]	improved SOM	F	T	T	F	T	T	T
[137]	SOM-SC	T	T	T	F	T	T	T

Note: consider (T), no consider (F), Muti (muti-objective).

6. Conclusions

Path planning is a hot and complex area of research in the USV field and a key technology to ensure its autonomy in the marine environment. This paper mainly reviews and analyzes the literature related to USV path planning in recent years. Path planning is divided into four aspects: global path planning, local path planning, hazard avoidance with approximate responses, and path planning under clustering. The advantages and disadvantages of different algorithms for global and local path planning are analyzed in Sections 2 and 3. As shown in Table 5, global path planning can cover the whole map and find the global optimal solution. However, the computational complexity is high, the operation efficiency is slow, and it cannot cope with dynamic environmental changes. The local path planning algorithm is simple, responsive, and can adapt to the dynamic environment. However, it cannot handle the dead ends and obstacles in global path planning, and it is easy to fall into local optimal solutions. In addition, most of the improved algorithms for global and local path planning ignore the influence of real environments, such as ocean currents and wind waves. Modeling USV kinematics and dynamic constraints is important for path planning when considering the effects of wind, waves, and currents on safe avoidance and energy consumption. Using USV clusters with multi-task assignment and path collaboration strategies is also the right choice when facing multiple tasks.

Table 5. Global path planning strategy and local path planning strategy contrast and analysis.

Characteristics	Global Path Planning Strategy	Local Path Planning Strategy
Information	known and all	sensor acquisition
Function	global optimization search	local optimization search
Calculation volume	complex and slow	simple and quick
Application scenario	static environments	dynamic environments

7. Prospects

Firstly, the future direction of current path planning optimization algorithms is multi-algorithm fusion, based on traditional optimization algorithms combined with deep reinforcement learning, digital twins, and other artificial intelligence optimization algorithms, which holds the promise of dynamic and online path planning.

Secondly, the evaluation system of the path and the collision risk assessment models need to be further improved. Under the premise of ensuring safety, it considers factors such as path length, smoothness, time consumption, and efficiency, and also further considers the motion performance of angular velocity, acceleration, and turning angle of the ship.

Finally, USV cluster collaboration for path planning is also a hot research topic for the future. How to make USV clusters perform multiple tasks while performing integrated obstacle avoidance is also a problem that needs to be tackled in the future. Improving the perception of the navigation area and the autonomous decision-making capability is a significant part of the solution to this problem.

Author Contributions: Conceptualization, B.X.; methodology, B.X., M.Y. and B.L.; writing—original draft preparation, M.Y., B.X. and Z.L.; writing—review and editing, B.X., M.Y. and B.L.; supervision, B.X., Y.S., Z.L. and Y.T.; project administration, B.X., Y.T. and Y.S.; funding acquisition, B.X. All authors have read and agreed to the published version of the manuscript.

Funding: This research was funded by Shanghai Science and Technology Committee (STCSM) Local Universities Capacity-building Project (No. 22010502200).

Institutional Review Board Statement: Not applicable.

Informed Consent Statement: Not applicable.

Data Availability Statement: The data are available on request.

Acknowledgments: The authors would like to express their gratitude for the support of Fishery Engineering and Equipment Innovation Team of Shanghai High-level Local University.

Conflicts of Interest: The authors declare no conflict of interest.

References

1. Yu, W.; Low, K.H.; Chen, L. Cooperative path planning for heterogeneous unmanned vehicles in a search-and-track mission aiming at an underwater target. *IEEE Trans. Veh. Technol.* **2020**, *69*, 6782–6787.
2. Zhou, C.H.; Gu, S.D.; Wen, Y.Q.; Du, Z.; Xiao, C.S.; Huang, L.; Zhu, M. The review unmanned surface vehicle path planning: Based on multi-modality constraint. *Ocean Eng.* **2020**, *200*, 107043. [CrossRef]
3. Meng, J.E.; Ma, C.; Liu, T.H.; Gong, H.B. Intelligent motion control of unmanned surface vehicles: A critical review. *Ocean Eng.* **2023**, *280*, 114562.
4. Liu, X.; Ye, X.M.; Wang, Q.B.; Li, W.G.; Gao, H.L. Review on the research of local path planning algorithms for unmanned surface vehicles. *Chin. J. Ship Res.* **2021**, *16* (Suppl. 1), 12–14.
5. Wu, G.X.; Xu, T.T.; Sun, Y.S.; Zhang, J.W. Review of multiple unmanned surface vessels collaborative search and hunting based on swarm intelligence. *Int. J. Adv. Robot. Syst.* **2022**, *19*, 17298806221091885. [CrossRef]
6. Cheng, C.X.; Sha, Q.X.; He, B.; Li, G.L. Path planning and obstacle avoidance for AUV: A review. *Ocean Eng.* **2021**, *235*, 109355. [CrossRef]
7. Ülkü, Ö.; Melih, A.; Tarık, A. A review of path planning algorithms in maritime autonomous surface ships: Navigation safety perspective. *Ocean Eng.* **2022**, *252*, 111010.
8. Liu, L.X.; Wang, X.; Yang, X.; Liu, H.J.; Li, J.P.; Wang, P.F. Path planning techniques for mobile robots: Review and prospect. *Expert Syst. Appl.* **2023**, *227*, 120254. [CrossRef]
9. Liu, J.; Wang, J. Overview of obstacle avoidance path planning algorithm for unmanned surface vehicle. *J. Comput. Appl. Softw.* **2020**, *37*, 1–10+20.. (In Chinese) [CrossRef]
10. Abdulsaheb, J.A.; Kadhim, D.J. Classical and Heuristic Approaches for Mobile Robot Path Planning: A Survey. *Robotics* **2023**, *12*, 93. [CrossRef]
11. Lin, S.; Liu, A.; Wang, J.; Kong, X. A Review of Path-Planning Approaches for Multiple Mobile Robots. *Machines* **2022**, *10*, 773. [CrossRef]
12. Gul, F.; Mir, I.; Abualigah, L.; Sumari, P.; Forestiero, A. A Consolidated Review of Path Planning and Optimization Techniques: Technical Perspectives and Future Directions. *Electronics* **2021**, *10*, 2250. [CrossRef]

13. Gan, L.; Yan, Z.; Zhang, L.; Liu, K.Z.; Zheng, Y.Z.; Zhou, C.H.; Shu, Y.Q. Ship path planning based on safety potential field in inland rivers. *J. Ocean Eng.* **2022**, *260*, 111928. [CrossRef]

14. Dijkstra, E.W. A Note on Two Problems in Connection with Graphs. *J. Numer. Math.* **1959**, *1*, 269–271. [CrossRef]

15. Borkar, P.; Sarode, M.V.; Malik, L.G. Acoustic signal based optimal route selection problem: performance comparison of multi-attribute decision making methods. *J. Trans. Internet Inf. Syst.* **2016**, *10*, 647–669.

16. Sekaran, J.F.; Kaluvan, H. Path Planning of Robot Using Modified Dijkstra Algorithm. In Proceedings of the 2018 National Power Engineering Conference (NPEC), Madurai, India, 9–10 March 2018.

17. Guo, Q.; Zhang, Z.; Xu, Y. Path-Planning of Automated Guided Vehicle Based on Improved Dijkstra Algorithm. In Proceedings of the 2017 29th Chinese Control And Decision Conference (CCDC), Chongqing, China, 28–30 May 2017.

18. Kang, W.H.; Xu, Y.C. Hierarchical Dijkstra algorithm for node-constrained shortest paths. *J. South China Univ. Technol.* **2017**, *45*, 66–73.

19. Zhuang, J.Y.; Wan, L.; Liao, Y.L.; Sun, H.B. Research on global path planning of surface unmanned boats based on electronic nautical charts. *J. Comput. Sci.* **2011**, *38*, 211–214.

20. Wang, H.L.; Mao, W.G.; Eriksson, L. A Three-Dimensional Dijkstra's algorithm for multi-objective ship voyage optimization. *J. Ocean Eng.* **2019**, *186*, 106131. [CrossRef]

21. Koenig, S.; Likhachev, M.; Furcy, D. Lifelong planning A*. *J. Artif. Intell.* **2004**, *155*, 93–146. [CrossRef]

22. Yu, J.B.; Yang, M.; Zhao, Z.Y.; Wang, X.Y.; Bai, Y.B.; Wu, J.G.; Xu, J.P. Path planning of unmanned surface vessel in an unknown environment based on improved D*Lite algorithm. *J. Ocean Eng.* **2022**, *266*, 112873. [CrossRef]

23. Hart, P. Nearest neighbor pattern classification. *J. IEEE Trans. Inf. Theory* **1967**, *13*, 21–27.

24. Sang, H.Q.; You, Y.S.; Sun, X.J.; Zhou, Y.; Liu, F. The hybrid path planning algorithm based on improved A* and artificial potential field for unmanned surface vehicle formations. *Ocean Eng.* **2021**, *223*, 107809. [CrossRef]

25. Yang, J.M.; Tseng, C.M.; Tseng, P.S. Path Planning on Satellite Images for Unmanned Surface Vehicles. *Int. J. Nav. Archit. Ocean Eng.* **2015**, *7*, 87–99. [CrossRef]

26. Bayili, S.; Polat, F. Limited-damage A*: A path search algorithm that considers damage as a feasibility criterion. *Knowl. Based Syst.* **2011**, *24*, 501–512. [CrossRef]

27. Singh, Y.; Sharma, S.; Sutton, R.; Hatton, D.; Khan, A. A constrained A* approach towards optimal path planning for an unmanned surface vehicle in a maritime environment containing dynamic obstacles and ocean currents. *Ocean Eng.* **2018**, *169*, 187–201. [CrossRef]

28. Campbell, S.; Abu-Tair, M.; Naeem, W. An automatic COLREGs-compliant obstacle avoidance system for an unmanned surface vehicle. *J. Eng. Marit. Environ.* **2014**, *228*, 108–121. [CrossRef]

29. Wang, Y.L.; Yu, X.M.; Liang, X. Design and Implementation of Global Path Planning System for Unmanned Surface Vehicle among Multiple Task Points. *J. Int. J. Veh. Auton. Syst.* **2018**, *14*, 82–105. [CrossRef]

30. Holland, J. Adaptation in Natural and Artificial Systems. In *Applying Genetic Algorithm to Increase the Efficiency of a Data Flow-Based Test Data Generation Approach*; University of Michigan Press: Ann Arbor, MI, USA, 1975.

31. Tsai, C.C.; Huang, H.C.; Chan, C.K. Parallel elite genetic algorithm and its application to global path planning for autonomous robot navigation. *IEEE Trans. Ind. Electron* **2011**, *58*, 4813–4821. [CrossRef]

32. Xiao, C.H.; Zou, Y.Y.; Li, S.Y. Path planning of multi–target points for multi UAV in obstacle area. *Space Control Technol. Appl.* **2019**, *45*, 46–52.

33. Yao, Z.; Ma, L. A Static Environment-Based Path Planning Method by Using Genetic Algorithm. In Proceedings of the 2010 International Conference on Computing, Control and Industrial Engineering, Wuhan, China, 5–6 June 2010.

34. Huang, S.Z.; Tian, J.W.; Qian, L.; Wang, Q.; Su, Y. Unmanned aerial vehicle path planning based on improved genetic algorithm. *J. Comput. Appl.* **2021**, *41*, 390–397.

35. Hao, K.; Zhao, J.L.; Li, Z.S.; Liu, Y.L.; Zhao, L. Dynamic path planning of a three-dimensional underwater AUV based on an adaptive genetic algorithm. *Ocean Eng.* **2022**, *263*, 112421. [CrossRef]

36. Hu, G.K.; Zhong, J.H.; Li, Y.Z.; Li, W.H. UAV 3D path planning based on IPSO-GA algorithm. *J. Mod. Electron. Tech.* **2023**, *46*, 115–120.

37. Nazarahari, M.; Khanmirza, E.; Doostie, S. Multi-objective multi-robot path planning in continuous environment using an enhanced genetic algorithm. *Expert Syst. Appl.* **2019**, *115*, 106–120. [CrossRef]

38. Li, K.R.; Hu, Q.Q.; Liu, J.P. Path planning of mobile robot based on improved multi-objective genetic algorithm Wireless. *Commun. Mob. Comput.* **2021**, *2021*, 8836615.

39. Mcculloch, W.S.; Pitts, W. A logical calculus of the ideas immanent in nervous activity. *Bull. Math. Biophys.* **1943**, *5*, 115–133. [CrossRef]

40. Chen, Y.; Liang, J.; Wang, Y.; Pan, Q. Autonomous mobile robot path planning in unknown dynamic environments using neural dynamics. *Soft Comput.* **2020**, *24*, 13979–13995. [CrossRef]

41. Praczyk, T. Neural anti-collision system for autonomous surface vehicle. *Neurocomputing* **2015**, *149*, 559–572. [CrossRef]

42. Xu, Q.; Zhang, C.; Zhang, L. Deep Convolutional Neural Network Based Unmanned Surface Vehicle Maneuvering. In Proceedings of the 2017 Chinese Automation Congress (CAC), Jinan, China, 20–22 October 2017.

43. Liu, Y.; Zheng, Z.; Qin, F.Y.; Zhang, X.Y.; Yao, H.N. A residual convolutional neural network based approach for real-time path planning. *Knowl. Based Syst.* **2022**, *242*, 108400. [CrossRef]

44. Sperduti, A.; Starita, A. Supervised neural networks for the classification of structures. *IEEE Trans. Neural Netw.* **1997**, *8*, 714–735 [CrossRef]

45. Waikhom, L.; Patgiri, R. A survey of graph neural networks in various learning paradigm: methods, applications, and challenges. *Artif. Intell. Rev.* **2022**, *56*, 6295–6364. [CrossRef]

46. Kipf, T.N.; Welling, M. Semi-supervised classification with graph convolutional networks. *arXiv* **2017**, arXiv:1609.02907.

47. Zhao, L.; Song, Y.; Zhang, C.; Liu, Y.; Wang, P.; Lin, T.; Deng, M.; Li, H.F. T-GCN: A temporal graph convolutional network for traffic prediction. *IEEE Trans. Intell. Transp. Syst.* **2019**, *21*, 3848–3858. [CrossRef]

48. Luo, C.; Yang, S.X. A Real-Time Cooperative Sweeping Strategy for Multiple Cleaning Robots. In Proceedings of the IEEE Internatinal Symposium on Intelligent Control, Vancouver, BC, Canada, 30–30 October 2002.

49. Xu, P.F.; Ding, Y.X.; Luo, J.C. Complete Coverage Path Planning of an Unmanned Surface Vehicle Based on a Complete Coverage Neural Network Algorithm. *J. Mar. Sci. Eng.* **2021**, *9*, 1163. [CrossRef]

50. Khatib, O. Real-Time Obstacle Avoidance for Manipulators and Mobile Robots. *Int. J. Robot. Res.* **1986**, *5*, 90–98. [CrossRef]

51. Mielniczuk, M. A Method of Artificial Potential Fields for the Determination of Ship's Safe Trajectory. *Sci. J. Marit. Univ. Szczec.* **2017**, *51*, 45–49.

52. Azmi, M.Z.; Ito, T. Artificial potential field with discrete map transformation for feasible indoor path planning. *J. Appl. Sci.* **2020**, *10*, 8987. [CrossRef]

53. Lazarowska, A. Discrete Artificial Potential Field Approach to Mobile Robot Path Planning. *IFAC-PapersOnLine* **2019**, *52*, 277–282. [CrossRef]

54. Wu, P.; Xie, S.; Luo, J.; Qu, D.; Li, Q. The USV path planning based on the combinatorial algorithm. *Rev. Tec. Fac. Ing. Univ. Zulia* **2015**, *38*, 62–70.

55. He, Z.H.; Chu, X.M.; Liu, C.G.; Wu, W.X. A novel model predictive artificial potential field based ship motion planning method considering COLREGs for complex encounter scenarios. *ISA Trans.* **2023**, *134*, 58–73. [CrossRef]

56. Lavalle, S.M. Rapidly-Exploring Random Trees: A New Tool for Path Planning. *Res. Rep.* **1998**. Available online: https://cir.nii.ac.jp/crid/1573950399665672960 (accessed on 22 May 2023).

57. Karaman, S.; Frazzoli, E. Optimal Kinodynamic Motion Planning Using Incremental Sampling-based Methods. In Proceedings of the 49th IEEE Conference on Decision and Control (CDC), Atlanta, GA, USA, 15–17 December 2010.

58. Feng, N. *A Path Planning Method for Autonomous Mobile Robot Based on RRT Algorithm*; Dalian University of Technology: Dalian, China, 2014.

59. Jeong, I.B.; Lee, S.J.; Kim, J.H. Quick-RRT*: Triangular inequality-based implementation of RRT* with improved initial solution and convergence rate. *Expert Syst.* **2019**, *123*, 82–90. [CrossRef]

60. Kuffner, J.J.; LaValle, S.M. RRT-connect: An efficient approach to single-query path planning. *IEEE Int. Conf. Robot. Autom. Symp. Proc.* **2000**, *2*, 995–1002.

61. Ding, J.; Zhou, Y.X.; Huang, X.; Song, K.; Lu, S.Q.; Wang, L.S. An improved RRT* algorithm for robot path planning based on path expansion heuristic sampling. *J. Comput. Sci.* **2023**, *67*, 101937. [CrossRef]

62. Ma, G.J.; Duan, Y.L.; Li, M.Z.; Xie, Z.B.; Zhu, J. A probability smoothing Bi-RRT path planning algorithm for indoor robot. *Future Gener. Comput. Syst.* **2023**, *143*, 349–360. [CrossRef]

63. Qureshi, A.H.; Ayaz, Y. Potential functions based sampling heuristic for optimal path planning. *Autom. Robots* **2016**, *40*, 1079–1093. [CrossRef]

64. Zhang, X.; Zhu, T.; Du, L.; Hu, Y.; Liu, H. Local Path Planning of Autonomous Vehicle Based on an Improved Heuristic Bi-RRT Algorithm in Dynamic Obstacle Avoidance Environment. *Sensors* **2022**, *22*, 7968. [CrossRef]

65. Fiorini, P.; Shiller, Z. Motion planning in dynamic environments using velocity obstacles. *Int. J. Robot. Res.* **1998**, *17*, 760–772. [CrossRef]

66. Zhuang, J.Y.; Zhang, G.C. Obstacle avoidance method for USV. *J. Southeast Univ.* **2013**, *43*, 126–130.

67. Cho, Y.; Han, J.; Kim, J.; Lee, P.; Park, S.B. Experimental validation of a velocity obstacle based collision avoidance algorithm for unmanned surface vehicles. *IFAC-PapersOnLine* **2019**, *52*, 329–334. [CrossRef]

68. Wang, J.; Wang, R.; Lu, D.; Zhou, H.; Tao, T. USV Dynamic Accurate Obstacle Avoidance Based on Improved Velocity Obstacle Method. *Electronics* **2022**, *11*, 2720. [CrossRef]

69. Song, L.F.; Su, Y.R.; Dong, Z.P.; Shen, W.; Xiang, Z.Q.; Mao, P.X. A Two-Level Dynamic Obstacle Avoidance Algorithm for Unmanned Surface Vehicles. *J. Ocean Eng.* **2018**, *170*, 351–360. [CrossRef]

70. Du, K.J.; Mao, Y.S.; Xiang, Z.Q.; Zhou, Y.Q.; Song, L.F.; Liu, B. A dynamic obstacle avoidance method of USV based on COLREGS. *J. Ship Eng.* **2015**, *3*, 119–124. (In Chinese)

71. Bareiss, D.; Berg, V.D. Generalized reciprocal collision avoidance. *J. Robot. Res.* **2015**, *34*, 1501–1514. [CrossRef]

72. Huang, Y.M.; Chen, L.Y.; Gelder, V.P. Generalized velocity obstacle algorithm for preventing ship collisions at sea. *J. Ocean Eng.* **2019**, *173*, 142–156. [CrossRef]

73. Su, Y.M.; Luo, J.; Zhuang, J.Y.; Song, S.Q.; Huang, B.; Zhang, L. A constrained locking sweeping method and velocity obstacle based path planning algorithm for unmanned surface vehicles in complex maritime traffic scenarios. *J. Ocean Eng.* **2023**, *279*, 113538. [CrossRef]

74. Fox, D.; Burgard, W.; Thrun, S. The dynamic window approach to collision avoidance. *IEEE Robot. Autom. Mag.* **1997**, *4*, 23–33. [CrossRef]

75. Brock, O.; Khatib, O. High-Speed Navigation Using the Global Dynamic Window Approach. In Proceedings of the 1999 IEEE International Conference on Robotics and Automation, Detroit, MI, USA, 10–15 May 1999.

76. Lin, X.G.; Fu, Y. Research of USV Obstacle Avoidance Strategy Based on Dynamic Window. In Proceedings of the 2017 IEEE International Conference on Mechatronics and Automation (ICMA), Takamatsu, Japan, 6–9 August 2017.

77. Yu, X.Y.; Zhu, Y.C.; Lu, L.; Ou, L.L. Dynamic window with virtual goal (DW-VG): A new reactive obstacle avoidance approach based on motion prediction. *J. Robot.* **2019**, *37*, 1438–1456.

78. Zhang, Y.Y.; Qu, D.; Ke, J.; Li, S.M. Dynamic obstacle avoidance of unmanned surface boats based on velocity barrier method and dynamic window method. *J. Shanghai Univ.* **2017**, *23*, 1–16.

79. Wang, N.; Gao, Y.; Zheng, Z.; Zhao, H.; Yin, J. A Hybrid Path-Planning Scheme for an Unmanned Surface Vehicle. In Proceedings of the 2018 Eighth International Conference on Information Science and Technology (ICIST), Cordoba, Granada, and Seville, Spain, 30 June–6 July 2018.

80. Han, S.; Wang, L.; Wang, Y.T.; He, H.C. A dynamically hybrid path planning for unmanned surface vehicles based on non-uniform Theta* and improved dynamic windows approach. *Ocean Eng.* **2022**, *257*, 111655. [CrossRef]

81. Wang, Q.; Li, J.; Yang, L.; Yang, Z.; Li, P.; Xia, G. Distributed Multi-Mobile Robot Path Planning and Obstacle Avoidance Based on ACO–DWA in Unknown Complex Terrain. *Electronics* **2022**, *11*, 2144. [CrossRef]

82. Wang, N.; Jin, X.Z.; Er, J.M. A multilayer path planner for a USV under complex marine environments. *Ocean Eng.* **2019**, *184*, 1–10. [CrossRef]

83. Sethian, J.A. A fast marching level set method for monotonically advancing fronts. *Proc. Natl. Acad. Sci. USA* **1996**, *93*, 1591–1595. [CrossRef]

84. Wang, A.G.; Zhi, P.F.; Zhu, W.L.; Qiu, H.Y.; Wang, H.; Wang, W.R. Path Planning of Unmanned Surface Vehicle Based on a Algorithm Optimization Considering the Influence of Risk Factors. In Proceedings of the 2021 4th IEEE International Conference on Industrial Cyber-Physical Systems (ICPS), Victoria, BC, Canada, 10–12 May 2021.

85. Zhao, W.; Wang, Y.; Zhang, Z.; Wang, H. Multicriteria Ship Route Planning Method Based on Improved Particle Swarm Optimization–Genetic Algorithm. *J. Mar. Sci. Eng.* **2021**, *9*, 357. [CrossRef]

86. Yin, S.; Mao, J.; Li, B. AUV 3D Path Planning Based on Improved Empire Competition Algorithm in Ocean Current Environment. In Proceedings of the 6th International Conference on Computing, Control and Industrial Engineering (CCIE 2021), Hangzhou, China, 15–16 January 2021; pp. 36–49.

87. Sun, B.; Zhang, W.; Li, S.Q.; Zhu, X.X. Energy optimised D* AUV path planning with obstacle avoidance and ocean current environment. *J. Navig.* **2022**, *75*, 685–703. [CrossRef]

88. Yu, J.B.; Chen, Z.H.; Zhao, Z.Y.; Yao, P.; Xu, J.P. A traversal multi-target path planning method for multi-unmanned surface vessels in space-varying ocean current. *Ocean Eng.* **2023**, *278*, 114423. [CrossRef]

89. Li, B.; Mao, J.; Yin, S.; Fu, L.; Wang, Y. Path Planning of Multi-Objective Underwater Robot Based on Improved Sparrow Search Algorithm in Complex Marine Environment. *J. Mar. Sci. Eng.* **2022**, *10*, 1695. [CrossRef]

90. Zhang, R.B.; Tang, P.P.; Yang, G.; Li, X.Y.; Shi, C.T. Convergence analysis of adaptive hazard avoidance decision process for surface unmanned boats. *J. Comput. Res. Dev.* **2014**, *12*, 2644–2652. (In Chinese)

91. Sivaraj, S.; Rajendran, S.; Prasad, L.P. Data driven control based on deep Q-network algorithm for heading control and path following of a ship in calm water and waves. *Ocean Eng.* **2022**, *259*, 111802. [CrossRef]

92. Rohit, D.R.S. Sanjeev, K.; Md Shadab, A.; Abhilash, S. Deep reinforcement learning based controller for ship navigation. *Ocean Eng.* **2023**, *273*, 113937.

93. Bhopale, P.; Kazi, F.; Singh, N. Reinforcement learning based obstacle avoidance for autonomous underwater vehicle. *J. Mar. Sci. Appl.* **2019**, *18*, 228–238. [CrossRef]

94. Vibhute, S. Adaptive Dynamic Programming Based Motion Control of Autonomous Underwater Vehicles. In Proceedings of the 2018 5th International Conference on Control, Decision and Information Technologies (CoDIT), Thessaloniki, Greece, 10–13 April 2018; pp. 966–971.

95. Li, P.J.; Yan, T.W.; Yang, S.T.; Li, R.; Du, J.F.; Qian, F.F.; Liu, Y.T. Energy-optimal Path Planning Algorithm for Unmanned Surface Vessel Based on Reinforcement Learning. *J. Unmanned Undersea Syst.* **2023**, *31*, 237–243. (In Chinese)

96. Huang, Y.M.; Chen, L.Y.; Chen, P.F. Ship collision avoidance methods: State-of-the-art. *Saf. Sci.* **2020**, *121*, 451–473. [CrossRef]

97. Melih, A.; Petter, S.; Tor, A.J. Collaborative collision avoidance for Maritime Autonomous Surface Ships: A review. *Ocean Eng.* **2022**, *250*, 110920.

98. Zhang, X.Y.; Wang, C.B.; Jian, L.L.; An, L.X.; Yang, R. Collision-avoidance navigation systems for Maritime Autonomous Surface Ships: A state of the art survey. *Ocean Eng.* **2021**, *235*, 109380. [CrossRef]

99. Zhu, Z.; Lyu, H.; Zhang, J.; Yin, Y. Decision supporting for ship collision avoidance in restricted waters. *Int. J. Simul. Process Model.* **2020**, *15*, 40–51.

100. Dong, J.X. Area Coverage Path Planning of UAV Based on Deep Reinforcement Learning. *J. Ind. Control Comput.* **2021**, *34*, 80–82.

101. Yang, L.; Xing, B.W.; Zhang, Z.Y.; Li, L.H. An Algorithm of Complete Coverage Path Planning Based on Improved DQN. *Lect. Notes Electr. Eng.* **2022**, *845*, 3728–3738.

102. Xing, B.W.; Wang, X.; Yang, L.; Liu, Z.; Wu, Q. An Algorithm of Complete Coverage Path Planning for Unmanned Surface Vehicle Based on Reinforcement Learning. *J. Mar. Sci. Eng.* **2023**, *11*, 645. [CrossRef]

103. Rong, H.; Teixeira, A.P.; Guedes, S.C. Ship collision avoidance behaviour recognition and analysis based on AIS data. *J. Ocean Eng.* **2022**, *245*, 110479. [CrossRef]
104. Zhang, L.; Meng, Q. A Big AIS data based spatial-temporal analyses of ship traffic in Singapore port waters Transportation Research. *Logist. Transp. Rev.* **2019**, *129*, 287–304. [CrossRef]
105. Gao, P.; Zhou, L.; Zhao, X.; Shao, B. Research on ship collision avoidance path planning based on modified potential field ant colony algorithm. *Ocean Coast. Manag.* **2023**, *235*, 106482. [CrossRef]
106. Li, M.; Mou, J.; Chen, L.; He, Y.; Huang, Y. A rule-aware time-varying conflict risk measure for MASS considering maritime practice Reliab. *J. Eng. Syst.* **2021**, *215*, 107816.
107. Zhou, X.Y.; Liu, Z.J.; Wang, F.W.; Ni, S.K. Collision Risk Identification of Autonomous Ships Based on the Synergy Ship Domain. In Proceedings of the 2018 Chinese Control And Decision Conference (CCDC), Shenyang, China, 9–11 June 2018; pp. 6746–6752.
108. Ning, J.; Chen, H.; Li, T.; Li, W.; Li, C. COLREGs-compliant unmanned surface vehicles collision avoidance based on multi-objective genetic algorithm. *IEEE Access* **2020**, *8*, 190367–190377. [CrossRef]
109. Jin, Q.B.; Tang, C.N.; Cai, W. Research on dynamic path planning based on the fusion algorithm of improved ant colony optimization and rolling window method. *IEEE Access* **2022**, *10*, 28322–28332. [CrossRef]
110. Zhang, C.G.; Xi, Y.G. Robot rolling path planning based on locally detected information. *Acta Autom. Sin.* **2003**, *29*, 38–44.
111. Gao, M.; Tang, H.; Zhang, P. Current status of research on robot cluster path planning technology. *J. Natl. Univ. Def. Technol.* **2021**, *43*, 127–138.
112. Hidalgo-Paniagua, A.; Vega-Rodríguez, M.A.; Ferruz, J.; Pavón, N. Solving the multi-objective path planning problem in mobile robotics with a firefly-based approach. *J. Soft Comput.* **2017**, *21*, 949–964. [CrossRef]
113. Su, M.M.; Cheng, Y.M.; Hu, J.W.; Zhao, C.H.; Jia, C.J.; Xu, Z.; Zhang, J.F. Combined optimization of swarm task allocation and path planning based on improved ant colony algorithm. *J. Unmanned Syst. Technol.* **2021**, *4*, 40–50. (In Chinese)
114. Dorigo, M.; Blum, C. Ant colony optimization theory: A survey. *Theor. Comput. Sci.* **2005**, *344*, 243–278. [CrossRef]
115. Liang, J.H.; Lee, C.H. Efficient collision-free path-planning of multiple mobile robots system using efficient artificial bee colony algorithm. *Adv. Eng. Softw.* **2015**, *79*, 45–56. [CrossRef]
116. Karaboga, D. *An Idea Based on Honey Bee Swarm for Numerical Optimization; Technical Report;* R. Computers Engineering Department, Engineering Faculty, Erciyes University: Kayseri, Türkiye, 2005.
117. Liu, Y.; Zhang, X.; Guan, X.; Delahaye, D. Adaptive sensitivity decision based path planning algorithm for unmanned aerial vehicle with improved particle swarm optimization. *Aerosp. Sci. Technol.* **2016**, *58*, 92–102. [CrossRef]
118. Gen, Q.; Zhao, Z. A Kind of Route Planning Method for UAV Based on Improved PSO Algorithm. In Proceedings of the 2013 25th Chinese Control and Decision Conference (CCDC), Guiyang, China, 25–27 May 2013; pp. 2328–2331.
119. Chen, P.; Shi, G.Y.; Liu, S.; Zhangi, Y.Q. Decision Support Based On Artificial Fish Swarm For Ship Collision Avoidance From Ais Data. In Proceedings of the 2018 International Conference on Machine Learning and Cybernetics (ICMLC), Chengdu, China, 15–18 July 2018; pp. 31–36.
120. Li, X.L.; Shao, Z.J.; Qian, J.X. Optimizing method based on autonomous animats: fish-swarm algorithm. *Syst. Eng. Theory Pract.* **2002**, *22*, 32–38. (In Chinese)
121. Zhang, Y.; Hua, Y.H. Path planning of mobile robot based on hybrid improved artificial fish swarm algorithm. *Vibroeng. Proc.* **2018**, *17*, 130–136. [CrossRef]
122. Zhao, Z.; Zhu, B.; Zhou, Y.; Yao, P.; Yu, J. Cooperative Path Planning of Multiple Unmanned Surface Vehicles for Search and Coverage Task. *Drones* **2023**, *7*, 21. [CrossRef]
123. Mirjalili, S.; Gandomi, A.H.; Mirjalili, S.Z.; Saremi, S.; Faris, H.; Mirjalili, S.M. Salp Swarm Algorithm: A bio-inspired optimizer for engineering design problems. *Adv. Eng. Softw.* **2017**, *114*, 163–191. [CrossRef]
124. Long, Y.; Liu, S.; Qiu, D.; Li, C.; Guo, X.; Shi, B.; AbouOmar, M.S. Local Path Planning with Multiple Constraints for USV Based on Improved Bacterial Foraging Optimization Algorithm. *J. Mar. Sci. Eng.* **2023**, *11*, 489. [CrossRef]
125. Long, Y.; Su, Y.; Shi, B.; Zuo, Z.; Li, J. A multi-subpopulation bacterial foraging optimisation algorithm with deletion and immigration strategies for unmanned surface vehicle path planning. *Intell. Serv. Robot.* **2021**, *14*, 303–312. [CrossRef]
126. Qu, X.Q.; Gan, W.H.; Song, D.L.; Zhou, L.Q. Pursuit-evasion game strategy of USV based on deep reinforcement learning in complex multi-obstacle environment. *Ocean Eng.* **2023**, *273*, 114016. [CrossRef]
127. Wang, L.L.; Zhu, D.Q.; Pang, W.; Zhang, Y.M. A survey of underwater search for multi-target using Multi-AUV: Task allocation, path planning, and formation control. *Ocean Eng.* **2023**, *278*, 114393. [CrossRef]
128. Venturini, F.; Mason, F.; Pase, F.; Testolin, A.; Zanella, A.; Zorzi, M. Distributed reinforcement learning for flexible and efficient UAV swarm control. *J. Cogn. Commun. Netw.* **2021**, *7*, 955–968. [CrossRef]
129. Mou, Z.Y.; Zhang, Y.; Gao, F.F.; Wang, H.G.; Zhang, T.; Han, Z. Deep reinforcement learning based three dimensional area coverage With UAV swarm. *IEEE J. Sel. Areas Commun.* **2021**, *39*, 3160–3176. [CrossRef]
130. Wang, J.; Yang, Y.X.; Li, L. Mobile robot path planning based on improved deep reinforcement learning. *Electron. Meas. Technol.* **2021**, *44*, 19–24.
131. Shi, Z.; Zhang, Q. Improved Q-Learning Algorithm Based on Pheromone Mechanism for Swarm Robot System. In Proceedings of the 32nd Chinese Control Conference, Xi'an, China, 26–28 July 2013.
132. Zhu, X.M.; Wang, L.L.; Li, Y.M.; Song, S.D.; Ma, S.Y.; Yang, F.; Zhai, L.B. Path planning of multi-UAVs based on deep Q-network for energy-efficient data collection in UAVs-assisted IoT. *Veh. Commun.* **2022**, *36*, 100491. [CrossRef]

133. Liu, Y.C.; Song, R.; Bucknall, R.; Zhang, X.Y. Intelligent multi-task allocation and planning for multiple unmanned surface vehicles (USVs) using self-organising maps and fast marching method. *Inf. Sci.* **2019**, *496*, 180–197. [CrossRef]
134. Kohonen, T. The self-organizing map. *Proc. IEEE* **1990**, *78*, 1464–1480. [CrossRef]
135. Garrido, S.; Moreno, L.; Blanco, D.; Martin, F. FM2: A Real-Time Fast Marching Sensor-Based Motion Planner. In Proceedings of the 2007 IEEE/ASME International Conference on Advanced Intelligent Mechatronics, Zurich, Switzerland, 4–7 September 2007.
136. Liu, Z.X.; Zhang, Y.M.; Yu, X.; Yuan, C. Unmanned surface vehicles: An overview of developments and challenges. *Annu. Rev. Control* **2016**, *41*, 71–93. [CrossRef]
137. Peng, Y.; Lou, Y.T.; Zhang, K.M. Multi-USV cooperative path planning by window update based self-organizing map and spectral clustering. *J. Ocean Eng.* **2023**, *275*, 114140.
138. Du, P.Z.; Yang, W.C.; Chen, Y.; Huang, S.H. Improved indirect adaptive line-of-sight guidance law for path following of under-actuated AUV subject to big ocean currents. *J. Ocean Eng.* **2023**, *281*, 114729. [CrossRef]

MDPI
St. Alban-Anlage 66
4052 Basel
Switzerland
www.mdpi.com

Journal of Marine Science and Engineering Editorial Office
E-mail: jmse@mdpi.com
www.mdpi.com/journal/jmse